W9-BMS-882

DESIGN AND EQUIPMENT FOR RESTAURANTS AND FOODSERVICE

A MANAGEMENT VIEW

SECOND EDITION

Costas Katsigris

Chris Thomas

WILEY

JOHN WILEY & SONS, INC.

This book is printed on acid-free paper. ♾

Published by John Wiley & Sons, Inc., Hoboken, New Jersey
Published simultaneously in Canada

For general information on our other products and services or for technical support, please contact our Customer Care Department within the United States at (800) 762-2974, outside the United States at (317) 572-3993 or fax (317) 572-4002.

Wiley also publishes its books in a variety of electronic formats. Some content that appears in print may not be available in electronic books. For more information about Wiley products, visit our web site at www.wiley.com.

Library of Congress Cataloging-in-Publication Data:

Katsigris, Costas.
Design and equipment for restaurants and foodservice : a management view / Costas Katsigris, Chris Thomas.—2nd ed.
p. cm.
Includes index.
ISBN 0-471-46006-0 (cloth)
1. Food service management. 2. Food service—Equipment and supplies. 3. Restaurants—Design and construction. I. Thomas, Chris. II. Title.
TX911.3.M27K395 2005
647.94'068'2—dc22 2004023810

Printed in the United States of America

10 9 8 7 6 5 4 3 2 1

CONTENTS

CHAPTER 20: LINENS AND TABLE COVERINGS | 513

PREFACE

There are hundreds of details involved in running a foodservice business, and the ultimate goals—good food, happy customers, and profitability—cannot be achieved without attending to them. This book began as an attempt to make these details easier to understand and cope with. The idea grew from the authors' needs—both as academics and foodservice professionals—to compile all the basic information for teaching, training, and troubleshooting a wide range of design and equipment-related topics.

This text is intended primarily for students in culinary or food and hospitality management programs. It is also useful for anyone taking a course in foodservice equipment, layout and/or facilities management. In short, the wealth of information can be used by just about anyone who aspires to open his or her own restaurant.

ORGANIZATION

The text is organized according to the typical steps you'd take to open a restaurant—from planning, to design, to equipment selection. Chapter 1 discusses concept development, site selection, budgeting, and market research. Design concepts and space allocation guidelines for both the kitchen and dining areas are then introduced (Chapters 2 through 4). A thorough discussion of utilities follows (gas, water, electricity), along with information on energy management and alternative energy sources. Chapter 7 covers individual components of indoor atmosphere, including lighting and color, sound, ventilation and temperature control. There is a great deal of emphasis on meeting safety and health requirements, as well as workers' safety, sanitation and waste management (Chapter 8).

The last dozen chapters, 9 through 20, focus on acquiring and using specific pieces of foodservice equipment in all areas of a restaurant, from the kitchen to the tabletop. The complexities of purchasing, installing, operating, and repairing foodservice equipment are explained in clear, simple terms. Individual chapters are devoted to equipment selection, purchasing and installation, storage equipment, the cook/chill process, ranges and ovens, broilers and griddles, fryers, steam cooking equipment, dishwashing and waste disposal, smallware, tableware, and linens. Perhaps these subjects are not as glamorous as other aspects of foodservice, but they're just as important as the design of your logo or the way the food tastes.

This industry is nothing if not dynamic, and since the first edition, we've added plenty of new material and reorganized much of the rest. The whole space planning process has become highly computerized, with designers able to use technology to create dining and kitchen layouts on-screen, preventing potentially expensive mistakes in the earliest planning stages. Outdoor dining and so-called display kitchens, where chefs work in full view of the guests, have gained great popularity—and both have their unique design challenges, which we discuss in Chapter 3.

Energy conservation has become a hallmark of well-run foodservice businesses, and an expensive drain on those that are inattentive to the topic. The text includes extensive new information (in Chapter 5) about energy efficiency, the impact of deregulation, alternative power sources, and planning for emergency power outages. We also discuss new types of lighting and HVAC technology (in Chapter 7) that can save money. Hundreds of cities, and more than a dozen states, have passed smoke-free regulations that eliminate "smoking sections" in restaurants; for the rest, basic ventilation knowledge has become even more critical. Other major health-related concerns, addressed in detail here, are safe food handling, use of safe but effective cleaning products, and how to deal efficiently with the mountains of waste that are generated by a foodservice business.

The underlying purpose behind all of these discussions is to demystify the myriad decisions a new restaurateur or foodservice manager will be faced with, which are as wide-ranging as where to put a laundry room, how many place settings you should order, how to lower utility bills, how to buy a walk-in cooler and how big it should be, how to create a pleasing tabletop presentation—even how the air conditioning system and water heaters work. The focus is on useful, updated information about equipment, procedures, technology, techniques, safety, government and industry regulations, and terms of the trade.

FEATURES

To help students process this wealth of information, several pedagogical features have been incorporated into the text.

Introduction and Learning Objectives appear at the beginning of each chapter. The introduction sets the scene for what will be covered, while the learning objectives—arranged in a bulleted list—alert students to the concepts they should focus on as they read.

The discussions in each chapter are supported by a multitude of real-world examples and useful hints, many excerpted from leading industry publications. These examples appear in the form of boxes that are inserted into the text at various points. The boxes have been organized into five distinct types:

Planning the Dining Atmosphere looks at ways to enhance the look and feel of the business from the customer's standpoint—that is, at the "front of the house." This includes ideas for creating a restaurant concept (Chapter 1), the use of artwork and murals on walls, and details about how LED lighting works (Chapter 7).

Budgeting and Planning refers to the bottom-line financial aspects of running a business.

In the Kitchen concentrates on design solutions for the kitchen and storage areas of a foodservice operation, with discussions as diverse as the safety rules for electricity and gas (Chapter 8), and suggestions for sizing a kitchen (Chapter 4).

Building and Grounds focuses on the maintenance and engineering of foodservice facilities, including recommendations for designing an energy-efficient building (Chapter 7) and negotiating lease agreements (Chapter 1).

Foodservice Equipment takes a closer look at specific pieces of equipment, with topics as wide-ranging as how to buy a refrigerator (Chapter 10) and the best way to dry linen (Chapter 20).

Photos and illustrations have been added to support the text and clarify concepts which might otherwise be difficult to visualize. Some of the technical information and code information is arranged as **tables** for easier comprehension and reference.

Each chapter ends with a concise **Summary** that highlights the most important points discussed within the chapter.

Study Questions appear at the end of each chapter to test students' recall of the material and serve as an aid in their studying.

A **Glossary** at the end of the book defines and explains all the key terms introduced in the chapters.

A Conversation with . . . To enhance the text's practical, real-world focus, and make the subject "come alive" for students, we have included several conversations with chefs, design consultants, and other foodservice professionals. A short biographical sketch precedes each conversation. There are ten conversations, placed throughout the book between chapters. We are indebted to the following individuals for sharing their best advice and insights with us. In alphabetical order, they are:

Mark Buersmeyer
General Manager, Business Foodservice,
Marriott Corporation
Boise, Idaho

Kathy Carpenter
Hospital Foodservice Consultant
Grand Junction, Colorado

Christophe Chatron-Michaud
General Manager, Aureole at Mandalay Bay
Las Vegas, Nevada

Mike Fleming
Tabletop Designer, US Foodservice
Portland, Oregon

Larry Forgione
Proprietor and Chef, An American Place
New York, New York

Jim Hungerford
Co-owner, Boise Appliance and Refrigeration
Boise, Idaho

Allan P. King, Jr.
Foodservice Design Consultant
Reno, Nevada

Joseph Phillips
Technical Manager, Food Equipment Program,
NSF International
Ann Arbor, Michigan

Kathleen Seelye
Design Principal, Thomas Ricca Associates
Denver, Colorado

Alice Waters
Chef and Owner, Chez Panisse
Berkeley, California

An **Instructor's Manual** (ISBN: 0471-69607-2) to accompany this textbook is also available from the publisher. The instructor's manual features outlines and suggestions for teaching from the book in a classroom setting, answers to the Study Questions, and chapter quizzes with answer keys. The Instructor's Manual is available for downloading to qualified adopters at www.wiley.com/college.

ACKNOWLEDGMENTS

We are indebted to many individuals who offered their guidance, suggestions, and insights as this book was taking shape. Foodservice educators from across the country reviewed early drafts of this manuscript and suggested ways to enhance its value as a learning tool. They include:

Andrew O. Coggins, Jr.
Pamplin College of Business, Virginia
Polytechnic Institute and State University

Ken Engel
The Art Institute of Washington

Pamela Graff
Minneapolis Community and Technical College

William Jarvie
Johnson & Wales University

Diane Miller
Liberty University

Stephani Robson
Cornell University

Donald Rose
Oklahoma State University

Kathleen Schiffman
St. Louis Community College

Andrew Tolbert
Texas Tech University

Carol Wohlleben
Baltimore International College

Compiling this type of book is truly a group effort, and we received a great deal of help and advice from individuals involved in the professional side of foodservice. John Trembley, a retired district manager of Hobart Corporation, read several chapters in draft form and provided us with helpful advice. Equipment repair experts Jim Hungerford and Roy McMurtry were always willing to explain the intricacies of refrigeration. The late Bill Hines of Lone Star Gas Company first made us aware of the many ways natural gas plays a role in foodservice, and the staff at the American Gas Association has also been most helpful. Kenneth Hutchison of the Electric Foodservice Council and Jeanette Bowman of Idaho Power Company offered up-to-date information on electric use and commercial rates.

Dennis Easton of the U.S. Food and Drug Administration's Dallas office clarified the most recent FDA guidelines for the Food Establishment Plan Review Guide. The University of Florida Department of Food Science and Human Nutrition, under the direction of Dr. Douglas Archer, provided the most current information on food safety and foodborne illness. Tabletop designer Mike Fleming enthusiastically offered us his insights and knowledge. Nancy Beavers of Beavers and Associates, and Norman Ackerman, FCSI, of Ackerman-Barnes Consulting, directed us to many of the food equipment dealers and manufacturers. The staff at Standard Restaurant Supply in Boise, Idaho, good-naturedly let coauthor Chris Thomas spend hours poring over their huge equipment catalog selection, and offered assistance no matter how busy they were.

The North American Association of Food Equipment Manufacturers (NAFEM) holds a biannual educational exhibition, which we'd recommend to anyone who wants to gather the latest information about equipment innovations. We'd also like to thank the editors of two first-rate industry publications, *Foodservice Equipment and Supplies* and *Foodservice Equipment Reports*, for their extensive coverage of equipment-related innovations. It seems that, from each and every issue, we gleaned something to add to the book about the latest restaurant-industry technology.

We are especially indebted to the late Joe Berger. As a retired principal of Fabricators, Inc., Joe taught the Foodservice Equipment course at El Centro College in Dallas, and urged Gus Katsigris to put all these ideas into textbook form more than a decade ago. As the original inspiration for this text, we hope the new edition honors his memory. Writer Lark Corbeil conducted some of the conversations that appear in the book, and many of the line drawings are the work of two talented artists, Richard Terra and Thomas Verdos.

Regina Gowans and her staff at El Centro Community College in Dallas helped Gus design the various charts and tables found in this second edition. And special thanks to Evelyn Katsigris, Gus' wife, who patiently and methodically transcribed hundreds of pages of handwritten instructor's notes that became the very first edition of this textbook, back in the 1990s. (We are pleased to report that Gus has learned to use a computer since then . . .)

Thanks to our intrepid editors at John Wiley & Sons, JoAnna Turtletaub, Julie Kerr, Melissa Oliver, and Tzviya Siegman. Finally, we thank the equipment manufacturers and trade associations—too numerous to mention—who graciously supplied photos.

With the completion of this Second Edition, Gus and Chris look back on a collaboration that has lasted more than ten years and produced two successful textbooks (so far!). Whether we are sitting in Dallas, in New Orleans, or in Leonidion, Greece—a few of the places we've shared dinners, laughs, and debates about the content of our textbooks—we've been committed to creating useful, readable information that may just inspire you to embark on a foodservice career. If we've succeeded, the journey will have been worth every merciless deadline. Best of luck!

Costas "Gus" Katsigris and Chris Thomas
February 2005

CHAPTER 1

ECONOMICS OF SITE SELECTION

INTRODUCTION AND LEARNING OBJECTIVES

Making all the decisions that go into the early planning of a restaurant is a process called ***concept development.*** The dictionary defines a ***concept*** as a generalized idea, but in the restaurant business, that definition hardly seems adequate. In the twenty-first century, restaurants and other foodservice operations must have a clear-cut identity. Today's sophisticated guests want to know the kind of food to expect, the mood of the place, and the menu's price range.

Your concept should encompass everything that influences how a guest might perceive the restaurant. Concept development is the identification, definition, and collection of ideas that constitute what guests will see as the restaurant's ***image.*** This process challenges the owner and planner to create a restaurant in which every component is designed to reinforce the guest's favorable perception and to make that guest a repeat customer.

Part of that is the job of mapping out a thorough business plan that will be profitable for the people who risk the money and effort to put this plan into action.

In this chapter, we will discuss:

- Concepts for restaurants, and what goes into choosing a concept
- Research that must be done to determine whether a concept fits an area or a particular location
- Site selection: the advantages, or possible problems, in choosing your location
- Major considerations when deciding whether to buy or lease your site
- Common factors and advice for negotiating a lease

1-1 CHOOSING A CONCEPT

To better understand the idea of concepts, the dining market may be divided into three basic subgroups:

Quick-Service Restaurants (QSR), formerly known as fast food; midscale, and upscale. The National Restaurant Association has developed lists of food types and/or concepts that fit each broad category:

QUICK SERVICE	MIDSCALE	UPSCALE
Burgers	Cafeteria	High check average
Chicken	Casual dining	Moderate check average
Seafood	Family-style	Hotel restaurant
Pizza	Hotel restaurant	Casual dining
Mexican food	Steakhouse	Steakhouse
Sandwiches	Specialty cuisine (seafood, ethnic, etc.)	Specialty cuisine (seafood, ethnic, etc.)
Ice cream	Eclectic menu	Eclectic menu

As you can see, concept groups are not clear-cut—the lines between them blur often in today's market. That's why it is so important to determine who your potential customers will be—what market segments you will serve—before you settle on a concept.

At the end of 1999, food professionals from around the U.S. met at a "Concepts of Tomorrow" conference in Acapulco, Mexico. Paraphrased from *Restaurant Hospitality* magazine's December 1999 issue, these were their findings about the market segments and dining trends for a new century:

1. The largest segment of the U.S. population is now in its fifties, and is looking for a different type of dining experience than they had in their twenties.
2. The nation is more culturally diverse than ever, with more diners who expect their native ethnic cuisines when they eat out, and others who are willing to try them.
3. Value is very important to today's families. They may have two paychecks coming in, but they are always feeling pinched and expecting to get what they pay for—or more.
4. There are more single-person households, who will be interested in home meal replacement (trends like gourmet carry-out, personal chefs, etc.) and in being accommodated at restaurants—either feeling welcome when dining alone, or comfortable that they can socialize and meet new people.
5. We are all crunched for time, so convenience is tantamount. Nearly 25 percent of U.S. adults say they eat at least one meal per week behind the wheel of their vehicles. Working people have the most interest in kitchen skills but no time to cook; retirees have cooked for years and now want to relax; and Generations X and Y never made cooking a priority in the first place!
6. People are increasingly better informed and expect dining experiences to be stress-free. They demand quality and expect freshness and plenty of interesting options from a menu—even when it's fast food.

The good news appears to be that there are plenty of people with plenty of reasons to eat out. You've just got to decide which ones you want to satisfy. It will be helpful for you to begin your concept search as many successful businesses do: by writing a ***mission statement,*** a few sentences that explain what your restaurant will be to future diners. Make the mission statement as simple and straightforward as possible.

One component of this statement that should help bring your idea to life is the name that is chosen for the restaurant. Its name should indicate the type of place it is (casual, upscale, etc.) or the type of food that is served. Suggested names should be researched, to make sure they're not already a registered name or trademark of some other business. When a name is chosen, immediate steps should be taken to register this name with your city or county, so no one else can use it. If it's a name that was used in the past (but not lately) and is still registered, your attorney can write to the original "owner" of the name to ask if you might use it now. Don't underestimate the importance of doing this name research. It could save you an expensive legal battle in the future.

The next crucial step is menu development. When you think about it, you'll realize you cannot even pick a location until you have decided on a menu. So many other factors—kitchen size and equipment, price range, skills of your potential workforce—all vary depending on the foods you will be serving.

How many dining concepts have you seen come and go over the years in your own area? Try listing them, briefly describing each one. Do their names reflect their "personalities"? Can you sum up their concepts in a single sentence? Truthfully, not all restaurants have a clear-cut concept. And of those that do, most are neither huge winners nor dismal failures. In fact, few of them handle all the variables equally well. So the successful concept is one containing a mix of variables that are, overall, better than those offered by its competitors.

As you create or select a concept, ask yourself: What do diners want in their restaurants?

Illustration 1-1 shows another way to check all the components of your concept. When laid out in circular form, it's easy to see how important each is to the other. When you insert your own concept information, be really critical. Do the "parts" complement each other? Is each compatible with the overall concept? Are they geared to satisfying the market segments you plan to serve?

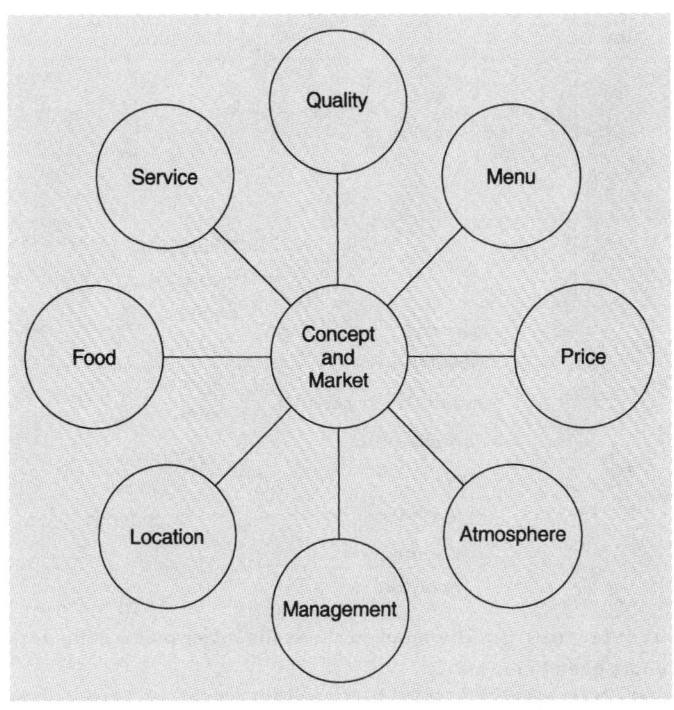

ILLUSTRATION 1-1 **A restaurant begins with a concept and potential market. All other components revolve around these two important considerations.**

SOURCE: *THE RESTAURANT FROM CONCEPT TO OPERATION,* 4TH EDITION, BY JOHN WALKER AND DONALD LUNDBERG. THIS MATERIAL IS USED BY PERMISSION OF JOHN WILEY & SONS, INC. © JOHN WILEY & SONS, INC., 2005.

Interestingly, as you learn this process, you may see some glaring inadequacies in the existing restaurants you visit. Certainly, everyone knows of at least one "hole-in-the-wall" that serves terrific Tex-Mex food. (Just don't think about its last health department inspection!) Or there may be a bistro that's considered incredibly romantic because it has outdoor tables and a beautiful fountain, not because the food is great. The point is that few restaurants manage to combine all the elements *and* add enough USPs to consistently attract loyal crowds of patrons. But that doesn't mean you shouldn't try!

The process from concept development to opening day may take as long as two years. It will also include the acquisition of space, the design process, building and/or remodeling, and placement of equipment. Illustration 1-2 shows a timeline of development for a typical restaurant.

1-2 THEMES AND CONCEPTS

If a concept is a generalized idea, a ***theme*** takes the idea and runs with it, reinforcing the original concept over and over again. A generation ago, the American dining public was bombarded with a wave of so-called theme restaurants. It became the norm to dine out in a Roman ruin, a Polynesian village, an English castle, or a train car. These illusions were costly to create, distracting at best and downright boorish at worst. Today, this type of recurring theme is viewed with some skepticism, perhaps because more attention was often paid to the trappings than to the quality of food served. As recently as 1998, a front page *New York Times* article chronicled "the demise of theme restaurants," suggesting that elaborate decor may be impressive, but it's also distracting and doesn't make up for mediocre food. A year later, in *Food Arts* magazine, writer Michael Whiteman contended that it's not the themes themselves that fail the *"eatertainment"* venues, but the fact that they too often depend on nonfood themes—and that today's more sophisticated diner gets bored with too much "tainment" and not enough "eat," so to speak.

All this nay-saying doesn't mean the theme restaurant has disappeared. It means that, to be successful, the theme must be driven as much by the menu as the clever setting or costumed wait staff.

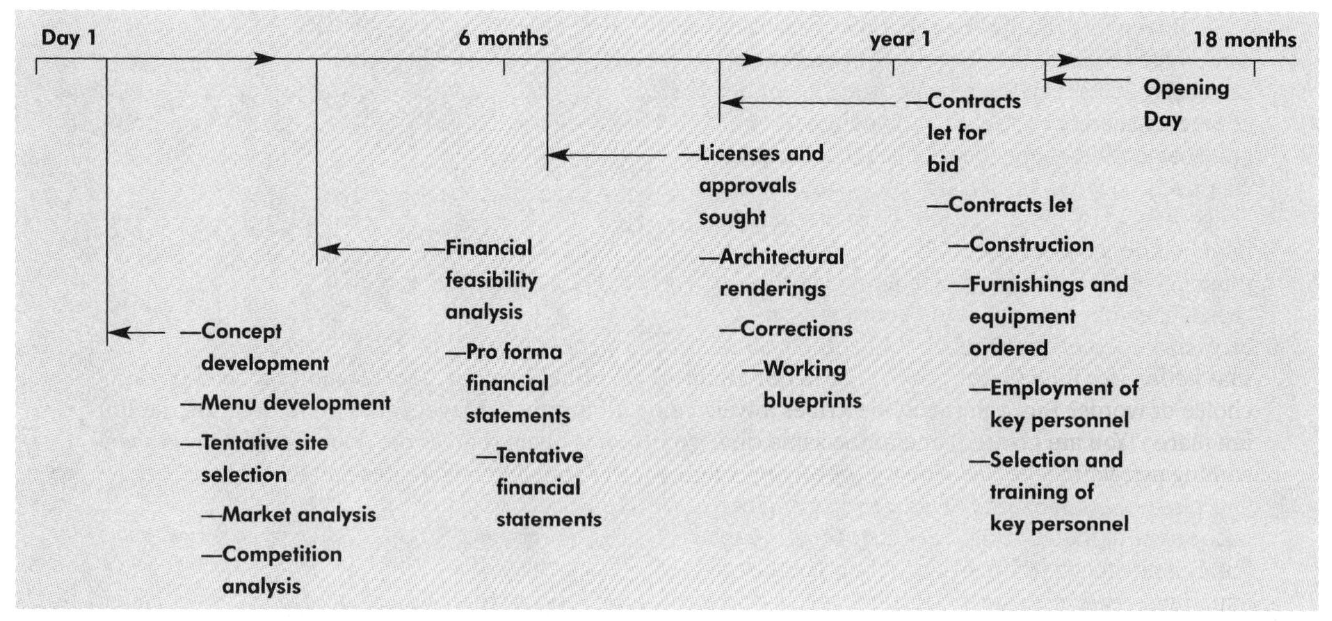

ILLUSTRATION 1–2 **This timeline shows the major points in the restaurant planning process. It may be two years or more before its doors open to the public!**

SOURCE: *THE RESTAURANT FROM CONCEPT TO OPERATION,* 4TH EDITION, BY JOHN WALKER AND DONALD LUNDBERG. THIS MATERIAL IS USED BY PERMISSION OF JOHN WILEY & SONS, INC. © JOHN WILEY & SONS, INC., 2005.

PLANNING THE DINING ATMOSPHERE

CREATING A CONCEPT

What are the variables in creating a concept? Here are the four big ones:

FOOD. What type of food will be offered? What is the style of preparation? How extensive will the menu be? What will the price range be?

SERVICE. How will the food be made available to the guest? Self-service? Counter service? Or full service, where the guest is seated and a wait staff takes orders? In each of these situations, the overall aim is that all guests feel reasonably well cared for by the employees who serve them.

DESIGN/DECOR. There are as many options here as there are restaurants! In general, however, the building's exterior should be inviting. Its interior should be comfortable and clean. The noise level should reflect the style of eatery. Very important: No matter how cavernous the room, seated guests must feel a sense of intimacy, of being able to watch the action without feeling "watched" themselves. This is a major component of most people's basic comfort and safety needs.

UNIQUENESS. In marketing, you'll hear the term "unique selling proposition," or "USP." A USP is like a signature. Everyone's is a little bit different, and the difference makes it special in some way.

A good restaurant concept will have USPs that enable it to attract and retain patrons. Some examples: A restaurant relies on homemade crackers and luscious cream soups as its hallmark, or serves the town's biggest cut of prime rib, or offers a selection of fresh pies that most moms don't have time to make anymore. Eating establishments with the best USPs provide instantly noticeable differences, which distinguish them from their competitors.

Take a look back at the restaurants you just listed. Can you easily determine their USPs?

Take a look at the amusing time line shown in Illustration 1-3. Remember these concepts? And now we're in the "Great Beyond" at the top of the chart, in the twenty-first century! If there's a theme so far in the 2000s, it is adaptability, as U.S. restaurateurs create unique niches to help them maintain a competitive edge. Some redesign interior or exterior spaces to match new menu items. Will an exotic Polynesian theme work in Idaho or Iowa? Maybe. If not, be prepared to transform it into something else that does.

One thing is certain: Successful restaurant concepts exude an attitude of ease and conviviality. It helps if you visualize your concept as a stage upon which to practice the art of hospitality. Your eatery must first be a welcoming place; only then can it take the next steps to becoming a great restaurant.

Take your cues from the places you like best. An Italian restaurant in Milan was the inspiration for New York's venerable Gramercy Tavern. As tavern owner Danny Meyers put it, "I've never been in a place that had better food, better service, or better kindness, all rolled into one." Isn't "kindness" an interesting choice of words? But it perfectly describes a welcoming atmosphere. Meyers goes on to describe the Italian place: "You are greeted, and at the same time, your coat is taken right at the door. I think that is a welcoming act; you feel taken care of. It's an opportunity for hospitality that creates goodwill."

In discussing concept development, restaurateurs must understand that food choices in an affluent society are highly personal, based on one's family background, income level, work environment, living conditions, and the particular social occasion, to name a few. There are so many variables to the dining experience that a hundred people can sit at the same table (not at the same time, of course), order the same menu item, be served by the same waiter, and pay the same amount, and yet it will be a different experience for each of them.

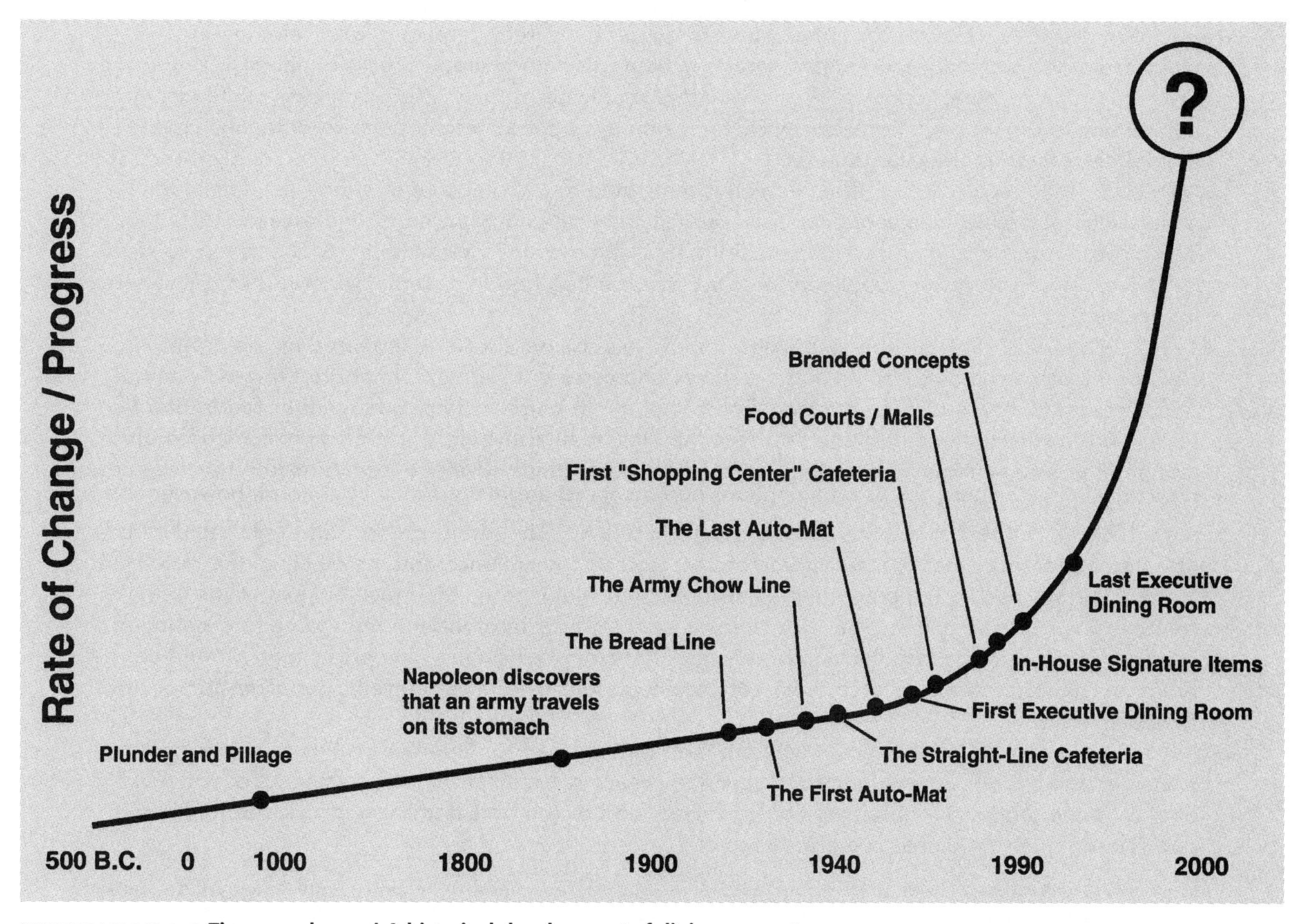

ILLUSTRATION 1-3 **Time marches on! A historical development of dining concepts.**

SOURCE: FOODSERVICE MANAGEMENT.

Therefore, you cannot escape the application of marketing principles to ensure a satisfactory meal experience. Consider a group of people who have deliberately decided to eat at a certain restaurant. Sure, they thought about the types of food and beverages that will be served; you can almost take that for granted. Ultimately, however, they chose the place because it suited their mood, their dress, their time frame, and so on. A National Restaurant Association study pinpoints just a few of these needs (see box below).

Nontraditional Concepts

An idea that has caught the attention of the foodservice industry in recent years is that of the "nontraditional" restaurant site. It is, perhaps, more an expansion strategy than a concept, but it has the potential to be a lasting phenomenon. The idea is simple: Instead of building a restaurant from the ground up and then trying to attract customers, you take the food to places where the customers already are. The possibilities are almost endless, including airports, convenience stores, retail stores, highway travel plazas, college campuses, shopping malls, sports stadiums, supermarkets, and hospitals.

PLANNING THE DINING ATMOSPHERE

DINING DECISIONS

HAVING A FUN TIME. The consumer wants to go out for a treat, a reward, or a celebration. The plans are made in advance, and table-service restaurants with average checks of above $10 are favored. (Authors' note: Today, the figure would probably be "above $20.") Concepts might include upbeat decor and theme, attentive service, food with eye appeal, a tableside presentation, experimental dishes, and perhaps live music.

HAVING A NICE MEAL OUT. The decision is motivated by the pleasure of eating out. It is made impulsively, with expectations of good food, quality, service, hearty portions, and exceptional value. Typical restaurants are table-service establishments with average checks below $10 and self-service cafeterias. The focus is the completeness of the meal rather than the event. Efficient, constant service is important.

SATISFYING A CRAVING. This is the most impulsive decision. It can be triggered by the aroma or sight of a food. It is the key to the home-delivery and carry-out segment. Important aspects are the price, freshness, and quality of the food. Foods that might be craved are pizza, ethnic foods, and barbecue. The restaurant operator can stimulate the craving by using signs, advertising, cooking aromas, and food displays. People with cravings most often satisfy them with food from a restaurant carry-out, food delivered to the home, or food from a supermarket or deli.

MAKING SURE EVERYONE HAS SOMETHING TO EAT. The family group with fragmented schedules and preferences makes this decision on the basis of convenience and the ability of the restaurant to meet various needs. The consumer is not concerned with a complete meal, but just wants to make sure everyone gets something. Such consumers want familiar fare and are not willing to experiment. The service needs to be fast. Restaurants should offer carry-out service, low prices, and promotional programs geared toward children. Such consumers go to fast-food restaurants, get carry-out, or have food delivered.

DOING THE EASY THING. People who are tired and pressed for time want a hassle-free experience while obtaining food. Convenience and speed of service are critical. To attract this group, you must offer a simple process for obtaining the food. Fast-food restaurants dominate, and restaurant carry-out, supermarkets, and delis are also popular.

Source: Study results courtesy National Restaurant Association, Chicago, first published in *Cornell Hotel Restaurant Association Quarterly,* December 1993.

The idea requires that the restaurateur enter into a symbiotic, if not parasitic, relationship with a partner. There are pros and cons to this type of arrangement. First, the potential advantages:

- You sell to an existing customer base, the present customers of the host establishment (baseball fans, airport travelers, theme park guests, etc.).
- It's faster to set up and open your business, because the basic construction may already have been done.
- The host has already done much of the paperwork, getting the necessary zoning, permits, and so on.
- You can be added to (and pay for your portion of) the host's utilities and insurance policies, so you don't have to get your own.
- In some cases, employees of the host company may be the ones implementing your concept, eliminating the need to hire your own daily workers.
- If the concept is already well established (a national chain like Starbucks, Kentucky Fried Chicken, etc.) you gain an instantly loyal, built-in customer base familiar with the brand and products.

However appealing this may sound, industry experts urge extreme caution in striking a deal between a foodservice establishment and a host site. The following caveats appeared in the August 1995 issue of *Restaurants USA,* the trade publication of the National Restaurant Association, and they are still appropriate today:

- You are only as good as your host site. Don't rely on your host's traffic projections; conduct your own customer counts and studies first.
- You may be judged by the company you keep. Even if the cash flow projections look promising, consider the marketing implications of the site for your overall image. Think of it this way: You wouldn't build a really nice house in a less-than-desirable neighborhood, because your surroundings would negatively impact your property value.
- Insist on quality control. In some nontraditional deals, employees of the host company are responsible for handling and selling your product. If it's a service station and your products are fine French pastries, do you see the potential incompatibility? Be sure your wares fit not only the environment but the prospective sales force.
- Be ready for the rush. Customers of a nontraditional "express" location have the same high expectations for good service, good value, and good food. They'll remember poor service, and they'll also remember your company's name when visiting your remote locations.
- Protect your product. If you're selling a signature item, insist on contract language that protects or restricts its sale elsewhere in that particular market area without your written permission. Carefully targeting, and even restricting, others' sales of your products can be critical to your success.
- Bid carefully. Read the fine print with an attorney, and be prepared to walk away from the contract if you are not absolutely sure you can make it work.

Both knowing your customers and knowing your capabilities are two basic tenets that particularly apply to nontraditional sites. Beyond these basics, it will be necessary to determine the minimum number(s) of items that can be served to meet customer demand (while preventing waste), as well as how you will make use of your limited space. Can ingredients be cross-utilized to minimize the need for storage and inventory? Can a few pieces of equipment make multiple menu items? What is the maximum amount of food your crew can turn out in its busiest fifteen-minute period? Finally, how few staffers will it take to provide prompt service and adequate cashiering duties as well?

In short, there is a lot more to the nontraditional idea than expanding handily into already existing buildings with already hungry customers.

The quick-service restaurant also has a new, nontraditional sibling/competitor—a hybrid that combines fast-food service with higher-end cuisine. The so-called ***fast casual or quick-casual restaurant*** focuses on fresh, healthful dining choices (geared to adults rather than children) and fast but attentive service. Menu prices range from $5 to $10 per item. Decor is upscale, airy, and comfortable. Unlike a QSR format, it encourages guests to linger and chat, even when service is speedy. At this writing, the fast casual segment accounts for only 4 percent of the total foodservice industry—but overall number of locations grew 23 percent in 2002 from the previous year. Good examples of fast casual include La Madeleine and the Corner Bakery restaurants.

Second Tier Concepts

Another popular expansion strategy is being used by many large national chains crowding the midscale restaurant segment. They open so-called *second tier* restaurants to keep their primary or flagship concept from saturating the marketplace. A second tier brand is similar to wineries' secondary label concepts—Robert Mondavi also bottles wine under the Woodbridge label; Frei Brothers Vineyards have Redwood Creek, and so on. The difference is that secondary labels are less expensive in the wine business. In the restaurant world, the second tier may compete at the same price point or higher than the flagship brand.

Chili's Bar and Grill has been a flagship brand for Brinker International, but its second tier Macaroni Grill restaurants represent a higher-end concept. Brinker also owns the Corner Bakery and On the Border franchises. Darden Restaurants, Inc.—with a flagship brand of Red Lobster—has the second tier concept Bahama Breeze. The Lone Star Steak House and Saloon represents the best-known but most casual concept in that Kansas-based company, with a check average of $18.50. Its "sister" eateries are Sullivan's Steakhouses (check average $40) and Del Frisco's Double Eagle Steakhouses (check average $60). If you're a national chain, to corner more "share of stomach," you have to expand—not just in numbers of restaurants, but in numbers of concepts to increase your appeal to more segments of the dining public.

Changing Concepts

Restaurateurs may come to the conclusion that they're in the right place—with the wrong concept. *Repositioning* and *reengineering* are the classy terms for updating a concept. Whether it is simple decor change or a menu revision, a concept change involves two basic parts: defining what the problem is with the current concept, and correctly addressing it. How do you know there's a problem in the first place? Lagging sales, fewer visits from repeat customers, higher employee turnover, sagging morale, or the general feeling that the restaurant is looking "tired" or "dated" are just a few of the signals. If you're doing your research—which should be a continuous process, even after the doors open—you'll see them.

You can ask employees and customers for ideas, hire a professional designer, employ a "mystery shopper" to rate the dining experience, and spend some time at your most successful competitors' locations studying what they are up to. Check all of your processes, from seating to order taking to garnishing—could anything be done faster, more efficiently, or more attractively for the guests? And finally, do some research about what might have changed about your surroundings. It may be a temporary economic slump, or may signal a demographic change in the overall area—local families' kids have grown up, for instance, and left empty-nester parents whose dining needs are different. Using what you have learned, update your mission statement and business plan along with your concept and menu. Chapter 2 offers more details on updating existing space in the section "The Redesign Dilemma."

1-3 STUDY YOUR MARKET

If you have a clear vision of where and what your restaurant will be in the future, you have a much better chance for survival. Why? You know where you are going, and have thoroughly researched potential advantages and pitfalls. In determining the potential success of your concept, you have to see if it will:

- **Work in the particular location you have chosen**
- **Generate sufficient sales to realize a profit**
- **Have a certain amount of "staying power" no matter what the economy does**

Any potential investor will most definitely want to see the proof that you have thought through these items thoroughly, and put them in writing. The written document is your ***feasibility study,*** the research you have done to justify the implementation of your concept.

There are two basic types of feasibility studies, and you should do them both. A ***market feasibility study*** defines the target customer, analyzes the competition, and also looks closely at the trade area around the restaurant. A ***financial feasibility study*** covers the money matters—income versus outgo—plus the costs of getting started.

Market Feasibility Study

How you go about this research depends on whether you have a site in mind already, or whether you have a concept and are searching for the best place to locate it. Site-specific research will focus on data for the immediate and nearby neighborhoods; research to pick a location will probably include data for an entire city, to be narrowed down as your site choices narrow.

Potential Customers. If your concept has been decided first, this is the time to define it so well that you can convince investors it is worth financing. The study's goal is to pinpoint who the average, most frequent guest at your proposed restaurant will be. To do this, you need ***demographics*** on these folks: their age, sex, income per household, level of education, number of kids, ethnic group, religious affiliation, and so on. Categorize them by lifestyle and see how much you can find out about them. Also consider the *life cycle* of the potential population. Singles marry and have children, or not. Traditional families may end up as two single-parent households after divorce. Empty nesters eventually retire, and become affluent, middle-income, or low-income senior citizens, active or inactive. Gauging these life cycle trends can help you fine-tune a concept that won't lose its appeal because its primary group of customers is dwindling.

This type of research can be time and labor intensive, and it may be easier to purchase data on some topics, so build in a modest budget for it. Most investors understand that market research is an investment that will not pay off for many months. However, they also understand that it is absolutely necessary.

Where do you find demographic information? Think about which government agencies in your city or county would want to know these same things, and ask if they'll share their statistics with you. You will find not only information about the customers themselves, but much about the area in which you want to locate.

The public library is another promising place to start. If nothing else, look there for a reference book called *The Insider's Guide to Demographic Know-How,* which lists more than 600 sources of demographic information and tells you how to analyze the data when you get it. A San Diego-based company, National Decisions System, offers market reports of all types: financial, consumer, crime, retail sales, and more, at www.sitereports.com. Its "Segmentation Report" groups every U.S. household into one of 50 unique market segments. The U.S. Census Department tracks statistics in about 20,000 different categories, and says only about 10 percent of them are requested by marketers and researchers. So there are plenty of numbers to crunch.

In addition to numbers, the personal habits of your guests are key, especially their eating and spending habits and family structure (empty nesters, single parents, etc.). This kind of information is known as ***psychographics,*** and it can be harder to dig up than demographics. Your area or state restaurant association may be a good starting point. Also, right on your own school campus there may be a marketing or advertising department with students who are willing to help research. Present them with a specific project or need, and ask if they will assist you in exchange for school credit or for a small fee.

Demographics and psychographics have multiple uses for the restaurateur. First, you can use them to determine if a project will be continued or scrapped. Then, if it is continued, this information will help determine how to price, how to provide service, what atmosphere will please the customers you have targeted, and so forth.

Trade Area and Location. This research begins after site selection is well under way. Can the location support your concept? Here, we evaluate the strength (or weakness) of the local economy. How much industrial, office, or retail development is going on? In what shape are nearby houses and apartments,

BUDGETING AND PLANNING

RESOURCES FOR DEMOGRAPHICS

WHERE TO ASK	WHAT TO ASK FOR
Department of Economic Development Chamber of Commerce City Planning Office	Maps, employment statistics (average income, unemployment rates, plans for upcoming commercial and residential developments, data about retail food and beverage sales, shopping habits and patterns, major employers and industries.
Building Department Planning and Zoning Department Commercial Realtors®	Area master plans, residential occupancy and housing statistics, property values, urban renewal projects, zoning information about property, laws about parking, signage, building height, permits, and any other construction-related restrictions.
Local publications (newspapers, magazines)	Dining guides, advertising rates, restaurant reviews, some business demographic data.
Transportation Department	Proposed road improvements, types and routes of public transportation, traffic count data by roadway.
Department of Revenue and/or Taxation	Property taxes by address (or neighborhood averages), breakdowns of real estate taxes, income and sales tax figures
Utility Companies	Past pattern of gas or electric bills at an address, information about typical rates, seasonal usage, conservation discounts.
Convention and Visitors' Bureau	Tourist information—data on numbers, spending habits, annual events and average attendance, dates and sizes of upcoming conventions and meetings.

Source: Neal Gersten, Partner, Citrin Cooperman & Co LLP for Myriad Restaurant Group. First appeared in *Sante* magazine, November 2002.

and what is the vacancy rate? How much is property worth? What's the crime rate? How often do businesses and homes change hands? In short, you're learning about the overall stability of the area.

This part of the study should also include details about a specific site: its visibility from the street, public accessibility to the driveway or parking lot, availability of parking, city parking ordinances or restrictions, and proximity to bus or subway lines. Ask about your ability to change the structure, if you decide to add a deck, a porch, or a second floor.

Don't forget to add details about your proximity to a museum, park, hotel, sports facility, college, military base—anything that would serve as a regular crowd generator for you. Starting from the potential location of your restaurant, the 5-mile radius around the site is your prime market for customers. For QSRs, the radius is a little smaller; for table-service restaurants, a little larger. You'll want to get to know this 5-mile circle as well as you know your own home.

Competition. You will be eating out a lot to do the research for this section! In your 5-mile radius, you must find out, in great detail, what other types of restaurants exist. Classify them as self-service or table-service, then zero in on any restaurant that has a concept even remotely similar to yours. These will be your *direct competitors.* Take notes as you observe their seating capacity, menu offerings, prices, hours of operation, service style, uniforms, table sizes, decor—even the brand of dishes they use is valuable information. Will your concept stand up to their challenge? In your market feasibility study, you might classify the competition in one of two ways:

1. The existing direct competitors seem to have more business than they can handle, so there's room for you.
2. Even though they're direct competitors, they have distinct weaknesses—outdated decor, overpriced menu, limited parking—that give you a viable reason to enter the market. If their concept is poorly executed, test your own skills by figuring out why.

Financial Feasibility Study

The idea of opening or working at a new restaurant is fun and exciting, but don't let the mood carry you away before you do the paperwork. You may be encouraged by the results of your market study. Unfortunately, however, in this business, romantic notions often determine failure rates.

It is going to take a solid financial analysis to make your dreams into reality. Your financial feasibility study should detail the following elements: your projected income and projected expenses, including both food and labor costs.

Projected Income. The first requirement is to project sales levels for each day, each week, and for the entire year. However, if you haven't even opened yet, how in the world do you calculate *that*? You must first determine:

1. The number of guests you plan to serve each day
2. The average amount that each guest will spend (known as the *average check* or *check average*)

On page 12 is a simple example of these calculations.

Understandable? Just remember, you can't get accurate projections unless you have realistic numbers to start with. And in putting your final plan together, it is always smart to slightly underestimate sales. After all, every seat will probably not be filled for every meal, every day. Leave yourself a buffer.

Projected Expenses. Before you decide you'll be a millionaire restaurateur, you must also calculate your expenses. The three major (meaning "most depressing") costs are food, beverage, and labor, which collectively are known as ***prime costs.***

The National Restaurant Association (NRA) conducts extensive annual studies to compile the expenses and percentages for many different types of restaurants. Because they use national averages, they are only estimates. Taking a look at them, however, will be very helpful in making your own calculations.

After studying the most recent NRA Restaurant Industry Operations Report, Chez Ralph owners have decided that their food costs will total no more than 30 percent of the restaurant's income, and beverages will cost 10 to 12 percent of its income. (How much food will people buy versus how much beverage? Chez Ralph figures the check will break down to about 85 percent food and 15 percent beverage.) We'll add labor costs to that in a moment. But, first, using the numbers we have, let's calculate the daily food and beverage costs for Chez Ralph (see box on page 13).

This type of "guesstimating" may seem a bit overwhelming at first, but it is really a simple, logical process based on your own estimates of how many people will walk into your eatery and how much money they will spend there.

Labor Costs. You can't decide how many people to hire or what to pay them until you've figured your labor costs. From the same NRA surveys, you can get an accurate picture of what restaurants spend on payroll, and compare this to the local wage and employment figures you get from the city. Of course, labor costs are always higher the first few months of business, since it takes extra time to train your staff and you may hire more than you need at first. At this writing, a restaurant spends 30 percent of its income on labor costs. This does not include employee benefits (some of which are optional), which add another 2 to 4 percent to overall costs.

BUDGETING AND PLANNING

FIGURING AVERAGE ATTENDANCE AND AVERAGE CHECK

"Chez Ralph" has 140 seats. In most restaurants, each seat is occupied two times for lunch and one time for dinner. (We're going to keep this simple, and assume that Chez Ralph is open daily for both lunch and dinner.) So, the daily customer count is:

Lunch 140 × 2 = 280
+ Dinner 140 × 1 = 140
Total guests per day = 420

- How many guests per week? Multiply 420 by 7 (days per week) for a total of 2940.
- How many guests per month? Multiply 420 by 30 (days per month) for a total of 12,600 guests.
- How many guests per year? Multiply 12,600 by 12 (months per year) for a total of 151,200.

Okay, now let's look at how much money these people spend at Chez Ralph. This is a very simple estimate, based on the price of the average entree. Since you're probably not far enough along to have created and priced a menu of your own, you could average the prices of your direct competitors. In Chez Ralph's market feasibility study, the direct competitors' average check is $10 per person.

DAILY SALES:

Average check = $10.00
× Daily number of guests = 420
Sales per day = $4200

WEEKLY SALES:

Average check = $10.00
× Weekly number of guests = 2940
Sales per day = $29,400

MONTHLY SALES:

Average check = $10.00
× Monthly number of guests = 12,600
Sales per month = $126,000

ANNUAL SALES:

Average check = $10.00
× Annual number of guests = 151,200
Sales per year = $1,512,000

Chez Ralph estimates labor costs will be 33 percent of its total sales. This means:

Daily labor cost

$3360 (daily total sales) x 33% (0.33) = $1108.80

Monthly labor cost

$1108.80 (daily labor cost) x 30 (days per month) = $33,264

Annual labor cost

$33,264 (monthly labor cost) x 12 (months per year) = $399,168

BUDGETING AND PLANNING

CALCULATING FOOD AND BEVERAGE COSTS

FOOD SALES

\$3360 (daily total sales) × 85% (0.85) = \$2856

FOOD COSTS

\$2856 (daily food sales) × 30 (0.30) = \$856.80

BEVERAGE SALES

\$3360 (daily total sales) × 15% (0.15) = \$504

BEVERAGE COSTS

\$504 (daily beverage sales) × 12 (0.12) = \$60.48

Multiply these totals by seven to obtain your weekly cost estimates:

FOOD

\$856.80 (daily cost) × 7 (days per week) = \$5997.60

BEVERAGE

\$60.48 (daily cost) × 7 (days per week) = \$423.36

Multiply these totals by 30 to obtain your monthly cost estimates:

FOOD

\$856.80 (daily cost) × 30 (days per month) = \$25,704

BEVERAGE

\$60.48 (daily cost) × 30 (days per month) = \$1814.40

Multiply the monthly totals by 12 to obtain your annual estimates:

FOOD

\$25,704 (monthly cost) × 12 (months per year) = \$308,448

BEVERAGE

\$1814.40 (monthly cost) × 12 (months per year) = \$21,772.80

Now do you see why the restaurant business is not for the faint-hearted? Sure, you may make a million dollars, but you'll spend almost \$730,000 of it on food, drinks, and employees to prepare and serve them. And did you notice that labor costs total more than food and beverage combined?

Other Expenses. There are other expenses that need to be estimated, from rent to utility cost to legal fees, taxes, and equipment payments.

When expenses are all gathered and estimated, you are ready to prepare your projected income statement and projected cash flow statement.

1-4 SELECTING A SITE

Be aware that in site selection, there are two important designations: whether your concept is ***convenience oriented*** or ***destination oriented.*** A convenience-oriented restaurant, like a fast-food franchise, depends primarily on a nearby base of customers to be "drop-ins," for often unplanned visits. Because there is likely to be similar competition throughout the area, customers probably would not drive a long distance to visit this particular site, unless it happens to be a convenient part of their daily commute or errand route.

Destination-oriented restaurants attract guests often because of their unique concepts. Customer visits in this case are planned ahead of time, and may involve driving ten or more miles, depending on the attractiveness and availability of the concept. A destination restaurant is more likely to be the choice for a special occasion or fancier meal. Ideally, you'll attract both types of customers: some drop-ins and some who planned their visits.

Site Selection Research

Unfortunately, most customer research seems to include only demographic information, when location-related background can be equally important—for example, frequency of use, distance traveled, where the trip originated, reasons for the visit, whether the visit was in tandem with other activities, proximity to home and/or work, and how often key competitors are also frequented.

Luckily, there is now site selection information specific to the foodservice industry—sales in existing eating and drinking establishments, market share of QSRs, and data known as the Restaurant Activity Index (RAI) and Restaurant Growth Index (RGI). The Restaurant Activity Index is an indication of a population's willingness to spend money eating out instead of cooking at home. The Restaurant Growth Index is a statistical prediction of cities where a new restaurant has the "best" chance of succeeding. In both surveys, the number 100 is used as the benchmark or "national average"—in other words, a score of above 100 indicates better than average prospects, and lower scores indicate poorer prospects. RGI is not a gauge of where restaurants *are* but where it is best for them *to go*. That's why well-established cities already known for their lively dining scenes may not fare as well in these surveys as yet undiscovered markets.

Lest the entire class now decide to move to Myrtle Beach, South Carolina, to open restaurants, the rest of the "Top 15" cities for high per-capita dining expenditures as of 2002 were: Flagstaff, Arizona; Anchorage, Alaska; Orlando, Florida; Honolulu, Hawaii; Barnstable–Yarmouth, Massachusetts; San Francisco, California; Santa Fe, New Mexico; Panama City, Florida; Atlantic City–Cape May, New Jersey; Naples, Florida; Wilmington, North Carolina; Billings, Montana; Bloomington–Normal, Illinois; and Portland, Maine.

And just for the heck of it, let's look at the "Bottom 10" cities—those where annual dining-out expenditures total less than $800 a year per person: Houma, Louisiana; Greeley, Colorado; Yuba City, California; Johnstown, Pennsylvania; Steubenville–Weirton, Ohio-Pennsylvania; Brazoria, Texas; Visalia–Tulare-Porterville, California; Vineland-Millville–Bridgeton, New Jersey; Provo–Orem, Utah; and Merced, California.

There's even a third chart, called the "Menu Potential Index," which estimates the success potential of a particular type of cuisine or dining spot in a particular market. For instance, in the market of Atlantic City–Cape May, New Jersey, a QSR donut shop, a sandwich and hot dog shop, a midscale Italian restaurant, or a casual seafood eatery are the four most likely concepts to gain acceptance. Restaurants that "make it" are those that meet a need, and research like this certainly pinpoints these needs by area. Finding out an area's fast-food market share is another valuable key because it provides an insight into the food preferences there, in terms of both price range and level of sophistication.

The sandwich giant Subway uses a special computer program for its site selection needs. Onscreen, a map shows each existing Subway store, with population rings around it that indicate the surrounding market potential for that store. Competitor locations have also been programmed into the system and are shown on the same map. This shows Subway where there are "holes" without restaurants. Then the population in those "hole" areas is studied for demographics, and compared to the customers of a successful Subway nearby. By the time the store development team visits potential sites in person, they have truly done their homework.

Technology does not ensure a successful restaurant. It is simply another useful tool that makes it easier and faster to analyze statistics on many levels, and to take advantage of research that's already available on the Internet. Specialized computer programs are expensive, but there are a variety of packages (and Internet sites with programs and research) that may be purchased for an annual fee, or on a by-site basis.

Guidelines for Site Selection

The location of a restaurant is the bridge between your target market and your concept. And yet, as long as there's a vacant building and a willing investor, there will be hot debate about the true importance of site selection. Some experts feel that as long as the restaurant is located in the "right" area of a city, with a strong economic base, the actual site doesn't matter as much—that unique food and beverage offerings and good service will attract customers, even to a less obvious location. Others are not so optimistic. For the time being, let's agree that, at the very least, the restaurant's concept and location must "fit" each other. This also means that, as certain things about the location change, the concept must often change to adapt.

Where does our fictitious Chez Ralph "fit"? The location selected will have an impact on:

- The type of customer
- Construction or remodeling costs
- Investment requirements by lending institutions
- Local ordinances, state and federal laws
- Availability of workers
- The option to sell alcoholic beverages
- Parking availability and accessibility
- Occupancy costs—rent, taxes, insurance, and so on

Study your own neighborhood, and you'll probably find a few restaurants that seem to thrive despite less than optimum locations. Most information, however, indicates the reverse: A good location is crucial to survival and success.

At this point, we will list some important guidelines for selecting a site. If any one of these conditions exist, proceed with extreme caution. Your business survival may be at risk if they are not handled properly:

Zoning Restrictions. Zoning ordinances must allow your specific type of operation to do business at this location, and must also permit adequate parking on or near the property. Do not waste time and money attempting to get a ***variance,*** which is a long and costly procedural struggle with your city or county.

Likewise, if there are restrictions on the sale of alcohol, look elsewhere. Unless your concept warrants absolutely no need to serve beer, wine, or drinks (during regular hours or for private functions), you most definitely restrict your profits when you limit yourself with a liquor prohibition.

Small or Oddly Shaped Lot or Building. If the parking area isn't big enough or easily accessible, you're asking for trouble. Can the lot meet the basic city parking requirements? (They are typically stringent, as discussed in detail in Chapter 4.) Can people drive in and out of the area as easily as they can park? Do garbage trucks have enough space to empty your dumpster? Also avoid narrow frontage to the street. You want people to be able to see the place!

There is no "perfectly shaped" space for a restaurant, but some configurations do work better than others for maximizing the total number of seats for the square footage. Even the way the building is placed on the lot determines whether you'll constantly be opening and closing window coverings at certain times of day because of glare or direct sunlight annoying the customers. Get professional design advice before you commit to an unusual space or site.

Short-Term Lease. Do not lease for less than five years, with one five-year option after that. Generally, short-term rental agreements prohibit most restaurants from realizing their potential. Also, beware the shopping center pad site, which requires that you pay rent, plus a percentage of profit based on monthly sales, to the owner or management company of the center. Can you handle the added expense?

Low Elevation. This is seldom a consideration, but it should always be. Elevation affects gravity, and gravity affects drainage away from the building. A good site is environmentally friendly. It will minimize your sewage backup, plumbing, and grease trap problems. It does not require extraordinary modifications to get good water pressure throughout the building. Also, find out if the property is located in the flood plain of an adjacent river or creek, which may make insurance difficult or expensive to obtain.

Utility Requirements. Amazingly, some people put down earnest money on a piece of property, only to find out that gas lines or sewer lines are not available there or are inadequate for commercial use. Your real estate agent or local utility companies can check this for you. And, as we'll cover in Chapter 5, electrical power requirements are complex for restaurants and may mean you need more power than now exists in the space. Be prepared to negotiate with the landlord for the cost of converting or modernizing the electric service.

Along the same lines, the HVAC (heating, ventilation, and air conditioning) system works hard in a dining establishment. Usually the landlord provides this service, so have the system inspected to ensure that it meets current city codes for ventilation. Add an inspection of the hood exhausts, or if there are none yet, determine where they will be located. Modern building codes require exhaust systems be located as much as 10 feet from any door, window, or fresh-air intake. In a multistory structure, hood installation involves building a shaft through other stories to the roof, which is very costly. Better to know early if it's even possible—or that the landlord will agree to do it.

Urban Challenges. Along those lines, there are some immediate concerns when locating in a downtown area as opposed to a roomier suburb. Exhaust vents can't simply be punched through a wall—as we mentioned, they may have to snake up several floors to the roof of the building. Sewer and grease trap requirements are often more complicated. Licensing can be more expensive, and regulations stricter overall, in an urban area.

Speed of Traffic. The roadway in front of your proposed site provides an important clue to its future success. Do cars whip by so quickly that the motorists never see your place? Drivers traveling at thirty-five miles an hour or less will be best able to read your signage, and to spontaneously turn into your parking lot without causing a traffic mishap.

Is the nearest intersection so busy that most people would think twice before trying to cross the street to patronize your business? Does the outside traffic hinder or help you? Stand outside the location and watch both automotive and foot traffic in the area. Ask yourself if anyone, coming from any direction, would be frustrated by the sheer hassle of getting to your site. This also applies to restaurant space in high-rises, shopping malls, and other locations that aren't necessarily at intersections.

Proximity to Workforce. Almost as important as being accessible to the public is the need to be located close to a potential labor pool. Are your employees able to live close by? Do the routes and hours of public transportation systems mesh well with your business?

Previous Ownership. Was this site previously a restaurant in its "past life"? If so, take great care to determine why it closed. If it is appropriate, talk with the previous tenants and neighboring businesses about the pluses and minuses of the location. Putting up an "Under New Management" sign is not enough to guarantee success where others have failed. A bankruptcy in the building's past may make it more difficult to secure financing for another venture there. Today's investors are cautious.

Now that we've discussed the possible drawbacks, how about those characteristics that are almost always attributes? Since we're speaking in general terms, no single factor will "make or break" a business. A combination of them, however, will certainly enhance the desirability of your site.

Visibility. If people cannot find you because they cannot see you, you're almost always in trouble. Exceptions exist, but they are rare. If possible, your building should be visible from both sides of the street, as far away as 400 feet. This also means checking local signage laws. Cities, counties, and even your neighborhood merchants' group may have restrictions on the types and sizes of outdoor signs allowed. If you're looking at a strip-style shopping center, go for the ***end cap,*** or end location; in other situations, consider a corner lot versus a middle-of-the-block lot.

Parking. In suburban locations, local ordinances specify the ratio of parking spaces to the size of the building. For restaurants, the general rule is one space for every three seats. Busy places may need more. Again, we describe the specifics in Chapter 4.

Closely examine parking availability during peak hours—at lunch, dinner, and on weekends—and query neighboring businesses. Will the dour dry cleaner next door tow your customers' vehicles before the main course has been served? (We know of one instance in which a law firm next door to an Italian restaurant has diners' cars towed from its lot even after dark and on weekends, when there's nary a lawyer in the office.) Don't assume that "after hours" it'll be okay to use others' parking lots without a specific written agreement. Parking fights among businesses can be nasty, protracted, and counterproductive.

If parking is that tight, study the option of offering valet parking. Will your guests pay the added cost for convenience or security? Or will you offer complimentary parking and foot the bill yourself? Finally, where will your employees park?

Accessibility. Make it easy for guests to enter and leave your parking lot and your building. Check the locations of traffic lights or stop signs, which may affect foot traffic. One-way streets or speed limits of more than forty miles an hour may make your place a little tougher for cars to get to.

Your city planning and zoning department will be able to provide any recent surveys of vehicle or foot traffic, as well as details of future plans for the street. A prolonged construction project that restricts access could be deadly to business.

Today, accessibility also includes compliance with the federal Americans with Disabilities Act (ADA), which was passed in 1992. Businesses are now required by law to provide commonsense amenities—such as handicapped parking spaces, wheelchair ramps, and restrooms—that enable physically challenged persons to be customers, too. (We detail a few of the major requirements of the ADA in Chapter 4.)

The first year the ADA went into effect, almost 1800 complaints were filed against businesses with the Department of Justice, 60 percent of them for physical barriers. Contact a charitable organization that assists disabled persons and ask for guidelines to help you make them loyal customers, instead of litigants.

Traffic Generators. Large, natural gathering places will affect your ability to draw people into your establishment. In some cases, when events are held near mealtimes and yours is one of the only restaurants in the area, they provide "captive audiences." In other situations, a crowded spot may be a drawback if the area gets a reputation for being inaccessible ("You can never find a parking space," etc.). Only you can decide if being near a sports arena, museum, department store, school, hospital, or mall will help your business. On the positive side, you can use its proximity as part of your theme or concept, and cater specifically to those patrons. However, there are some drawbacks.

Let's say you're considering a site near a high-rise office with 2500 employees. It sounds promising at first, and you're imagining a bustling lunch business. If the office building already has its own cafeteria, executive dining room, and ground-floor sandwich shop, however, the location may be less than ideal.

Also, ask if the area has an active merchants' association, which may sponsor business builders like parades, street fairs, or holiday events.

Design Flexibility. Your own ability to adapt to an attractive space when you find it is important. For instance, Starbucks has four different prototype store designs, knowing it will encounter promising locations that may be differently sized or shaped. Adaptable designs for both an urban and a suburban setting might be the way to go if you're open to either type of location—or if you're simply planning ahead for expansion.

Restaurant Cluster. There are streets in most cities that seem to be lined with eating places. Such areas are commonly known as ***restaurant clusters.*** People tend to congregate in these areas, which helps your place become a regular destination for some. For all their advantages, however, clusters have life cycles. Make sure you choose one that is not headed for a downturn in economics or popularity. Do not try to duplicate any of the other concepts already at work in your cluster, and remember that the national, casual dining chains seem to work better in a cluster than fancy, white-tablecloth establishments.

Guests, Both Regular and Infrequent. You want 50 percent of your guests to fall into the category of "regulars," who visit your restaurant three to five times per month. The market segments that can typically provide these diners are:

- Singles
- Young families
- Retirees
- Affluent empty nesters
- Office or professional crowd

These folks must live or work nearby. When you visit your competition to do market research, notice who eats out and when. Observe what they order and how long they stay. What is it about this particular site that seems to attract them?

However, do not overlook the infrequent guest as an attribute. Having a strong base of infrequent guests—tourists, conventioneers, regional salespeople who come to town occasionally—helps insulate a restaurant against the impact a new competitor may have on the whims of the regulars. A regular clientele may suddenly evaporate, at least temporarily, as they all rush to try a new eatery, while the infrequent guest is not prey to these whims.

PLANNING THE DINING ATMOSPHERE

HOW BAD LOCATIONS ARE CHOSEN

1. Insist on the most convenient location for *you*, ignoring demographics or site problems.
2. Select a weak strategic position in a highly competitive market.
3. Overlook the importance of drive-by features, like access and visibility.
4. Select a site that can be easily "outpositioned" by a major competitor.
5. Be one of the last concepts to enter a market that is already saturated.
6. Don't bother to find out why there are no competitors near the site you're considering.
7. Locate in an office area where the businesses close earlier than your normal hours.
8. Overlook demographic information that suggests the closest neighbors aren't really the lifestyle types who will enjoy your concept.
9. Select an out-of-the-way location because it's quaint—when there's little foot traffic.
10. Overlook vehicle traffic patterns around the site.

1-5 OWNING OR LEASING SPACE

Once you have identified a location where your concept will fit, where you can find employees, and where members of your target market will visit on a regular basis, your work has just begun. Will you lease the site or purchase it? Remodel or build from the ground up?

You may consider hiring a commercial real estate broker to assist in your property search, but it is smart to do your concept and target customer research before you do so. Some property owners don't want to pay the broker's commission fee to make a sale, but the broker brings real estate negotiation experience to the table, as well as knowledge of multiple locations and market demographics. They're not in the restaurant business—but then again, you're not in the real estate business—so they may add some valuable insight to the selection process.

Buying Land and Building on It

Even if it is raw land, the same attributes (and problems) of site selection in general hold true here, too. Your first step should be to find a Realtor® who will help with the myriad details of land purchase. You'll want to know if the selling price is fair, based on other recent sales in the area.

Sales prices in urban areas are calculated in dollars per square foot, and in undeveloped areas, dollars per acre. If you can afford to pay cash for the land, you'll bring the cost down considerably. Most people, however, buy land for businesses just like they buy houses—with a down payment and a mortgage of monthly payments with interest charges. Will the person selling the land be willing to finance the mortgage ("carry the note") or will you get a bank loan? In either case, the land itself will be used as *collateral* to guarantee the loan. If you don't make payments, you lose the land and, most likely, anything you've built on it.

When making a land purchase for a restaurant site, pay close attention to two long-term factors:

1. The ***floating interest rate*** is the rate of interest you will pay. It is one or two percentage points higher than the bank's *prime rate* or the *Consumer Price Index* (CPI).

2. The ***payout period*** is the number of years it will take to pay back the amount borrowed. In most situations, this will be ten to fifteen years.

When it's time to build on your land, you'll use your projected income statement to tell you the upper limits you should spend on construction. This may sound cynical, but only you can decide if you'll have enough guests coming in the door to support massive construction debts in addition to land costs. So, before you decide on the "buy land and build" strategy, you would do well to meet with a restaurant or foodservice consultant and an architect. Together, after listening to your needs, they should be able to present a preliminary budget that covers all aspects of construction. In addition to design fees, it must include:

- Electrical
- Mechanical
- Plumbing
- Painting
- Heating/air conditioning
- Interior finish-out (for specific areas: kitchen, dining room, bar)
- Special features (glass, doors, etc.)

It is impossible to generalize about these costs, but professionals can break them down into an approximate *cost per square foot* figure that you can plug into your budget. Costs in 2004 to build a restaurant ranged from $200 to $400 per square foot in Dallas, Texas.

The time to get an architect or foodservice consultant involved is now, before the lease is signed. Yes, their services can be expensive, but it is worth it to have a realistic idea of whether the space can become what you intend it to be. Stay on good terms with your designers and contractors by being completely honest as your estimates are being compiled. It is time consuming and expensive to ask them to draw up preliminary plans, then continually downsize them because your budget does not fit the project.

One way to save money in constructing your own restaurant is to rehabilitate an existing structure. Some call this ***adaptive reuse.*** The Internal Revenue Code details the potential savings for renovation work, since Congress has decided it is generally better to save old buildings than demolish them. Usually, if the building was constructed before 1936, you can receive a tax credit of 10 percent of all qualified expenditures; in refurbishing some historical buildings, the tax credit can be as high as 20 percent. The credit is deducted directly from actual taxes owed. For instance, if qualified rehabilitation expenses total $100,000, that means a $10,000 tax credit is allowed. The actual cost of purchasing the structure, however, is not included. An accountant and/or tax attorney should be consulted if you're undertaking historical renovation.

Leasing an Existing Space

Most first-time restaurant owners lease space because it is the least expensive option. In some instances, the site is simply not available for purchase. Again, a good real estate broker will save time in site selection and lease negotiations. In every city, there are real estate professionals who specialize in restaurant locations or in specific trade areas, and their expertise is worth their fee. In fact, the fee (which usually amounts to 4 percent of the total lease amount) is often paid by the property owner. Make sure to check on this before you begin the negotiating process, to avoid misunderstandings. Even if you will not own the space, professional design or architectural assistance is also a good idea, long before a lease is signed.

As with a purchase, a lease price is referred to in cost per square foot. Since you're planning on a successful long-term operation, ask for a lease of at least five years. That way, you know the exact monthly amount to budget for, and you know it won't increase before then.

In negotiating for space, you should ask about:

- The exact size of space available, including parking
- The exact price, in dollars per square foot
- The amounts of finish-out allowances, and exactly what is included in them

A ***finish-out*** or ***build-out allowance*** is money for a new tenant to complete the interior of the space with paint and fixtures, usually chosen by the tenant. If the landlord provides a finish-out allowance, the rent figure will be higher than if the tenant opts to pay for finish-out. Who should pay? Your cash flow situation should determine that and, frankly, most new restaurants need their cash to sustain them in the first lean months before they see a profit. So try to let the landlord assume finish-out costs.

Or, if you're quite a negotiator, test your skills by offering to do the finish-out in exchange for a much lower rent for the first six months to one year. We also know of situations in which restaurateurs have done their own finish-out in exchange for *free* rent for the first three to six months. It all depends on how badly the landlord needs a tenant in the space.

The other time period to ask for leniency in paying rent is during the time just after move-in, when improvements are taking place. Many property owners will grant a 60- to 90-day grace period before the first rent check is due, knowing that the place is not yet open. Even with this reprieve, budget in rent costs anyway. You never know when overruns and delays will slow your progress. If you're technically in the space, you should be able to pay the rent.

Specifics of Most Restaurant Leases

The annual cost of for leased space is calculated per square foot, and is known as the ***base rate.*** Chez Ralph is considering a space of 5000 square feet, at $20 per square foot. The annual rent would be:

5000 (size in square feet) × $20 (cost per square foot) = $100,000

The monthly rent would be:

$100,000 (annual rent) ÷ 12 (months) = $8333.33

On average, total rent costs should be about 7 percent of yearly gross sales. In many lease agreements, there is also a ***percentage factor*** added to the base rate, usually 6 to 8 percent. This means that the more profit the restaurant makes, the more money it must pay to remain in the space. Let's say the percentage factor in this particular lease is 8 percent. How much money can Chez Ralph make without an automatic rent hike? It's easy to calculate:

$100,000 (monthly rent) × 1.08 (rent plus 8%) = $108,000.00

This lease states that Chez Ralph's rent will remain at $8333.33 per month, as long as its *net sales do not exceed* $108,000.00. Be sure that this figure is based on net sales, *not* gross sales. Also, specify what the net sales figure will *not* include: items like liquor taxes or license fees and other expenses (credit card discounts, sales tax on food, etc.). All parties must agree on how sales for each month will be verified. Who will "drop by" to look at your books, and when? Whenever a percentage factor is part of a lease, your attorney should go over it carefully with you before you sign anything.

The reason percentage factors are controversial is simply that they ask you to pay an additional price for success. Let's say, after 7 months in business, Chez Ralph has a terrific month: $112,000 in net sales. Well, this also means more rent to pay. How much more?

$112,000 (net sales) – $108,000.00 (rent base plus the 8% factor) = $4,000.00

$4,000.00 (additional sales) × 0.08 (8% factor) = $320.00

Some restaurateurs shrug this off as the cost of doing business; others see it as a subtle disincentive to strive for bigger bucks.

Term of Lease. Most foodservice business leases are for a period of five years, with two more five-year options—a total of fifteen years. In addition to rent and percentage factors, it is not unusual to have an ***escalation clause*** in the lease, detailing a "reasonable" rent hike after the first five-year term. The increase may be based on the Consumer Price Index (CPI) or the ***prevailing market rate*** (what similar spaces are being rented for at the time the lease is negotiated). Make sure the basis for any rent hike is clearly spelled out in the lease agreement.

Financial Responsibility. Early in the lease negotiations, you should cover the touchy topic of who will be responsible for paying off the lease in case, for any reason, the restaurant must close its doors. If an individual signs the lease, the individual is responsible for covering these costs with his or her personal assets. If the lease is signed as a corporation, then the corporation is legally liable. As you can imagine, it makes more sense to pay the state fees to incorporate before signing a lease.

It may well be your landlord who goes broke, not you. Your lease should stipulate that, in case of the landlord's financial default (or sale to a new owner), your business cannot be forced out and the new owner must abide by the terms of the existing lease until it expires. This is sometimes called a ***recognition clause.***

Within your corporation, multiple partners must have specific agreements about their individual roles in running the business. You should probably also outline how a split would be handled if any partner decides to leave the company. Your attorney and accountant fees will be well worth the peace of mind to have these important contractual agreements written and reviewed.

Maintenance Agreement. Another important part of a lease is the complete rundown of who is responsible for repairs to the building. Some leases specify that the tenant takes full responsibility for upkeep. Others give the landlord responsibility for structural and exterior repairs, such as roofing and foundation work, while tenants handle interior maintenance, such as pest control service or plumbing and electrical repairs. These items are easy to gloss over if you have your heart set on a particular site. Remember, however, that all buildings need maintenance, and the costs can really add up. How much are *you* willing to do—and pay for?

Insurance. Generally, the tenant is responsible for obtaining insurance against fire, flooding, and other natural disasters, as well as general liability insurance for accidents or injuries on the premises. The lease must specify how the policy should be paid—monthly or yearly are the most common stipulations—and also the amount of coverage required. Both tenant and landlord are listed as the insured parties, so the landlord should be given copies of all insurance policies for his or her records.

Real Estate Taxes. Each city and county decides on the value of land and buildings, and taxes an address based on its *assessed value.* These taxes are typically due once a year, in a lump sum, but most landlords will ask that the taxes be prorated and paid monthly, along with rent and insurance. A ***triple net lease*** is the term for a lease that includes rent, taxes, and insurance in one monthly payment.

Municipal Approval. Just because you sign a lease doesn't mean you'll ever serve a meal at this site. Cover your bases by insisting in writing that this lease is void if city or county authorities do not approve the location to operate as a restaurant (or bar, or cafeteria, or whatever you're planning). Potential roadblocks: Do you intend to serve alcohol? Is your concept somewhat controversial—scantily clad wait staff, for instance? You'll save yourself a lot of time and money if your lease allows these items in writing, and if you also obtain permission from the county or city first. Politely inquire about all necessary licenses and permits before you begin finish-out work on the site.

Moving On. Always allow for contingency plans should you decide that this is not the right location for you. Insert a clause in the lease that says you do not have to pay the remainder of the lease in full if you decide to close and/or relocate. If there *is* a "lease buyout" amount, it should decrease with the age of the lease. Also include your right to *assign and sublease*—that is, to lease the premises (along with the lease itself) to another business without having to pay the landlord for the unexpired term of your lease. The landlord will usually require his or her prior written consent, which is understandable. So insert this statement in the lease: "Consent will not be withheld unnecessarily." That way, when you find a reasonable and financially sound tenant to take over your lease, the landlord can't refuse them without a very good reason.

The Infamous "Other." There are many unusual or unforeseen circumstances that might hamper your progress, as there are different types of cuisines. It may be an antiquated town ordinance, a surprise deed restriction, or an old lien on the property that has resurfaced in your research of the site. Whatever the case, check carefully at city hall to make sure there is nothing unexpected on the horizon. Talk with the landlord, the Realtor® and with other businesses in the area to piece together a complete history of the site you are about to lease. Think of it as a marriage of sorts. Go in optimistically, but go in with your eyes open.

As an example, a restaurant will normally need a greater than normal number of roof penetrations to accommodate kitchen and bathroom exhaust systems, fresh air intakes, and so on. The landlord should be aware of this and provisions should be made in the lease for the original roofing company to make these modifications and seal the holes. This should keep the roof's warranty intact.

In the "Building and Grounds" box below, we reproduce some advice for tenants from *Restaurants USA,* the monthly publication of the National Restaurant Association. These hints were first published in the November 1993 issue and still apply today.

Lease-Purchase Options

A final word on the question of owning versus leasing space. Many successful restaurateurs have included in their leases an option to purchase the space at a predetermined time and price. Should your location prove to be successful, we urge you to exercise the option to buy it. From personal experience, we feel that owning the land brings you much additional freedom to operate your business. Of course, it can also cause you many sleepless nights!

BUILDING AND GROUNDS

ADVICE FOR TENANTS

BE A SMART NEGOTIATOR

1. Research rents and common area maintenance charges (CAMs) in your area—research anything that will help in the negotiations.
2. Set limits on what you'll pay based on your break-even point and your plans.
3. Leave a lot of time for negotiation, so you'll have room to dicker. Smart negotiators give themselves months to conclude favorable agreements, rather than being pressured into making hasty decisions that could mean trouble later.
4. Tell the landlord what you want. If you don't ask, you won't get it.
5. State your case, giving the facts and their impact on your restaurant. Armed with information, a restaurateur can tell a landlord what market rent and CAMs are, and ask that the rent be reduced by several dollars a square foot to be more competitive.
6. Recognize that all leases pose risks. Decide which are acceptable to take and which aren't. Understand that the landlord is also taking a risk (investing thousands of dollars) and wants some assurance that he or she will be properly compensated.
7. Use longevity in business as a bargaining chip. It brings clout to the negotiating table.
8. Include as many options and escape clauses as possible in your lease to increase flexibility.
9. Seek lower rent in the first few years of the lease, to give your restaurant a fighting chance.
10. Be prepared to make concessions for concessions. If a landlord lowers the rent, you might add a few years to the lease. One landlord lost $45,500 in future rent from a tenant, who moved out when they couldn't close a $3,000 gap during negotiations.
11. Get professional help before signing a lease. In fact, bring in an attorney well before you sign. It can produce benefits during negotiations.
12. Make final decisions yourself. Don't leave it to lawyers. They should advise you, not decide your course of action.
13. If you can't make the deal work, walk away from negotiations.

Source: Reprinted from *Restaurants USA* with the permission of the National Restaurant Association.

SUMMARY

At the heart of any successful foodservice establishment is a creative, marketable concept. The idea will differentiate you from your competitors, and create a mix of variables that will encourage your customers to return.

Who are today's customers? They are aging; they are culturally diverse; they expect quality and variety; and they are always pressed for time.

How can your restaurant serve these people, and how will it be different? Begin by creating a mission statement that defines your concept and target market. Then outline specifics: the name of the establishment, menu, pricing, location, design, and decor. To fine-tune your concept, visit restaurants that you enjoy, as well as those that will be your competitors. In your visits, try to pinpoint exactly what the owners do well and what needs improvement.

You will need to develop both market and financial feasibility studies for potential investors. A market study analyzes the overall area, the competition within a 5-mile radius, and the demographics and psychographics (physical and mental characteristics) of potential customers. Sophisticated types of research are now available that can tell you how much money customers in your market spend dining out, and how saturated the market is with any particular concept. The financial study projects how much money it will take to start the business, and what kind of profit can be expected.

One of the most important considerations in the future success of your restaurant is site selection. Pay attention to zoning restrictions, access to adequate utilities, proximity to both a customer base and a potential workforce, and previous ownership. Also examine both lease and purchase options for the site. Hire experts (designers, architects, foodservice consultants) before mortgage or lease paperwork is ever signed, for their opinions on whether the space can be adapted to what you have in mind, and what that will cost.

STUDY QUESTIONS

1. In your own words, define the term "concept." Give examples of restaurant concepts you have visited in your own city.
2. Using examples from your own dining-out experiences, list and briefly discuss four components of a restaurant's concept.
3. When doing market research, what's the difference between demographics and psychographics?
4. Let's do some simple figuring. At your own "Chez Restaurant," we have 200 seats. If each seat is occupied two times at lunch and one time at dinner, and the average check is $10, how much money do you collect in one day?
5. Why is it important to do both financial and market feasibility studies? What are you trying to prove when you do them?
6. What is a restaurant cluster? Is there an example of one in your town?
7. Name three of the long-term factors to consider when buying a restaurant site and working with designers and building contractors.
8. List two of the positive—and two of the negative—factors in restaurant site selection. What makes them good or bad?
9. If you had the choice, would you lease or buy a site to start your foodservice business? Explain your decision.
10. Define the terms "base rate" and "percentage factor" in a restaurant lease agreement. Give examples of each.

A Conversation with . . .

Larry Forgione

PROPRIETOR/CHEF, AN AMERICAN PLACE
NEW YORK, NEW YORK

Larry Forgione has made his culinary reputation as a creator and champion of "New American" cuisine. Among his many awards: the Culinary Institute of America's 1989 Chef of the Year, and an inductee into the Fine Dining Hall of Fame. Forgione began his career at London's Connaught Hotel. After winning medals for innovative cooking in both Britain and France, he held several executive-chef positions at New York's *El Morocco, Regine's,* and the *River Café* in Brooklyn.

In 1983, at age thirty, he opened *An American Place* in New York City and, eight years later, the *Beekman 1766 Tavern* in Rhinebeck, New York. Forgione is also a frequent contributor to *Cooks* magazine, a culinary adviser to American Airlines, and cofounder of American Spoon Foods, a specialty foods company.

Q: You graduated from the Culinary Institute of America before you started your career. Do you think it is critical that chefs today go to the CIA or have a so-called "orthodox" educational background in cooking?
A: I'm not sure it's critical that they do. I think it would certainly be a help to anybody who could afford the time and money to get the experience of going to a structured culinary school. You get to see a very broad picture of what the industry is about, what kitchens and cooking and table service is about, and how it all works together. But I certainly don't think you have to go to a culinary school to be a fine chef. You can also work your way through kitchens, learning completely on the job.

Q: You began your career in London and traveled quite a bit, and many other famous American chefs have also worked in Europe. How important is it to have international experience?
A: Again, I don't think it's absolutely necessary. But all experience helps build the foundation for your life's pyramid. I would recommend that anybody who has the opportunity to go to another country to experience the food and culture should certainly do it. I was very happy that I did. I learned a great deal while honing the skills I learned at culinary school.

Q: So much of creating a successful restaurant involves the design details—like where the utilities should be placed, whether to install a pass-through window, or where to put the restrooms. How do you make all these "nonfood" types of decisions?
A: First, you have to decide on the concept of the restaurant. Generally, it's influenced by neighborhood or location. Is it an area with offices where people will be coming down for cocktails before going home, or is it in a neighborhood where people are returning home from their offices? All of those little factors come into play determining your concept.

You mentioned the pass-through window. I prefer it, because whenever you can create less traffic, it's better. Here at An American Place, you have to come into the kitchen to pick up the food, but we have all the food fed to one spot on the pickup line

instead of five or six places so people are not running into each other all the time, and also to ensure that the chef and expediter see everything that's coming out of the kitchen.

Deciding where to put utilities or equipment is all part of the process of being actively involved in laying out your kitchen. You need to speak with engineers who know how to lay it out, how much power you need, and what's the proper flow. I never had formal training in utilities, but it would have come in handy. It behooves you to figure out how to fix a salamander on your own without waiting for the repair person to get there, or at least figure out a way to work with it until it can be fixed! It's helpful to know as much as possible about the tools of your trade.

Q: If those classes had been available to you in school, would you have bothered to take them? Do you think it's necessary to have formal training in these areas?

A: It depends on whether you have designs or plans to open your own restaurant. I mean, there are certainly engineers you can work with in kitchen design who have all that information and can take care of all that for you. You need to understand it, certainly, but not necessarily be the expert.

Q: You are sometimes called the "Godfather of American Cuisine." How did you come up with the concept for An American Place?

A: An American Place promotes the use of American products and highlights the bounty of America, and that was the concept—to create a restaurant and a style of food based on American ingredients and some sound American culinary concepts. This idea has been my life's work for the past twenty years. It came from a sense I got growing up on my grandmother's farm, being familiar with farmland and the wonderful, farm-fresh foods. Traveling abroad, I noticed there was so much more emphasis on that than in the United States. I felt it was wrong that there wasn't more emphasis put on the growing and the love of food, so I decided that was something I wanted to do when I came back [to the United States].

Q: You've had a successful long-term partnership with your investors. How do you handle disagreements, and what do you need to watch out for when you go into business with somebody?

A: When people go into partnerships and take on investors, the deal has to be sensible for everybody. If it's not sensible or is unreasonable in the beginning, it's never going to become reasonable. Barring unforeseen problems which arise naturally through life, good partnerships from the beginning usually remain good. The best way to formulate a business partnership is to base it on sound business principles. I'd add that sometimes when you get involved with family or friends, the decision-making process becomes too personal. Business decisions should always be based on business.

Q: What do you need to know to negotiate a lease?

A: You have to have a fairly good business plan and keep within a budget concept. Not all leases work. If rent is set at 6 to 8 percent of your gross sales and it turns into 12 to 15 percent of your gross sales, you're going to have a problem with your bottom line.

Q: Are there any design or aesthetic elements that you think are particularly important in foodservice?

A: There are design elements I personally think are important, but I can't imagine anyone believing that if certain things exist, there's a sort of guaranteed success. I really like dining rooms to be open, airy, or spacious, with clean-looking lines. For me that means high ceilings, nice sculptured columns, and polished wood finishes.

Q: How do you keep the energy and love of cooking alive when you have to do it to such a degree that it might become drudgery?

A: You just keep changing and creating, doing different menus. There is a part of cooking in a restaurant that is focused on getting the food out on time, and making sure that things are the way you want them, and you have to learn an excitement from that to make it fun. When the restaurant is busy and the kitchen and dining room are clicking, the energy that rises in you becomes a stimulus rather than a burden. You're in the wrong profession if you don't get excited by that energy—the energy and controlled chaos of being busy, pans everywhere, people picking up and ordering. All the things that would be overwhelming to an outsider are the things that make it exciting to a person who is part of it.

CHAPTER 2

RESTAURANT ATMOSPHERE AND DESIGN

INTRODUCTION AND LEARNING OBJECTIVES

All the physical surroundings and decorative details of a foodservice establishment combine to create its ***atmosphere,*** or ***ambience,*** its overall mood. When people enter *your* new restaurant, do you want them to feel excited? Relaxed? Important?

In some cases, it is enough that a restaurant be clean and uncluttered. In this very competitive market, however, most merchants realize they have to sell a little "sizzle" along with the steak. Why all the fuss about design? The components of a well-designed restaurant do not go unnoticed by its customers. Although they may not be able to pinpoint specifics, any diner will tell you that attractive surroundings seem to make a meal better. Even in large, industrial cafeteria settings, there are small but significant touches that can contribute to a warm and inviting feeling—greenery, fresh produce displays, suggestion boxes for customer comments, inventive bulletin boards. No matter what your theme or price range, the idea is to make people feel welcome, safe, and cared for. In this chapter, we will discuss:

- Individual details that contribute to the atmosphere of a restaurant
- Hiring and working with design professionals
- How restaurant space is planned and subdivided
- Guidelines for restrooms
- "Facelifts" for older restaurants
- Guidelines for deciding whether to include a bar in your establishment
- Guidelines for choosing chairs and tables
- Mobile dining options—carts, kiosks, and the like

2-1 CREATING AN ATMOSPHERE

Marketing and design experts insist "atmosphere" is a key ingredient when people decide where to dine out. However, it is one of those terms that seem to defy precise definition. Think about it! Everything that is seen, heard, touched, tasted, or smelled is part of the atmosphere of any location. When you put it that way, even the other customers are part of the package. And how can you match the atmosphere to the customer if the customer requires a different "feel" for different meals?

One good rule for developing atmosphere is to provide a change of pace. When we go to the movies, for instance, we almost "shift gears" as we enter the lobby, then the theater, then prepare to watch the show. You should already have researched the needs and interests of potential customers in your site evaluation. Now ask yourself: What would be a welcome change of pace for *them*?

Some examples: Lunch in a bright, casual café provides a respite from the everyday office cubicle. Outdoor dining on an umbrella-covered patio beats the heck out of a windowless skyscraper. And a glowing fireplace in a corner of the room is incredibly inviting when the weather is cold. Can you think of others?

The development of a successful design begins with a firm concept for the restaurant. You have to actually have an image before you can decide how to convey and promote that image. This applies whether you are opening a brand new eatery or repositioning an existing concept to build business. If you're unclear how to proceed, reread Chapter 1.

Today's dining designs focus on making a space comfortable, inviting, and entertaining, with a consistent theme throughout the experience. Efficiency, value, and convenience are the hallmarks of modern restaurants, reflected in their design. More kitchen activity is "out front"—with display kitchens, wood-burning ovens, and sauté stations that showcase the chef and his or her staff members for diners to watch. Restaurants with "celebrity chefs" have really learned to show them off. Fresh ingredients are displayed, and sensory details are not overlooked. The message? "Our food is fresh and we care about this!" It is no surprise that the dollar volume generated by a display kitchen is higher than a traditional, "hidden" kitchen. The guest experience is enhanced, the food quality improves, and profitability increases.

There are two distinct paths to successful dining design. One is to create the latest "hot spot," by making a trendy statement that makes the place a "must-visit" for sophisticated clientele. In this scenario, nothing is unimaginable if the budget will allow it. The second path is to create a dining experience that is unique but still has some "staying power." Design in this case will include simple, classic details that will still look good in five years. Regardless of the path, the new watchword in restaurant design seems to be "simplicity." Instead of cluttering walls and shelves with memorabilia, today's destination restaurants are more like public buildings, clean and unobtrusive. Furnishings are attractive, comfortable, and functional. Use of light, fabrics, and forms has replaced the clutter, putting the emphasis on the food rather than the decor.

Another influential factor is the target market for the concept. Will it be a gathering place for singles? The ultimate power lunch spot for the business crowd? A haven for families? An escape for empty nesters? The design should be developed to suit its target market. Examples: The choice of large, round tables always leaves room for a latecomer to join his or her dining party. A wide aisle, stretching from the entryway to the bar, spotlights people as they enter and creates an impression of bustling movement. By elevating the bar or raising the tables at the perimeter of the room, guests are provided with a better view of other tables and activity.

The menu is an important tool for use by the designer. The food selections, and even the type of paper used, can complement the decor of the restaurant.

Working with a Design Team

The design process is very much a team effort. The composition of the team will vary depending on the capabilities (and budget) of the owner and the requirements of the project. For our discussion, the team will consist of:

- An architect (with engineering, mechanical, electrical, and HVAC professional advisers)
- An interior designer
- A kitchen designer
- A foodservice consultant
- A general contractor, with construction crew (builders, plumbers, painters, electricians, roofers, etc.)

More complex jobs may include other types of specialists (lighting, landscaping, acoustics, kitchen exhaust consultants, etc.).

The foodservice consultant guides the owner through the conceptual development of the restaurant, defining the concept, researching its applicability and financial viability in a chosen market, and determining menu and wine list needs. The architect then coordinates the building plans to execute the concept. The designer provides color schemes and suggests materials for floors, walls, and ceilings, as well as choices of fixtures and furniture.

A foodservice consultant should be hired if there's any uncertainty about how to proceed with a project. Often, if a person has money and wants to be involved in the restaurant industry but doesn't have any hands-on experience, the consultant is hired to help brainstorm concepts, conduct research, analyze sites—a sort of "personal trainer" for the would-be owner. However, the owner should not abdicate the decision-making responsibilities, but rather should listen carefully to the advice for which he or she is paying. When hiring any kind of consultant, be sure you understand the charges for their services, and exactly what is (and is not) included in the fees.

Like any good partnership, the team members should support one another's strengths and curtail one another's excesses. For the owner/manager, this may mean reminding the designer of the real-life intricacies and practicalities of running a restaurant, and nixing those ideas that simply don't lend themselves to a smooth-running operation. The owner/manager is also far more aware of budget constraints and can prioritize the team's list of ideas, deciding which can be afforded now and which must wait until later.

The professional team contributes creativity, expertise, knowledge of industry trends, and familiarity with the construction and decorating crews it will take to get the job done. They can save time, and often get better deals on furnishings, fabrics, and other items from suppliers. However, the team's most important contribution may be objectivity. Often, the new restaurateur is so captivated by an idea that he or she loses the ability to judge accurately what will work best for the dining public, relying instead on personal tastes or convictions.

Together, they can manage to incorporate both "big picture" elements and "small picture" touches that reflect the personalities of the owner and/or chef—but only if there is accurate and timely communication among the team members, who all must be working under the same assumptions and heading in the same direction. In the interest of running the team smoothly, some written direction is required.

The ***design program*** is a document that details all the criteria and assumptions on which the restaurant design will be based. From entryway to receiving dock, it should specify the equipment and flow patterns necessary to operate the business. The design program will change over time, as the project progresses.

A *time schedule* is key to successful completion of this type of project. Start dates and deadlines, as well as assignment of responsibility for each task, should be put into writing and shared with all team members. *Planning meetings* should be part of the schedule, each with a preset agenda to keep everyone on task. Not every team member must attend every meeting, but accurate and detailed minutes of the meetings should be sent to all.

A separate contract for each company or individual consultant involved in the project, specifying the terms of their involvement and the fees they charge, is a requirement. (This makes an attorney an ancillary, but very necessary, part of the design package.) During construction, most projects seem to take on a life of their own. Inspectors don't show up when they're supposed to. Weather causes delays. Deadlines aren't met and costs exceed budgets. There are all manner of hurdles and challenges. Communication, coordination, trust, and respect for the other team members—and all agreements in writing—will get you through the rough spots.

Because this interaction is so intense and important, choose your design team members carefully. Visit other restaurants they have worked on, and talk with other owners if you can. The designer's personality is especially important. Choose one who listens to what you are saying and is excited about the project. Of course, just as there are celebrity chefs, there are celebrity architects and designers who specialize in creating dining spaces. You might hire a newer firm, or someone who has been known for home decor, to get professional results with lower fees than those of big-name design stars.

When all is finally in place—even down to the chairs, tables, and linens—it is helpful to stand back periodically and evaluate the final design. Ask other trusted associates to do the same. Will it impress the guests the first time they come in? Is it comfortable enough to prompt them to return? Why not ask the guests themselves? The design process is all-consuming at first, but it should not be shelved after opening day. Constant evaluation and modification are part of a good design. Keep in touch with your design team and ask them to come in periodically with recommendations for "updates" and news about trends. Look at your competitors and listen to your customers.

A few restaurateurs prefer to tackle the design process themselves, without the aid of a professional. They believe a restaurant's personality is best communicated to the guest by the owner. Those who do it themselves say the chief advantage is that they control the process. And there are feelings among some long-time foodservice operators that designers "decorate," with little concern for comfort or functionality; that they're more concerned with how something *looks* than how it *works*.

One caution for do-it-yourselfers: If a menu item—or even a whole menu, for that matter—doesn't work out, it can be changed relatively easily. But what about wall treatments, carpeting, or furnishings? If they don't work out, change is expensive and inconvenient. It may be easier in the long run to pay for the services of a designer. Just resolve to stay as involved as possible in the process, on every level. Remember, the architect creates, the contractor builds, and the designer designs the restaurant—once. The operator must run it for years to come.

Creating a Comfortable Atmosphere

There is some debate about exactly how much impact the look and feel of a restaurant actually have on its success as an eating place. After all, the most important elements are food and service. However, the look and feel of the restaurant are what invite people to go in.

Often, you'll hear the word "comfort" associated with atmosphere. Of course, this means different things to different people. The temperature of the room, the style or padding of the chairs, personal preference for booths instead of tables, a sense of privacy or openness, proximity to other diners who smoke, and accessibility for patrons with physical disabilities are just a few common concerns. One thing is certain: Discomfort is somehow attributable to a poorly designed atmosphere.

Let's take privacy as an example. Is it important that your potential customers "see and be seen"? If this is a priority, then seating them in a dark corner booth will make them feel slighted. Do you want people to choose your place for intimate, romantic getaways? In that case, a brightly lit table in the middle of a noisy room, with other diners at each elbow, won't be satisfactory. Perhaps you see the need for both options in the same restaurant. Without answers to crucial questions about what your customers want from their dining experience, it will be tough to manipulate your space to meet those needs.

ADVICE FROM CELEBRITY DESIGNERS

ADAM TIHANY **(Le Cirque 2000 and Jean Georges, New York City): Tihany believes design can play the most important role in a restaurant by setting the stage for food and service, and that the three elements are inseparable.**

TONY CHI **(Aqua, Las Vegas and Asiate, New York City): His passion is to integrate the local culture into the room design; the "human element" is an integral part of all dining designs.**

PAT KULETO **(Farallon, San Francisco and Buckhead Diner, Atlanta): Kuleto's resume reads like a guide to San Francisco's best restaurants. His goal is to match the atmosphere to the food. A diner should be influenced by the design of the space. Kuleto calls it "grand culinary choreography." Atmosphere should be timeless, yet comfortable.**

MICHAEL BOUDIN **(Myriad Restaurant Group, New York City): Great service begins with great design. This means creating a floor plan that saves steps and reduces stress for the wait staff. Elements like roomy service stations, more space for pastry chefs, some extra seating, and a dish room configuration that keeps waiters from colliding on busy evenings are all important to keeping things running smoothly.**

Another consideration is changing demographics. For instance, by the year 2010, about 38 percent of the U.S. population will be over fifty years of age. For these diners, setting and food quality will replace novelty in importance. Changes in patrons' eyesight and hearing should be reflected in design—one of older diners' most common complaints about restaurants in general is that they're "too noisy." Buffets are popular with the older generation; so are "doggie bags," since leftovers are perceived as a good value. Nutritional choices (and information about them) will be requested. The senior diner prefers the privacy of a booth to a table, and would rather wait on a busy evening in a spacious lobby or waiting area than a bar. The population is also becoming more ethnically diverse, with more African-American and Hispanic customers in most markets.

Don't underestimate the theatrical nature of the dining business! Putting wares on display leads to impulse buying, thus increasing sales. Whether it's wine, appetizers, or decadent desserts, think of ways to show them off. Speaking of impulses, one of the strongest is vanity—which has an impact on dining room design. Many diners like to see and be seen, especially in upscale eateries. Having a chef's table right in the kitchen can be fun and exciting for them. Offering a communal table for singles to meet and mingle is another option.

The bottom line with any of these factors is how they affect the guests of the restaurant. When you take them all into account, you see that the square footage of a room is meaningless as a measure of space. There is so much more to it!

The Redesign Dilemma

There will come a time when any restaurant's design becomes a bit tired and dated. This is a tough call for many owners, since they spend every day there and often overlook the telltale signs that a redesign is in order, to freshen up the place and make great food and service even better. Decor is usually not a priority, even in places that pride themselves on other aspects of business. And when a redesign is discussed, there are still logistics to be worked out. Sometimes the owner plods on rather than think about whether and how long to close the restaurant to do the work, as well as how to pay for it. Soon, sales start to erode. Experienced restaurateurs know how difficult and costly it is to get those customers to come back to an "older" restaurant when there are fresh, new ones to be tried down the street.

PLANNING THE DINING ATMOSPHERE

ATMOSPHERE AWARENESS

VISION. Exterior signage and facade; high or low light levels; bright or subdued colors; use of mirrors or partitions (either portable or fixed) to expand or reduce space; height of ceilings; menu design; artwork on walls; window coverings; positioning of tables.

TOUCH. Floors of marble, tile, carpet, or wood; chairs of wood, metal, leather, or fabric; seats cushioned or not cushioned; table linens or bare surfaces; chunky or dainty glassware; baskets or plastic plates, earthenware, or fine china; plastic tableware, stainless steel, or silver flatware; paper on which the menu is printed.

SOUND. Loud or subdued conversation; type and loudness of music, live or on the sound system; dishes being bused; kitchen or bar noise; hum of central heat or air-conditioning system; cash registers; street noise from outside.

SMELL. Aromas of baking or spices; rancid odor of fryer oil that needs changing; colognes of guests and staff; wood in fireplace; "new" smells of carpet, fabrics, or linens; restroom air fresheners.

TASTE. A cool drink; a fluffy soufflé; a crisp onion ring; a perfectly cooked steak; a hot curried dish.

TEMPERATURE. The thermostat setting of a room; body heat of guests and staff; heat from the kitchen or coffee station; breeze created by ceiling fans; breeze when seated directly above (or below) a vent or open window; direct sunlight or use of window coverings; hot food served hot; cold food served cold.

MOTION. The effort it takes to get to a table or chair; the way wait staff negotiate the dining room with full trays; the waiting line as perceived by passersby (and people still dining); activity within the dining room as viewed through windows; outside activity as viewed by diners.

How do you avoid this dilemma? Stay "in training," just as good athletes do. Some experts suggest a redesign every five to seven years, with an encouraging payoff—sales can increase an average of 15 percent!

Components of a relatively inexpensive makeover may include:

- Painting walls—both interior and exterior; include interior ceilings and restrooms
- Changing design elements, such as wall hangings and artwork; adding faux finishes, mirrors, brass, or copper accents
- Reupholstering chairs and booths
- Retrofitting with new light fixtures
- Replacing the carpet, tile, or other flooring
- Changing window treatments
- Adding revenue-generating space—patio, atrium, bar area, gift shop
- Adding live plants (or, at least, clean and rearrange the silk ones!)
- Adding big-screen televisions
- Changing the tabletop surfaces or linen colors

One of the first things to decide before an update is simple: Do you want to retain the regular customers, or attract new guests? These are two very different goals.

More elaborate redesigns focus on some element of display or merchandising, because customers react positively to the smells and sights of the cooking process. The idea of "mingling" with the kitchen appeals to them, since research tells us that even those people who enjoy cooking spend less and less time in their own kitchens. But the job doesn't have to be as extensive as creating a display kitchen. You might use the opportunity to correct a persistent problem that you didn't realize when you first opened the doors, by moving a wait station or rearranging tables.

It is easier and more affordable to clean, repaint, reupholster, and spruce up what you've already got than to move to a new space and start over. Again, a professional designer can offer several options, and a price range for each of them, using that all-important set of "new eyes" to look at the place. Ask the designer to pay special attention to guests' comfort and security.

The question often arises about how much to do at once, especially for an existing business that may not be able to afford to close its doors for a number of weeks to complete extensive renovations. A gradual upgrade over time, with projects budgeted in annual installments, is a practical alternative. However, the danger in this approach is that it can amount to applying a stick-on bandage to a gaping wound if you're not working from a coherent, long-term plan for overall facility improvement.

Rather than tend to whatever need appears to be screaming the loudest at the time, a restaurateur has to anticipate the inevitable "down cycles" of the business, and even undertake changes just when everything seems to be going right. You don't have to think big—as long as you think continuously, to keep competitors from chipping away at your business. If someone else is doing something better than you, your guests will notice.

2-2 TODAY'S TRENDS

Here are just a few of the ways today's restaurants can increase profitability and use space efficiently.

Home Replacement Meals. The "HRM" can be a lucrative option if, ironically, your target market is too busy to sit down and eat at your restaurant! The take-home counter, with a menu that focuses on easily packaged items for consumers to heat and eat at home, is growing in popularity as an "extra" offered by the traditional restaurant. It works well for both upscale and home-style cuisine.

HRM isn't much different than a supermarket, with diners stopping in briefly on their way home, so it must offer similar convenience in terms of location and accessibility, with ample parking close to the entrance and perhaps some of those "ten-minute" spaces for people popping in and out. The customer's ability to make a right turn into and out of the parking lot is a necessity, also for convenience.

Inside, food should be located within 30 to 50 feet of the entrance, attractively displayed to take the guesswork out of decision making and allow easy viewing of multiple choices. There should be no cross-traffic between the HRM and the regular sit-down dining area, since their two purposes and moods are so different.

Private Dining Rooms. These have become part of many midscale to upscale restaurants, and they're replacing people's homes or apartments as the site of intimate dinner parties and celebrations. This target customer wishes to entertain somewhat privately, and can afford to do so without the fuss of setup and cleanup. The host has all the pleasure of choosing a menu and inviting guests, with none of the mess or stress of cooking and cleaning! Unlike commandeering a large table in the main dining room, a private room gives the host control of virtually every aspect of the event—a personalized menu, selection of table settings and linens, music, wine choices, and so on.

The host probably won't consider this, but the restaurateur should—the private dining area is also a benefit to the rest of the guests who happen to be in the restaurant that night. Anyone who has been seated next to a large, loud, boisterous party knows how uncomfortable it can be. It's better to corral them in their own space.

Private dining rooms are profitable, too. When the menu is planned, there is a particular entrée, or perhaps two or three, for guests to choose from. The food is preordered and prepaid, so the kitchen knows how much to prepare and the manager knows how much labor to schedule. If a party is too small to make the private room profitable, consider adding a flat fee for the room rental.

Experts recommend the private room be a space in the dining area that can be walled off separately, or at least adjacent to the main dining room. They say rooms on other floors—a basement wine cellar or a second-floor balcony—don't do as well because people can't see them, so they require more marketing and word-of-mouth to stay booked. A few restaurants can afford to buy adjacent buildings to house their special-event rooms.

ILLUSTRATION 2–1 **The wine room at Mortimer's Idaho Cuisine, Boise.**
COURTESY OF GREG SIMS, TRI-DIGITAL GROUP, LLC, BOISE, IDAHO.

You must also consider the feasibility of operating a separate room. If there is not a separate kitchen, what happens when the private party's dinner order arrives, all at once, in addition to preparing meals for the rest of the dining room? Will service suffer in one room or the other? Is there adequate parking for private events? Can the building itself accommodate the noise level?

One terrific example of a successful private dining room is at Eleven Madison Park in New York City. This separate space—not tucked away, but on a second floor with a full-length glass wall—can seat up to 55 guests, who can see both the dining room and the outdoor park beyond (see Illustration 2-2). The private dining room has its own kitchen, bar, restroom, and elevator, a virtual "restaurant within a restaurant." It also features heavy sliding wooden panels that can split the room into two halves to hold separate, smaller functions.

Multipurpose Spaces. A trend in front-of-house design is to stretch the usefulness of every square foot of space beyond the rather limited scope of one or two day-parts (that is, lunch or dinner, or lunch *and* dinner), making the facility inviting in ways that prompt guests to visit anytime, day or night. Design experts break the mealtimes down in terms of guests' needs: relative solitude in the morning; speed and convenience for lunch; comfort in midafternoon; value in the evening; and a "scene" to participate in for late-night. The Globe Restaurant in New York City's Gramercy Park is a fine example of maximum functionality:

- The day begins with self-service breakfast items like muffins and coffee, ordered at a circular steel counter.

ILLUSTRATION 2-2 **The private dining room at Eleven Madison Park in New York City.**
COURTESY OF ELEVEN MADISON PARK. PHOTO BY ROGER SHERMAN.

- At lunch, the same counter becomes a buffet bar where cooks dish up soup, panini sandwiches, and pastas.
- At dinner, the counter is transformed once again, this time into a raw bar where shrimp, clams, and oysters are displayed with bottles of chilled white wine.
- Late at night, the lights are turned down, bar stools emerge from on-site storage, and the counter area becomes a coat-check stand. The expediting kitchen and shelves that held soft drinks and prepared salads earlier in the day are hidden from view behind vertical blinds and shades.

Outdoor Patios. There's nothing more pleasant than dining outside on a lovely day, and adding a patio has a number of benefits. First, it can expand seating capacity (for at least part of the year) and it can be a good way to use otherwise wasted space. If it can be seen from the street, the patio adds liveliness and action to an otherwise sedate storefront. And it can attract a different type of customer, the more casual drinks-and-snacks and after-work crowd.

Patios bring their own challenges as well—notably weather, traffic noise, temperature control, and of course, insect control. Weather is the real wild card. Large tabletop umbrellas can shelter guests both from rain and too much sun, but if it's a real downpour, be prepared to reseat indoors.

Adjusting to overly warm days, put up the big umbrellas once again; install an awning (retractable or not) over the space; or outfit the patio covering with misters that spray a fine mist of water on guests. Depending on how well the misters are adjusted, this can be very refreshing and fun on a hot day—or it can make mascara run and guests feel as though they are fresh produce on a supermarket shelf! For cold weather, there are a number of attractive patio heaters (see Illustration 2-3) that use natural gas or liquefied petroleum gas (LPG). A trendy little wine bar in Boise, Idaho, The Grape Escape, puts monogrammed wool blankets at its outdoor tables on chilly spring and fall mornings, for guests to wrap around their shoulders or legs.

When it comes to pest control, outdoor diners have to learn to live with nature to a certain extent. Bees, flies, mosquitoes, and birds not only can be bothersome, but pose a sanitation risk. Two ways to cope are by keeping food covered as long as possible, and using decorative citronella candles on tables. Outdoor spraying is only marginally successful.

ILLUSTRATION 2-3 **To keep outdoor dining spaces warm during chilly nights, there are lots of patio heaters to choose from. This one is from Homestead Heat-Seekers in Glendale, California. Even if they are used strictly outdoors, always check with your local fire department in case of specific rules about their use, minimum distance from guests, and so forth.**
COURTESY OF TEECO PRODUCTS, INC.

Design Mistakes

A good summary of what *not* to do comes from writer Howard Riell, in the fall 1998 issue of the trade publication *The Consultant*. Here, we summarize Riell's excellent roundup of the most common restaurant design gaffes he has seen.

1. **Inconsistent ambience. Theme, space, layout, and flow patterns. Colors, materials, lighting, graphics, and artwork.**

Uniforms, linens, menus, dishes, and signage. All should complement each other. When they do, it implies great attention to detail. The big restaurant chains seem to have grasped this "centrality" of design continuity, but small restaurants, probably in their quest to cut costs, seem to treat visual aspects of the dining experience as an unnecessary expense.

2. Too many people involved in the decision-making process. It is good to have candid input and a variety of opinions in the design process, but there must be one person who drives the design program. Otherwise, you end up with a mishmash of ideas and skyrocketing costs as you experiment with them. Also, Riell adds, be sure to include the chef in decision making.
3. The target market is forgotten. This happens when the owner or manager defines the tone, menu, and concept based on what he or she wants, instead of what truly fits the location. This mistake is often made in suburban areas, where the customer base can afford an upscale concept but the restaurateur opts for a middle-of-the-road concept.
4. Inadequate space between tables. It is not unusual to underestimate the level of intimacy in a given concept, but you should leave at least 3.5 feet between tables so servers can do their jobs comfortably. Casual dining can cut the space a bit, to a range of 24 to 36 inches, but for fine dining, there should be a full 4 feet between tables.
5. Traffic patterns are overlooked. This refers to the movement of people within the space, both guests and employees. Major "sins" include placing the kitchen too far from the dining room, and restricting movement into and out of the kitchen. Good service and comfort are the result of uncluttered circulation patterns, or *flow,* which is also discussed in Chapter 3.
6. Unrealistic budgets. There are no fixed budget guidelines for restaurants, which results in too much variation. It can cost anywhere from $50 to $60 per square foot to design a casual eatery, while a formal dining room may run $400 per square foot! Unrealistic budgets come back to haunt a project. Most major expenses are covered, but what about the so-called "soft costs"—insurance, permit fees, advertising? Some consultants suggest a contingency budget of 10 to 20 percent, just to cover possible cost overruns.
7. Cutbacks in nonrevenue-producing space. It is important to balance the idea of more seats in the dining room with sufficient space in the kitchen to produce the foods the menu promises. Municipal fire and building codes and ADA requirements (see Chapter 4) must be met.
8. Poor lighting. Lighting is much more than the need not to have darkness. A lighting mistake can actually be offensive to guests if the results are too harsh, too dim, or at angles that make the atmosphere uncomfortable. Pick a lighting style and stick with it, rather than use four or five different types of fixtures to achieve everything from walkway lighting to enhancing food presentation to making diners look good. And never light directly at eye level.
9. Offensive colors. The use of color impacts the way the food looks as well as the mood of the dining room. Blue feels "frozen in space"—warm hues, like reds or browns, are generally more inviting. Consider colors that either invigorate or relax the dining space.
10. Forgetting the future. Restaurants need a plan for growth, so they can respond to it without major expense. You can only do things like expand the dining area, add a banquet room, or build a private dining room if the original design was created to allow for this at a reasonable cost.

Riell's additional advice: Don't neglect storage as a space need. Think about overhead storage for some items, or storing them in another location entirely and bringing them on-site as needed. Don't procrastinate about making decisions—especially about items like furniture that must be ordered in quantity and well in advance. And finally, if you're going to hire a consultant, be open-minded and willing to listen to his or her advice—otherwise you're wasting the consultant's time and your money.

2-3 FOLLOW THE GUEST

Now that we know something about the components of atmosphere, let's follow a hypothetical guest, Mr. Smith, as he approaches a restaurant. His reasons for choosing this particular place, even though they may be "spur of the moment," require some serious judgments on his part, as you'll see. They may include:

- **Convenient location (it's close to work or home; he was so hungry he didn't want to drive any more)**
- **Outside signage (it's clearly a Tex-Mex place, and he loves Tex-Mex food; lunch special sign indicates a bargain or sale price)**
- **Parking (an empty space right up front; on the convenient side of the street for him; doesn't look too crowded and he's in a hurry)**
- **Architecture (resembles a fast-food Tex-Mex place he already likes; building and grounds look clean and inviting)**

This impromptu lunch search is almost like wandering around a library after deciding you want to read something. The cover of the book will entice a reader to pick it, out of rows and rows of other books!

First Impressions

Both quick-service and table-service chain restaurants rely heavily on exterior appearance as marketing tools. McDonald's "golden arches" provide instant recognition. Steak and Ale restaurants or TGI Friday's, although not all alike, have similar exteriors that make it impossible to confuse them with any other chain.

For independent restaurants, the only warning is that it is often difficult to inherit a once well-known location and mask its former identity without costly reconstruction. To simply repaint the distinctive, pointed roof of an International House of Pancakes, for instance, and open a Thai restaurant there instead is not a wise decision.

Mr. Smith approaches the front door. Remember, he already has plenty of impressions from what he's seen in the parking lot: the other customers' cars, front signage, the landscaping, whether the lot is clean or littered. The next thing he encounters is the door itself. Is it heavy wood with beveled glass insets? Clean, clear glass? A sleek, colorful laminate? Even the doorknob or handle is a key to what awaits him inside. Whatever the case, the door should be easy to open. Mr. Smith enters the restaurant.

Entryway Etiquette

The entryway provides another important clue to the character of the eatery. Although some restaurants have eliminated reception areas and waiting rooms because of space constraints, others use them wisely. Here's how:

- **Pay phones, newsstands with free local publications, or cigarette machines are placed here.**
- **Cash register and hostess stand are located here.**
- **In quick-service, the front counter is here.**
- **Menus and daily specials are displayed here.**
- **Waiting diners sit or stand here.**
- **Wines or prepared desserts are displayed here.**
- **Raw foods are displayed, perhaps in glass cases: whole salmon on ice, fresh meats or pasta, lobsters in a bubbling tank.**
- **If a waiting or serving line is the norm, make it very clear where this line begins and ends, with signs, railings, and partitions.**

Whether it is the prelude to an elegant dining experience or the start of a fast-food line, the waiting area should be very clean, properly lit, and temperature controlled for the guests' comfort.

Oops, we almost forgot about Mr. Smith. There he is, waiting (rather impatiently) for a table. He says he wants the "best seat in the house." Don't they all?

Dining Room Layout

Restaurateurs like to think every seat is the "best," of course. When designing the dining area, a well-planned scheme carefully shapes the customer's perception with the following components:

- Table shapes, sizes, and positions
- Number of seats at each table
- Multiple floors, steps, or elevated areas of seating
- Paintings, posters, or murals
- Type and intensity of lighting
- Planters, partitions, or screens
- Attention to sight lines, to block any undesirable view (restroom, kitchen, service areas)
- Muffling of distracting noises (clattering dishes, outside traffic, or construction)
- Placement of service areas (coffee stations, dirty dish bins, etc.)

Each of the preceding considerations plays a role in creating the ***flow*** (or *flow pattern*) of the restaurant. The flow pattern is the process of delivering food and beverages to customers. That is, it involves the methods and routes used to transfer items from the kitchen, to the dining tables, and, finally, to the dishwasher. How well can the waiters manage their full trays of food? How far is it from the kitchen to the dining area? Does anyone have to hike up and down stairs? How hard is it for Mr. Smith to negotiate a path to his table on this busy day? When seating guests, does the hostess sometimes seem more like a traffic cop?

When planning flow patterns, customer safety should be paramount. Remember that most of us tend to walk to the right of oncoming persons. Look at both the maximum seating and the average time customers will spend at a table. Generally, the faster the turnover, the greater the need for clear flow patterns that do not cross. Conversely, if dining is to be leisurely, flow should be designed mostly to make the wait staff seem as unobtrusive as possible.

When laying out your dining room plans, you might think of the room as a neighborhood and the flow as the major streets in that neighborhood. Avoid traffic congestion, and everybody likes living there.

One important design rule is that tables of different shapes and sizes should be mixed to create visual harmony. A lineup that's too orderly evokes more of a military mess hall image. And how close is too close when determining table position? Generally, allow 15 square feet per person. More specific seating guidelines are listed later in this chapter.

Restroom Facilities

Back to our fictitious Mr. Smith. At this point, he has been seated, has admired the surroundings, and has ordered his lunch. Now he goes into the restroom to wash up. The single most critical public perception of a restaurant is that if the restroom is clean, so is the kitchen—and, of course, the reverse is also considered true. All too often, especially during busy times, the restrooms are neglected because no one on staff has been given specific responsibility for checking on them.

Although customers don't spend more than a few minutes in the restroom, this short time impacts the rest of the dining experience. So why does the concept of atmosphere seem to be left outside the restroom doors of so many restaurants? Your guests' expectations are simple: cleanliness, privacy, and comfort. So why not pay the same attention to decor here as you do in the dining area, with plants or attractive wallpaper or artwork reflecting your theme or mood? Temperature control and lighting are critical here, too—select warm, incandescent lighting or color-corrected fluorescents that are bright without being harsh.

Placement of restrooms in your building usually depends on the location of your water lines, so they are often near the kitchen. However, guests should never have to walk through the kitchen to use the restroom. Another pet peeve of some diners is having to wait outside a locked restroom door, in a narrow hallway or in view of other guests. If at all possible, divide your restroom into a small waiting area with sink and mirror and at least one stall with a locking door. Most restaurants can't spare enough space for restrooms to be lavish, but they should at least be roomy enough that guests don't feel uncomfortably cramped.

Local health ordinances may require a specific number of toilets and urinals, depending on your square footage or total seating capacity. Guidelines to ensure that your restroom facilities are accessible to persons with physical disabilities are covered in Chapter 4.

Luckily, a few cosmetic touches can add to the perception of cleanliness in a restroom. The use of easy-to-clean tile (smaller tiles on the floor, larger ones on the walls) is one option. Stark white is not a good color for walls because it's hard to keep clean; conversely, dark colors are not perceived as clean, even if they are, so it is best to pick a lighter paint shade. The stall partitions don't have to be dull gray metal when there are several sturdy options in attractive, solid colors that discourage scratched-on graffiti.

Other amenities to consider: Exhaust fans should operate whenever the toilets are occupied; soft incandescent lighting is preferable to harsh fluorescent lighting. Make sure the stall doors properly close and latch. Provide hooks on stall doors, to hang coats or handbags. Install high-quality, undistorted mirrors, including full-length mirrors. Provide privacy screens between men's urinals. Provide well-stocked (and working) vending machines for women's sanitary products. Complimentary mouthwash, hand lotion, or hairspray can sometimes be found in the restrooms of fine-dining establishments. If smoking is allowed, provide ashtrays or cigarette urns to prevent fires in other waste receptacles.

In family-oriented eateries, consider diaper changing tables in restrooms for both sexes. (The String Bean, a family-style restaurant complete with playroom in Richardson, Texas, provides spare diapers and wipes in its restrooms.)

Finally, assign employees to check the restrooms regularly, to wipe up spills on sinks and countertops, empty trash cans, and replenish supplies of hand soap, toilet paper, and paper towels.

2-4 MOBILE FOODSERVICE OPTIONS

Being part of a food court, a street vendor, or an outdoor-event food provider is one way an existing restaurant or institutional foodservice company can expand economically into nontraditional locations where potential customers congregate.

Whether it's called a *mobile merchandiser, vending cart, kiosk, food center,* or *concession trailer,* it is designed to take food to customers when time constraints and/or distance prevent them from patronizing the primary dining facility. The first mobile merchandisers were little more than tables on wheels from which prepackaged foods and beverages were sold. Today, they can be custom-ordered with food-holding compartments, fryers, warmers, induction cookers, microwave or pizza ovens, refrigerators or refrigerated display cases, beverage dispensers, cash registers, storage space, power and water lines, exhaust fans, and vented hoods.

Your mobile merchandising choice (and cost) will depend on answers to a few questions:

- Will service take place on one, two, or three sides of the cart?
- How extensive will the menu be? How many hot (and cold) items?
- Will the operation require a staff, or be self-service?
- Where will money change hands, and cash be safely stored?
- Where will the cart be stationed—mall, indoor arena, ballpark, carnival, food court?
- Which meals (and what times of day) will the cart serve?
- How will wares be displayed—photos and menu boards? Display cases?

Carts vary in length from 3 to 10 feet. Most fold down to some degree, making them relatively easy to secure after hours and to store. A handy note: If you order your unit with a pop-out heater or cooler

ILLUSTRATION 2–4 **There is a mobile merchandising option for almost any type of food preparation!**
COURTESY OF SUPREME PRODUCTS, WACO, TEXAS.

in the base, it can do double duty as a warmer or refrigerator, depending on your needs. Other experts recommend you purchase carts that will produce as much heat as possible, since doors of the heated compartments will be opened and closed frequently.

Some carts are modular and can be purchased in fit-together units—a cold-food module, a hot-food module, or a coffee module that can work together. These can often be custom-made to fit a particular space.

Indoor carts are usually electric, running on 125/250-volt power; outdoor carts may be equipped with rechargeable batteries or a small propane generator. Propane is the fuel choice for many fire marshals, who consider it the safest, but it is also the most expensive. A 22-volt, 50-amp propane generator can cost as much as $4,000, and a single propane tank provides about three hours of heat. You must decide early in the buying process whether you want your unit to be completely self-contained or a plug-in.

What makes the carts mobile are the wheels, and here too there are options. For indoor use, smaller carts should have hard neoprene casters that won't mar floors. For outdoor use, operators can select foam-filled or air-filled tires. Foam-filled works well for carts that move often, and keeps the tires from getting flat spots in them; air-filled tires work best on rough surfaces and are appropriate for carts that remain stationary most of the time.

Other construction details: Awnings should be retractable so the person pushing the cart can see where he or she is going with it, and the awning should be tall enough (8 to 10 feet) so that tall customers won't hit their heads on it. Durability of awning fabric is important, since it can dry out, fade, and tear over time.

A cart put together with screws will cost more than one put together with rivets, but it's worth the extra expense. A loose screw can easily be tightened; a loose rivet has to have a new hole drilled. Be certain there is enough storage space on the cart—a good rule is 4 cubic feet of storage for each 5 feet of cart length. And the ability to lock the space is also important, since being able to secure the contents reduces time and labor as well as theft potential.

A kiosk is set in a fixed location. It is more like a small structure than a cart, with the employee typically sitting inside the structure. Security is a major concern, since the kiosk cannot be rolled behind a locked door after hours. Make sure yours can be covered and/or locked when not in operation, or you'll need to make arrangements to have it guarded. Kiosks come with the same range of equipment options as carts and are large enough to accommodate ventilation for more elaborate restaurant functions, like grilling. They can be fitted with hoods, conveyor ovens . . . just about anything you'd need to turn out a specialty product.

One notable observation: Kiosks have a way of seducing their operators into adding a wide variety of items for sale, to capture more customers. Remember, this also requires more highly skilled labor to staff them. Some restaurants use a cart within the primary dining facility, to sample and test new menu items, to serve a particular type of food (such as desserts or specialty coffees), or to handle overflow crowds. When tastes and/or traffic changes, the cart or kiosk can be moved.

At this writing, the major chains are likely to use carts in places like airports, colleges, and hospitals, while independent operators seem to choose kiosks more often. One is not necessarily better than the other; they meet different foodservice needs.

A final option, for vendors who serve food at ballparks, county fairs, concerts, and other outdoor venues, is the concession trailer. Riding on 15-inch tires, it can be pulled behind a vehicle and set up with stabilizing jacks at each site of operation. Like any other trailer, a concession trailer has its own brake and signal lights, axles, and brakes. Inside, the equipment can run on either gas or electricity.

In most jurisdictions, there are strict rules for mobile foodservice that apply to safety, cleanliness, and appearance. Many cities require that mobile units be dismantled daily for cleaning and sanitizing. Some communities (Los Angeles and Orange County, California, for instance) require that mobile units have 50 percent more space for wastewater than fresh water, to prevent overflows. There must be a safe way to drain the water created by melting ice rather than keep food sitting in it. Other sanitation requirements include a separate hand-washing sink—and for food preparation, a three-compartment sink deep enough to submerge all utensils—so a water supply is critical. The cart can carry its own water, or be capable of hookup to a remote water supply. The cooking equipment on the cart must meet NSF International and Underwriters Laboratory standards for safety and sanitation (more about these in Chapter 8).

Food safety is a major consideration. If you store raw meats or vegetables, or serve perishable items or even cream-filled pastries, refrigeration is a requirement. Heated compartments and warming lamps usually require an electric power source. Some carts can be plugged in long enough to heat the food, then unplugged to move around, with storage in an insulated compartment that keeps the food safely warm for several hours. Excellent insulation should be one of the things you look for in mobile equipment purchases. A couple of food thermometers and knowledge of safe temperatures should be requirements.

Getting food to and from your remote location is another potential dilemma. You may need to invest in extremely rugged coolers, made of high-impact plastic or stainless steel. When making this decision, look at how easy the cooler is to clean, carry, and repair. With use, the gaskets around the doors and lids wear out and need replacing. Some coolers can be purchased with wheels, which is an advantage if they have to be hauled very far. Some have their lids on the top, others on the side. If the coolers will be stacked, side access is preferable. There's a portable option for every need.

Now let's consider the costs involved in cart and kiosk operation. Prices for standard or custom-made carts vary according to their sizes and the equipment they contain.

The most basic mobile merchandising equipment ranges in price from $1,800 to $2,400 for a simple setup to sell bottled drinks and packaged snacks. (In some cases, food manufacturers even supply the carts free of charge to restaurateurs who will sell their wares.) Higher-end models can cost as much as $50,000 or $60,000 for virtual kitchens on wheels that can be linked to others to form a larger unit. However, sales can top $1 million for profitable cart locations. The promise of high rewards with relatively low investment and low risk has led some national quick-service chains to "go mobile" outside the United States. In international cities where real estate costs have soared out of sight, a cart or kiosk is a low-cost alternative that seems to work well.

2-5 TO BAR OR NOT TO BAR?

For many adults, a bar is their preferred gathering place—after work, on weekends, to mingle with friends or watch a major sporting event. Even if a bar is considered chic or upscale, it is still expected to be a friendly and comfortable place, with a style that promotes interaction between patrons, servers, and bartenders. Today's customer doesn't go to a bar "to drink." Instead, the combination of atmosphere, entertainment, service, food, and beverages makes it a pleasant place "to be."

In some restaurants, the bar also serves as a waiting area for diners. Most restaurants with bars place them near the entrance, to avoid bar patrons having to maneuver through the dining area to get there. Chili's, the Dallas-based casual dining chain, has its bar areas front and center, with frozen margaritas in stout, frosted mugs to beckon thirsty guests.

You'll also have to think about where to put those guests who don't drink or who don't wish to sit in a bar while they wait for tables. You certainly don't want to offend them or make them feel pressured to buy something, but a menu of nonalcoholic drinks is a good idea.

In quick-service and cafeteria settings, having a bar is not usually a consideration. However, the growing trend is to provide a beverage area, away from the food counter, where customers can fill their own drink cups. This process speeds up service, by allowing the counter personnel to concentrate on food sales and preparation. How big will your beverage area be and where will it be located?

As for a full-fledged cocktail lounge, make the decision—whether "to bar or not to bar"—based on how much profit you expect liquor sales to generate compared to the rest of your business, and how much space you have to achieve the atmosphere you want to provide (live bands, a deejay, big-screen televisions, or other types of entertainment).

At T.G.I. Friday's, the cocktail area is a casual gathering place, attracting a young crowd around its central, island-style bar. Even on a slow dining-out night, the bar area may be hopping. In a hotel, there might be different expectations of the bar—or more than one bar for more than one type of patron.

Today's upscale bars are much like upscale restaurants—their theatrical touches set them apart from competitors and attract the "see and be seen" crowd. For instance, anyone who's been to Las Vegas has probably heard of Charlie Palmer's restaurant and bar Aureole, located in the Mandalay Bay Hotel and Casino. The showpiece at Aureole is the "wine tower," a glass-encased, temperature- and humidity-controlled wine storage area that rises from the restaurant's bar to a height of 50 feet (see Illustration 2-5). Circular staircases allow guests to meander around the perimeter of the tower as they make their way from the ground-floor bar area to the restaurant below. Wines are retrieved in the tower by "wine angels," lady wine stewards who are literally attached by cables to four pulley systems. They are propelled up and

ILLUSTRATION 2-5 **Aureole "wine tower" in Las Vegas.**

COURTESY OF AUREOLE AT MANDALAY BAY RESORT AND CASINO, LAS VEGAS, NEVADA.

PLANNING THE DINING ATMOSPHERE

PLANNING A BAR

The four prime considerations in planning bar space are customers, service, atmosphere, and efficiency.

CUSTOMERS. Who are they? The local office happy-hour crowd? Visiting businesspersons? Why are they here? To watch televised sporting events? For quiet, after-dinner conversation? Again, research your market.

SERVICE. Do people have to walk to a central bar and order their own drinks or are there cocktail servers? Will you offer a full menu, or appetizers, or just peanuts or popcorn in tabletop baskets?

ATMOSPHERE. Big and bright or small and cozy? Your choices of lighting and seating will determine whether the bar is used primarily as a waiting area or as a destination in and of itself. Will there be cocktail tables and overstuffed chairs or only a few bar stools? Is it located at the front of the restaurant or tucked discreetly in a back room?

EFFICIENCY. Can a bar coexist with other parts of your business? What are the flow patterns from bar to kitchen to dining area? Are there convenient outlets for utilities, storage spaces for bar equipment and glassware, adequate places for trash disposal? Will the music or TV noise in the bar distract restaurant customers? And is there enough room to set up "the perfect bar"?

down the tower's seven levels using computerized joysticks, like videogame players or jet fighter pilots! The tower holds 9,000 bottles of wine, with more housed in a separate cellar and in off-premise storage. The entire operation is controlled by a two-way radio system, not in view of guests.

The Perfect Bar

Every bar, no matter where it's located, how big it is, or how it is shaped, has three interrelated parts:

- The front bar
- The back bar
- The under bar

Following is a brief description of each.

The Front Bar. The front bar is where customers' drinks are served. The space is 16 to 18 inches wide, topped by a waterproof surface of treated wood, marble, or laminated plastic. Some bars have a 6- to 8-inch padded armrest along their front edge. The recessed area, nearest the bartender, is known as the ***rail*** *(glass rail, drip rail)* or *spill trough.* That's where the bartender mixes the drinks.

The typical bar is 42 to 48 inches tall—the optimum height for the bartender's work, as well as for leaning against. All bar-related equipment is designed to fit under a 42-inch bar.

The vertical front panel that supports the front of the bar is called the ***bar die.*** It shields the under bar from public view. If you're sitting at the bar and kick it, what your foot hits is the bar die. On the guest side, there is usually a metal footrest running the length of the bar die, about 1 foot off the floor. It's sometimes called a ***brass rail,*** a remnant of Old West saloons.

When selecting a bar, avoid the straight-line, rectangular model in favor of one with corners or angles, which prompts guests to sit opposite each other and visit instead of staring straight ahead at the wall. (Illustration 2-6 should give you one idea.)

Of course, the ultimate "conversation bar" is the island, an oblong bar in which the bartender occupies the center. People can sit all the way around it and see each other easily. However, island bars take up a lot of space. They are also the most expensive to build.

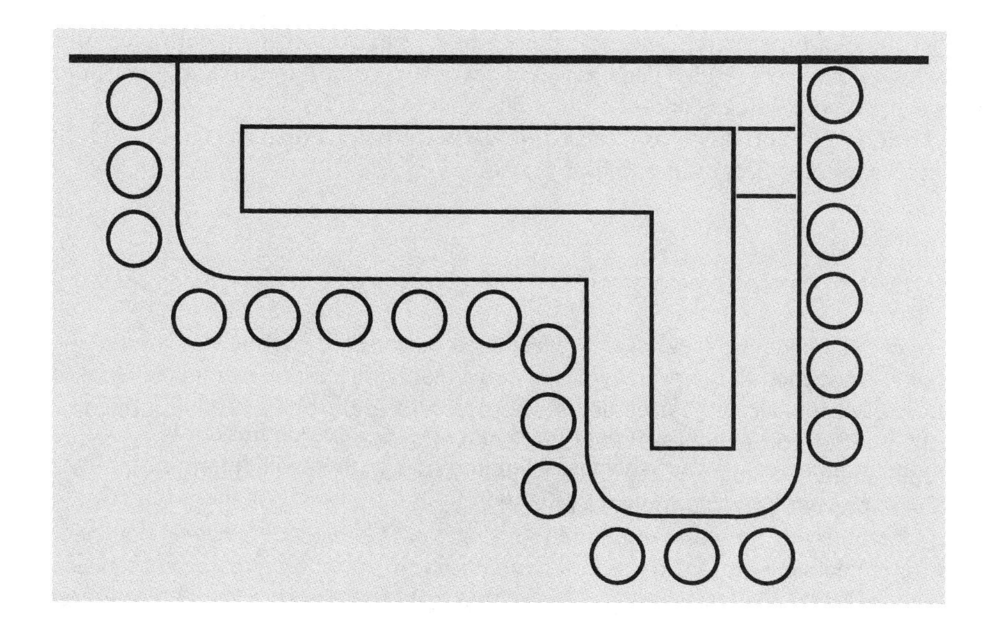

ILLUSTRATION 2–6 Think about what your customers have to look at as they sit at a bar. Designs with corners or angles prompt more conversation and allow people to "see and be seen."

A word about bar stools. Choose them for height and comfort, and allow 2 feet of linear space per stool. Because they're high off the ground, it is more comfortable to provide additional rungs for footrests. As with any type of chair, you've got plenty of choices!

The Back Bar. The ***back bar*** is the wall area behind the bar structure. It serves a dual function, providing both decorative display and storage space. Sparkling glassware and all the various brands of spirits are neatly arranged, usually highlighted by a mirror. This display is certainly a form of merchandising, but it's also the handiest place to store the bartender's most often-used supplies. The mirror adds depth to the room, while allowing guests to view others at the bar, as well as the action behind them. Smart bartenders learn to use the mirror to observe the scene without being noticed.

The base of the back bar is an excellent place for refrigerated storage space. The cash register or point-of-sale system is usually found at the back bar, sometimes built in. There are only 3 feet separating the back bar from the under bar, so when designing this space, choose doors that won't block the bartender's movements when open. Make sure there's an opening wide enough to accommodate the largest piece of movable equipment, in case it needs to be repaired or replaced. Otherwise, you may have to lift it out!

Back-bar design requires specific plumbing and electrical considerations. You need outlets where you'll be placing appliances, and plumbing for water supply, ice making, and drainage.

The Under Bar. The ***under bar*** is the heart of the entire beverage operation and deserves careful attention to design. Too often, a bar is selected because it looks good from the customer's point of view, without regard to the bartender's practical considerations. Is there plenty of storage? Will all supplies and fountain valves be easy to access and easy to restock? Will the bar area be easy to clean? Equipment and supplies must be arranged in a compact space, but being able to use them speedily is the primary concern.

The central point of the under bar is the ***pouring station,*** where you'll find the automatic dispensing system for carbonated beverages and juices. The system has lines running from individual tanks, through a ***cold plate*** at the bottom of the ***ice bin,*** to the push-button dispensing head (commonly referred to as a ***gun***).

Also in the pouring station are ***bottle wells*** and a ***speed rail***—both places to store the most often-used liquors and mixers. Blenders and other small bar tools are located in the under bar, too. Every bartender sets it up a little differently for speed and ease of use.

Storage needs include chilling space for some wines and refrigeration for beer, both in bottles and in kegs. Ice supply is another major concern. An ice maker is usually installed in the under bar, near a double sink where ice cubes can be stored to allow the ice maker to refill itself. Another option is to put the ice maker elsewhere and keep the ice bins behind the bar refilled. This is more labor intensive, but it frees up the space to install a small sink or dishwasher behind the bar.

You'll also require a couple of other sinks. The typical *bar sink* is 7 to 10 feet long, with four compartments, two drain boards, two faucets (which each swivel to cover two compartments), and bar drain overflow pipes. Finally, a separate *hand sink* is another plumbing need to consider.

Clean glasses may be stored upside down on a *glass rail,* on drain boards near the ice bin, on the back bar, or in overhead racks, grouped according to type and size.

Portable Bars

If your goal is to offer space for banquets, meetings, or receptions, a portable bar may be a workable alternative to a full-sized one. These modular bars are typically 4 to 6 feet long, built on casters with brakes. At private functions, you can offer the bar and bartending services for a fixed time period at a set price.

If your banquet space is large, you may need more than one portable bar unit. Two of them can be arranged side by side at right angles to create a larger bar, or you can locate them on different sides of a room to divide the crowd and prevent long waiting lines for drink service.

Portable bars should have service shelves, laminated tops that can't be stained by water or alcohol, and a stainless steel bin for cubed ice, with a drain and a large reservoir to hold water (from the ice as it melts). Decide where you will store the bar when not in use, then choose a durable model that is easy to move and easy to clean.

Also note that, especially when you're on a budget, a portable bar is a luxury. Many caterers actually prefer to use banquet tables instead of lugging a bar unit around town. They say a skirted table is roomier and offers more storage space beneath than some portable bars. A 6-foot table (30 inches wide) will work for a single bartender; for two bartenders working side by side, use two 6- or 8-foot tables. Two bartenders can share ice and chilled items, in coolers placed on the floor between them.

2-6 CHAIRS AND TABLES

Most people have distinct seating preferences at movies, plays, ballparks, and restaurants. Whenever possible, give your customers a choice of where they want to sit: booth or table, near the window or away from it, smoking or nonsmoking section (if your state still allows smoking in restaurants at all), and so forth. A diner's displeasure with seating arrangements can color the rest of the meal (if he or she even stays long enough to order).

As for the furnishings themselves, there are countless styles from which to choose. As a chorus, industry experts tell us to select seating and tables "from the customer's eye." What does this mean? Well, let's take a family-oriented Italian restaurant as an example, as seen from the eyes of parents with two children who will be dining there. You want them to appreciate the fact that you went to some trouble to carry the "Italian" look and feel through in your decor. You also want them to be comfortable dining out with their kids. They may need high chairs and booster seats—how many should you have for a busy night? Is there enough space between tables to accommodate a high chair and allow the server to do a good job? Does the upholstery clean up well after being slathered with marinara sauce? Is there a place with a bit more privacy for some diners—say, the parents who have hired a babysitter—and who, for once, don't *want* to be seated with a bunch of families? As you can see, design and decor factor into a lot of related decisions.

Since we're on the topic of seating, let's continue with it. Popular types of seating include chairs, stools, booths, and banquettes. Booths offer a certain feeling of privacy or intimacy, but tables and chairs are more adaptable since they can be moved around as needed. A ***banquette*** is an upholstered couch fixed to the wall, with a table placed in front of it. Banquettes are a hybrid of booth and table—more adaptable than a booth, but they still must hug the wall. Banquettes happen to be very fashionable at the moment. Not only can they be upholstered in any number of stylish fabrics, but they maximize seating by filling up corners and allowing more guests to be seated than would fit at individual chairs. Bar stools, either at bars or taller cocktail-style tables, are the most casual seating option.

Because seating is a major investment that will probably be in use for a long time, great care should be taken to choose seats that are comfortable, durable, adaptable, and appropriate to the type of dining you will offer. The typical restaurant chair has a life of five years, but the best ones may last

PLANNING THE DINING ATMOSPHERE

"DRESSING FOR SUCCESS": RESTAURANT FURNITURE AND FASHION SENSE

Some eateries borrow from different time periods, cultures, and themes when selecting their furnishings. Here are a few examples of trends.

- Theme: Bar stools that are Western saddles, beach chairs, tables that look like spaceships. They're all in use somewhere!
- Cultural and historic inspirations: Middle Eastern fabrics, 1950s vinyl-covered bar stools, rattan tables, leather banquettes, wrought-iron railings. All are strongly linked to the past, but can be adapted to contemporary tastes.
- Residential: Armchairs, cozy couches, and table lamps in bar areas, even in restaurants that are quite upscale, make customers feel at home.
- Eclectic: Mismatched antiques, casually grouped wingback chairs, and overstuffed ottomans create a classy sense of informality. The idea here is to make the environment artsy and stimulating, but still relaxing.

Source: Adapted from designer Donna Oetzel in *Restaurants USA* magazine, December 1998. Used with permission of the National Restaurant Association.

ten or more. Chairs are part of the overall design of the room, so the style you select should be in line with the image and ambience of the room. You may hear the term *scale* used in chair selection. The scale of an object is your visual perception of its size. For instance, look at the two chairs in Illustration 2-7. When you compare the captain's chair to the Windsor chair, they're actually about the same size, but the Windsor chair appears lighter and more delicate. This illustrates the difference in scale between the two.

Once you've selected a style, examine the technical aspects of chair construction. Less expensive chairs may be glued or even stapled together—not the optimum for durability. Upkeep and maintenance are important, including whether the manufacturer will keep spare parts available over the years.

Frames can be made of metal, wood, or plastic. They can be stained, dyed, painted, or lacquered—stained and dyed frames are the easiest to maintain, and dying allows for an endless choice of colors. Seats may or may not be upholstered. The chairs can have arms, but only if your tables are roomy enough to accommodate them. Ask about protective laminate finishes for wooden chairs, which would otherwise chip and dent easily.

Check for design flaws that would be troublesome in a public setting: does clothing catch or snag on edges? Are any of the edges sharp enough for someone to scrape or cut themselves accidentally (guests or the wait staff)? Are the legs wobbly? How will the chair hold up when someone quite overweight sits in it? If you have many female guests, can the chair hold a handbag or jacket slung over the back without these items sliding to the floor?

It is wise to order samples of several chairs and test them for a week or two. Here are some specifics that may help in chair selection:

- A 15-degree angle for the chair back is recommended.
- The depth of the seat, from edge to chair back, should be 16 inches.
- The height of the chair, from the floor to the top of the chair back, should be no more than 34 inches. Anything higher impedes the servers.
- The standard distance from the seat to the floor should be 18 inches.

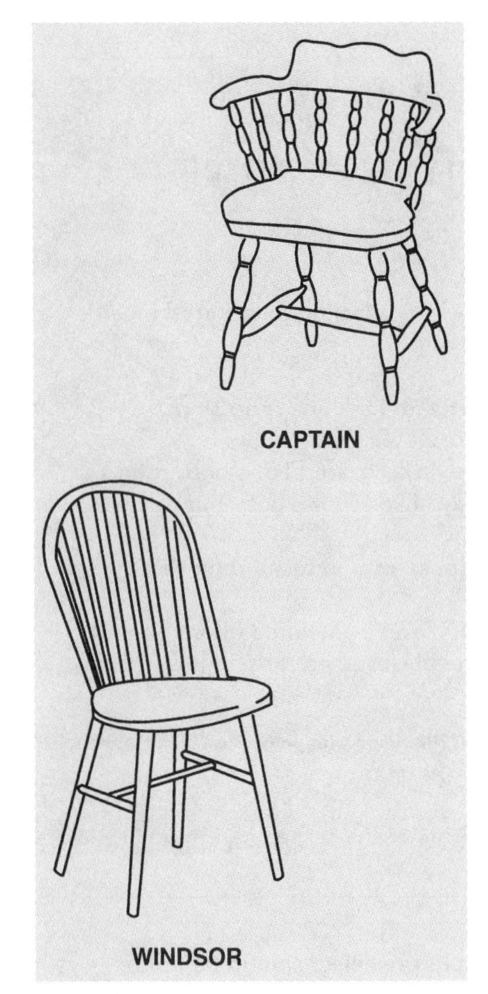

ILLUSTRATION 2-7 **These two chairs are similar in size, but their styles give them very different looks.**

- The distance between the seat and the tabletop should be 12 inches.
- Allow 24 to 26 inches of space for each chair at a table; 28 inches for armchairs.
- For bar or countertop seating, allow 24 to 26 inches per stool.

Consider how the chairs or stools work when empty, too. Do they fit under the tables or the bar armrest? Can they easily be pushed in when not in use, to make the aisles roomier? Are they stackable? Are they easy to clean and easy to move when the floor needs cleaning?

Booth seating is another common choice. You used to see booths only in bar lounges and casual restaurants, but they also look very smart in upscale eateries, where they afford a sense of privacy and romance. Booths can save space, taking up as little as 8 square feet per person. On the other hand, booths are more labor-intensive than tables, since it's harder to clean beneath and around them, and they can't be moved to accommodate various sizes of dining parties.

Deciding on your overall dining atmosphere will help greatly in table selection. Space usually dictates how many tables you will need, and, in most cases, you can get more square tables than round ones in the same square footage. Research shows square tables also seem to produce faster turnover, while round ones prompt guests to linger a bit longer. Attention to aesthetics may require that you blend both square and round tables in your dining area, arranging them at different angles to avoid that military mess hall look mentioned earlier.

The more upscale a restaurant is, the more "elbow room" you allocate for each customer. A fine-dining establishment should allow 15 to 18 square feet per guest; a moderately priced restaurant, 12 feet per guest; for banquets, a minimum of 10 square feet per person.

When purchasing tables, check for sturdy construction. You want long use and solid service from them! Self-leveling legs or bases allow you to adjust for wobbles and also permit the table to glide easily along the floor if it needs to be moved. Think about whether you will cover the tables with linens, butcher paper, or nothing at all. Tabletop linens and accoutrements are discussed in Chapter 20, but you must make an early decision about the type of finish you want on your tables, especially if you won't be using tablecloths. There's a world of choices, from marble, wood, and ceramic to the durable plastic laminates such as Formica and Corian, which are stain resistant and easy to maintain. Nowadays, they come in many patterns, including faux marble, which would work even in an upscale restaurant setting.

No matter what you decide, a table should have a waterproof top, and its ***base*** should be placed to give your customers a comfortable amount of legroom beneath. If lighting will be low and your tables will be draped, a simple pedestal-style base is appropriate. However, if the dining space is airy and open and the tables will not have cloths, the style of the table base can be part of your design. Tables should be chosen in tandem with chairs, since they will be used together. For example, tables that are 26 inches tall work best with chairs that measure 16 inches from seat to floor; 30-inch tall tables work best with chairs that measure 18 inches from seat to floor.

Table bases don't just come in chrome, brass, or black enamel anymore. Today's trends range from fire engine red and deep evergreen to plated finishes of copper, pewter, or bronze.

Those little feet at the bottom of the base that hold the table steady also come in different styles. You'll probably choose between the so-called four-pronged *spider base* and the cylindrical *mushroom base.*

The table, from floor to tabletop, should be 30 inches in height. Here are some basic table sizes and uses:

- For one or two guests: a 24-by-30-inch square table, also known as a ***two-top*** or ***deuce***.
- For three or four guests: a 36-by-36-inch square table; a 30-by-48-inch rectangular table, commonly called a ***four-top,*** or a 42-inch-diameter round table. (There are 36-inch-diameter round tables, but they're a tight fit for four persons.)

- For five or six guests: a two-top and four-top can be joined to create seating for up to six or use a 48-inch-diameter or 54-inch-diameter round table.
- For seven or eight guests: two four-top tables can be joined, or use a 72-inch-diameter round table.
- For cocktail lounges: a 20-by-20-inch square table; or a 20-inch-diameter round table.

2-7 SPECIAL BANQUET NEEDS

Banquet facilities, catering companies, hotels, and country clubs may have additional needs for folding tables. These are typically 30 inches wide and come in two lengths: 72 inches (seats six) and 96 inches (seats eight). There are also narrower tables, from 15 to 18 inches wide, designed to seat people on one side only. This is known, for obvious reasons, as ***classroom-style*** seating.

There are a few exotic shapes in the banquet table universe: arc-shaped *serpentine* tables, which can be placed end to end to create an S shape; graceful, rounded *ovals; trapezoids* that create a solid hexagonal (six-sided) table when two are placed together; ***half-rounds*** and ***quarter-rounds,*** which can be placed at the ends of rectangular tables to "round them out" for buffets or extra seating space. Take a closer look at them (see Illustration 2-8) and the ways they can be arranged.

For maximum legroom at banquet tables, choose tables with ***wishbone-style legs*** (see Illustration 2-9). When setting up rows of oblong banquet room tables, leave 5 feet between them for comfortable back-to-back chair space. For round tables, allow 52 inches between them. In classroom-style setups, allow 30 inches between tables and 30 inches of aisle space between table edges and walls.

Banquet tables are probably the hardest-working pieces of furniture in the business! They should be made with rugged steel underpinnings and reinforced with angle iron rails that run the length of the table. Each leg should have its own brace to lock it in place, as well as a self-leveling device sometimes known as a ***glide.***

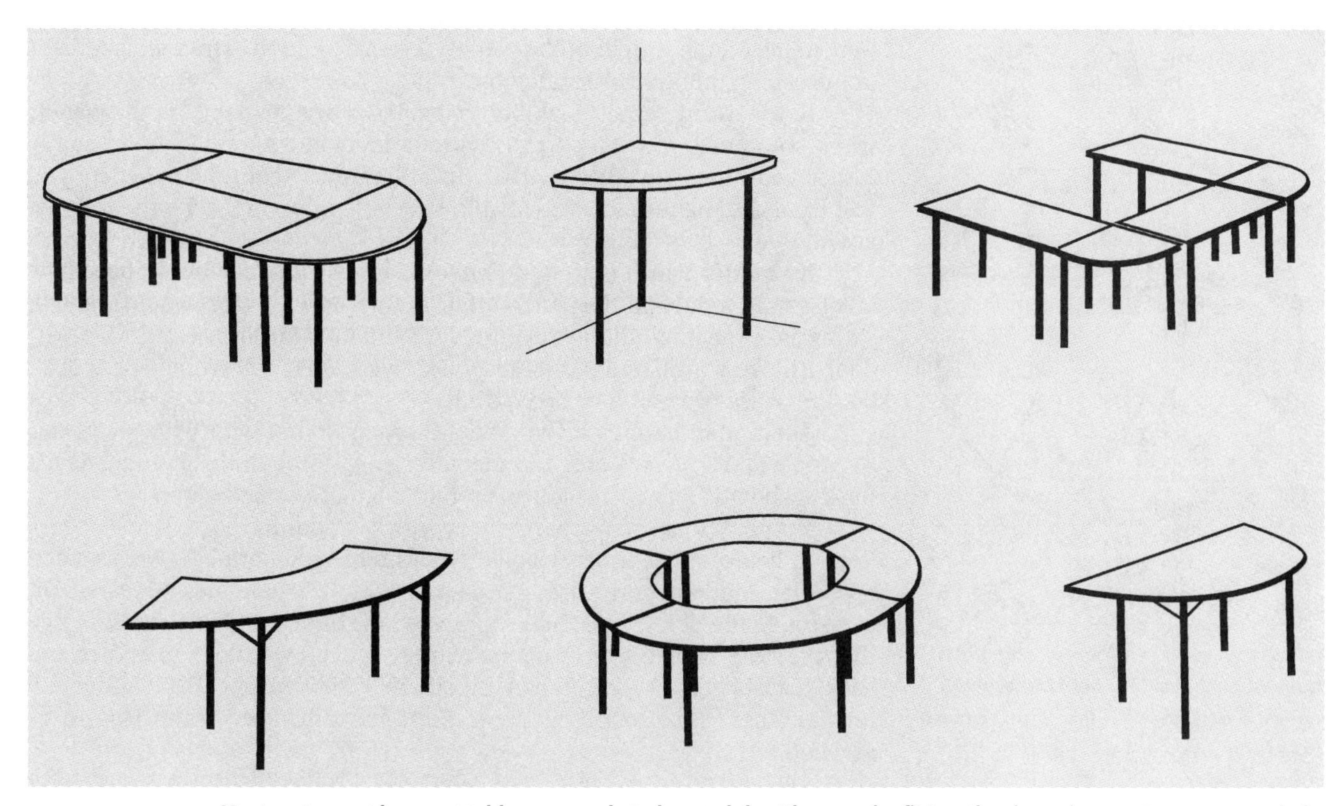

ILLUSTRATION 2-8 **Most restaurant banquet tables are made to be modular. They can be fit together in various patterns, as needed.**

ILLUSTRATION 2-9 **When selecting banquet tables, try sitting at them, especially near the legs, to see if they'll be comfortable for guests.**

The lightweight plywood tabletop may seem flimsy in comparison to its underpinnings, but this makes the banquet table easier to transport, handle, and store. If metal edging rims the table, it should be firmly riveted to the wood as a safety feature and for ease of storage.

Almost everybody knows that seating comfort at a banquet is not exactly top priority. Indeed, the most popular large group seating choices are folding chairs and stackable chairs, selected primarily for ease of storage and handling. If you're in the market to buy them, use the same guidelines you would for other chairs: comfort, durability, and adaptability. This means that folding chairs should fold and unfold easily, have strong locking mechanisms, and be made of sturdy construction. Stack chairs, as they're often called, should be easy to stack as high as a person can reach without marring other chairs beneath them. Also ask for ***wallsaver legs,*** which extend beyond the chair back and prevent the chairs from rubbing or marring the wall they are stacked against.

Risers or *folding platforms* can also be useful in banquet situations. They can be used to elevate the head table, a speaker, or a performance above the crowd. Schools find them useful for band and choir practices and events in their cafeterias. These platforms must be far sturdier than tabletops, with locking mechanisms to bolt the legs firmly in place when the platforms is in use. Dual-height platforms have two sets of legs attached to them, so you can make one side of the platform shorter, the other taller, for a step-up effect.

Typical platform heights are 8 to 10 inches; anything over 12 inches is not recommended for safety reasons.

Portable partitions will be handy if you're going to boast maximum flexibility for your meeting and banquet rooms. You've got two basic options, depending on the ways you plan to divide the room. If it will always be divided at the same spot, you can install partitions that operate on a track attached to the ceiling, or you can use a stiff, accordion-folded panel that splits a single, large room into two equal halves (see Illustration 2-10). Depending on the weight of the panels, this may require extra reinforcement in the ceiling, plus basic installation costs.

If you want more flexibility and have storage room for the panels when not in use, you can purchase freestanding partitions. They're usually 3 feet wide and no taller than 12 feet. However, consider whether you will have the manual labor available to set them up and take them down at will.

No matter which option you choose, ask the manufacturer about how much sound reduction the panels offer. There's nothing more annoying than sitting in a meeting and competing with the noise from the meeting next door. The best insulated panels may cost twice what the average ones cost, but the peace of mind may be worth it.

Additional features include pass-through doors, with window panels so people don't hit each other accidentally going through them; attachments for chalkboards or screens; louvers for air circulation; and more.

Turn any room into a party room with the addition of a *portable dance floor.* In lieu of a fixed, hardwood area of flooring, a portable floor is made up of flat, modular units that interlock to form a dance area of just about any size. Made of tough, durable hardwood, its outer rim has a slanted edge that prevents people from tripping as they move from floor to carpet and back. Portable floors come in panels that are 3 feet square or 3-by-6-foot rectangles. There are also roll-out (or roll-up) floors made of jointed, flat strips of wood.

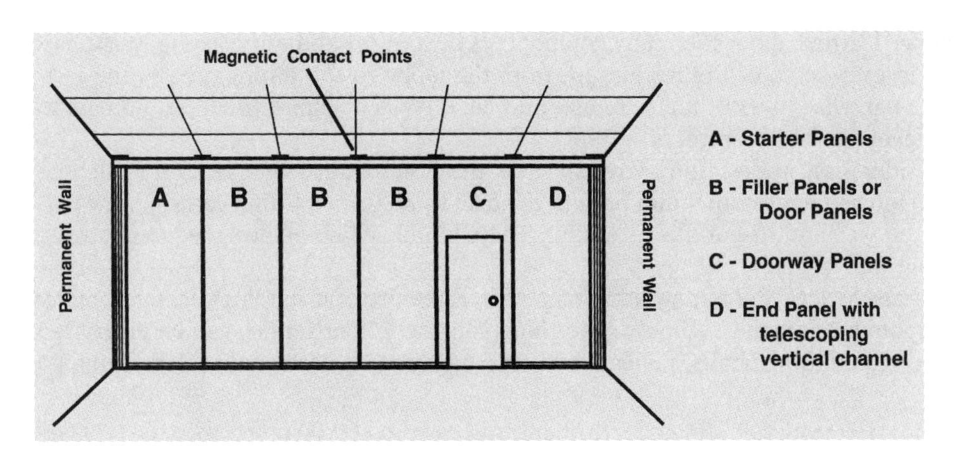

ILLUSTRATION 2-10 **Portable wall panels, available in different sizes and with or without doorways, can divide one 600-seat room into three 200-seat rooms, or other configurations.**

2-8 A FINAL WORD ABOUT ATMOSPHERE

Atmosphere alone cannot guarantee long life to a restaurant, or any other type of foodservice establishment. However, attention to design and detail can give it a competitive edge, attracting customers and reinforcing the statements its owner wants to make. As restaurateur Wolfgang Puck puts it, "You have an idea, and you try to bring it to life. The menu is the last thing. It's made of paper and you can throw it away. The floors and columns are made of cement. You make it look right, and work right, first."

In the foreword to *Dining by Design* by Eddie Lee Cohen and Sherman R. Emery, restaurant guru George Lang suggests that the restaurant designer has the responsibility to "fulfill every guest's personal fantasy in a single interior." Lang says that while it is true that there is a market for every restaurant, one that is "incorrectly conceived and designed" won't last long enough to find its own market. If he were writing a Holy Bible of design, Lang jokes that the first chapter, first verse would be: "In the beginning, there was the feasibility study!"

Many experts feel that restaurants have a lifespan, of sorts, of five to seven years. Also in *Dining by Design,* Brad Elias advocates an understated theme for this very reason. Within five years, fashions change, and many of the interior fixtures and fabrics have simply worn out with use. Elias's advice? "Make a statement, and then let the customer alone," focusing your time on food, ambience, and comfort.

Finally, restaurant designer Spiros Zakas says do not forget the food itself, and the way it is presented, as a crucial design element. Zakas suggests displaying fresh produce, hors d'oeuvres, and desserts as decorative elements. Its underlying message to the customer, he says, "is that the food is fresh, and that it is real." Anything that looks too carefully or blatantly "designed," Zakas fears, squelches the spontaneity that is the hallmark of many chic and successful restaurants. Instead, he strives for "an air of informality that promotes table-hopping and casual friendliness among patrons."

What will *your* place be like?

SUMMARY

Restaurant atmosphere is developed by offering diners a change of pace from their normal environment, using what you know about your target market and what they find comfortable. You can hire a professional designer to create your design and draft the plans or do it yourself. If you hire a designer, be sure to check the person or firm's credentials, and be absolutely honest with the designer and contractors about your budget. The same rules apply to older foodservice buildings that may need to be redesigned and/or modernized.

In this chapter, you've learned about the many components of comfort and atmosphere. Customer perceptions are affected in every area of the restaurant, from the lobby to the dining area to the restrooms. Whether to have a bar, what sizes of tables to use, and what types of chairs you'll buy all help to create the right comfort level for your customers.

One important consideration in designing your space is the flow pattern that is used to deliver foods and beverages from kitchen to customer and back to the dish room. Like a traffic pattern, flow patterns vary depending on the type and size of the restaurant. They should always be designed to minimize backtracking, for efficiency and safety.

This chapter also discussed specialty dining situations, such as the use of a vending cart for mobility or setting up a banquet room for flexibility. Almost everything you use in foodservice can be purchased to be moved and/or reconfigured, from chairs, tables, risers, and partitions to carts and trailers equipped with cooking appliances.

STUDY QUESTIONS

1. What are three of the basic components of a restaurant's overall atmosphere?
2. Name three things that potential customers might observe outside a restaurant that might keep them from going inside.
3. What can a customer learn from standing in a restaurant entryway?
4. What is the single most important rule about restrooms?
5. How should a restaurant determine the size of its bar area?
6. What is a flow pattern?
7. Name three functions of the beverage area.
8. How do you choose chairs for your dining establishment?
9. You have 1240 square feet of dining area, 16 four-top square tables, and 10 two-tops. Arrange them to seat 86 people, making your arrangement as visually pleasing as possible. Briefly describe your layout.
10. Select a restaurant in your area that you believe needs a "facelift." Explain why you think so, and what you would suggest to improve its atmosphere and/or design.

A Conversation with . . .

Kathleen Seelye

DESIGN PRINCIPAL, THOMAS RICCA ASSOCIATES
DENVER, COLORADO

When you can catch her at home, Kathleen Seelye is in Denver, Colorado, but she spends most of her time traveling. Serving as the first woman president of the Foodservice Consultants Society International (FCSI) and principal in the foodservice and laundry design consulting firm of Thomas Ricca Associates doesn't leave her much time to relax. But she says it's actually easier to balance work and family as a designer than as a foodservice operator.

Kathleen began her career in 1975 in a Westin Hotel food and beverage department during college, then entered an apprentice program to work with equipment designers. She says instead of focusing strictly on architecture and engineering, she brings an understanding of actual production experience to the design assignments she undertakes.

Q: It seems like your career came to life during an apprenticeship. Are those programs still available?
A: There is no college degree in foodservice design or consulting. It doesn't exist. There are a number of hotel schools and colleges that provide very limited course work on foodservice design.

Q: So is your organization actively promoting this in the colleges?
A: Absolutely. FCSI has about 800 member consultants in the world, so it's a fairly good-sized field. One of its primary missions is to promote education at the university level and within the business practices of our own membership. We're aiming at developing a licensing or certification program in schools at the university level within ten years, similar to the American Institute of Architects.

Q: So many people work in foodservice for years and probably never realize there is a whole other world of consulting they could move into.
A: Knowing there is more to foodservice makes a big difference. There are truly a lot of good people who leave the hospitality industry because of the demands it places on your time, and the difficulty of balancing family life and other interests. Before I began consulting, I worked every Mother's Day, every holiday—every chef works every weekend—and there are some very high-quality people who leave because they don't want to work within that requirement. I think that's a critical issue for our industry. I don't mean to imply that consultants have "banker's hours," because our hours are difficult also and the travel is incredible. I'm out of my office 80 percent of the time, either nationally or internationally.

Q: If someone doesn't go through the official school channels, how would they know about these other opportunities?
A: There are a lot of different ways you can come into foodservice consulting. We have people who join our staff with strictly technical design backgrounds. They were taught at industrial or

technical schools and trained on CAD [computer-aided design]. As they develop plans, they work just as I did, under other principals who have already been trained in the design of food systems. Eventually, they're able to implement a full design process on their own.

Unfortunately, those people do not have operational experience. To me, the ideal person is someone who has had some education in food systems management, not necessarily a degree because I don't even have that. It's important to have worked in the food and beverage business and understand what it takes to produce food, to serve food that meets all the criteria demanded by the customer, and recognize what systems are the most efficient. It really takes a person who likes research, and has the desire to know what manufacturers are doing and to help them make their products even better. We're the ones who work with the chefs and operators on a daily basis, and we can best communicate to the manufacturers what the problems are and how we can best answer them.

Q: What kind of person has these abilities?

A: Somebody who's in operations now, saying to themselves, "You know, I wish I could fix this facility. I think it would work better if I just move this. . . ." It's the kind of person who's thinking outside the box, thinking creatively instead of assuming they just have to work with what they've got. Typically, a good consultant can find ways of making things work much better, even with what you've got! But the person who's always out there searching for something better is the person who makes the best consultant.

Q: What is the future of foodservice consulting?

A: It's a very focused field. A number of people specialize in only certain areas. I handle a lot of different chains and resorts, as a production methodologist. We help large chains determine their production systems, and what systems are the most efficient.

It's really important to be able to develop a system specific to one person or one company's need that has nothing to do with anyone else's. You need to close the box on the project and not ever take information from one project to another, especially when you're working with chain accounts, because of confidentiality.

It's a very fine line that the consultant walks. Part of a consultant's ethics is to know when you've met that limitation, and convey that to your client. If someone has a proprietary system, it can't even be discussed with another client.

Q: So you always have to start from scratch. Maybe that ensures continued innovations in the industry.

A: Yes. Another important thing about working in a very focused field is that you can develop a very strong expertise, where there are only three or four people in the entire world who do exactly what you do. If you can become the very best in-flight catering foodservice design consultant, you can certainly narrow your competition. On the other hand, that market could go away for one reason or another—the airlines scaling back their meal service is a good example. I appreciate the specialists, but I do think it's important to develop knowledge about a number of different marketplaces.

Q: The foodservice industry seems to lend itself to change. Have you found that to be true, and how do you feel about it?

A: The turnover in operations is very high, but it's much lower in the consulting field. We don't have anyone in my firm who's been here less than ten years. But the entrepreneurial spirit drives this whole industry. We see chefs all the time who say, "I'm going to open my own restaurant," and I say great, but we see a very high exposure to loss. They simply are not able to make it because too many of them are trained in operations and meeting customer needs, but not in business. The schools do not emphasize enough how to make your business succeed by using better business practices.

Consultants are no less entrepreneurial, so we suffer from the same thing. We're working in the FCSI to help correct that as well. But it is much less risky to become a foodservice consultant and work with somebody on an apprentice level before going out on your own, than it is to invest in the high-capital expense of opening a restaurant. I think that's part of the fun of this industry, though. It has a high level of enthusiasm and entrepreneurial spirit that you don't find in a lot of other industries.

Q: What are some of the biggest or most surprising mistakes you've seen people make in foodservice?

A: Not being flexible enough to meet customer demands. It's very similar to the medical industry, where everybody is becoming a specialist—chicken rotisserie chains, for example—but they're finding out if they don't diversify, they won't be able to hold on to their market share. They become so focused on a particular item or style of menu that they're not flexible enough to shift quickly.

Really, the most surprising thing is for the operator, who discovers his lack of knowledge about the process of implementing a design to meet his needs. Many times I hear from a client that the process is a lot harder than they expected, and they didn't know how involved they'd have to be.

CHAPTER 3

PRINCIPLES OF KITCHEN DESIGN

INTRODUCTION AND LEARNING OBJECTIVES

Owners and managers of restaurants are constantly bombarded with new ideas, concepts, and plans for using their kitchen space wisely by reducing operating costs or increasing productivity. Appliances, gadgets, and space savers of all sorts may be tempting, but if you don't have a kitchen that is well planned in the first place, you may be throwing money away by purchasing them.

The three basic kitchen-related costs (and, in parentheses, the ways to reduce them) are:

1. Labor (increased productivity)
2. Utilities (increased energy efficiency)
3. Food (menu flexibility and planning)

What does kitchen design have to do with these? Well, careful space planning gives the restaurant owner or manager the best possible environment and tools with which to accomplish these three critical cost controls.

In this chapter, we will discuss:

- Trends in modern kitchen design
- How to budget for the kitchen you want
- Where to put your kitchen within your facility
- How to create flow patterns that make the service system and work centers run smoothly
- Food safety considerations when designing a kitchen
- Guidelines for placement of equipment
- The unique design needs of service areas and each part of the kitchen

Design refers to overall space planning; it defines the size, shape, style, and decoration of space and equipment in the kitchen. The ***layout*** is the detailed arrangement of the kitchen floor and counter space: where each piece of equipment will be located and where each ***work center*** will be. A work center is an area in which workers perform a specific task, such as tossing salads or garnishing plates. When several work centers are grouped together by the nature of the work being done, the whole area is referred to as a ***work section***: cooking section, baking section, and so forth.

It's smart to design the maximum amount of flexibility into any foodservice setting, and there are different types of flexibility to consider: multiple uses for equipment, and how that may impact the design of the work sections; mobility of the equipment within the kitchen space; operational flexibility and labor flexibility. Shortsighted decisions early in the planning process impede flexibility later.

The natural tendency in kitchen design is to try to fit the equipment into the space available, instead of making the space work to fit the specific needs of the operation. As foodservice professionals, we look at preliminary drawings of a space and notice its shortcomings—insufficient electrical capacity, not enough room for adequate waste disposal, and so on. Instead of living with these, why not tell the architects what we require, and where we want it? If it's going to be *your* space, you must be aggressive about whether, and how, it can be made to work for you.

An important early consideration may be optimistic, but necessary nonetheless—and that is the need for eventual expansion. In addition to your own square footage, look at whatever space is located next door. Would it suit you if you needed to use it? Would it be expensive to modify? Would expansion be possible into that space without too much disruption of business? Even the locations of your building's exterior walls are important. If you're expanding into an existing parking lot, that is possible. If it's a garden area, you might be able to design an expansion that works with it, to create outdoor dining space. But if it's a street on the other side of the wall, expansion is simply not possible in that direction.

3-1 TRENDS IN KITCHEN DESIGN

Industry professionals say now more than ever, kitchen design is driven by consumer demands and economic factors. The fact is, we are seeing smaller and more efficient kitchens. Experts see this trend as the result of three things: a shortage of qualified labor, an ever-increasing battle for space in general for business uses, and budget constraints, including the demand for an increased return on investment. Here are just a few examples of what's in vogue in the commercial kitchen world.

The Display Kitchen

A ***display kitchen*** is where much of the food preparation is done in full view of customers. Being able to watch a busy kitchen staff at work really is interesting to most of us. It whets the appetite and gives the feeling that the guest is being catered to, with a meal that is freshly prepared as he or she looks on. For today's more sophisticated diner, the perceptions of quality, freshness, and presentation are just as important as how the food tastes. A well-functioning display kitchen also accentuates the sense of showmanship that certainly is part of the culinary arts. It enhances the total dining experience by being part of the atmosphere and the evening's entertainment value. It presents opportunities for the culinary staff to interact with guests. Of course, this may impact the type of person you hire as a staff member! Not everyone is good at, or comfortable with, conveying such a "public" image. But for chefs who enjoy the limelight, something magical happens when they can see the patrons, and vice versa.

One nice design detail is to install half-walls, in what is sometimes called a *semi-open kitchen*. The staff can be seen preparing food "from the waist up," without a view of the inevitably messy and unsightly aspects of cooking—soiled pots and pans, stacks of plates, the dirty floor, and so on. Nothing should go on in a display area that indicates any type of "volume cooking." The emphasis is on individually prepared dishes.

Food preparation in view of the guests also addresses another modern-day concern: food safety. Most guests believe that when food is prepared in full view, the staff is more conscious of safe food handling practices than they would be closeted away at a prep station in the back of the house.

IN THE KITCHEN

CHEFS ON DISPLAY: CHINA BEACH

This is the display kitchen in Lucaya Beach Golf Resort, in the Grand Bahamas Hotel. The designers of this Pan-Asian concept, called China Beach, actually placed the kitchen directly in the center of the dining room! China Beach has a prep kitchen behind closed doors, where all meats, vegetables, and fruits are washed and sliced. Most of the dry storage and the walk-in coolers are also located in the prep kitchen, and are sized to hold three days' worth of supplies. This allows room in the display kitchen for additional equipment. There is some storage in the display area: pull-out drawers and compartments to hold vegetables and spices.

China Beach

To prep kitchen

Our Lucaya's China Beach found a different way to include an open kitchen. By putting the kitchen in the middle, guests get a better view of the chefs in action. It also gives the restaurant revenue opportunities by letting guests customize their entrees.

Dining area

Service pick up

Noodles-Dim Sum-Cold pantry

Sushi

Asian Fusion

Chinese cooking line

Service pick up

Dining area

ILLUSTRATION 3-1 **Our Lucaya's China Beach put an open kitchen in the middle of the restaurant, allowing customers to watch the chefs prepare food.**

SOURCE: *HOTEL MAGAZINE*, FEBRUARY 2003.

Today, as the cost of restaurant space keeps climbing, there is some financial urgency behind such a multitask environment. The modern restaurateur must maximize profit per square foot of space, and risks failure by using space extravagantly and having to pay the higher costs of heating, cooling, and insuring it. Combining at least part of the kitchen with the dining area is one way to conserve space. On the other hand, the display kitchen is generally more expensive: up to $360 per linear foot, compared to $115 for a regular back-of-house commercial kitchen. When it is in public view, everything from equipment to walls to preparation surfaces has to look good.

A final word about display or semi-open kitchens: They should only be considered as an option when the menu and food preparation techniques actually lend themselves to display—pizza dough being twirled overhead, steaks being flame-broiled over an open grill, the intricacy of sushi preparation.

Appliances "on Display." In terms of equipment, we've noticed massive brick wood-burning ovens (or gas-fired counterparts) as display kitchen staples. They're not exactly portable, weighing up to 3,000 pounds, but they're attractive, energy-efficient, and quite functional—they can turn out a pizza in three to five minutes.

Induction range tops have found their way into display kitchens, since they are sleek-looking, easy to clean, and are speedy and energy-efficient. Induction cooking works by creating an electromagnetic field, which causes the molecules (in this case, of a pan) to move so rapidly that the pan—not the range top—heats up, in turn cooking the food inside. The magnetic field only prompts other magnetic items (i.e., metal cookware) to heat, while its ceramic surface stays cool to the touch. Not every metal pan is well suited to induction cooking, but there are specific, multi-ply metal pans made for this purpose. Cleanup is as simple as wiping off the cooktop surface; there are no spills seeping into burners, and no baked-on messes. A 2.5-kilowatt induction burner puts out the equivalent cooking power of the 20,000 BTU burner on a typical sauté range.

Yet another display kitchen requirement is the rotisserie oven or grill. We usually think of whole chickens, browning perfectly in a glass-front rotisserie cabinet, but there are now attachments that allow you to bake pasta, casseroles, fish, vegetables, and more. From countertop units no more than 30 inches wide, to floor models six feet in width, rotisseries may be purchased as gas-, electric-, or wood-fired. Ease of cleaning should be a consideration when choosing a rotisserie unit, because they are in view of patrons.

And finally, the *cooking suite* is a real boon to today's hard-working "chef on display." A cooking suite (or *cooking island*) is a freestanding, custom-built unit into which just about any combination of kitchen equipment can be installed. Instead of a battery of heavy-duty appliances against a wall, a cooking suite allows workers to man both sides of the island. (Illustration 3-2). It is a way to effectively concentrate the cooking activities, improving communication because appliances and personnel are centrally located. Often, a cooking suite requires less floor space than a conventional hot line, with shorter electrical and plumbing connection lines. Menus that include mostly sautéed, grilled, or charbroiled items would be well served by a cooking suite. You'll find more information about cooking suites later in this chapter, in the section "Production Areas."

The Marche Kitchen. If you can walk up to a stand-alone counter, place an order, and get fresh food, cooked to order, as you wait, you are in a ***Marche kitchen*** (pronounced mar-SHAY). A display-style retail concept with European origins, this is different than a display kitchen, since the diners stand and

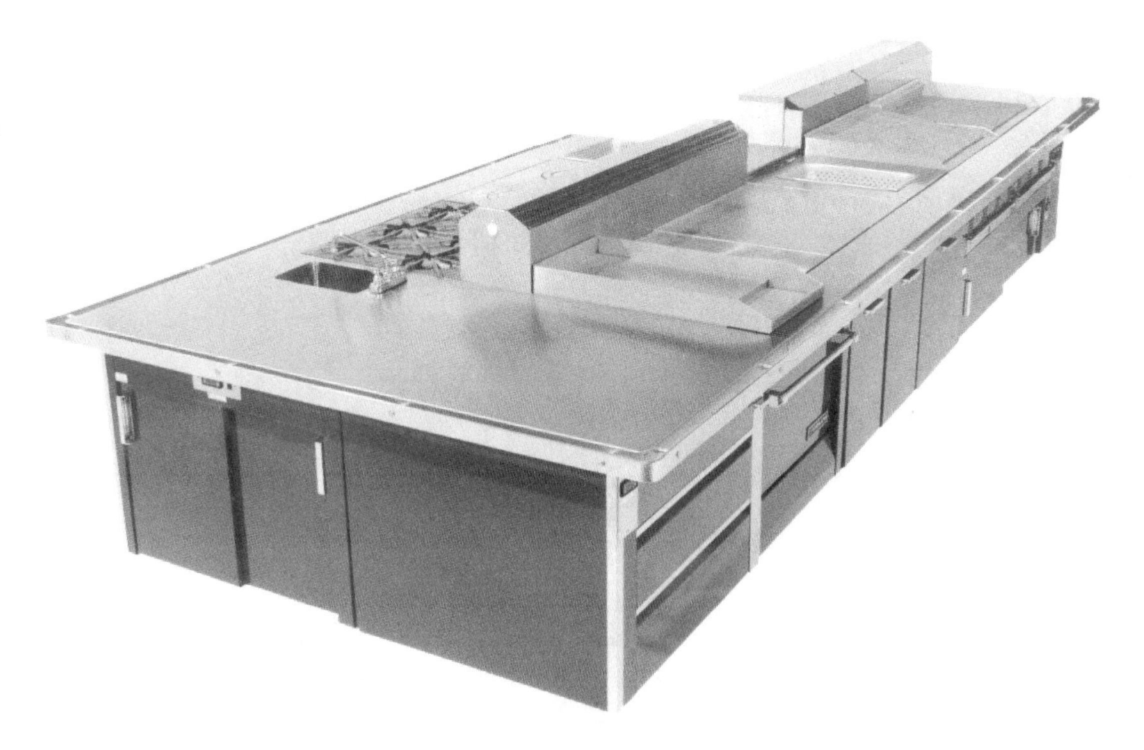

ILLUSTRATION 3-2
A cooking suite or cooking island.
USED WITH PERMISSION FROM THE GARLAND GROUP, A WELBILT COMPANY, FREELAND, PENNSYLVANIA.

watch the action instead of being seated and waited on. The novelty and excitement comes from the combination of freshness and the commercial equipment moved from back of the house, right up front! It's almost as though the wall between dining room and kitchen has been peeled away, and you're practically standing inside a clean, attractive, well-lit kitchen full of griddles, fryers, broilers, refrigerators, steam-jacketed kettles, and all the trappings of "big deal cooking." Food is prepped, cooked, and served in one place, which saves steps for the workers and doesn't require the heavy hauling of ingredients from kitchen to serving line.

Most Marche kitchens are designed with some flair—blonde wood, tiled pedestals and warmer trays, faux finishes on ventilation hoods and equipment to make them more attractive. In some of them, customers pay for each course separately as they receive it on the serving line.

The Small, Technologically Superior Kitchen. Another idea taking hold today is simply a small kitchen with carefully orchestrated work centers, all designed with ergonomics and efficient labor utilization in mind. It is created to use the fewest possible number of people to perform each task, and outfitted with the highest-quality equipment to minimize downtime. Greater use of technology (touch screens, programmable appliances) improves customer service and saves time. Commonsense touches—trash receptacles built into the counters, compact storage for individual areas to minimize wasted steps having to go fetch supplies—all mean the crew is able to work smarter, not harder. "Plug and use" (portable) equipment combinations (steamer/ovens; refrigerated space beneath range tops) that are easy to service, with surfaces that are easy to clean, also are popular.

3-2 MAKING THE NUMBERS WORK

Deciding how much money to spend on design, construction, and equipment is a crucial first step—not just a total figure, but an estimate of how funds are to be allocated and how much time each step will take. First, does the design budget realistically match the concept in size and scope? The budget should reflect the type of market in which the finished restaurant will operate. There's a tendency to overspend at the front end of a project, so much so that the finished facility can't be profitable for many months while construction and design costs are being paid off. So early on, set your ego aside and, likewise, refuse to accept grandiose plans from consultants who may have good ideas but don't seem to be thinking about your cost parameters.

In fact, the lack of early budgeting and planning is the undoing of many a promising concept. It's true that it is difficult to decide months in advance what to spend on design, construction, decor, furniture, fixtures, equipment, china, flatware, glassware, table linens, smallware, publicity, and other preopening expenses, but a good budget should include all of these and more, plus a contingency amount of about 20 percent to cover any cost overruns. For renovations or adaptive reuse projects in older structures, the contingency amount may even be higher than 20 percent, since problems often crop up that were not anticipated due to the ages or conditions of buildings, utilities, and fixtures, as well as the special care and additional permits often required to refurbish them. It is also more difficult to budget for a restaurant that is part of a larger project—say, a hotel restaurant—than a freestanding restaurant, because you are often bound by the design decisions of others.

Underestimating your budget will not impress your banker or your investors. It is more likely to compromise your concept and misrepresent your ideas to the contractors you hire to work with you.

All your costs will vary widely, and will depend on many factors in your market: economic conditions, location, the type of restaurant you want, and so on. The key to controlling them is planning. Let's take an equipment purchase as an example. Invest some time with the salespeople or manufacturers' representatives. Pay attention during their equipment demonstrations. Don't order something that is more, or bigger, than what you need. Check Chapter 9 of this text for guidelines on equipment specifications, and what to look for in a warranty or guarantee.

Price new equipment before you look at used—that way, you have an idea of what is truly a "good deal." Look for equipment that can perform more than one function—a combi oven, for instance, can hold, bake, or steam. If you order it custom-made, give the manufacturer enough time to do it right. Then, well before it arrives, be sure the preliminary utility connections (called ***rough-ins***) are in the correct

IN THE KITCHEN

THE CHEF-DESIGNED KITCHEN

As professional chefs assume near-celebrity status, their input as part of the design team is indispensable. What you really have with a highly qualified chef is a person who understands food and the important details of a functional kitchen. Combining that know-how with the technical expertise of engineers or designers means a greater range of workable possibilities.

Such is the case with Tru, a Chicago restaurant that is a joint effort of chef Rick Tramonto, pastry chef Gail Gand, and Richard Melman, a longtime restaurateur and chairman of "Lettuce Entertain You" Enterprises. Their 200-square-foot kitchen produces 800 to 1,000 meals nightly, with a staff of no more than 23 at peak periods (see Illustration 3-3).

A cooking suite solves the traffic problem, with a central cooking area surrounded by garde manger, fish and meat prep, and saucier stations along the walls. A special station handles the Tru house specialty, amuse bouchées, bite-sized predinner palate-teasers. A separate pastry room includes granite counters, mixers, convection ovens, and cold storage.

To ensure freshness, Tru accepts deliveries twice daily—with separate walk-in storage areas for meat, fish, and produce. The kitchen preps and vacuum-packages these items in single-serve portions for quick later use. Refrigeration beneath the cooking surfaces keeps food at optimum temperatures until it is needed.

ILLUSTRATION 3-3 **Executive Chef Rick Tramonto and Executive Pastry Chef Gale Gand designed the kitchens at Tru Restaurant in Chicago.**

COURTESY OF TIM TURNER PHOTOGRAPHY. TRU KITCHEN DESIGNED BY EXECUTIVE CHEF/PARTNER RICK TRAMONTO AND EXECUTIVE PASTRY CHEF/PARTNER GALE GAND.

places. Check widths of doors and the intended space where the equipment will be installed to be certain they are wide enough. These are all small steps, and many of them don't specifically have to do with budgeting. But they all add up to one huge factor: making a smart investment that you can live with.

In creating a budget, it is reasonable to expect cost estimates up front from any designer, architect, or consultant you may hire, after they have received enough information from you to prepare their own fee estimates. And remember, "costs" and "fees" are different animals.

One factor that has major cost implications is time. If work crews don't show up on schedule or materials you've ordered don't arrive promptly, the delays can become costly. Contractors may charge expensive overtime, and there'll be no time to dispute any details you may not be satisfied with. In some cases, you may have to delay your opening. So, along with your budget, you must develop a time line, which will help you pinpoint when you have to spend certain amounts. If your concept is fully developed and you've already obtained the site for your foodservice facility, it is reasonable to assume:

- The design phase should take no more than 16 to 18 weeks. This includes the time to get necessary permits from authorities and the bidding process to hire contractors.
- The construction and/or remodel process should not take more than 16 to 24 weeks.

Having to make "change orders"—requests to alter your requirements after the initial agreements with contractors—can increase costs. However, there are situations in which modifications are needed that were not anticipated. In this case, the details should be presented (by you, in writing) to the contractor. This document is called a ***bulletin,*** and it usually includes a drawing and/or specifications for the exact requirements.

In Illustration 3-4, you'll find a checklist for a budget to construct a "theoretical" restaurant. It's a freestanding eatery of 6,900 square feet, which will have 175 seats and 40 barstools.

To whom should you turn for assistance with kitchen design? You can rely on the advice of consultants or the expertise of equipment manufacturers. Either way, you should take the following steps:

- In addition to defining the concept itself, define the goals you expect to achieve from it. Decide on your menu before the design process begins.
- Be very clear on how much money you intend to spend. Trust the professionals who pay attention to your budget, not the ones who consistently try to talk you into spending "just a little bit extra."
- Make a commitment to yourself to choose the most energy-efficient equipment in your price range.
- Consider future growth that may require additional space and utilities.
- Consider equipment and work spaces (tables, shelves) that are movable. Think "maximum flexibility."
- Separate the stages of food production so that raw materials can be prepared well in advance.
- Look for equipment that has a good reputation and a proven performance record. As much as possible, each piece should be multifunctional, simple to operate, easy to maintain, and energy efficient.
- Control costs where the customers won't notice and the kitchen crew won't be affected.

Some experts suggest that the overall cost of your kitchen equipment should not exceed one-third of the total investment for the entire facility.

3-3 KITCHEN PLACEMENT

Placement of the kitchen in the building is the first critical step. We view the kitchen as the "heart" of the house. Like a human heart, its job is to pump and circulate life-giving blood throughout the rest of the

"BACK OF HOUSE"

Electrical ($ per square foot of kitchen space)	________
Plumbing ($ per square foot of kitchen space)	________
Mechanical ($ per square foot of kitchen space)	________
Fire protection/sprinklers	________
Kitchen walls, ceilings (ceramic, tile, etc.)	________
Kitchen equipment	________
Kitchen smallware	________
SUBTOTAL	________

Note: To determine the total cost PER SQUARE FOOT, divide the total number of dollars on this part of the list by the number of square feet for the back of the house.

"FRONT OF HOUSE"

Millwork ($ per square foot)	________
Finishes ($ per square foot)	________
Includes drywall, paint, tile, carpet, window treatments, etc.	________
Specialties ($ per square foot)	
Includes door glazing, etc.	________
Lighting (ceiling and wall fixtures, dimmers)	________
Sound (systems and treatment)	________
Furniture (tables, booths, chairs, stools)	________
Accessories (includes artwork)	________
Plateware	________
Glassware	________
Flatware	________
Linens	________
SUBTOTAL	________

Note: To determine the total cost PER SQUARE FOOT, divide the total number of dollars on this part of the list by the number of square feet for the front of the house.

FEES

Architect	________
Engineering firm	________
Mechanical contractor	________
Designer and/or consultant	________
Legal fees (includes permits, licenses, etc.)	________
SUBTOTAL	________

FEES

Facade treatment	________
Signage	________
Parking lot, curbs, sidewalk construction	________
Exterior lighting (includes parking lot)	________
Landscaping (design and installation)	________
Sprinkler system for landscaping	________
SUBTOTAL	________

ADD ALL FOUR SUBTOTALS TOGETHER TO ARRIVE AT YOUR TOTAL PRELIMINARY BUDGET.

ILLUSTRATION 3–4 **Budget checklist.**

operation. Kitchen placement will affect the quality of the food, the number of guests who can dine at any particular time of day, the roles of the servers, utility costs, and even the atmosphere of the dining area. Consider each of these factors for a moment. Can you see why they are impacted by kitchen location? And remember, each also figures in the overall profitability of the business.

A poorly designed kitchen can make food preparation and service more difficult than it should be, and it can even undermine the morale of the staff. So, if a new restaurateur has little cash to spend on professional designers, that cash is probably best spent planning the location and design of the kitchen—the one area where equipment, ventilation, plumbing, and general construction costs combine for a major investment.

We can also learn to plan wisely by thinking of the restaurant kitchen as a manufacturing plant: With a combination of labor and raw materials, it turns out product. The unique aspect of foodservice is that this finished product is sold in an outlet that is attached to the factory!

Like any other type of manufacturing plant, productivity is highest when the assembly lines and machinery are arranged in a logical, sequential order to put the components together. In foodservice, this includes everything from tossing a salad to turning in orders so that no guest is left waiting too long for a meal.

3-4 THE SERVICE SYSTEM

A major issue that must be addressed before deciding on a kitchen design is the way in which food will be delivered to guests. This is known as the ***service system.*** A large operation, such as a hotel, can have more than one service system at work simultaneously: elegant tableside service, room service, and casual bar service. At the other end of the spectrum, quick-service restaurants employ service systems that emphasize speed and convenience, including takeout service and the fast-food option of standing at the same counter to order, pay for, and wait for a meal served within minutes.

Each service system has subsystems; together, they encompass every aspect of the progression of food from kitchen, to table, and back to the dishwashing area. This progression is called ***flow,*** much like the traffic flow of a busy street grid. There are two types of flow to consider when planning your kitchen design: product flow and traffic flow. ***Product flow*** is the movement of all food items, from their arrival at the receiving area, through the kitchen, to the guests. ***Traffic flow*** is the movement of employees through the building as they go about their duties. The ideal, in both types of flow systems, is to minimize backtracking and crossovers—again, to make sure the "streets" don't get clogged.

There are three basic ***flow patterns*** in every foodservice operation:

1. **The raw materials to create each dish have a "back to front to back" flow pattern. They arrive at the "back" of the restaurant, in the kitchen, where they are prepared.**
2. **Next, they travel to the "front" of the restaurant, to be served in the dining area.**
3. **Finally, they return to the "back" again, as waste. The movements are so predictable that they can be charted, as shown in the functional flow diagram of Illustration 3-5.**

Within the kitchen, there is also a flow unique to each cooking section. It could be a pattern of steps the chefs follow to put each dish together or the methodical way the dishwashers scrape, sort, and wash dishes and dispose of waste.

The third type of traffic pattern is the flow of the service staff as waiters pick up food, deliver it to the guests, and clear the tables. On a busy night, the whole system really does resemble a busy freeway. So, as you might imagine, there's always the possibility of disaster if somebody makes a wrong turn. The key to managing these three types of flow is that each should not interfere with the others.

The service systems and flow patterns of your business should guide your kitchen design. An operation with huge numbers to feed in short time periods will differ from one that also feeds large numbers but in a longer time period. Can you see how?

The distance from the kitchen to the dining area is one important consideration, and kitchen designers have devised numerous strategies to cope with it. You may have noticed that, at some restaurants, the waiters are expected to do quite a few food-related tasks outside the kitchen, at wait stations

ILLUSTRATION 3–5
Typical kitchen product/traffic flow.
SOURCE: EDWARD A. KAZARIAN, *FOODSERVICE FACILITIES PLANNING*, JOHN WILEY & SONS, INC.

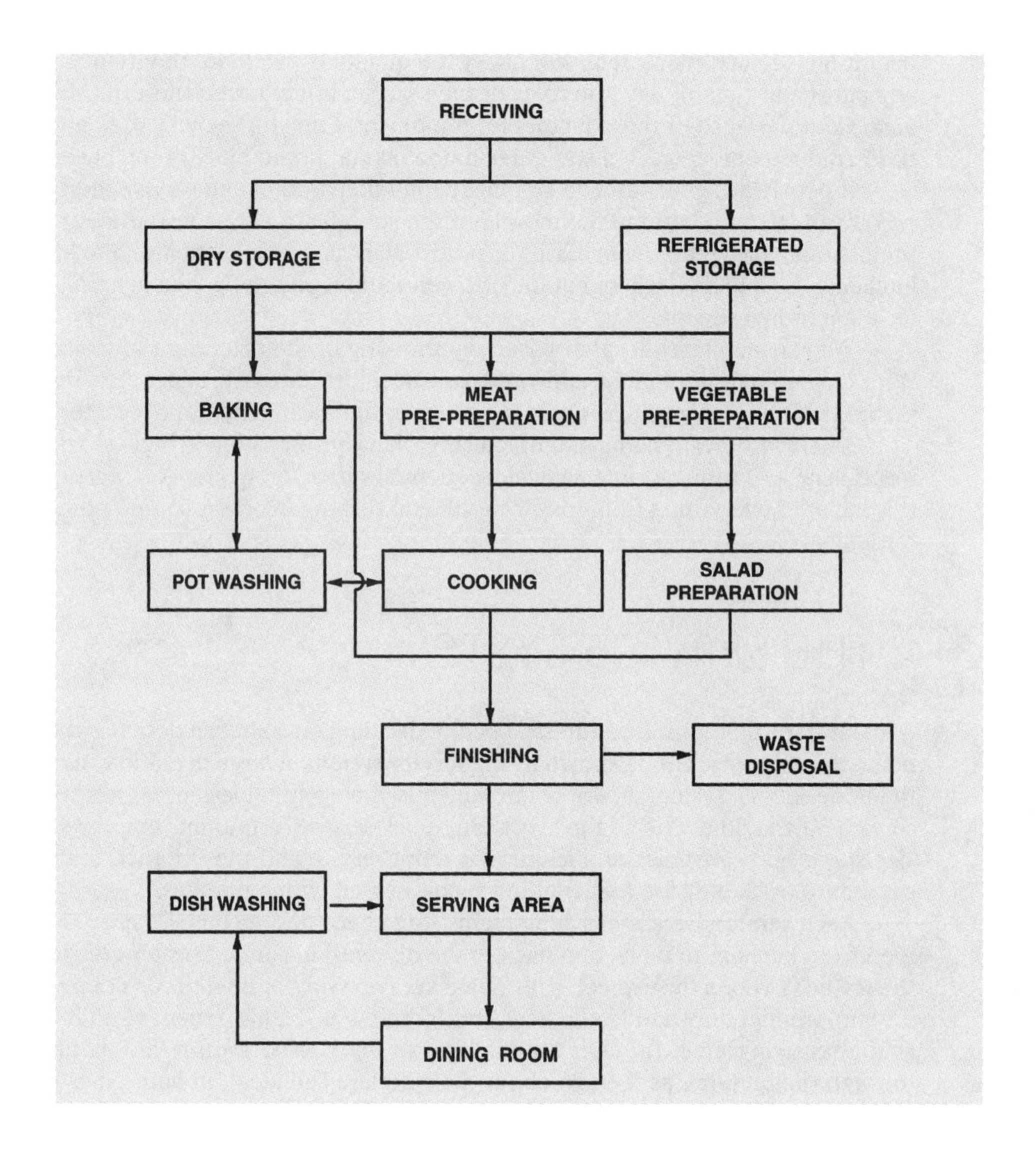

closer to the guests. They might slice and serve bread, ladle soup, arrange and dress salads, or pour beverages themselves. The idea is to speed service and preserve the (sometimes inadequate) kitchen space for actual cooking tasks.

Another critical decision to be made early in the design process: Should the wait staff come into the kitchen to pick up food, or should it be handed to them through a ***pass window*** between kitchen and dining area? Although the pass window is considered more informal, it can still be used in a fancier restaurant, perhaps masked from public view by a wall or partition.

Each of these items—distance and kitchen access—help determine your flow patterns. In a perfect world, flow patterns would all be straight lines that do not intersect. However, this ideal is rarely achieved. One simple rule is that the faster you want your service to be, the more important it is that your flow patterns do not cross. In a fast-service scenario, the flow lines must be short and straight. The next time you're standing at a fast-food counter, notice how few steps most of the workers have to take to pour your soft drink, pick up your burger, and bag your fries. Speed is the desired outcome.

The reverse is true in a fine-dining establishment, where the work may all be done in the kitchen specifically to enhance the feeling of a leisurely dining experience. No clattering plates, no bustling wait stations here.

Now that we've looked at the flow of people as they perform their restaurant duties, let's follow the *food flow line*: the path of raw materials from the time they enter the building to the time they become leftovers.

The ***receiving area*** is where the food is unloaded from delivery trucks and brought into the building. Most restaurants locate their receiving areas close to the back door. Our next stop is *storage*—dry storage, refrigerated storage, or freezer storage—where large quantities of food are held at the proper temperatures until needed.

Food that emerges from storage goes to one of several *preparation* or "prep" areas for vegetables, meats, or salad items. Slicing and dicing take place here, to prepare the food for its next stop: the *production* area. The size and function of the prep area varies widely, depending mostly on the style of service and type of kitchen.

When most people think of a restaurant kitchen, what they imagine is the ***production line.*** Here, the food is given its final form prior to serving: boiling, sautéing, frying, baking, broiling, and steaming are the major activities of this area. The food is plated and garnished before it heads out the door on a serving tray. And that's the end of the typical food flow line.

There are several kitchen work centers that are not included in the typical food flow sequence, but are closely tied to it. For instance, storage areas should be in close proximity to the preparation area, to minimize employees' walking back and forth. In some kitchens, there is a separate ***ingredient room,*** where everything needed for one recipe is organized, to be picked up or delivered to a specific workstation. Storage is much more useful when it's placed near the prep area than near the receiving area, saving steps for busy workers. The bakery is usually placed between the dry storage and cooking areas, because mixers and ovens can be shared with the cooking area. A meat-cutting area is also essential. It should be in close proximity to both refrigerators and sinks for safety and sanitation reasons, as well as for ease of cleanup. Remember, however, that some kitchens are simply not big enough to accommodate separate, specialized work centers. Kitchen space planning becomes a matter of juggling priorities, and it is a continuous compromise.

As you juggle yours, think about each task being done in each work center. How important is it to the overall mission of the kitchen? Are there duties that might be altered, rearranged, or eliminated altogether to save time and/or space? Some of the ideas that should be discussed here are frequency of movements between various pieces of equipment, the distance between pieces of equipment, allowing space for temporary "landing areas" for raw materials or finished plates to sit until needed, putting equipment on wheels so it can be rolled from one site to another, and making "parking space" for the equipment when it's not being used.

Simply stated, if work centers are adjacent to one another, without being cramped, you save time and energy, and if people who work in more than one area have handy, unobstructed paths between those areas, they can work more efficiently.

One work center that is often misplaced is the pot sink, which always seems to be relegated to the most obscure back corner of the kitchen. True, it's not the most attractive area, but think of the many other work centers that depend on it! The typical kitchen generates an overflow of pots and pans. Why isn't the pot sink placed closer to the production line to deal with the mess?

And, speaking of pots, think carefully about where to store them. Both clean and dirty, they take up a lot of space and require creative storage solutions. Often, pot/pan racks can hang directly above the sink area, giving dishwashers a handy place to store clean pots directly from the drain board. (Remember that anything stored near the floor has to be at least 6 inches off the floor for health reasons.)

3-5 FLOW AND KITCHEN DESIGN

Let's take a look at some common flow plans for food preparation that you'll find inside the kitchen. The most basic, and most desirable, flow plan is the ***straight line,*** also called the *assembly line flow.* Materials move steadily from one process to another in a straight line (see Illustration 3-6). This type of design minimizes backtracking; it saves preparation time and confusion about what's going out of the kitchen and what's coming back in. The straight-line arrangement works well for small installations because it can be placed against a wall and adapted to the cooks' duties.

Where there is not enough space to arrange food preparation in a straight line, a popular and efficient choice is the ***parallel flow.*** There are four variations of the parallel design:

1. **Back to Back.** Equipment is arranged in a long, central counter or island in two straight lines that run parallel to each other (see Illustration 3-7). Sometimes, a 4- or 5-foot room divider or low wall is placed between the two lines. It's primarily a safety precaution, which keeps noise and clutter to a minimum and prevents liquids spilled on one side from spreading onto the other. However, placement of a wall here also makes cleaning and sanitation more difficult.

The back-to-back arrangement centralizes plumbing and utilities; you may not have to install as many drains, sinks, or outlets, since both sides of the counter can share the same ones.

A back-to-back arrangement in which the pass window is parallel to (and behind one of) the production areas is sometimes known as a "California-style" kitchen.

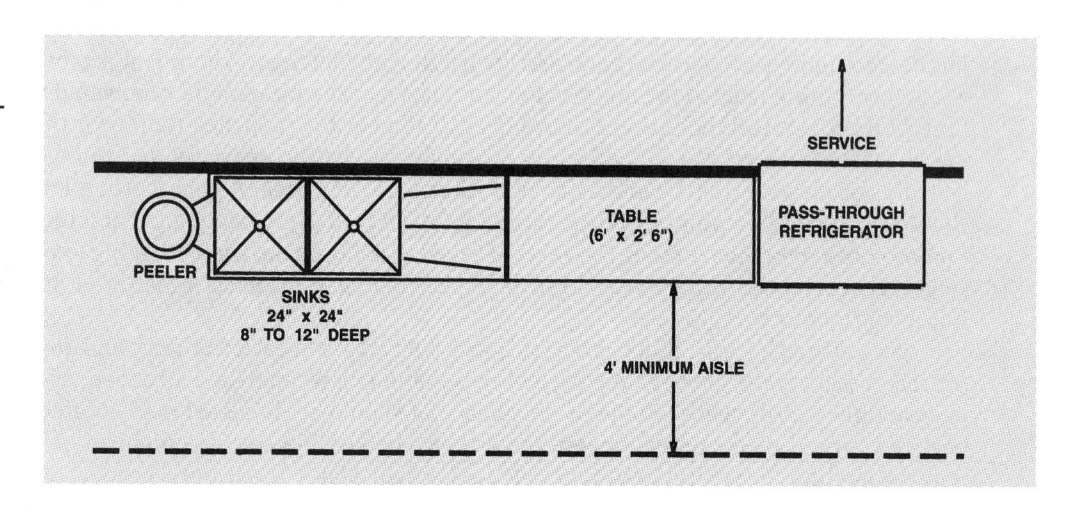

ILLUSTRATION 3-6 **Straight-line arrangement. In a straight-line or assembly line kitchen, food and materials are passed from one work center to another in a straight line.**
SOURCE: ROBERT A. MODLIN, EDITOR, *COMMERCIAL KITCHENS,* 7TH ED., AMERICAN GAS ASSOCIATION, ARLINGTON, VIRGINIA.

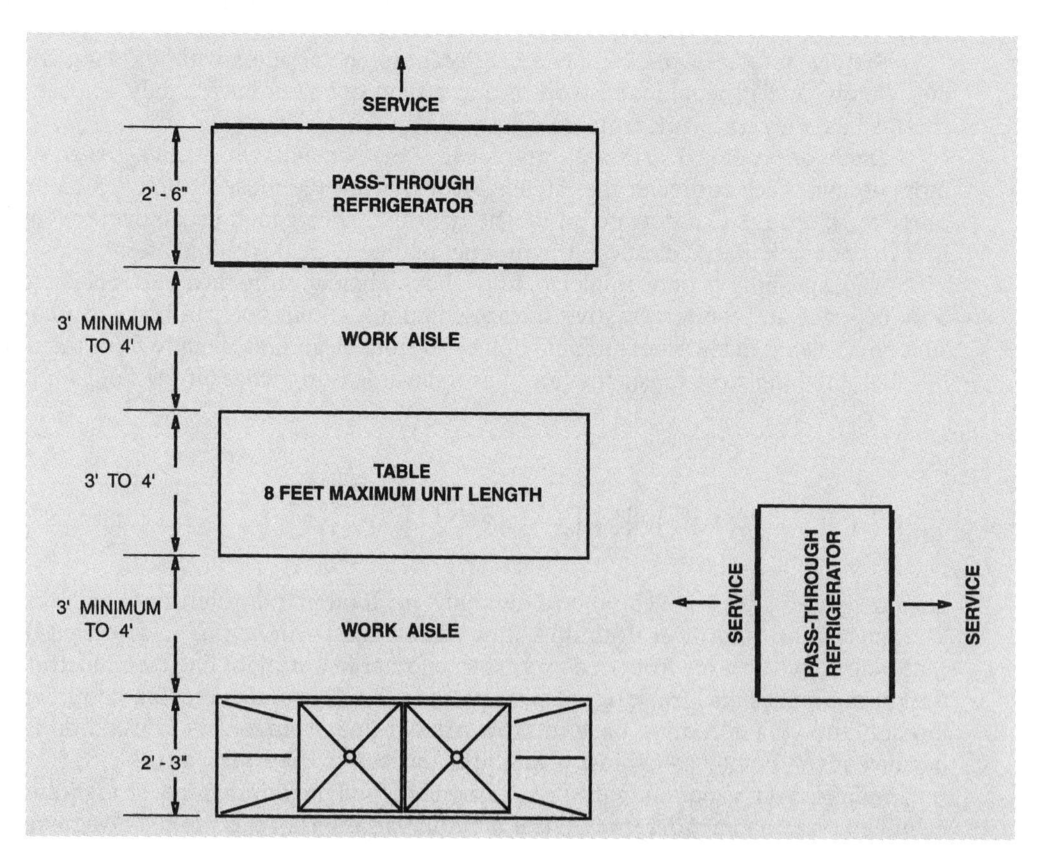

ILLUSTRATION 3-7 **In a parallel-flow configuration, equipment and work centers are arranged to save space and allow efficient movement of product and workers.**
SOURCE: ROBERT A. MODLIN, EDITOR, *COMMERCIAL KITCHENS,* 7TH ED., AMERICAN GAS ASSOCIATION, ARLINGTON, VIRGINIA.

When the pass window is located perpendicular to the production line, it may be referred to as a "European-style" kitchen design. The advantage of the European design is that each cook on the line can see the progression of multiple dishes that make up one table's order.

2. **Face to Face.** In this kitchen configuration, a central aisle separates two straight lines of equipment on either side of the room. Sometimes, the aisle is wide enough to add a straight line of worktables between the two rows of equipment. This setup works well for high-volume feeding facilities like schools and hospitals, but it does not take advantage of single-source utilities. Although it is a good layout for supervision of workers, it forces people to work with their backs to one another, in effect separating the cooking of the food from the rest of the distribution process. Therefore, it's probably not the best design for a restaurant.
3. **L-Shape.** Where space is not sufficient for a straight-line or parallel arrangement, the L-shaped kitchen design is well suited to access several groups of equipment, and is adaptable for table service restaurants (see Illustration 3-8). It gives you the ability to place more equipment in a smaller space. You'll often find an L-shaped design in dishwashing areas, with the dish machine placed at the center corner of the L.

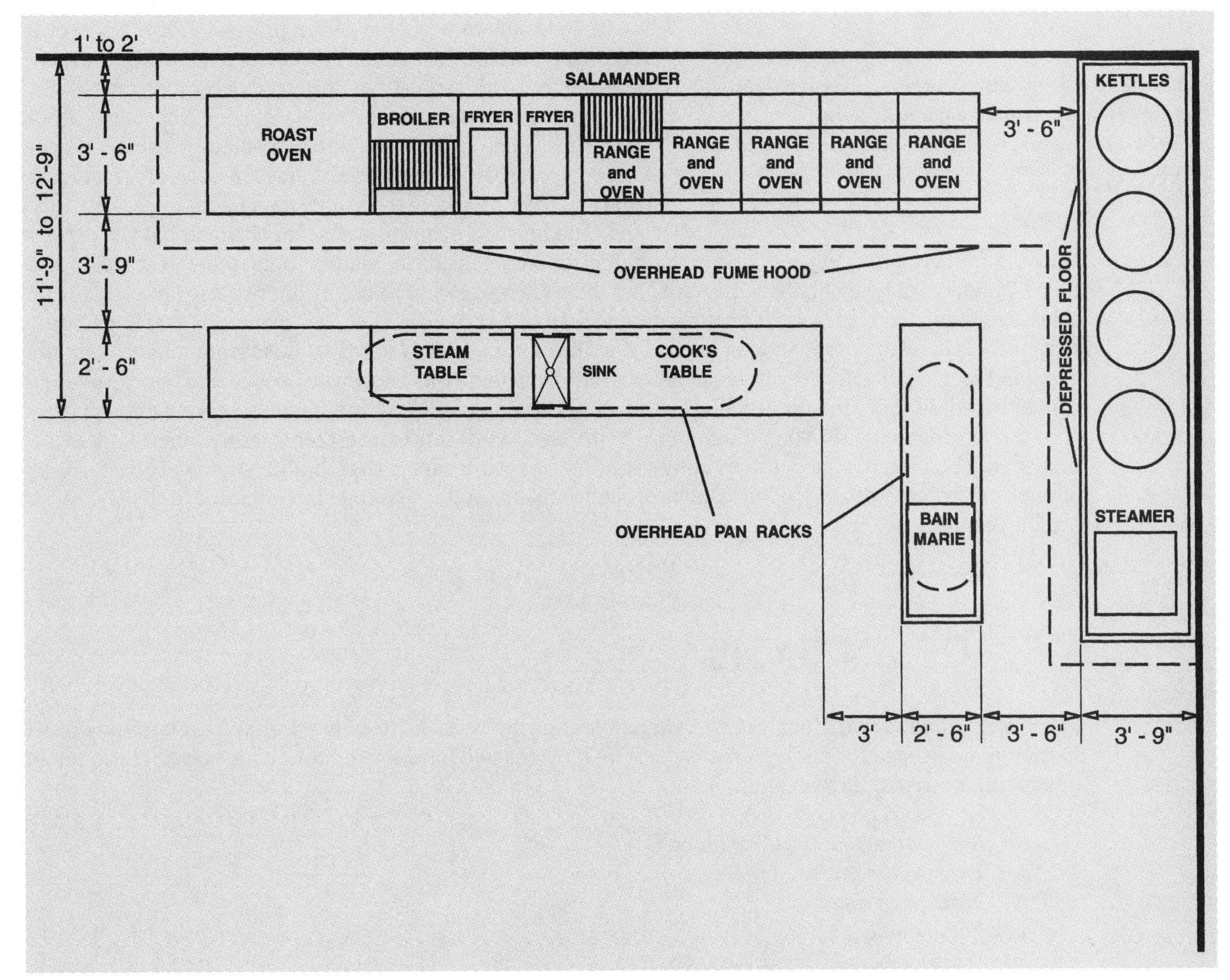

ILLUSTRATION 3-8 **The L-shaped kitchen layout works well in long, rectangular spaces and allows more equipment to be placed in a smaller area.**

SOURCE: ROBERT A. MODLIN, EDITOR, *COMMERCIAL KITCHENS*, 7TH ED., AMERICAN GAS ASSOCIATION, ARLINGTON, VIRGINIA.

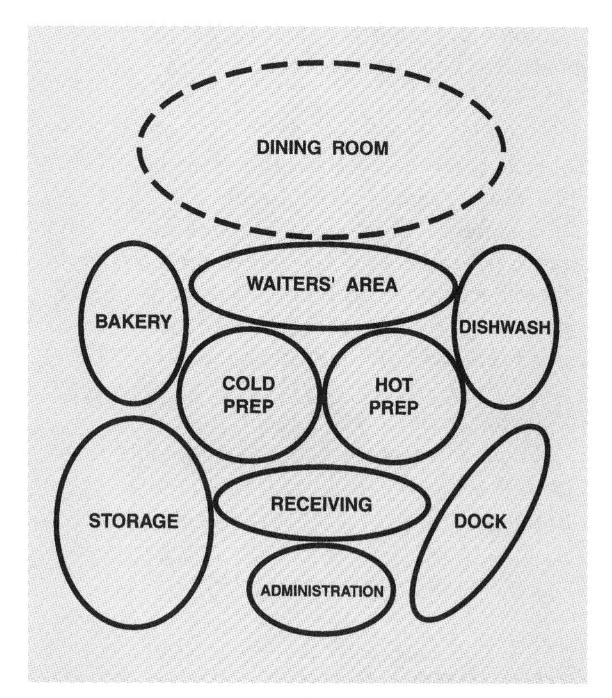

ILLUSTRATION 3–9 **Bubble diagrams are rough sketches that show the relationship among different areas of a restaurant, or just the kitchen or dining area.**
SOURCE: *LODGING,* THE MAGAZINE OF THE AMERICAN HOTEL AND LODGING ASSOCIATION, YARDLEY, PENNSYLVANIA.

4. **U-Shape.** This arrangement is seldom used, but it is ideal for a small space with one or two employees, such as a salad preparation or pantry area. An island bar, such as the ones in T.G.I. Friday's restaurants, is another example of the U-shape at work.

There are also circular and square kitchen designs, but their limited flow patterns make them impractical. Avoid wasted space if you can, by making your kitchen rectangular, with its entrance on one of the longest walls to save steps.

The more foodservice establishments you visit, the more you will realize that the "back of the house" is a separate and distinct entity from the rest of the business, with its own peculiar problems and unique solutions. Correct flow planning sometimes means breaking each kitchen function down into a "department," of sorts, and then deciding how those departments should interact with each other. They must also interact with the other, "external" departments of the facility: your dining room, bar, cashier, and so on.

A good way to begin the design process—both for the overall business and for the kitchen—is to create a ***bubble diagram.*** Each area (or workstation) is represented as a circle, or "bubble," drawn in pencil in the location you've decided is the most logical for that function. If two different workstations will be sharing some equipment, you might let the sides of their circles intersect slightly, to indicate where the shared equipment might best be located. The finished diagram will seem abstract, but the exercise allows you to visualize each work center and think about its needs in relation to the other centers. Illustration 3-9 shows this type of bubble diagram, and there are other ways to use it. You can also lay a kitchen out using a diamond configuration, situating the cooking area at one point of the "diamond" shape, and other crucial areas in relation to it at other points, as in Illustration 3-10. Notice that this layout minimizes confusion (and accidents) with a separate kitchen entrance and exit. This allows the people who bus the tables to deliver soiled dishes to the dishwashing area without having to walk through the entire kitchen to do so.

An alternative to drawing diagrams is to list each work center and then list any other work center that should be placed adjacent to it. Conversely, list any work center that should *not* be next to it. For example, it's probably not a good idea to have the ice maker and ice storage bin adjacent to the frying and broiling center.

3-6 SPACE ANALYSIS

In the next chapter, we'll analyze the kitchen section by section, to help you approximate how much space will be needed for each. Of course, not all kitchens will have a need for all areas we mention here. In general, however, they are:

- Service areas and wait stations
- Preparation areas
- Production areas
- Bakery area
- Warewashing (dishwashing) area
- Storage areas: dry, refrigerated, cleaning supplies, dishes and utensils
- Receiving area
- Office
- Employee locker rooms, toilets

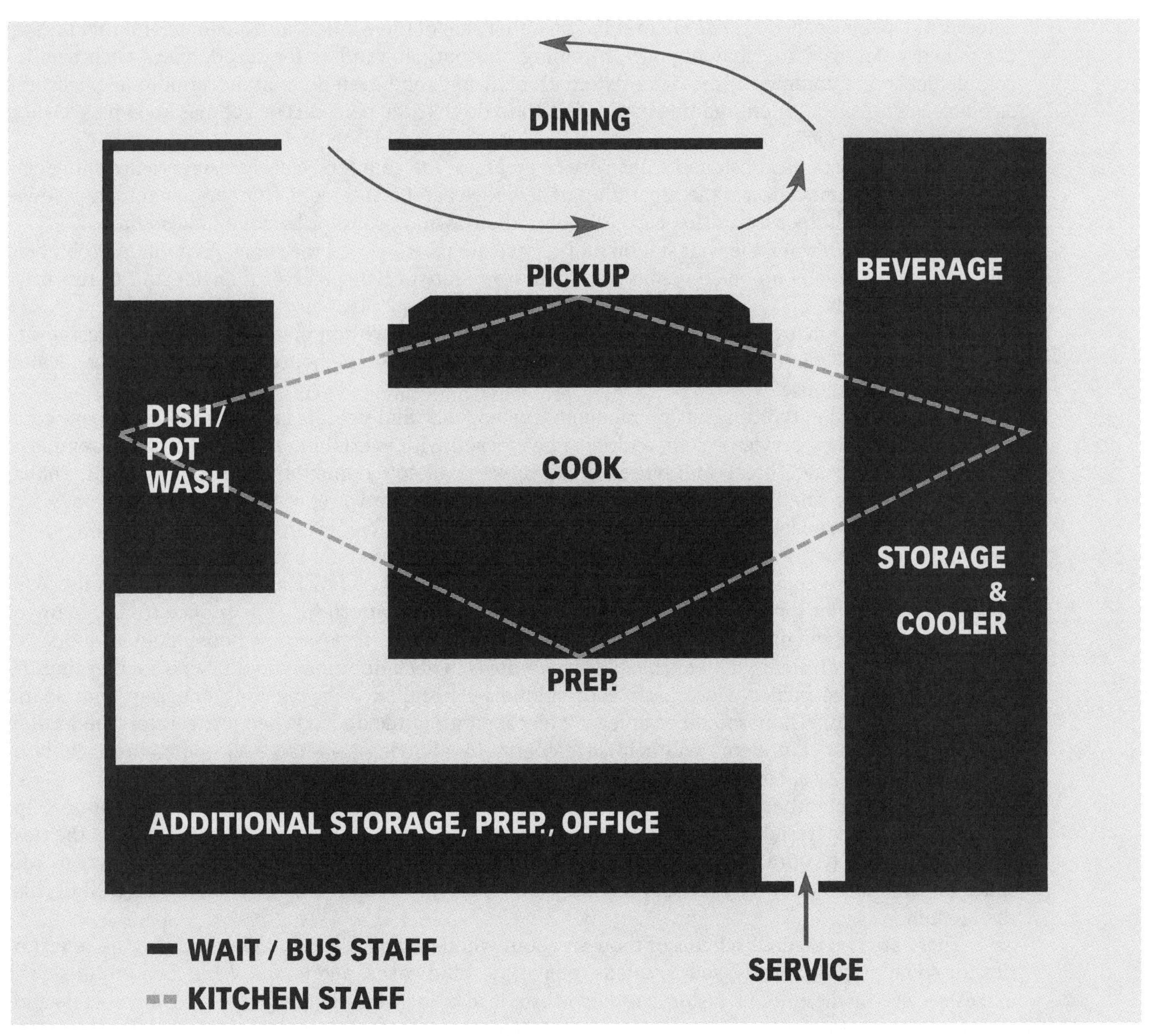

ILLUSTRATION 3-10 **The "diamond" shape can also be used to create a diagram that shows the four primary work areas (prep, cooking, dishwashing, and storage) in relation to one another and unobstructed paths for kitchen and wait staffs.**
SOURCE: *RESTAURANT HOSPITALITY* MAGAZINE, OCTOBER, 1998.

You'll learn more about warewashing, storage and receiving areas, offices, and employee locker rooms in Chapter 4 of this text. For now, let's briefly discuss the serving and cooking functions of the operation, and the unique space requirements of each.

Service Areas/Wait Stations

These are probably the most difficult areas to describe, since there are as many types of service areas as there are different restaurants! The service area is the busy zone between the kitchen and the guests, in which the food production staff and wait staff use the most efficient means to get food out of the kitchen and to the customers.

The teamwork on both sides of the pass window evolved into today's service area. At a small table-service restaurant, it may be an area no larger than 4 to 6 feet long that houses the pass window. Many eateries hire people called ***expediters*** to stand on either side of the window and facilitate ordering and order delivery. An expediter may organize incoming and outgoing orders for speed, check each tray for completeness and accuracy of the order before it's delivered, and even do a bit of last-minute garnishing as plates are finished. When standing in the kitchen to do this, the person is sometimes known as a *wheel person* or *ticket person*.

In quick-service establishments, the service areas are the counters, clearly visible to incoming customers. A recent innovation is the separation of beverages and condiments from prepared foods. Nowadays, customers fill their own drink cups, dispense their own mustard or ketchup, and so on.

In some restaurants, the wait station and service area are one and the same. A common problem in this area is that usually no one remembers to plan for plenty of flat space for the wait staff to rest trays or set up tray holders.

Generally, no matter what it is called, this area is viewed with apprehension by many restaurant designers. And no wonder. It is actually an extension of the back of the house that happens to be located in the front of the house.

The wait station typically has no inherent "eye appeal," and yet it is an absolutely necessary component of an efficient service system and must be stocked with everything the wait staff uses regularly. A likely list will include bread and butter, bread baskets, all coffee-making and serving paraphernalia, assorted garnishes, salad dressings and condiments, dishes and flatware, water pitchers and glasses, ice, linen napkins and tablecloths, bins for soiled linens, tray holders, credit card imprinters, computers, calculators, counter space, and so forth. Get the idea? Things are always hopping here.

If your service area is immediately adjacent to your kitchen, a few simple design precautions will make the space safer for employees and provide a better flow pattern for your service traffic. Many of these areas have swinging doors that link them to the kitchen. If possible, choose double doors, 42 inches in width, with clear, unbreakable plastic windows in each door that should be no smaller than 18 inches high and 24 inches wide. Each door should be installed to swing only one way and clearly marked "in" or "out." There should be at least 2 feet separating the doors. When refrigerators are located in the service area, glass doors will help save energy by allowing servers to see the contents without opening the door. Also, consider sliding doors instead of doors that swing open.

The floor is another important consideration. Make it nonskid, just like the kitchen floor. (Options for flooring materials are detailed in Chapter 8 of this text.) And, because there's always the possibility of a spill, be sure there are adequate floor drains. Finally, because of the intense use of this space and the interaction of your servers and production line people, this area should be well lit, like the kitchen.

There are two distinct schools of thought about whether wait stations should also be the repositories for some prepared foods: soups, salads, precut pies and cakes, and so on. While one group says it makes service of these items faster, another insists it also increases pilferage by staff members—and, therefore, increases costs.

Where should your wait station(s) be located? The shorter the distance between wait station and kitchen, the better off you'll be. Long hikes between the two actually increase labor costs, as waiters spend their time going back and forth. It is also harder to keep items at proper temperatures if they must be carried for longer distances.

If space does not permit your wait station to be adjacent to your kitchen, there are a couple of options. The use of metal plate covers will keep food warm for an additional two to four minutes as it is delivered to guests. Or purchase ***lowerators,*** spring-loaded plate holders that can be temperature controlled to provide preheated plates.

In so many situations, food is now prepared in full view of the guests. Therefore, it is more important than ever that designers create work centers that are well organized and can be kept looking clean. Take advantage of undercounter storage space to minimize messiness. Provide a hand sink for hand washing and a utility sink to rinse kitchen tools and utensils, right at the work center. Select surfaces that are easy to clean, from wall paints to countertop materials. Think carefully about where to dispose of waste. Ventilation should receive professional attention, to minimize the grease buildup that naturally comes with cooking and to avoid the possibility of smoke wafting into the dining room.

Cafeteria Dining. For large group-feeding facilities—cafeterias, hospitals, prisons—the service area takes on a complexity rarely seen in a table-service or fast-food eatery. An institutional kitchen may need as

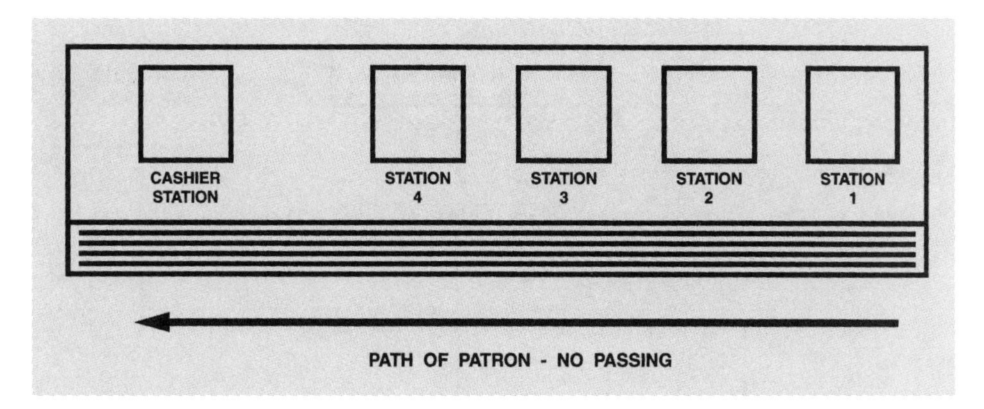

ILLUSTRATION 3-11 **Most cafeteria serving lines are set up as straight-line patterns, although this is often the slowest-moving option on a busy day.**

SOURCE: ROBERT A. MODLIN, EDITOR, *COMMERCIAL KITCHENS,* 7TH ED., AMERICAN GAS ASSOCIATION, ARLINGTON, VIRGINIA.

much as 2000 to 3000 square feet of service area, since this is where serving lines are set up in a multitude of combinations:

1. The straight serving line
2. The shopping center system
3. The scramble (or free-flow) system

The ***straight line*** is exactly what its name implies. In terms of speeding customers through the food line, it is the slowest-moving arrangement, since most guests are reluctant to pass slower ones in front of them. However, single or double straight lines are still the most common design in commercial cafeterias, because they take up the least space and the average guest is comfortable with the arrangement. Since customers must walk by all the food choices, they are also more likely to make an impulse purchase (see Illustration 3-11).

The ***shopping center*** (also called a *bypass line*) is a variation of the straight line. Instead of being perfectly straight, sections of the line are indented, separating salads from hot foods and so on. This makes it easier for guests to bypass one section. In serving lines where burgers, omelets, or sandwiches are prepared to individual order, the bypass arrangement keeps things moving.

The *free-flow* or ***scramble system*** is designed so that each guest can go directly to the areas he or she is interested in. (Once in a while, you'll hear it referred to as a *hollow-square system.*) Food stations may be laid out in a giant U-shape, a square with islands in the middle, or just about any shape the room size will permit (see Illustration 3-12.) It can be attractive but is often confusing for first-time customers. So you're most likely to find this layout in an industrial cafeteria, where employees eat every day and soon become familiar with it. Scramble systems offer fast service and minimal waiting. They also allow for some types of exhibition cooking, including items grilled, stir-fried, or sliced to order.

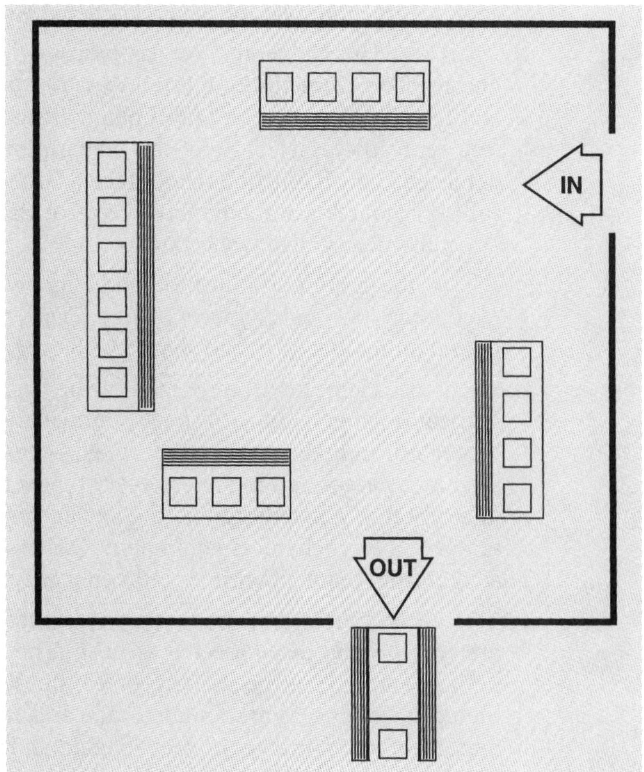

ILLUSTRATION 3-12 **The hollow-square serving arrangement lets cafeteria diners wander between stations.**

SOURCE: ROBERT A. MODLIN, EDITOR, *COMMERCIAL KITCHENS,* 7TH ED., AMERICAN GAS ASSOCIATION, ARLINGTON, VIRGINIA.

Airline foodservice kitchens seem to have the largest and most complex service areas. Several dozen workers line a system of conveyor belts, assembling meal trays for as many as 70,000 passengers a day. To produce this kind of quantity, the

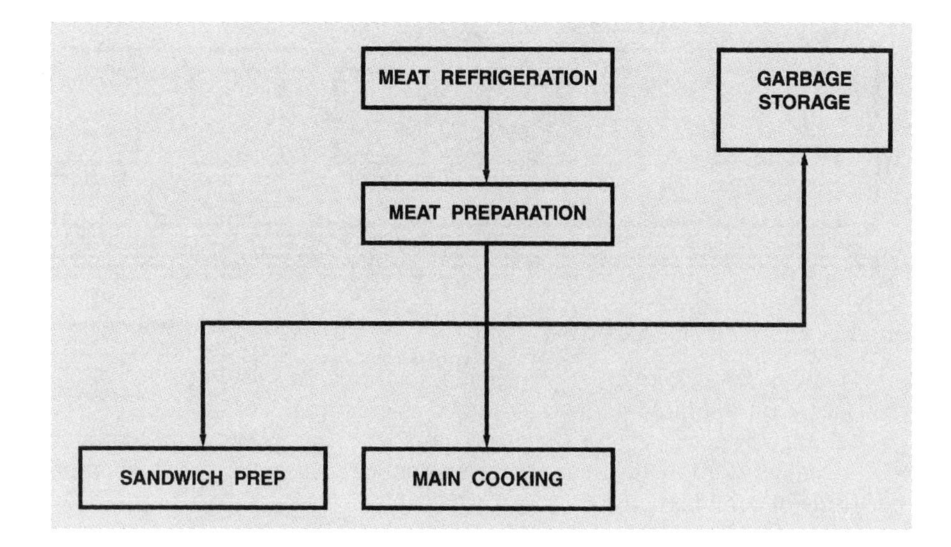

ILLUSTRATION 3-13 A flowchart is the next step after a bubble diagram in planning the movement of materials through your kitchen.

SOURCE: ROBERT A. MODLIN, EDITOR, *COMMERCIAL KITCHENS*, 7TH ED., AMERICAN GAS ASSOCIATION, ARLINGTON, VIRGINIA.

prepared food is held in hot carts, and much of the preparation is done ahead of time to make the assembly process go quickly.

Preparation Areas

We'll use the term *preparation areas* to refer to food preparation. After all, a lot more activities than cooking are crammed into most kitchens! Here are some of the major work sections you may find.

Fabrication. This is where raw (or processed) foods begin their journey to their final destination—the guest's plate. Sometimes referred to as ***pre-prep,*** it is here that we break down prime cuts of beef, clean and fillet fish, cut up chickens, open crates of fresh produce, and decide what gets stored and what gets sent on to the other parts of the preparation area. In planning for each area, start with a flowchart to determine which functions should be included. Illustrations 3-13 and 3-14 show the flowchart and the resulting layout for a butcher shop, one of the most common types of pre-prep sites for restaurants that don't buy meats already custom-cut.

If the restaurant plans to handle its own meat-cutting duties (and many do to save money), you'll need room for a sink, a heavy cutting board, portion scales, meat saws, grinders, and slicers. Some can be placed on mobile carts and shared with other areas of the kitchen.

Preparation. Here, foods are sorted further into individual or batch servings. The loin we trimmed in the fabrication area is cut into steaks, lettuce and tomatoes are diced for salad assembly, shrimp is battered or peeled. Ingredients are also mixed—meat loaves, salad dressings, casseroles. Salad and vegetable prep areas are found in almost every foodservice setting. They are busy places, and their focus must be on efficiency. When designing the layout, remember the need for worktables, compartment sinks, refrigerators, and mechanical equipment. Order some worktables with food and condiment "wells" that are cooled from beneath with ice, allowing easy accessibility.

A prep area with unique requirements is the ***garde manger,*** a term that encompasses both food preparation and decoration or garnish. The garde manger area is the source of cold foods—chilled appetizers, hors d'oeuvres, salads, pâtés, sandwiches, and so on. Obviously, refrigeration is of paramount importance here, as are knife storage and room for handheld and small appliances: ricers, salad spinners, graters, portable mixers, blenders, juicers. Color-coded bowls, cutting boards, knives, scrub brushes and even kitchen towels all help to avoid cross-contamination among different types of raw foods.

Production. Yes, it's finally time to do some cooking, in the production area! This is divided into hot-food preparation, usually known as the ***hot line,*** and cold-food preparation, called the ***pantry.*** Production is the heart of the kitchen, and all the other areas are meant to support it.

Holding. As its name suggests, this is the area in which either hot or cold foods are held until they are needed. The holding area takes on different degrees of importance in different types of foodservice operations. Basically, the larger the quantity of meals produced, the more critical the need for holding space. For banquet service and in cafeterias and hospitals, food must be prepared well in advance and stored at proper temperatures. In fast-food restaurants, the need is not as great but it still exists.

Assembly. The final activity of the preparation area is assembly of each item in an order. At a fast-food place, the worktable is where hamburgers are dressed and wrapped and fries are bagged. At an à la carte restaurant, the cook's side of the pass window is where steak and baked potato are put on the same plate and garnished. Again, in large-scale foodservice operations, grand-scale assembly takes up more room.

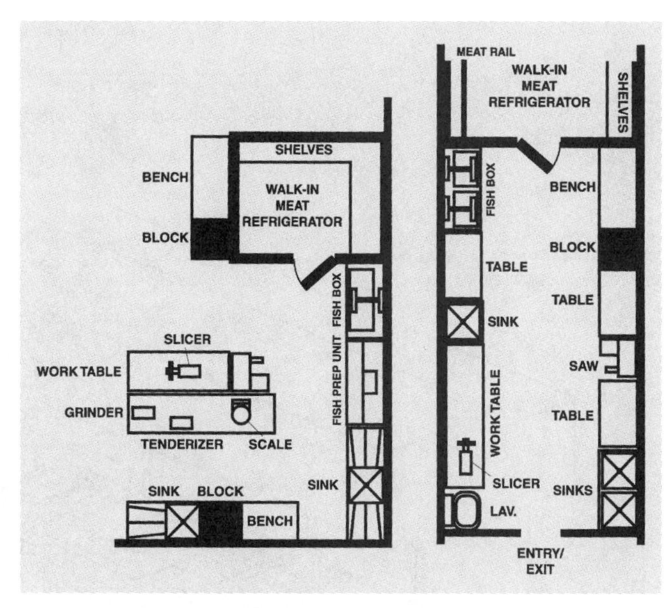

ILLUSTRATION 3-14 **After you've completed a flowchart, it is easier to draw the actual layout of a work center.**

SOURCE: ROBERT A. MODLIN, EDITOR, *COMMERCIAL KITCHENS,* 7TH ED., AMERICAN GAS ASSOCIATION, ARLINGTON, VIRGINIA.

The menu and type of cooking you do will determine the makeup of your production area. Will you need a fabrication area at all, if yours is a fast-food franchise that uses mostly prepackaged convenience foods? Conversely, cooking "from scratch" will probably require a lot of space for preparation, baking, and storage. ***Batch cooking,*** or preparing several servings at a time, will also affect your space allocation. Finally, the number of meals served in a given time period should be a factor in planning your space. Your kitchen must be able to operate at peak capacity with plenty of room for everyone to work efficiently. For a hotel with banquet facilities or for an intimate 75-seat bistro, this means very different things.

Production Areas

Now let's go into greater detail about the hot line and the pantry areas and what goes on in each. Much of today's line layout can be traced back to the classical French brigade system, in which each cook or chef stood in a row and was responsible for preparing certain food items—one fish, another vegetables, and so on.

On the hot line, you'll need to determine placement of equipment based on the cooking methods you will use. There are two major ways to cook food: "dry heat" methods (sauté, broil, roast, fry, bake) and "moist heat" methods (braise, boil, steam). The difference is, of course, the use of liquid in the cooking process.

A kitchen that is well laid out may position its volume cooking at the back and its to-order cooking at the front. In large-volume food production, deck ovens, steam-jacketed kettles, and tilting braising pans will be found together. In smaller (batch or à la carte) kitchens, fryers, broilers, open-burner ranges, and steam equipment are grouped together.

In kitchens that require both types of cooking (large and small batches), the large-volume, slower-cooking equipment is placed behind a half-wall under an exhaust hood (see Illustration 3-15). On the other side, or in front of the wall, is the equipment used for smaller-volume or quick-cooking dishes. Because the smaller dishes require more attention, putting this equipment at the front of the wall allows them to be watched more carefully.

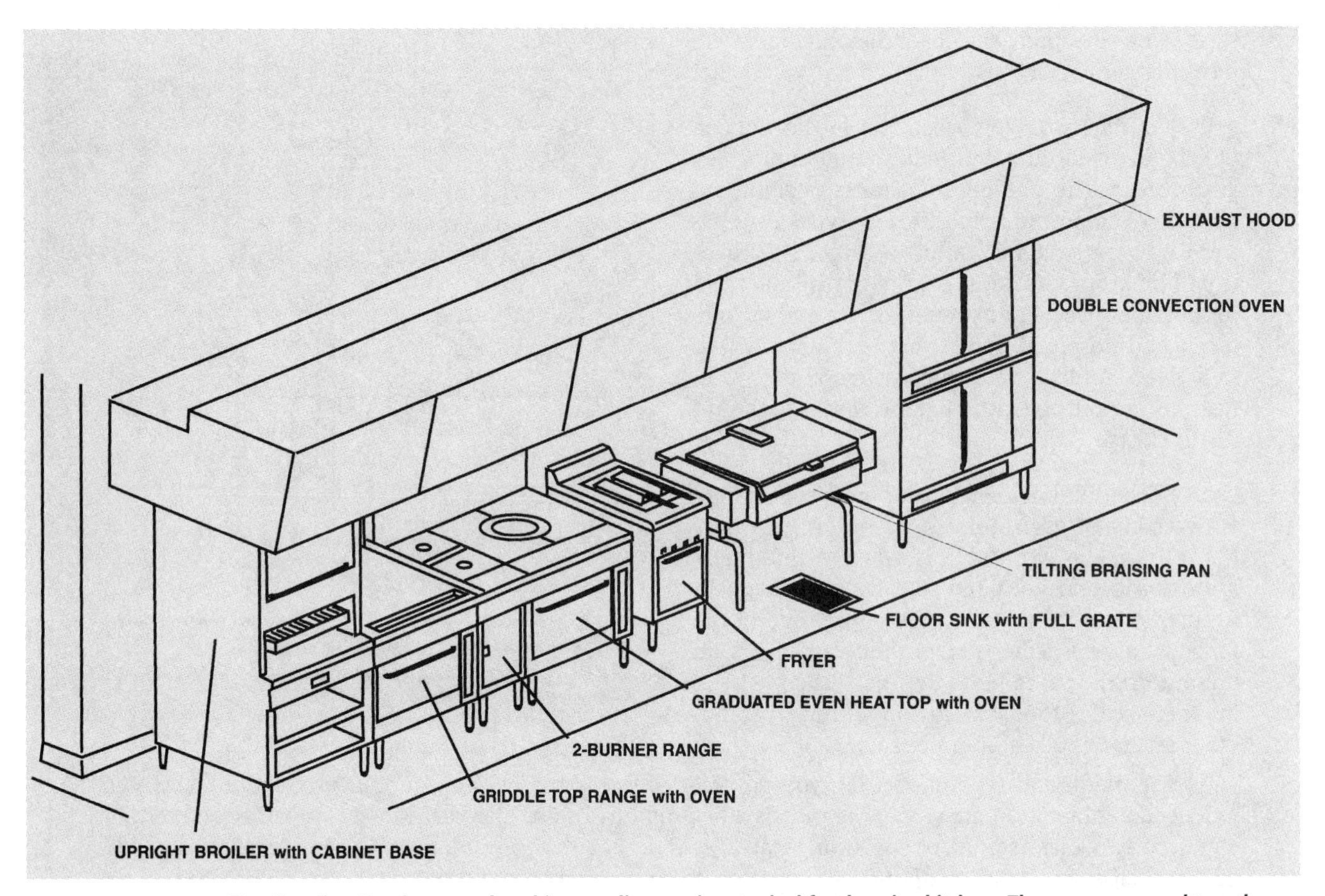

ILLUSTRATION 3–15 **The "hot line," or battery of cooking appliances, in a typical foodservice kitchen. They are arranged together under an exhaust hood system.**

SOURCE: CARL R. SCRIVEN AND JAMES W. STEVENS, *MANUAL OF EQUIPMENT AND DESIGN FOR THE FOODSERVICE INDUSTRY.* THOMSON LEARNING, 1989.

In most communities, it is a law that all heat- and/or moisture-producing equipment must be located under ventilating hoods. Also, all surface cooking equipment (range tops, broilers) must be located so that the automatic fire-extinguishing (sprinkler) system can reach them. Ventilating hoods are very expensive, so the tendency is to crowd a lot of cooking equipment into the small space beneath them.

Here are some of the "stations," or sections, of the typical hot line.

Broiler Station. A single or double-deck broiler is found here.

Griddle Station. The width of the griddle determines its capacity. It can be either a floor-standing model or a tabletop griddle that sits atop a counter. The tabletop model is a smart use of the space below it, where the cook can store supplies. There are even refrigerated drawers, so that meats can be kept chilled until they are ready to be cooked.

Sauté Station. Here, we find three types of ranges: the flat-top sectional range, with individual heat controls for each section; the ring-top range, with rings of various sizes that can be removed to bring flames into direct contact with the sauté pan; and the open-top range, also sectional, with two burners per section. Ovens can be placed tabletop in this area or they can be installed below the range. Above the range is often a mini-broiler known as a ***salamander,*** used for quick duties like melting butter or as a holding area.

Sauce Station. Not every kitchen has room for this station, where soups, sauces, and casseroles are prepared on range tops or in steam-jacketed kettles.

Holding Station. Once again, we have an area designated for holding finished food before it is assembled on plates or put onto trays. Any of the individual stations can have their own holding areas or there can be one central area. Either way, they often include a dry or wet steam table or hot-water baths for keeping sauces warm.

Utility Distribution Center. In high-volume production kitchens, such as banquet and catering operations, the installation of a ***utility distribution system*** is recommended. A utility distribution system is designed to provide all the necessary services (gas, electricity, hot and cold water, and steam) to the cooking equipment placed under the exhaust canopy (see Illustration 3-16). Connected directly to the canopy, it's like having a huge fuse box that controls all the utilities, with a single connection for each of them, plus emergency shutoffs and inspection panels to allow quick access to (and easy cleaning of) all components. As cooking equipment is added, or moved, the system can be adapted, rearranged, or expanded.

We'll discuss utilities more in Chapters 5 and 6, but there are a couple of terms you should be familiar with in discussing a utility distribution system. Most systems are shaped like a big H. The two vertical, upright posts are called ***risers.*** The larger one, which houses the main utility connections, is called the *primary riser;* the smaller one is the *secondary riser,* which serves mostly for balance. The horizontal section between them is known as the ***raceway.*** Here, you'll find the outlets that connect the system to the cooking equipment.

Pantry. The pantry is where cold foods are prepared for serving. Preparation responsibilities here may include salads, sandwiches, cold appetizers and entrées, and desserts. Sometimes it is referred to as the "cold kitchen." Illustration 3-17 shows the components of this complete workstation.

A two-compartment sink is a must for this area, since salad greens must be washed and drained here. Worktables with cutting boards should be adjacent to the sink, and it is helpful if the sink itself has corrugated drain boards.

Refrigeration is also necessary, since cold storage is required for many ingredients, as well as a holding area for cold prepared foods. In restaurants, this could be a pass-through setup, with sliding glass doors on both sides so that pantry employees can slide finished items into the refrigerator and servers can remove them as needed from the other side. In banquet or other high-volume foodservice operations, a walk-in refrigerator may be more appropriate. Inside the walk-in, mobile carts can store finished trays of plated salads for easy access. If reach-in carts are used, make sure they have tray slides on them that will hold large sheet pans.

For sandwich-making, the pantry needs slicers and other types of cutting machines, as well as mixers for making dressings. Because these are often items that are shared with the fabrication area, the equipment should be placed on rolling carts.

Another unique need is keeping breads enclosed so they do not dry out. And, if recipes require toasted bread, at least one commercial toaster is needed.

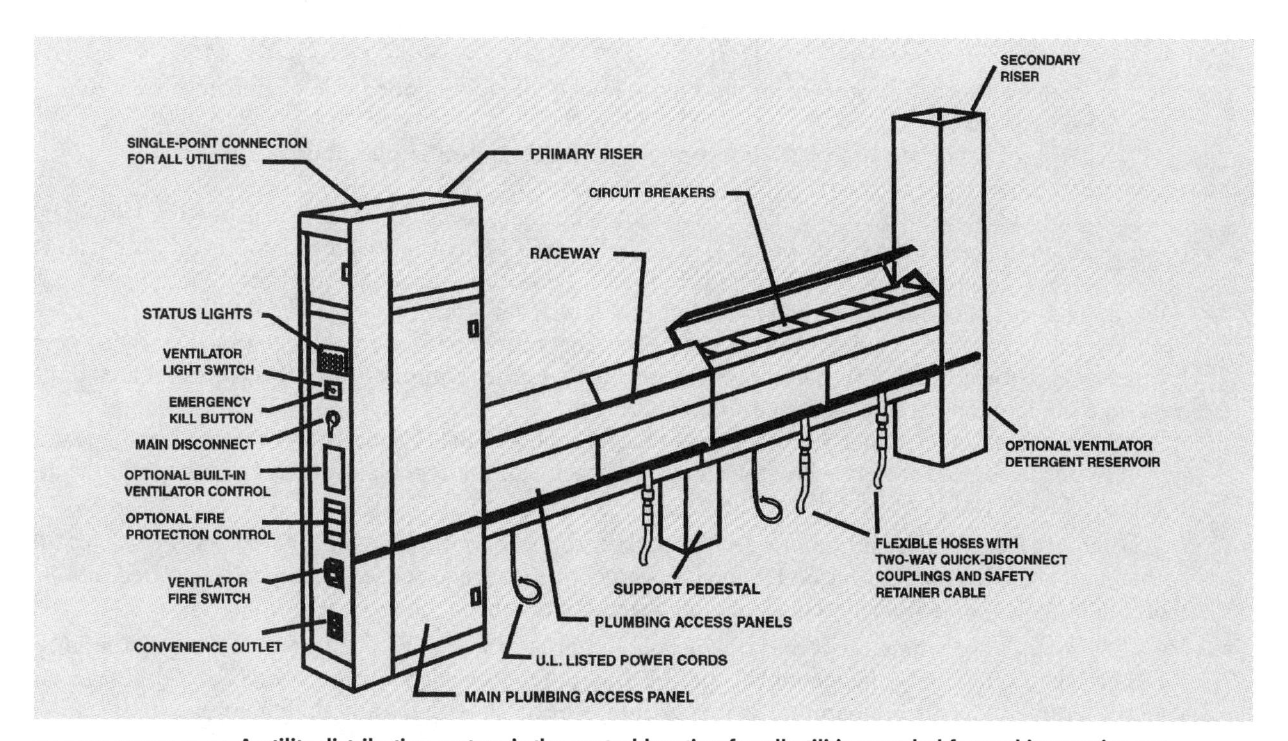

ILLUSTRATION 3-16 **A utility distribution system is the central location for all utilities needed for cooking equipment.**

ILLUSTRATION 3-17 **A pantry area layout.**

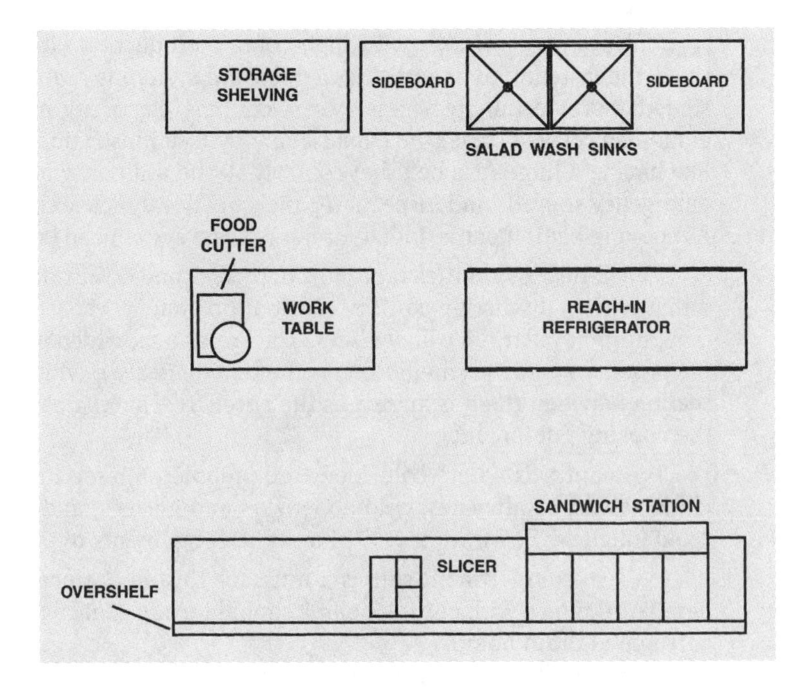

When the pantry is also the site of dessert preparation, additional refrigeration and counter space are needed. You must find room for a freezer and an ice cream chest with its special water well to store scoops. The latter must meet local health ordinance requirements. Soft-serve machines may also be needed. Again, mobile carts equipped with shelf slides can be helpful, both to store and to transport products.

Bakery Area

We use the term "baking" to refer to foods that start as batter or dough. Breads, desserts, and pastries are the output of the typical restaurant bakery area, which has its own unique characteristics and needs.

Before designing a bakery area, ask yourself:

1. **What baked goods will we prepare here, and what baked goods will we purchase from outside sources?**
2. **Will we bake "from scratch" or use premixed items and/or frozen doughs?**

Often, to bake or not to bake on site is strictly a business decision, since it does require additional room, equipment, and skilled personnel. There are dozens of hybrid arrangements in foodservice: A restaurant may decide to purchase all breads but make desserts in house, or vice versa. Pizza operations must do baking on site, but other than that, there are lots of variations.

Within the kitchen, a bakery generally takes on one of two basic forms. It is either a compact area located next to the hot line, like the sample layout shown in Illustration 3-18, or it is a distinct area of its own, like the larger bake shop shown in Illustration 3-19.

When it is a compact area, the setup may be as simple as a baker's table with storage bins beneath, which shares ovens and mixers with other areas. A larger, stand-alone baking area has other, unique attributes, some of which we will discuss next.

Mixing Station. Here, you'll find an array of large floor or tabletop mixers and their accessories. This, of course, requires electrical power and a generous amount of storage space. Here, you'll also find a table with scales, where ingredients are weighed, measured, and mixed.

Proofing Station. This is where the mixed dough is held to rise, with controlled temperature and humidity. Proofing boxes (enclosed cabinets for this purpose) require electricity, a water source, and drainage capability. Mobile racks also roll around in this section, storing sheet pans and dough.

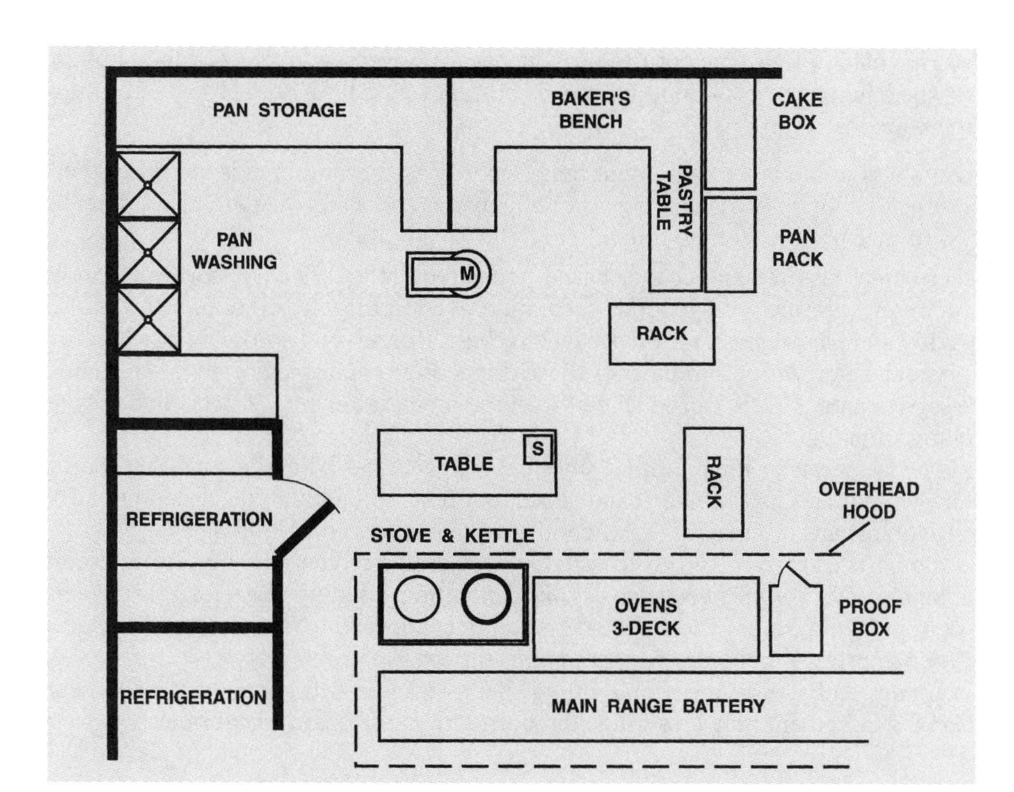

ILLUSTRATION 3–18 The bakery area is an integral part of many kitchens. Here is a layout for a small corner site.

SOURCE: ROBERT A. MODLIN, EDITOR, *COMMERCIAL KITCHENS,* 7TH ED., AMERICAN GAS ASSOCIATION, ARLINGTON, VIRGINIA.

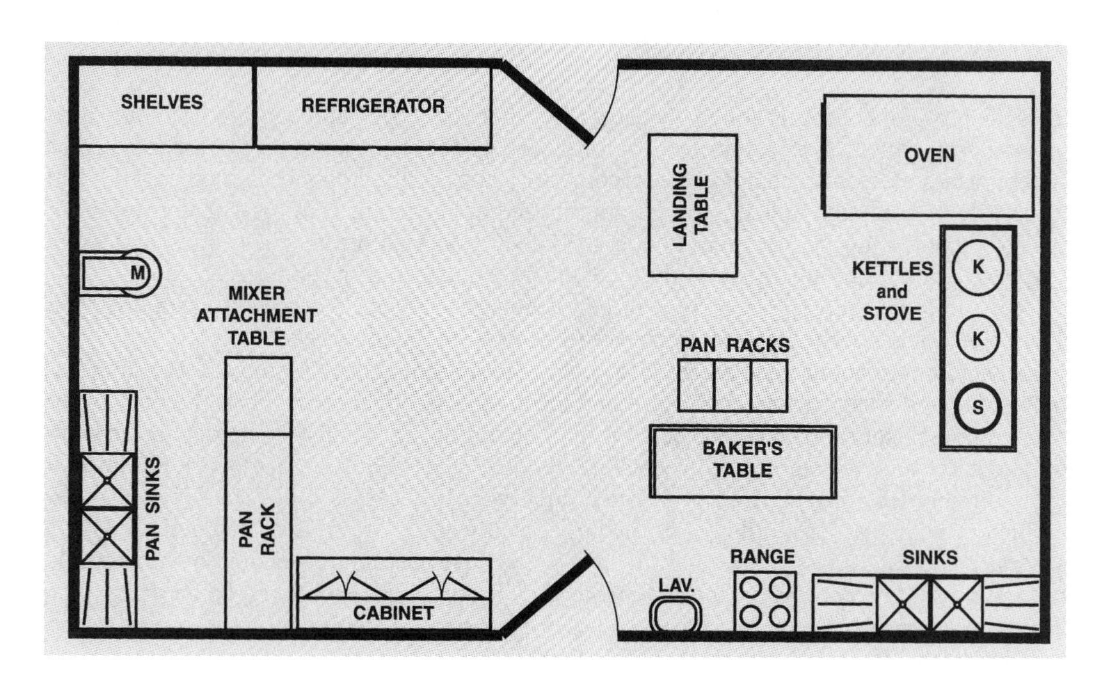

ILLUSTRATION 3–19 A layout for a large, high-volume bakery.

SOURCE: ROBERT A. MODLIN, EDITOR, *COMMERCIAL KITCHENS,* 7TH ED., AMERICAN GAS ASSOCIATION, ARLINGTON, VIRGINIA.

Forming Area. This is where dough is shaped into rolls, pies and cakes are put in pans, and so forth. In small kitchens, the mixing station must sometimes double as the forming area. The tasks performed here require bakers' tables with tops approved by local health authorities, and mechanical dough dividers and shapers. The same mobile racks from the proofing station can stop here on their way to the oven.

Baking Station. Ovens and exhaust hoods take up most of the space here. The types of ovens depend entirely on what's being baked. If there is only one type of product, in large quantities, a carousel or reel oven is preferable. The combination steamer/oven can add moisture to the baking process to make harder crusts for some types of bread. Deck ovens are stacked on top of each other, each individually temperature controlled, for baking two types of products at once. Convection ovens circulate hot air inside the ovens during the baking process—good for breads but not so great for some types of pastries and cheesecakes. So there are many options, depending on your menu needs.

Finishing Station. A baker's worktable is a requirement here, where pastries and breads are given their final form prior to serving. This includes slicing, decorating, glazing, and so forth. You'll need room to stand the mobile racks, so that product can be rolled in and out of this station.

In creating a bakery area, also remember that baked goods require a variety of storage environments, ranging from room temperature with good air circulation to constant refrigeration. If this is also the general dessert-making area, leave room for the ice cream chest, freezer, and soft-serve machine.

What can a bakery area share with other parts of the kitchen? Refrigerator space might be shared with the pantry or the service area. Mixers can be shared with the preparation area, especially if they're tabletop models on rolling carts.

The bakery should be located near the pot sinks, since baking generates a lot of dirty dishes. Some bakery areas have their own pot sinks, as well as hand sinks, too. It's also handy to be close to the dry storage and walk-in refrigerator areas, since that is where the raw materials come from.

Worktable space should be no less than 6 feet per employee, and adequate lighting is important because of the delicate kinds of work done here, such as cake decorating. Landing space for baked goods is needed, especially in front of the ovens. There should be plenty of room for sheet pans, which measure 18 by 26 inches. Proofing boxes should also be placed near the ovens.

In preparation of fillings and syrups, a portable tilting steam-jacketed kettle may be needed near the ovens. Another special requirement may be a range top burner to melt sugar or heat sauces.

Warewashing Area

Warewashing is the term for collecting soiled dishes, glasses, flatware, pots, and pans and scraping, rinsing, sanitizing, and drying them. It is by far the most necessary, and least exciting, part of the foodservice business. Whether dishes are done by hand or by machine, or a combination of both, it's a messy job with high temperatures, high humidity, and slippery floors that require constant caution. For the restaurant owner, this is also one of the most costly areas to operate. Labor (and high turnover), utilities, and equipment are all expensive, and breakage can account for 10 to 15 percent of dish room expenses.

Cost is not the only reason warewashing is of major importance. Quite simply, it has a direct impact on public health. Done poorly, you jeopardize your business on a daily basis.

You'll learn much more about mechanical dishwashing requirements in Chapter 16. For now, let's cover the way bussing and washing impact the size and location of the dish room. First, the dirty dishes have to get there. Some restaurants have bus persons or wait staff bring the dishes directly to the warewashing area; others try to save some steps by stacking the dishes in bus tubs on carts located throughout the dining room and bringing them to be washed only when the cart is full. (The biggest type of mobile cart, which can carry at least a dozen of the large, oval waiters' trays, is nicknamed the Queen Mary.) Quick-service and cafeteria-style restaurants may opt to have customers bus their own tables, depositing waste and trays at receptacles near the doors as they leave. Emphasis should be on designing a system to make sure every plate, cup, or fork has its own flow pattern, to get it in and out of service quickly and efficiently.

A smooth, short flow pattern from dining room to dish room will keep breakage to a minimum and lower labor costs. You'll need to find the shortest route that minimizes the natural noise and clatter of working with dishes, and also keeps them out of sight of the guests as much as possible. In institutional settings, the dish room should be placed in the flow pattern of outgoing guests, so as not to interfere with the incoming ones. A conveyor belt or placement of walls or partitions can minimize contact with dish room sights and sounds.

Dishes must be scraped and stacked when they arrive at the warewashing area. Sometimes, the scraping and stacking is the responsibility of the bus persons or waiters who bring in the dishes. This method is known as ***decoying,*** and it allows the dishwashers to concentrate on rinsing the dishes and loading them into machines. No matter who scrapes, you'd better figure out where the scrapings are going to end up. A trash can with plastic liner? A garbage disposal?

Glasses and flatware have special sanitation needs. As used flatware is taken off a table, it should be placed immediately into a presoaking solution—either in a sink or a bus tub. When heading into the dishwasher, flatware should be placed, facing up, in perforated, round containers, much like those you're familiar with in home dishwashers. Glasses and cups are arranged face down on special racks, not placed on regular dish racks along with the plates. If stemware is used, like tall wine or water glasses, there are special racks with individual, high compartments to prevent chipping the glass bases. There must be adequate room for storage of these specialized containers, both full and empty.

Neither flatware nor glassware should be towel-dried upon removal from the dishwasher. Instead, place flatware on large, absorbent towels and sort it as it dries; allow glasses to air-dry before removing them from racks. Again, this requires space. Most restaurants store their glasses in the racks until they're used again, which saves storage space and time.

There are a number of ways to organize the dish room. In most cases, you will set your equipment in a straight line, an L-shape, or a hollow square, with the equipment making the sides of the square and employees standing inside. Illustration 3-20 shows sample dish rooms in each of these three common configurations.

In determining the size of your dish room, consider the size of the dining room. Remember that each person served will generate six to eight dirty dish items. Calculate how many full dish racks your dish machine can process in one hour. Finally, realize that the best dish rooms are, on average, working at 70 percent efficiency.

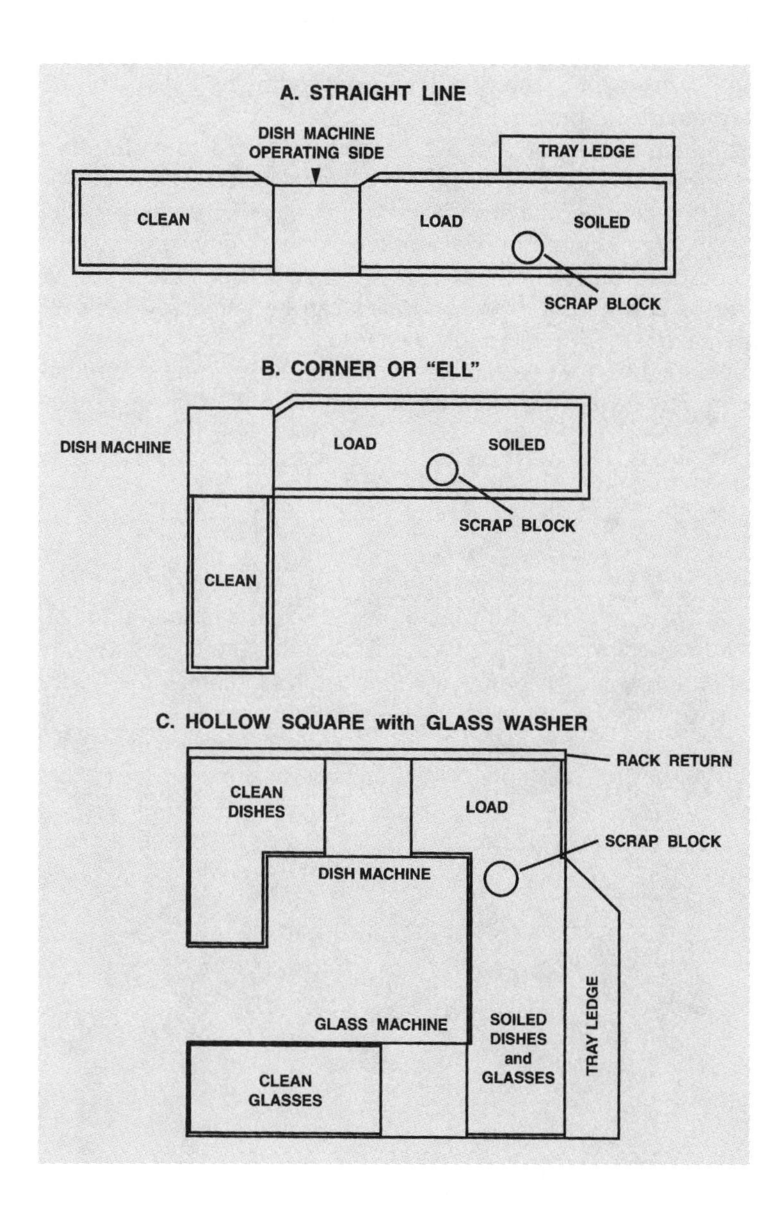

ILLUSTRATION 3-20 **Dish room configurations.**

SOURCE: ROBERT A. MODLIN, EDITOR, *COMMERCIAL KITCHENS*, 7TH ED., AMERICAN GAS ASSOCIATION, ARLINGTON, VIRGINIA.

Inside the dish room, scraping, spraying, lifting, and carrying mean greater labor costs and the potential for disaster without special attention to safety. Nonslip floors and adequate drainage are necessities. Usually, city ordinances also specify lighting requirements (70 to 100 foot-candles of brightness) and govern the ventilation system. Dish machines generate so much heat and humidity that proper ventilation is necessary for employee comfort; as a bonus, ventilation also helps clean dishes dry faster. Acoustical tile for walls and ceilings is important in this area, to keep dish room noise from spilling into the dining area.

Another item that must be carefully chosen is the type of door that links the warewashing area to the rest of the facility. Swinging double doors are not a good idea here, for the same reason we discourage them between the kitchen and service areas, unless each door is installed to swing only *one* way, there is at least 2 feet between the doors, and each door is clearly labeled "IN" and "OUT."

Other Considerations

As long as we're on the subject of doors, what is a "good" restaurant kitchen door? First, it should be lightweight. It should open easily, since it will be opened often by people who have their hands full, carrying trays of food, bus tubs, and other heavy items. Prolong every door's useful life by making sure it has a metal ***kick plate*** or *scuff plate* over the bottom area on both sides—the spot most likely to be kicked open by scurrying feet. Finally, kitchen doors should always contain eye-level windows, so employees don't barrel through and hit each other accidentally.

These kinds of doors come in standard sizes, with a variety of looks, depending on the size of the kick plate, the color, and whether you want guards for the door jambs.

Every kitchen needs work surfaces, which you know by now are often referred to as ***landing spaces.*** These surfaces may be mobile or permanent. They are usually made of stainless steel. Illustration 3-21 shows some basic worktables. As you can see, they can be ordered with shelves, sinks, slots for sheet pans, drawers, and even overhead pot racks for additional storage. As you study your options, think about safety. Can anyone catch a pocket or button on drawers? Make sure they have recessed handles. If you'll be doing cooking classes or demonstrations, worktables can be purchased with removable, adjustable overhead mirrors or chalkboards. There's a lot to choose from!

Another decision to make is what type of edges your worktables will have. Edges can be rounded (called a rolled edge), curved upward (a marine edge), raised, straight, and so on. If your employees

ILLUSTRATION 3-21 **Stainless steel worktables are important in any kitchen.**

COURTESY OF INTERMETRO INDUSTRIES, WILKES-BARRE, PENNSYLVANIA.

must slide heavy containers from one table to the other, you'll need different kinds of table edges than if your goal is simply to prevent spills.

In Chapter 8, you will learn more about the correct heights for work surfaces, to ensure the comfort of your kitchen staff. Worktables also come in standard widths (usually 30 inches) and lengths (from 24 inches up, in 1-foot increments). Tables longer than 72 inches require underbracing for additional support. As you design your kitchen space, remember that long tables are often more inefficient because of all the steps it takes to walk around them. If you can do the same job with two shorter tables and a bit of space between them, do it. Better yet, use tables with ***casters*** that allow them to roll. Locking casters act as brakes, so you can keep them from rolling if you wish. Casters are capable of carrying heavy loads without squeaking, rusting, or corroding in the intense heat and humidity of a kitchen environment. The heavier the load, the larger the wheels you will need as casters. The bearings in each wheel allow them to roll with ease under weight; roller bearings carry more weight than ball bearings, but ball bearings roll more easily.

Placement of sinks, water supply, and electrical outlets should be priorities in designing any kitchen. We'll focus more on specific utility needs in Chapters 5 to 7.

Floor and wall materials should be damage resistant, with easy-to-clean surfaces. Ceramic glazed tile on walls will withstand both heat and grease. For floors, quarry tile that contains Carborundum™ chips is an excellent option that provides natural slip resistance. There should be a floor drain in front of every sink in the preparation area, and a floor drain for every 6 linear feet of your hot line. One smart alternative to individual drains is to cut a 4-inch deep trough along the hot line floor containing several drains, covered by a metal grate.

We have talked mostly about placement of equipment and worktables, but do not overlook the importance of adequate aisle space. After all, that's where you'll "store" your people! In addition to maximum foot traffic, they will be rolling carts and equipment around constantly. Where space is at a minimum, you can declare some narrow aisles "one-way" and work this right into your flow pattern. Overall, however, kitchen aisles should be at least 36 inches wide, or wider if they carry "two-way" traffic or mobile cart traffic.

Finally, you may not think of it as equipment, but you'll need plenty of trash receptacles—and places to put them—throughout the kitchen. Trash cans should be lightweight plastic, covered, and on dollies so they can roll around. Always use trash-can liners to make them easier to empty.

3-7 SERVICE OR BANQUET KITCHENS

If your restaurant plans include space for private dining—meeting rooms or separate catering areas—you may also need a ***banquet kitchen*** to properly service these areas. This kitchen will probably not see daily use, but when it is needed, it is a labor-intensive, production-oriented place that requires powerful, reliable, and multifunctional equipment. If your banquet and special events business becomes successful, you will stretch your main kitchen resources awfully thin without an "extra" banquet kitchen, and you'll run your wait staff ragged if the meeting rooms are located far from the main kitchen. Banquet kitchens are sometimes called ***service kitchens.***

Think of your service kitchen as an extension of your main kitchen. The purpose of the banquet kitchen is to make only the final food preparation before serving to the banquet or meeting crowd. Only modest, one-day storage is needed here and—since most foods will be delivered partly or completely prepared, directly from the main kitchen—there is no need for separate prep areas in the banquet kitchen. All cleaning, peeling, slicing, and butchering can take place in the main kitchen. You do not need a separate dishwashing area, either.

If you make the financial commitment to a service kitchen, it is true that it is making money only when it is in operation. However, don't think of it as a waste if it is not constantly busy. Remember, when the lights are out and nobody's using it, it is still depreciating as an asset, not actively costing operating dollars or shrinking your bottom line.

How much space will you need for these additional kitchen facilities? It depends mostly on how many people you can seat in your banquet rooms. In the Sofitel Hotel in midtown Manhattan, an 800-square-foot production kitchen in the basement also functions as the prep kitchen for the hotel's on-site restaurant and a second, 700-square foot kitchen adjacent to the hotel's banquet rooms is where the meals are heated, finished, and assembled.

A service kitchen of 75 square feet can accommodate seating of 50 to 100; for 1000 seats, you'll need as much as 500 square feet of kitchen space. The general rule is 50 square feet of kitchen space for every 75 to 100 seats. In addition, you will need about ½ square foot per seat for storage of banquet tables and chairs when they are not in use.

In designing a layout for your service kitchen, look at portable equipment you can use to hold and serve food. It will be the most adaptable. If the banquet rooms are close to the main kitchen, you may be able to assign a certain amount of main kitchen space as your "banquet area." At its simplest, it could be a few stainless steel tables on which banquet food is plated for wait staff to pick up and deliver. This area may also include a separate beverage dispenser and ice machine, so waiters from banquet and main dining areas won't trip over each other using the same facilities. Think carefully about traffic flow for busy times of day.

IN THE KITCHEN

DESIGNING A SERVICE KITCHEN

TOP PRIORITY FOR CATERING/SERVICE KITCHENS

1. Stainless steel tables for plating food
2. Combi oven/steamer
3. Cook-and-hold oven
4. Hot food holding boxes
5. Steam table
6. Mixer
7. Tilting kettle
8. Braising oven or tilting braiser
9. Salamander
10. Range top
11. Reach-in and walk-in refrigeration
12. Sink, with hot and cold water
13. Beverage containers
14. Ice bin or (better) ice machine
15. Electrical outlets for all portable equipment
16. Storage for linens, plateware, flatware, glassware
17. Storage for tables and chairs

"NICE TO HAVE" BUT NOT ABSOLUTELY NECESSARY

1. Portable steam table
2. Portable salad bar
3. Fryer, broiler, griddle
4. Three-compartment sink
5. Dishwashing machine

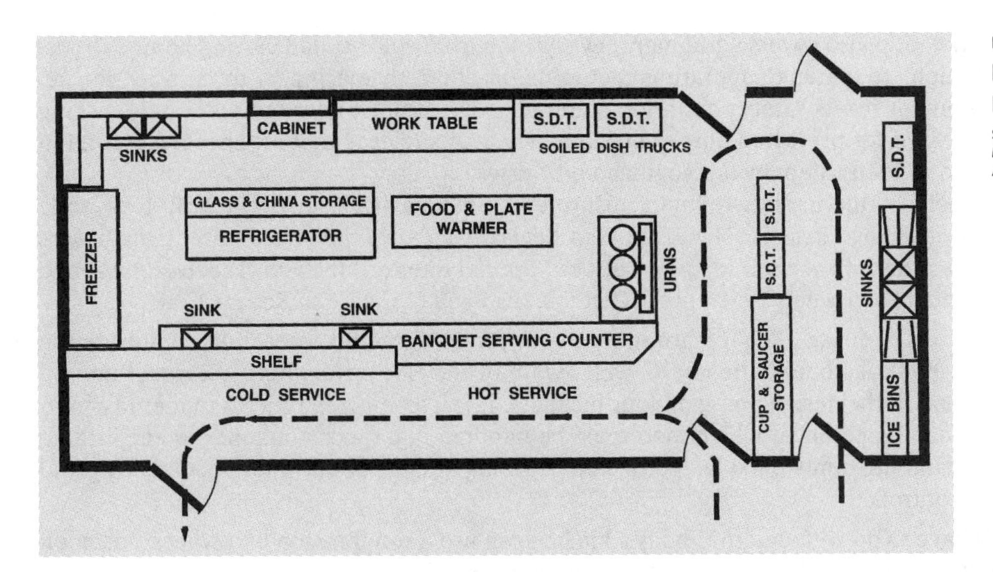

ILLUSTRATION 3-22 The banquet kitchen requires its own separate layout and flowchart.

SOURCE: ROBERT A. MODLIN, EDITOR, *COMMERCIAL KITCHENS*, 7TH ED., AMERICAN GAS ASSOCIATION, ARLINGTON, VIRGINIA.

Depending on how busy your banquet/catering business becomes, there may be a need for additional cooking equipment on the main hot line or perhaps expansion of other stations to meet the demands. Although this does enable you to keep the food preparation all in one space, it has a downside. The additional equipment is needed only when there is a banquet. At other times, it either sits idle or is used by the staff simply because it's sitting there, which is an inefficient way to use your appliances. The real solution is to separate the service kitchen from the main kitchen from the start.

Illustration 3-22 shows a large banquet department, which is located adjacent to a main kitchen.

3-8 FOOD SAFETY AND KITCHEN DESIGN

This part of the kitchen design process is enough to tarnish the stainless steel dreams of any chef! It concerns food safety, a topic no one likes to talk about but which makes headlines nonetheless, any time patrons become ill when dining out. The U.S. Centers for Disease Control estimate there are 76 million cases of food-borne illness annually in the United States alone! The causes of about 95 percent of them are never pinpointed; however, if a pattern is established that points to a certain product, supplier, or restaurant, the chance for legal liability—or at least, very bad publicity—is major.

The U.S. Food and Drug Administration (FDA) developed a preliminary food safety system known as a Food Establishment Plan Review Guide. The idea is to complete the Plan Review and submit it to the local health department before a food-related business opens its doors. This way, work does not have to be redone to comply with the health regulations—they are followed from the very beginning of construction.

A plan review is not a requirement, but it certainly is a good idea for new businesses in the design phase. The checklist and guidelines cover many food buying, handling, and preparation concerns. The initial questionnaire is quite extensive, but it's an excellent reason to sit down and pin down all the particulars of the new operation: numbers of seats and meals to be served, type of service, staffing, and so forth. Required documents to submit with the questionnaire are proposed menus; a site and floor plan showing aisles, storage spaces, restrooms, basements, and any areas where trash may be stored; and manufacturers' specifications for the equipment that's been ordered.

A flow plan is another requirement, including the flow patterns for food, dishes, utensils, and waste. Also necessary is a list of foods that will be prepared more than 12 hours in advance, and a safety plan (based on the HACCP system; see below) for each of these food categories. This includes the methods by which they will be thawed, cooked, cooled, chilled, kept hot until service, and so on, including temperatures. For cold storage, the plan must list how various types of foods stored in the same refrigerated space will be protected from cross-contamination.

The Plan Review Guide also covers equipment: everything from color-coded cutting boards to prevent cross-contamination, to water temperature requirements for dish machines, to certification requirements that equipment meets safety sanitation standards. There must be plenty of hand-washing sinks, conveniently located for use by all employees; and the FDA suggests that at least one person in the kitchen be certified in food safety by a recognized authority.

Using these guidelines, the famous Culinary Institute of America (CIA) in Hyde Park, New York, modified one of its 38 teaching kitchens. The FDA's Plan Review Guide can be obtained on the agency's website. The main page is http://vm.cfsan.fda.gov. Click on "Special Interest," then on "Federal/State Food Programs," then on "Retail Food Safety Reference." That will lead you to the "Plan Review Code" page.

Design and HACCP Compliance. You'll learn more about food safety and inspection compliance in Chapter 8 of this text, but you should already be well aware of the *Hazard Analysis of Critical Control Points,* or HACCP system, in the design phase of your business. It is a seven-step process to identify "critical control points," meaning points at which food must be handled in a specific manner to keep it safe for consumption; determining limits for safe and unsafe handling at each point; then documenting and complying with your controls.

What does this have to do with design? Today's kitchen can use a combination of advanced technology and an intelligent layout to minimize many contamination risks. For example:

- Install reach-in coolers in every prep area, so employees don't have to trek to the walk-ins all the time. This keeps the walk-in temperature steadier, and also keeps batch ingredients at their optimum temperatures until use.
- Make ovens, fryers, rangers and even storage racks mobile so they can be moved away from walls more easily for cleaning around them.
- Place hand-washing sinks closest to the stations that will need them most. There should be separate hand-washing facilities for people who handle produce, versus those who handle raw meats. Electronic sensors on the sinks will turn water on and off and dispense hand soap so employees don't touch the faucets.
- Store raw and finished foods in separate refrigerators.

In a HACCP-compliant kitchen, everything flows forward. Ingredients move in a single direction, without backtracking, from receiving, to storage, to pre-prep and prep, to holding and/or serving. What isn't served, or can't be safely held, is discarded. Equipment also moves without backtracking—it arrives in an area clean and sanitized, remains there until its job is done, and is cleaned and sanitized again before being used in another area.

SUMMARY

Kitchens are organized into work sections, and each work section is composed of work centers where certain tasks are performed. Layout and equipment placement are determined by the duties of each section and which appliances are needed to perform the tasks. Kitchen design should always allow for flexibility in case the menu or concept changes.

Flow patterns are just as critical in designing a kitchen as they are in the layout of your dining area. Decide how workstations relate to each other by drawing a bubble diagram, with each bubble (circle) representing a task or area. Let the circles intersect to indicate shared functions or equipment needs. In this chapter, examples of flow patterns are shown for cafeteria dining, food preparation and storage areas, and dish rooms.

You will also need to determine your service system, the way the food will be delivered to the guests. Distance and kitchen accessibility are the keys to creating workable flow patterns. A major component of your service system is the wait station, which should be designed for efficiency. Placement of wait stations is critical in your design and in the guest's perception of the dining experience. Even the amount of aisle space you allow is important in space planning, because it impacts guest comfort (in the dining area), ease of equipment operation (in the kitchen), and safety (in both areas.)

The biggest modern-day concern of health departments is the ability of commercial kitchens to comply with food safety and sanitation requirements. This has many design implications—everything from where food is stored (individual refrigerators for each work section; separate storage for separate products, to minimize cross-contamination), to where trash is discarded, to where hand-washing stations are placed and how many there should be, to mobility of equipment. It is easier to make these decisions "on paper" in the design phase than to figure them out—and deal with health inspectors' concerns—after the business is open.

STUDY QUESTIONS

1. How do kitchen location and type of service system impact each other in a restaurant?
2. What is a flow pattern?
3. What is the difference between a fabrication area, a preparation area, and a production area?
4. Name three types of space that are often overlooked in kitchen design.
5. What equipment would you install in a separate service kitchen for banquets? What equipment would you share with your main kitchen?
6. Do quick-service (fast-food) restaurants have service areas?
7. What factors do you consider when deciding how much space to allocate for your kitchen production area?
8. City ordinances require that all cooking equipment on the hot line be located near two things. Explain which two, and why.
9. What makes washing glasses and flatware different from washing plates or pots?
10. What are the safety rules for using doors that lead in and out of kitchen and dishwashing areas?

CHAPTER 4

SPACE ALLOCATION

INTRODUCTION AND LEARNING OBJECTIVES

There are two major considerations in deciding on an appropriate amount of space for a foodservice facility, and they should be constant guidelines for you:

- The exact purpose and use of the space
- The cost of building and/or renting the space

There are so many variables in space planning. It is not as simple as deciding how many guests you're going to serve and making sure there's enough room for them. In fact, the guests themselves have some interesting perceptions about dining space. In the early 1990s, the trade publication *Restaurants and Institutions* surveyed 1000 restaurant customers about what space-related improvements they would make at their favorite eateries. Most said they'd like bigger dining areas so they wouldn't have to wait for a table, or more restroom stalls so they wouldn't have to wait there, either.

Space planning begins by examining all the parts of the foodservice facility, even beyond the front and the back of the house.

In this chapter, we will discuss:

- **Exterior:** The parking lot, patio seating, sidewalk tables. Restaurants located in malls or large office buildings must also account for food-court-style seating or any public area that surrounds their business.
- **Interior (Front of the House):** Entryway, dining area, wait stations, beverage service area, restrooms.
- **Interior (Back of the House):** Hot-food preparation area, cold-food preparation area, serving/plating area, bakery, dish room.
- **Kitchen Auxiliary:** Receiving area, dry storage, cold (refrigerated) storage, employee locker rooms, office space.

Not every foodservice operation will need space for each of these areas, and the significance of each area depends entirely on the type of restaurant. Using a combination of square footage and meals served, you can actually chart the amount of space that is needed, as shown on the graphs in Illustration 4-1. These graphs indicate individual space requirements for storage, preparation, cooking, and serving of food, plus dishwashing and staff facilities. As you can see, space requirements do not increase in the same ratio as the number of guests served.

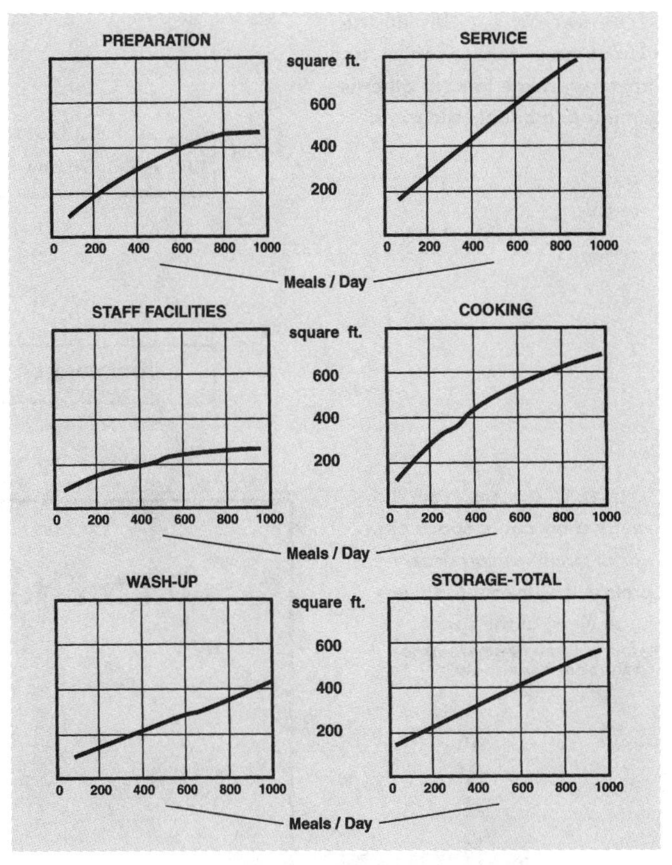

ILLUSTRATION 4-1 **Space requirements for a foodservice operation. As you can see, space requirements do not increase in the same ratio as the number of guests served.**

4-1 WHERE TO START?

There are two fact-finding stages for space allocation and planning. The first requires that you decide what your end product will be: What are the quantities of food you will turn out? This includes numbers of meals, sizes of portions, amount of seat turnover, and type of service. To a lesser degree, you must also consider the surroundings—both kitchen and dining area. What do you have to work with?

The second stage is to decide on your methods of food preparation. Why? Because you must decide which pieces of equipment your space must accommodate. Equipment takes up a lot of space, and you must figure out which pieces will meet your volume needs, as well as what times of day each piece will be used. In this step, you will also be forced to consider related space needs, such as storage areas, serving areas, and utility availability. Yes, it's a lot to consider.

For each piece of equipment you think you will need, draw a rough sketch of the work center where it will sit. As you learned in Chapter 3, a work center is where a group of closely related tasks are done, such as the mixing center of a bakery area. The average work center encompasses about 15 square feet. It contains the machines needed to accomplish the tasks, or room for them if they are movable; space for incoming materials; utilities; storage space for finished products; and enough room for the workers in the area.

You've already read in this text about how each area, or service center, of the restaurant is interconnected with the others. Use these relationships as a starting point to plan your layout. Which one makes sense located nearest to others? You can start with a diagram, as shown in Illustration 4-2, and work toward an overhead view of the space itself, as shown in Illustration 4-3. This is known as your *preliminary space plan*.

No great amount of detail is needed for the preliminary space plan. Like the bubble diagram we mentioned in Chapter 3, it is an exercise on paper, to help you discover whether food and supplies can move logically, and with the least effort, through your operation. The idea is to minimize wasteful backtracking. All this sketching may be time consuming. But wouldn't you rather make mistakes, changes, and improvements on paper than later, with expensive, detailed equipment plans and architectural renderings in hand?

As you work, remember that the restaurant market is so volatile that chances are good you will change something—your menu, concept, or size—within three to five years. Try to give yourself the room and flexibility to grow.

Looming over all these questions is the availability of resources and the ability and willingness of your investors to allow you to create the restaurant you truly want.

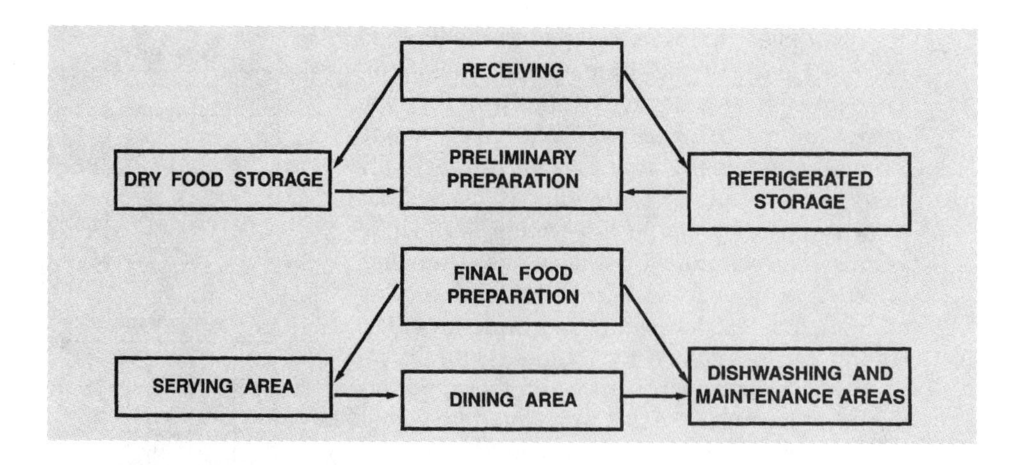

ILLUSTRATION 4-2 This illustration of major service centers and how they relate to each other is similar to a bubble diagram.

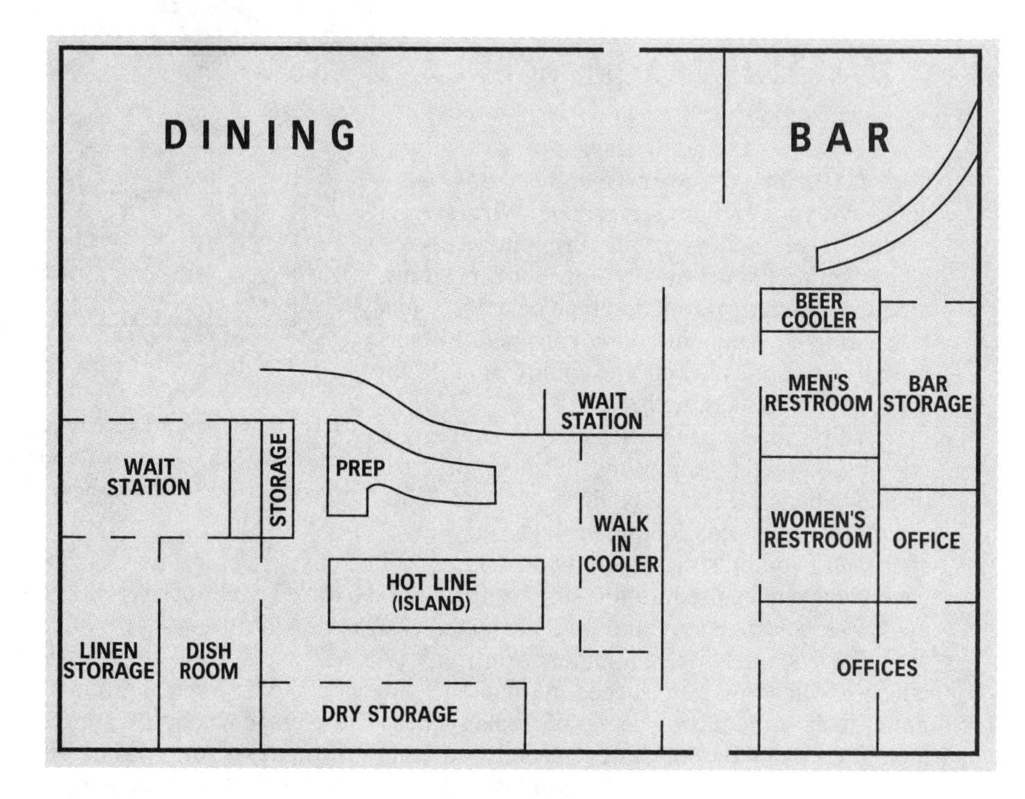

ILLUSTRATION 4-3 From the rough diagram, a space planner or architect can draw a preliminary layout (an overhead view) of the space.
LAYOUT COURTESY OF THOMAS VERDOS, TEVALIA DESIGN BOISE, IDAHO.

4-2 PLANNING PUBLIC AREAS

There are several other major benefits of these preliminary sketches: You'll see if the site you've selected can actually fit your needs. Or, if no site has been selected yet, it will give you a guideline to decide how much property is needed. This means you must determine the number of seats necessary to produce a satisfactory sales volume and, in turn, sufficient return on your investment.

Keep in mind that customer comfort should be very high on your space allocation priority list. As you visualize the dining area, bar or waiting area, lobby, and restrooms, here are some points to consider:

- **Assign the "best" or most convenient area to tables and chairs, booths, or banquettes. "Best" means different things in different settings: proximity to the parking lot, the best view, the least noise, or the most foot traffic. In a "see-and-be-seen" restaurant, the best spots will be quite different from those in a small, romantic bistro or a busy family-friendly café.**

- No surprise here—people usually eat at mealtimes. This means most restaurants are busy only 20 to 25 percent of the hours they are open. Despite all the clever attempts to attract off-hours crowds, it is seldom accomplished. Provide ample aisle space and a comfortable number of square feet per guest so, no matter how crowded it gets, you don't annoy guests by bumping into them or forcing them to bump into each other. Decide where you will store items such as booster seats, high chairs, and extra chairs.
- In planning your seating area, be aware that your guests have some predetermined dining-out habits that influence the efficiency of various table sizes and groupings. Especially at lunch, half of your customers will arrive as twosomes, another 30 percent will be singles or parties of three, the remaining 20 percent will be parties of four or more. Review the discussion of tables and booths in Chapter 2 for layout suggestions.
- If customers must line up to wait for tables or place orders, make adequate provisions in your layout so the line won't interfere with other guests or the wait staff. Also, make plans to control the waiting line, so customers can't ignore it and wander into the dining area to seat themselves at a table that hasn't been cleared.

Some of these sound like small, perhaps even petty, concerns. However, they are best addressed in your initial layout—not on opening week, when you'll have enough to worry about!

In fact, during the space planning process, it is wise to make more than one plan (see Illustration 4-4). Realize that your preliminary drawings can always be altered to improve the final results.

4-3 OUTDOOR AREAS

Parking Areas

The question of adequate parking should be considered very early in the planning process, for two reasons. One: whether you're a freestanding restaurant or in the middle of a shopping mall, ease and availability of parking is a critical issue that can make or break a restaurant. Two: the parking area may be both the first, and last, impression the guest has of your place. We'll talk in a moment about what makes a parking lot attractive, safe, and convenient. First, let's discuss the legal requirements for making parking available in the first place.

In most cases, a city ordinance will clearly spell out the space needs for parking. For freestanding buildings, a widely accepted standard is the need for *one parking space for every 100 square feet of space covered by a roof.* This includes patio dining covered by permanent awnings, for instance, but excludes areas where umbrellas are used. The space "under roof" includes *all* space: hallways, closets, kitchens, bathrooms, and so on. So if we have a total of 3000 square feet "under roof," we need 30 parking spaces.

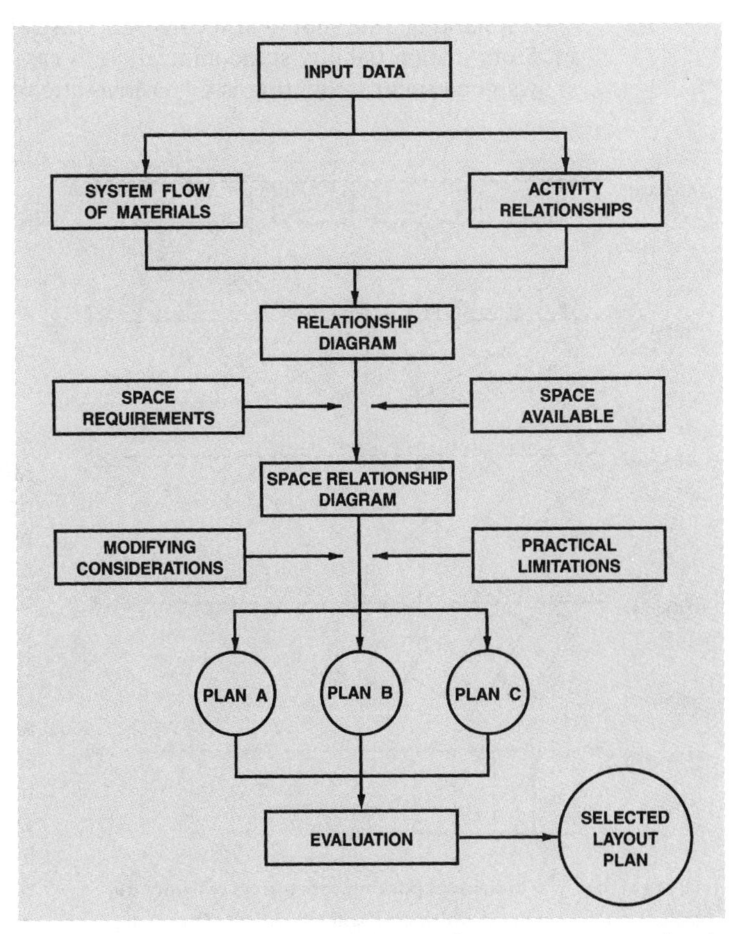

ILLUSTRATION 4-4 **The steps required to go from concept to completed layout. Note that it is a good idea to design more than one layout plan.**

Further, your city will probably spell out how much space to allocate per car. Most ordinances specify that a single parking space must measure 7½ to 10 feet wide and 16 to 18 feet in length. The shorter dimensions are for compact cars, and a parking lot may generally allocate up to 35 percent of its space for compacts. Roughly, this means that a space 40 feet wide is needed to park a car at a 90-degree angle to the building and to provide a traffic lane; slightly less space is required if the car is parked at an angle.

Because parking requirements are quite technical, consult your city zoning authorities to be sure there is adequate space for parking. Illustration 4-5 shows common layout methods for a 60-foot parking lot.

If there is not adequate parking space on the property, don't give up hope. Most cities allow off-premise parking, but only with a signed *parking agreement* with owners of space adjacent to the restaurant. Valet parking is another option, if the valet lot is located within 600 feet of the restaurant. And, if you are located in a historic district or a busy downtown area, there are often different, less stringent standards that require fewer spaces (as few as one space for every 2000 square feet under roof) and allow on-street parking.

Asking for a variance in local parking laws should be your last resort. The process is expensive, time consuming, and loaded with competing (and political) interests that could impact your business.

Now that you've got a parking lot, make it a good one! Its appearance should be consistent with the image you wish to create and, to a certain extent, the parking area also serves as a buffer between your business and the neighbors. The lot should always be kept free of litter, and someone on your staff should be assigned to check it on a daily basis. If there are islands that might be attractively planted with small trees or easy-care shrubs, do a bit of landscaping, or at least put a layer of bark or stones in the island rather than leave it as dirt and weeds. Chicago is one city that encourages businesses to landscape parking areas to help reduce the baking summer heat that radiates from large, flat stretches of cement and asphalt.

A parking area should also convey a sense of safety, keeping criminals out as it draws customers in. From a legal liability standpoint, there is case law (about crimes committed in parking lots) that shows good security lighting is a "positive attempt" on the part of a business to provide a safe environment. There are three basic goals for parking lot lighting:

- To provide a specified intensity throughout the lot
- To provide good visibility (with minimal glare) for customers and employees
- To accomplish the first two goals without annoying the neighbors

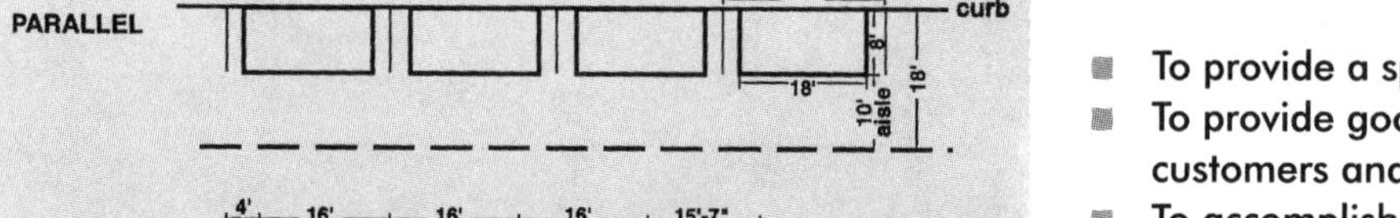

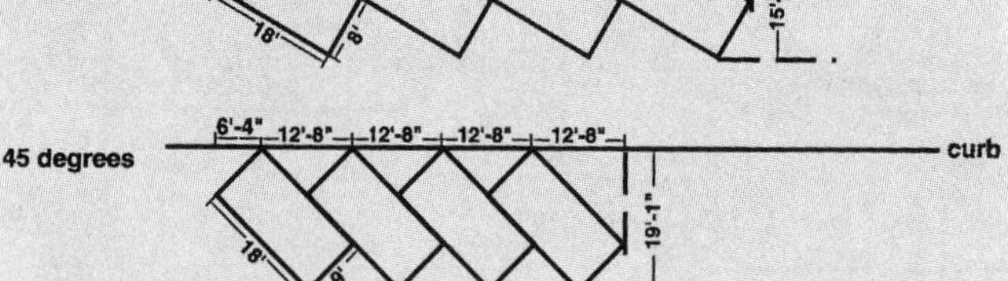

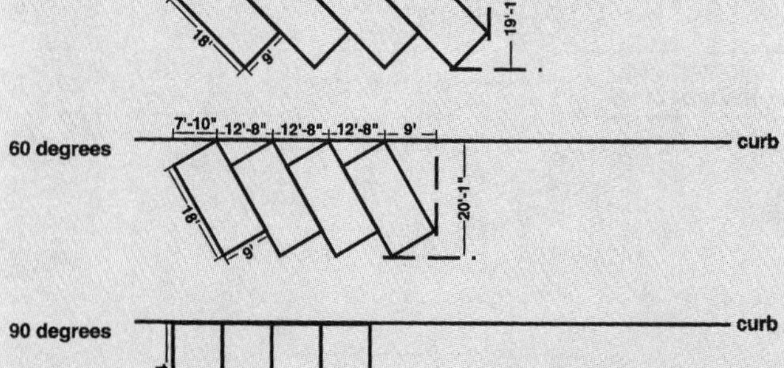

ILLUSTRATION 4-5 **Individual parking space sizes will vary depending on the angle at which cars will be parked. These sketches are based on guidelines from the Traffic Control Department, City of Dallas, Texas.**

The biggest challenge in lighting a parking area is the corridors created by the parked cars. To minimize these, it is recommended that illumination be provided from at least two, and preferably four, light poles that are at least 20 feet tall. (As you landscape, remember that low shrubs and bushes are best, since they won't diffuse overhead light like mature trees will.) The optimum light level for an open parking lot (measured at the pavement) is 3 or 4 foot-candles. (In Chapter 7, you'll find more information about light and how it is measured.) Supermarkets average 3.45 foot-candles in rural areas, and almost 5 foot-candles in urban areas. For the entrances and exits, the lanes within the lot, and any loading zones, the illumination level should be twice the level of either the lot itself or the adjacent street—whichever is greater.

Then again, too much lighting creates another problem. Quick-service restaurants, service stations, supermarkets, and car dealerships have prompted some nasty neighborhood battles with their intensely illuminated lots. Too much light—meant to create safety—may actually result in a hazard as drivers' eyes adjust from nighttime conditions to the intense glare. All the more reason to get bids from outdoor lighting specialists to handle this task correctly.

And finally, depending on your service style and clientele, one of today's parking trends is to reserve (and specially mark) several close-in spaces for specialty use. We've seen spaces for "to-go" pickups, for senior citizens, and for guests with small children.

Patio Dining Areas

Outdoor dining is very popular, but it is not without its space planning problems. In many areas, outdoor dining is only permitted if you have the parking to accommodate it, but the rules sometimes depend on whether (and how) the outdoor dining area is covered: with a tent or umbrella or a rolled or permanent awning. Again, check with the zoning authorities.

The rule for creating an outdoor dining area is to blend it comfortably into the space that surrounds the restaurant. Find out early if it can be fenced or cordoned off in some way, if necessary, and decide whether the existing doors of the building will work for outdoor service: Is there a separate entrance for your outdoor wait staff or must they use the front doors to deliver food to outdoor tables? When things get crowded, is it still workable? Do the indoor guests get blasted by the wind and rain because the doors are constantly swinging?

Outdoor dining also has unique space considerations: an additional outdoor service area, the purchase of outdoor furniture, and the means to secure it from theft or during inclement weather.

Adjacent Public Areas

To maximize public awareness and build traffic, choose a building that will give your restaurant its own outside entrance. If this is not possible, look for space that allows people in the hallways of the building to notice the activity within the restaurant. Some buildings will allow restaurant tables in the hallway, a sort of indoor "outdoor" dining!

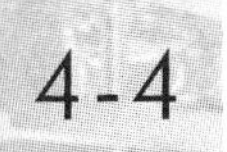

4-4 THE FRONT OF THE HOUSE

Entry Areas

We've discussed this area in detail in Chapter 2. The space allocated to an entryway is a function of the type of restaurant and the owner's overall feelings about the need for this space. In a fast-food facility, the space allocated for an entryway will be a large portion of the whole facility, because it is here that guests line up, place and pay for orders, and pick up their food.

On the other hand, table-service restaurants have entryways that vary widely. They reflect the concept of the facility and provide a small waiting area for overflow crowds. A suggestion here for a casual place is to allow 3 square feet of entry space for each dining room seat, and 5 or 6 square feet per seat for an upscale dining room. The larger entryway sets the mood for a more "grand" entrance into the dining room.

Take-Out Space. The recent upsurge in popularity of take-out foods (the chic new term is "portable dining") and Home Replacement Meals means a different type of entryway experience. The rules are "keep 'em moving," and "keep 'em busy," so it feels like progress is being made—they order, and wait a bit; then pay, and wait a bit; then receive their food. Here are the major considerations in planning for this type of space:

- Speed: The faster you can serve "in-and-out" customers, the better. Drive-through has become the norm, and not just for national quick-service chains, and this requires a different kind of design altogether.
- Space: How many people will be lining up at one time? Do you have room for two ordering, payment, and pickup stations for busy times? What can you do to guide people into lines?
- Menu: Offer only those items on your "to-go" menu that can be easily packaged and transported. Consider ways the guests can help themselves, such as soft drink and condiment selection, or reach-in refrigerators of prewrapped salads and desserts.

Dining Areas

Did you know that in a table service restaurant, customers spend up to 25 percent of their time just waiting? They wait to order, wait for the food to arrive, and so on. They don't feel comfortable if the space is too crowded, and they don't feel comfortable if the space is too empty.

To a large extent, the question of how many seats can fit into a particular dining space defines the experience a guest will have at the restaurant. The trend in major metropolitan areas seems to be to squeeze in "just one or two more" tables, even if this means the difference between convivial and catastrophic. But New Yorkers, for instance, are accustomed to tiny apartments and crowded subways, so

IN THE KITCHEN

SPACE AND SIZING GUIDELINES

1. The restaurant kitchen is approximately one half the size of the dining room.
2. Sizing by seat count:

DINING ROOM

Deluxe—15 to 20 square feet per seat
Medium—12 to 18 square feet per seat
Banquet—10 to 15 square feet per seat

BACK OF THE HOUSE (KITCHEN)

Deluxe—7 to 10 square feet per seat
Medium—5 to 9 square feet per seat
Banquet—3 to 5 square feet per seat
(Add the banquet requirement to the kitchen)

3. Food prep is approximately 50 percent of the back of the house.
4. Storage is approximately 20 percent of the back of the house.
5. Warewashing is approximately 15 percent of the back of the house.
6. Waitstaff circulation is approximately 15 percent of the back of the house.

Source: *Lodging*, American Hotel & Lodging Association, Yardley, Pennsylvania.

perhaps they tolerate the jostling and closeness better than others. It is true that "packing" a restaurant with guests contributes to an overall air of excitement, which designers realize and sometimes exploit in their layouts.

Perhaps it is exciting, but is it safe? As long as we mentioned New York City, the fire codes there do not specify (or limit) a number of seats based on total square footage. Instead, the codes require that restaurants maintain clear corridors, 3 feet wide, leading to fire exits. Restaurants with 74 seats or more are required to have two fire exits; those smaller than 74 seats must still have unobstructed aisles, but to a single exit.

Most people think about how to accommodate crowds; but in larger spaces, you might also want to consider how to separate your dining area into smaller spaces (or spaces that *appear* smaller) when things are slow. Many eateries seat in only one of two or three dining rooms at off-peak times.

A good guideline is to allow 15 square feet per seat. This figure includes aisles and waiters' service stations, but generally excludes the entryway and restrooms. Of course, your figures may be modified by the shape or size of the dining room and the sizes of the tables and chairs.

Various types of facilities have "industry-accepted" standards for dining space allocation, as shown in the "In the Kitchen" box on page 90.

Generally, the lowest space-per-seat allocation is in a school cafeteria, with 8 to 10 feet; the highest is in restaurants with a high average check, at 15 to 18 feet.

One factor to take into account is the ***seat turnover,*** or *seat turn,* which is the number of times a seat is occupied during a mealtime. How much turnover you have depends on the method of service, the time or day, the type of customer, the type of menu and atmosphere, and even the availability of alcoholic beverages at the restaurant.

Arranging Tables and Booths. After the overall dining space is agreed on, you must consider how tables or booths will be arranged within that space. The Floor Plan Guide (see Illustration 4-6) compares different types of table arrangements, with suggested table sizes for different types of dining facilities. Notice that chairs are also drawn, as well as tables, to ensure there'll be enough room to pull them out and seat people comfortably.

We've included lots of suggestions because there are so many variables in table arrangement. Notice that none of the diagrams takes into account items like columns, doorways, and architectural features (unusual wall placement) that often exist.

And what if you have an extremely small dining area and want to get the most number of tables into it? Illustration 4-7 shows how less can be more.

Another parameter in determining seat turnover is that, even when your dining room is "full," all the seats may not be! A party of two may occupy a table that could seat four, and so on. This partial vacancy rate can be as high as 20 percent in table service restaurants or 10 to 12 percent at cafeterias or coffee shop counters. Vacancy rates don't apply to facilities in which meals are all eaten at the same time, such as prisons and military mess halls. For most eateries, however, table sizes can help control the vacancy rate. Arrange tables for two (***deuces,*** or *two-tops*) so that they can be easily pushed together to create larger tables if necessary. Quick-service establishments can also try stools and countertops or classroom-style seats with ***tablet arms*** to accommodate people eating alone.

Arranging Banquet/Meeting Space. One major decision you must make early on is whether to offer round or rectangular tables for banquet seating. It is especially important when planning this space to allow enough room for aisles, since the wait staff will truly be bustling (with full trays) in this environment. Illustration 4-8 includes suggested aisle and wall space measurements. With the right tables and good plans on paper, the same room can take on different personalities for every occasion, as shown in Illustration 4-9.

There is a handy formula for calculating banquet seating: If you are using standard rectangular tables, divide the square footage of the room *by eight* to find out how many seats the area will accommodate. For instance, a 500-square-foot area, divided by eight, will seat 62 or 63 persons.

When using round tables (of any standard circumference), divide the square footage of the room *by ten.* A 500-square-foot area with round tables will seat 50.

This formula allows room for chairs as well as space for aisles. Its use is only limited by columns, entrances, or service doors that would require fewer people sitting in those particular areas.

FLOOR PLAN GUIDE

All suggestions given are approximate and minimum. It should be pointed out that no rule of square feet per person can be exact, because too many variables exist. The space consumed by entry and kitchen door aisles, for example, are almost equal in a room of 800 square feet and a room of 1600 square feet, but the percentage of the room used is less in the latter. Seating capacities can only be determined by a final layout, but the approximate capacity of a room can be determined by this rough guide:

Banquet or institutional seating: 10-12 square ft. per person
Cafeteria or lunchroom seating: 12-14 square ft. per person
Fine dining: 14-16 square ft. per person

SUGGESTED TABLE SIZES:

	Banquet Institutional	Lunchroom Cafeteria	Fine Dining
2 persons	24" x 24"	24" x 30"	24/30" x 30/36"
4 persons	30" x 30"	30" x 30"	36" x 36" or 42" x 42"
4 persons	24" x 42"	24/30" x 48"	30" x 48"
6 persons	30" x 72"	30" x 72"	48" diameter
8 persons	30" x 96" or 60" diam.	30" x 96"	60/72" diameter
10 persons	72" diameter	30" x 120"	96" diameter

Note: In self-bussed tray service cafeterias, tables should be of adequate size to accommodate the trays.

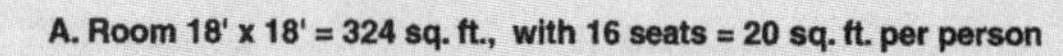

SUGGESTED MINIMUM AISLE DIMENSIONS:

	Customer Access Aisles	Service Aisles	Main Aisles
Institutional Banquet	18"	24/30"	48"
Lunchroom Cafeteria	18"	30"	48"
Fine Dining	18"	36"	54"

Note: Allow 18" from edge of table to back of chair in use. For diagonally spaced tables allow 9" more between corners of tables than needed for the type of aisle needed (e.g. for 30" service aisle, allow 39").

As rough rules of thumb, remember that tables laid out diagonally will increase seating capacity, and a smaller quantity of tables with greater seating per table increases seating capacity but reduces flexibility.

TYPICAL SEATING LAYOUTS

SQUARE SPACING

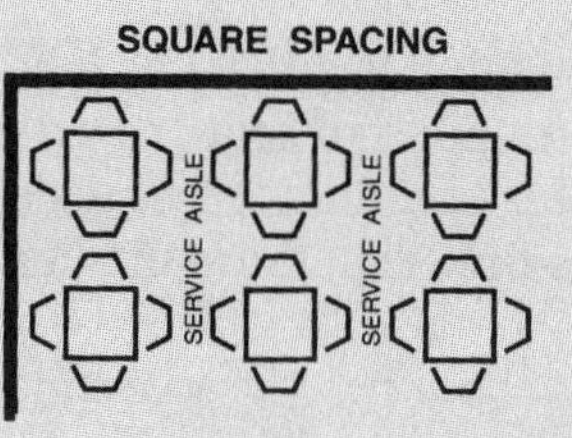

DIAGONAL SPACING

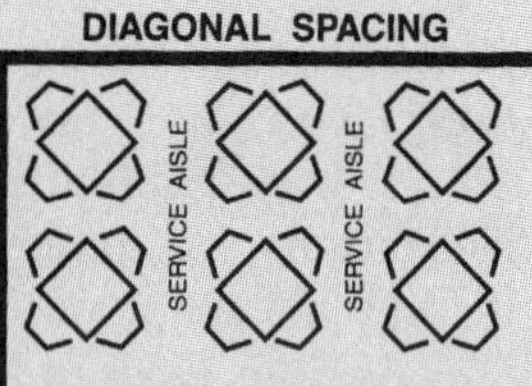

DEUCES

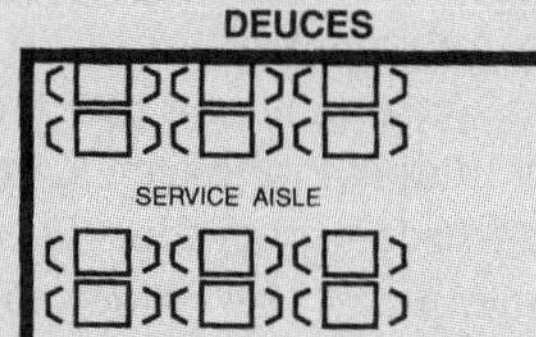

FOUR-SEATERS

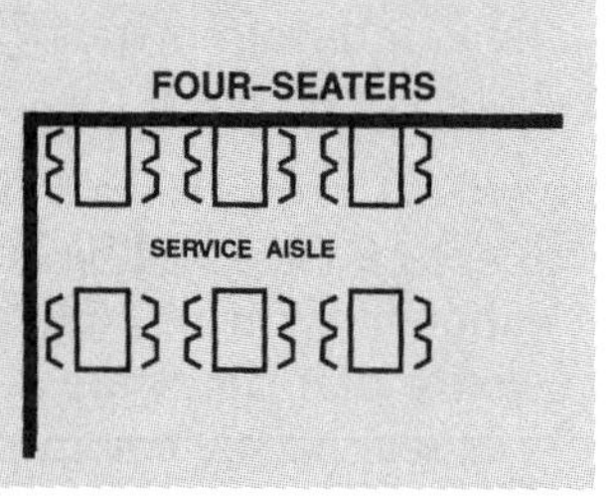

ILLUSTRATION 4-6 **Floor plan guides.**

SOURCE: CARL R. SCRIVEN AND JAMES W. STEVENS, *MANUAL OF EQUIPMENT AND DESIGN FOR THE FOODSERVICE INDUSTRY,* THOMSON LEARNING, 1989.

A. Room 18' x 18' = 324 sq. ft., with 16 seats = 20 sq. ft. per person

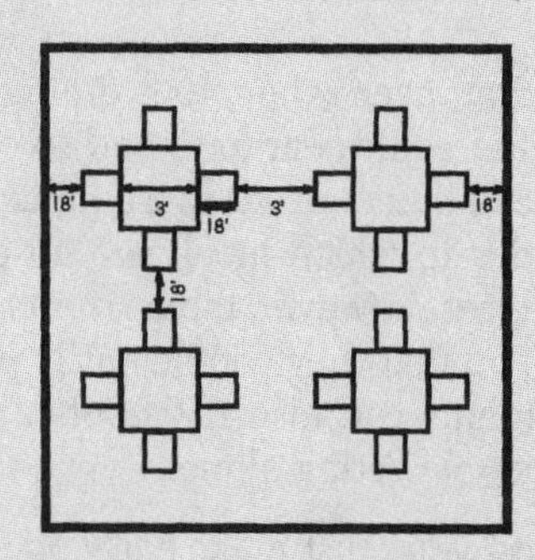

B. Now reduce room size to 16' x 15' (dotted line) Same 16 seats in 240 sq. ft. = 15 sq. ft. per person

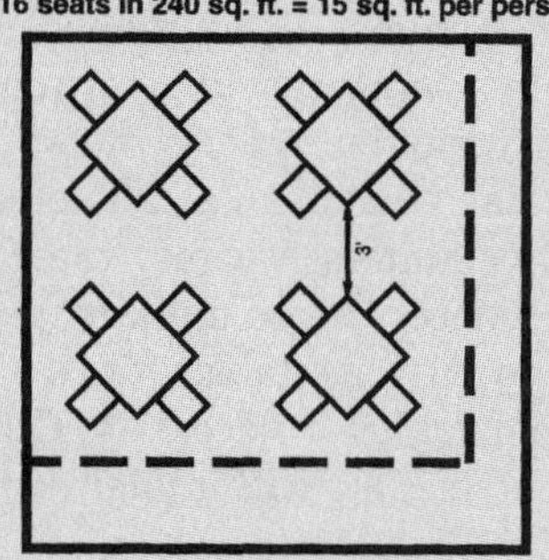

C. Same room size (18' x 18'), with 12 seats on booths, 8 on tables 20 seats in reduced room (16' x 16') = 256 sq. ft., or 12.8 sq. ft. per person

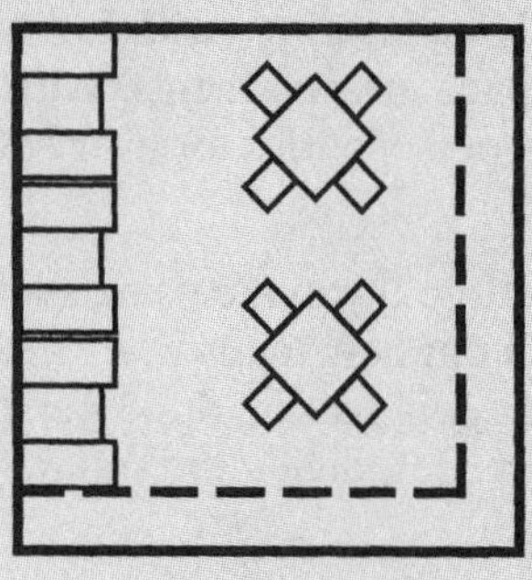

D. 24 seats on booths, 8 on tables, for 32 seats Room size (18' x 16') = 288 sq. ft., or 9 sq. ft. per person

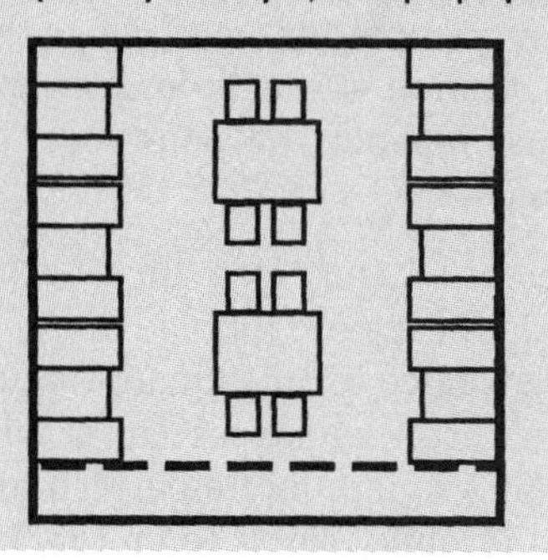

ILLUSTRATION 4-7 **Examples of typical table arrangements. A safe and comfortable amount of aisle space is another important consideration in dining area layout.**

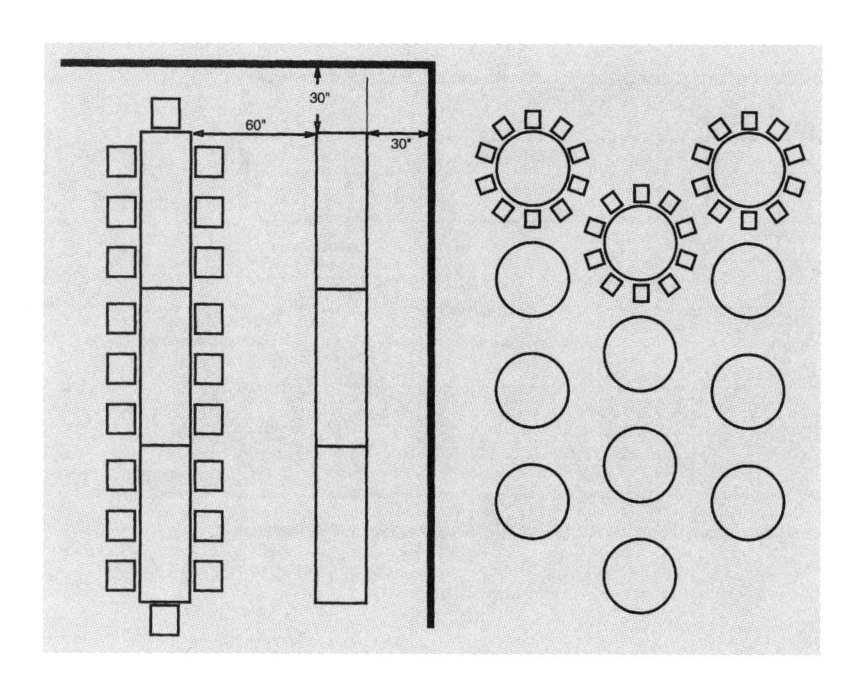

ILLUSTRATION 4-8 **Common banquet table setups with aisle space requirements.**

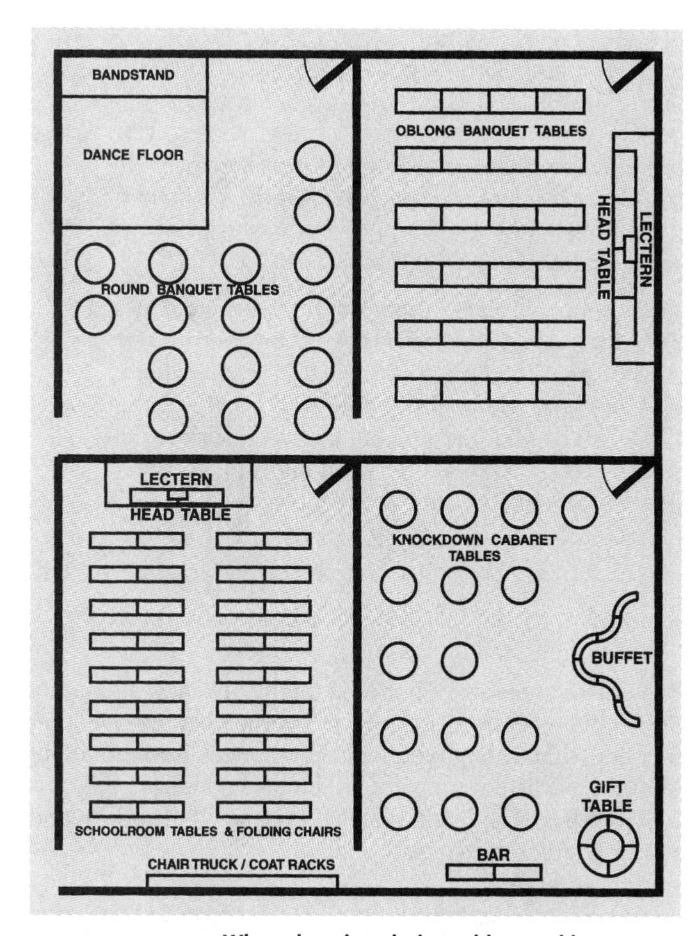

ILLUSTRATION 4-9 **When choosing chairs, tables, and banquet accessories, the goal is to make the room as versatile as possible for different occasions. This illustration shows four different layouts in the same space.**

Service Areas

Finally, you must make room for service areas or wait stations. Generally, you should provide a "small" wait station (20 to 36 inches, square or rectangular in shape) for every 20 to 30 seats, or a "large" one (as large as 8 to 10 feet long and 24 to 30 inches wide) for every 50 to 75 seats.

The need for a large wait station depends mostly on how far away it is from the kitchen and food pickup area. Placement of wait stations also depends on the availability of utilities, since they need electricity and water. Also, design them with counter space, work surfaces, and overhead shelving in mind. This, of course, requires separate, detailed drawings of the station itself, as shown in Illustration 4-10.

Beverage Areas

These are treated differently depending on whether alcohol is served. As a rule, restaurants that do not serve alcohol incorporate beverage dispensing into their kitchen service area or wait stations. The exception is the fast-food operation, where guests receive a cup and help themselves at an ice and beverage dispenser.

For bar service, the requirements are considerably more complex. Often, the space allocated to the bar depends on the importance of

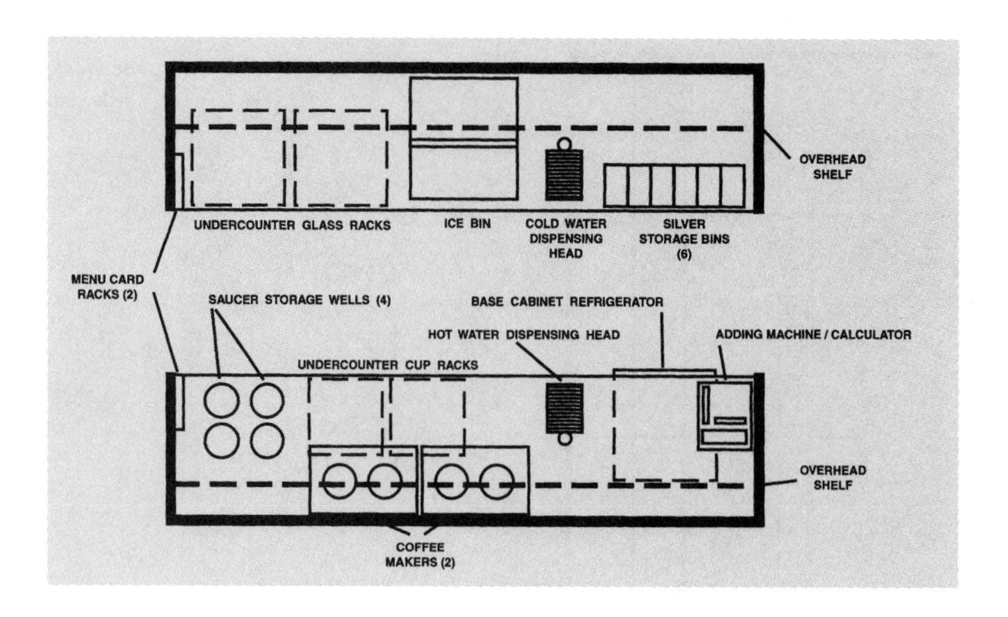

ILLUSTRATION 4–10 **Even an area as small as a wait station should begin with a layout on paper.**

PLANNING THE DINING ATMOSPHERE

STAIRCASES AND SECOND-FLOOR DINING

Upstairs dining spaces used to be opened only for private functions, banquets, and the occasional overflow crowd—and they were never considered "prime" seating. Today, rising rents have led a surprising number of restaurants to look upward for additional room, and the profit it provides. Second-floor dining spaces are commonplace, and there is often only one way to get there—up and down the stairs.

Staircases in some restaurants are elegant, providing an impressive vista of the ground floor and a chance for diners to make grand exits. In other eateries, where every inch counts, function overrules form. Either way, check the building codes about public stairways. If you have at least 75 seats upstairs, your staircase has a host of "minimums" to meet: 44 inches wide, risers of no more than 7 inches, treads of no more than 11 inches, and a handrail. Lighting should be plentiful, and nonskid surfaces are mandatory (think rain, melting snow, or anything that is spilled). One misstep, and a guest (or server) becomes "unscheduled entertainment" for the diners below. Ouch!

beverage sales to the total restaurant income. A single bar may serve both the dining area and a separate bar or lounge; or there may be separate bars, one for dining customers, one with lounge seating. Whether it's a rolling cart, a liquor closet, or a long, ornate bar with seating, you will need to decide where to put it, and how to secure it when not in use to prevent the persistent problem of employee theft.

An overhead layout of the bar is necessary. Sketches of individual areas, such as wait stations and bars, should be more detailed at this point than the original drawings.

Here are some size guidelines for your consideration.

A full-fledged lounge, with tables and chairs, will take up a little less than 5 percent of the total restaurant square footage. Each seat will require 10 square feet of lounge space.

A standard bar, which is 20 to 24 feet in length, will use up to 3 percent of the total restaurant space. The standard-size bar accommodates 10 to 12 bar stools; allow 2 feet per seated guest at the bar. Also, consider the customers who stand behind the seated ones. The area directly in back of the bar stools will hold three times the number of seated guests. So you'll need a 72-square-foot space behind the bar stools for them to comfortably stand, sip, and chat. Bars that have dance floors will require an even greater amount of space.

Where the bar is an important part of the overall concept, the standard guidelines may be too conservative. Build in room for extra wait stations and supplies if yours is a "destination" type of bar, which tends to become crowded during happy hours or on weekend evenings. Additional space is needed to set up serving tables if you offer hors d'oeuvres at happy hour. Finally, if the bar and dining room are in close proximity, separate the two areas with a rail so that seated customers don't get accidentally bumped or spilled on while eating. Draft initial sketches before you spend any design money.

Restrooms

The size of your restroom will depend on the seating capacity of your restaurant. Among experts, there are two very different schools of thought about restroom placement. One group thinks they should be located near the entrance, so that guests can freshen up before dining; the other thinks they should be nestled discreetly at the back of the dining area. Suit yourself. Realistically, restroom locations are most likely a function of where your plumbing lines are, and these are usually near the bar and/or kitchen. Minimum restroom space requirements based on the number of guests in your restaurant at any one time are spelled out in city ordinances. These include space for water closets (the common legal name for toilets) and the lavatory or washbasin for hand washing. Basic requirements of the 2003 Uniform Plumbing Code are listed in Table 4-1. The code is updated every three years.

For the small business, with up to 50 seats, a 35- to 40-square-foot area is the absolute minimum for a toilet and wash basin. Table 4-2 lists another set of requirements, taken from *Restaurants, Clubs and Bars* by Fred Lawson.

Lawson suggests that a facility with up to 70 seats should allow for 75 to 80 square feet of restroom space. He also believes the fancier the restaurant (the higher the check average), the roomier the restrooms should be—one urinal for every 15 guests, for instance.

In our experience, the number of toilets for female guests should be higher than either of these charts—perhaps three or four toilets for every 100 female guests. Statistically, women spend more time in the restroom and expect enough spaciousness for a modicum of privacy. Our own survey of restaurants in a major U.S. city with a population greater than one million found that the average 140-seat restaurant provided the following restroom space: for men, 110 to 140 square feet, containing

TABLE 4–1

RESTROOM REQUIREMENTS

Guests	Water Closets	Lavatories	Urinals
Males 1–50	1 fixture	1 fixture	1 fixture
Females 1–50	1 fixture	1 fixture	—
Males 51–150	2 fixtures	1 fixture	1 fixture
Females 51–150	2 fixtures	1 fixture	—
Males 151–300	3 fixtures	3 fixtures	2fixtures
Females 151–300	4 fixtures	3 fixtures	—

Source: 2003 Uniform Plumbing Code, International Association of Plumbing and Mechanical Officials (IAPMO), Ontario, California.

TABLE 4–2

RATIO OF RESTROOM FACILITIES PER GUEST

Fixture	Male Guests	Female Guests
Toilet	1 for every 100	2 for every 100
Urinal	1 for every 25	—
Wash basin	1 for every toilet or 1 for every 5 urinals	1 for every toilet

Source: Fred Lawson, *Restaurants, Clubs and Bars,* Butterworth Architecture, Linacre House, Oxford, Great Britain.

two toilets, three urinals, and two washbasins; for women, 100 to 150 square feet, containing three toilets and two washbasins.

Two additional legal requirements govern your restroom space. One is that, in most cities, places that serve alcoholic beverages *must* provide separate restroom facilities for men and women; typically, no "unisex" toilets are permitted where alcohol sales exceed 30 percent of total sales. The other, which we will cover later in more detail, is the Americans with Disabilities Act of 1990, which mandates accessibility measures and space requirements to accommodate guests with physical limitations.

It is advisable to have separate restroom facilities for staff and customers; however, this is not always possible. We've noticed that most restaurants with separate staff restrooms offer a 30- to 40-square-foot unisex facility, with one toilet and one washbasin.

4-5 THE BACK OF THE HOUSE

Kitchen

Many of the considerations for kitchen space allocation are detailed in Chapter 3. Briefly, design your kitchen based on the type of menu you will serve, the number of guests to be fed at one time (or one meal), the type of cooking you will do (individual dishes versus large batches), and the service system you will use to deliver the food.

An important step in your space allocation plan is to take your rough sketches a step further: Pencil in the footprints of various pieces of kitchen equipment. Some people number each piece of equipment on this floor plan, which is commonly referred to as an ***equipment key*** or *equipment schedule* (see Illustration 4-11).

One type of space that is often overlooked is the aisles. Aisles are not just for human traffic and rolling carts, but for opening oven doors and refrigerators and carrying stockpots and sheet pans. Aisle width should be determined by how many people use the space simultaneously. Where one person works alone, 36 to 42 inches is sufficient. Where workers must pass each other, a width of 48 to 60 inches is necessary. Main traffic aisles, for guests and wait staff, should be at least 60 inches wide.

Storage considerations include overshelves, undershelves, drawers, and cabinets, as well as those all-important parking or landing spaces where shared equipment is stored when not in use.

Most experts suggest a conventional kitchen take up 40 to 50 percent of the total restaurant space. Those that use more convenience foods and have more frequent deliveries to minimize storage space can squeak by with 25 to 35 percent of the total restaurant space. In most cases, tradeoffs are necessary. However, the steadily rising costs of construction have dramatically changed space requirements. The

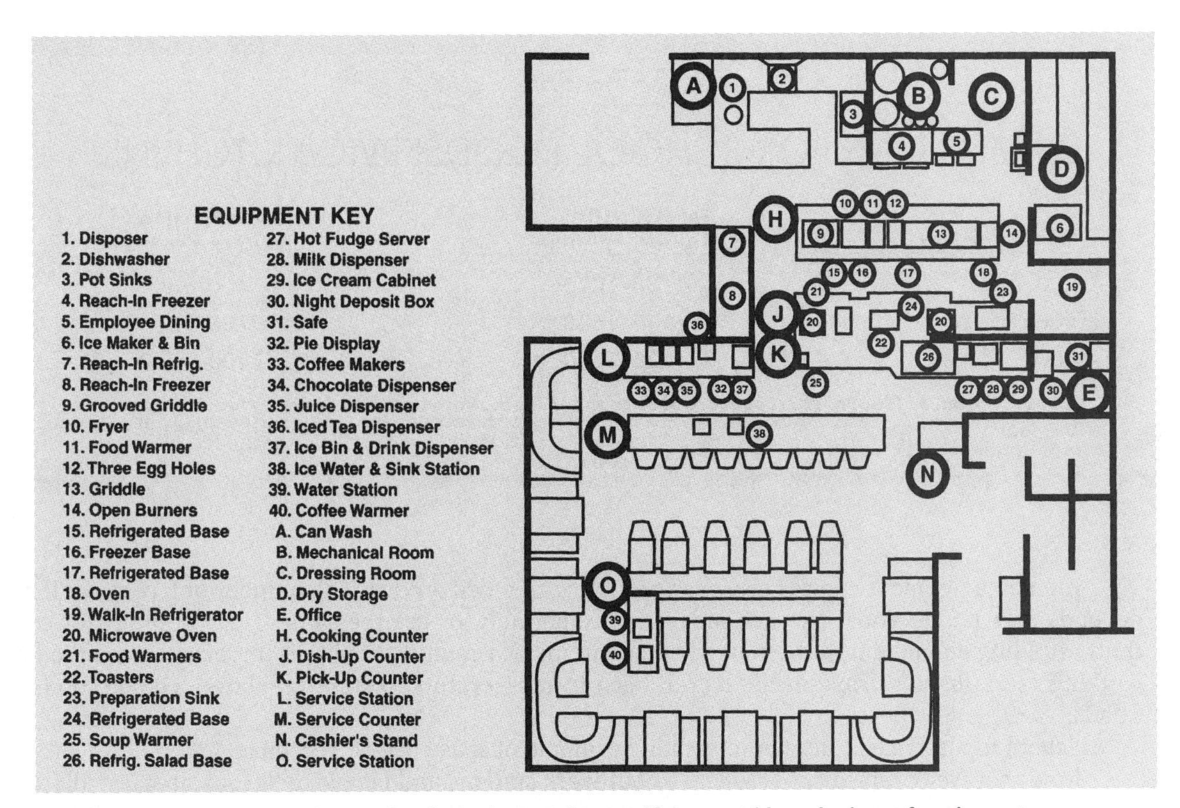

ILLUSTRATION 4–11 **An equipment key is the layout of a specific area, with each piece of equipment numbered and positioned where it will fit into the finished room.**

SOURCE: COPYRIGHT © 1989. ELECTRIC POWER RESEARCH INSTITUTE, CU-6702-V1, *THE MODEL ELECTRIC RESTAURANT, VOLUME 1: RESTAURANT SUBSYSTEM ANALYSIS AND EVALUATION.* REPRINTED WITH PERMISSION.

trend is to consolidate the back-of-the-house functions so more dollars can be spent on the front of the house, where guests are served and money is made. Table 4-3 is reproduced from the American Gas Association's guidebook *Commercial Kitchens.*

Quick-Service Considerations. If yours is a "fast-food" operation, or has a major take-out component to its menu, remember that functions like the food delivery and payment windows will be located in the back of the house. Having the cash separated from the food pick-up offers the advantage of keeping the customers "busy" so they won't perceive their wait as long. In high-crime areas, the take-out windows require special touches like bulletproof glass, but in most areas, the main decisions will be whether they are sliding or swing-out, and whether they are operated manually or with a foot switch or pedal.

TABLE 4–3

DIMENSIONS FOR COMMERCIAL FOODSERVICE KITCHENS

Type of Service	Kitchen Square Footage per Dining Room Seat	Total Square Footage in the Back of the House per Seat
Cafeteria/commercial	6–8	10–12
Coffee shop	4–6	810
Table service restaurant	5–7	10–12

Source: Robert A. Modlin, ed., *Commercial Kitchens,* 7th ed., American Gas Association, Arlington, Virginia.

TABLE 4–4

DIMENSIONS FOR SCHOOL FOODSERVICE KITCHENS

Meals per Day	Total Facility Square Footage	Main Kitchen Square Footage
200	730–1015	400–500
400	1215–1620	700–900
600	1825–2250	1100–1300

Source: Robert A. Modlin, ed., *Commercial Kitchens,* 7th ed., American Gas Association, Arlington, Virginia.

Kitchen equipment choices also change a bit with quick service. The number of fryers you'll need depends on whether you'll fry each menu item separately, or use the same fryer (and oil) for everything. Holding equipment, to keep newly made items warm until serving, is important, as is the hard-working fountain soft drink machine and reach-in refrigerators to store preplated, chilled foods like salads.

School foodservice is another interesting example of space needs (see Table 4-4).

Now that we've talked about where to put work stations and people, let's look at space allocation in a different light—from the point of view of the equipment. The industry norm is that up to 30 percent of the floor space in a commercial kitchen is occupied by machinery; another 10 percent is taken up by work surfaces, tables, carts, and the landing spaces referred to previously.

Warewashing

Again, the size of this area depends on the number of guests to be fed during peak periods. This might be the number of guests the dining room can accommodate at one time, or the number of seat turns per meal period.

The simplest space allocation method is to bring the soiled dishes from the dining room into a straight-line configuration, where they arrive at soiled dish tables; are scraped, stacked, washed, and sanitized; and come out sparkling at the other end of the line. Just scraping and stacking the things waiting to be washed takes up 60 percent of dish room counter space; clean dishes take up the remaining 40 percent. Space should also be allocated for a three-compartment sink for more efficient scraping and rinsing.

Between "dirty" and "clean" is the dishwashing machine. The smallest space for a single-tank dishwasher is 250 square feet, in an area approximately 20 feet by 12 feet. Add to that the space needed for dish carts full of bus trays, empty racks for loading dishes, and room to wash those unwieldy pots and pans. All of this will require a minimum of 40 square feet, which includes a 4-foot aisle that separates the pot sink from any other equipment.

If the dish room space is too short for a straight-line setup, the L arrangement is popular. Two tables, one for dirty dishes and one for clean, are placed at right angles to each other, with the dish machine located at the corner of the L. The table width should be no more than 36 inches, with a comfortable working height of 36 to 38 inches. The length, of course, will depend on the available space.

You'll probably need to decide on the room configuration before you order your dishwasher. Sizes vary from 30 to 36 inches wide for single-tank machines to 30-foot-wide commercial flight-type machines, which include a long conveyor belt to carry dishes into the washer.

Ponder for a moment Table 4-5, which includes some suggestions for dish machine capacity and space requirements.

TABLE 4–5

SPACE DIMENSIONS FOR DISHWASHING EQUIPMENT

Type of Dish System	Dishes per Hour	Space Required
Single-tank dishwasher	1,500	250 square feet
Single-tank conveyor system	4,000	400 square feet
Two-tank conveyor system	6,000	500 square feet
Flight-type conveyor system	12,000	700 square feet

Source: Carl Scriven and James Stevens, *Food Equipment Facts,* John Wiley and Sons, Inc., New York.

Receiving Area

Vigilant cost control begins here, in the space set aside to receive food and beverage shipments and other deliveries. It is in the receiving area that you count and weigh items, check orders for accuracy, and refuse incorrect orders or those that don't meet your quality standards.

In creating this space, consider these three factors:

- Volume of goods to be received
- Frequency of delivery
- Distance between receiving and storage areas

Some floor and counter space should be allocated for temporary storage, where things can be stacked until they're properly checked in. Weight scales, dollies, or carts will be located here, and sometimes there will be a sink and drain board for a quick rinse and inspection of incoming produce; remember space for waste disposal, too.

The absolute minimum receiving area space is about an 8-foot square, which allows room for a receiving table, a scale, a dolly or cart, and a trash can. In large operations, gravity rollers or conveyor belts move materials from receiving into storage.

Most product delivery is done by truck, which means you will need a loading dock outside the receiving area. Delivery drivers have tough enough jobs without adequate room for backing up and proper ramps to use on the dock. The minimum loading dock space should be 8 feet wide and 10 to 15 feet long. Space for two trucks—from 80 to 120 square feet—is preferable and prevents congestion if you require that all deliveries be made in a short time period. In *Food Equipment Facts: A Handbook of the Food Service Industry,* Carl Scriven and James Stevens, suggest the receiving area space allocations for various types of facilities, as shown in Table 4-6.

Note that in hospitals and other group feeding situations, dock space is shared with other departments that also have daily deliveries. In these cases, consider adding efficient transport equipment to move supplies quickly to their storage destinations.

We've seen businesses at both ends of the receiving area spectrum. Some don't even *have* receiving areas, which seems to invite cost control problems. If there is no place to "check" materials in and out logically, there is a tendency to be lax about keeping track of shipments overall. This is when inferior products, incomplete orders, and pilferage start taking their tolls. On the other hand, a receiving area that is too large will end up being the dumping ground for empty boxes, broken equipment, and anything that doesn't fit anywhere else.

TABLE 4–6

SPACE DIMENSIONS FOR RECEIVING AREAS

Restaurants

Meals Served per Day	Receiving Area Square Footage
200–300	50–60
300–500	60–90
500–1000	90–130

Health Care Facilities

Number of Beds	Receiving Area Square Footage
Up to 50	50
50–100	50–80
100–200	80–130
200–400	130–175

Schools

Meals Served per Day	Receiving Area Square Footage
200–300	30–40
300–500	40–60
500–700	60–75

Source: Carl Scriven and James Stevens, *Food Equipment Facts,* John Wiley & Sons, Inc., New York.

General Storage

There is a tendency to think of storage space as unproductive and to downplay its importance. However, more and more cities are mandating specific ratios of storage space to kitchen space.

How much storage you need depends on how much stuff you want to store. Simple? Actually, it's not. During the planning phase, you should design storage areas that can increase in capacity without increasing in size, through the use of additional shelf space. A system of grids, movable shelves, and accessories can attach directly to your wall studs. Some shelves can be hinged to swing up when needed and down when empty.

Other storage systems are movable, rolling shelf units that are attached to an overhead track installed on the ceiling. As shown in Illustration 4-12, systems like Top-Track have drop-in baskets, mats, and dividers to accommodate different sizes of containers. A track-style system, with only one aisle, can give you up to 40 percent more storage space.

In foodservice, you must determine not only the number of products you'll be storing, but the number of days they'll be there before you use them. This is known as your ***inventory turnover rate.*** Experts suggest a restaurant should have at least enough space to store one to two weeks' worth of supplies. Believe it or not, they've even calculated exactly how much storage space is needed per meal served per day, when deliveries are made every two weeks: 4 to 6 square inches! (Get out your ruler and start measuring those shelves!) Of course, if your location is more remote—a hunting lodge, a small island resort, a rural café with infrequent product deliveries—you will need more storage room for extended time periods.

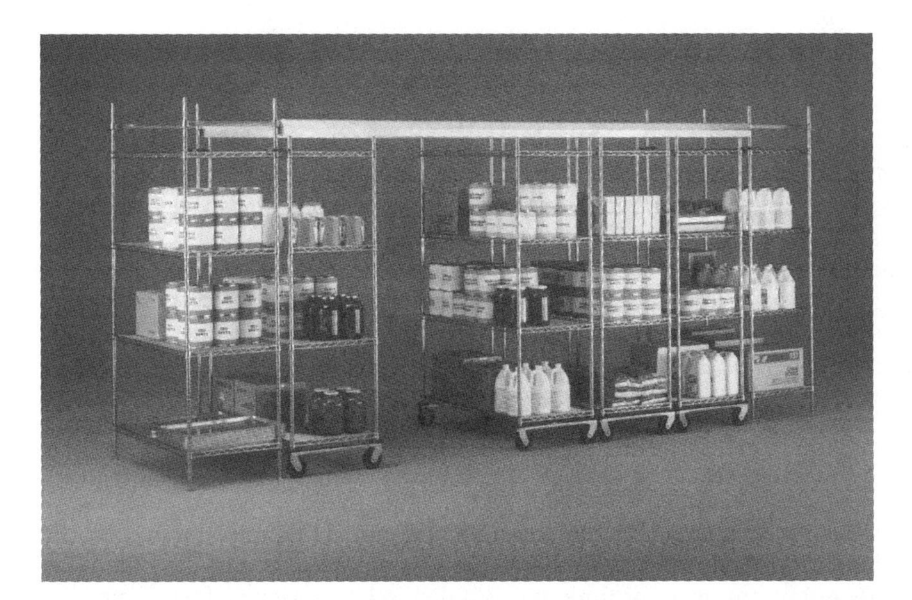

ILLUSTRATION 4–12 **Top-Track is one of the high-density storage systems. Its components are attached to an overhead track installed on the ceiling.**
COURTESY OF INTERMETRO INDUSTRIES, WILKES-BARRE, PENNSYLVANIA.

From this, it is clear that storage space depends on how management handles ordering and receiving.

Common fixtures you must consider in setting up storage areas include shelving, racks, bins, pallets, carts, and dollies. Remember to plan for aisle space of at least 3 feet in all storage situations. Make sure racks and carts have wheels (called ***casters***), to enable you to roll them when they're fully loaded.

A pallet is a (usually wooden) platform that sits 4 or 5 inches above the floor and measures anywhere from 36 to 48 inches square. Boxes and cases can be stacked on pallets, and the full pallets can be stacked atop each other. Their portability and stackability make them good, flexible storage options.

IN THE KITCHEN

STORAGE AT SEA

To illustrate the importance of space efficiency in foodservice, let's dive into the watery deep with the U.S. Navy. On its submarines, food has to be stored for more than 100 days—with no deliveries—in extremely limited space. The galley (kitchen area), including storage, cooking, and dining, is no more than 300 feet. Yet it turns out three meals a day for up to 200 crew members!

You can't walk into a walk-in cooler on a submarine. Instead, there are detailed storage diagrams to follow. For instance, the frozen green beans may be stored on the right side of a completely packed freezer compartment, two rows back. The person retrieving the beans knows exactly what else must be moved to get to them.

The benches on which crew members sit are actually chests in which canned goods are stored. Pots, pans, and utensils are nestled in special racks that hold them in place. Despite the space limitations, a submarine kitchen includes most of the same equipment you'd find on shore: a range, oven, griddle, broiler, grill, dishwasher, and trash compactor are all standard. The compacted trash, incidentally, is weighted and launched into the sea through the torpedo tubes.

Source: ***Restaurants and Institutions Magazine,*** **a division of Reed Business Information, Oak Brook, Illinois.**

Dry Storage

The standard restaurant storeroom is at least 8 feet wide, and its depth is determined by the needs of the operation. If the storeroom must be wider than 8 feet, increase it in multiples of seven. This will allow for two rows of shelves, each a standard 21 inches wide, plus the 3-foot aisle. Some restaurants purchase movable shelving, which is usually 27 inches wide and about 5 feet in length.

The storeroom door should open out, into an aisle, to maximize storage space within the room.

Dry storage is "robbed," in a way, by the different sizes of cans and bottles that must be stored. Minimize this problem by storing same-size items together on shelves of adjustable heights. For storage space requirements, we defer again to *Food Equipment Facts* by Scriven and Stevens (see Table 4-7).

Refrigerated Storage

There are three types of refrigerated storage space: the "reach-in" type of refrigerator or freezer, the walk-in cooler, and the walk-in freezer. Storage space for these appliances is calculated in cubic feet. The expert guideline for an average restaurant open for three meals a day is 1 to 1½ cubic feet of refrigerated space per meal served. For fine dining, it increases to 2 to 5 cubic feet per meal served.

As an example, let's examine the refrigeration needs of an "average" restaurant (140 seats) that serves three meals daily. Using our rule of thumb, the place will need between 420 and 630 cubic feet of refrigeration.

TABLE 4–7

SPACE DIMENSIONS FOR DRY STORAGE

Restaurants

Meals Served per Day	Dry Storage Square Footage
100–200	120–200
200–350	200–250
350–500	250–400

Health Care Facilities

Number of Beds	Dry Storage Square Footage
50	150–225
100	250–375
400	700–900

Schools

Meals Served per Day	Dry Storage Square Footage
200	150–250
400	250–300
600	350–450
800	450–550

Source: Carl Scriven and James Stevens, *Food Equipment Facts,* John Wiley & Sons, Inc., New York.

Management has been considering a combination of a 392-cubic-foot cooler and a 245-cubic-foot freezer, and is surprised to learn that this will not be sufficient. Why not? Well, only half of the total 637 cubic feet is *usable* storage space—not insulation, aisles, motor, evaporator, or fans that are part of every refrigerated unit. Divide 637 by 2, and you get only 318.5 cubic feet of space—which is inadequate for this restaurant's needs.

Tables 4-8 and 4-9 summarize the storage capacities for some standard sizes of reach-ins and walk-ins.

In determining the useful space of a walk-in unit, remember that from one-third to one-half of it will be taken up by the aisle(s); still more, by the evaporator and fans. No wonder, in our opinion, that most foodservice facilities allocate too little refrigeration space and then have to find ways to live with it.

Employee Locker Rooms

Unfortunately, providing even the most basic amenities for employees seems to be an afterthought in the minds of most restaurant owners. It is our view that "employee-only" areas deserve a higher priority in the planning process, because they have a direct impact on sanitation, morale, productivity, and security. Once again, this area should be sketched from the beginning, along with storage and receiving areas (see Illustration 4-13).

A locker room area, well lit and properly ventilated, provides secure storage space for personal belongings and a place to unwind at break times. Tall, single lockers or smaller, stacked ones should have perforations to allow for air circulation. Benches in front of the lockers are a good idea. If a table or two

TABLE 4–8

FULL-DOOR REACH-INS

Number of Doors	Height (inches)	Width (inches)	Depth (inches)	Cubic Feet
1	78	28	32	22
2	78	56	32	50
3	78	84	32	70–80

Source: Carl Scrivner and James Stevens, *Food Equipment Facts,* John Wiley & Sons, Inc., New York.

TABLE 4–9

WALK-INS (ALL 7′6″ HEIGHT)

Size of Unit	Square Footage	Cubic Feet
5′9″ x 7′8″	35.7	259.9
6′8″ x 8′7″	47.4	331.8
7′8″ x 7′8″	49.0	340.2
8′7″ x 11′6″	86.4	604.8

Source: Carl Scriven and James Stevens, *Food Equipment Facts,* John Wiley & Sons, Inc., New York.

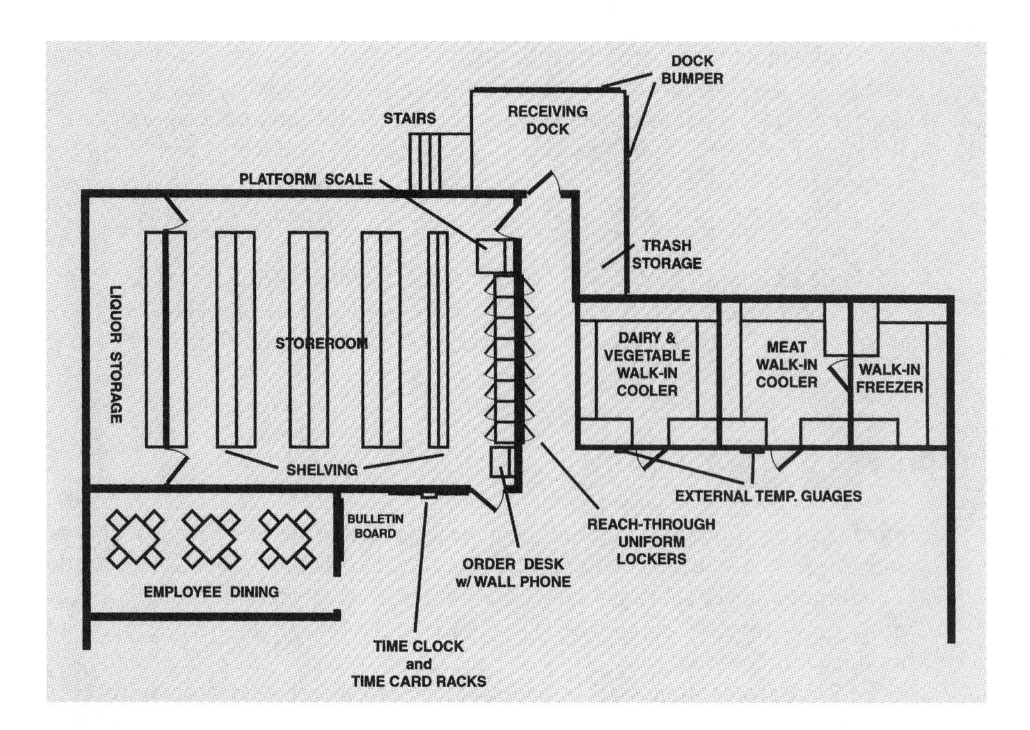

ILLUSTRATION 4-13 A layout for a receiving and storage area that includes a separate employee break room.

and some chairs fit in this area, it will prevent workers from sitting around in the dining area during their breaks—a practice that, though understandable, still looks unprofessional.

In larger facilities, the locker room is the site of any uniform exchange system, with a supply of clean uniforms and laundry hampers for soiled ones. The size of the employee locker room/restroom area depends on the size of your staff; some guidelines are given in Table 4-10.

Toilets and hand sinks for employee use only are another smart idea, with enough shelf or counter space so that employees can place personal toiletry items. Install a full-length mirror in the restroom or locker room, too. The general rule for restroom fixtures is one hand sink and one toilet for every 8 employees. The ratio moves upward to two of each for every 20 to 25 employees, and three of each for every 25 to 35 employees. The minimum size for a single toilet and stall is 4½ square feet, and your city health department will regulate the size and sanitation of employee restrooms just as stringently as the public ones. The same handicapped-accessible guidelines of the Americans with Disabilities Act also apply.

Office Area

There are two types of restaurant offices. The ones we usually think of are tucked strictly in the back of the house, generally pretty messy and dingy. At country clubs, hotels, or catering companies, however, there may be offices where the guests interact with the staff to plan functions, get price quotes, and so on.

For the "public" office, you should plan a minimum of 150 square feet, with an attractive desk and several arm chairs. Space for filing cabinets, computer equipment, bookshelves, and display space for photos or albums of previous events will all be needed—especially the latter, because potential clients will want to see examples of your handiwork.

A back-of-the-house office is more likely used for accounting, paperwork storage, and one-on-one employee meetings. It can be as small as 60 square feet, up to a relatively roomy 120 square feet. If accounting is done on-site, we recommend it be housed in its own separate office. It is much easier that way to keep track of specific accounting paperwork.

Purchasing supplies and planning menus takes time and concentration. The office should be designed (and its filing system kept tidy enough) to allow the owner, manager, or chef to perform these tasks efficiently.

TABLE 4–10

SPACE DIMENSIONS FOR LOCKER ROOM

Number of Employees at a Single Location	Locker Room Square Footage
1–10	90
10–20	150
20–40	250

We have occasionally noticed that restaurants with bars locate their offices directly behind the back bar, with a two-way mirror that allows management to watch the bar area. We've even seen a chef's office built so that, almost like an airport tower, it is raised and glass-enclosed to allow easy observation of all activity in the busy kitchen.

Overall, restaurant office space does not have to be big or fancy. Like so many other areas, you must decide in advance exactly what you want to use the office for, then plan the space around those needs.

4-6 SPACE PLANNING AND THE AMERICANS WITH DISABILITIES ACT

When planning your dining area, you must become familiar with the Americans with Disabilities Act (ADA) of 1990. Although it is a U.S. law, and this textbook will presumably be read in other countries, the ADA guidelines can be considered in any nation as a call for sensitivity to the needs of persons with physical limitations. As of 1992, the ADA also included requirements that companies with 15 or more employees cannot fire or refuse to hire people with disabilities, unless the impairment prevents the individual from performing the job.

Much grumbling resulted when the ADA was enacted, mostly by business owners who felt the law was ambiguous, that it was being too broadly interpreted, and that it was costing them money by requiring expensive modifications to their facilities. Lawsuits and complaints under the ADA range from customers with impaired mobility asking for wheelchair ramps to employees asking for preventive programs and compensation for on-the-job back injuries.

So, for many reasons, it is wise for any business owner to stay abreast of ADA guidelines and developments. From the parking lot to the restrooms, they will most definitely affect your space planning.

Parking. Among other things, the ADA mandates a certain number of "handicapped" parking spaces. (Although we dislike the use of the term, it seems to be the most common description of these spaces.) The numbers range from a single parking space for a lot with only 25 total spaces to nine spaces for lots with more than 400 spaces. These spaces must be those closest to the public entrance, and they must be clearly marked (usually with the universal symbol of accessibility) so that the markings cannot be obscured by other parked vehicles. Most of the figures were developed to ensure wheelchair access. Each of these special spaces must be at least 8 feet wide, with an adjacent "aisle" of at least 60 inches. At least one van-accessible space is required, which also requires a wider (72-inch) aisle.

From the vehicle to the building entrance, the path of travel must be at least 36 inches wide. For stairs, handrails between 34 and 38 inches above the stairs themselves are required. Wheelchair ramps must slope gently, with a height ratio of 1:20. This means that, for every 20 inches in length, the ramp height increases by only 1 inch. If the ramp is longer than 6 feet, it must be equipped with handrails with the same height requirements as those for stairs.

Entrances. At least 50 percent of the entrances to a foodservice facility must be accessible to disabled persons, including emergency exits. At nonconforming doors, signs must be posted indicating the location of the accessible entrances. If the door does not open electronically, it must be 32 inches wide, with a door handle that is no higher than 48 inches from the floor. "Loop"- or "lever"-style handles are preferable to doorknobs. Next to the handle of a pull-open door, there must be 18 inches of clear wall space. If there are double doors, the requirement is a 30- to 40-inch clear floor space, not counting the space the door would normally require to swing open. If there are revolving doors, an adjacent handicapped-accessible door is also required.

Public Areas. Once inside the restaurant, the disabled guest requires an aisle width of at least 36 inches. If counter service is offered, a 5-foot portion of the counter must be as low as 28 to 34 inches from the ground, to facilitate ordering from a seated position. Tabletops must meet the same height requirements as counters. In banquet situations, if people will be sitting at a raised head table, for instance, a ramp or platform lift must be provided.

Self-serve items must be positioned such that they can be reached by someone in a seated position—cups stored horizontally, for example, instead of being stacked vertically. On salad bars, cold pans may be tilted so all products are visible and may be reached easily; reach-in cooling units may have "air screens" instead of doors. Sneeze guard heights may have to be adjusted.

Aisles that lead to restrooms, and to individual toilet fixtures, must also be 36 inches in width. At restrooms that don't meet the rules, signs must be posted with directions to accessible ones. Doorways to both restrooms and individual stalls must be at least 32 inches wide. The sizes of accessible toilet stalls are also regulated—they must be at least 5 feet square.

Lavatories (washbasins) require clear floor space around them to accommodate the wheelchair-bound patron. This means a 30-by-48-inch space, with the rim of the basin no more than 34 inches from the floor. The bottom edges of mirrors must hang no higher than 40 inches from the floor. Soap, towel, and toilet paper dispensers should be no higher than 54 inches from the floor. Finally, whether they are placed in restrooms, lobbies, or elsewhere in the restaurant, telephones should not be mounted higher than 54 inches from the floor.

Kitchen Area. Two pieces of equipment that require special ADA consideration are the hand sink and the worktable. Neither can have obstacles underneath that would prevent a wheelchair-bound employee from getting close enough to safely use them, and both should be of wheelchair-friendly height.

Overall, the hospitality industry has been exemplary in recognizing the need for accessibility, because we are in a business that focuses intently on customer service. Most of us recognize that making these adjustments is a commonsense way to cater to an important part of our clientele. The ADA has helped formalize many policies and practices that, before 1990, existed informally in many locations. In recent years, the U.S. Justice Department (which enforces ADA requirements) released a document you may find useful. It was written for the hotel industry, but much of the information is applicable in foodservice settings. Access the "ADA Checklist for New Lodging Facilities" at www.usdoj.gov/crt/ada/lodgesur.htm

SUMMARY

When deciding how much space you'll need for various parts of your restaurant, consider how much food you plan to serve, what types of meals you will prepare, and the costs to either build or rent the space.

Draw a preliminary space plan to examine the potential flow patterns—both for food and customers—and where to locate the major work areas. Assign the most favorable spots to customers, and allow for some room to expand if you need to. At the same time, pay attention to how the public area may look when it is not crowded. Your space requirements and ambience will also be affected by whether you want a high turnover or whether you want your customers to be able to linger over a meal.

Outdoor facilities may include dining area, front entryway, and parking lots. City ordinances specify the particulars for these outdoor areas, including the minimum number of parking spaces you should have for your size of building.

After the overall space plan is firmed up, try planning the table arrangement. Be sure there is adequate room for wait stations, restrooms, a bar, or banquet room if you want to include them. Think about the "nonpublic" areas, too: storage and receiving, dishwashing space, employee lounge or locker rooms, and office space.

Familiarize yourself with the Americans with Disabilities Act, which will affect many aspects of your space planning. Work with experts, if necessary, to ensure your eatery is truly accessible to the public, and to prospective employees with disabilities. This includes handicapped parking, entrance ramps, and restroom facilities, to name a few.

STUDY QUESTIONS

1. Why put your preliminary layout and estimate your space needs on paper?
2. List two advantages, and two disadvantages, of having an outdoor dining area. How would you decide if the advantages outweigh the disadvantages?
3. In what way does the entry of the building set the mood for the rest of the dining experience?
4. List three ways you can increase turnover in your restaurant, and explain why you think they work.
5. How do you decide how large your restrooms should be?
6. What are the three main parts of a dishwashing area?
7. Why is it important to have a receiving area? What are the few basic items that should be part of a receiving area, even if space is tight?
8. List and explain three ways you can make better use of storage space, both dry and refrigerated.
9. Do you think separate locker rooms and restrooms for employees are necessary? Why or why not?
10. In what ways does the Americans with Disabilities Act affect space planning?

A Conversation with . . .

Alice Waters

OWNER/CHEF, CHEZ PANISSE
BERKELEY, CALIFORNIA

Alice Waters has authored eight cookbooks (so far!) and won numerous awards, including Best Chef in America in 1992 by the James Beard Foundation, and Best Restaurant in America for Chez Panisse. She also has a second eatery, Café Fanny, named after her daughter. Waters has developed an edible garden pilot project for schoolchildren, to teach them to grow, harvest, and prepare their own foods at school.

Waters seems to have it all, but when we interviewed her at the idyllic Berkeley restaurant where she supervises 114 employees, she said her dream kitchen would be organized like those of medieval castles: "a fireplace at one end, a cooking line at the other, and copper pots going up the middle." She feels for a restaurant to be successful, it must begin with a good bakery.

Q: A lot of restaurant buffs claim they can tell if a place is going to make it just by walking in the door. How can you tell?
A: If people are sitting in the chairs, that's how! Seriously though, the bottom line is the food, which needs to be nourishing as well as delicious. I'm convinced that has been the basis for the success of my restaurant. We have never compromised on that; everything else is secondary. I mean, Chez Panisse started out in a house that was remodeled, bit by bit. The kitchen was tiny to begin with, and we used to grill outside the restaurant, much to our neighbors' horror! So anything is possible if you are concerned with the nourishment of the customers, first and foremost. If you're in it just to make a living and money is number one, or if you think it might be fun to have a restaurant or hospitality business, then I think you're in it for the wrong reasons.

Q: What's important to you as a restaurant customer?
A: I've seen a lot of little restaurants. I've eaten in hundreds and hundreds, and I'm fascinated by them. The ones that are really successful have very delicious food, and they smell good! I'm always concerned about the way the restaurant smells. I want to walk in and just enjoy the smell. As far as the tables go, we've always been a little small, so the tables are quite close together. But again, I think if they didn't get good food and they had to sit that close together, they'd be completely unhappy.

Q: How do you get that good smell consistently? You can't bake bread all the time, and potpourri probably won't do the trick!
A: Good food! We have a pizza oven in the restaurant, and that's a big part of it. Also, we have a fireplace with oakwood burning, so you have those aromas. As far as a block away, you can sense there's a fire. Once in a while, we'll burn a branch of rosemary in the fire, and you can get it with good garlic and things sautéed in different herbs, good fish cooking. It really comes from the food you prepare.

Q: How did you pick the site for Chez Panisse?
A: I wanted a place that felt like home. I didn't want a commercial restaurant, so I chose a house very specifically because I wanted

the feeling that you were in somebody's house and that somebody was home, cooking for you. I like really small restaurants and I think we lose something when they get too big. I like a 50-seat restaurant. If it's much bigger, you can't feel connected with all the people who come in.

Q: Do you think that location matters?
A: I have never thought that it did, and I still don't think that it does. Of course, you have to be sensitive to what's around you, and try not to be ever so rigid that you can't adapt to people's likes or dislikes, or the weather, or the ingredients. You have to be enormously flexible. I think somebody with a very strong idea, and not wanting to change, in the wrong spot could have a lot of problems. But I think if you are basically interested in hospitality, people will come to a lot of different types of places.

Q: Do you have any theories about kitchen equipment?
A: Sometimes it seems to me the equipment gets in the way of what one is trying to do, and I think you can stock a kitchen with too much equipment. I think the kitchen space should be big enough and appealing enough to inspire the cooks there, because that's what it's all about. You can't put them in some little corner full of soulless machines and expect to have something really tasty coming out.

Q: How do you make decisions about the layout of the kitchen?
A: I guess there are some basic rules. You should have the dishwashing area easily accessible to the waiters, but not too close to the dining room, in terms of noise. I don't think the dish room should ever be a real back room, though. I think the dish washers need air and light and communication with everyone else in the kitchen. It's a very important part of the restaurant!

It's also very hard to have areas that need to be wet near areas that need to be dry. For instance, you can't spray down every little part of the wet area when it's close to the linens and glasses. It's very, very important to me that every little part of the kitchen is clean. I mean really clean. That's vital! You need floor drains and really washable walls. We have tiles up to about 4 or 5 feet, which help enormously. You have to be able to spray behind everything and underneath as well, all the way down to the floor, every inch of the place. You'd be surprised where food will go with heavy use! We are constantly washing and painting the ceiling. Our kitchen is open to the public, but we would do it anyway.

The walk-ins need to be spotless. Again, the kind of shelving in the walk-ins is very important so that you can clean behind everything. It's great to have separate walk-ins and that's one extravagance I really approve of and would spend money on. That way you can keep the right temperature for vegetables, for meat. If you can afford to, have another one for fish. It really helps to keep things clean and organized at their correct temperatures. You don't want much moisture in a meat fridge; you don't want vegetables and fish stored together. I don't believe much should be in freezers except ice cream, and it's nice to have one that keeps the right temperature for that, too.

And finally, it's great to have big, big sinks to wash things in; and of course, I think a fireplace in the kitchen, for cooking, is indispensable. It's the kind of centerpiece for the kitchen.

Q: Is there anything else you'd like to suggest?
A: Overall, think about making your kitchen into a place where you'd like to be, for long periods of time. You don't always have to opt for the stainless steel, either—there are lots of other materials that are beautiful and cleanable. We did our meat butchering room with Tunisian tiles, which changes the whole feeling and mood of the room to have those blue tiles. Before, it was just a little cubbyhole—now, it's a great place to be.

CHAPTER 5

ELECTRICITY AND ENERGY MANAGEMENT

INTRODUCTION AND LEARNING OBJECTIVES

If your business is going to be profitable, you've got to control your operating costs. By learning more about energy management, you can cut your utility bills—saving money, as well as the earth's limited natural resources.

In this chapter, we will discuss:

- How to determine your annual electricity costs
- How to get and understand an energy audit of your site
- Basic terms you should know to track your own energy use and decipher your utility bills
- Basic electrical principles: how power flows into your building and equipment
- Tips for choosing energy-efficient equipment
- Tips for saving energy in all phases of a foodservice operation

This chapter focuses primarily on electricity. Other utilities—gas, steam, and water—and plumbing issues are covered in Chapter 6.

5-1 UNDERSTANDING ENERGY USE

Until the early 1970s, when the U.S. experienced its first energy crisis since World War II, most restaurateurs were simply not focused on cutting utility bills. When budgets needed to be trimmed, the emphasis was on curtailing labor costs, insurance costs and—as always—food costs. It seemed downright miserly to fret over when to turn on an oven, or whether to adjust the air conditioner a few degrees warmer. Today, however, restaurants from Maine to California are finding ways to save money by saving energy. It's not only a cost-saving measure; it is considered forward-thinking to conserve resources by operating more efficiently. At times, this requires making financial investments in equipment, mechanical or electrical systems, and in the building itself. For the typical restaurant, these make sense only if the initial expense can be recouped within five years.

Energy use per square foot in restaurants is greater than in any other type of commercial building—more than triple what a hospital uses per square foot and at least six times what an office building uses per square foot (see Illustration 5-1). Illustration 5-2 gives you the basic guidelines for how energy is used in a restaurant. On your profit-and-loss statement, utility costs will be only 4 to 7 percent of your total operating expenses. However, National Restaurant Association studies suggest they are expenses that can be cut by as much as 20 percent with smarter energy consumption; utility companies claim you can realize savings as high as 30 percent. This has a major and direct impact on your bottom-line profit. Table 5-1 charts the potential savings, which vary (of course) with your electric costs and amount of savings. And remember, because its energy use is so high, this type of business is especially vulnerable to fluctuations in energy costs.

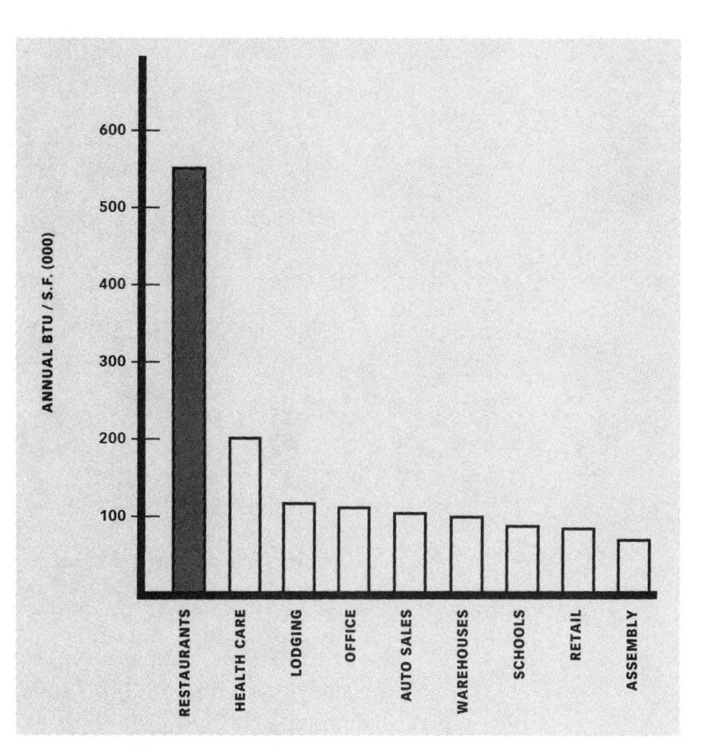

ILLUSTRATION 5-1 **Businesses that use the most energy.**

SOURCE: *THE ECONOMICS OF GAS AND ELECTRIC COOKING*, ELECTRIC COOKING COUNCIL, FAYETTEVILLE, GEORGIA.

The best way to save both energy and money is to plan and implement an Energy Management System, which consists of six components:

1. **Energy accounting:** A monthly tabulation of energy use and costs will allow the owners of a business to track this data, season to season and year to year. Put it on a standard form or spreadsheet. Include the information from your power bills—total costs, total amount of consumption, and demand charges. (The latter will be explained shortly in this chapter.)
2. **Energy audit:** This is explained in detail in the next section of this chapter.
3. **Retrofitting:** In foodservice, about half of energy conservation comes from retrofitting to make existing appliances, systems, or buildings more energy-efficient. This includes everything from increasing the insulation in the building, to insulating the hot water tanks, to installing timers on outside lighting and climate control systems. A retrofit type of project usually only makes sense when you know it will end up paying for itself in cost savings within a few years.
4. **Low-cost and no-cost ideas:** Before you do the retrofitting, which

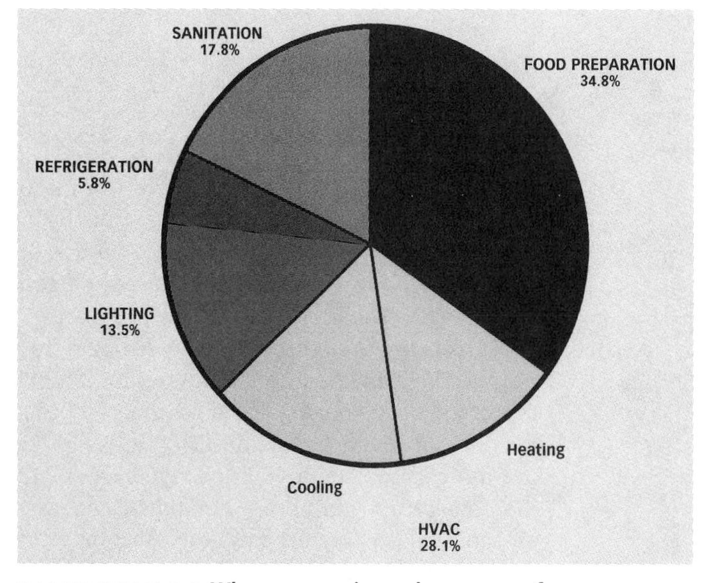

ILLUSTRATION 5-2 **Where energy is used: a survey of seven restaurants.**

SOURCE: *THE ECONOMICS OF GAS AND ELECTRIC COOKING*, ELECTRIC COOKING COUNCIL, FAYETTEVILLE, GEORGIA.

TABLE 5–1

YEARLY VALUE OF ELECTRICITY SAVED

Your Electricity Cost Per kWh	Value of the Electricity Saved if the Number of Kilowatt-Hours Used per Year Is Reduced by					
	500	1,000	2,500	5,000	10,000	25,000
5 cents	$25	$50	$125	$250	$500	$1,250
7 cents	$35	$70	$175	$350	$700	$1,750
9 cents	$45	$90	$225	$450	$900	$2,250
11 cents	$55	$110	$275	$550	$1,100	$2,750
13 cents	$65	$130	$325	$650	$1,300	$3,250
15 cents	$75	$150	$375	$750	$1,500	$3,750

***Source: How to Reduce Your Energy Costs*, 2nd ed., Center for Information Sharing, Boston, Massachusetts.**

costs money, try changing the "habits" of your staff and their routines. Turn off lights in unoccupied areas; don't leave appliances on when they are not being used; ask your staff members for energy-saving ideas—then put them to work.

5. **Capital projects:** Like home improvements that add value to a house, there are similar improvements that add value to a business. New technology may have a high price tag, but consider its long-term usefulness. Computerizing the HVAC system to turn on and off automatically and hold predetermined temperatures can result in huge savings over time. Cogeneration is an idea discussed near the end of this chapter, adding new equipment that captures "wasted" heat from appliances and uses it to heat water or generate steam. When appliances wear out and need to be replaced, look for the newest energy-saving features.
6. **Continued surveillance:** In order to make any of the other five steps effective, they must be monitored and their importance must be communicated to the staff. Soon, you'll be documenting savings instead of dreading each month's utility bills!

5-2 ENERGY AUDITS

The good news is, because foodservice uses so much gas and/or electricity, you're automatically a good customer—and an important one—to your utility companies. You can use this clout to buy the utility's expertise, which can help you manage your business. Utility companies often provide cash rebates, low-interest loans for equipment updates, and free design and technical advice. You might as well get to know them! In fact, you should have a contact person at every public utility (or, in deregulated markets, your Retail Electric Provider) with which you deal.

An ***energy audit*** is a great way to begin. In fact, you won't really know where savings can be obtained until you have an energy audit of your building. Most utility companies routinely perform energy audits at the request of customers, and there are two basic kinds: the *walk-through audit* and the *analysis audit.*

Walk-through audits are usually free of charge. The service provider's representative will already have information about your average bills when he or she comes out to inspect the property, look at your food preparation and storage equipment, ask about habits and procedures, and recommend improvements you could make. Some suggestions (such as replacing an appliance or adding insulation) may cost money. Of course, whether you act on the audit results is up to you.

There's usually a fee for an analysis audit, since it requires gathering more detailed information about the types of heating and cooling systems and appliances, and even the illumination levels of the lights in your dining room. The data is input into a computer, which might recommend:

- Structural or design modifications to your building
- Replacing or retrofitting some equipment
- A target electric rate that is "best" for your particular business

The recommendations are designed to give you lots of options. For example, let's say you have recessed lighting in the dining room ceiling, and the audit suggests it is inefficient. You could:

- Change from regular light bulbs to more efficient reflector (R) or ellipsoidal reflector (ER) bulbs
- Paint the room a lighter color, which would be more reflective and therefore require less artificial light
- Install more light switches or dimmers, to better control individual areas or rooms
- Lower the ceiling height
- Install skylights or light tubes

Who performs these audits? A call to your utility company or Retail Electric Provider (REP) should get you started. One advantage the utility has is that it can track the history of energy use at your site even before you got there. If it used to be another restaurant, this could be extremely helpful. There are also private firms listed in the telephone directories under *Energy Management and Conservation Consultants.* When selecting a private contractor, it's important to ask:

- What they charge
- Exactly what the fee includes
- Whether they represent a variety of equipment manufacturers (so they're not just trying to sell you their own brand of system or appliance)
- If they are qualified as Certified Energy Managers (CEMs) by the Association of Energy Engineers
- For sample reports and references

Often, equipment manufacturers' representatives will offer a no-cost energy audit as part of their "introductory service." Remember, although their advice might be free, they are trying to sell you their products.

Some restaurateurs challenge their own employees to self-audit, finding the best ways to save energy. A walk-through self-audit form has been developed by TU Electric, the electric utility that serves Dallas/Fort Worth, Texas. Not every line applies to a restaurant or foodservice business, but it's a thorough starting point.

Involving your staff in energy management is a very wise move, not only because it is profitable, but because it is a responsible way to do business. Make it the topic of some staff meetings. Give employees an incentive, and let them try their own walk-through and report on their findings, as often as every six months. You might be surprised at what they come up with!

Just remember, especially with a brand-new business, there's no substitute for the expertise of a utility company or contractor. If there are changes to be made, a thorough energy audit will help you prioritize them.

5-3 BECOMING ENERGY PROACTIVE

Utility providers are not necessarily run by a higher power—no pun intended. They might make mistakes. They may misread a meter or estimate a bill without checking the meter at all. This is why foodservice managers should know how to read their own meters and to compare those readings with the bills when they arrive. Not only will you keep the utility companies honest, you'll be able to chart usage and see if the energy conservation measures you've implemented are working, week to week.

In a well-run operation, meter readings are taken on a regular basis, usually weekly or every other week. Another purpose of these readings is to diagnose any energy use problem. It is wise to designate one person as the "energy specialist" in your business, so that he or she can take responsibility for this and for negotiating with utility providers for greater savings. In multiple locations, such as a chain restaurant, it may be wise to hire an energy management coordinator who can handle these affairs on behalf of all locations. There are also independent companies that monitor and check utility bills for a percentage of the savings obtained. Bills should be checked for two types of errors: calculation errors (incorrect addition or subtraction) and classification errors (being charged a residential rate, for instance, instead of a commercial rate).

There are thousands of electric utility providers. Almost every one of them has a slightly different way of converting electrical usage into an electric bill; and almost every company will bill a restaurant differently than they bill a residence. However, there are two major parts to any restaurant's electric bill:

- The ***energy charge*** (consumption charge) is based on the total amount of electricity used over the billing period.
- The ***demand charge*** (capacity charge) is based on the highest amount of electricity used during peak use periods.

Demand is the amount of power that is required at a point in time to operate the equipment connected to the electrical system (also known as the ***power grid***) of your business. It is measured in kilowatts. Demand should not be confused with electrical energy, which refers to power usage over a period of time and is measured in kilowatt-hours. To illustrate the difference, a piece of electrical equipment that is rated at 30 kilowatts (KW) requires 30 KW of power when it is operating. If it operates for one hour, it uses 30 kilowatt-hours (kWh) of electricity. If it operates for 20 hours, it will use 600 kWh of energy—but it's still rated at 30 KW.

Now let's look at the difference demand charges make in a utility bill. The power company must have enough generating capacity to produce whatever amount of electricity the customers need—not the number of kilowatt hours, but the amount of demand that may be put on the system at any given moment.

Let's pretend our fictitious "Power Company, Inc." has only two customers. Customer A has a 50 KW load that is required 24 hours a day. Their total usage is 1200 kilowatt-hours (kWh), or 50 multiplied by 24. To serve Customer A, the Power Company must have the capacity to generate, transmit, and distribute 50 KW of electricity.

Customer B also uses a total of 1200 kWh of electricity, but this usage is only six hours a day—200 KW for six hours equals 1200 kWh. To serve Customer B, Power Company, Inc. must have the capacity to generate, transmit, and distribute 200 KW, or four times the load required for Customer A. Thus, the overall cost to serve Customer B is much greater than for Customer A. It's not fair to overcharge Customer A for the greater peak demands of Customer B, so Customer B pays more in demand charges than Customer A.

To bill customers accurately, both usage and demand must be measured, and this is done by meters. A meter is an electric motor with a large, circulating disk that rotates as energy passes through it. There are four types of meters. On three of them, you will see a row of dials that resemble clock faces. The dials record the number of rotations made by the disk.

You are billed for electricity in one of two ways: by the ***kilowatt-hour (kWh)*** or by the ***kilovolt-ampere (kVA).*** A kilowatt is 1000 watts, so a kilowatt-hour is the time it takes to use 1000 watts. A kilovolt is 1000 volts, so a kilovolt-ampere is the "push" it takes to propel 1000 volts through your wires. A meter keeps track of either kilowatt-hours or kilovolt-amperes.

On a ***kilowatt-hour meter,*** there's only one row of dials. This type is used when the rate schedule is based only on kilowatt-hours used. You probably have this type of meter at your home.

The second type of meter is called a ***demand meter.*** As you just learned with Customers A and B, utility providers monitor (and can charge higher rates for) what they consider "excessive" use of electricity for businesses during peak time periods. The demand meter may only measure for random 15- or 30-minute intervals. However, it has one hand that stays in the maximum-demand position to record the most energy that was used, and another hand that moves as power needs increase or decrease. It is reset, using a key, by the meter reader on his or her normal rounds. Ask your meter reader exactly how to read your demand meter, since there are many different models. Some are zeroed after every reading; others are cumulative.

Your meter may have two rows of dials. The top row is a kilowatt-hour meter; the bottom row is the demand meter. In commercial buildings, you may have two separate meters or this type of combination meter.

The ***digital meter*** is the most modern, as well as the easiest to read. It looks like (and is read the same way as) the mileage indicator on a car speedometer.

No matter what kind of meter you have, subtract the first reading you take (at the beginning of the month, for instance) from the next reading (say, two weeks later) to get the total number of kilowatt-hours used during that time. Some meters have these words on them: "Multiply all readings by ——." If so, you should multiply the readings you get by that particular number or they won't be accurate.

When learning to read an electric meter, the thing that can be confusing is that the hands on the dials rotate in opposite directions from each other. To figure out which way each hand rotates, look at the dial itself. "Zero" will always be at the top of the dial. If the number to the *right* of "zero" is "one," we know this dial rotates clockwise, like a regular clock. If the number to the right of "zero" is "nine," we can assume the dial rotates counterclockwise.

Why are there several dials in a row? The one on the far right represents amounts from 0 to 10 kilowatt-hours; the one to its left, from 10 to 100 kilowatt-hours; the next one to the left, from 100 to 1000 kilowatt-hours; and so on. The hand on the dial at the extreme right will complete one full circle, but the one to its immediate left advances only one number at a time—like the minute and hour hands of a clock. The same relationship exists between the other adjacent dials. (See Illustration 5-3.)

If the hand is pointing between two numbers, read it as the lower of the two numbers.

Most utility providers require installation of a demand meter when the maximum energy use is 10 kilowatts or higher during their peak time period, which is usually from noon to 8:00 P.M. weekdays. This type of energy usage is called ***peak loading.***

Whether you are in a state where the electric industry has been deregulated or not, any provider varies its pricing structure somewhat, especially the demand charges. Some demand charges vary according to the time of year. Others use a formula based on the relationship between the level of demand and the actual amount of energy used. Still others will impose a minimum demand charge based on a high-energy-use period (midwinter in a very cold climate) and make customers pay it each and every month. This practice is known as a ***ratchet clause,*** and it is understandably unpopular, because high demand during one month can result in a higher bill during the rest of the year.

Because energy costs can vary drastically from season to season, a ***tariff analysis*** should be prepared by the utility for every restaurant, using its actual energy consumption as a guideline. This is a kind of overview of the billing, rates, and rules offered by the provider. The tariff analysis should be reviewed (and modified, if necessary) when major changes occur in the restaurant's operations or the utility provider's rate structure. Major changes include undertaking an energy conservation program or extensive remodeling. Illustration 5-4 is a sample tariff analysis. You can see how helpful it might be to have all these details on paper.

Another item on your electric bill that should be reviewed periodically is the sales tax imposed by your local city or state. You may be able to apply for tax-exempt status for your utility bill, if you use electricity (or gas) for "processing" foods. "Processing" is defined, for these purposes, as performing an operation in which a physical or chemical change is brought about. In a restaurant, this would include most types of preparation and cooking: baking, broiling, chopping, slicing, frying, and mixing. It would not include keeping foods hot or cold (since no chemical change is happening to them at that point), heating or air conditioning, dishwashing, or lighting.

There are no separate meters for exempt and nonexempt utility uses, so it is up to you, as the restaurant owner, to prove that more than half of your utility use is for processing (exempt) purposes. If you can, the entire bill may qualify for a

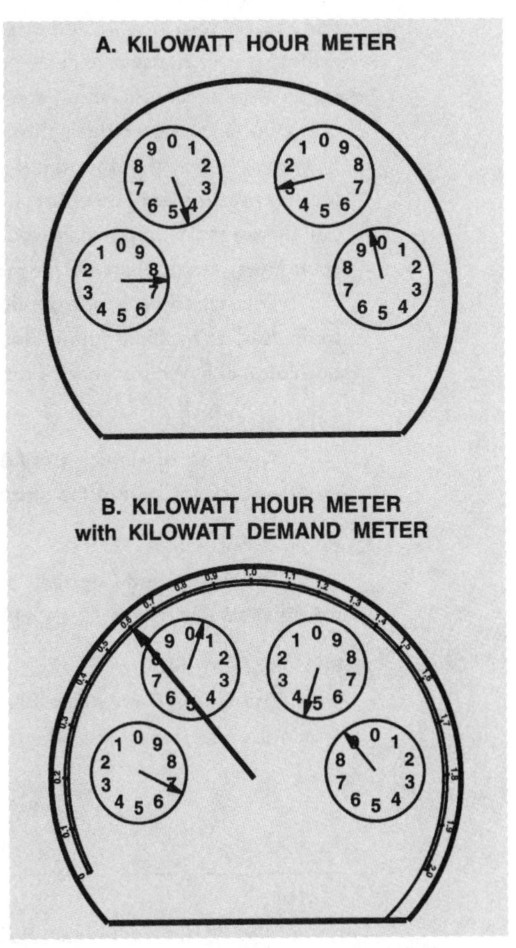

ILLUSTRATION 5-3 **(a) A kilowatt-hour meter (multiply all readings by 10 kilowatt hours) and (b) a kilowatt demand meter (to get kilowatt hours, multiply demand reading by 10).**

Idaho Power Company | IDAHO PUBLIC UTILITIES COMMISSION
Approved Effective
I.P.U.C. No. 27, Tariff No. 101 Original Sheet No. 9-1 June 1, 2004 June 1, 2004
Jean D. Jewell Secretary

SCHEDULE 9
LARGE GENERAL SERVICE

AVAILABILITY

Service under this schedule is available at points on the Company's interconnected system within the State of Idaho where existing facilities of adequate capacity and desired phase and voltage are adjacent to the Premises to be served and additional investment by the Company for new transmission, substation, or terminal facilities is not necessary to supply the desired service.

APPLICABILITY

Service under this schedule is applicable to firm Electric Service supplied to a Customer at one Point of Delivery and measured through one meter. This service is applicable to Customers whose metered energy usage exceeds 3,000 kWh per Billing Period for a minimum of three Billing Periods during the most recent 12 consecutive Billing Periods and whose metered Demand per Billing Period has not equaled or exceeded 3,000 kW more than twice during the most recent 12 consecutive Billing Periods. Where the Customer's Billing period is less than 27 days or greater than 33 days, the metered energy usage will be prorated to 30 days for purposes of determining eligibility under this schedule. Customers whose metered energy usage does not exceed 3,000 kWh per Billing Period on an actual or prorated basis three or more times during the most recent 12 consecutive Billing Periods or whose metered demand equals or exceeds 1,000 kW per Billing Period three times or more during the most recent 12 consecutive Billing Periods are not eligible for service under this schedule and will be automatically transferred to the applicable schedule effective with the next Billing Period. New customers may initially be placed on this schedule based on estimated usage.

This schedule is not applicable to standby service, service for resale, or shared service, or to individual or multiple family dwellings first served through one meter after February 9, 1982, or to agricultural irrigation service after October 31, 2004.

TYPE OF SERVICE

The type of service provided under this schedule is single and/or three-phase, at approximately 60 cycles and at the standard service voltage available at the Premises to be served.

BASIC LOAD CAPACITY

The Basic Load Capacity is the average of the two greatest non-zero monthly Billing Demands established during the 12-month period which includes and ends with the current Billing Period.

BILLING DEMAND

The Billing Demand is the average kW supplied during the 15-consecutive-minute period of maximum use during the Billing Period, adjusted for Power Factor.

IDAHO
Issued Per IPUC Order Nos. 29505 and 29506
Effective – June 1, 2004

Issued by IDAHO POWER COMPANY
John R. Gale, Vice President, Regulatory Affairs
1221 West Idaho Street, Boise, Idaho

ILLUSTRATION 5-4 A typical tariff analysis provided by an electric company. It is a summary of all the types of charges for which a customer may be billed.
COURTESY OF IDAHO POWER COMPANY, BOISE, IDAHO.

Idaho Power Company

I.P.U.C. No. 27, Tariff No. 101 Original Sheet No. 9-2

IDAHO PUBLIC UTILITIES COMMISSION
Approved June 1, 2004 Effective June 1, 2004
Jean D. Jewell Secretary

SCHEDULE 9
LARGE GENERAL SERVICE
(Continued)

FACILITIES BEYOND THE POINT OF DELIVERY

At the option of the Company, transformers and other facilities installed beyond the Point of Delivery to provide Primary or Transmission Service may be owned, operated, and maintained by the Company in consideration of the Customer paying a Facilities Charge to the Company.

Company-owned Facilities Beyond the Point of Delivery will be set forth in a Distribution Facilities Investment Report provided to the Customer. As the Company's investment in Facilities Beyond the Point of Delivery changes in order to provide the Customer's service requirements, the Company shall notify the Customer of the additions and/or deletions of facilities by forwarding to the Customer a revised Distribution Facilities Investment Report.

In the event the Customer requests the Company to remove or reinstall or change Company-owned Facilities Beyond the Point of Delivery, the Customer shall pay to the Company the "non-salvable cost" of such removal, reinstallation or change. Non-salvable cost as used herein is comprised of the total original costs of materials, labor and overheads of the facilities, less the difference between the salvable cost of material removed and removal labor cost including appropriate overhead costs.

POWER FACTOR

Where the Customer's Power Factor is less than 85 percent, as determined by measurement under actual load conditions, the Company may adjust the kW measured to determine the Billing Demand by multiplying the measured kW by 85 percent and dividing by the actual Power Factor. Effective November 1, 2004, where the Customer's Power Factor is less than 90 percent, as determined by measurement under actual load conditions, the Company may adjust the kW measured to determine the Billing Demand by multiplying the measured kW by 90 percent and dividing by the actual Power Factor.

SUMMER AND NON-SUMMER SEASONS

The summer season begins on June 1 of each year and ends on August 31 of each year. The non-summer season begins on September 1 of each year and ends on May 31 of each year.

MONTHLY CHARGE

The Monthly Charge is the sum of the Service, the Basic, the Demand, the Energy, the Power Cost Adjustment, and the Facilities Charges at the following rates.

IDAHO
Issued Per IPUC Order Nos. 29505 and 29506
Effective – June 1, 2004

Issued by IDAHO POWER COMPANY
John R. Gale, Vice President, Regulatory Affairs
1221 West Idaho Street, Boise, Idaho

ILLUSTRATION 5-4 *Continued*

Idaho Power Company | **IDAHO PUBLIC UTILITIES COMMISSION**

I.P.U.C. No. 27, Tariff No. 101 **Original Sheet No. 9-3**

Approved June 1, 2004 **Effective June 1, 2004**

Jean D. Jewell Secretary

SCHEDULE 9
LARGE GENERAL SERVICE
(Continued)

SECONDARY SERVICE	SUMMER	NON-SUMMER
Service Charge, per month	$5.60	$5.60
Service Charge, per kW of Basic Load Capacity	$0.37	$0.37
Demand Charge, per kW of Billing Demand	$3.00	2.73
Energy Charge, per kWh	2.8903¢	2.5784¢
Power Cost Adjustment*, per kWh	0.6039¢	0.6039¢

***This Power Cost Adjustment is computed as provided in Schedule 55.**

Facilities Charge
None.

Minimum Charge
The monthly Minimum Charge shall be the sum of the Service Charge, the Basic Charge, the Demand Charge, the Energy Charge, and the Power Cost Adjustment.

IDAHO
Issued Per IPUC Order Nos. 29505 and 29506
Effective – June 1, 2004

Issued by IDAHO POWER COMPANY
John R. Gale, Vice President, Regulatory Affairs
1221 West Idaho Street, Boise, Idaho

ILLUSTRATION 5-4 ***Continued***

sales tax exemption. Illustration 5-5 is a list compiled by the Texas Restaurant Association to advise its members what that state's comptroller considers reasonable types of processing equipment that use energy, as well as those that do not qualify in determining your sales tax exemption.

Your power company or Retail Electric Provider probably has written information on this topic, as well as pamphlets explaining utility rates and how to read meters. Ask for similar information from gas and water suppliers.

To be truly energy proactive, it is important for the foodservice manager to know that there are state and local codes governing the use of electricity, the calibration of meters, installation techniques for electrical equipment, and licensing requirements for electricians.

The most widely accepted standard is the *National Electrical Code,* which specifies how wiring must be done and what safety precautions must be taken. At this writing, the newest edition is dated 1990. (For people who are building new restaurants, note that it includes many changes from the previous code about the sizes of wires, conduit, and main electrical panels. These changes allow you to use smaller sizes than would have been allowed under the old code, which will probably cost less, so be sure your architects and designers are familiar with the newer code.) The code can also come in handy if you are considering adding equipment to an existing kitchen, as shown in the following "In the Kitchen" box, provided by the Electric Power Research Institute.

Most local codes are adopted from the National Electrical Code, which is part of the U.S. Occupational Safety and Health Act (OSHA) and therefore has the clout of federal law. Equipment manufacturers also use the National Electrical Code when designing their products.

The following pieces of equipment are representative of the types of equipment that the comptroller has determined will qualify as processing and those that are considered nonprocessing.

Prepare a list of equipment that you use to process food. You may include all of the following listed equipment:

Rotisseries	Malt Machines
Ovens	Soft-Drink dispensers
Ranges (stoves)	Ice Cream Machines
Microwave Ovens	Ice Machines
Fryolators	Heat Strips*
Toasters	Steam Table*
Food Choppers	Mixers/Grinders
Slicers	Broilers, Charbroilers
Grills	Proof Boxes
Coffee Makers	Chicken and Fish Breaders

***Heat strips and steam tables cannot qualify if used to maintain a warm temperature after preparation, but may qualify if used to bring food from a cold or frozen stage to serving temperature; or, in the case of a heat strip, to keep food warm during preparation.**

Your list may NOT include the following equipment:

Cash Registers	Air Conditioning
Coffee Warmers	Lighting
Cream-O-Matics	Exhaust Fans
Milk Dispensers	Music, Signs
Heat Lamps	Dishwashers
Refrigerators	Hot-Water Heaters
Freezers	Ice Cream Holding Cabinets

While this list is not all-inclusive, it will by way of example help you understand the comptroller's definition of processing.

ILLUSTRATION 5-5 **Most states have strict rules about what types of foodservice equipment do (and do not) qualify for sales tax exemptions.**
COURTESY OF TEXAS RESTAURANT ASSOCIATION, AUSTIN, TEXAS.

Another agency involved in the safe use and installation of electrical equipment is *Underwriters' Laboratories (UL)*. This organization establishes test procedures and tests equipment for compliance with its own rigorous set of standards. The UL certification label means the equipment meets these standards and is safe to use under normal conditions. However, be sure when you see the label that the entire piece of equipment has been certified—not simply one component, such as the fan of a convection oven.

5-4 UNDERSTANDING AND MEASURING ELECTRICITY

Electricity is a source of power made up of billions of individual electrons flowing in a path, much like water in a stream. The stream is the *electrical current.* Current flows through *wires,* sometimes called **conductors.** Wires are coated with a layer of *insulation,* which prevents shocks and ensures that the full amount of power reaches its destination. The type of insulation may vary, depending on the type of appliance or whether it will be in a wet or dry location. When several wires are bound together with additional

IN THE KITCHEN

ADDING EQUIPMENT TO EXISTING RESTAURANTS

As menus change, it is often desirable to add a new piece of electric cooking equipment to do the job. However, a major concern of owners is whether this addition will also require an increase in the size of the existing electrical service. The new code provision does not address this.

However, there is an existing provision in the code (Section 220-35) for "Additional Loads to Existing Installations." This is sometimes overlooked. With a year of electric bills from your files, or from your local electric company, it is easy to calculate whether additional service capacity is required. Following is an example of how to do the calculations (taken from a real restaurant). Be sure to do these calculations before contracting for any increase in electric service.

THE PROBLEM

A restaurateur wishes to install two new EPRI/Frymaster electric deep-fat fryers—14 kW each. Can it be done without the added cost of enlarging the electric service entrance and main electrical panel?

THE FACTS

- Existing service entrance (PF = 0.9): 800 amps—288 kVA—259 kW
- Current connected load (PF = 0.9): 798 amps—287 kVA—259 kW
- New load (2 fryers: PF = 1.0): 78 amps—28 kVA—28 kW
- 12-month billing history (from the local electric provider)

Month	kWh	kW Demand	Month	kWh	kW Demand
Jan	26000	73	Jul	26000	78
Feb	27760	70	Aug	29040	97
Mar	23360	73	Sep	27120	85
Apr	28400	68	Oct	27120	89
May	23280	76	Nov	26400	80
Jun	24320	119	Dec	25920	75

(System voltage is 208Y/120.3O)

Consult the National Electric Code (Section 220-35) *220-35. Optional Calculations for Additional Loads to Existing Installations.* For the purpose of allowing additional loads to be connected to existing feeders and services, it shall be permitted to use actual maximum kVA demand figures to determine the existing load on a service or feeder when all the following conditions are met:

1. The maximum demand data is available in kVA for a minimum of a one-year period.
2. The existing demand at 125 percent plus the new load does not exceed the ampacity of the feeder or rating of the service.
3. The feeder has overcurrent protection in accordance with Section 240-3, and the service has overload protection in accordance with Section 230-90.

THE SOLUTION

- Highest recorded 12-month demand: 119 kW
- Highest demand + 25%: 149 kW
- Highest demand + 25% + new load (28 kW): 177 kW
- Existing panel capacity: 259 kW

Yes! Both new electric fryers can be added with capacity to spare.

insulation, you have a ***cable.*** Cables in foodservice, heavy-duty but flexible, are wrapped in steel and often referred to as BX cables. When the wires are encased in nonflexible steel pipe, they're called ***conduits*** instead of cables.

The complete path of a current, from start to finish, is known as a ***circuit.*** Any circuit has three basic components: the *source,* or starting place; the *path* along which it flows; and the *electrical device* (the appliance) it flows into. A common type of circuit is found in the fluorescent light bulb (see Illustration 5-6).

A complete, working circuit is called a *closed circuit.* An *open* (or *broken) circuit* is one that has been interrupted so that no electricity flows. A ***short circuit*** occurs when current flows in an unplanned path, due to a break in a wire or some other reason. Short circuits are unsafe because they can cause fire and/or other types of equipment damage.

Fortunately, there are a couple of safety devices to prevent short circuits. A ***fuse*** is a load-limiting device that can automatically interrupt an electrical circuit if an overload condition exists. Fuses protect both appliances and humans. More about them in a moment.

A ***circuit breaker*** looks like an electrical switch. It automatically switches off when it begins to receive a higher load of electricity than its capacity. As the load decreases, the breaker may be manually switched on again without damage to electric lines or appliances.

In addition to kilowatt-hours and kilovolt-amperes, there are four basic terms to learn in measuring electrical energy. You need to know them, because most kitchen equipment is "rated" (for output or energy efficiency) based on the following terms.

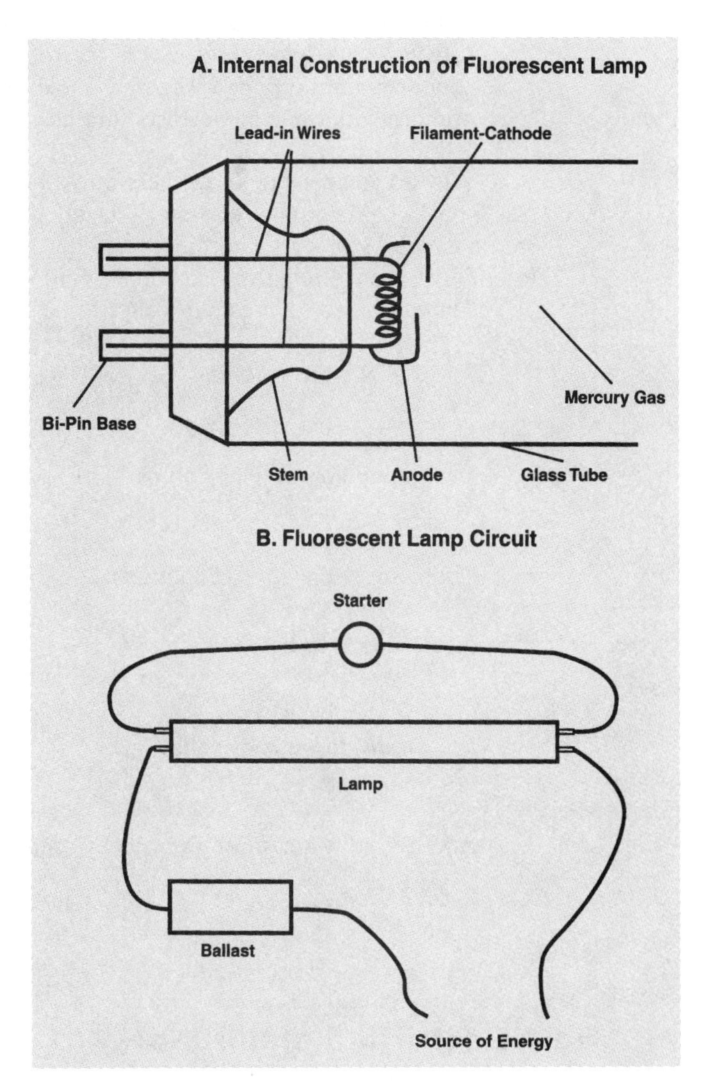

ILLUSTRATION 5-6 The inner workings of (a) a fluorescent lamp and (b) a simple fluorescent lamp circuit.

SOURCE: FRANK D. BORSENIK, *MAINTENANCE AND ENGINEERING FOR LODGING AND FOODSERVICE FACILITIES,* THE EDUCATIONAL INSTITUTE OF AH&MA, LANSING, MICHIGAN.

AMPERE. Commonly known as an ***amp,*** this is a term for how much electric current flows through a circuit. A 100-watt light bulb, for instance, requires 0.832 amps of electricity to light up. The larger the diameter of an electric wire, the more amperes it can safely carry, so it is vitally important that a licensed electrician determine the correct wire sizes for your needs. The National Electrical Code sets the trade standards for wires. The minimum size for copper wire, for instance, is #14; for aluminum wire, #12.

Circuit breakers are also rated in amperes, most often 15 amps (a single circuit breaker) or 30 amps (a double). The typical home electrical outlet provides 15 amps to plug-in appliances, lamps, and so on. As few as 0.02 amps can render an electrical shock. In electrical formulas, the letter *I* indicates amperes.

VOLT. The ***volt*** is the driving force that pushes an ampere through an electrical wire. One volt is the force required to push 1 amp of electricity for 1 second. For appliances, common voltages are 110–120 and 208–240. This means the amount of voltage necessary to operate the appliance efficiently; most appliances have a minimum requirement of 120 volts. In electrical formulas, the letter *E* indicates volts.

WATT. The ***watt*** represents actual consumption of electrical energy, the amount of power in a circuit. One watt equals the flow of 1 amp of electricity, at a pressure of 1 volt. Many electrical appliances are rated in terms of both watts and volts. In electrical formulas, the letter *W* indicates watts.

OHM. This is a less familiar term; an ***ohm*** is a unit of electrical resistance. It refers to whether a substance is a good conductor of electricity or an insulator that restricts electric flow. Examples of conductors are copper, silver, and drinking water. Examples of insulators are dry air, wood, rubber, and distilled water. In electrical formulas, the letter *R* (for "resistance") stands for ohms.

Why do you need to know these terms? You may need to figure amps, volts, or watts to help you decide what equipment to buy, to check the efficiency of existing appliances, or to see if you're about to overload a circuit by plugging in something new. Now that you know the terminology, there are some simple formulas for converting the data you *have* into the energy measurement you *need* for a given situation.

There are two ways to find volts:

1. Divide watts by amps

$$E = \frac{W}{I}$$

2. Or multiply amps by ohms

$$E = I \times R$$

There are two ways to find amps:

1. Divide volts by ohms

$$I = \frac{E}{R}$$

2. Or divide watts by volts

$$I = \frac{W}{E}$$

And to find watts, multiply volts by amps:

$$W = E \times I$$

How Electrical Systems Work

The Edison Electric Institute estimates that 80 percent of all power-related equipment problems occur on the customer's side of the meter. When your repair person shows up, you'll be thankful for your basic knowledge of electrical systems as you explain the problem and listen to his or her recommendations.

A building's electrical system uses either ***alternating current (AC)*** or ***direct current (DC).*** *Alternating current* is by far the most common type of electrical current. It actually reverses its flow at regular intervals. Alternating current can be transmitted over long distances with minimal loss. The equipment costs to use AC power are lower than those for DC, and AC doesn't tend to overheat equipment like DC does. The most common AC energy in the United States is 60-cycle energy, which means the electrical current "cycles" 60 times every second. Why is this important? The number of AC cycles varies from country to country, and if you're purchasing equipment made in another country, it may run on 50-cycle power instead of 60-cycle power. If that's the case, you'll need to buy a converter to enable it to work in your facility. The nameplate on the appliance should state how many cycles it uses to operate.

There are two main types of alternating current, single phase and three phase. ***Single-phase current*** is one current of alternating voltage. Most household appliances run on single-phase power. ***Three-phase current*** combines three different streams of current that "peak" in a regular, steady pattern. Heavy-duty commercial appliances and industrial equipment run on three-phase power.

The common battery-powered flashlight is a good example of DC power. It has a constant flow of amps and volts, it doesn't travel a long distance, and it can overheat equipment. It's not widely used in the United States, but it has limited applications such as lighted exit signs, elevators, and security systems.

Your utility company or Retail Electric Provider will recommend what source of energy is preferable for your site. Often only one source will be available anyway, and it must be brought into the building and channeled to the various points of use. Because different voltages and types of electrical circuits are available, it's important to know exactly what you have in your building before you order equipment or anything with a motor.

Voltage Combinations and Load Factors

Electrical equipment manufacturers design appliances with combinations of voltage and phase—that is, the force with which the amperes are pushed through the line and the number of alternating voltages available. Voltage and phase specifics should be part of any new appliance's service information, and are found on the nameplate of the appliance. As you will see, the more wires there are, the greater options for different levels of maximum voltage.

You need to be somewhat familiar with these options because so many large appliances require ***dual voltage***—more than one type. Perhaps the controls or motors need 120 volts, at the same time a heavy-duty heating element requires 208 or 240 volts. Therefore, you must be sure your building's wiring can handle the challenges, whether it's a brand-new facility or a remodel or even just replacement of an old piece of equipment with a newer one. The voltage supplied to each electrical outlet will directly affect the performance of the appliance plugged in there, so it must be correctly specified when ordering the equipment. The most common voltage combinations are:

1. Single phase: 110–120 volts. This system includes two wires, one "hot" and one "neutral." Together, the maximum amount of voltage they can generate is 125 volts.
2. Single phase: 220–240 volts. In this two-wire system, both are hot wires. The maximum voltage between them is 250 volts.
3. Single phase: 110–120/220–240 volts. This three-wire system includes two hot wires and one neutral wire. Between either of the hot wires and the neutral wire, the maximum voltage is 125 volts, but between the two hot wires, the maximum is 250 volts.
4. Three phase: 110–120/208 volts. This four-wire system is made up of three hot wires and one neutral wire. The voltage between any single hot wire and the neutral wire is 125 volts, but the voltage between any two hot wires and the neutral one is 208 volts (single phase) and the voltage between all three hot wires without the neutral is 208 volts (three phase).
5. Three phase: 220–240 volts. All three wires in this system are hot. The voltage between any two of them is a maximum of 250 volts.
6. Three phase: 440–480 volts. This is a three-wire system in which the voltage between any two of the wires is a maximum of 480 volts.

7. Three phase: 110–120/220–240 volts. This is a four-wire system, three hot and one neutral. The voltage between any one of the hot wires and the neutral wire is a maximum of 125 volts; the voltage between any two hot wires is a maximum of 250 volts.
8. Three phase: 440–480/277 volts. Finally, this is a four-wire system, three hot and one neutral. The maximum voltage between each hot wire individually and the neutral is 277 volts; between any two hot wires, a maximum of 480 volts; and between all three hot wires, a 480-volt maximum.

Now, let's put all this terminology into perspective. Generally, as you have learned, single-phase circuits are used for most small or light-duty appliances, such as a two-slice toaster. Heavy-duty appliances with large power and heating loads are usually wired at the factory for three-phase, three-wire electricity. They sometimes include smaller fuses and circuit breakers for different functions within the appliance. A complete wiring diagram must be supplied by the manufacturer and should always be kept with the appliance's service records. Countertop cooking equipment usually requires a two-wire, single-phase circuit. In short, the requirements will be very specific.

It is critical that the equipment voltage matches the service voltage of the building. To determine your restaurant's electric service voltage, you can:

- Call your electric utility provider and ask! (This is the surest and safest method.)
- Look at the electric meter. Most of them are labeled to indicate if the meter itself is a Delta (120/240 volts) or a Wye (120-208 or 277-408).
- Check the nameplates on your cooking equipment for the electrical ratings.
- Ask an electrician to check the large electrical wall outlets with a voltage meter.

The other key piece of information you should learn is the restaurant's ***load factor.*** Expressed as a percentage, it is the amount of power consumed (in kilowatt-hours), divided by the worst-case (highest) consumption over a stated period of time (number of hours). The formula is:

$$\frac{\text{Kilowatt hours}}{\text{Kilowatts} \times \text{Hours}} \qquad \frac{(\text{KWh})}{\text{KW} \times \text{h}} = \text{Load Factor (\%)}$$

The business should have a target load factor. Then, by tracking the load factor weekly or monthly, you can see how well the business is doing—or how well it *could* be doing. Luckily, with today's computer-based metering systems, this kind of analysis is possible. It's easy to calculate a target load factor:

1. Take into account all the hours the business is open (not just the business hours, but the after-hours times that employees are there). If a business site is being used 14 out of 24 hours a day, the target load factor is 14 divided by 24, which equals 58.3%. If you're calculating the target load factor for a full month, divide the total number of hours you're open that month by the total number of hours in the month. (In a 30-day month, there are 720 hours.) A restaurant open 15 hours a day, 7 days a week is open 450 hours a month. Divide 450 by 720 and you get a target load factor of 62.5%.
2. Now you're ready to compute the *actual* load factor. You'll need these figures for the month:
   The total amount of power consumption (in kilowatt-hours)
   The demand meter reading (in kilowatts)
   The number of hours you're open in a day (not the whole month, just one day)

Let's plug them into the formula and see what happens:

$$\frac{\text{(Total consumption) } 2100}{\text{(Demand reading} \times \text{hours in operation) } 180 \times 15} = \frac{2100}{2700} = .7777\text{, or } 77.77\%$$

In the case of the restaurant open 15 hours daily, its actual power use is 15.27 percent higher than its target load factor. In the case of the business open 14 hours, actual power use is 19.47 percent higher than the target. In either case, some energy-saving measures are overdue.

Tracking your load factors will allow you to create a ***load profile*** for the business—that is, an accurate snapshot of how much power you need, and what times of day you need the most. Later in this chapter, when we discuss deregulation of the electric industry, you will see that having a load profile is necessary to get the best price for the electricity you buy.

How Appliances Heat

In electric equipment, the heat is generated through a heating element. Wires inside the heating element accept the incoming electric current. It may be microchrome wire (round, or flat "ribbon"-style wire), fiberglass-coated, or mica-coated. Fiberglass withstands temperatures up to 400 degrees Fahrenheit, and more expensive mica-coated wire withstands up to 1,000 degrees Fahrenheit. Heating elements are also categorized by their ***watt density,*** or number of watts that they can produce per square inch. The higher the watt density, the greater the appliance's ability to provide uniform heating.

If a piece of equipment doesn't have a simple on-off switch, it has a ***thermostat*** to control the heat. A thermostat also turns the equipment on and off, allowing it to heat to a desired temperature and then cycle on and off to hold that temperature for as long as needed. The cycle is often controlled by magnetic force. Magnets within the thermostat serve as contacts, keeping the power connected until the correct temperature is reached; then a bimetal control causes the magnets to snap open. This releases the power connection and the heat stops—until it cools down enough to cool the bimetal, which causes the magnets to snap shut and complete the power connection, and the cycle begins again. Solid-state controls—which use microprocessors rather than moving parts or heated filaments to turn equipment on and off—are now available in many types of kitchen equipment.

Kitchen appliances should also be equipped with their own set of cords, which can be easily disconnected for cleaning and servicing. Neoprene cords are popular, because neoprene is a tough but flexible substance that can withstand both water and grease. However, neoprene will soften under conditions of extreme heat. Rubber-coated cords are not recommended, because they soften and deteriorate under grease and heat.

Today, many heavy-duty commercial cooking appliances can be ordered (at extra cost) for direct hookup to a 460-volt system. However, 460-volt equipment does not contain internal fuses or circuit breakers to isolate trouble. A malfunction in any part of the equipment will cause the whole thing to shut down completely until the problem is located and repaired.

When new equipment arrives in your kitchen, always read the nameplate one more time before connecting it. If the electrical requirements are not followed, you will notice problems. If the operating voltage is lower than it's supposed to be, the appliance will heat more slowly or not heat fully, no matter how long it is left on. For instance, a 12-kilowatt fryer connected to a 208-volt power supply will be 25 percent less efficient and take longer to preheat.

If the operating voltage is higher than recommended, the appliance will become hotter, heat faster, and perhaps burn out the elements more quickly than it should. The same 12-kilowatt fryer, when wired to a 240-volt power outlet, will heat faster, but this will shorten the useful life of the appliance because all its parts are working harder than they were designed to work.

If you know some of the energy-related information for a particular appliance but not all of it, remember there are formulas that can be used to calculate whatever you need to know. If an appliance is rated in kilowatt hours (KW), for instance, you can convert this rating to watts (*W*) and figure out how many amps (*I*) it will require. Let's try it. Let's say we have a single-phase 13.2 KW appliance. It's rated at 208 volts, but we need to know how many amps of electricity it will need. First, we need to convert the kilowatts to watts. This is a standard calculation, because we know that one kilowatt equals 1000 watts:

$$13.2 \text{ KW} \times 1000 = 13{,}200 \text{ watts}$$

Then, we use the formula we learned previously to convert watts into amps. For this appliance, that means:

$$\text{Amps} = \frac{\text{Watts}}{\text{Volts}} \quad \text{or} \quad \text{Amps} = \frac{13{,}200}{208} = 63.5$$

This appliance requires 63.5 amps of electricity.

The Electric Foodservice Council has calculated the energy requirements of several different types of appliances, which you'll see in Table 5-2. This chart shows the approximate amounts of time and electricity required to preheat and maintain some commonly used pieces of cooking equipment.

In this chapter, we focus mainly on electricity as a power source, but don't forget your gas-fired appliances when adding up energy costs. For example, a single gas pilot light (a necessary part of every cooking device approved by the American Gas Association) consumes about 750 Btus of gas per hour. Multiply that by 24 hours a day, and that's 18,000 Btus per day, or 18 cubic feet (gas is purchased in cubic feet). If a cubic foot costs 5 cents, this equals 90 cents per day. The kitchen that has a dozen gas appliances will spend almost $4000 a year just to keep the pilot lights on! More about gas use in Chapter 6.

The Electrical Service Entrance

Electricity is generated (converted into electricity from another source like gas, coal, sunlight, or wind), then transmitted (moved in bulk to a wholesale purchaser), and finally distributed (delivered to end users, like homes and businesses.)

The physical point at which a building is connected to the power source is called the ***electric service entrance.*** Before you add new pieces of electric cooking equipment, you must find out whether this will also require an increase in the size of the electric service entrance and/or the main circuit panel, a considerable expense.

The electric service entrance may be overhead or underground, easily visible or not. This very important place has eight basic components, and, as shown in Illustration 5-7, you're already familiar with a couple of them. If there's a power outage, fire, or malfunction, you'll want to know enough to correctly describe it in these terms:

1. **Connection Point.** The exact site at which the wires touch the building.
2. **Electric Meter.** Records the kilowatt-hour consumption of the building.
3. **Demand Meter.** Records the maximum demand for kilowatts.
4. **Master Service Switch.** Controls the flow of energy to the building and will automatically shut off as a safety precaution if there's an overload or excessive demand. It's usually located on the outside of the building, and it is locked.
5. **Transformer.** As its name suggests, it changes incoming alternating current into whatever voltage is usable for a particular facility. Not all buildings have (or need) a transformer; most very large ones do.

TABLE 5–2

ENERGY USE OF COMMERCIAL KITCHEN EQUIPMENT

Type of Appliance	Kilowatt Input	Minutes to Preheat to 450°F	Watts to Maintain at: 300°F	350°F	400°F
All-Purpose Oven	6.2	36	531	649	767
Baking Oven	7.5	90	660	807	953
Pizza Oven	7.2	45	410	507	599
Convection Oven	11	10 or 11 (to 350°F.)		1800	
Electric Fryer (28-lb. Fat Capacity)	12	5 (to 350°F.)		770	

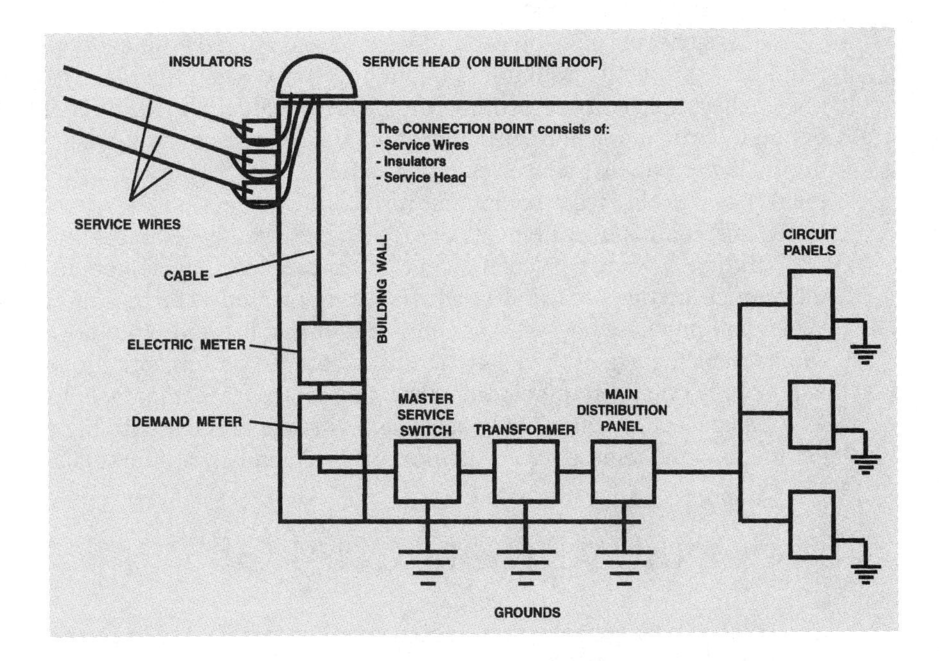

ILLUSTRATION 5-7 **The electrical service entrance of a building.**

SOURCE: FRANK D. BORSENIK, *MAINTENANCE AND ENGINEERING FOR LODGING AND FOODSERVICE FACILITIES*, THE EDUCATIONAL INSTITUTE OF AH&MA, LANSING, MICHIGAN.

6. **Main Circuit Panel.** Also called a distribution panel, this is the equivalent of the fuse box in your home. Separate switches, with circuit breakers, direct power to different areas of the building or pieces of equipment. Each panel also has a master fuse or circuit breaker that controls power to the entire panel.
7. **Secondary Circuit Panel.** Local codes may require separate circuits, in separate locations, for one particular appliance or function in the building. This provides a way to shut off the air conditioner, for instance, or all heavy equipment, without interfering with other power use. It's especially handy when repairs have to be made.
8. **Ground Rod.** One low-impedance connection to the earth, which serves as a conducting body to which an electric circuit can be connected. A "ground" is necessary to operate electronic equipment.

Take an Electrical Inventory

The more equipment you buy and maintain, the harder it may be to keep track of it all if there is a problem. Problems may include:

- Voltage transients (short, fast, high-voltage pulses of electricity that may damage equipment immediately or over time; also called *spikes*)
- Momentary surges or sags in voltage
- Momentary power losses
- Electrical noise
- Harmonic distortion (an imbalance in one wire of a single-phase AC line that can create an overload; seen primarily in fluorescent lighting systems and computers)

Some types of equipment are more likely to be the victims of power problems (phone systems, credit card scanners, or personal computers), while others are more likely to cause problems for more sensitive equipment. Your heating or air-conditioning system, on a high-demand day, is an example of this. Solving a problem may be as simple as making sure the "victims" and "culprits" are not on the same power circuit so they can't affect each other, or upgrading wiring so that it is large enough to handle a larger operating load.

The smart business owner will become acquainted with every piece of electrical equipment in the building, by walking around and taking inventory when the place is closed. From the largest motor to the smallest mixer, list everything. If a piece of equipment has an identification number or model number on it, write it down. Note the location of each piece of equipment and which electrical panel is feeding it. If the panels themselves aren't labeled, now is the time to do that, too, by shutting off each breaker and seeing what equipment turns off. One caution: *Do not* turn the large breakers (controlling HVAC and computer equipment) on or off unless you have an electrician with you.

If there are any particular symptoms or problems, make note of them. Be sure to ask employees if they notice anything unusual—flickering, surging, and so forth—and what time of day it seems to occur. When you purchase new equipment, put it on the list and write down the date it was put into service. (Sometimes the age of the equipment is a factor in its susceptibility to power problems.) This way, you will be fully equipped to help the electrician or electric company help you.

Most electric utilities have specialists who are familiar with the needs of foodservice businesses, so get to know the representative in your area. Remember, restaurants are their best customers!

5-5 CONSUMPTION CHARGES

Now that you know something about reading meters and converting different quantities of energy, you're ready for the most daunting test—reading the electric bill!

Most foodservice establishments aren't large enough to qualify for lower, industrial rates. Instead, they are billed on a "commercial" or "small general" rate schedule. They are rarely charged a single rate for every hour of electricity they use. A single rate, or ***uniform charge,*** is more likely to be seen on a residential power bill.

Instead, for commercial customers, you'll probably find a *step-rate* or *declining-block schedule*: that is, a graduated schedule of rates that decreases as consumption increases. For instance, the first 50 kilowatt-hours may cost you $0.100, but the next 50 will only cost $0.050, and the 100 after that will only cost $0.030 per kilowatt-hour.

A ***time-of-day option*** allows you to be billed based on what time of day your business is open. More power is used during weekdays, so that's when rates are highest. However, businesses that operate mostly nights, holiday, or weekends can qualify for lower, off-peak rates by choosing a time-of-day plan.

Another type of commercial rate is the ***demand charge.*** Remember the demand meter? The demand charge multiplies that reading—the average peak demand—by a charge per kilowatt-hour. The demand charges may also be a declining step rate, depending on how much power is used; and if your facility exceeds its normal demand, there will also be an additional charge for consuming more than usual.

The utility provider may also impose additional charges on customers with low power factors. Your ***power factor (PF)*** is the ratio between what the demand meter says (*real power*) and what the utility company says is the total capacity available to you (*apparent power*). Most foodservice businesses don't run the kinds and amounts of equipment that would result in a low-power-factor penalty. If it is a problem, however, capacitors can be installed to bring your voltage and current levels into closer sync with each other, increasing your power factor and lowering your power bill.

Fuel adjustment charges (sometimes called *energy cost adjustments*) will show up on your bill, too. Because energy costs fluctuate constantly, utility companies in nonderegulated states are authorized to pass on these fluctuations to their customers. Sometimes it's a charge—other times it's a credit!

Finally, don't forget *late charges,* which will inevitably be tacked on if you don't pay the bill on time.

See why you need to keep a very close eye on your power bills? In fact, now let's examine one, line by line.

Charting Annual Utility Costs

In addition to regular meter readings, there are two useful methods that foodservice operations can use to track energy consumption. You may find them especially useful in preparing cost estimates to open a new facility. They are utility costs as a percentage of total sales and an ***energy index*** of Btus per meal served.

BUILDING AND GROUNDS

READING AN ELECTRIC BILL

Every manager should understand the information on utility bills. This will enable a comparison of his or her own meter readings to those of the utility companies, and an awareness of how the billing figure is arrived at. Although utility bills come in a wide variety of shapes and forms, they usually display the same type information shown on the sample electric bill.

LOCAL ELECTRIC POWER COMPANY

THE HOTEL
50 States Avenue
City, State Zip

ACCOUNT NUMBER			
FOLIO	S	ROUTE	RG
123	4	5678	9

(1)

Meter Reading by Rate Schedule			Service		Meter Reading x Constant		Constant	Meter Number	
SCH.	KW Demand	KW Hours	From	To	Previous	Present			Overdue After Jul 15, 1999
GS4	615	171,400	5/07	6/06	710400	881800	100	GS7890	
(2)	(7)	(6)	(3) (11)		(5)		(4)		
* Fuel Cost Per KWH When Applicable .00087480									

Billing Data by Rate Schedule			Energy Charge	Demand Charge	Tax		Amount Due Now
SCH.	KW Demand	KW Hours					
GS4	615	171,400	5,848.11	1,410.00	223.74		7681.85
	(9)	(10)	(12)	(13)	(14)		(15)
			Control Demand If Applicable	Environmental Surcharge	County Energy Tax Surcharge		
			(8)	(16)			7681.85

ILLUSTRATION 5-8 **A sample electric bill and key for interpreting it.**

NOTES FOR SAMPLE BILL

1. Your account number. Use it when communicating with the utility.
2. The rate schedule that applies specifically to your property. The utility company will furnish you with a copy.
3. Dates on which the meters were read. They represent the beginning and end of a billing period.
4. A constant by which the meter reading must be multiplied to get the actual usage.
5. The meter readings multiplied by the constant. If no constant is shown on your bill, this entry will be the same as the meter reading.
6. Kilowatt-hours of electricity used.
7. The maximum kilowatt demand that has been recorded during a demand interval in this billing period. The demand interval is measured in periods of 15, 30, or 60 minutes and is shown in the rate schedule.
8. Control demand would be shown here if it is part of the rate schedule. In general, it is based on the maximum demand of the previous 12 months. The rate schedule will show how it is determined and its effect on the computation of the electric bill.

BUILDING AND GROUNDS

READING AN ELECTRIC BILL (CONT.)

9. Kilowatt demand for billing purposes.
10. Kilowatt-hours for billing purposes, usually the same as item 6.
11. Fuel adjustment cost per kilowatt-hour. It varies as the utility company's fuel cost changes from a fixed "normal" cost.
12. Charge for kilowatt-hours used based on the rate schedule and the fuel adjustment cost.
13. Charge for maximum demand based on the rate schedule.
14. Local tax.
15. Total amount of the bill.
16. Surcharges that may be added to the bill to offset expenses of local governments.

The same type information is shown on bills for other types of utilities, with the obvious differences due to consumption in the applicable units, such as cubic feet (gas), gallons or cubic feet (water), and pounds (steam).

Source: ***Guide to Energy Conservation for Food Service.*** **U.S. Department of Energy, Washington, D.C.**

Let's look at them one at a time. First, to determine the percentage of utility costs, we simply add up all the utility bills (except the phone bill) for the month or year and divide the total by our total sales for the same period. (The total sales figure is found in the income statement.) Although this is a very simple method, it certainly doesn't give much detail about conservation measures.

To develop an energy index, we must convert all the different types of energy used into a similar form, so the figures can be added together. This is where the *Btu* comes in handy. The abbreviation stands for ***British thermal unit,*** which is the amount of heat needed to raise the temperature of 1 pound of water 1 degree Fahrenheit. This measurement is useful because all energy sources can be easily converted to Btus. Here's how:

ELECTRICITY.	Multiply kilowatt-hours by 3413.
NATURAL GAS.	Multiply cubic feet by 1000.
OIL.	Multiply gallons by 140,000.
STEAM.	Multiply pounds by 1000.

When the resulting figures are added together, the total is the amount of energy (in Btus) consumed in the restaurant overall. Don't be shocked—it will be in the millions!

Make your own copy of Energy Tracking Worksheet 1 (see Illustration 5-9) and write the Btu total on it for the month. These worksheets were developed by the Federal Energy Administration, the predecessor of the U.S. Department of Energy. You will need two more monthly totals:

- The total amount you paid for utilities
- The total number of customers you served

Write all three totals on Energy Tracking Worksheet 2 (see Illustration 5-10). Now divide the number of Btus by the number of customers to get the number of *Btus used per customer.* And divide your energy costs by the number of customers to get the *energy cost per customer.*

ELECTRICITY

Month	$	kWh	multiplied by	equals	divided by	equals
Example	**513**	**23,666**	× 3413	80,772,058 Btu	1,000,000	80.8 million Btu
January			× 3413	Btu	1,000,000	million Btu
February			× 3413	Btu	1,000,000	million Btu
March			× 3413	Btu	1,000,000	million Btu
April			× 3413	Btu	1,000,000	million Btu
May			× 3413	Btu	1,000,000	million Btu
June			× 3413	Btu	1,000,000	million Btu
July			× 3413	Btu	1,000,000	million Btu
August			× 3413	Btu	1,000,000	million Btu
September			× 3413	Btu	1,000,000	million Btu
October			× 3413	Btu	1,000,000	million Btu
November			× 3413	Btu	1,000,000	million Btu
December			× 3413	Btu	1,000,000	million Btu

NATURAL GAS

Month	$	kWh	multiplied by	equals	divided by	equals
Example	223	208,000	× 1000	208,000,000 Btu	1,000,000	208 million Btu
January			× 1000	Btu	1,000,000	million Btu
February			× 1000	Btu	1,000,000	million Btu
March			× 1000	Btu	1,000,000	million Btu
April			× 1000	Btu	1,000,000	million Btu
May			× 1000	Btu	1,000,000	million Btu
June			× 1000	Btu	1,000,000	million Btu
July			× 1000	Btu	1,000,000	million Btu
August			× 1000	Btu	1,000,000	million Btu
September			× 1000	Btu	1,000,000	million Btu
October			× 1000	Btu	1,000,000	million Btu
November			× 1000	Btu	1,000,000	million Btu
December			× 1000	Btu	1,000,000	million Btu

OIL

Month	$	kWh	multiplied by	equals	divided by	equals
Example	85	**243**	× 140,000	34,020,000 Btu	1,000,000	34.0 million Btu
January			× 140,000	Btu	1,000,000	million Btu
February			× 140,000	Btu	1,000,000	million Btu
March			× 140,000	Btu	1,000,000	million Btu
April			× 140,000	Btu	1,000,000	million Btu
May			× 140,000	Btu	1,000,000	million Btu
June			× 140,000	Btu	1,000,000	million Btu
July			× 140,000	Btu	1,000,000	million Btu
August			× 140,000	Btu	1,000,000	million Btu
September			× 140,000	Btu	1,000,000	million Btu
October			× 140,000	Btu	1,000,000	million Btu
November			× 140,000	Btu	1,000,000	million Btu
December			× 140,000	Btu	1,000,000	million Btu

PURCHASED STEAM

Month	$	kWh	multiplied by	equals	divided by	equals
Example	2,000	291,667	× 1000	291,667,000 Btu	1,000,000	291.7 million Btu
January			× 1000	Btu	1,000,000	million Btu
February			× 1000	Btu	1,000,000	million Btu
March			× 1000	Btu	1,000,000	million Btu
April			× 1000	Btu	1,000,000	million Btu
May			× 1000	Btu	1,000,000	million Btu
June			× 1000	Btu	1,000,000	million Btu
July			× 1000	Btu	1,000,000	million Btu
August			× 1000	Btu	1,000,000	million Btu
September			× 1000	Btu	1,000,000	million Btu
October			× 1000	Btu	1,000,000	million Btu
November			× 1000	Btu	1,000,000	million Btu
December			× 1000	Btu	1,000,000	million Btu

ILLUSTRATION 5-9 **Energy Tracking Worksheet 1.**

SOURCE: U.S. DEPARTMENT OF ENERGY, WASHINGTON, D.C.

Month	Energy Cost	Cost/Customer	Energy Usage (million Btu)	Energy Usage/Customer
Example	$821	$0.09	322.8	33,979
January				
February				
March				
April				
May				
June				
July				
August				
September				
October				
November				
December				

ILLUSTRATION 5-10 **Energy Tracking Worksheet 2.**
SOURCE: U.S. DEPARTMENT OF ENERGY, WASHINGTON, D.C.

Using these charts, Chez Ralph owners discovered they had spent the following amounts on energy in April:

Electricity	$513
Gas	223
Oil	85
Total	$821

From their utility bills, they calculated they had used 323 million Btus of energy. From their daily customer tallies, they knew they had served 9,500 customers. How much did they spend per customer on energy? Nine cents.

A final note about electricity costs: Any experienced restaurateur will tell you they're tough to track under the best conditions. Unusual weather and rate increases are just two of the factors that make forecasting difficult. In our experience, it is still preferable to monitor and pay your own utility bills than to agree to include utilities in a lease agreement and let the landlord keep track of them. Leases that include utility payments leave little or no incentive for conservation, and leave the door wide open for rent increases based on "higher utility costs" you have no way of confirming.

5-6 ENERGY CONSERVATION

There are thousands of ways to conserve energy, but most can be grouped into four simple categories:

- Improve efficiency of equipment.
- Reduce equipment operating time.

- Recover otherwise "wasted" energy.
- Use a cheaper energy source.

Cooking efficiency means maximizing the quantity of heat that is transferred from the cooking equipment into the cooked product rather than the surrounding environment. This keeps your energy bills lower, and also keeps your kitchen temperature more comfortable for employees.

The primary causes of inefficient cooking are:

- Preheating equipment too long
- Keeping equipment turned on when it is not being used
- Using higher temperatures than necessary
- Opening appliance doors frequently, letting heat escape

Sometimes, the piece of equipment itself is at fault: it is poorly insulated; its temperature has been incorrectly calibrated; perhaps it hasn't been kept clean enough to work at maximum efficiency. Energy-saving features should be considered in any equipment purchase, as it is almost impossible to retrofit an appliance to make it more energy efficient.

For equipment, energy efficiency is measured in ratios (by percentage), and can be calculated any of three ways:

1. *Combustion efficiency* refers to fuel-fired equipment that uses gas, oil, or propane to produce heat. The key here is to control both the heat that is lost "up the flue" of the appliance, and the radiant heat released into the atmosphere. There's a formula for this:

$$\text{Combustion Efficiency} = \frac{\text{Available Heat (the total amount of energy the appliance can produce)}}{\text{Purchased Heat (the amount of energy purchased for cooking with this appliance)}}$$

Typical combustion efficiency ranges from 40 to 70 percent for some gas-fired equipment.

2. *Heat transfer efficiency* means the amount of heat that is transferred into the cooked product, compared to the total amount of available heat the appliance can produce. This percentage depends in part to the cooking process that is being used, the design of the equipment, and how efficiently it is being operated. Again, the formula:

$$\text{Heat Transfer Efficiency} = \frac{\text{Heat Transferred to Food (the amount of energy used cooking)}}{\text{Available Heat (the total amount of energy the appliance can produce)}}$$

Equipment design can increase this type of efficiency by adding insulation, or recirculating exhausted air. Smart use of the equipment—keeping oven doors closed, putting the cover down on a tilting braising pan—also minimizes heat losses.

3. Overall energy efficiency is the ratio of heat that is transferred to the cooked product in comparison to the amount of energy purchased for that appliance. The formula:

$$\textbf{Energy Efficiency} = \frac{\textbf{Heat Transferred to Food}}{\textbf{Purchased Energy}}$$

This takes into account heat losses due both to combustion inefficiency and the surroundings.

Another way to shop for appliances, including computer equipment, is to look for the "Energy Star" rating. This program, developed by the U.S. Environmental Protection Agency (EPA), began in the early 1990s as an effort to save energy and/or reduce pollution by "rating" items and offering a sort of "seal of approval." An appliance with an Energy Star rating typically scores among the top 25 percent in its product category when it comes to energy efficiency. An Energy Star-approved reach-in refrigerator, for instance, can save an average of $140 per year in energy costs over a model that did not receive the rating. And here's something to ponder—the EPA claims that if all the current reach-in refrigerators and freezers in the U.S. were replaced with Energy Star-approved models (which contain more modern refrigerants), we'd eliminate as much air pollution as taking 475,000 cars off the road!

Reducing operating time is another way to save energy. In most kitchens, the hot water heater is a perfect example of this. When used primarily for hand washing, the water is comfortable at a temperature of 105 degrees Fahrenheit or less. However, the most common setting is 140 degrees Fahrenheit. Do you know why? This is the original factory setting, and most people just never bother to change it after it is installed—an expensive oversight! The water heater works harder to maintain the higher temperature, even during the hours when the business is closed.

For foodservice, it is more economical to install separate water heaters for high-temperature uses such as dish-washing or laundry and low-temperature ones for hand washing. As you'll learn in Chapter 6, there are also boosters that can be installed near the hot-water appliances. As its name suggests, a *booster* can boost the temperature of the water as it enters the appliance, lowering the heat loss that would occur if it had to be piped in from a more remote source. And how about checking for leaks and using spring-loaded taps in sinks, to minimize wasteful dripping?

Another typical restaurant oversight is the fryer. Today's fryers take ten minutes or less to heat to full temperature, but in many foodservice settings, they are on and fully heated all day. This not only wastes energy, it minimizes the life of the fryer oil. As you'll learn in Chapter 12, it is best to turn an unused fryer down to a lower temperature (if not off altogether) when it is not needed. Griddles and hot tops (see Chapter 11) can be purchased with heat controls in sections—heat only what you need. There are plenty of ways to save energy by using just "as much appliance" as you need, and no more.

One way to minimize the demand charges discussed earlier in this chapter is to stagger the preheating of equipment. By sticking to a start-up schedule and not turning everything on to preheat at once, the peak demand reading will be lower on your electric meter. How much difference can this make? Take a look at the charts (Tables 5-3a and 5-3b) from the Electric Foodservice Council. Remember, it's not the overall amount of energy used that varies—it's the peak, or maximum demand for which you are also being charged.

In foodservice, energy conservation goals must also be seen from the point of view of the guest. If you're only looking at the bills, you're not looking far enough. Nowhere is this more evident than in your heating, ventilation, and air-conditioning (HVAC) system. If your guests are complaining that they're uncomfortable, your lower utility bill may actually be having an adverse impact on business. Set the

TABLE 5–3A

SIMULTANEOUS START-UP

15 min. Intervals	15 min.	30 min.	45 min.	60 min.	75 min.	90 min.
Oven 6.2 KW	36 min. (ON through ~36 min.)					
Fryer (1) 12 KW	5 min. (then IDLE to 45 min.)					
Fryer (2) 12 KW	5 min. (then IDLE to 45 min.)					
Griddle 6.5 KW	7 min. (then IDLE to 45 min.)					
EST. DEMAND	18.7 KW	8.7 KW	5.4 KW			

Start-Up Time–36 Minutes
Maximum Demand–18.7 KW
Total Energy Used–7.7 KWH

ON ——

IDLE – – –

thermostat at 68 degrees Fahrenheit in winter and 76 degrees Fahrenheit in summer, and you combine pleasant temperatures with savings of up to 18 percent in winter and up to 30 percent in summer.

One big HVAC consideration is whether you allow people to smoke. Having a smoking area requires enough airflow to "change out" the "old" air 20 times per hour. In a nonsmoking area, only six to eight air changes per hour are necessary. This definitely impacts energy costs as well as wear and tear on the system itself.

Lighting can be another power drain. Fluorescent lights are certainly the most economical, but they may also be the least flattering unless you investigate some of the more modern types, which we discuss in greater detail in Chapter 7. Also remember that lighting creates heat; therefore, when you reduce lighting, you lose heat, and if you lose heat, your central heating system must work harder (and use more energy) to make up for it. You will learn more about both HVAC and lighting basics in Chapter 7.

Little wonder that, in the typical 24-hour coffee shop, nearly 60 percent of the energy that is used flows through the kitchen. We realize it has become popular to sear food quickly or to prepare dishes individually. From a power consumption standpoint, however, you realize the most savings by:

- Cooking in the largest possible volume at a time
- Cooking at the lowest temperature that can still give satisfactory results
- Carefully monitoring preheating and cooking temperatures
- Reducing ***peak loading,*** or the amount of energy used during peak demand times (usually between 10 A.M. and 8 P.M.)
- Using appliances (ovens, dishwashers) at their full capacity
- Venting dining room air into the kitchen to meet kitchen ventilation requirements
- Using a heat pump water heater, which heats water as it cools (or dehumidifies) the kitchen
- Using an evaporative cooling system
- Recovering heat from equipment—refrigerators, the HVAC system, even kitchen vents—for reuse
- Increasing the hot-water storage tank size
- Keeping equipment clean and serviced

Later in this text, in the chapters on individual types of equipment, we'll discuss how to maintain them to maximize energy efficiency.

TABLE 5–3B

STAGGERED START-UP

15 min. Intervals	15 min.	30 min.	45 min.	60 min.	75 min.	90 min.
Oven 6.2 KW	36 min.					
Fryer (1) 12 KW				5 min.		
Fryer (2) 12 KW					5 min.	
Griddle 6.5 KW						7 min.
EST. DEMAND	6.2 KW	6.2 KW	2.9 KW	5.1 KW	5.9 KW	5.7 KW

Start-Up Time – 82 Minutes
Maximum Demand – 6.9 KW
Total Energy Used – 7.6 KWH

ON ———
IDLE – – –

BUILDING AND GROUNDS

THE ENERGY-EFFICIENT McDONALD'S

In 2000, McDonald's received the Edison Electric Institute's National Accounts Award for Innovation and Technology. "The Energy-Efficient McDonald's" (or TEEM, for short) is a project by this giant quick-service restaurant chain to impact its utility bills. You think yours are high? McDonalds' power bills exceed one-half billion dollars per year in the U.S. alone! The TEEM research has shown that not every type of technology "fits" in every location. Each site is unique, and therefore uses only the techniques that will be of benefit there. However, the list to choose from is extensive. Here's just a sample:

1. **Electronically controlled ballasts:** These dim fluorescent lights up to 50 percent during daylight hours in the dining areas, when natural light is often sufficient.
2. **Light pipes:** Using the same idea as skylights, these bring daylight into interior spaces, from kitchens to storage areas.
3. **Two-speed exhaust fans:** Over the cooking equipment, these control the volume of exhaust and reduce the amount of make-up air in the kitchen area, which saves fan use as well as heating and cooling energy.
4. **High efficiency HVAC systems:** A unit with an efficiency ratio of 10.5 or higher saves the company 5,000 to 10,000 kilowatt-hours per year, depending on location.
5. **Infrared-controlled hand dryers:** In restrooms, these save electricity and contribute to a more sanitary, "touch-free" environment for guests and employees.
6. **Occupancy sensors:** When walk-in storage and refrigerated areas are unoccupied, the lights go off.
7. **CO_2 monitors:** This type of system controls the carbon dioxide levels of indoor air, regulating the amount and flow of outside air brought into the space. Outside air, of course, must be heated and/or cooled when it gets indoors.
8. **Overall energy monitors:** With all these sophisticated systems, you need a system to alert employees if something isn't working right. An energy monitoring system provides automatic notification when any system fails. This includes a real-time accounting of utility use, which can analyze the data by time of day, weather conditions, or unit sales.
9. **Turbo-generators:** This type of power generator has a price tag of about $40,000 per store, but it pays for itself within five years by saving on peak loading and providing emergency power in case of service interruptions.

Source: Adapted from *Foodservice Equipment Reports,* a Gill Ashton Publication, Skokie, Illinois, February 2001.

5-7 CONSTRUCTING AN ENERGY-EFFICIENT BUILDING

There are two major advantages to building a new structure instead of moving into an existing one:

1. You are free to select the most energy-efficient systems and designs on the market (as long as you can pay for them.)
2. You can design the building itself to minimize its energy requirements, using daylight and other lighting techniques to minimize the need for heating and cooling.

By designing and selecting wisely, your future energy bills can be 30 to 50 percent lower than they would otherwise have been! In fact, you might even take your city or state's minimum standards for construction and go beyond them, adding a bit more insulation, self-closing valves on water faucets, or any number of "extras" to enhance your savings even more. Before construction begins, look at your blueprints and designs and check the following list of items. Are you saving as much energy as you could be?

- Plenty of daylight, minimizing the use of electric lighting
- Lots of light switches and/or dimmers, allowing for flexibility in turning off lights that are not needed
- The most efficient types of lamps and fixtures; energy-efficient ballasts for fluorescent lights
- High-pressure sodium lights in parking areas
- Efficient exit signage
- Timers, computerized or photoelectric controls for indoor/outdoor lighting
- Occupancy sensors for storerooms
- Glazing for windows to reduce incoming heat and increase daylight penetration
- Use of sufficient insulation for roof and walls
- Use of light colors, both inside and outside
- Positioning of building so that, if possible, trees or sloping land provide an insulating shield from wind and weather
- Awnings or overhangs to shield windows from direct sunlight
- Use of "spectrally selective" window film that cuts incoming heat in hot-weather areas
- Adjustable shades or blinds and, if appropriate in your climate, windows that open
- Caulking and weather stripping around doors and windows
- Double doors or revolving doors at entrances
- Energy-efficient hot-water system, with tank located near main point of use and insulated pipes
- Low-flow and dripless faucets
- Efficient HVAC system, organized in zones so that only areas in use are heated or cooled, with programmable wall thermostats and adjustable vents
- Locking covers on wall thermostats
- Heat pumps, where appropriate
- Restroom exhaust fans wired to go on/off with lights
- Installation of a computerized Energy Management System for optimum energy control

Many of the items on the preceding list can also be used in improving the efficiency of existing buildings.

Think of the building itself as a shell that is the primary barrier separating a controlled, temperate indoor environment from the often harsh and unpredictable outdoor conditions. Better yet, think of the building as a filter, and use it to allow selected bits of the outdoors inside to make the indoor environment more comfortable: light, fresh air, and humidity. If the building works well as both a barrier and a filter, you're on the right track to using energy wisely.

Insulation and Air Quality

Air that leaks out of your building and air that leaks in have one thing in common: They both place an additional burden on your heating and cooling systems. So it's your job to minimize these losses and gains by sealing and insulating. Doors and windows are prime culprits, but dampers, skylights, and utility and plumbing entrances also must be secured.

Weather stripping your doors is a good example of an often-ignored insulation priority. Let's say you have double doors at the exterior entrance to your cafeteria. Where the doors meet, there's an opening of about ¼ inch. It's not an eyesore, and you're busy, so you never quite get around to doing the weather stripping. However, on a pair of 80-inch doors, this quarter-inch gap adds up to a 20-square-inch opening! Most homes and offices are full of these kinds of seemingly insignificant leaks, which truly can become energy drains. Illustration 5-11 gives examples of where to caulk and weather-strip, valid for any type of building.

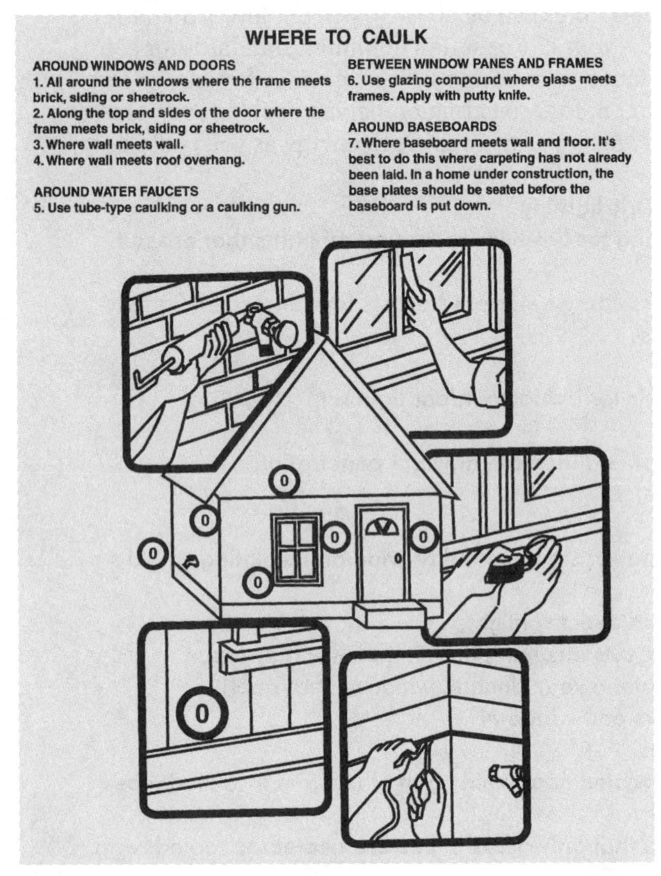

ILLUSTRATION 5-11 **Small weatherizing details such as caulking and weather stripping can save energy and money.**

Insulation minimizes the transfer of heat by some of the same principles as cooking, conduction, and convection. We know that heat flows naturally from a warmer environment to a colder one. All materials conduct heat, but some conduct it so poorly that they actually resist the heat. Insulation is measured by its heat resistance, or ***R-value.*** The higher the R-value, the more resistant (better insulator) it is. Your city building codes will probably specify minimum levels of insulation, most commonly R-19 for walls and R-30 for roofs.

Even through layers of insulation, buildings "breathe." That is, they take in a certain amount of fresh air and vent stale air back outside, through windows, fans, and ducts. The amount of fresh air delivered is often dictated by health ordinances or building codes. The current recommendation by the American Society of Heating, Refrigeration and Air-Conditioning Engineers (ASHRAE) is 15 to 20 cubic feet of air per minute, per occupant of the building.

Especially in older structures, air quality may suffer at first when the building is "tightened" to make it more energy efficient. Humidity may condense on walls or windows, and existing ventilation systems may no longer be adequate. As you work at energy savings, ask about ways to do it without compromising air quality.

Measuring Your Progress

You've heard elsewhere in this text about the need for keeping good records, and nowhere is it more important than in tracking utility bills and savings. You can set goals, complete surveys, determine priorities, and embark on numerous impressive conservation efforts, but how do you know how well they have really worked, unless you can compare the "before" and "after" figures?

Ideally, you should begin by charting costs and consumption figures from the utility bills for the previous twelve months. If you have not been in business that long, but the building was occupied by others, contact your utility companies for energy use records for the address. Setting an accurate base figure or starting point is critical to your success. How well you do from here is completely up to you!

5-8 POWER FAILURES

After all this discussion about how important electricity is, now we've got to focus on what happens when it is suddenly not available. The U.S. government's Electric Information Administration reports at least two dozen "major disturbances and unusual occurrences" per year that interrupt electrical service. Many of them are weather-related, lasting from two hours to several days. The EIA publishes an interesting monthly report about power prices, usage and disruptions that covers both electricity and gas. It's available on-line at eia.doe.gov/cneaf/electricity.

Being prepared for a power outage seems to be a necessary part of the restaurant business in the twenty-first century. Ask anyone in California, after the "rolling blackouts" of 2000, 2001, and 2002. Whether caused by man or nature, a disruption in the power supply can have costly consequences if it lasts more than two hours. Most potentially hazardous foods (known as "PHFs" in food safety jargon) can be kept out of the temperature "danger zone" for that long, but no longer.

Of course, during an outage you will have no way of knowing how long it will last. Managers on duty should take the following precautions:

1. Record the time the outage began.
2. Take periodic temperature readings (every 30 to 45 minutes) of the food that was prepared and being held at the time.
3. Keep all refrigerator and freezer doors closed as much as possible. (This is why, as you'll learn in Chapter 10, it's handy to have a thermometer that measures its internal temperature on the outside of the unit.)
4. Cover any open display cases in which food is iced or chilled.
5. Used canned fuel (like Sterno®) to keep hot foods hot. If the temperature of hot food on a steam table, for instance, drops below 140 degrees Fahrenheit for more than two hours, it should be discarded. If the outage doesn't last that long, and the food has dropped to a temperature of less than 140 degrees F., reheat it rapidly to 165 degrees F., then hold it at 140 degrees F. and serve as quickly as possible.
6. Discontinue cooking and serving if it is not possible to wash, rinse, and sanitize utensils.
7. Discontinue cooking if there is not sufficient light to allow for safe food preparation, including cleaning and sanitizing food contact surfaces.
8. Don't cook if there is no hot water or not enough water pressure.
9. Don't assume that cooking cold food stored at room temperature for too long will kill any microorganisms and make it safe to eat. That is not the case. The rule is always, "When it doubt, throw it out."

In areas prone to weather-related power outages, the purchase of a power generator is an attractive option. If a foodservice business experiences more than two or three major power interruptions a year, having a backup generator makes sense. The generator can also be used for so-called ***peak shaving,*** curtailing power use or shifting to another source to avoid peak loading and the higher costs it entails. The generator can be tied into the primary power grid, and programmed to "kick in" not just when there's a power failure, but whenever demand charges are highest. In other words, when the electric or gas company is charging its highest rates, you generate some of your own power and reduce dependency on the utility company, at least temporarily.

TABLE 5–4

GUIDELINES FOR HANDLING POTENTIALLY HAZARDOUS FOODS (PHFs) THAT HAVE BEEN STORED IN REFRIGERATION UNITS DURING A POWER OUTAGE

	Cold Food Temperatures		
Duration of Power Outage	**45°F or below**	**46°F to 50°F**	**50°F or above**
0–2 hours	PHF can be sold.	Immediately cool PHF to 45° F or below within 2 hours.	PHF cannot be sold.
2–3 hours	PHF can be sold, but must be cooled to 41° F or below within 2 hours.	Immediately cool PHF to 45° F or below within 1 hour.	PHF cannot be sold.
4+ hours	Immediately cool PHF to 41° F or below within 1 hour.	PHF cannot be sold.	PHF cannot be sold.

Source: Developed by Lacie Thrall, Director of Safety Management Services for Food Handler, Inc.

5-9 POWER OUTLOOK FOR THE FUTURE

The power industry is one of the largest in the U.S., with annual revenues of more than $200 billion. The U.S. uses about 40 percent of the world's oil, and 23 percent of its gas and coal. Table 5-5 shows where we get our energy—these figures are from 1998, but they're the most current at this writing.

Some updates from the U.S. Department of Energy: In 2003, power generated by nuclear plants was down 2 percent, and natural gas power generation was down 10 percent—but hydroelectric power (from dams) and coal generation were up slightly, and petroleum-fired generation increased a substantial 41 percent. So the percentages may change slightly from year to year, but it is clear that we are still depending mostly on fossil fuels to heat and cool us, and cook our food.

The Impact of Deregulation

At this writing, at least two dozen states have either enacted legislation or issued a regulatory order to deregulate utility companies. (The best source of information about what's happening in your area is the U.S. Department of Energy's Energy Information Administration website, which lists and describes restructuring activity by state: eia.doe.gov/cneaf/electricity/chg_str). Deregulation signals a huge and unprecedented change in the way power is sold to the end users. Some say that's a good thing, but not everyone agrees.

The idea behind deregulation is to break up an industry of big systems into smaller, more competitive ones. Instead of a single entity that generates, distributes, and sells electric power, what we know as the "power company" or "utility company" will only generate the power, and handle repairs and emergencies. ***Retail Electric Providers*** (REPs) buy the power, and sell it to business and residential customers. The REP provides such functions as customer service and billing. In deregulated markets, there are also ***Affiliate Retail Electric Providers*** (AREPs). The difference between a REP and an AREP is that the REP is an independent business; an AREP is an offshoot of the original utility company, now in a separate business to do the selling. Customers who choose not to shop around and/or switch providers will end up with their local utility's AREP. Interestingly, the AREP is required to offer a standard rate for electric service—literally called the "Price to Beat"—which is set by the state's utility commission. If a REP can beat the standard rate, the customer is free to switch service providers at any time.

TABLE 5-5

U.S. ENERGY PRODUCTION, 1998

Commodity	Production (in quadrillion Btus, which means add 15 zeros!)	Percent of total production
Oil	13.2	18.1
Natural Gas	19.5	26.7
Coal	23.8	32.6
Nuclear	7.2	9.8
Renewable (wind, solar)	6.7	9.2
Hydroelectric	3.4	4.7

Source: U.S. Geological Survey, 1998.

A group of customers, such as a trade association, business complex, restaurant chain, or neighborhood, can also hire an ***aggregator*** to represent them. This is a person or company that conducts research about prices, availability, and contract terms, then negotiates for electricity at a group rate based on the services the group requires. The risk with using an aggregator is that, even if you're doing everything possible to conserve energy, you may be penalized for the inefficiencies of others in your group when you're all lumped together.

The increased competition that deregulation sparks, in any industry, is supposed to be good for the end users—but it can also be confusing. Traditionally, a business paid a flat rate for power based on how much was consumed and billed in numbers of kilowatt-hours. Now, the business will have several options, not only for pricing, but for "package deals" that include different rates at different times of day, as well as other types of services. This is why it becomes even more important for a business or building owner to know the facility's load profile, which shows how and when power is used over a given time span. Knowing usage patterns will make it easier to negotiate more favorable rates from a power supplier and forecast demand when creating a budget. Predictably, REPs will offer the best deals to customers with consistent, predictable energy use patterns—those with "better" load profiles.

A business that keeps up with its load factors will also be able to shift a portion of the demand back and forth to match the supplier's lowest-cost hours, saving money. Some start-up companies in the early 2000s have created computer programs to help track the information.

When deregulation was first proposed, its backers promised savings of 25 to 35 percent. In reality, the hot competition for your power business has not been so hot. For the first few years, although venture capitalists poured $300 million into the industry in 2000 alone, very few start-up firms emerged. Those that did offered savings that often amounted to less than 10 percent. Critics claim sales is now the core objective of the REPs—not reliable, inexpensive energy and certainly not a focus on new, cleaner, and more environmentally safe technology.

The second factor is that people are accustomed to dealing with their old public utility companies. The idea of comparing rates, bidding for service, and switching to a new energy provider can be overwhelming or, at least, a big hassle for an already harried business owner. A lot more complexity for minimal savings? We'll see.

Alternative Energy Sources

No matter how it is generated, it is generally agreed that electricity costs will continue to rise, which has prompted more research into renewable energy sources like solar and wind power. The U.S. Department of Energy throws a little money—the experts agree not nearly enough—at developing alternate ways to generate energy. The amounts have ranged from $296 million in 1997, to $375 million in 2001, dropping drastically during the George W. Bush administration to a low of $237 million in fiscal year 2002. It's ironic, then, that the White House has installed solar panels that provide a token 1100 kilowatt-hours per month.

The lack of support does not appear to be stopping power generation pioneers. Between 1994 and 1999, annual worldwide shipments of solar photovoltaic (PV) cells and modules almost tripled, and the solar energy industry is growing at about 20 percent a year. Japan and Germany are the main competitors of the U.S. in solar research and technology.

Solar PV systems make electricity like this: a solar cell is made up of semiconducting material that absorbs sunlight. The cells are grouped into modules, and the modules are mounted in PV "arrays." In larger installations, the arrays collect and focus sunlight with mirrors to create a higher-intensity heat source. Solar power is as adaptable as electricity generated any other way—it can be used for heating water, heating or cooling air, and so forth.

In making a decision to "go solar," try to determine the life cycle or life expectancy of the system before you buy, as well as the potential tax benefits. Solar systems currently used to heat dishwashing water qualify for a federal energy-saving tax credit of 15 percent, plus a 10 percent business investment credit—a total tax write-off of 25 percent. Fifteen states also offer rebates of some sort for renewable energy projects. However, as your accountant will surely explain, you must first spend the money to be

BUILDING AND GROUNDS

A "REP" OR AN "AREP"?

Here are some issues to clarify when making the decision to stay with your "old" utility company's "new" Affiliated Retail Electric Provider . . . or to choose a third-party competitor, a Retail Electric Provider.

- What is the price per kilowatt-hour, by day-part? Are the prices fixed or variable and, if they're variable, based on what? Can the prices change without notice to customers? Are there other charges in addition to the kilowatt-hour rate? What, and how much? Are additional taxes included in the bills?
- How long is the service contract? This is a volatile market, so you don't want to lock in for too long if it will prevent you from taking a better deal if you find one. Keep shopping around.
- Is the provider certified by the state's Public Utility Commission? How long has the company been in business, and where? What are the top executives' qualifications to do this?
- Are there up-front fees to begin service? Must you pay for a new or different type of meter? Is there a set-up charge?
- Who will actually perform each of the critical functions in the process: who generates the power? Who reads the meters? Who provides maintenance? Who consults if you need an energy audit or other types of advice? Who do you call if there's an overcharge, an outage, a problem?
- Does the company encourage use of renewable energy sources, like wind or solar power? Will they assist, or offer price breaks, for these?

Source: Adapted from the Public Utility Commission of Texas, Austin.

able to write off the expense or receive the rebate. Learn more about solar energy technology at the website of the American Solar Energy Society, ases.org

Wind power is another renewable energy source, which any farmer with a windmill has known for years. A single windmill produces from 1 to 5 kilowatts, just enough to pump water for livestock or home use, but new technology allows the progressive farmer to retrofit an existing pump system to generate 300 to 500 kilowatts—enough to power the property's irrigation system. They're not called "windmills" anymore, but wind towers and windfarms. Windfarms now power the equivalent of 7.5 million average American homes, worldwide.

In the U.S., federal tax breaks of 1.8 cents per kilowatt-hour have prompted more interest in wind power. California leads the U.S. in wind generation, with almost 2000 megawatts a year. Texas and Minnesota are the runners up, but America accounts for only 15 percent of the world's wind power-generating capacity. We're being outdone by Germany and Spain. Denmark now gets 20 percent of its power from wind generation. Windfarms in Europe are generating more than 20,000 megawatts per year.

What does this have to do with your restaurant? According to the American Wind Energy Association, during the 1990s, the costs of wind turbines dropped by 75 percent, and their reliability improved from 60 to 95 percent. On the downside, some people complain that the giant towers and turbines are noisy and unattractive, especially in large numbers, as well as hazardous to birds. Wind power advocates are making technical breakthroughs to bring down costs and make production more reliable, but natural-gas-fired plants can produce electricity at 3 cents per kilowatt-hour or less, while windfarms produce power at about 5 cents per kilowatt-hour. The newest studies say this cost could be reduced enough to compete with gas-fired plants if windfarm developers could obtain the same favorable financing terms as utility companies, or if the utilities would be willing to own the windfarms.

You can learn more about the costs and feasibility of wind power at awea.org, the website of the American Wind Energy Association. The organization has European (ewea.org) and Canadian (canwea.org) counterparts.

BUILDING AND GROUNDS

SOLAR POWER FOR MAUNA LANI

Rising energy costs and an electric shortage in Hawaii callled for some bold initiatives. The Mauna Lani Resort in Kohala Coast, Hawaii, teamed up with a manufacturer of commercial-scale solar electric products and devised a long-range energy plan in four parts.

PHASE 1: An 80-kilowatt solar power system was installed on the resort's roof. It instantly supplied 3 percent of the resort's electric needs.

PHASE 2: 110-kilowatt solar panels were installed in the golf course area. They supply 100 percent of the electricity used in the clubhouse, pro shop, and golf warehouse. And, since these areas are used only in the daytime, the panels produce excess energy, which is routed to provide 50 percent of the power needed for an adjoining restaurant.

PHASE 3: Solar panels were installed on the roofs of 120 golf carts. They reduce the carts' battery consumption by one-third, and also reduce the time it takes to recharge the carts' batteries.

PHASE 4: A 250-kilowatt power tracking system was installed. It rotates the solar panels on the buildings' roofs, following the seasonal patterns of the sun to maximize their exposure. This increases the effectiveness of the panels, allowing them to produce 30 percent more power.

The cost of the entire four-phase project is expected to be paid off in about six years with the savings in electricity bills.

Source: *Hotels* magazine, July 2002.

Cogeneration and Heat Recovery

One of the most intriguing ways to save energy dollars is to recapture and reuse heat that would otherwise be wasted. This is called ***heat recovery,*** *heat reclamation,* or *cogeneration* (the latter because it uses the same power for two jobs). Waste heat can also be recaptured from HVAC systems, refrigeration compressors, and computers, and the idea is adaptable enough that:

- Hot exhaust air can heat incoming cold air.
- Hot exhaust air can heat water.
- Hot refrigerant can heat air or another liquid.
- Heat from a boiler can be used to make steam or to heat air or water.

Both gas and electric utility companies have done extensive research on cogeneration. The Gas Research Institute conducted a study of 1000 restaurants' utilities in 28 U.S. cities, suggesting a cogeneration system in each case. This particular system, which impacted both gas and electric appliances in the restaurant, cost $128,000 to install. However, the average utility cost savings was an impressive $22,000 per year, meaning the system would pay for itself within five or six years.

Another study was conducted by the Electric Power Research Institute in conjunction with Pennsylvania State University. Reuse possibilities in four types of restaurants (cafeteria, full menu, fast food, and pizzeria) were assessed in cities with a variety of climate conditions, and the study revealed that each of the following technologies reduced energy consumption by at least 10 (and as much as 25) percent.

Reuse of Dining Room Air for Kitchen Ventilation. Since dining room air is already conditioned, it can be used to meet hood ventilation requirements above the hot line. This reduces the amount of (naturally hotter) kitchen air that must be conditioned. Air can be transferred from the dining area to the kitchen using a single-duct system or transfer grills. After use in the exhaust hood, the heated air can even be used to preheat incoming makeup air! However, the cost of this method makes it feasible only in restaurants that use the largest amounts of hood ventilation air.

Heat Pump Water Heaters. A ***heat pump*** is a mechanical device that moves heat from one area to another.

In all restaurant types, this was cited as an effective way to use less energy to heat water. An additional benefit of heat pump water heaters is the cool air from the evaporator, which can be routed to spot-cool the hottest areas of the kitchen. The heat pump contains an evaporator, a compressor, a condenser, and an expansion valve (see Illustration 5-12), much like the refrigeration systems you'll learn about in Chapter 10. Instead of cooling, however, the heat pump recovers heat from an area and uses it to heat water.

The location of a heat pump water heater is important. It should be in a space where the ambient temperature is between 45 and 95 degrees Fahrenheit to work at its best. If operating properly, and if its air intake filters are kept clean, this handy appliance is capable of producing 30 to 70 percent of the hot water needs of an average restaurant.

Recovery of Refrigerator Heat. Hot air released by refrigerators and walk-ins can be used to heat water. However, because refrigerators and walk-ins run twenty-four hours a day, storage is necessary to retain the heat during the hours when there's no demand for hot water. This is possible with a two-tank system that consists of a preheating tank that keeps water warm using air from the refrigerator, then feeds it into a conventional water heater tank, which doesn't use as much energy because it doesn't have to work as hard. A side benefit of this system is that in warm months, the building's HVAC system doesn't have to work so hard to cool the space, either. The heat is being captured and used for other purposes.

HVAC Heat Recovery. When the restaurant is being cooled (air-conditioned), the heat from the cooling process can be captured to heat water elsewhere in the building. Typically, more heat is generated than needed, so correctly sizing the storage tank is critical to the success of this method. This method also uses a two-tank system like the one mentioned previously.

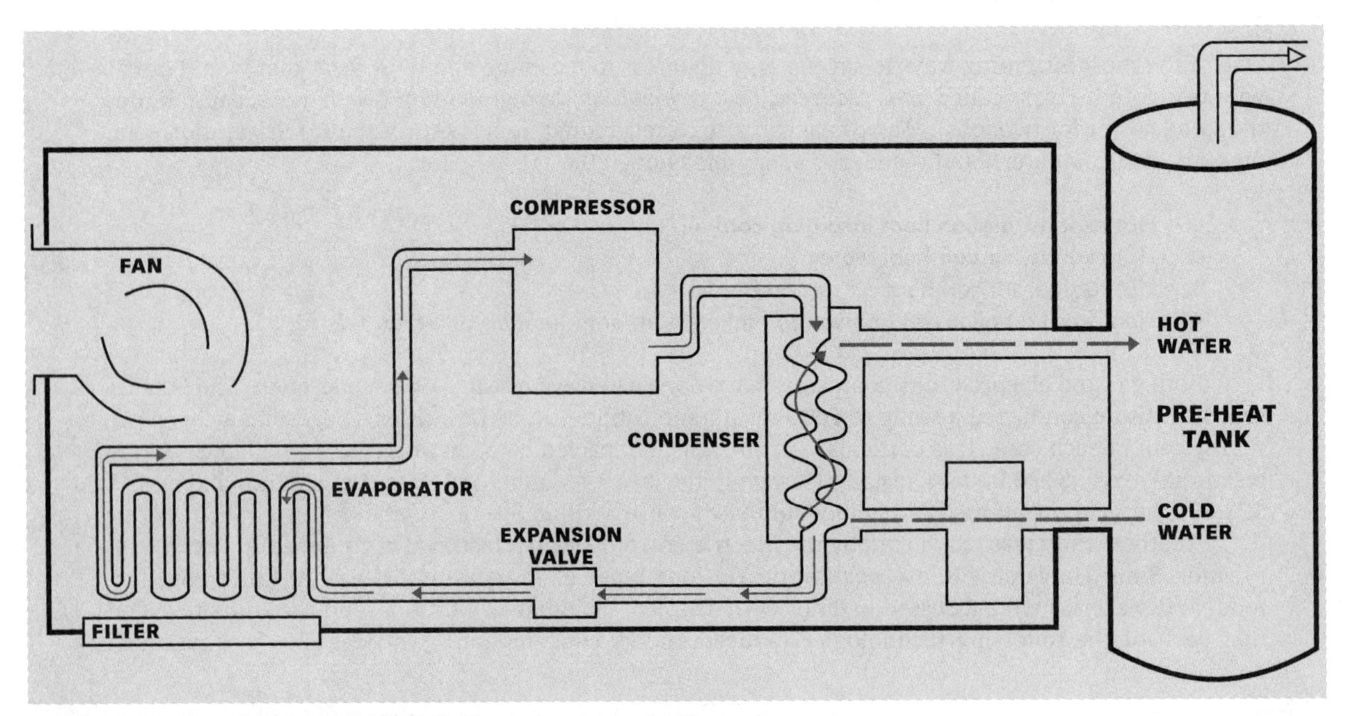

ILLUSTRATION 5-12 The inner workings of a heat pump water heating system.
COURTESY OF THE ELECTRIC FOODSERVICE COUNCIL, FAYETTEVILLE, GEORGIA.

SUMMARY

Energy use per square footage in restaurants is greater than in any other type of commercial building, but research indicates that utility costs can be trimmed by 20 to 30 percent with an eye on energy conservation.

Having an energy audit is a good way to begin. There are two types of energy audits: a walk-through audit, usually free of charge from your utility company; and an analysis audit, which is a more detailed summary of every energy-use item in your building and how it might be improved.

Smart restaurateurs will learn how to read not only their electric meters, but gas and water meters, too. You should have a contact person at each local utility to discuss questions or problems. If your state or area has deregulated its utility industry, you can negotiate with a different provider other than a traditional utility company, and sign a contract with them to provide your power, plus billing and customer service.

This chapter explains how electricity is measured (in amps, volts, watts, and ohms) and what materials are good or bad conductors of electricity. There are two different types of electrical current: alternating current (AC) and direct current (DC). Alternating current is the most common. Smaller appliances use single-phase electrical wiring; larger or more heavy-duty equipment uses three-phase electrical wiring. Combinations of voltage and phase will be listed in specifications for every piece of electrical equipment you purchase. You should be able to determine if an individual circuit has enough electrical capacity to accommodate the appliances you will plug into it.

You can chart your utility costs annually or even break them down per meal. Energy conservation methods can be grouped into four simple categories: improve equipment efficiency, reduce equipment operating time, recover energy that would otherwise be wasted, or use a cheaper energy source. The more you know about your utilities, the more you can track down and correct water, gas, or electricity wasters.

STUDY QUESTIONS

1. List three questions you would ask if you were giving an energy audit to a restaurant.
2. What is the difference between a *kilowatt-hour* and a *kilovolt-ampere*? Why do you need to know that?
3. Let's say your state has just decided to deregulate its electric industry, and you now have the opportunity to pick a new service provider. What do you think are the two most important things to consider in making this decision—and why?
4. List three reasons it is necessary to be familiar with terms such as *amps, volts,* and *ohms* as they relate to foodservice.
5. Explain briefly why a business might be billed differently for different hours of electricity use.
6. What is an *energy index,* and what does it do for you as the owner of a foodservice business?
7. What is the *fuel adjustment charge* on your electric bill?
8. List four ways you can save energy in a commercial kitchen.
9. In case of emergency, is it okay to pull or flip the *master service switch* of your electrical system? Why or why not?
10. Do you think solar or wind power is practical for a restaurant or other foodservice business? Why or why not?

A Conversation with . . .

Mark Buersmeyer

GENERAL MANAGER, BUSINESS FOODSERVICE, MARRIOTT CORPORATION BOISE, IDAHO

At the time of this interview, Mark Buersmeyer was in charge of all foodservice operations at Hewlett Packard's largest manufacturing facility in the United States, in Boise, Idaho. Working for Marriott Corporation's contract foodservice division, Mark's responsibilities included running a 24-hour cafeteria for 5000 employees, catering meetings, providing coffee service in remote locations throughout the plant, and stocking all food-related vending machines. Marriott earns its money based on sales in this type of arrangement, so Mark's goal was to increase sales.

A graduate of the Cornell University Hotel and Restaurant School, he began his career in hotels before working for a cafeteria chain that was bought by Marriott.

Q: Since you got the prestigious four-year degree, why did you choose this line of work instead of becoming a chef and naming a restaurant after yourself?

A: Well, the salesmanship of the person who hired me was compelling! He convinced me that you could have a Monday through Friday schedule, not having to work weekends, nights, or holidays. In the food business, that's unheard of. The salaries may be a little less in the contract end of foodservice, but the environment is very comfortable.

I also like the fact that, at a young age, you can assume more responsibility. After Cornell, I found myself at age twenty-one managing a hotel bar and restaurant. It sounds impressive, but I was not in charge of the food, the bartenders, the laundry facility, the cash accounting functions; I didn't do any ordering. It was like I was a glorified maître'd, and yet I was called a "manager."

But when I got into contract foodservice, in a smaller account you can have complete control of the menu, the hiring and scheduling, total accountability. You operate as your own boss, and in many cases your boss is in another city. So you're much more independent and you have much more control over what you do, and I like that.

Q: Do you still think the hotel business is a good way to get started in the industry?

A: Yes, because there are so many different departments that you can dabble in, and so many entry-level supervisory positions that they can put you in. You can get the same thing in a contract foodservice account, if it's big enough. I worked at Federal Express for three years doing employee feeding and we had eleven managers.

There's all kinds of contract foodservice, from the cafeterias and break rooms at the big auto manufacturing plants to executive dining rooms at banks or insurance companies. We also have vending, which is not a glamorous part of the business but certainly a profitable part.

Q: What kind of business do you do here, at the Hewlett Packard Boise site?

A: We do about $4 million in volume here, and about $1 million of it is just catering, off site and on site. Another two-thirds of a million is vending; coffee service is about half a million. So the

cafeteria itself, although it seems like the biggest piece of the pie, isn't always. We serve a total of about sixteen buildings, and three other cafeteria locations. It's kind of exciting. We feed about five thousand people, and every one of them expects service, right now. The bigger the account, the more the demands there are.

Q: How important is it to know things like the electrical capacity of your kitchen, how to set appliances up correctly, and the other technical details of foodservice?
A: The more experience I've gotten, and the bigger account I'm dealing with, the more important it has become. When you've got a central location and satellite locations like this one, somebody has to get things installed and deal with salespeople. They do offer advice, and there are also consultants, but when you do a project, you really have to understand the pitfalls. Things like electrical, plumbing, and drains become very important as you try to expand.

There are always communication issues and if you don't understand the language the service person is using, it's tough. For instance, I end up talking to electricians all the time about our remote coffee service sites: what kind of plug, what kind of amperage, what kind of wiring.

Of course, Marriott can hire a consultant if necessary, but the client will have to pay for it. That's expensive, and the client doesn't always understand why they have to pay for that service.

Q: If you had it to do all over again, what do you wish you had studied in school that would help you on the job today?
A: Well, I used to sleep through some of the psychology and human resources courses, and now I wish I hadn't! I guess as I look back, one of the most important skills is to get people to work for you, to know how to make things run smoothly with employees. People who warm up to other people naturally have a real advantage in this field. And in our case, the client can always switch and go with another foodservice contractor if they're not comfortable with the people who run it.

Q: I noticed a suggestion box and a bulletin board in your cafeteria of customer questions and answers. Why did you decide to do that?
A: I started that when I got here, and I think I do it better than most because I try to make it a point to answer every question, and to post the tough questions as well as the easy ones. Sometimes the customers are antagonistic and the demands are unrealistic, and you've got to post their comments and hope that when other people read them, they also feel the demand is unreasonable. And here's a tip: When you put a suggestion box by the dish return you don't get as good a level of activity as when you put it at the front, where people first come in. That way, they have their whole lunch period to do what they want, to browse and read the bulletin board and fill out a comment card. If they don't see it until they're done eating, they are usually running back to work and not as likely to write a question or opinion.

Q: Tell us a little bit about the design and traffic flow of a cafeteria.
A: This one is a scatter system, the type I prefer. It's a series of independent stations where people don't move in a straight-line pattern. Being able to wander around and order what they want usually helps them get through the line quicker and not be stuck behind someone else. The stations here are fixed, but you can buy modular display cases and salad bars that can be moved around as needed.

Q: How do you build warmth and ambience into a cafeteria setting?
A: In a new facility, that is where I would rely on the opinion of my architect. They typically try to blend style into a place within the physical structure. In an existing building, when I first take over an account, I try to find people who have a talent for decorating. It isn't really a strength of mine, so I try to encourage the employees to use seasonal themes and fresh foods, think about displaying and marketing the place as a fresh-food place.

We also make sure people see that nutritional awareness is important to us. The bulletin boards have nutritional displays and information on them. I get a lot of the information on the computer. There's good nutrition software available as well as the Internet.

Q: How do you know what people want to eat?
A: We do surveys and ask for their feedback. We offer two main entrees a day, so it is no big deal for us to try new things and track their success. And I'll tell you how important display is—if the food looks bad, it just flops.

Q: Do you have advice for people just starting out in foodservice?
A: Don't just go to school. Use this time to work in the industry and see what you like to do. It's a huge and very diverse field, so don't just look at restaurants or chain restaurants for work opportunities. You might even start with two years of general foodservice courses and job experience, then use the next two years to specialize in the area that interests you most.

CHAPTER 6

GAS, STEAM, AND WATER

INTRODUCTION AND LEARNING OBJECTIVES

Just when you thought you'd learned a lot about the utility needs of a restaurant . . . this chapter focuses on utilities other than electricity. And there are plenty of them! Gas and steam provide power for many of the major kitchen appliances you will use, and may also run your heating and/or cooling system. A basic knowledge of water and plumbing is also necessary.

In this chapter, we will discuss:

- The uses of gas and steam in foodservice
- The basics of how gas and steam equipment works
- Energy-saving use and maintenance tips
- Water quality issues and how to deal with them
- Basic foodservice plumbing requirements
- Installing and maintaining a drainage system
- Hot-water needs, and how water heaters work

6-1 GAS ENERGY

Gas has many uses in foodservice. You may use it to heat or cool your building, to heat water, to cook food, to chill food, to incinerate waste, and/or to dry dishes or linens.

It's called *natural gas* because, indeed, it is not manmade. It was formed underground several million years ago by the decay of prehistoric plants and animals, and is now pumped to the earth's surface for use as a fuel. Illustration 6-1 shows how natural gas is extracted from wells, then processed and transported through a series of pipelines to its final destination—your restaurant.

The American Gas Association credits the Abell House, a stagecoach stop in Fredonia, New York, as the first commercial establishment to use natural gas for cooking. Back in 1825, the "pipes" were

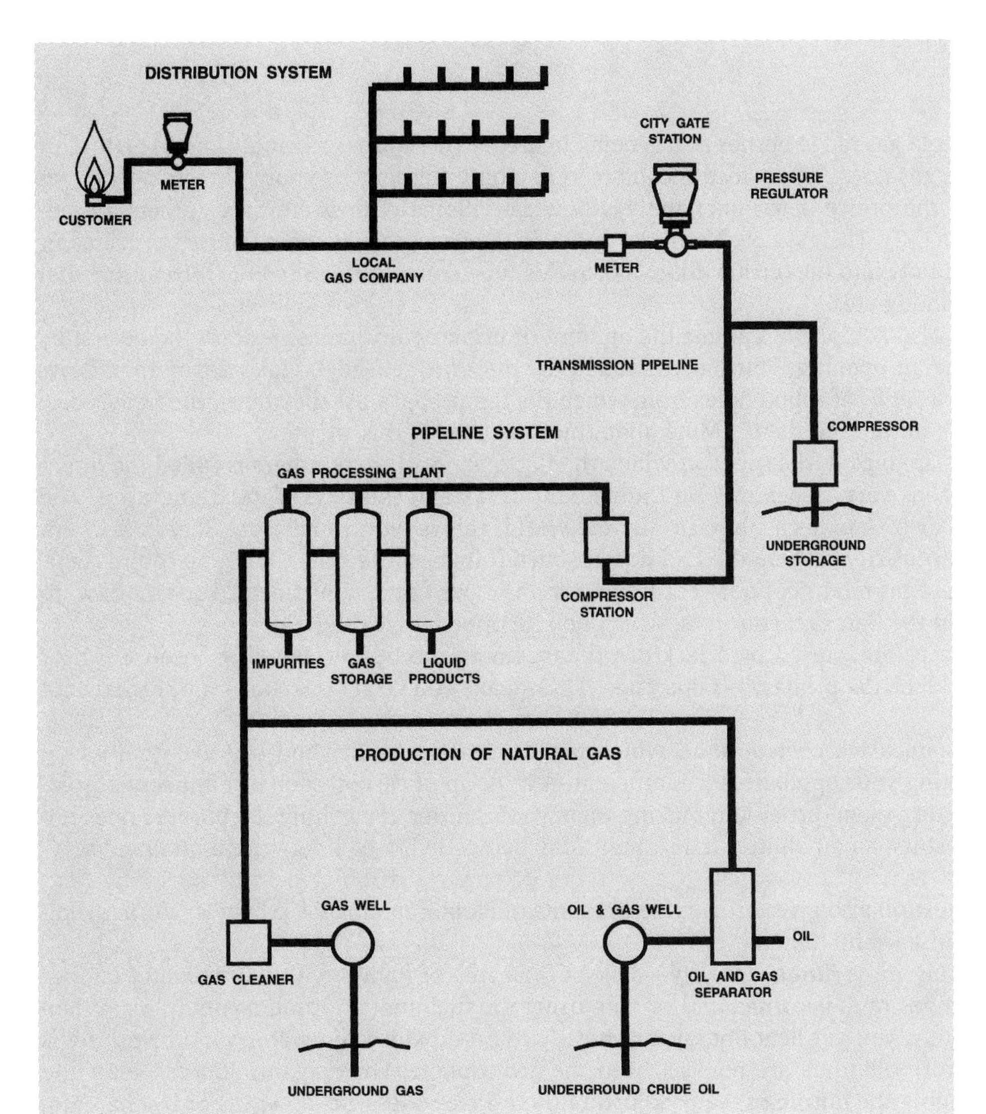

ILLUSTRATION 6-1 How natural gas gets from the ground to the customer.

SOURCE: ROBERT A. MODLIN, EDITOR, *COMMERCIAL KITCHENS,* 7TH ED., AMERICAN GAS ASSOCIATION, ARLINGTON, VIRGINIA.

hollowed-out logs! Gas was propelled through the logs into the building, to a single-flame stove with a reflector plate. We've come a long way since then.

There are different types of gas for different uses. The one most commonly known as natural gas is mostly ***methane.*** When it is highly compressed for storage, under incredibly cold conditions (below -260 degrees Fahrenheit), it becomes ***liquified natural gas (LNG).***

When it is manufactured—in a process that mixes methane with hydrogen and carbon monoxide—it is known as *synthetic gas.* And there are other gas combinations—propane, butane, isobutane—which may be called *liquified petroleum gas, LP gas,* or *bottled gas.*

In the U.S., restaurants use natural gas to operate as much as two-thirds of their major cooking equipment. Its chief benefit is its instant ability to provide intense heat. Chefs like cooking with gas because it allows them to work quickly and to use different types of burners to direct heat where it is most needed. Modern-day foodservice ranges may feature ***power burners,*** high-input gas burners that burn twice the amount of gas (and can supply twice the heat) of a conventional gas burner.

Other appliances that may be gas fired include broilers, fryers, griddles, steamers, coffee urns, and ovens. These appliances may have automatic *pilot lights,* which stay lit and indicate the gas is ready if you turn on the appliance, or you may have to light the pilot manually with a match when you need it.

Gas Terminology

Let's take a look at a single gas range burner as it is being lighted. Every burner assembly has an ***orifice,*** or hole through which the gas flows. The orifice is contained in a unit that may be known as a hood, cap, or spud. The diameter of the orifice determines the gas flow rate. Some orifices are fixed; others are adjustable (see Illustration 6-2). The gas flows from the orifice in the form of a ***jet,*** which causes a rush of air (called ***primary air***) to flow into the ***burner tube.*** It is in this tube (sometimes called the *mixer tube*) that the gas mixes with air and ignites.

Most burners are also equipped to adjust the amount of primary air that comes into the burner by adjusting the size of the air opening. This piece is called the ***air shutter.*** The air-gas mixture then flows through ***burner ports,*** a series of round holes from which the flames burn. As they burn, the flames also use ***secondary air***—the air around them. (More about burners later in this chapter.)

When air and gas are mixed, the speed at which the flame shoots through them is called the ***burning speed.*** Burning speeds vary, depending on the amount of air and the type of gas being used. The ideal ratio is 10 parts air to 1 part gas, but, in the real world, this is hard to achieve. So you may encounter ***incomplete combustion,*** where the fuel doesn't burn fully because something's not quite right: not enough air in the air-gas mixture, poor ventilation, or improper flame adjustment. You can spot incomplete combustion in the flames themselves, which will be tipped with yellow.

Yellow-tipped flames are caused by a lack of primary air to the burner, possibly because lint or grime has collected to block the primary air openings. This means you've got to clean your burners, not adjust them.

Yellow flames create black ***carbon soot,*** which makes cleanup harder and can eventually clog vents and orifices, making your appliances less efficient. A buildup of carbon soot will impact your exhaust canopies too, getting them dirtier and making them work harder. If cleaning the burners does not give you the desired results, an air shutter adjustment may be needed, which requires a qualified service person.

Incomplete combustion also gives off varying amounts of carbon monoxide, which is odorless, colorless, and tasteless, but harmful nonetheless.

What you're looking for is ***flame stability***—a clear, blue ring of flames with a firm center cone—indicating that your air-gas ratio is correct and you are using the fuel under optimum conditions. When gas fuel burns completely, you get heat energy, harmless carbon dioxide, and water vapor. Nothing is wasted, and no harmful pollutants are released into the atmosphere. Once again, you can alter the flame stability by changing the burning speed (adjusting the orifice so less gas flows in) or by changing the primary airflow into the mixture (adjusting the air shutter).

Don't confuse yellow-tipped flames with the red or orange streaks you sometimes see in a gas flame. These colored streaks are the result of dust in the air that turns color as it is zapped by the flame, and should not be a problem. Also remember you are wasting gas when you use high flames that lick the sides of your pots and skillets. In fact, when a completely unheated pan is placed on the gas range, it is

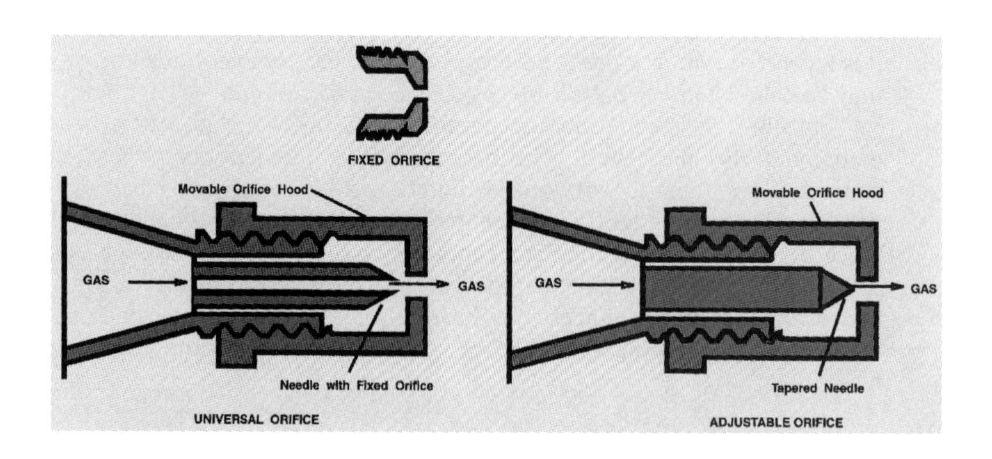

ILLUSTRATION 6-2 The orifice, or point at which gas flows into the range burner, may be fixed or adjustable.

best to begin heating it on medium heat, so the tops of the flames do not touch the surface of the pan. Carbon monoxide and soot are produced when intense heat hits the cool metal surface. Increase the heat only after the pot or pan has had a chance to warm up.

The Gas Flame

The flame of the gas burner represents the ultimate challenge: to mix gas and oxygen in just the right amounts to produce combustion, giving us controlled heat with minimum light. The simplest, most effective example of this is the old-fashioned *Bunsen burner.* This type of burner premixes air and gas prior to reaching the flame, resulting in a highly efficient flame that burns intensely, but with a clean, smokeless flame.

The shape and size of the burner are the two factors that place the flame exactly where direct heat is needed most. In a toaster or broiler, for instance, the gas flame is directed at a molded ceramic or metal screen, which is heated to a deep red color and emits infrared heat rays that penetrate the food being cooked. The latest innovation in the industry, as mentioned earlier in this chapter, is the high-input gas burner, which burns twice the amount of gas (for greater intensity of heat) as a conventional burner of the same size.

Amazingly, the natural gas–air combination can produce a stove temperature of up to 3000 degrees Fahrenheit! So one of the functions of the gas-fired appliance is to limit and distribute the available heat, to reach the correct temperature to cook foods properly.

You'll notice a gas burner sometimes lets out a "whoosh" or roaring sound as it lights. This is called ***flashback,*** and it happens because the burning speed is faster than the gas flow. This type of flashback occurs more often with fast-burning gases such as propane.

Another type of flashback happens when the burner is turned off, creating a popping sound that is known as the ***extinction pop.*** Sometimes it's so pronounced that it blows out the pilot light flame. You can usually correct both types of flashback by reducing primary air input to the burner. If you're unsure about how to do it yourself, remember that burner adjustment is a free service of many gas companies.

Flashback is not hazardous, but it is annoying. It creates soot and carbon monoxide and often means you have to relight your pilot. Inside the burner, repeated flashback occurrences may cause it to warp or crack. It makes more sense to get the burner adjusted than to live with it.

Several other conditions may require professional attention and adjustment (see Illustration 6-3). You may notice that the flames seem to lift and then drop on some parts of the burner head at irregular intervals, as though some unseen hand were playing with the control knob. This burner may seem a bit noisier than the others, making a roaring sound whenever the flames increase. ***Flame lift,*** as this is sometimes known, is not a stable, normal burner condition and should be corrected immediately.

Incomplete combustion may also cause ***floating flames*** that are lazy looking and are not shaped as well-defined cones. This is a dangerous condition, and you'll usually notice it in the first minute or two that a burner has been turned on, before it achieves the proper airflow. If the flames don't assume their normal, conical shapes quickly, have the burner checked.

Finally, the most serious condition is ***flame rollout.*** When the burner is turned on, flames shoot out of the combustion chamber opening instead of the top of the burner. Flame rollout is a serious fire hazard and must be repaired immediately. The burner may not be correctly positioned or something may have obstructed its inner workings. Either way, call the service person—fast.

Gas Burners

The burner assembly is the basic unit in a gas-fired piece of equipment that mixes gas with oxygen and thus produces the heat required for cooking. Today, there are many different types of burners, each designed to meet the demands of a particular appliance. All burners have ports, a series of round holes

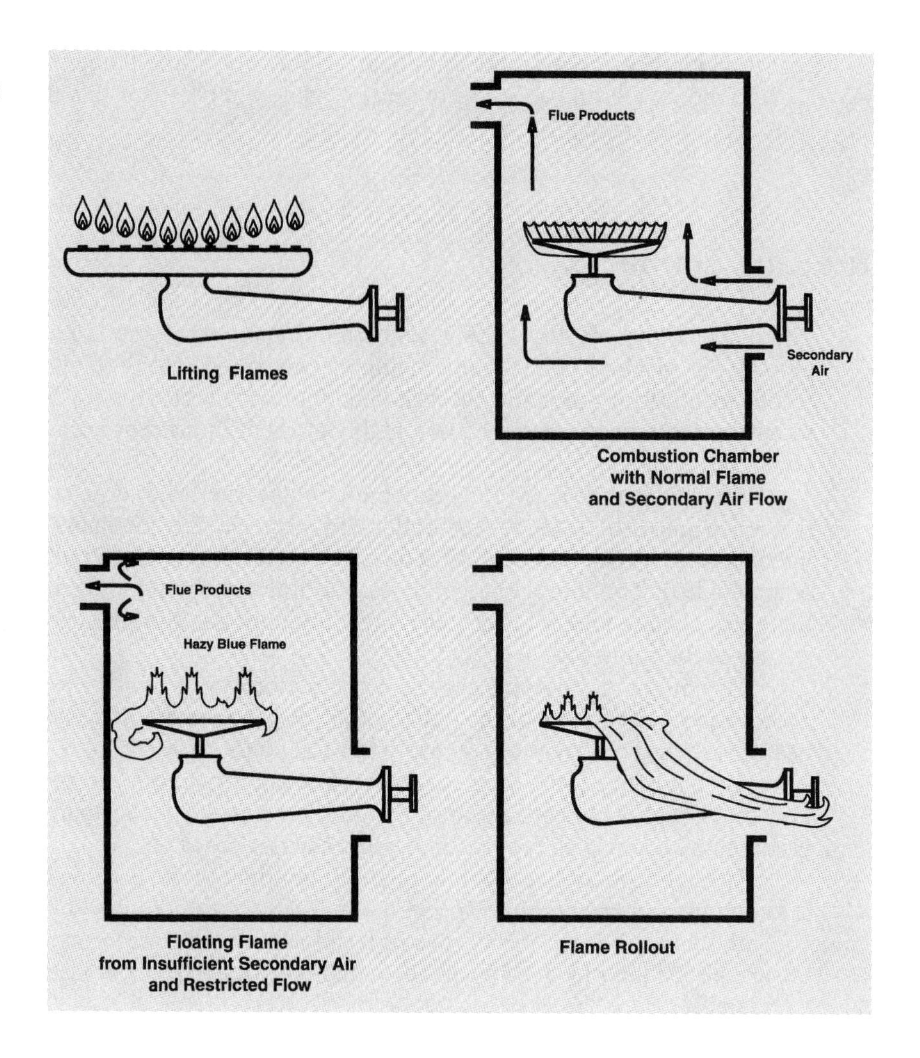

ILLUSTRATION 6–3 **The four most hazardous gas flame conditions, all of which require adjustment.**
SOURCE: *COOKING FOR PROFIT* BY ANN MARIE JOHNSON, ROBERT HALE, LTD., 1991.

from which the flames burn. There are wide and narrow ports, and they can be arranged in various patterns (see Illustration 6-4). A few of the most common burner types are discussed next.

Pipe Burner. This is a pipe (usually made of cast iron) with two or more rows of ports drilled along its length. You'll find pipe burners in ovens, griddles, and broilers. Although most pipe burners are straight, there are also loop burners, in which the pipe has been bent into a circular or oval shape.

Ring Burner. These are widely used on range tops, steam tables, boilers, and coffee urns. The standard ring burner has one or two rows of ports arranged in a circle. It's made of cast iron and comes in a variety of sizes. To increase the capacity of a burner, several ring burners of different diameters can be nestled one inside the other so that one, some, or all of them can be turned on as needed. Some ring burner ports face sideways instead of straight up, so that food isn't spilled into them accidentally.

Ring burners are sometimes known as atmospheric burners, because the secondary air comes from the atmosphere around them. However, the flame tips often suck in too much oxygen, making them less efficient and requiring frequent adjustment of the primary air–gas mixture.

Slotted Burner. This is really a type of pipe burner, because it is made of the same cast-iron pipe and can be straight or circular. The ports in a slotted burner are all aimed in the same direction to form a single, large flame. Some have only one wide slot, with a corrugated ribbon of heat-resistant metal alloy located inside the slot to allow a wider (but still safe) path for the flame. Slotted burners are typically found in hot-top ranges and deep-fat fryers.

Flame Retention Burner. This is a slotted burner with additional ports drilled into the pipe. It allows more heat to flow into the burner, improves combustion, and reduces flashback. Flame retention burners are considered very efficient, with a wide range of heat settings and the ability to fine-tune your primary-air adjustment.

Radiant (Infrared) Burner. The usual infrared burner is a set of porous ceramic plates, with about 200 holes per square inch on its surface. Air and gas flow through these holes and burn very hot (about 1650 degrees Fahrenheit), which makes this type of burner ideal for broilers. They can be located at the sides of a fry kettle for maximum heat transfer or suspended inside a protective cylinder located at the bottom of the fry tank.

Fryers with infrared burners boast 80 percent energy efficiency, compared to about 47 percent for conventional fryers. Their heat recovery time (the time it takes to return to optimum cooking temperature after a new batch of cold food has been loaded into the kettle) is less than two minutes. The same benefit—heat intensity—makes the infrared burner popular for griddles. A sturdy, 1-inch-thick griddle plate can be used instead of a thinner one that is not able to retain heat as well.

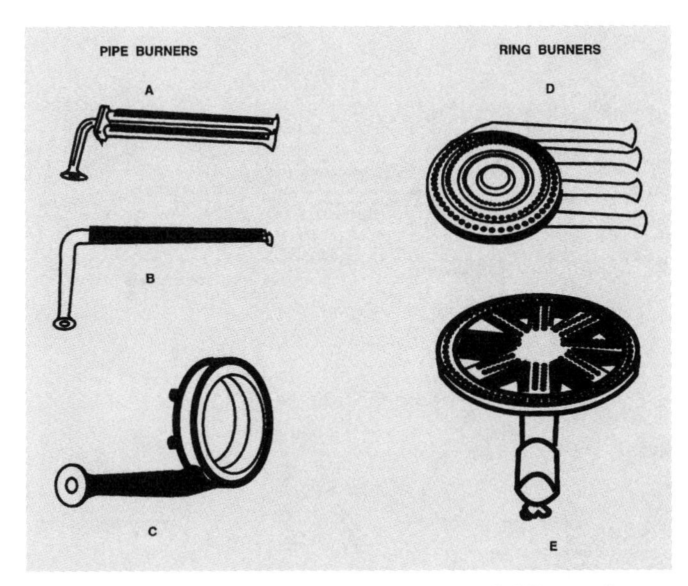

ILLUSTRATION 6-4 **Gas burners come in several different shapes and sizes.**

Infrared burners work so well because they use ***radiant heat,*** and the best example of radiant heat is the sun. Have you noticed how the sun can warm your face on a winter day, even though the air around you is cold? In the same way, the infrared burner sends its high-frequency waves directly from the heat source to the food. The rays only turn into thermal (heat) energy when they hit the food; they do not heat the air. When you're cooking, you get the most energy efficiency from an appliance that heats the food—not the air that surrounds it.

Range Top Power Burner. This burner premixes gas and combustion air in correct proportion to produce high heat and efficiency. Unlike the standard ring burner, the power burner does not rely on the atmosphere to supply its secondary air. The burner head is enclosed in a sealed metal ring through which no excess air can enter. Flames are spread evenly through the ring over the bottom of the cooking pot, so less heat is wasted, more heat is delivered to the food, and the kitchen stays cooler. The burner head acts as a shutter mechanism, readjusting the premixed air and gas whenever the controls are turned up or down.

A recent appliance development is the power burner range, with two (front) power burners and two (back) conventional burners, the front ones for speedy cooking and the back ones for keeping food warm.

Infrared Jet Impingement Burner. The "IR jet," as it's nicknamed, is a new type of high-efficiency burner that also uses less gas than conventional burners. It is a power burner that premixes gas and air in a separate chamber before burning. This mixture is fed into the burner by a blower and ignited at the burner surface. The perforated ceramic burner plate holds the flames in place and allows them to impinge (hit hard) on the bottom surface of the pan.

Pilot Lights and Thermostats

The pilot light is an absolute necessity in the gas-fired commercial kitchen. There are several different kinds of pilot lights, some automatic and some manual, and most gas appliances make it possible for you to easily adjust the pilot light, if necessary, with the turn of a screw. Illustration 6-5 shows the inner workings of the two most common types of pilot lights.

The pilot light should be about three-quarters of an inch high. This small flame is located next to the main burners of the appliance. In some cases, there's a separate pilot for each burner; in others, a single pilot light is used to light more than one burner.

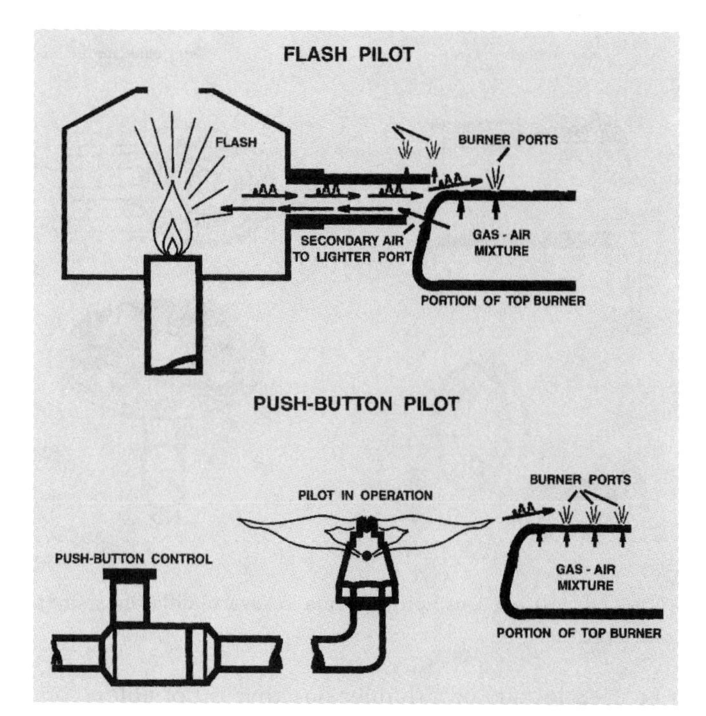

ILLUSTRATION 6-5 The main parts of a typical pilot light system. The flash pilot lights on its own; the push-button pilot requires someone to push the button.

SOURCE: ROBERT A. MODLIN, EDITOR, *COMMERCIAL KITCHENS*, 7TH ED., AMERICAN GAS ASSOCIATION, ARLINGTON, VIRGINIA.

If the pilot light goes out, it is a signal that the gas has been shut off. Like other gas flames, pilots can also burn yellow, which means dirt or lint may be blocking its opening. This can be removed by brushing it clean again.

Pilot lights have safety features, the most common of which is the ***thermoelectric control.*** When a junction of two metal wires (called a ***thermocouple***) is heated by the pilot flame, a very low electric voltage is generated—just enough to fire an electromagnetic gas valve and hold it in an open position. If the pilot light fails, the thermocouple cools, the electric current stops, and the gas valve is closed by spring action. To resume the flow of gas, the pilot must be manually relit.

The ***thermostat*** is the control used on most gas-fired equipment to maintain the desired burner temperature. By far, the most common thermostat is a knob or dial called a *throttling* or *modulating control.* It allows the flames to rise or fall quickly, then regulates the gas flow to keep the burner temperature constant.

Some types of cooking appliances, such as deep-fat fryers, require quick heat recovery, in less than two minutes. In these cases, a ***snap-action thermostat*** is used, which opens fully to permit maximum heating until the desired temperature is reached. Then it shuts off just as quickly.

Remember that the function of a properly working thermostat is to turn down, or shut off, the supply of gas as soon as the burner reaches the desired temperature. Should you need to reduce its temperature, turning down the thermostat will not be sufficient. There's already heat stored in the appliance, and it will take time for it to dissipate and cool down. On a gas oven, for instance, you should set the thermostat to the new, lower level and then open the oven doors to allow it to cool more quickly.

When you need to raise the temperature (e.g., starting the oven when it's cold), many people have the mistaken impression that the appliance will heat more quickly if you blast it immediately to "High," and then turn it down. In fact, it won't heat any faster, and you risk forgetting to turn it down and damaging the food by cooking it at an excessive temperature.

Maintaining Gas-Powered Equipment

Preventive maintenance is as important with gas appliances as any others to prevent equipment malfunction. We've adapted these maintenance suggestions from an article in *Equipment Solutions* magazine's September 2002 issue:

- Perform routine checks and/or maintenance both weekly and monthly, and document what you've done.
- Weekly tasks include cleaning burner parts and orifices; checking the primary blower speeds; vacuum-cleaning the entire blower system.
- Monthly tasks include adjusting and cleaning air inputs and pilot lights; checking and calibrating thermostats; checking the burner valves. If the latter are difficult to turn or move, they need to be lubricated (sparingly) with high-heat valve grease. Also, check monthly for a good balance of exhaust and makeup air. Remember, gas-operated equipment depends on a uniform exchange of "new" air to replace the air used in the combustion process.
- Whenever another piece of equipment is added to the cooking line, have a flow test performed on the main gas line (by a professional repairperson or gas company representative) to ensure

that there is sufficient pressure to your hot line when all the cooking equipment is at peak gas consumption levels. This, in turn, will ensure the efficient operation of each appliance.

Reading Gas Meters and Bills

If you have gas appliances, you will also have gas meters, which look like Illustration 6-6. The single, upper dial on a gas meter is used only to test the meter and is not part of your reading. The dials are sometimes called ***registers.***

Although gas is measured in cubic feet instead of kilowatt-hours, the meter dials work the same way (and are read the same way) as electric meter dials.

In large commercial buildings, the gas meter may be more sophisticated, with two sets of dials much like the combination electric meter. Called a compensating meter, this meter also adjusts ("compensates") when gas pressure or temperature at the location varies from normal conditions. Of the two sets of registers, one will be marked "Uncorrected" or "Uncompensated" and the other will be labeled "Corrected" or "Compensated." Read the "Corrected" meter to determine your gas usage; the "Uncorrected" dials are used by the utility company for checking the meter.

Illustration 6-7 is a sample gas bill. In all instances, the measuring of gas consumed is in cubic feet. However, the rates may be based on therms. (One therm equals 100,000 Btus or approximately 100 cubic feet.) In our sample bill, the cubic feet are converted to therms to determine the cost of the gas fuel. The following notes describe items appearing on the bill:

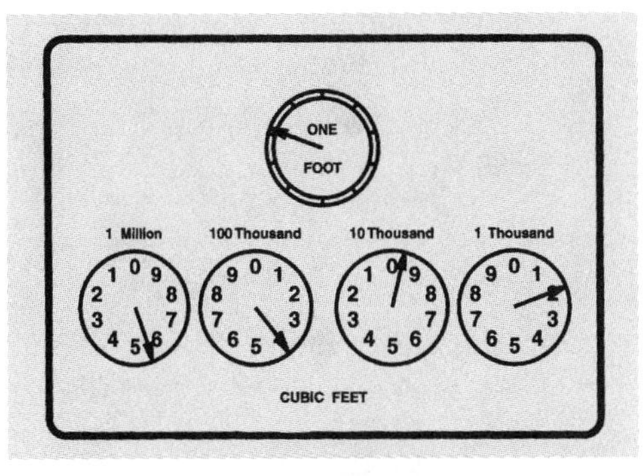

ILLUSTRATION 6-6 Gas meter dials are read the same way as an electric meter, but they measure cubic feet instead of kilowatt-hours.

1. **Your account number is to be used when communicating with your gas company.**
2. **These are the dates on which the meters were read and represent the billing period.**
3. **This is the rate schedule that applies specifically to your account. While only one rate is shown on this sample bill, differing rates may apply to gas used for heating, cooling, or other special applications.**
4. **This matches the number on the meter at your restaurant.**
5. **These are the meter readings at the beginning and end of the billing period expressed in hundreds of cubic feet of gas.**
6. **The total amount of gas consumed in hundreds of cubic feet is arrived at by subtracting the previous meter reading from the current meter reading (67390 − 63710 = 3680).**
7. **This factor converts cubic feet to therms. The particular gas distributed by this company contains 102,200 Btus per 100 cubic feet, from which the factor of 1.022 is developed.**
8. **The therms used are determined by multiplying item 6 by item 7 (3680 × 1.022 = 3761). Items 7 and 8 do not appear on gas bills if rate schedules are based on cubic feet.**
9. **Charge for gas consumed.**
10. **State tax is shown. County or other municipal taxes may also appear.**
11. **Total bill.**

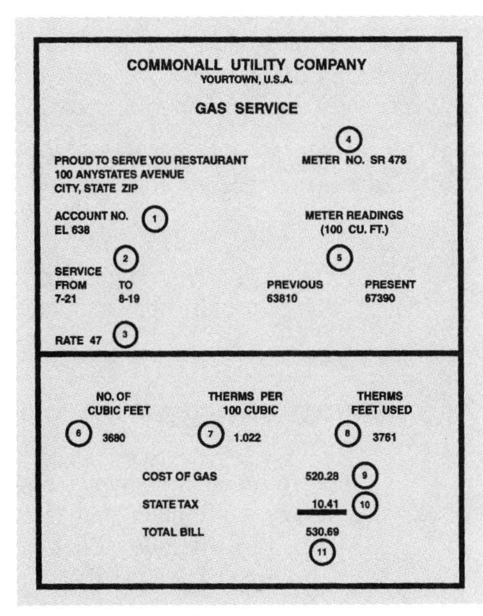

COMMONALL UTILITY COMPANY
YOURTOWN, U.S.A.

GAS SERVICE

(4) METER NO. SR 478

PROUD TO SERVE YOU RESTAURANT
100 ANYSTATES AVENUE
CITY, STATE ZIP

ACCOUNT NO. (1)
EL 638

METER READINGS (100 CU. FT.) (5)

SERVICE (2)

FROM	TO	PREVIOUS	PRESENT
7-21	8-19	63810	67390

RATE 47 (3)

NO. OF CUBIC FEET	THERMS PER 100 CUBIC	THERMS FEET USED
(6) 3680	(7) 1.022	(8) 3761

COST OF GAS	520.28	(9)
STATE TAX	10.41	(10)
TOTAL BILL	530.69	(11)

ILLUSTRATION 6-7 A sample gas bill.

SOURCE: REPRODUCED WITH THE PERMISSION OF THE NATIONAL RESTAURANT ASSOCIATION.

FOODSERVICE EQUIPMENT

MAINTENANCE TIPS FROM THE PROS

Frank Murphy of Gas Consumer Services, Inc., supplies these maintenance tips for specific types of equipment:

BROILERS AND CHARBROILERS

Most problems with these are the result of grease and food particles clogging the burners, pilots, and/or shutters. These components need to be cleaned weekly, and to be properly adjusted to restore optimum operation.

A good way to prevent this buildup is to make sure that broiler grates are positioned to direct the excess grease flow for burn-off, or collect it in grease drawers. Clean the broiler grates at the end of each day, which is easy to do—just place the grates flat on the broiler and set the gas valve on High for 45 minutes. Then turn off the broiler and allow the grates to cool. Remove them from the broiler when they've cooled down, and clean them (both top and bottom surfaces) with a wire brush, damp cloth, and mild detergent.

The grate channels and burner radiants should be thoroughly cleaned as well. Brush the burner's heat reflector to remove and dust or debris, and clear all the burner portholes at least once a week.

FRYERS

A major maintenance problem for the gas-fired fryer may be its location in the kitchen. It is essential that there are no restrictions for "new" air entering the burners or blower motor. If the airflow is restricted, the fryer's sidewalls and internal control components will be abnormally high. This will cause the electrical controls to overheat, and soon the equipment's performance will diminish.

To prevent this problem, make sure the gas connections to the fryer are tight. Be sure that enough fresh, makeup air is available. And clean the hood filters each day, checking again for airflow restrictions there.

Another maintenance concern for fryers is slow temperature recovery, which is related to having a reliable and controllable heat source. A fryer that takes too long to recover its temperature when cold food items are dropped into the kettle is losing its capacity to conduct or radiate heat efficiently. In high-efficiency fryers, the burner seals may be leaking, the blower motor speed may be too low, or a broken

A "purchased gas adjustment" may also appear on your utility bill. In addition, a potential late charge usually appears if payment is made after a prescribed date.

Saving Energy with Gas

There are many simple, practical ways to take full advantage of the instant heating power of gas. Choose equipment that is enclosed and insulated, keeping the energy within the appliance (or absorbed by the food). Cook at the lowest temperature, or in the largest volume, possible. Especially for solid-top ranges, use flat-bottomed cookware that makes full contact with the cooking surface. Curves and dents in pots and pans end up wasting money. The bottoms of the cookware should be about 1 inch wider than the diameter of the burner.

Although there are occasions for "big" flames and kitchen showmanship, for most cooking duties it is sufficient that the gas flame tips barely touch the bottoms of the cookware and do not lap up over the

FOODSERVICE EQUIPMENT

MAINTENANCE TIPS FROM THE PROS (CONTD.)

temperature probe could be the problem. In tube-heated fryers, internal heat-baffle wear causes recovery problems. Your owner's manual should offer some guidance in these situations.

CONVEYOR OVENS

The cooling fan exhaust grilles should be wiped clean on a daily basis. Check to see that the cooling fan is turning when the oven is operating. Every month, the entire conveyor should be removed so you can clean the jet-air "fingers," being careful to replace all components in their original positions. And every two or three months, clean the combustion motor's blower air intake, which is usually located behind a closed panel.

RANGES

Most maintenance problems with gas-operated ranges happen because bits of food and greasy particles settle into the gas lines or ports, preventing a smooth flow of gas and air to the pilot lights and burners. To identify this condition, check the burner pilot lights and flames for clear, even combustion. Clean the burner grate surfaces every day using a wire brush, damp cloth, and mild detergent. Clean and adjust the range pilots monthly—they should sit level in their mounts, and be situated so the pilot light flame can easily ignite the burner. Check the range burners at least once a month for heat flow. If foods are taking too long to cook, ask a factory-authorized service person to check the gas pressure and flue.

GRIDDLES

Every month, check the underside of the griddle to make sure grease is not running where it's not supposed to go—like the air vent for the gas pressure regulator. If the vent is blocked, the griddle's heating performance will become erratic. Each griddle section and its burner/orifice must be kept clean to maintain a consistent flame and heat correctly. Also be sure that the thermostat mechanism is securely in place.

Source: Reprinted with permission of *Equipment Solutions* magazine, September 2002.

sides. Burners should be adjusted accordingly. Don't keep pots at a boil when simmering them would be sufficient, and cover them to hold in heat.

A common tendency is to turn equipment on early to let it "heat up." Again, this is a waste of fuel and time. For open-top ranges, preheating is simply not necessary; for griddles, low or medium flames are sufficient for just about any kind of frying. Broilers don't require much, if any, preheating; gas ovens, solid-top ranges, and steamers can be preheated, but no more than ten minutes.

Energy saving is another good reason many ranges and griddles are built as adjoining, temperature-controlled multiple burner sections. During slow times, learn to group food items on the least possible number of sections, which eliminates the need to keep the entire cooking surface hot.

Regular cleaning and maintenance of the appliances are two important keys to wise use of natural gas, but there are also energy-saving innovations in the works. One is a concept called ***heat transfer fluids (HTF).*** The idea is to power several pieces of equipment with a single burner, using a series of pipes and a heated fluid that runs through the pipes to different appliances. (The heated fluid can't be water, because its pressure would become too high and create steam.) The fluid may also be run through a heat exchanger, if necessary, to boost its temperature along the way.

A major hotel chain testing an early HTF system uses the same heating fluid to do such disparate tasks as drying laundry and frying chicken! At this writing, researchers are looking for a completely nontoxic fluid, because a leak or accident might release some of it into the food. On the drawing boards, however, is an entire integrated HTF kitchen, all heated by a single, closed loop of hot fluid and piping.

Gas Pipes

Natural gas for commercial kitchens flows through large pipelines at pressures of 600 to 1000 pounds per square inch (psi). This high pressure is reduced by a series of valves, to arrive at your gas meter at about 25 psi. Both the size and the quality of pipes used are critical in setting up a gas system for your business.

By totaling the amount of Btus required when all gas equipment is on, you can estimate the total amount of gas required and calculate the size of the pipes needed. Divide the total number of Btus needed per hour by 1000. This figure is the total number of cubic feet of gas needed. Let's say your place would use 400,000 Btus per hour. That's 400 cubic feet per hour. Then, use the pipe sizing table (see Illustration 6-8) to estimate the diameter of pipe to install. You'll notice that this depends, in part, on how far the gas has to travel from the meter to the kitchen. For our example, let's estimate that the distance from meter to kitchen is about 40 feet. The table says a 1¼-inch pipe should be more than adequate. We might install a 1½-inch pipe, to compensate for any additional equipment installed in the future.

Most gas utility companies have commercial representatives who will help determine proper pipe sizing, and their consulting service is usually free.

Because the gas pipes will be a permanent part of the building, they should be durably constructed of wrought iron. It is always preferable to install gas pipes in hollow partitions rather than solid walls, to minimize their potential contact with corrosive materials. Gas pipes should never be installed in chimneys or flues, elevator shafts, or ventilation ducts.

Gas appliances are attached to their gas source with a ***gas connector,*** which is a flexible, heavy-duty brass or stainless steel tube, usually coated with thick plastic. One end is fastened permanently to the building's gas supply, the other to the back of the appliance. The appliance end should be a ***quick-disconnect coupling,*** an easy shutoff device that instantly stops the flow of gas with an internal assembly of ball bearings and a spring-loaded plug (see Illustration 6-9). Shutoff occurs automatically when kitchen air reaches a certain high temperature (as in a fire) or when an employee disconnects the coupling to move or clean

Nominal Iron Pipe Size (in inches)	Internal Diameter (in inches)	Length of Pipe (in feet)									
		10	20	30	40	50	60	70	80	90	100
½	0.622	132	92	73	63	56	50	46	43	40	38
¾	0.824	278	190	152	130	115	105	96	90	84	79
1	1.049	520	350	285	245	215	195	180	170	160	150
1¼	1.380	1050	730	590	500	440	400	370	350	320	305
1½	1.610	1600	1100	890	760	670	610	560	530	490	480
2	2.067	3050	2100	1650	1450	1270	1150	1050	990	930	870
2½	2.469	4800	3300	2700	2300	2000	1850	1700	1600	1500	1400
3	3.068	8500	5900	4700	4100	3600	3250	3000	2800	2600	2500
4	4.026	17500	12000	9700	8300	7400	6800	6200	5800	5400	5100

ILLUSTRATION 6-8 **Pipe sizing table. This shows the maximum capacity of pipe in cubic feet of gas per hour (based on a pressure drop of 0.3-inch water column and 0.6 specific gravity gas).**

SOURCE: STEVEN R. BATTISTONE, *SPEC RITE: KITCHEN EQUIPMENT,* FOOD SERVICE INFORMATION LIBRARY, CINCINNATI, OHIO.

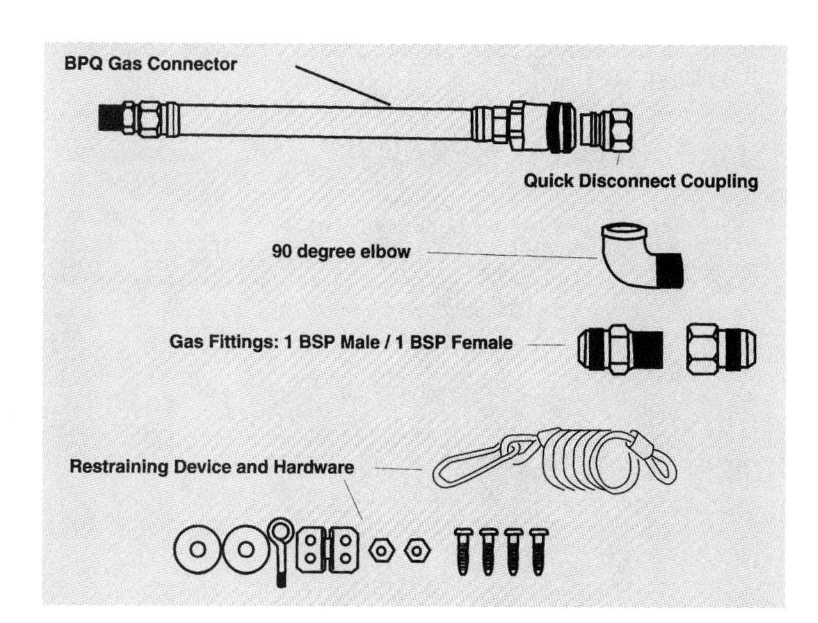

ILLUSTRATION 6-9 The components of a connector kit, used to attach an appliance to its gas source.

SOURCE: DORMANT MANUFACTURING COMPANY, PITTSBURGH, PENNSYLVANIA.

behind the appliance. Quick disconnects are not only safety precautions, they are also very handy when rearranging and cleaning the kitchen or servicing the appliances. Like pipes, quick disconnects come in different lengths and diameters. You'll order them based on the connection size, the gas pressure, and the length needed. There are also quick disconnects available for steam and water appliances.

6-2 STEAM ENERGY

Steam is water vapor, which occurs when water molecules are suspended in air by the heat added to them. Steam molecules carry large quantities of heat, and they return to their original form (condense) when they come into contact with a cooler surface.

When we discuss steam and its uses in foodservice, the terms *heat* and *temperature* (sometimes used interchangeably) take on completely different meanings:

- ***Heat* is the total amount of energy contained in steam or water at a given temperature.**
- ***Temperature* is used to describe how hot a particular object is.**

It takes 180 Btus to heat 1 pound of water from freezing (32 degrees Fahrenheit) to boiling (212 degrees Fahrenheit). However, to change this same pound of water to steam requires an additional 790 Btus. This means steam contains about six times the energy of boiling water.

The temperature of steam is generally related to its pressure. In short, the hotter the steam is, the higher its pressure, as shown in Table 6-1. The higher its pressure, the more steam molecules it contains. As these molecules condense, most of the Btus they contain are transferred quickly to the food being cooked, and the condensation creates room in the airspace for even more steam molecules to take the place of the ones that just condensed, in a cycle that continues until the heat source is turned down or off.

Steam is simple, clean, and quick, and it has been around longer than either electricity or gas as a heat source. In foodservice, steam is used extensively in the dish room, to heat water and sanitize and dry dishes. In cooking, steaming is a healthful alternative to range-top cooking that holds in nutrients and can be done quickly. Most foods can be cooked in a steam appliance with three significant advantages: greater control over the food quality; less energy use than other types of cooking equipment; and minimal handling, since the food can often be prepared, cooked, and served in the same pan. Steam is also a more efficient way to thaw frozen foods, instead of immersing them in boiling water.

Steam is a major component of the following popular kitchen appliances:

- The ***steam-jacketed kettle*** is a large "bowl within a bowl" used for making sauces, soups, and stocks. The kettle has a sturdy outer layer. Between the two bowls is an area about 2 inches wide

TABLE 6–1

STEAM PRESSURE AND TEMPERATURE

Steam Pressure (pounds)	Temperature (degrees Fahrenheit)
0	212
1	215
2	218
4	224
8	235
15	250
20	259
25	267
40	287
45	292
50	298

into which steam is pumped, which provides high but uniform cooking temperatures. The water used to create the steam can be heated with either gas or electricity (see Illustration 6-10).

- A ***steamer*** is a rectangular-shaped ovenlike appliance with an insulated door, which can be used for steaming vegetables, braising meats, cooking rice, thawing frozen food—any process that would benefit from the addition of moisture (see Illustration 6-11).
- ***Convection steamers*** contain a fan or *blower* that circulates the warm, moist air for quicker, more even cooking.
- ***Steam tables,*** often seen in cafeterias or on serving lines, hold food above a reservoir of hot water to keep it warm; similarly, a *bain marie* is a hot-water bath in which an urn of gravy or a delicate sauce is immersed, also to keep it warm.

ILLUSTRATION 6–10 **A steam-jacketed kettle is among the most versatile pieces of cooking equipment.**
COURTESY OF VULCAN-HART, DIVISION OF ITW FOOD EQUIPMENT GROUP, LLC, LOUISVILLE, KENTUCKY.

Steam systems and appliances work in one of three ways:

- ***Steam generators*** use electricity to heat water and make their own steam. Small generators, called *boilers,* can be located right in the kitchen, under or near the steam equipment.
- ***Heat exchangers*** take steam already made from one source, circulate it through a series of coils to clean it, and use it to heat another source. The steam from a building's heating system could, for example, be captured and "recycled" by a heat exchanger, and then be used to heat the same building's hot-water tanks.
- ***Steam injectors*** shoot pressurized steam directly into an appliance to produce heat. This is the least efficient way to use steam, because it's a one-time use. Condensation is drained away, not reheated and reused.

Steam equipment can further be classified into pressurized and unpressurized. The amount of pressure in a steam appliance is related to the temperature of the steam: the higher the temperature, the higher the pressure. Pressurized steamers cook food quickly, because the steam can be superheated and comes into direct contact with the food. Unpressurized steamers are not as efficient. Unpressurized steam may not become as hot, and, as it touches the colder foods or cookware, its temperature is lowered even more. Eventually, it condenses back into water and is vented away into a drain.

ILLUSTRATION 6-11 Any type of cooking or thawing that requires moist air can be done in a steamer.

COURTESY OF VULCAN-HART, DIVISION OF ITW FOOD EQUIPMENT GROUP, LLC, LOUISVILLE, KENTUCKY.

Steam Requirements for Equipment

We measure steam in boiler horsepower (BHP). As a general rule, 1 BHP creates 34.5 pounds of steam per hour, and is equivalent to about 10 kilowatts of electricity. The ***boiler*** is the piece of equipment that boils the water to make it into steam. If a boiler is rated by the manufacturer as producing 5 BHP, this means it produces $5 \times 34.5 = 172.5$ pounds of steam per hour.

To calculate the size of boiler you'll need for a whole kitchen, you must find out how much steam flow is needed by each piece of equipment; then add them and divide by the 34.5 figure. For instance, a kitchen with eight pieces of steam equipment may require a total steam output of 187.5 BHP. Divide 187.5 by 34.5, and you discover you will need a boiler with an output of at least 5.43 BHP.

However, you should know more than how much steam will be produced. You must also know how much force, or pressure, the steam will have. In most English-speaking countries, steam pressure is measured in *pounds per square inch,* or ***psi.*** Once again, remember that the temperature of the steam goes *up* when the pressure goes *up.*

The other factor that impacts steam pressure is your altitude—not drastically, but the equivalent of a 2- or 3-degree drop in temperature for every 1000 feet above sea level.

Finally, both steam temperature and pressure are impacted by the distance the steam must travel to get from the boiler to the appliance. Heat loss is determined by the number of feet of pipe traveled, plus every valve and fitting through which the steam must flow.

Foodservice industry research indicates that the most expensive way to set up a kitchen is to install individual boilers for each piece of equipment, so it is ironic that that's the most common way it is done. Self-contained boilers have higher maintenance costs than a single, large unit, and they add more heat to the already sweltering kitchen environment.

Like other appliances, boilers also have efficiency ratings to consider. A boiler that requires 140,000 Btus and has a 50 percent efficiency rating will deliver 70,000 Btus of heat to its water supply to make steam. (This is the equivalent of a little more than 2 BHP.)

Steam can be a very economical energy source, especially if your building already has a ***clean steam*** system built in. (When steam is referred to as "clean," it means it is pure and has not been contaminated by chemicals.) If the building is not already fitted with steam pipes, you must decide if you will be using enough different steam appliances to justify the expense of installing them.

Steam Terminology

Here's the way a steam system works. Steam is made by boiling water in the boiler, which may also be called a *converter.* It is then piped to the appliance where it will be used. At the appliance, the steam hits a ***coil*** (coiled copper or stainless steel tubing), which condenses the steam and transfers its heat to be used in the appliance. As this transfer occurs, the steam cools and becomes water again. This ***condensation*** is removed from the appliance through a ***steam trap.*** The condensation usually returns to the boiler through another set of pipes, called the ***return piping,*** to be reheated and made into steam again (see Illustration 6-12).

The steam trap is one of the most vital parts of a steam system, because it helps regulate the overall pressure of the system. Oddly enough, steam traps are not placed directly in the main lines of a steam system. Steam does not flow through the steam trap. Instead, the trap is placed near the "ends" of the steam lines. It operates almost like an overflow valve, opening now and then to discharge water without affecting the rest of the steam or the steam pressure. Your steam appliance (or system) will have one of two kinds of steam traps: an *inverted bucket* or a *thermodynamic disc.* Either way, the results are the same. Condensation, plus air and carbon dioxide, is collected in the trap, then discharged as the trap becomes full.

Valves control the steam pressure based on the amount of flow, size of pipe, and intensity of pressure needed for the appliance or system. There are several types of valves: *Electric solenoid valves* control the steam flow; *pressure-reducing valves* regulate steam pressure within the main supply lines; *test valves* allow you to test steam pressure at a point close to the appliance; and *manual shutoff valves* ensure the steam can be safely turned off by hand any time the system needs to be serviced.

When purchasing or installing steam equipment, you'll need to decide if the unit is adequate for the job; and, if the unit does not generate its own steam, can the steam system in your building make enough steam to operate it?

As you've already learned, the length and diameter of the pipes can affect the performance of the equipment. Both friction and condensation in the pipeline will naturally cause a drop in pressure, and you've got to allow for that, too. The basic steps for sizing pipe are:

1. Determine the steam requirements of the equipment.
2. Determine how much the pressure can drop between the steam source and the equipment. To do this, substract the required amount of pressure for the appliance from the total amount of available pressure. This number is called the *allowable pressure drop.*
3. Calculate how long the pipe will be. This includes not only the actual length, but also the *equivalent length* that you must allow for pipe fittings, elbow joints, valves, and so on. Determine this length in hundreds of feet. This is called the *effective length* of the pipe.

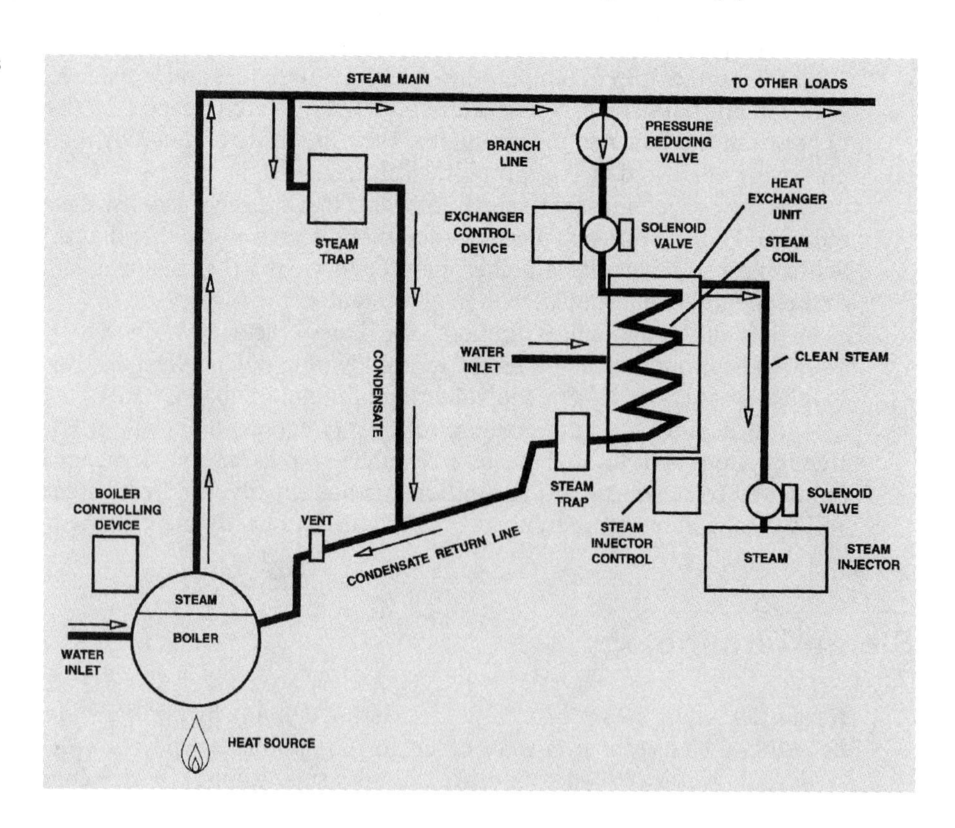

ILLUSTRATION 6-12 The basics of steam system operation.
SOURCE: *STEAM IN PERSPECTIVE,* HOBART CORPORATION, TROY, OHIO.

4. Divide the allowable pressure drop by the effective length of the pipe, then divide that number by 100. This will give you the *acceptable pressure drop* per 100 feet of pipe.
5. Refer to the correct chart (see Illustration 6-13) to determine if the pipe size you have is adequate for the job.

At this point, you may be wondering, "Why do I need to know all this technical information? Can't I hire someone to handle this?" Of course you can, and most equipment sales representatives are very familiar with the particulars of installation. However, you still need to know the basics, when replacing an old machine, purchasing a new one, or troubleshooting a steam-related problem.

Common Problems and Diagnoses

Safety is the main consideration in dealing with steam equipment malfunctions. The mandatory first step is to shut off the steam supply and depressurize the steam line before attempting to disassemble the equipment.

The problem may be as simple as inadequate steam pressure. Take pressure readings as close to the appliance as possible, while steam is flowing through the line. Trace the steam flow through the entire appliance, by checking all valves, strainers, and coils for visible leaks or damage.

On most steam equipment, you will find evidence of hard-water mineral deposits, as the traces of minerals dissolved in the water settle and form *scales* inside tanks and pipes or lime buildup inside boiler tanks. Even if chemical additives are used periodically to control scale buildup, you should disassemble equipment now and then and remove scales manually. Excessive or frequent buildup is a sign that the problem is not being properly treated, and perhaps a professional water treatment expert should be consulted. We'll talk more about water quality problems later in this chapter, but it's safe to say that most cities' water supplies are hard enough to cause significant problems in commercial steam use for foodservice. Steam equipment manufacturers cover themselves in these situations by specifying a minimum water hardness acceptable for their appliances. Equipment failure caused by unacceptable water quality is not covered under their warranties. They may also recommend the installation of a water-softening system or, at least, a filter to remove silica and chlorine from water used to make steam. A water-softening method known as zeolite is often recommended for hard-water areas, which specifically attracts and filters minerals and salts out of the water. Some manufacturers offer their own descaling kits.

6-3 YOUR WATER SUPPLY

Safe, plentiful water is often taken for granted by most guests, employees, and managers in foodservice—which is ironic when you consider what a truly scarce resource it is. In fact, salt water makes up about 97 percent of all the water on Earth. Another 2 percent is inaccessible, frozen in remote ice caps and glaciers. More than half of the single percentage that remains worldwide is now diverted for human use, and yet the combination of increased population, industrial technology, and irrigation has pushed people to use an amazing 35 times more water than our ancestors did just three centuries ago.

Luckily, the last 20 years of the twentieth century brought a big wake-up call when it comes to water conservation. For instance, since the 1990s, you've probably noticed more restaurants advising guests that they will only be given a glass of water if they request it. As some of these tabletop notices remind guests, for every glass of water on their table, it takes as many as five more gallons to wash, dry, and sanitize it. Hotels are notorious water-guzzlers—according to a study by the School of Hotel Administration at Cornell University, the average hotel room requires 144 gallons per day. Now, many have installed low-flow showerheads and water-saving toilets; they don't change bed linens or replenish towels as often, to cut down on laundry volume. The U.S. Environmental Protection Agency began its "Water Alliances for Voluntary Efficiency" program (WAVE) for hotels, restaurants, and other businesses. Slowly but surely, these efforts are working. From 1980 to 1995, water use in the U.S. decreased by 20 percent per person.

(a) STEAM PIPE CAPACITY IN POUNDS PER HOUR AT 15 PSI

Pipe Size (inches)	Pressure Drop Per 100 Feet of Pipe Length					
½	5	8	11	14	16	23
¾	13	18	26	32	37	52
1	27	38	53	65	76	110
1¼	59	83	120	140	160	230
1½	91	130	180	220	260	360
2	180	260	370	450	520	740
2½	300	430	600	740	860	1,210
3	560	790	1,110	1,360	1,570	2,220
3½	830	1,180	1,660	2,040	2,350	3,320
4	1,180	1,660	2,350	2,880	3,330	4,700
5	2,180	3,080	4,350	5,330	6,160	8,700
6	3,580	5,060	7,150	8,750	10,120	14,290
8	7,450	10,530	14,880	18,220	21,060	29,740
10	13,600	19,220	27,150	33,250	38,430	54,270
12	21,830	30,840	43,570	53,370	61,690	87,100

(b) STEAM PIPE CAPACITY IN POUNDS PER HOUR AT 60 PSI

Pipe Size in (inches)	Pressure Drop Per 100 Feet of Pipe Length						
½	8	12	17	21	25	35	55
¾	20	28	40	49	57	81	128
1	40	57	81	100	115	163	258
1¼	89	127	179	219	253	358	567
1½	139	197	279	342	395	558	882
2	282	400	565	691	800	1,130	1,790
2½	465	660	930	1,140	1,318	1,860	2,940
3	853	1,205	1,690	2,090	2,410	3,410	5,400
3½	1,275	1,800	2,550	3,120	3,605	5,090	8,060
4	1,800	2,550	3,610	4,462	5,100	7,220	11,400
5	3,320	4,710	6,660	8,150	9,440	13,300	21,100
6	5,475	7,725	10,950	13,420	15,450	21,900	34,600
8	11,360	16,100	22,800	27,900	32,200	45,550	72,100
10	20,800	29,400	41,500	51,000	58,900	83,250	131,200
12	33,300	47,100	66,700	81,750	94,500	133,200	210,600

(c) EQUIVALENT LENGTH OF PIPE TO BE ADDED FOR FITTINGS

	Length to be Added to Run				
Pipe Size (in inches)	Standard Elbow	Side Outlet Tee	Gate Valve*	Globe Valve*	Angle Valve
½	1.3	3	0.3	14	7
¾	1.8	4	0.4	18	10
1	2.2	5	0.5	23	12
1¼	3.0	6	0.6	29	15
1½	3.5	7	0.8	34	18
2	4.3	8	1.0	46	22
2½	5.0	11	1.1	54	27
3	6.5	13	1.4	66	34
3½	8.0	15	1.6	80	40
4	9.0	18	1.9	92	45
5	11.0	22	2.2	112	56
6	13.0	27	2.8	136	67
8	17.0	35	3.7	180	92
10	21.0	45	4.6	230	112
12	27.0	53	5.5	270	132

** Valve in full open position*

ILLUSTRATION 6-13 a, b, c **Charts a and b are samples of pipe capacities for two different levels of steam pressure (15 and 60 psi, respectively). Chart c indicates the amount of pipe length to add for various sizes of fittings.**

SOURCE: *STEAM IN PERSPECTIVE,* HOBART CORPORATION, TROY, OHIO.

The other major water-related problem, at least in the United States, is the delivery system itself. Many of our water mains and pipes are more than a century old and, long neglected, are reaching the end of their useful life. There are more than 237,000 water main breaks in the U.S. annually. When pipes break, water pressure drops and dirt and debris are sucked into the system and jeopardize water quality. So at this writing, there is widespread agreement among experts that the nation's water system is in need of an enormous and expensive overhaul, and fixing it may change the way Americans use, and pay for, water.

In recent years, desalination (removal of salt from salt water to make it drinkable) has received much attention. Most current plants employ a technique called ***multistage flash distillation,*** which removes contaminants from seawater by boiling it, then condensing (distilling) the steam.

Another technique is called ***reverse osmosis (RO).*** Highly pressurized seawater is pumped through a semipermeable membrane that allows only the freshwater molecules to flow through, leaving the mineral ions behind. In foodservice, reverse osmosis equipment is becoming popular as a way to purify water for steam, drinking, cooking, and humidification. RO technology can address the problems of both hard-water scaling (caused by calcium, magnesium, and manganese salts) and soft-water scaling (caused by sodium and potassium chloride) in water pipes. Because it can remove solids better than normal filtration, RO offers the advantages of reduced water- related maintenance and better equipment life in addition to improvements in water quality. Illustration 6-14 is a diagram that shows the major components of a reverse osmosis water filter.

Water is a major expense, and water in foodservice establishments is given much more scrutiny than you'd probably ever give your home tap water. Samples are checked for bacteria, pesticides, trace metals, alkalinity, and chemicals. Before you lease or purchase a site, a water test is in order. And before the water is inspected in a laboratory setting, your own senses can offer clues to a few important basics.

Taste or Odor. Sometimes, you just happen to live in an area where—there's no other way to put it—the water tastes or smells funny. The locals may be accustomed to it, but visitors to the area notice it right away. It can affect the flavor of coffee, hot or iced tea, and any beverage in which

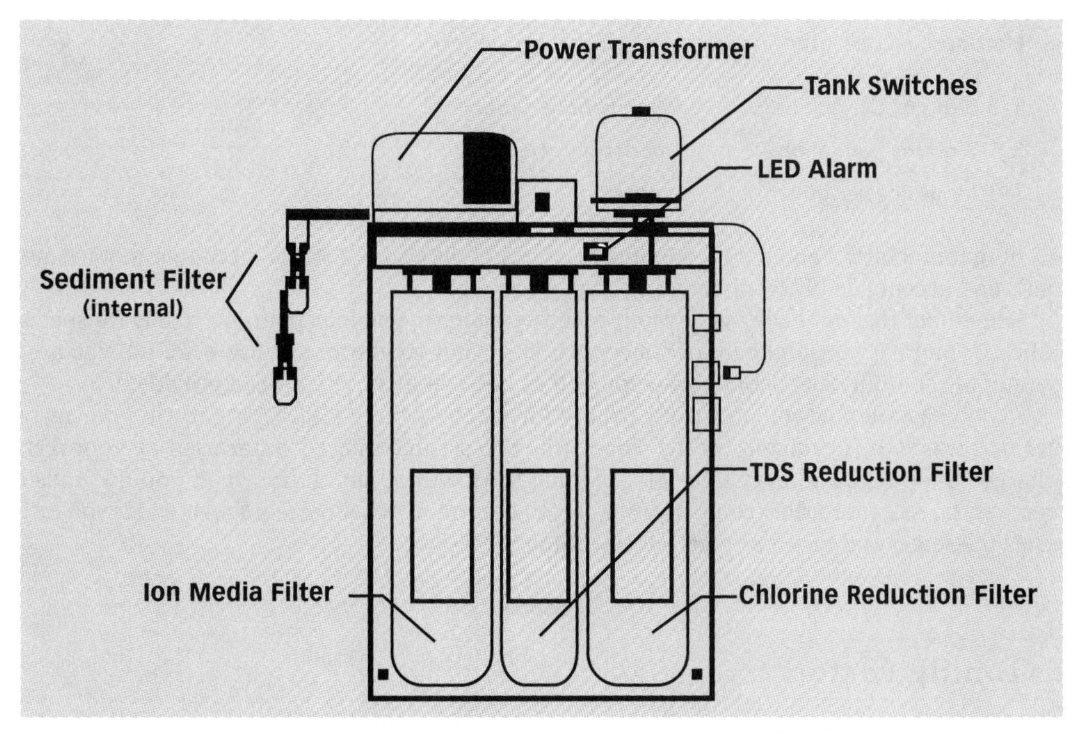

ILLUSTRATION 6-14 **A reverse osmosis water filtration system includes a prefilter for sediments (not shown); a filter to reduce chlorine; a filter for Total Dissolved Solids (TDS) that cause scale formation in pipes; an ion media filter that replaces the minerals with non-scale-forming particles; and a storage tank for the filtered water.**

you place ice cubes. Taste and odor problems are typically caused by the presence of organic materials in the water. You may need to find an outside source of ice to purchase, make ice and beverages with (more expensive) bottled water, or install crushed carbon filters to minimize customer contact with off-tasting water.

Color. Expert water quality advice is needed for this one. Iron or manganese in the water supply can result in odd colors, which, through clear water glasses, look unpalatable. Filtering may help, but getting rid of this condition is a surprisingly technical problem.

Turbidity. When solids are suspended in water, it looks cloudy or murky—a definite turnoff in foodservice. Filtration, with a water-softening system, is a reasonably priced, low-maintenance alternative.

There are two other common water quality concerns:

Corrosion. This condition is the result of oxygen or carbon dioxide becoming trapped within the water supply, and is often caused by the level of acidity in the water system. It affects the useful life of pipes and equipment, and can be corrected by installing a filtration system. What kind, and how extensive the problem is, can be determined by a water quality specialist.

Water Hardness. "Hard" water contains a high proportion of minerals and/or salts. As you're learned, this condition causes an eventual buildup of scales on equipment, which requires constant preventive maintenance to prevent clogging of tanks and water lines and malfunction of the equipment.

Hard water is also troublesome because it encourages the formation of soap scum and makes it more difficult for the *surfactants* (foaming agents) in soap to produce lather. In a hard water area, you have to compensate by using more dish detergent. You may also notice that dishes or cooking utensils washed in hard water become slightly discolored over time. Illustration 6-15 is a chart by the Hobart Corporation, which lists common maladies and possible causes. Notice how many of them are related to water quality.

Hardness is measured in grains (of solids) per gallon:

1–3.5 grains per gallon:	slightly hard water
3.5–7 grains per gallon:	moderately hard
7–10.5 grains per gallon:	very hard

In manufacturers' equipment specifications, you will also see solids measured in parts per million (ppm), and acceptable pH factors may also be mentioned.

Remember that in dealing with water quality problems, you may not have to pay for specialized assistance if you take your problems or concerns first to your local water utility. After all, you are a paying customer of the utility and should avail yourself of any expertise its staff can provide.

Also be aware that you aren't just paying for water—you're also paying to rid your restaurant of water and waste. It is common for the water utility to assume that all water used by your restaurant is discharged as sewage and to charge you accordingly. However, not all the water you use ends up in the sewer system. Ask your utility company to help you determine what percentage of water you buy actually reaches the sewer and to adjust your bill accordingly.

Water Quality Factors

Water quality is not always a major concern. Indeed, in the United States, modern purification techniques have virtually eliminated such waterborne illnesses as cholera, typhoid, and dysentery. However, our water supplies are not without problems. Day-to-day human activities—farming, construction, mining, manufacturing, and landfill operation—impact water quality, affecting wildlife and marine life as well as humans.

STEAM DISH MACHINES	SOME COMMON CAUSES														
SOME COMMON SYMPTOMS	LOW STEAM PRESSURE	STEAM TRAP OPEN	STEAM TRAP CLOSED	BACK PRESSURE in CONDENSATE RETURN	STEAM PIPE TOO SMALL	MINERAL BUILDUP ON COIL	CONTROLS MALFUNCTION	LINE STRAINER FOULED	STEAM INJECTOR BROKEN OR FOULED	SOLENOID VALVE FOULED	PRESSURE RELIEF BAD	INCONSISTENT STEAM PRESSURE	BLOWER DRYER VENTS MISADJUSTED	IMPROPER CHEMICAL ADDITIVES TO TREAT STEAM	
LOW TANK TEMPERATURE	X	X	X	X	X	X	X	X	X	X	X	X	X		
HIGH TANK TEMPERATURE							X						X		
VIBRATION				X			X		X	X					
LEAKING STEAM PIPES											X			X	
FREQUENT VALVE FAILURE										X	X			X	
SCALE/CORROSION in TANK														X	
SCALE/CORROSION in COILS														X	
HIGH FINAL RINSE TEMP.							X			X					
LOW FINAL RINSE TEMP.	X	X	X	X	X	X	X	X		X	X	X			
LOW BLOWER DRYER TEMP.	X	X	X	X	X	X	X	X		X		X	X		

ILLUSTRATION 6-15 **Common problems and symptoms when dish machines malfunction.**
COURTESY OF HOBART CORPORATION, TROY, OHIO.

Strange but true: Most water available for drinking is unfit for consumption before it is treated. Illustration 6-16 shows the typical journey from source to end user, through a water treatment plant. In 1974, Congress enacted the Safe Drinking Water Act, which authorized the federal government to establish the standards and regulations for drinking water safety. The Environmental Protection Agency (EPA) now sets and implements those standards and conducts research. State governments are primarily responsible for implementing and enforcing the federal mandates. The EPA reports that about 90 percent of community water systems comply with its standards. This figure is always controversial, because some experts assert the EPA's standards are not tough enough. They claim there are as many as 1000 different potential pollutants, and the EPA rules have established enforceable limits for only 100 of them.

Water quality debates are often in the news, especially when a system is found to have higher-than-normal lead levels. But there are plenty of other contaminants and pollutants making headlines:

- Arsenic occurs naturally in some groundwater, and is also a residue of mining and industry. At low doses, it is linked to cancer and diabetes; at high doses, it is poisonous.
- Pathogens are bacteria that can cause gastrointestinal illnesses. In news reports, you've heard some of their names: *Escherichia coli (E. coli O157:H7)*, *cryptosporidium*, and *yersinia enterocolitica* are just a few. Farm waste runoff and sewage discharge can result in their accidental introduction to a water system.
- MTBE is a fuel additive designed to reduce air pollution. But when spilled or leaked from storage tanks, it can contaminate water and cause liver, digestive, and nervous system disorders.
- Perclorate is used in making fireworks, weapons, and rocket fuel. It interferes with thyroid function in humans.
- Trihalomethanes (THMs) are among the most common groundwater contaminants. They form when chlorine reacts with organic material, something as simple as leaves. THMs may contribute to miscarriage risks and bladder cancer.

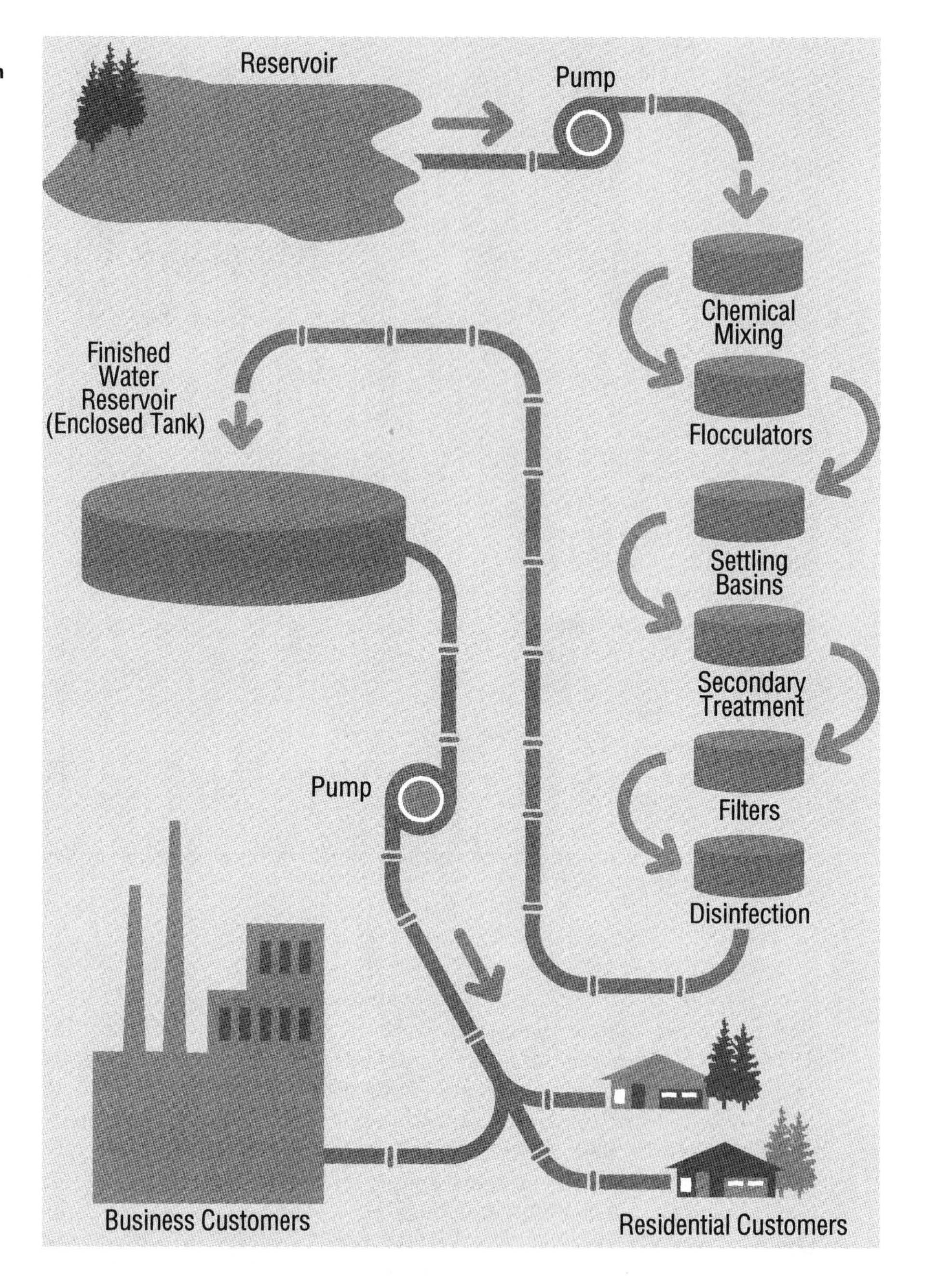

ILLUSTRATION 6-16 A flow diagram of how water gets from its source, through a treatment plant, and to homes and businesses.
COURTESY OF DALLAS WATER UTILITIES, DALLAS, TEXAS.

The water treatment process itself doesn't sound all that appetizing! Chemicals and gases—including lime, ferric sulfate, chlorine, ammonia, carbon, polymers, ozone, carbon dioxide, and fluoride—are added to drinking water to remove impurities, kill harmful viruses and bacteria, eliminate "off" tastes and odors, and even help prevent tooth decay. These substances are mixed into the water, which is then sent through huge basins called flocculators, where large paddles mix the water while the additives do their various jobs and prompt the "bad" particles to group together and sink to the bottom of the tank. After this, the water passes into a settling basin, flowing slowly for four to eight hours as the enlarged particles continue to settle to the bottom. There's a secondary treatment phase—more chemicals, more mixing, more settling—that removes most of the chemicals that were originally put in, not just the "bad" stuff. The water is then filtered through anthracite coal, sand, and gravel, a process that catches any remaining particles. And finally, it is disinfected to kill bacteria.

As a health precaution, or in areas where the local water has persistent mineral content that results in taste or odor problems, many restaurants opt to filter their own water. You can purchase different types of filters to counteract different problems: a carbon filter for odor and taste problems, an integrated UV-plus activated carbon filter to kill viruses and remove particles. There are filters for chemical absorption, turbidity reduction, and heavy-metal reduction.

Capacities of water filtration systems can range from a small, single-cartridge unit that can be attached to a single machine like a coffeemaker, to a multicartridge system capable of filtering all the water that enters a building, more than 100 gallons per minute! No matter what their size, the principle is simple: Cold water enters the filter at an inlet valve, where it is directed through the internal filter chamber. It exits the body of the filter through an outlet connection, generally at the bottom of the chamber. Changing the internal filter element is usually as easy as opening the chamber, removing the old element, and putting in a new one. The manufacturer will recommend the frequency with which filters should be changed.

The effectiveness of filtration systems probably varies as much as their manufacturers' claims, but all of them should adhere to two important NSF International Standards: #42 ("Aesthetic Effects," which governs taste, odor, chlorine content, and particular reduction) and #53 ("Health Effects," which governs turbidity, Giardia cyst content, and asbestos reduction). Properly filtered water can extend the life of your most expensive equipment, like steamers, combi ovens, dish machines, ice machines, and beverage dispensers, by eliminating scale and slime buildup. Better energy efficiency and fewer maintenance calls can translate into cost savings. And of course, using filtered water for customers—in beverages, ice making, and cooking—is also a plus.

Yet another technique has been developed for identifying hazardous particles in water. Laser beams are shot through a stream of water to check for microorganisms. They can detect anthrax, *E. coli,* or any other particle not previously identified in a particular water supply. Each type of microorganism looks different, and the lasers are precise enough to differentiate them. So far, this type of system scans for live organisms (like bacteria), but still sees chemicals only as "unidentified" particles. It can also detect and report any type of increased particle activity. The newly patented technology was used as a security device to monitor water safety at the 2003 Super Bowl football game, in case of terrorist attack.

An increasing number of people have decided they distrust their municipal water supplies, and purchase bottled water for drinking. Those who expect it to be purer than tap water may be wasting their money. Legally, bottled water does not have to be any cleaner than tap water—it's covered by the same Safe Drinking Water Act provisions—and yet it costs, on average, 625 times more!

Mineral water is exempt from the Safe Drinking Water Act, because it contains a higher mineral content than allowed by the U.S. Food and Drug Administration (FDA) regulations. "Manufactured" waters, such as club soda and seltzer, are also exempt from the act, because they are considered soft drinks. For other types of bottled water, however, the FDA now requires additional labeling on individual bottles to identify better the source of the water. Common terms include:

SPRING WATER. Collected as it flows naturally to the surface from an identified underground source, or pumped through a bore hole from the spring source

ARTESIAN WATER. Tapped from a confirmed source before it flows to the surface

WELL WATER. Tapped from a drilled or bored hole in the ground

MINERAL WATER. Collected from a protected underground source; contains appreciable levels of minerals (at least 250 ppm of total undissolved solids)

SPARKLING WATER. Contains the same level of carbon dioxide in the bottle as it does when it emerges from its source

PURIFIED (DISTILLED) WATER. Produced by distillation, deionization, or reverse osmosis

Buying and Using Water

Water is purchased much the same way we purchase electricity. A meter measures the number of gallons that enters the water system, either in cubic feet or in hundreds of gallons. The meter, which isn't equipped to record the huge numbers used by most foodservice locations, will show a basic number.

The meter reader takes that number and multiplies it by a constant figure, known as the ***constant multiplier.*** For instance, if the meter shows 1200 and the multiplier is 100, we have consumed 120,000 gallons.

When you turn on a sink in your kitchen, the water rushes out at 50 to 100 psi. This pressure is more than enough to get it from the city's pipes into the building, which only takes up to 20 psi. So the excess pressure is used to move water into numerous pipes throughout the facility. This is called the ***upfeed system*** of getting water. In fact, 50 to 100 psi is strong enough to supply water to the upstairs area of a building four to six stories high! If your facility is in a taller building than that, you will probably need water pumps to boost the pressure and flow.

Pumps can be used to increase water pressure; valves (called ***regulation valves***) can be used to decrease it. Your goal is to control the water coming into your facility to avoid fluctuating pressure or an uneven flow rate. And whenever there's a possibility that contaminated water could backflow into the potable water system, a *backflow preventor* should be installed. ***Backflow*** might occur whenever a piece of equipment, such as a commercial dishwasher, is capable of creating pressure that is greater than the incoming pressure of its supply line.

The segments of the typical upfeed system are as follows:

WATER METER. The device that records water consumption. It is the last point of the public water utility's service. Anything on "your" side of the water meter, including all pipes and maintenance, is your responsibility, not the water company's.

SERVICE PIPE. The main supply line between the meter and the building.

FIXTURE BRANCH. A pipe that carries water to a single fixture. It can be vertical or horizontal and carry hot or cold water.

RISER. A vertical pipe that extends 20 or more feet. It can carry hot or cold water.

FIXTURES. The devices (faucets, toilets, sinks) that allow the water to be used by guests and/or employees.

HOT-WATER HEATER. The tank used to heat and store hot water. (To be discussed later in this chapter.)

PIPES. The tubes that are fitted together to provide a system for water to travel through. They can be copper, brass, galvanized steel, or even plastic. Building codes determine what materials are acceptable for different uses. Copper is the most expensive type of pipe, but it's considered easy and economical to work on. Plastic pipes are only allowed for limited, special uses. The most common type of plastic pipe is made of ***polyvinyl chloride (PVC)***. It is inexpensive, corrosion resistant, and has a long life, if you're allowed to use it.

FITTINGS. The joints of the pipe system. They fit onto the ends of pipes, allowing them to make turns and to connect to each other and to other appliances or fixtures. Some of their names describe their shapes, and the most popular fittings include the bushing, cap, coupling, elbow, plug, and tee. Some fittings have threads (either internal or external) to be screwed into place; others are compression type (see Illustration 6-17).

VALVES. Valves control water flow, and are made of brass, copper, or cast iron. Use of the correct valve minimizes plumbing problems. *Gate valves* are used to vary water flow and allow water to go in either direction. *Check valves* allow water to flow only one way. They are marked with an arrow indicating the direction of flow. *Safety valves* are spring-loaded valves that are operated by temperature or water pressure, to relieve excess pressure if they sense a buildup.

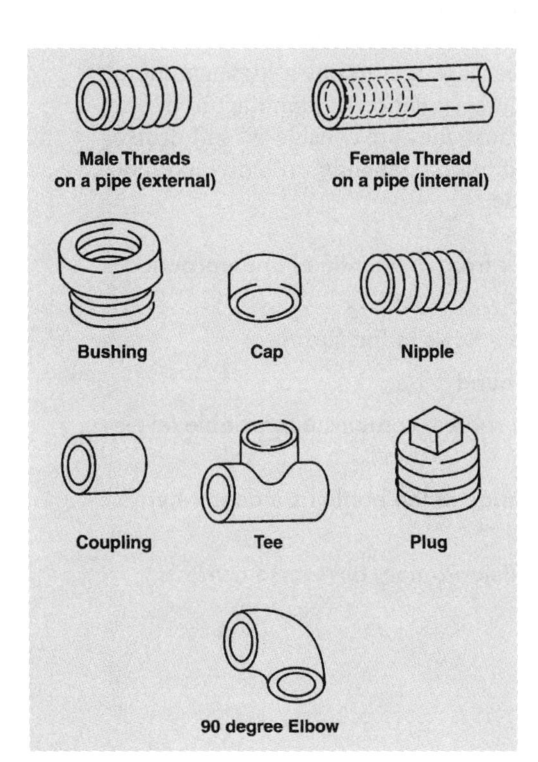

ILLUSTRATION 6-17 **Examples of the most common pipe fittings. Fittings allow pipes to connect and bend.**

Reading Water Meters and Bills

Like gas, water consumption is usually measured in cubic feet, but occasionally you will see a meter that measures in gallons. If so, that will be printed on the meter. There are two common faces on these meters: One has a simple readout in the center that indicates the number of cubic feet that have been used (see Illustration 6-18A). The hand that makes its way around the dial is used only to indicate that water is flowing through the meter.

The other type of meter has a series of small dials, which are read like gas and/or electric meter dials. The 1-foot dial is not part of the reading; it only indicates whether water is flowing through the meter. Start your reading with the 10-foot dial (see Illustration 6-18B).

There will be several different charges on your water bill. General use is billed at a flat rate for every 1000 gallons you use. However, there's also a charge for sewer services, also billed in 1000-gallon increments. As shown in the sample Dallas Water Utilities rate schedule (see Illustration 6-19), you can often get a small break on your bill by paying early. In some cities, there may also be a variety of nonmetered charges, including a fee for maintaining your town's firefighting equipment, a water line repair service fee ("leak insurance"), or a water treatment or "water quality" fee. Illustration 6-20 is a sample water bill.

You might ask your utility company if your business can install a submeter, a separate meter to track water that does not go into the sewer. Use of a submeter is not common, but it allows you to subtract the submeter count from your total gallons used and, therefore, pay less for your sewer bill. A resort hotel, for example, uses water to fill its swimming pools or a irrigate golf course—uses that do not flush water into the sewer system.

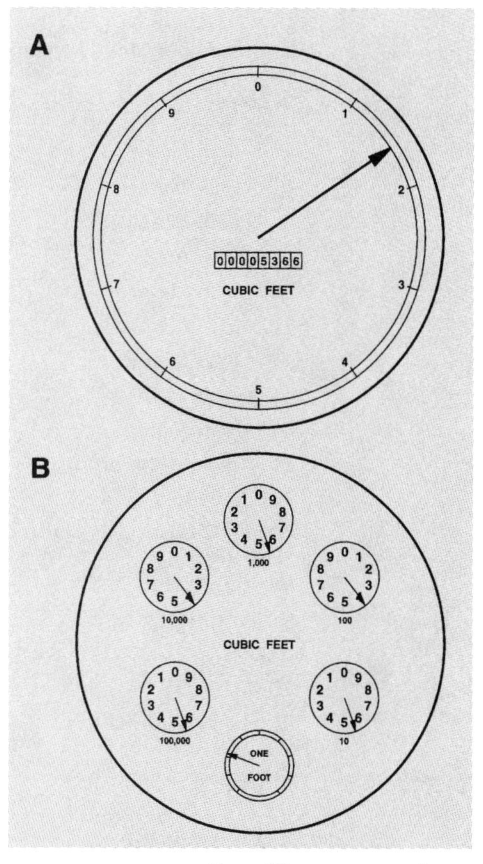

ILLUSTRATION 6-18 **Two different types of water meters:**
(a) a simple readout and (b) a dial meter.

Water Conservation

All our best efforts aside, Americans still use 35 billion gallons of water a day. Restrooms, kitchens, and landscaping are the three most water-intensive areas—most restaurants have all three! You may be surprised at how many water utility companies offer water conservation tips, often on Internet sites, so there is a mountain of information out there for the restaurateur who wants to train employees to practice conservation. Most suggestions are simply common sense; a few are truly inventive. These, for instance, were adapted from the Massachusetts Water Resources Authority in Boston:

KITCHEN AND SERVICE AREAS

- Turn off the "continuous flow" feature of drain trays on coffee/milk/soda beverage islands. Clean them thoroughly, as needed.
- Check the manufacturer's instructions to see if the dish machine spray heads can be reduced to lower-flow ones.
- Do your food thawing, and utensil presoaking, in tubs or basins of water, not running water. Better yet, allow for longer thawing times in refrigeration, so you won't have to use water to speed up the process.
- Adjust the ice machines to make and dispense less ice when less is needed.

Dallas Water Utilities Monthly Prompt Payment Rates Effective October 1, 2003

Customer Charge	Water	Sewer	Combined
5/8 Inch Meter	$3.34	$2.65	$5.99
3/4 Inch Meter	3.88	2.88	6.76
1 Inch Meter	5.58	3.62	9.20
1 1/2 Inch Meter	10.30	5.70	16.00
2 Inch Meter	15.92	8.20	24.12
3 Inch Meter	46.45	17.93	64.38
4 Inch Meter	69.21	29.04	98.25
6 Inch Meter	131.72	56.84	188.56
8 Inch Meter	215.90	94.36	310.26
10 Inch Meter or larger	322.73	143.00	465.73

Usage Charge per 1,000 gallons*		
Residential		
Up to 4,000 gallons	1.13	3.06
4,0001 to 10,000 gallons	1.87	3.06
10,001 to 15,000 gallons	2.50	3.06
Above 15,000 gallons	3.24	3.06
General Services		
Up to 10,000 gallons	1.20	1.74
Above 10,000 gallons	1.48	1.74
Above 10,000 gallons & 1.4 times annual average monthly usage	2.60	1.74
Optional General Services		
1st million gallons or less (minimum)	1,000.00	1.65
Above 1 million gallons (per 1,000 gallons)	1.22	1.65
Sewer Metered Separately		1.83
Untreated Water		
Uninterruptible	0.3868	
Interruptible	0.2413	

The above Prompt Payment Rates apply if payment is received on or before the due date shown on the bill. These represent a 5% discount from the Standard Rates.

* Sewer Charges for residential accounts are calculated on an average of the water billed in December, January, February and March (40,000 gallons maximum) or the actual month's water consumption, whichever is less. Sewer charges for general services and optional general services accounts are based on the month's water consumption unless sewer is metered separately. Industrial wastewater discharges containing concentrations of BOD and/or Suspended Solids greater than 250 milligrams per liter are assessed sewer surcharges. Certain commercial users such as restaurants, bars/lounges, small food processors and equipment service facilities are assessed standard surcharges. These surcharges are included as part of the monthly bill.

ILLUSTRATION 6–19 **A sample list of water charges, called a rate schedule. Your water utility should be able to supply these figures for your area.**

COURTESY OF DALLAS WATER UTILITIES, DALLAS, TEXAS.

CITY WATER UTILITY

1234 Sample Avenue
City, State Zip

THEN WEST, LLC
C/O XYZ RESTAURANT GROUP
1234 MCKINNEY AVENUE
DALLAS, TX 75123-4567

**TO QUALIFY FOR PROMPT PAY AMOUNT, PAYMENT MUST BE RECEIVED IN OUR OFFICE BY THE DUE DATE OF JULY 14, 2004

PLEASE RETURN, THIS PORTION WITH YOUR PAYMENT
PLEASE KEEP THIS PORTION FOR YOUR RECORDS

COMMERCIAL

Customer: THEN WEST, LLC
Service Address: 1234 MCKINNEY AVE

Account Number:
001-12345678-010

Billing Date: 06-30-04
Water Used This Month: 268,700 Gallons

Days Served: 31

SERVICE PROVIDED	METER NUMBER	READ PREVIOUS	READ 6/30/2004	USAGE IN 100 GALS	USAGE CHARGE	CUSTOMER CHARGE	TOTAL
Water	123456	89181	91868	2687	394 88	3.34	398.22
Sewer				2687	467 54	2.65	470.19
Surcharge				2687	209 69	0.00	209.69
					CURRENT CHARGES		$1,078.09

NOTES:
Water charged at $1.20 per 1000 gallons up to 10,000; $1.48 per 1,000 gallons above 10,000 gallons.
Sewer charged at $1.74 per 1000 gallons.
Industrial Wastewater charged at $0.78040 per 1000 gallons.

ILLUSTRATION 6–20 **A typical commercial water bill.**
COURTESY OF DALLAS WATER UTILITIES, DALLAS, TEXAS.

RESTROOMS

- Repair leaky toilets and faucets! One leaking toilet can waste 50 gallons of water a day; a dripping faucet can waste at least 75 gallons a week.
- Install aerators, spring-loaded valves, electronic sensors, or timers on all faucets.
- Replace worn-out fixtures with water-saving ones.
- Apply water conservation stickers on mirrors to remind both employees and customers not to waste water.

LANDSCAPING

- One inch of water per week is sufficient to sustain an established lawn or landscape. Gauge rainfall, and augment with only as much water as is needed to equal one inch per week.
- After a heavy rain, wait at least ten days to water again.

- Don't water on overly windy, rainy, or hot days, when more water evaporates than gets to your lawn!
- Investigate a drip irrigation system for flowers, shrubs, and new plantings. Drip irrigation saves 30 to 70 percent of the water used by an overhead sprinkler system.
- Sweep sidewalks, loading docks, and parking lots instead of hosing them down.

The water-guzzling capital of the U.S. is Las Vegas, with its massive (and highly landscaped) casinos, backyard pools, and green boulevards transforming what was once desert land. Las Vegas uses an estimated 325 gallons of water per person, per day. At that rate, the region is expected to run out by the mid-2030s, according to some experts. The city's hospitality industry is finally embracing conservation measures, from water-saving plumbing fixtures to lawn-watering restrictions, to new types of water purification technology. Do these individual efforts matter? Apparently so. Within one year, Las Vegas' water consumption had decreased by 13 percent.

6-4 CHOOSING PLUMBING FIXTURES

Your plumbing fixtures are among the hardest-working items in your business. Fortunately, there are many guidelines to assist you in selecting them. The Uniform Plumbing Code sets fixture requirements for the public area of your restaurant, primarily the restrooms. For the kitchen, and food preparation in general, NSF International (formerly the National Sanitation Foundation) has extensive guidelines. Here are some things to think about when selecting your fixtures and designing your restrooms.

Water Closet. Yes, that's the fancy name for "toilet." It should be made of solid, glazed porcelain, with a flush tank that discharges water when a lever or button is pushed. Another way to flush the tank is with pressure valves; however, they use more water. The toilet should have a self-closing lid.

The Uniform Plumbing Code will specify the number of toilets and urinals you must have for your restaurant; some cities require more if you serve alcoholic beverages. The general rule is two toilets for every 150 female guests and two urinals for every 150 male guests. The code requirements are listed back in Chapter 4, in Table 4-1.

Since 1994, when Congress mandated that new construction include water-saving toilets (which use 1.6 gallons per flush) instead of the traditional toilets (which use 4 to 7 gallons per flush), there have been attempts to repeal the law. Its detractors say the water-saving toilet just doesn't do the job with less than two gallons, so people flush two or three times! The law has even created a sort-of "black market" trade for toilets salvaged from older homes, or imported quietly from other countries. Luckily, most conservation methods are not so controversial.

Urinal. This companion fixture for men's restrooms should also be solid, glazed porcelain. There are stall, wall, and pedestal-style installations; the wall-mounted urinal is the best, because it makes cleaning easier beneath the urinal. The flush valve is the most common mode for flushing urinals.

Lavatory. The ***lavatory*** is also called a *hand sink.* The preferred material for this important part of every restroom is, again, glazed porcelain. The hand sink is required in most cities to supply both hot and cold water, with a common mixing faucet for temperature control. The sink should have an overflow drain. Other health code requirements include soap dispensers (not bar soap) and disposable towels for hand drying. Although heater-blowers can dry hands with warm air, they are not particularly energy efficient. For every 100 guests, you will need to provide one hand sink in each restroom. You must be certain that at least one sink is installed such that a person in a wheelchair can use it, to meet the guidelines of the Americans with Disabilities Act. We'll discuss hand sinks in the kitchens in just a moment.

Other Considerations. Generally, it is advisable to have one floor drain in each restroom stall and at least one in the urinal area of the men's room. If the restrooms are large, consider installing additional floor drains to make mopping easier, as well as to catch any potential plumbing overflows.

A working exhaust fan may be required by the local health code—even if it's not, it is a good idea, to circulate the restroom air. Install spring-loaded doors on restrooms to prevent people from leaving them open. Finally, another crucial consideration: Restrooms must meet both local and federal requirements of accessibility for physically disabled guests.

At the back of the house, the plumbing fixtures must withstand heat, grease, heavy-duty cleaning products, and all the rigors of cooking. They include sinks and drains, discharge systems, venting systems, and hot-water tanks. As a rule, the architectural drawings of your building will include plumbing, electrical, and mechanical connections: ask that the drawings be rendered in ¼-inch scale, and include a schedule of equipment to be plumbed.

Sinks and Hand-Washing Systems

Before we discuss the multiple types of sinks used in foodservice, let's talk for a moment about the particular importance of the hand sink. The U.S. Food and Drug Administration reports that 40 percent of all food-borne illness is the result of poor hand-washing practices by employees, and cross-contamination from touching the faucets or other surfaces that may not be clean. Proper hand washing is a combination of appropriate water temperature, the duration and type of scrubbing, and the use of soap. It's a matter of teaching employees the right way to do it and then, human nature being what it is, monitoring to make sure they do it right. *You cannot wash hands in the food sinks, or wash food in the hand sinks—it is absolutely against health codes!*

Food contamination is a serious enough problem that manufacturers now produce automated hand-washing systems or stations. Some can be installed using existing plumbing; others are self-contained units. For use in either restroom or kitchen, the unit is activated by motion, not touch, and leads the user through the wash (dispensing soap and warm water) and dry (with warm air) process in less than a minute. The unit can also "read" the hand-washing frequency of each individual. And as a child, you thought your mother was strict about this? Think again! Here's a typical computerized hand washing operation:

1. A "beep" in the food production area reminds you that it's time to wash your hands. It can be programmed either to sound at certain timed intervals, or after a task is completed.
2. You approach the sink and, without touching anything, water streams out at just the right temperature. Get your hands wet . . .
3. . . . and then, about seven seconds later, a built-in device dispenses soap. (Depending on the system, you may have to punch in your employee code number to get the soap.) The lathering and scrubbing is up to you, but it should take about 20 seconds.
4. At that moment, more water comes on for rinsing—again, about 20 seconds' worth.
5. Dry your hands using the hot-air dryer.
6. Some units provide an optional antibacterial spray for sanitizing.
7. The computer software "records" the fact that you accomplished a proper wash! In some systems your employee code number even qualifies you for a small gift for frequent washes.

The most sophisticated systems, most often used in hospitals, allow the hand-washer to insert the hands, past their wrists, into two separate cylinders. The machine provides a low-volume (but high-pressure) spray of water and sanitizing solution, from 12 to 20 seconds in duration. It requires electric power as well as standard plumbing connections.

Even if you have "just plain" hand sinks, most health departments have rules about how many, and where they must be located. (These requirements usually apply to bar areas as well.) At this writing, the norms are:

- A hand sink should be within 15 feet (in a straight line) of any food prep area.
- One hand sink is required for every five employees, or every 300 square feet of facility space.
- One hand sink is required for every prep and cooking area.

In addition to hand sinks, you will need several other types of sinks in your kitchen. In the dish room, there is the pot sink (for washing pots and pans), the warewashing or *scullery sink,* and the three-compartment dish sink. The three compartments are for washing, rinsing, and sanitizing. Elsewhere, there's the prep sink (for scrubbing and peeling vegetables), the utility sink (for mops and cleaning), and the bar sink (for the bar area).

Sinks should always be made of stainless steel, which is durable and easy to clean. Manufacturers use two types of stainless steel, known as Type 430 and Type 304. Both are approved for foodservice use, but Type 304 is considered more durable because of its content: 8 percent nickel, in addition to the standard 16 percent chromium. Rounded corners (called *coved* corners) make sinks easier to clean. You can also clean more thoroughly under the sinks if they are installed so that the water faucets come straight out from the walls (and the water pipes are located behind the walls instead of beneath the sinks themselves).

Other requirements are a swiveling, gooseneck-style faucet that can reach each compartment of the sink; an overflow drain for each compartment of the sink; and ample supplies of both hot and cold water. You can choose from many faucet types but aerators and stream regulators will save the most water.

For pot sinks, add a drain board to the list of requirements. If the drain board is more than 36 inches long, it will need its own, separate support legs. NSF International now requires that the drain boards be welded to the sink bowls. Here are some basic sink installation guidelines:

POT SINKS. The height of the sink edge should not exceed 38 inches. The sink itself should not be more than 15 inches deep (most are 12 to 14 inches deep) on legs or a pedestal no more than 24 inches tall. The depth of the sink, from front to back, should not exceed 28 inches (see Illustration 6-21).

WAREWASHING (SCULLERY) SINKS. Your local health codes will dictate the number of compartments or bowls these sink units must have; their backsplash height, water depth, drain board size, and so on.

DISH SINKS. These are used mostly in small, limited-menu operations. They are three-compartment sinks, with a minimum bowl size (for each compartment) of 16 by 20 inches, with a water-level depth of 14 inches. A dish sink also usually requires a double drain board—on each side of the far left and far right sink bowls. Illustration 6-22 shows a handy three-compartment design made to fit into a corner.

COMBINATION POT-DISH SINKS. Also in small restaurants, a three-compartment unit can be installed with slightly larger sinks to do double-duty for washing both pots and dishes. The minimum bowl size here would be 20 by 20 inches, with a water depth of at least 14 inches. You can also order these with 24 by 24-inch bowls and, if you'll be washing a lot of full-sized baking sheets, you'll want 24 by 28-inch bowls.

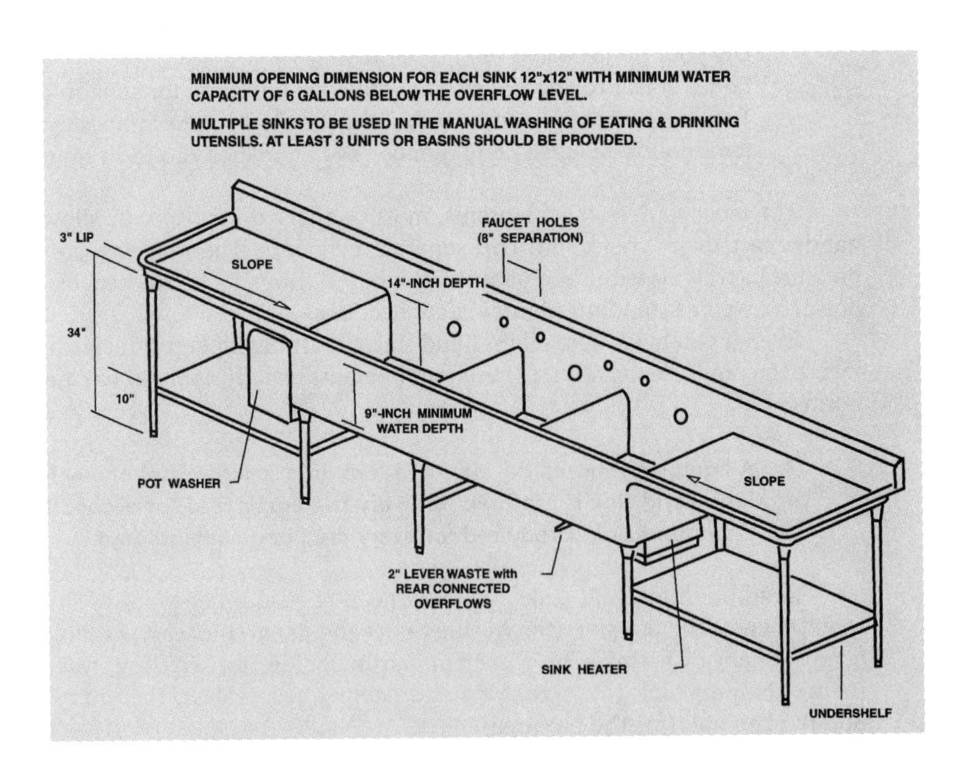

ILLUSTRATION 6-21 **The pot sink is organized to permit the addition of faucets, shelves, water heaters, and a pot washer.**
SOURCE: ADAPTED FROM CARL R. SCRIVEN AND JAMES W. STEVENS, *MANUAL OF EQUIPMENT AND DESIGN FOR THE FOODSERVICE INDUSTRY,* THOMSON LEARNING, 1989.

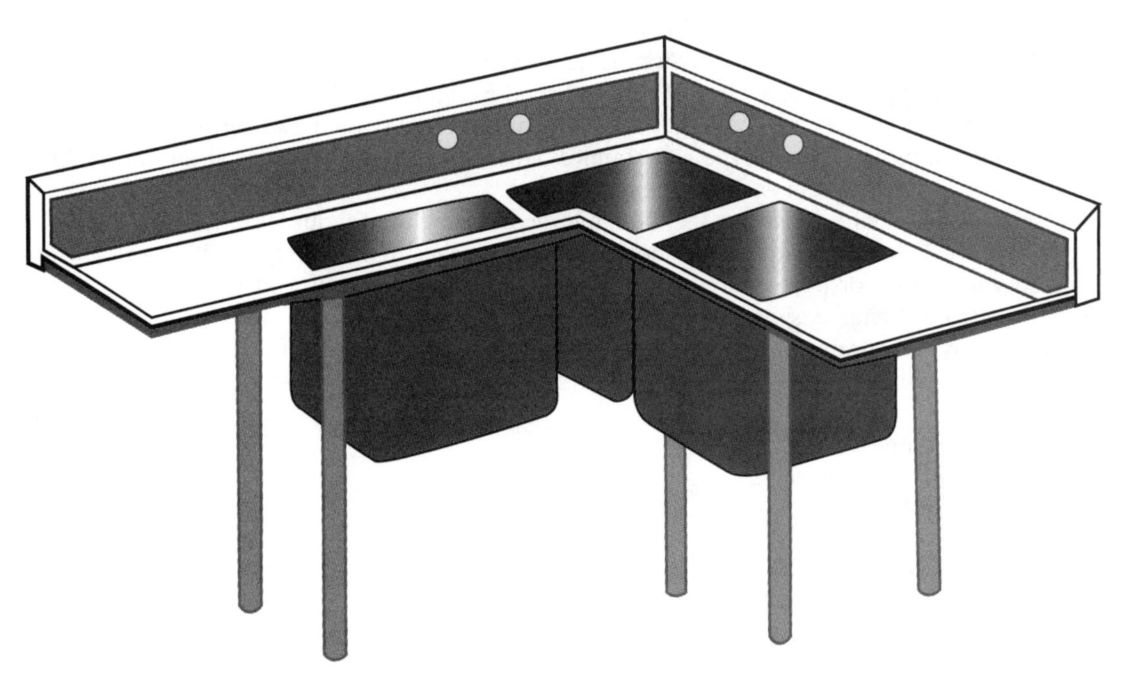

ILLUSTRATION 6–22 **For small spaces, there are three-compartment sink designs that fit into corners.**

HAND SINKS. The hand sink most often required by city ordinance is either wall mounted or pedestal style. Again, the wall-mounted sink is easier to clean beneath. The typical hand sink is 20 inches long by 16 inches wide, with a depth of 8 inches, but smaller sizes are also acceptable. In choosing one, remember that it will get a lot of use.

BAR SINKS. This is a four-compartment sink, either 8 or 10 feet in length. A minimum of 24 inches of drain board space is recommended on both sides of the bar sink. The bar sink is generally 18 to 24 inches wide and 1 foot deep. Most have a special overflow drain; make sure this drain is at least 1 inch in diameter.

PREP SINKS. This is usually a two-compartment sink unit, although some health codes mandate a third bowl. A heavy-duty garbage disposal may be installed in one of the compartments. Size will vary depending on the amount of prep work done in the kitchen; the most common size is a 20-inch square sink, with a 10-inch depth.

UTILITY SINKS. This is typically the big, deep, rectangular sink in the back of the kitchen that always looks so beat up and untidy! At least it is useful, if not attractive. A wall-mounted sink will allow storage (of buckets and the like) beneath it.

At least one manufacturer has introduced "mobile" sinks—moveable, on casters, with quick-disconnect water lines so the sink unit can be completely relocated (temporarily) so you can clean more thoroughly behind it. This is a most difficult area to reach, and over time can harbor lots of bacteria and other hazardous gunk. No tools or special skills are required to disconnect the lines.

Drains and the Discharge System

Your myriad sinks are drained into the drainage or ***discharge system,*** which receives the liquid discharge created by the food and beverage preparation area. The first component of the discharge system is on the sink itself: the ***trap.*** It is a curved section of pipe, where the lowest part of the pipe "traps" (or retains) some water. The trap is called a *P trap* when the drainpipes go into the wall, an *S trap* when the drainpipes go into the floor.

In addition to these traps, it's a good idea to have floor drains located directly beneath your larger sinks. The drains in a commercial kitchen must have a ***dome strainer*** (or *sediment bucket*) much like a perforated sink stopper that traps bits of dirt and food as liquids go down the drain. For the heaviest-duty jobs, a floor drain with a much larger strainer compartment (called a ***sump***) is recommended. The sump is at least 8 inches square. Type 304 stainless steel is the preferred material for drain fabrication, and coved corners make them easier to clean.

Drains should not be flush with the floor, but recessed slightly (about 1⁄16 of an inch) to prompt water to flow toward them. The drainpipe should be 3 to 4 inches in diameter, and its interior walls must be coated with acrylic or porcelain enamel that is both nonporous and acid resistant. A nonslip floor mat, with slats for drainage, should be a standard accessory beneath every sink.

How many floor drains should you have in your kitchen? Let's count the areas in which drains are a must to catch spills, overflow, and dirty water from floor cleaning:

- Hot line area
- Prep and pantry area
- By the pot sinks
- Dishwashing area
- Dry storage area
- Outside the walk-in refrigerator
- Wait stations/service areas
- Near steam equipment
- By the bar sinks
- Under the ice maker

The ice maker has another unique drainage requirement: a recessed floor.

One smart idea is to install several drains, in a trench that is about 2 feet wide and several feet long, covered with a rustproof metal grate. This is very effective along the length of the hot line area or in the constantly wet dish room.

When we talk about draining away waste, we're not just discussing water. The water often contains grease, and grease disposal is an enormous (and messy) problem in foodservice. A *grease interceptor,* more commonly known as the ***grease trap,*** is required by law in most towns and cities. Your area's building code will list which kitchen fixtures must be plumbed to the grease trap—typically, the water/waste output of the garbage disposal, dishwasher, and all sinks and floor drains must pass through the grease trap before it enters the sewer. Employee restrooms and on-premise laundry appliances generally do not have to be connected to the grease trap.

The role of the grease trap is to prevent grease from leaving the restaurant's drainage system and clogging the city sewer system. Foodservice wastewater is a big problem for sewers designed primarily for residential waste. Thus, fines and surcharges may be imposed on restaurants if their effluent (outflow) exceeds the local standards for its percentage of fats, oil, and grease ("FOG," in industry jargon).

As waste enters the grease trap, it separates into three layers: The heaviest particles of food and dirt sink to the bottom; the middle layer is mostly water, with a little bit of suspended solids and grease in it; and the top layer is grease and oil. The trap holds the top and bottom layers, while allowing the middle layer to flow away into the sewage system (see Illustration 6-23).

Grease traps come in different sizes, and you should choose one based on the gallons of water that can run through it per minute, the number of appliances connected to it, and its capacity to retain grease. Table 6-2 is a sample grease trap size chart from the Uniform Plumbing Code.

Cleaning the grease trap regularly is necessary because the bottom layer can clog pipes if allowed to build up, and the top layer can mix with, and pollute, the middle layer too much. Most restaurants hire a grease trap service company to handle this unpleasant task. It is a costly activity, and not without legal ramifications. The service company must be licensed to haul the grease waste to specially approved treatment areas. It's not enough anymore for a restaurateur to trust that the grease is being taken care of. The smart ones take a proactive approach. Once in a while, you'll see news reports about grease trap service companies that skirt the law by dumping waste into creeks or unapproved areas. You would be wise to thoroughly research your area's grease removal requirements and to interview several service companies. Ask for, and contact, their references.

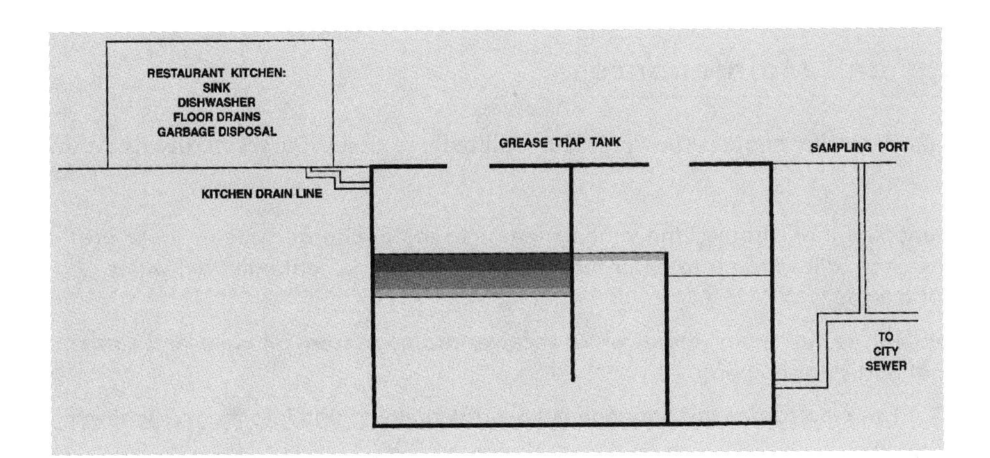

ILLUSTRATION 6-23 Diagram of a grease trap. The trap must be cleaned thoroughly, not just skimmed, to keep working properly.

TABLE 6-2

GREASE TRAP REQUIREMENTS

Total Number of Fixtures Connected	Required Rate of Flow per Minute (gallons)	Grease Retention Capacity (pounds)	Required Rate of Flow per Minute (liters)	Grease Retention Capacity (kilograms)
1	20	40	76	18.2
2	25	50	95	22.7
3	35	70	132	31.8
4	50	100	189	45.4

Source: 2003 Uniform Plumbing Code, International Association of Plumbing and Mechanical Officials (IAPMO), Ontario, California.

There are two types of grease-trap cleaning: "skimming" (removing the top layer), versus a full pump-out of the tank. For most foodservice operations, skimming is not sufficient. The heavy, lower layer of particles must also be filtered away. You might decide on a combination of services—frequent skimming, with a full pump-out at regular intervals. The types of foods you serve and your volume of business should be your guidelines, along with a scientific measurement of the effluent to see how much "FOG" and/or chemicals it contains. In some cities, the penalties are so strict that restaurateurs include a pretreatment step, adding fat-dissolving chemicals or filtering the waste before it even gets to the grease trap. Undercounter units operate using electricity to recover grease for discarding as trash, not sewage.

Outside installation of the grease interceptor is recommended, at a level that is several feet below the kitchen to use gravity in your favor in grease elimination. Building inspectors will seldom allow a grease trap to be located anywhere inside the building, but if it happens to be inside, it should be flush with the kitchen floor. Early in the building process, a call to your local plumbing inspector will provide the particulars for your city, and probably save you a lot of trouble.

We must also discuss the "dry" part of the discharge system, which is known as the ***venting system.*** Its main purpose is to prevent siphoning of water from the traps. Vents (called "black vents") on both sides of the grease trap equalize the air pressure throughout the drainage system, circulating enough air to reduce pipe corrosion and help remove odors. Vent pipes extend up and through the roof, for kitchens and restrooms.

Drainage Terminology and Maintenance

Drainage and vent pipes have specific names. Knowing them will make it easier for you to discuss your discharge system:

BLACK VENT. A vent pipe that connects the venting system to the discharge system. These are found near the grease trap, allowing air to enter the trap and preventing contaminated water from flowing out of the trap.

BUILDING DRAIN. This is the "main" drain, which receives drainage from all pipes and carries it to the sewer for that particular building.

BUILDING SEWER. This pipe carries the drainage beyond the building and into the public sewer.

DRAINPIPE. Any pipe that carries away discharge from plumbing fixtures. Usually, drainpipes are horizontal; if they're vertical, they are called *stacks.* Sometimes, drainpipes are referred to as *soil pipes.*

VENT. A pipe that allows airflow to and from the drainage pipes.

The Uniform Plumbing Code contains specifics about choosing sizes of pipes and vents. The general rule is that smaller-sized pipes flow into larger pipes—never the reverse.

The flooded kitchen floor is every restaurant manager's nightmare. Typically, the floor floods because the drains are clogged; and they're clogged because everybody thought it was somebody else's job to take care of.

Drainage systems require periodic maintenance to keep them open and working efficiently. Simply because they depend on gravity to work, they occasionally become clogged when debris blocks the natural flow of the system. You can use flexible metal rods, called ***augurs*** or *snakes,* inserted into pipes to break up the debris; or you can pour in chemicals, which are formulated to dissolve grease and soap buildup. Either way, the clogged material should be removed and not flushed back into the system. There are all types of augurs, including heavy-duty ones with gas-operated motors and 300 feet of line.

Drainage Problems

The two terms most frequently heard when there's a water backup in the kitchen are ***backflow*** and *back siphonage.* Backflow results when dirty water (or other unsafe materials) flows into the drinking water supply. Back siphonage occurs when negative pressure builds up and sucks contaminated water into the freshwater supply. Either way, you're in trouble.

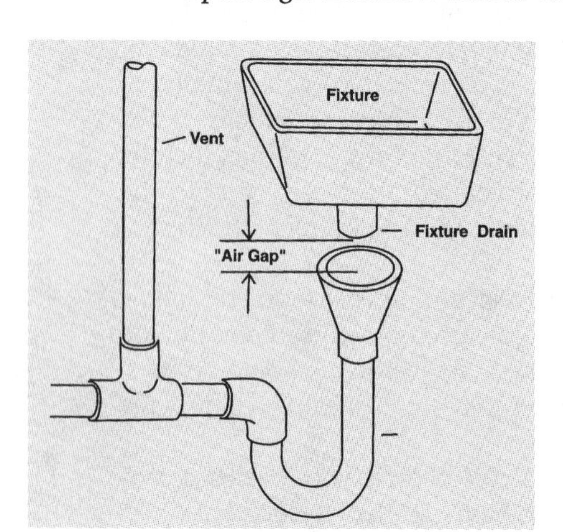

ILLUSTRATION 6-24 Proper sink installation will prevent backflow.

SOURCE: FRANK D. BORSENIK, *MAINTENANCE AND ENGINEERING FOR LODGING AND FOODSERVICE FACILITIES,* THE EDUCATIONAL INSTITUTE OF AH&MA, LANSING, MICHIGAN.

Avoiding backflow is not complicated if you've hired an experienced plumber who will plumb your system to include a space (never less than 1 inch) between each pipe and its drain, which prevents contaminated water from flowing back into the water supply (see Illustration 6-24). This space is often referred to as an ***indirect waste.*** Floor drains that receive condensate from refrigerators are also required to have an indirect waste.

An ***atmospheric vacuum breaker*** is another smart addition to your water line, especially on any hoses you use in the kitchen to clean the floor or flush out drains. The vacuum breaker (see Illustration 6-25) is a small shutoff valve that allows the water to drain completely after the faucet is turned off and minimizes the chances that fresh water can be contaminated by whatever the hose has touched.

Back siphonage is the term for reversing the normal flow of your water system because of a vacuum (or partial vacuum) in the pipes. This sometimes happens after firefighters use a hydrant in your area or the water system has been shut off temporarily for repairs. Like sipping a soft drink through a straw, when water comes back on after being shut off, it may create pressure to move it in the opposite direction than is desired. The vacuum breaker is, again, the most popular preventive measure.

Restaurants in areas with winter temperature extremes must also deal with frozen pipes, which create their own set of challenges for your staff. Heed this advice from the Restaurant and Hospitality Association of Indiana.

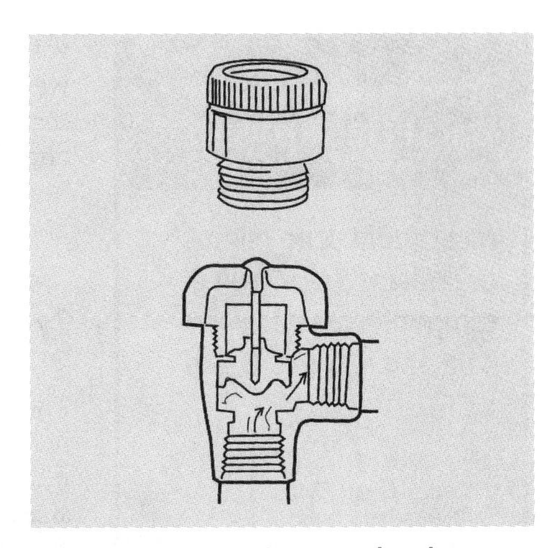

ILLUSTRATION 6-25 **A vacuum breaker works to drain excess water from a pipe or line and prevent back siphonage.**

6-5 HOT-WATER HEATING

Most hot-water heaters aren't given much attention—until there's no hot water. Because you'll spend more money heating water than you spend on the water itself, you probably should know a few basics to manage this valuable resource.

The average restaurant guest prompts the use of 5 gallons of hot water. This figure decreases in the fast-food arena, of course, where disposable utensils and plasticware are the norm. In a table-service restaurant, however, the 5-gallon figure includes water to wash, dry, and sanitize dishes, glassware, and utensils, plus pots and pans and serving pieces.

Water must, by health ordinance, reach certain temperatures for certain foodservice needs. They are listed in Table 6-3. These correspond to the "thermometer" chart in Illustration 6-26, which shows minimum temperatures for safe handling and storage of food. You'll learn plenty more about food safety in Chapter 8.

In a small restaurant, there may be a single hot-water tank with a temperature of 140 degrees Fahrenheit. Near the dishwasher, a second "booster" heater will be installed to raise the final rinse water to the required 180 degrees Fahrenheit. In larger operations, you may install two or more separate water heating systems for different needs.

To determine what size water heater you will need, take the 5-gallons-per-guest average and multiply it by the maximum number of guests you would serve at a peak mealtime. For example, if 200 guests are likely to be served, multiply by 5 gallons and you'll need 1000 gallons of hot water per hour. The other figure you will need to determine is the maximum amount the water temperature will have to rise to become fully heated. For instance, in the winter, the water may be as cold as 35 degrees Fahrenheit when it enters the building. Your water heater must work hard to get it to 140 degrees Fahrenheit. The "temperature rise" in this case is 140 minus 35, or 105 degrees Fahrenheit. Manufacturers' charts will tell you how much gas or electricity your water heater will consume for different temperature rises.

Even if your water heater is large enough and its output is hot enough, there is one more variable that impacts the availability of sufficient hot water—the way it is piped. If the pipes are too small, the

TABLE 6–3

REQUIRED WATER TEMPERATURES FOR FOODSERVICE OPERATIONS

Restrooms	110–120 degrees Fahrenheit
Pot sinks	120–140 degrees Fahrenheit
Dishwashers	140–160 degrees Fahrenheit
Final dish rinse	180 degrees Fahrenheit

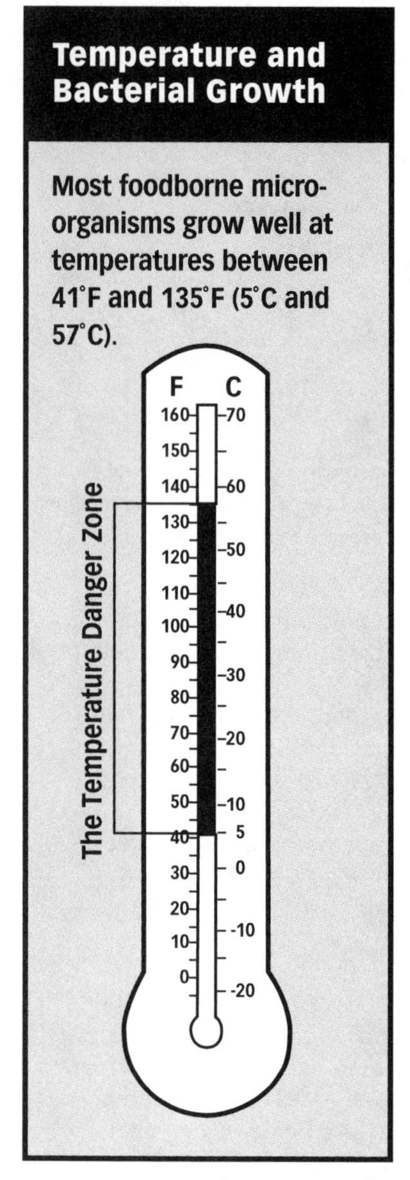

ILLUSTRATION 6-26 Temperatures for food safety.

SOURCE: NATIONAL RESTAURANT ASSOCIATION PUBLIC HEALTH AND SAFETY COMMITTEE, WASHINGTON, D.C.

water heater doesn't empty and refill fast enough. Also, heat is lost along the way when the water must travel long distances to reach appliances or faucets. Insulate the pipes against this heat loss. And, if the distance cannot be shortened between the source and appliances, you may need to install recirculation lines and a pump.

Types of Water Heaters

The most common type of water heater is the *self-contained storage heater.* It heats and holds water up to 180 degrees Fahrenheit, delivers on demand, and requires no external storage tank. This type of heater comes in a variety of sizes, ranging from 5 to 100 gallons. At a 100-degree temperature rise, the self-contained storage heater can heat 500 gallons per hour. A closely related type of water heater is the *automatic instantaneous heater,* designed to heat water immediately as it is drawn through the tank. Again, there is no external storage tank needed. Finally, the *circulating tank water heater* heats the water, then passes it immediately to a separate storage tank, using gravity or a pump.

There are some exciting technological developments in the field of water heating that may help restaurants save money. Some companies are experimenting with recovering ***waste heat.*** This means reusing heat given off by air conditioners or kitchen appliances (such as the big walk-in refrigerators) that is normally wasted, by capturing it with a heat pump and using it to heat water.

The ***heat pump water heater (HPWH)*** should be located wherever such waste heat is available. An air-conditioning system gives off as much as 16,000 Btus per hour of waste heat, so it's easy to recover the 3000 to 5000 Btus necessary to heat water. In hot-weather climates when the air conditioners run all day, it is not uncommon to be able to heat all the water you need at no cost. Because heat pumps can also capture heat from the outside air, they work best in locations where the temperature is more than 50 degrees Fahrenheit year round. Illustration 6-27 shows how the HPWH works. A survey of eight U.S. quick-service restaurant chains found their HPWHs saved from $851 to almost $4000 per year in water-heating costs, and the systems paid for themselves in four to twenty months' time. A side benefit of the system that siphons off waste heat is it makes places like kitchens and laundry rooms more comfortable and easier to cool.

The use of solar energy to heat water is also being introduced in foodservice. So far, it has been expensive to install, but the long-term savings potential should be considered. Generally, the solar water heating system is used only to preheat water; an electric or gas-powered water heater is still necessary to

ILLUSTRATION 6-27 A diagram of a heat pump water heater. Notice that its parts and their functions are not unlike refrigeration systems.

SOURCE: THE ELECTRIFICATION COUNCIL, WASHINGTON, DC.

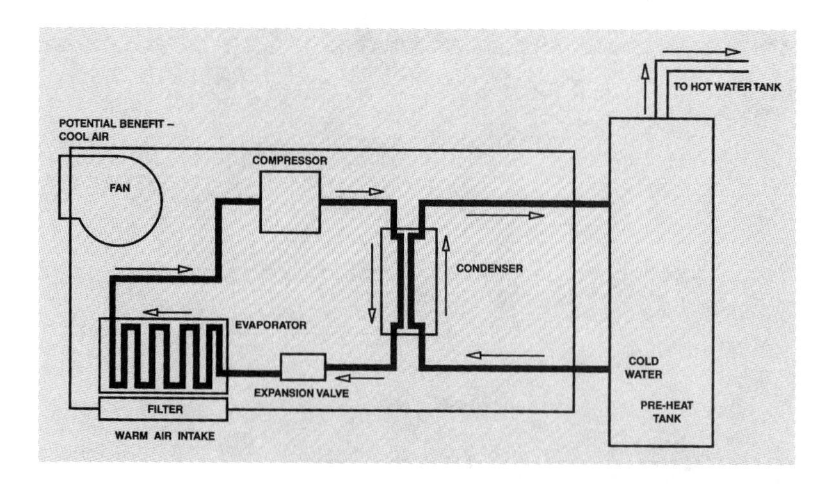

BUILDING AND GROUNDS

FROZEN PIPES? YIPES!

In the heart of winter most facilities need extra maintenance—wiping up wet floors and shoveling snow at entrances, for example. Don't let frozen pipes become another headache. Follow these prevention tips from maintenance experts.

1. In autumn, shut off valves that supply outdoor water pipes. But because some water might be trapped in the pipes, leave outdoor faucets on to prevent pipes from bursting.
2. Set the thermostat at a minimum of 65 degreees Fahrenheit. Higher is fine, but if your establishment is scheduled to close for an extended period, don't turn the heat down too far.
3. Periodically check the water flow in faucets. When the temperature drops to 0 degreees Fahrenheit or below, open wide both the hot- and cold-water taps on all faucets for 30 seconds. Then close them to a point where they drip slightly. This flow should reduce the threat of freezing in the line.
4. Heat hidden pipes. During a cold snap, open cabinet doors that house hidden bathroom and kitchen pipes to let heat in.
5. Raise the temperature to 75 degreees Fahrenheit if pipes freeze. Then open both the hot and cold taps so they can drip slightly and allow water a place to go when the pipe begins to thaw.
6. Take care when heating a frozen pipe. You can be electrocuted if you try to use a hair dryer on a pipe near standing water. Never use a butane torch or open flame on a pipe, either.

Source: *Newsletter of the Restaurant and Hospitality Association of Indiana*, Indianapolis.

bring the water up to acceptably high temperatures. And the amount of solar energy available depends mostly on the amount of sunlight your location receives.

SUMMARY

Gas energy has many uses in foodservice, from heating buildings to powering ranges to drying dishes. This chapter explains how a gas burner works and describes the potential problems if it is not kept clean and properly adjusted. There are several different types of burners, depending on cooking needs.

You also learned about the working parts and significance of the pilot light on gas appliances. There are manual or automatic pilot lights, and most can be adjusted easily by hand. Pilot lights also must be kept clean and properly adjusted.

Although many people assume it is necessary to turn gas equipment on early to let it "heat up," this is a waste of time and fuel. It is also wasteful to cook with large flames. In fact, the gas flame tips on a range burner should barely touch the bottom of your cookware and should not lap up over the sides.

Steam energy is water vapor that carries a large quantity of heat and can also be used to cook food, heat water, and more. The hotter the steam is, the higher its pressure. The size and capability of the boiler, and the sizes and lengths of pipes through which the steam must pass, all affect the output of the steam system. Water quality is another component in clean, efficient steam output.

Water quality is also critical for drinking, cooking, and dish washing. If there's something wrong with the taste or appearance of your water, try getting help first from the local water utility, which will be less expensive than calling in a private consultant. The watchword in this century is "conservation," both of water and of electricity, and there are lots of simple ways to do this that really pay off.

The National Uniform Plumbing Code contains guidelines for equipping restrooms, and specifies the proper sizes for fixtures, pipes, and vents. Hand-to-food contamination is responsible for nearly 40 percent of all food-borne illnesses, so providing enough of the correct types of hand-washing facilities is critical. There are now automated hand-washing systems to track employees' progress in this task.

In addition to several types of sinks for different purposes, no kitchen is complete without a grease trap, to prevent sewer blockage by intercepting grease and solids before they enter the sewer system. The grease trap must be cleaned regularly by a reliable company that will dispose of the grease correctly, and it is not enough to rely on them to do so. A business can be fined for putting out "too much" solid waste, and your locality may have strict enough laws that, to avoid the fines, you are required to pretreat restaurant waste before it even enters the grease trap.

STUDY QUESTIONS

1. Briefly describe how a burner on a gas range works. What two components mix to make the fire, and how do you adjust each of these components?
2. What is flashback, and how do you prevent it?
3. List three ways to save energy when cooking with natural gas.
4. How do you decide whether to use steam-cooking equipment in a commercial kitchen?
5. How do you compensate for hard water in a restaurant?
6. Why must you determine the length and diameter of pipe to use when installing steam equipment?
7. What's the difference between a P trap and an S trap?
8. List three features to consider when choosing sinks for your kitchen, and explain why they are important.
9. What is a grease trap, and how do you empty one?
10. How do you unclog a kitchen floor drain? Why might it become clogged in the first place?
11. When discussing water pressure, what is an upfeed system?
12. How do you determine the size of water heater you will need for your foodservice facility?

CHAPTER 7

DESIGN AND ENVIRONMENT

INTRODUCTION AND LEARNING OBJECTIVES

We've discussed many of the practical aspects of designing useful space for a foodservice facility. ***Environment*** is another one that cannot be ignored. For the guest, environment refers to the things that make him or her feel welcome, comfortable, and secure. For the employee, environment means a safe, comfortable, and productive work setting.

We've already made the comparison between fine dining and other art forms. It's just as much a "show," in many respects, as a movie or play, and many of the same elements used by a film or theater director can be adapted to make a foodservice business attractive and functional. A few of them involve safety as well as atmosphere. We'll cover most of the employee-related safety issues in Chapter 8. In this chapter, we will discuss:

- Lighting
- Color
- Noise levels
- Temperature and humidity
- Heating and air conditioning systems
- Ventilation and indoor air quality

7-1 LIGHTING

Lighting is the single most important environmental consideration in foodservice, but lighting rules are probably the most difficult to define. Correct lighting enhances the mood of a dining area, the appeal of the food, and the efficiency of a kitchen. And yet, in each of these situations, "correct" means an entirely different thing.

First, consider that no one actually sees light. Instead, what we see as "light" is a reflection off an object. Its intensity can be measured (in foot-candles, which you'll learn more about in a moment)—and yet the human eye cannot see a "foot-candle." What the eye can sense are differences, in brightness, color, and other characteristics. Try striking two matches, one in darkness and one in daylight. Their appearance is very different, and it's not because the two matches lit differently! It's because our perception of them differs with the surrounding conditions.

Seen in this light (so to speak), it's a complex sensory challenge to illuminate a dining area. Increasingly, lighting is seen as an integral part of creating ambience, with artsy fixtures and new technology that allows smaller fixtures to do "bigger" lighting jobs. A lighting design consultant should be on your planning team if the budget allows. If not, be sure to choose an interior designer with extensive experience in lighting. This person will be determining not only the placement and appearance of light fixtures, but their intensity, their direction, and the contrast of light levels in different parts of the dining area.

The lighting environment should match the type of facility. A fast-food restaurant, for instance, is brightly lit to help move guests through the ordering and pickup process and to discourage lingering (sometimes given the unflattering term "camping" by restaurant folk). However, the intimate eatery will require subdued light levels, to encourage a leisurely or romantic dining experience. An establishment that features fine artwork on the walls will want to play up its collection by spotlighting it. In most dining situations, experts try to think of each table as a single space and design the lighting accordingly. This results in more sources of light, but at lower levels.

Restaurants that are multipurpose—serving business lunches during the day and intimate dinners in the evening—should install dimmers on the lighting system to enable a change of light levels to reflect different moods. Dimmers also come in handy after closing time, because they allow you to crank up the lights for thorough cleaning. When dining areas also serve as meeting rooms, there are special needs: spotlight capability for speakers at podiums; perhaps track lighting with multiple circuits so the lights can be moved as needed for panel discussions, note taking, use of chalkboards, and so on. All fixtures that could interfere with the use of a projection screen (for slides or videos) should be dimmer controlled.

Restaurant Daniel, in New York City, is a case study of how lighting design can be both beautiful and useful. The dining room light sources here are, for the most part, dramatic bronze chandeliers with alabaster cups; but there are also wall sconces, hanging sconces, track lights, and so-called "downlights" directed at specific objects—a painting or mural, a stained-glass piece or flower arrangement (see Illustration 7-1).

Restaurant Daniel has divided its lighting grid into 16 different zones, controlled from a single panel. In the main dining room, there are four settings:

- "Full power," for cleaning and resetting tables when guests are nowhere in sight
- A lunch and daytime setting, when artwork and floral designs are lit
- An evening setting, when light levels are lowered but artwork and flower arrangements are highlighted more dramatically
- An intimate setting, with all light levels at their lowest

Even within a dining room, light levels will vary. An entryway, waiting room, cocktail lounge, and dining room each require separate treatment. This may require a ***lighting transition zone,*** the technical term for shifting people comfortably between two different types of lighting, giving their eyesight a moment to adjust to the change. A lighting transition zone is necessary in a lobby area, for example, when guests come from bright sunlight into a darker dining room.

Another common lighting design term is ***sparkle,*** which refers to the pleasant glittering effect, not unlike a lit candle, of a lighting fixture. Sparkle is a must in creating a leisurely or elegant dining atmosphere.

In lighting your dining area, think about making people feel good. Don't throw spotlights onto guests when you could surround them with softer lights. One caution that we add from experience: If yours is a truly dark dining room, be prepared for the occasional complaint that it is "too dark to read the menu," usually from older diners. They are not just being ornery—the average 60-year-old receives only one-third as much light into his or her eye as the typical 16-year-old. Equip the hostess or wait staff with a few pocket-sized flashlights, and they'll be able to offer immediate "enlightenment," so to speak, in these cases.

Light sources, and the level of light intensity, can help both the guest and the food to look good. Although every design expert has a theory about how to accomplish this, all agree that the "look" of incandescent light is the best way to achieve this ideal. If the building is being constructed, the architect can locate windows to avoid the glare of natural light during daytime serving hours or to take advantage of daylight if it benefits your overall look and concept.

If it is an existing structure, be sure to consider the effects of daylight on light and heat levels, and also any light that may seep in from nearby hallways, parking lots, or exterior street lights. Remember that many problems can be remedied by adding window blinds, shades, curtains, or awnings and that these, too, should be carefully chosen to blend with the mood of the facility and for ease of cleaning.

There are several different ways to use light. The most common are indirect and direct.

Indirect lighting washes a space with light instead of aiming the light at a specific spot. Indirect lighting minimizes shadows and is considered flattering in most cases. Wall sconces are one example of indirect lighting. Often, the light fixture itself is concealed.

Direct lighting aims a certain light at a certain place, to accent an area such as a tabletop. When that area is directly beneath the light fixture, it is referred to as ***downlighting.*** A chandelier is an example of direct lighting.

ILLUSTRATION 7–1 **The Lutron company was invited by Chef Daniel Boulud to consult on the Restaurant Daniel's lighting. Glass artist Dan Dailey created hand-blown wall sconces that he custom-designed for Daniel.**

COURTESY OF THE RESTAURANT DANIEL, NEW YORK, NEW YORK.

Measuring Light

Now we'll introduce you to some of the terms you will use when making lighting decisions.

Lumen is a measurement of light output; it is short for *luminous flux.* One lumen is the amount of light generated when 1 foot-candle of light shines from a single, uniform source.

Illumination (also sometimes called *illuminance*) is the effect achieved as light strikes a surface. We measure illumination in ***foot-candles***; 1 foot-candle is the light level of 1 lumen on 1 square foot of space. In the metric system, foot-candles are measured in terms of *lux,* or 1 lumen per square meter. A lux is about 1⁄10th of a foot-candle.

All light sources except natural light are classified in terms of their *efficiency.* Efficiency is the percentage of light that leaves the fixture instead of being absorbed by it, and efficiency is measured in *lumens per watt.* You may also hear the term ***efficacy,*** which refers to how well a lighting system converts electricity into light. Efficacy is a ratio of light output to power output, and is also measured in lumens per watt.

The ***color rendering index (CRI)*** is yet another measurement of light. The CRI is a numerical scale of 0 to 100, which indicates the effect of a light source on the color appearance of objects. It's not as confusing as it sounds! The index measures the "naturalness" of artificial light compared to actual sunlight. A higher CRI (from 75 to 100) means a better color rendering. Simply put, things look more true to their natural color when the light on them is strong enough to get a good look at them. So a CRI of 75 to 100 is considered excellent, while 60 to 75 is good, and below 50 is poor. Incandescent and halogen lamps (bulbs) have the highest CRI ratings; clear mercury lights usually have the lowest CRI ratings.

You can't discuss about lighting without deciding on the color of the light. Most of us are familiar with "warm" tones or "cool" tones, although we don't think much about why different types of lighting appear to be different temperatures. Interestingly, you may assume warm, reddish tones are the "hottest" colors on the spectrum, but scientists rank ***color temperatures*** from low (red) to orange to yellow and, finally, to bluish white, which is the highest or "hottest" color temperature. Whether a light source appears warm or cool is its *correlated color temperature (CCT)*, or ***chromaticity.*** Color temperatures, CCTs, and chromaticity are all measured in ***kelvins (K).*** Kelvins are named for Baron William Thomson Kelvin, a British physicist and mathematician in the 1800s who developed this color temperature degree scale based on the Celsius temperature scale. The lower the degrees on the Kelvin scale, the warmer and cozier the effect of the light. In hot climates, designers try to light with high Kelvin output to make the environment seem cooler!

There are also definitions for factors that take away from a light source's potential, called ***light loss factors.*** (The word *lamp,* in professional lighting terms, refers to what consumers usually call a "bulb" or "light bulb.")

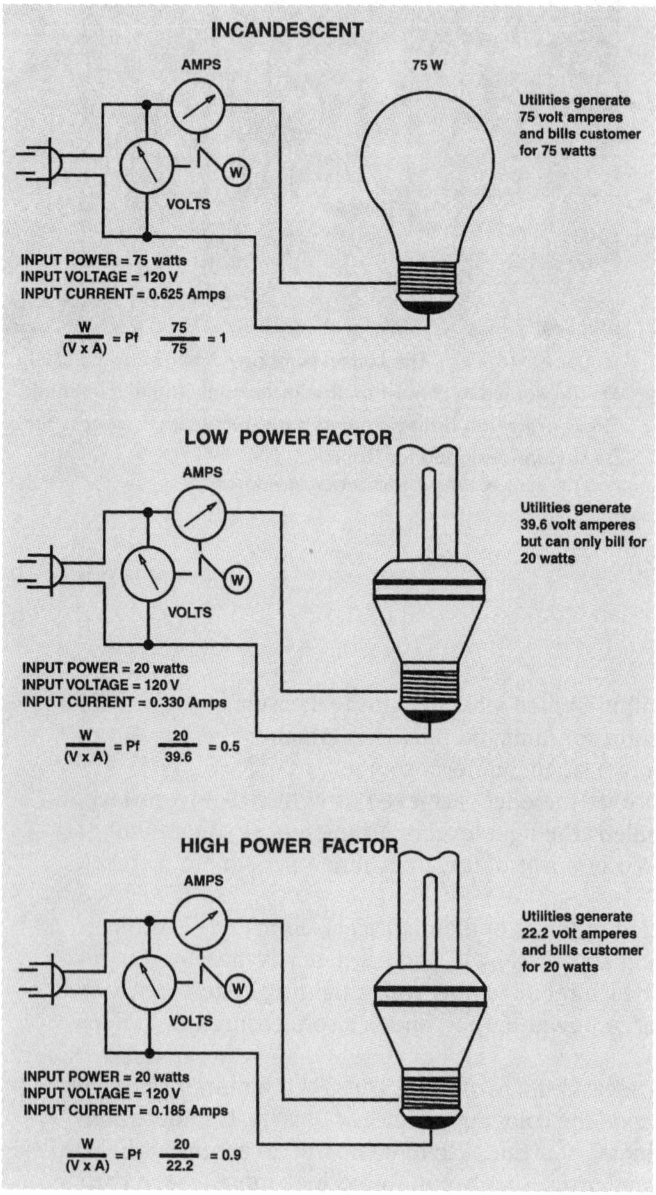

Lamp lumen depreciation is the gradual reduction in a bulb's light output as the bulb ages, losing some of its filament or phosphorous.

Lumen direct depreciation is the reduction of light output due to the accumulation of dirt, dust, or grease on the bulb.

The ***ballast factor*** explains the lamp's output using a given ballast. As you'll learn in a moment, some types of lamps require a separate piece of equipment called a ***ballast*** to be able to work, and some ballasts are more effective than others at allowing the lamp to emit its maximum light capability. The ballast factor is the ratio between the light output using a particular ballast and the maximum potential output of the light under perfect conditions. Most ballast factors are less than one.

Finally, the ***power factor (PF)*** is a measurement of how efficiently a device uses power. A lamp that converts all the power supplied to it into watts without wasting any in the process has a power factor of one. Often, lamps that require ballasts have a PF less than one (0.60 or 0.90, for example), because some of the electric current is used to create a magnetic field within the ballast, not to produce light.

Lighting devices are referred to as ***high power factor (HPF)*** or ***low power factor (LPF).*** This is important because sometimes utility companies will penalize customers when their electric loads have a low PF—that is, if you're using too much electricity to generate the amount of light and/or heat that you need. They may impose standards on the types of lamps you can buy to qualify for energy-saving discounts or rebates, requiring that they be rated HPF (see Illustration 7-2).

These factors are so critical that lighting experts generally recommend you use only one lamp manufacturer for reorders, instead of buying from various sources and getting mixed results.

Kitchen lighting is another animal altogether than dining area lighting. For employees' safety and comfort, and to allow them proper attention to detail

INCANDESCENT
AMPS
75 W
Utilities generate 75 volt amperes and bills customer for 75 watts
VOLTS
INPUT POWER = 75 watts
INPUT VOLTAGE = 120 V
INPUT CURRENT = 0.625 Amps

$\frac{W}{(V \times A)} = Pf \quad \frac{75}{75} = 1$

LOW POWER FACTOR
AMPS
Utilities generate 39.6 volt amperes but can only bill for 20 watts
VOLTS
INPUT POWER = 20 watts
INPUT VOLTAGE = 120 V
INPUT CURRENT = 0.330 Amps

$\frac{W}{(V \times A)} = Pf \quad \frac{20}{39.6} = 0.5$

HIGH POWER FACTOR
AMPS
Utilities generate 22.2 volt amperes and bills customer for 20 watts
VOLTS
INPUT POWER = 20 watts
INPUT VOLTAGE = 120 V
INPUT CURRENT = 0.185 Amps

$\frac{W}{(V \times A)} = Pf \quad \frac{20}{22.2} = 0.9$

ILLUSTRATION 7-2 **Utility companies prefer HPF products over LPF products. These diagrams show how power is used in each type of lamp.**

as they work, kitchen lighting must be bright and long-lasting, and should be selected to give off the least possible amount of heat. The most popular all-purpose kitchen lighting is still fluorescent lamps—inexpensive, durable, and bright. A new generation of fluorescents, known as T-5s, offer more output of lumens than their predecessor T-12s.

The trick in lighting a kitchen is to make things bright without causing unnecessary glare (and the resulting eyestrain). Recommendations from the Illuminating Engineers' Society of North America (IESNA) include matte or brushed finishes on countertops; and careful lighting around highly reflective surfaces like mirrors and glazed walls. In addition, kitchen light fixtures and lamps must be able to withstand the rigors of professional cooking and cleaning. Lamps that can withstand humidity for areas like dish rooms and walk-in refrigerators are called ***damp-labeled*** luminaries. Both the U.S. Food and Drug Administration and Occupational Safety and Health Administration (OSHA) now require shatterproof lamps. This so-called ***protective lighting*** usually includes a coating to prevent any bits of glass or chemical from flying out in case of breakage.

Artificial Lighting

The operation of your lighting system will account for about one-third of the electricity costs of your restaurant. When you include the costs of eliminating the heat produced by the lighting system, the figure rises to 40 percent. So look very closely at what you light, how you light it, and how much the lighting system will cost to operate. If you occupy an older building, consider retrofitting with more modern, energy-efficient lighting sources. According to the Electric Power Research Institute, the three areas for lighting improvement with the most potential cost savings are:

- Upgrading older fluorescent fixtures with new, improved components (better bulbs and ballasts, which we'll define shortly)
- Replacing incandescent bulbs with new, compact fluorescent lamps (CFLs), which last up to ten times longer
- Installing lighting controls: dimmers, occupancy sensors, or programmable lighting systems that turn on and off automatically

We are fortunate that technology keeps leaping ahead when it comes to lighting. There are three basic types of artificial lighting: ***light-emitting diodes*** (commonly known as *LEDs)*, ***incandescent lamps,*** and ***electric discharge lamps.***

LEDs are the newest type of lighting—a low-voltage system that is already revolutionizing the industry. First used in the 1990s to illuminate appliances, remote control devices, and digital alarm clocks, they were soon adapted for traffic lights and signage, and then for home and office lighting. They're now used in the brake-light systems of most automobiles. In 2002, the White House Christmas tree lights were powered for the first time with LEDs.

An individual LED looks like a tiny light bulb that fits into an electrical circuit (see Illustration 7-3). In fact, it works much differently. It contains a simple semiconductor, a microchip called a *diode.* In most applications, diodes don't give off much light at

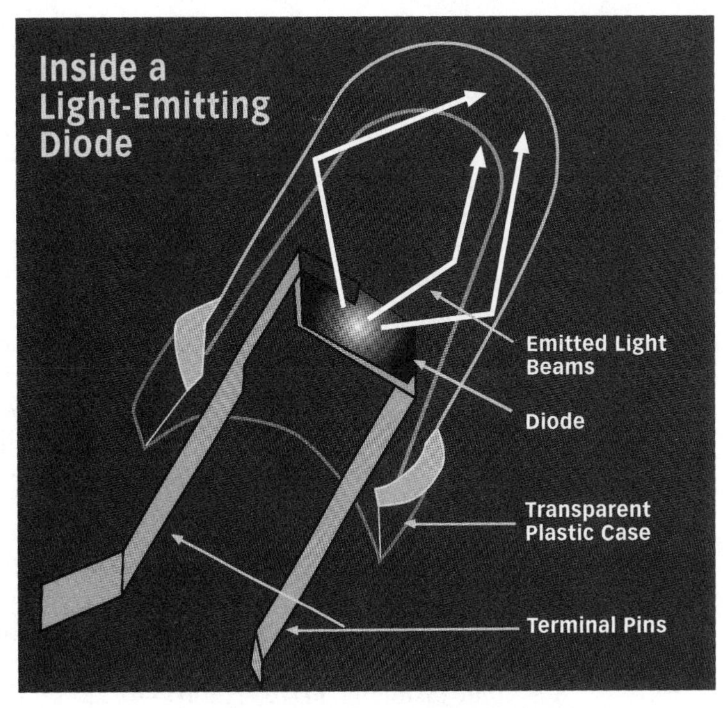

ILLUSTRATION 7-3 **Diagram of a light-emitting diode, or LED.**

all. But a light-emitting diode is specially constructed to release photons in large numbers, and the "light bulb" (made of plastic) that houses the diode concentrates or aims the light in a certain direction. By using different chemical compositions in the diode, the LED can be made to give off almost any color.

The advantages of LEDs are substantial. They put out brighter light than incandescent bulbs, but they consume up to 80 percent less energy and generate very little heat. They can be safely used indoors or outdoors, since they don't require high-voltage electricity. They last up to ten times as long as a regular lamp and, when they're ending their useful life, they gradually get dimmer instead of burning out with an unexpected "pop." The downside of LED use is that it is still more expensive than more traditional lighting methods. But, like so many other types of technology, the prices come down a bit every year. They're still not inexpensive enough to use, say, to light a parking lot or loading dock area, where lighting quality doesn't matter as much as brightness.

What most of us think of as a "traditional" light bulb is actually an ***incandescent lamp.*** It is a filament, encased in a sealed glass bulb. Electricity flows into the filament through the base of the bulb, and the glass is usually coated to diffuse the light. Most light bulbs have a relatively short life (up to 2000 hours), which means they have to be changed fairly often in a commercial environment. If they are used at a higher voltage than originally intended, their lives are even shorter. (The suggested voltage is usually stamped right on top of the bulb.) They also have rather poor efficiency (only 15 to 20 lumens per watt), which means they give off heat and it may cost a bit more in the long run to cool off an area in which they are used.

However, incandescent bulbs are relatively inexpensive, work instantly, and can be used with dimmers. Also, light from these bulbs is considered to be "warm" in hue, which is a desirable color rendition.

There are lots of variations of the incandescent lamp, including spotlights, floodlights, and elliptical reflector (ER) lamps. Their names often refer to the type of lighting they are used for. We'll talk more about ER lamps shortly.

The ***electric discharge lamp*** is one that generates light by passing an electric arc through a space filled with a special mixture of gases. That's why you'll sometimes hear them called *gaseous discharge lamps.* In this category, you find:

- Fluorescent
- Mercury vapor
- Halide or halogen
- High- and low-pressure sodium

Unlike the incandescent lamp, the electric discharge lamp cannot operate directly by threading or screwing it into a fixture. An additional piece of equipment, called a ***ballast,*** is required. The ballast is a current-limiting device that acts as a starting mechanism. It is generally mounted on top of the fluorescent light fixture and is replaceable. (Replacement costs are lower for separate ballasts than for those permanently attached to the fixture.)

There are two types of ballasts: the magnetic ballast and the electronic ballast. Electronic ballasts cost 60 to 70 percent more than magnetic ones, but the additional expense buys you a ballast that is more energy efficient and has less of the humming and flickering sometimes associated with fluorescent lighting. Electronic ballasts can also be used with dimmers; magnetic ballasts cannot. You can purchase rapid-start electronic ballasts that can operate up to four light fixtures at a time, with parallel wiring so that if one burns out, the other three continue to operate.

When electronic ballasts were first introduced, their life spans were very short, but recent improvements have increased them. However, buyers should still specify a warranty that covers replacement materials and labor for up to three years.

When comparing ballasts, look for those that are rated with the letters *P* (which indicates it is thermally protected and self-resetting) and *A* (which is the most quiet in terms of humming noise).

Both magnetic and electronic ballasts may generate ***harmonics,*** a distortion of power frequency that can create electromagnetic interference on power circuits. This wreaks havoc with sensitive electronic equipment, such as computerized cash registers or order-taking systems. Ballasts even have ***total harmonic distortion (THD) ratings***—the lower the percentage of THD, the less likely it is to distort the power line. "Low" THD is less than 32 percent, which is the recommendation of the American National

Standards Institute. Therefore, high power factors and low THDs are the rule for whatever type of ballast you buy.

An ***integral ballast lamp*** is a lamp and ballast in a single unit. It may be magnetic or electronic, and has a screw-in base that allows it to fit directly into almost any standard light fixture. Combining lamp and ballast makes for easy retrofitting. Replacement costs are higher, however, because the whole unit must be replaced. There are also adapters you can purchase that contain a ballast. These last through the lives of three or four lamps, and are generally more economical than the integral ballast lamp.

Fluorescent lamps are the most common form of electric discharge lamp used in the hospitality industry. Unlike an incandescent bulb's 2000 hours of life and 15 to 20 lumens per watt, the fluorescent tube will last from 7000 to 20,000 hours, with an efficiency of 40 to 80 lumens per watt. Most fluorescent tube-shaped bulbs come in three standard sizes: 24-inch, 48-inch, and 64-inch. However, there are many different sizes of *compact fluorescent lamps (CFLs),* which feature smaller-diameter tubes bent into twin tubes, quad tubes, or even circular shapes. The smaller diameter of the CFL tube makes it economical to manufacture these lamps with higher-quality phosphorus, which improves light output and makes the color seem more natural.

CFLs are the most popular type of fluorescent lamp, and for good reason. They typically are three to four times more efficient than regular light bulbs; they're available in several different color temperatures (from a warm 2700 Kelvins to a cool 4100 Kelvins) to achieve different effects; they don't emit heat, which saves on air-conditioning costs; and they don't have to be changed as often as other types of bulbs. Table 7-1 gives you a look at the potential savings when CFLs are selected over standard incandescents.

On the downside, most CFLs cannot be used with dimmers, in outdoor locations, or in any situation where they might get wet. They also work best in mild temperatures, not in cold weather, and their lives will be shorter if you use them in recessed or enclosed fixtures.

Another breakthrough in lighting technology is the ***E-lamp,*** which uses a high-frequency radio signal instead of a filament to produce light. Inside the sealed globe, the rapidly oscillating radio waves excite a gas mixture, which, in turn, gives off light. The light hits a phosphorous coating on the inside of the globe and glows.

TABLE 7–1

HOW TO SAVE $25 ON A LIGHT BULB

COMPARING TOTAL COSTS OVER 10,000 HOURS FOR A COMPACT FLUORESCENT AND AN INCANDESCENT

	Compact Fluorescent	Incandescent
Wattage	15 W	60 W
Light output (lumens)	900	890
Energy use	150 kWh	600 kWh
Energy cost @ 9 cents/kWh	$13.50	$54.00
Lamp life (hours)	10,000	1,000
Lamp replacements	0	9
Cost of lamp replacement	0	$6.75
Original lamp cost	$23.00	$0.75
Total cost for 10,000 hours	$36.50	$61.50

Source: Texas Utilities Electric Company, Dallas, Texas.

The E-lamp lasts even longer than the CFL—at least 20,000 hours—because it does not contain an electrode that can burn out. However, the phosphorous coating wears out, and the E-lamp gradually dims over time. E-lamps fit most fixtures, will operate whether they're installed upside down or sideways, and are not susceptible to cold climates like some CFLs.

The renewed interest in lighting technology happened back in the 1990s when Congress passed the 1992 National Energy Policy Act (EPACT), which mandated greater energy efficiency in certain bulbs. New labels were required on lighting products beginning in 1995, listing lumens, watts, and hours of life. The least efficient bulbs and fluorescent tubes were forced off the market by minimum performance standards they could not meet, and manufacturers were given just 12 months to find replacements and make all the labeling changes. A few types of specialty lighting (plant growth lights, colored lights, impact-resistant medical, stage and studio lighting) were exempted, but, overall, EPACT turned the industry on its ear.

For consumers, the changes were beneficial. Now that they undergo performance testing, some of today's bulbs, coated inside with a substance called rare earth (RE) triphosphor, achieve a color rendering index (CRI) of 70 to 89, compared to a CRI of only 50 to 65 for the older-style halophosphorous lamps. Rare earth lamps produce 5 to 8 percent more light with no increase in energy consumption, and they are a popular choice in retrofitting older fluorescent lighting systems. However, they cost up to $10 apiece—quite a difference from the $3 average cost of the conventional lamps.

Some interesting combinations have also been created, such as the *compact halogen lamp.* This is actually an incandescent bulb with halogen gas inside, which prolongs the life of the filament. Compact halogen lamps give off brighter, whiter light than regular incandescents and last longer, too. They are used in downlighting, accent lighting, and retail display.

Mercury vapor lamps are also known as *high-intensity discharge (HID) lamps.* They're used for lighting streets and parking lots, and their intensity varies from 15 to 25 lumens per square foot. The lifetime of these lamps is between 12,000 and 20,000 hours. There are clear mercury vapor lamps and white ones; the white ones have better color rendition. EPACT did not set any requirements for HID lamps, but the legislation suggested that the U.S. Department of Energy develop standards by 1999. These will probably result in restrictions on the use of mercury, in favor of metal halide and high-pressure sodium lamps.

In fact, the ***metal halide lamp*** is a variation of the mercury vapor lamp, with metallic halide gases added to improve color rendition and efficiency. These lamps have an efficiency level of 80 to 100 lumens per watt and lifetimes of 7500 to 15,000 hours.

Sodium pressure lamps are highly efficient light sources, used primarily in parking garages, driveways, hallways, and anywhere lighting is used as a security measure. Their color rendition is very poor, but these bulbs are very efficient, with 85 to 140 lumens per watt and a life of 16,000 to 24,000 hours. There are high-pressure sodium (HPS) and low-pressure sodium (LPS) types.

No matter what type of bulb you use, you must also make a choice of ***light fixtures*** or *luminaries* to hold the bulbs. A fixture is a base, with wires that connect to an electric power source and a reflective surface that directs the light to whatever surface you want to illuminate. A fixture can be highly visible and decorative, such as a chandelier or stained-glass tabletop lamp, or recessed into a wall or ceiling where it is virtually invisible. For recessed fixtures, you can purchase a special bulb called an ***elliptical reflector (ER) lamp.*** This is an incandescent bulb with a unique shape that focuses its light outside the lamp envelope. None of the light is wasted by "heating up" the inside of the fixture. Using ER lamps will save 50 percent of the energy you'd use with regular incandescent bulbs, but only if they are used in recessed fixtures.

For electrical discharge lamps, the fixture also holds the ballast in place. Before you buy fixtures, think about how easy (or hard) they will be to clean and replace. Also, consider the added costs of any replacement items, such as decorative globes or safety covers.

If you're retrofitting an existing lighting system, professional advice from a lighting consultant is not a waste of money. The price tags for state-of-the-art fixtures, ballasts, and lamps can be so high that you immediately assume you can't fit them into your budget when, after the initial expense, they could save you more than 40 percent on lighting costs month after month and replace old fixtures that were just about worn out anyway. A consultant can help you evaluate what you've got and decide if the improvements are worth the money.

Controlling Light Levels

How about the use of natural light? One interesting way to light your dining area is to install ***solar tubes*** on your roof (see Illustration 7-4). Each aluminum tube is a mere 13 inches in diameter (so small that it doesn't disturb the rafters or ducts between roof and ceiling) but can light a 10-square-foot indoor area, providing the equivalent of 1500 watts. One end of the tube is covered with a clear acrylic dome and peeks out several inches above the roofline, facing the best direction (usually southwest) for maximum sunlight exposure. The other end opens as a hole in the dining room ceiling. The tube is sealed with special flashing to prevent water leaks. Because aluminum is highly reflective, the tube can generate light even on cloudy days. From the inside of the building, the opening is sealed by a frosted globe, which diffuses the light and looks like just another light fixture attached to the ceiling.

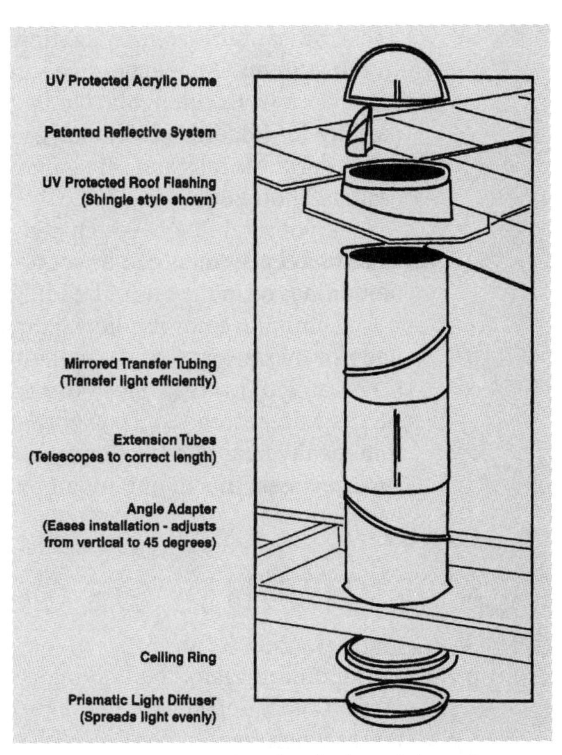

ILLUSTRATION 7-4 **Only a few inches of the solar tube's total length can be seen above the roofline.**
SOURCE: SOLATUBE, CARLSBAD, CALIFORNIA.

Reflection is one good way to make light go farther. A ***reflector*** is a shiny, custom-fit surface that mirrors the light that bounces off of it. Reflectors fit between the bulb and fixture. Because most are shaped like an inverted bowl, they also help aim the light, reducing glare and focusing the light downward. On the downside, walls can sometimes appear darker and it's harder to make illumination seem uniform throughout a room.

Reflectors cannot be used in most recessed light fixtures, but are often recommended as a retrofitting technique for other types of lamps because of their potential energy savings. By installing a reflector, you can remove up to half of the bulb in a fixture, saving half on your lighting costs and a little on your heating and cooling costs, too. For example, a four-bulb lamp operated with two bulbs and a reflector still has 60 to 75 percent of the light output it would have with all four bulbs burning. Also, for spaces that are too dark, use of a reflector with a full contingent of bulbs can actually boost existing light levels by 15 percent or more.

Most fluorescent fixtures use either a ***lens*** or a ***louver*** to cover the bulb itself and prevent direct viewing of it at most angles to minimize glare. In fact, there's a standard ***glare zone***—at just about any angle of more than 45 degrees from the fixture's vertical axis, most "naked" bulb lighting is uncomfortable to look at and causes reflections on work surfaces and computer screens. A sturdy, clear plastic lens will help diffuse glare and make the light more uniform.

Louvers do the same thing, in a slightly different way. Instead of covering the bulb, they are like small, partially open doors that aim the light to eliminate glare at certain angles. Louvers are more effective than lenses at reducing glare, but, because they "shade" some of the light, they allow less light output overall. If louvers are not properly adjusted, people may complain that ceilings look dark.

Lighting control systems can help save energy, but only if you think it through carefully before installing one. Many an expensive system has been disconnected because it caused too much inconvenience or resulted in false alarms.

An ***occupancy sensor*** can be used in a conference room, storage area, restroom, or walk-in refrigerator. It detects motion and activates a controller device that turns on light fixtures. Then, if no motion is detected within a specific period of time, the lights turn off until motion is sensed again. Compared to manual on-off switches, occupancy sensors can reduce the use of your lights by 15 to 30 percent. Most sensors allow for adjustments of motion sensitivity, time periods the lights stay on or off, and a bypass switch that overrides the system when necessary. They are also handy at dark back-of-the-building loading docks and entrances.

Occupancy sensor costs often include installation of mounting devices in walls or ceilings and wiring to hook them up to the building's power supply. In simple systems, regular light switches are merely replaced by sensors.

You can also control lighting with *timing switches,* light-sensitive ***photocells,*** and programmable ***sweep systems.*** An inexpensive timer can be used to turn lights on and off at the same times every day. They work well with lights that have predictable operation periods, from outdoor signage to corridors to parking lot floodlights. Similarly, outdoor lighting can be hooked to a photocell that turns the light on when darkness falls and off again when the sun rises. The only problem with using photocells is that occasionally on dark, rainy days, the outdoor lights will come on unnecessarily.

If your lighting needs change from day to day, a more complex programmable system is better. So-called sweep systems can be set up to turn lights on and off sequentially in one area, on a whole floor of a building, or in the entire building.

Dimming controls allow light output to be adjusted for the needs of a particular space, either manually or by sensors that detect light levels, daylight, or occupancy. Precise adjustment of lights is sometimes called ***tuning.*** Most dimmers are installed during new construction, but they can also be successfully retrofitted. Daylight sensors use photocells to detect light levels and will dim the fixtures automatically near windows or skylights. In a kitchen setting, a dimmer can be used to adjust light in work surface areas; in a dining room, lights can be adjusted to full output to assist the cleaning staff.

7-2 THE USE OF COLOR

In the dining room, the lighting system and color scheme must work together to enhance the environment. Colors depend on their light source because, as you know, the same color can look completely different when seen under different types or intensities of light. Color and light together can also be used to make space seem larger or smaller and more intimate.

Color can be used to convey a theme, a style, a geographic region, a way of life, or even a climate. The next time you're out studying your competitors, look at what colors their walls are. Much research has been done about the psychological impact of certain colors on people's moods—one example that is often cited in the restaurant business is that people don't linger long in a restaurant where red is the predominant color. Is the restaurateur trying to make guests uncomfortable or just trying to turn the tables faster? Is the premise true or just a supposition? We've never heard the definitive word on that one! However, authors and scholars Regina S. Baraban and Joseph F. Durocher assert there are *color cycles* in the restaurant industry—popularity trends that seem to last about eight years at a time. It would be smart to ask your decorating and design team what cycle you're in. As you make lighting decisions, remember that they will serve you (and the guest) in several ways: visually, psychologically, thermally, and socially as well as aesthetically. The food should look appetizing and wholesome; the dining area must look pleasant and appropriate to the image, price range, and type of cuisine.

The Use of Artwork

Making a real statement with your decor by using murals and other forms of art is an idea worth exploring. The walls aren't going to be bare, are they? What you choose to put there will add personality and flair to your public areas.

A *mural* is a wall-sized painting, often created directly on the wall itself. It's a big, gutsy artistic statement that usually becomes the focal point of a room, and it can impact your overall design in several ways:

- It can be a signature that communicates the restaurant's concept.
- It can be used to "personalize" a large dining space, adding color and motion to an area that might otherwise feel more like a warehouse than a restaurant.
- Conversely, a mural may be used to "open up" small dining spaces, by adding extra dimension to the wall space.
- A mural can add sophistication. It can be wild, vibrant, and ultramodern, or soft and subdued. Either way, it cannot help but make a statement.

Adding a mural is as big a step as the size of the mural itself, and the price tag for such a large original work can easily top $10,000, so it requires careful consideration. When hiring an artist, ask to see many photos of his or her work, and visit some of the other mural sites in person. The mural can be an impressive conversation piece, but if not properly planned and professionally painted, it can be distracting and downright annoying instead of mood enhancing. Either way, it's an investment that you, and your guests, must live with for a long time. Be certain you have a contract with the artist or designer that specifies the size and content of the finished work (nothing that might be considered offensive), including preliminary sketches, a time line for completion of the mural, the materials that will be used, the procedure and price if touch-ups or alterations are needed, and how payment is expected—usually some up front and the rest upon completion and approval.

A less permanent and less expensive way to add color to your dining area is to display the work of local artists, galleries, or art dealers. Depending on your arrangements with the art supplier, the works are priced and the restaurant may receive a small commission when one is sold. This allows you to change the decor with the seasons, by the month, or whenever you think the place needs a change. Whenever a new artist or selection is featured, you have a nice reason to invite guests and members of the local news media to an "opening" reception.

As with purchasing a mural, when selling the works of others it's a good idea to have a short, basic contract signed by both you and the artist or gallery representative. It should specify the number of days and/or months that the work will appear in the restaurant, with starting and ending dates; specifics about commissions; rules about who has responsibility for safely hanging and/or lighting the artwork, accepting money for the pieces that sell, and picking up the pieces that don't sell within a certain time frame (so you won't be stuck storing them). It should also contain a statement that waives your liability in case any of the work is damaged or stolen while in your possession. The gallery owner or artist should give you a neatly typed sheet that lists every piece being given to you for sale, along with the prices and a stack of their business cards. You'll need to have this information on hand in case prospective buyers ask for it.

It is not without risks, but the idea of showcasing local talent has a satisfying, community-minded spirit as a side benefit. It decorates your walls, boosts your business, and nurtures goodwill—all inexpensively.

ILLUSTRATION 7-5 **Gino's Italian Ristorante in Boise, Idaho, features huge, lush murals on almost every wall. The artist is Fred Choate, who designed each of them with "surprises" for the guests to discover as they study them.**

PHOTO COURTESY OF GINO VUOLO, GINO'S RESTAURANT, BOISE, IDAHO.

Finally, when thinking about artwork, don't forget the restrooms. Many restaurants decorate all other areas to the max, but leave their restroom walls completely bare. Why?

Kitchen Lighting and Color

When lighting a foodservice kitchen, trends and ambience most definitely take a back seat to efficiency. The idea here is to reduce employees' eyestrain by minimizing glare while making sure light levels are bright enough to allow a safe working environment. Planners must consider:

- Square footage of the area
- Ceiling height
- Contrast and colors of the products being processed at the workstation
- Colors of the surrounding walls

A back-of-the-house lighting system is usually fluorescent, with shields on the fixtures to protect food and workers from falling glass in case a bulb should break. Fluorescent fixtures may be placed parallel to workers' lines of sight, since this results in less glare. In storage areas, light fixtures should be located over the centers of the aisles for maximum safety. And we must also counteract the reflective properties of stainless steel surfaces, such as work tables and refrigerator doors. A ***satin finish*** instead of a shiny one will reduce the glare considerably.

Some kitchen tasks—cake decorating and garde manger duties (cold-food work, preparation of hors d'oeuvres, and so on)—are detailed enough to necessitate brighter lighting than other tasks. The Illuminating Engineering Society recommends a light intensity of at least 30 foot-candles throughout the restaurant, and 70 foot-candles at so-called "points of inspection," such as pass windows and garnishing areas, where the food gets a close look before being served. You may find that, in your area, the city or county government has already set minimum lighting requirements as part of your licensing or permit process. We know of some cities that require a minimum of 50 foot-candles throughout the kitchen.

Kitchen lights require considerable maintenance, because grease can build up on the fixtures and shields, so choose them for easy cleaning. In addition, fixtures that are labeled ***vapor proof*** (moisture resistant) must be used in exhaust hoods and dishwashing areas.

The colors of walls and furnishings don't seem to be a major concern in most kitchens. However, a well-designed combination of light and color can reduce eyestrain and increase worker efficiency, possibly even reducing accidents and boosting morale. Light, cool colors are probably the best choices. Using one lighter color and a darker version of the same hue is an option that creates a mild contrast that is easier on the eye.

Because white is highly reflective, it should be avoided for kitchen walls; however, an off-white for kitchen ceilings can be a smart choice. Its natural brightness helps boost your light levels.

The use of colors to signal special equipment and areas is also recommended. Danger is usually red, marked with paint or reflective tape on moving parts of equipment, swinging doors, and so on. Yellow is used to mark the edges of steps and landings. Green is used to identify first-aid kits or areas.

A final note about lighting: Local building codes, as well as your insurance company, will have very specific requirements for emergency lighting. Most of them come from either the National Electrical Code or the Life Safety Code. An emergency lighting system must be able to provide at least 1 foot-candle of brightness for a period of 90 minutes; and it must have its own power source, independent of the main power system, in case of an electrical failure.

7-3 NOISE AND SOUND CONTROL

Both the dining area and the kitchen pose numerous challenges when it comes to noise reduction. Unfortunately, they are not usually obvious until the space is occupied and somebody— either employees or guests—starts complaining.

Typical restaurant sounds are many and varied: people conversing; waiters reciting the day's specials or picking up orders; cleaning of tables and bussing of dishes; kitchen equipment grinding, whirring, and sizzling; the hum of the lights or the HVAC system; the strains of background music or a live band. If you're located on a busy street, add traffic noise; if it's an airport—well, you get the point. Music is also an integral part of many restaurant concepts. Where would the Hard Rock Cafés or Joe's Crab Shacks be without it?

It's interesting that, while too much noise can cause discomfort, an absence of noise is just as awkward. It feels strange to sit at a table and be able to hear every word of your neighbors' conversation, and to know that your own conversation is probably being eavesdropped on as well. In short, a restaurant's noise level should never be accidental. It is an important component of the environment and mood.

Some restaurant owners pump up the sound system volume in hopes of creating a sense of "happening" in the dining area. The challenge here is not to have a loud restaurant, but to achieve a sound level of high enough quality and volume that the guests will notice and enjoy it, yet can also comfortably converse through it. Juggling these priorities is harder than it sounds. Even if yours is a fabulous sound system, a stimulating noise level is bound to turn some customers away.

The Nature of Noise

In controlling noise, we need first to be aware that it has two basic characteristics: *intensity* and *frequency.* Intensity, or loudness, is measured in *decibels,* abbreviated dB. The lowest noise an average person can hear close to his or her ear is assigned the level of 1 decibel, while a 150-decibel level would cause pain to the average ear. A "noisy" restaurant averages 70 to 80 decibels, as shown in Table 7-2.

Frequency is the number of times per second that a sound vibration occurs. One vibration per second is a hertz (abbreviated Hz). Humans hear vibrations that range all the way from 20 per second (low frequency) to 20,000 per second (high frequency). In a restaurant setting, a high-frequency sound is more objectionable than a low-frequency sound. Even sounds that are low intensity may be objectionable to guests if they are also high frequency. Sound travels from the source to the listener and back again in very speedy fashion—about 1100 feet per second. Inside a building, sound can either be absorbed or reflected

TABLE 7-2

COMPARATIVE NOISE LEVELS FOR FOODSERVICE

Decibels	Noise Equivalent
10–20	Broadcast studio
20–30	Average whisper at 5 feet, very quiet residence
30–40	Average school, average residence, library reading room, museum
40–50	Quiet restaurant dining room, quiet office
50–60	Average office, noisy residence, quiet street
60–70	Average restaurant dining room, noisy office, automobile at 30 mph
70–80	Noisy restaurant dining room, automobile at 50 mph, restaurant kitchen
80–90	Noisy street, police whistle at 15 feet
90–100	Fire siren at 75 feet, subway train, riveter at 30 feet
100–120	Train passing at high speed, airplane propeller at 10 feet
130	Threshold of pain

Source: Reproduced with the permission of the National Restaurant Association.

by all the other things in that space: walls, ceilings, floors, furniture, or equipment. If it reflects or "bounces off" of surfaces (called ***reverberant sound***), the area can build up a sound level that is much higher than if the space was not enclosed.

It is not just the dining area where noise is a problem. At the back of the house, everything seems to conspire to *create* noise! Dishes clatter, chefs bark their orders, dishwashers whoosh, toilets flush, refrigerators hum, and the kitchen exhaust system drones on and on. It can become a real cacophony. Your goal is not only to keep it from reaching the guests, but to keep it from driving your own staff crazy.

A study on restaurant noise levels by the University of California at San Francisco raises the issue of potential health hazards to employees of consistently loud eateries. Although the findings are inconclusive, the study authors measured everything from a low of 50 decibels in a Chinese bistro, to almost 91 decibels in a busy microbrewery. A noise volume of 75 decibels or more requires that most people raise their voices to be heard in conversation—which, ironically, just creates more noise—and OSHA guidelines require employees in other fields of work to wear earplugs with noise levels of 90 decibels or more.

To determine problem areas, the primary concerns of the designer are those points at which noise is most likely to emerge from the kitchen. This means the pass window area, where expediters and/or wait staff are positioned to call out orders and pick up food. Computerized ordering systems have already improved and quieted this process. The other problem site is the dishwashing area, where doors that are too thin or opened too frequently permit the clatter to emanate into the dining room.

One simple noise control strategy, often overlooked, is to make equipment operate more quietly by keeping it properly maintained, as we discuss in more detail in Chapter 9. This includes careful examination of basic mechanical systems. The heating, ventilation, and air-conditioning (HVAC) system has ducts that can act as chambers for sound transmission, amplifying the motor noise from the system's fan and condenser and sending it throughout the building. Metal plumbing pipes also transmit noises, as rushing water makes them vibrate and trapped air creates knocking sounds.

Now let's look at ways you can choose "sound smart" alternatives, inside and outside of the kitchen. General sound control can be accomplished in two ways: Either suppress the sounds from their source, or reduce the amount of reverberant sound by cutting the travel direction of the sound waves. Sound engineers use *hard concave surfaces* to concentrate sound. The hard surface traps the sound, then sends it in the desired direction. On the other hand, if a *soft convex surface* is used, the sound is absorbed and deadened. These principles can be adapted to the ceilings, walls, and furnishings of any room.

Ceilings. The ceiling of a room is a natural choice for sound control treatment, because there is not much on the ceiling that would get in the way. Spray-on acoustic surfaces, acoustical tile, fiberglass panels padded with fabric, wooden slats, or perforated metal facings are less noisy alternatives to plain plaster or concrete ceilings. Panels and slats should be at least ¾ inch to 1 inch thick and suspended instead of being directly attached to the ceiling. Be sure not to paint acoustical tile or it will lose its sound-absorbent effect.

Walls. Covering walls with padding, fabric, or carpet helps a great deal to muffle sound. Organizing the dining room on different levels also helps somewhat. And the use of movable partitions, which can be purchased up to 5 feet in height, can contain sound as well as create intimacy by breaking up a large dining area into smaller spaces. Walls and partitions can even be decorative and see-through, to maintain the open feeling of a room.

Draperies and Furnishings. Fabric can be used in many parts of the dining area, chosen for its sound absorption qualities. Window coverings can muffle sound if they're made of heavy material. Tables can be padded and covered with cloth, minimizing clanking dish noise. Chairs can also be padded and covered; although a porous fabric allows sound waves to penetrate, it also absorbs dirt more quickly. High-backed booths absorb noise.

Even something as simple as choosing a slightly smaller table can have an impact on sound, as it prompts diners to talk closer and more quietly to each other.

Carpets. Floor coverings have a major impact on noise. If yours is carpeted, choose a carpet with high pile. It will be more expensive to purchase and maintain, but won't wear as readily as cheaper carpets. Learn more about carpet and flooring choices in Chapter 8.

The Use of Music

Finally, because music is so often viewed as a marketing tool, we can't overlook its role in the restaurant environment. Many casual and theme restaurants make a considerable investment in sound systems meant to entertain guests. The music serves the dual purpose of "drowning out" the noise levels of guests and employees, so never fail to think of it in terms of noise control, not just entertainment. Make sure you get expert advice in selecting the right-sized amplifier—that is, one designed to work with your space and sized (by watts and number of channels) to complement your number of speakers. The newest amplifiers are sophisticated enough to allow programming sound levels based on time of day and/or day of the week! Some have built-in microphones that sense the ambient sound level and automatically adjust the music level to fit the background noise level. Proper placement of speakers is another job for experts. In foodservice, ceiling-installed speakers are more common than wall-mounted ones.

Make music selections based on the demographic mix of your guests and your overall concept. A direct satellite music feed is probably the best way to ensure employees don't hear the same song several times during a work shift—a surprisingly common employee gripe! With 120 satellite channels to choose from, you can select different ones for different times of day and clientele. Some casual eateries opt to play a local radio station on their sound systems, but it's unwise. Why would you want customers to hear advertisements, for your competitors or anything else, while they eat?

No matter what your musical selections, you must pay an annual licensing fee to play almost any song in a public setting. The American Society of Composers, Authors and Publishers (ASCAP) and Broadcast Musicians Incorporated (BMI) are the two major music-licensing organizations in the U.S. They represent the people who create the music, ensuring that they receive royalties for its use. Music licensing is a multi-million dollar industry. The good news is, purchasing the annual ASCAP and BMI licenses entitles you to freely play millions of songs. The bad news is, licensing is expensive. From a minimum of about $300 to a maximum of more than $8,000 a year for each license, the cost is determined by a number of factors: the size of your establishment, whether you include dancing or a cover charge or use a jukebox, and others. Discounts may be given if your restaurant is a member of a recognized state restaurant association.

Music licensing is nothing to neglect. You can be fined and even sued for playing tunes without it. The licensing companies, which refer to themselves as "performing rights organizations," have enforcement people whose jobs entail going into stores, restaurants, bus stations, office buildings—anywhere music might be playing in the background—to check for current licenses and cite business owners who don't have them. Learn more about music licensing specifics at the websites of these companies: ascap.com and bmi.com.

Kitchen Noise Control

Controlling loudness in the kitchen begins with the idea of having doors that separate the kitchen from the dining room. Some kitchens don't have doors, but passageways, making it especially important to have acoustical treatments on both walls and ceilings in these transition zones.

Of course, there is a popular belief among some restaurateurs that having customers witness kitchen sights, sounds, and smells is part of the mystique and excitement of eating out. In the open or ***display kitchen,*** noise levels are part of the atmosphere. Romano's Macaroni Grill restaurant chain is one concept that uses the open kitchen and its associated noise level as part of the design and concept.

Inside a kitchen, consider the following noise abatement options:

- Installing acoustical tile ceilings
- Undercoating all work surfaces
- Putting the dish room in an enclosed area
- Installing all refrigerator compressors in a separate "mechanical room" or area outside the kitchen
- Properly installing and maintaining exhaust fans to reduce humming and vibration
- Using plastic or fiberglass dish carts and bus containers instead of metal ones

7-4 HEATING AND AIR CONDITIONING

Heating, ventilation, and air conditioning, commonly known as ***HVAC,*** are used to maintain a level of comfort for both guests and employees. Your building's HVAC system must be carefully selected, properly operated, and continuously maintained to do its job effectively. The key environmental comfort factors at work here are:

- Indoor temperature
- Humidity
- Air movement
- Room surface temperature
- Air quality

To modify and control the factors listed previously, the following types of equipment are part of most HVAC systems:

- Furnaces (to produce hot air)
- Boilers (to produce hot air)
- Air conditioners (to produce cold air)
- Chillers (to produce cold air)
- Fans (to circulate and remove air)
- Duct work (to move air)
- Filters (to clean air)

Individual comfort is a simple matter of balancing a person's body temperature with that of the surrounding environment. The body gives off heat in three ways: *convection, evaporation,* and *radiation.* An example of heat loss by convection is when air moves over a person's skin. The movement creates a temperature difference between skin and air. An example of heat loss by evaporation is perspiration, when heat causes liquid to turn to vapor. Heat loss by radiation happens when two surfaces of differing temperatures are placed right next to each other, like being seated by a window on a cold day. The right combination of temperature and relative humidity to make people comfortable is referred to by experts as the ***comfort zone.*** The parameters of this zone are highly subjective, but let's just say that, in finding your comfort zone, you're trying to balance the environmental conditions with the natural heat loss of the bodies of your guests, or workers. Comfort is important in the kitchen, too!

The challenge to foodservice managers and planners is to find the comfort zone while also paying attention to odors, noise, and air quality. The HVAC system must also be flexible enough to change if environmental demands change, whether that means a change in season, a change in crowd size, or a change in building size. Select a system that has a quick response time and can automatically be controlled and operated at low cost. These are challenges that can be met, but only with the help of engineers. Once the HVAC system is installed, several of your staff should always be aware of how the it operates.

When discussing an HVAC system, the air that runs through it is known by different names, depending on its location and use. In alphabetical order, common terms include:

CONDITIONED AIR. Air that has been cooled or heated mechanically (by HVAC) and released into the building's interior.

DESSICANT AIR. Dessicants are drying agents, which may be included in the HVAC system to reduce humidity.

EXHAUST AIR. Air, that must be removed from cooking sources (e.g., ranges and fryers) or enclosed spaces (e.g., restrooms). Once exhaust air has been removed from the building, it should not be reused. The volume of air removal (how fast it is removed) is measured in cubic feet per minute (cfm).

MAKEUP AIR. Air that must be supplied to an area to replace the exhaust air that has been removed.

OUTDOOR AIR. Air that is taken from outdoors.

RETURN AIR. Air that is removed from an interior space, then returned to the HVAC system for recirculation or exhaust. Sometimes called recirculated air.

SUPPLY AIR. Air that is delivered to an area by the HVAC system. It may be used for ventilation, heating, cooling, humidifying, or dehumidifying.

TRANSFER AIR. Air that flows from one part of a building to another. In a restaurant, transfer air comes from the dining area into the kitchen, which helps keep cooking odors from drifting back into the dining area.

Now, let's talk briefly about each part of an HVAC system and how it contributes to the overall heating and/or cooling process.

How HVAC Systems Work

The two major HVAC components that heat cold air are the heating plant and the heating system. The ***heating plant*** is where fuel is consumed and heat is produced. The ***heating system*** is the means by which the heat is distributed and controlled. The heating plant will use one of several heat sources: electricity, natural gas, liquefied petroleum gas (LPG), fuel oil, or steam. Your options will depend on the availability of the fuel, its cost, the cost of equipment and upkeep, environmental regulations, and safety concerns, some of which were discussed in Chapters 5 and 6.

Electricity is available everywhere, and, while it is safe and clean, it tends to cost the most. If the need for heat is low and other fuel sources are not reliable, electricity may be your first choice. Natural gas, fuel oil, and LPG all produce heat by combustion, a combination that requires special equipment such as boilers and furnaces. So, while fuel costs are lower, equipment and maintenance costs are higher. Also, the availability of these fuels is not as widespread in some areas as electric power.

Parts of the heating plant (shown in Illustration 7-6) are:

- The *preheater,* which takes completely cold air and begins the heating process.
- The *heater* itself, which is a type of motor that runs on fuel or electricity to create heat.
- The ***humidifier,*** which adds water to the air (often necessary because heat dries the air so much).
- The ***dehumidifier,*** needed in special circumstances when outside air brought into the building for heating (or cooling) contains too much moisture. ***Dessicants*** are drying agents used to dehumidify air.
- The *fan,* which blows the air from the motor or brings in air from outside. Fans have their own electric motors.
- The ***reheater,*** used in some systems with humidifiers and/or dehumidifiers. A reheater boosts the heat of air that is otherwise too "wet" to blow into a room to dry it out enough to send it through the ductwork.

Parts of the heating system are:

- *Ducts* and *vents,* which are the pipes and openings through which the air travels into parts of the building. Vents are also called ***diffusers;*** they are covered with ***grilles*** that can be manually opened and closed to restrict or encourage airflow.
- ***Dampers,*** which are used inside the ductwork to regulate airflow. They can either heat or cool the air that passes through them, depending on the season.

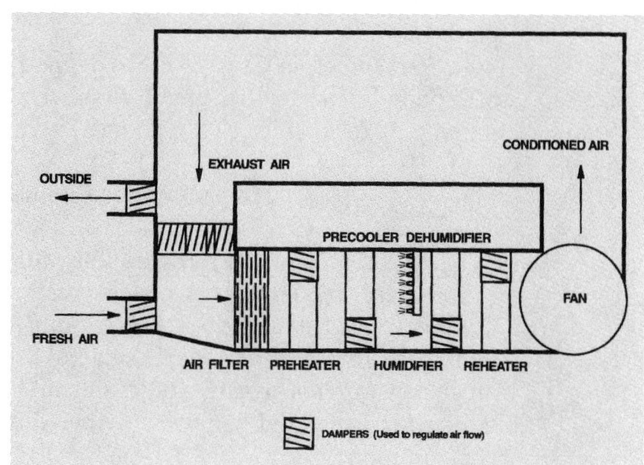

ILLUSTRATION 7-6 The layout of a year-round HVAC system.

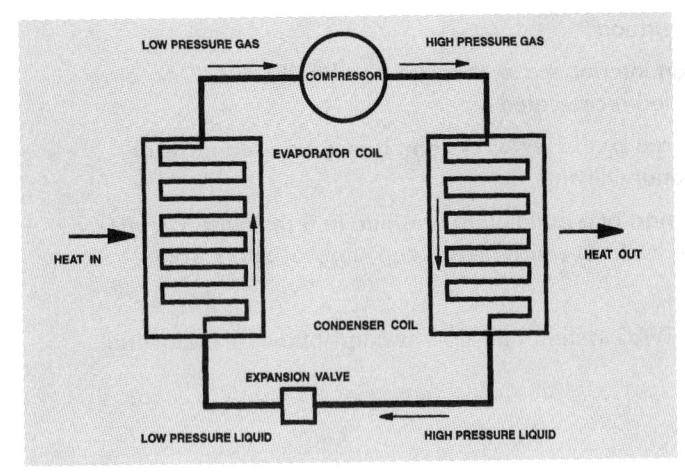

ILLUSTRATION 7-7 **How vapor compression works to refrigerate air.**

SOURCE: FRANK D. BORSENIK AND ALAN T. STUTTS, *THE MANAGEMENT OF MAINTENANCE AND ENGINEERING SYSTEMS IN THE HOSPITALITY INDUSTRY*, 4TH ED., JOHN WILEY & SONS, INC., NEW YORK.

- The *filter,* which purifies the air by catching dust particles that would otherwise circulate.
- The ***thermostat,*** which is the control unit that allows you to regulate the air temperature of a building or area.

Most climates require both heating and air conditioning. When you think of air conditioning, you probably think of cold air. However, the typical air-conditioning system does both cooling and heating, plus all the same filtering and humidifying that a heating system does. Mechanical cooling equipment works by extracting heat from air or water, then using the cooled air or water to absorb the heat in a space or building, thus cooling that space.

Refrigerated air conditioning is similar to commercial refrigeration in that air is cooled in the same way—in a room or inside a refrigerated cabinet. The components of the system may be assembled in several different ways, but they accomplish the same goal: to produce refrigerated air that cools a given space. The main components are a metering device, an evaporator, a compressor, and a condenser (see Illustration 7-7).

Here's how the system works. The ***refrigerant*** (which is, at this point, about 75 percent liquid and 25 percent vapor) leaves the metering device and enters the ***evaporator.*** This is called the "low side" (meaning low pressure) of the refrigeration system.

The refrigerant mixture moves through the coiled tubing of the evaporator. As its name suggests, the liquid evaporates as it moves along. By the time it leaves the evaporator, it should be 100 percent vapor. (This is known as the "saturation point" of the refrigerant.) A fan is typically used to cool the evaporator area.

The vapor is drawn into the ***compressor*** by the pumping action of pistons and valves. It is super heated by the time it leaves the compressor, or the "high side" (meaning high pressure) of the system. The hot vapor travels through more tubes to the ***condenser,*** where it begins to turn back into liquid. Another fan is used here to help the condensing process, cooling the hot gas to help it return to its liquid state. By the time the refrigerant reaches the end of the condenser coil, it should be 100 percent liquid again. From here, an *expansion valve* controls the flow of the liquid refrigerant. Its pressure drops as it is forced through the valve, which causes its temperature to drop. The cold, low-pressure liquid refrigerant then cycles back into the evaporator and the process begins again.

An air-conditioning system is chosen primarily based on the size of the space you intend to cool and on the overall climate of your area. You will purchase a unit that provides a certain number of ***refrigerated tons.*** A ton sounds like a lot, doesn't it? For measurement purposes, 1 ton of refrigeration is equivalent to the energy required to melt 1 ton of ice (at 32 degrees Fahrenheit) in a 24-hour period.

Remember, in Chapter 5, we defined a Btu as the amount of heat needed to raise the temperature of 1 pound of water 1 degree Fahrenheit. Well, the melting of 1 pound of ice absorbs 144 Btu—so the melting of 1 *ton* of ice (2000 pounds) would take:

144 (Btu) × 2000 (pounds per ton) ÷ 24 (hours) = 12,000 Btu per hour

How many tons of refrigeration your system requires depends on many factors, including the size of the space, the number of people you'll serve at one time, and the type of cooking you'll do. A refrigeration specialist will do a *heating and cooling load calculation* to determine the size of unit you will need. In the United States, federal law also requires manufacturers to label air conditioners with an ***energy-efficient rating (EER)*** so you'll know how much power it will use. The EER is figured by dividing the Btus used per hour by the watts used per hour. The higher the EER, the lower the energy consumption.

The most popular HVAC system is known as a ***packaged air-conditioning system,*** so named because it is self-contained. A fuel provides heat, and a refrigeration system provides cooling. Heating and cooling share the same ductwork, dampers, and thermostat.

Speaking of air ducts, they must be properly sized to allow free passage of air. The fan and motors in a system must work too hard if the ducts are too small.

HVAC Technology Advances

In addition to ducts, there are regions of the country where you will need ***heat pipe exchangers*** when summer weather brings both high temperatures and high humidity, especially if the doors to your eatery are opened and closed often. The heat pipe, pioneered by the National Aeronautics and Space Administration (NASA) to cool the electronic components of spacecraft, uses no electricity. Installed as a part of your HVAC system, it evaporates and condenses a fluid in a continuous loop that transfers heat from the supply air to the return air stream. That is, it precools warm air before it reaches the evaporator coil, takes some of the moisture out of that warm air, and discharges it into the return air. Air flows through the heat pipe simply because of where it is placed; no power is required to force it to work. When several Burger King restaurants in Clearwater, Florida, experimented with heat pipes, they decreased the relative humidity of their buildings from 85 percent to 60 percent, and saved more than 20 percent on their electric bills!

Yet another option is to use the earth's constant temperature to heat and cool air. The ***ground source heat pump*** may be the most energy-efficient, clean, and cost-effective HVAC system available, but ***geoexchange technology,*** as it is called, is still fairly new. In the U.S., less than a million systems are installed nationwide. It is a combination of traditional plumbing equipment and air conditioning ducts, combined with large ***ground loops,*** flexible piping filled with water and nontoxic antifreeze. The loops of pipe are buried four to six feet underground, either vertically or horizontally, where they adopt the natural underground temperature of 55 degrees Fahrenheit. A simple electric compressor circulates the liquid and cools or heats the air going into a home or other building. The advantage is that, because it's already at a temperate 55 degrees, it requires less energy to heat or cool the indoor air to the desirable "end result" temperature. Another plus: no noisy condensers, and no pollution. Geoexchange or geothermal technology can be expensive to install, but some utility companies provide grants or rebates to encourage use of this nonpolluting energy source.

In climates famous for their high humidity, everyone knows comfort involves more than temperature. It requires reducing the moisture in the air, which also helps eliminate humidity-related problems like mold, mildew, and fungus growth and their associated odors. For this purpose, ***dessicant HVAC systems*** were created. A blower, usually located on the roof of the building, pulls fresh air through a rotor coated with either silica gel or titanium silicate. The combination of these substances and the rotor motion dries the inbound air. From that point, the drier air passes through a heat exchanger (a pipe, or another rotor), to either heat it or cool it. The air is then forced through ducts into the building's interior "as is," or into a conventional air-conditioning system for further cooling or heating. At the end of the "chain," another blower sucks in more air, into which the excess moisture from the system is dumped and vented back to the outdoors. The system can maintain a relative humidity level of 60 percent or less, and has the added benefit of keeping coils and filters cleaner.

Dessicant HVAC is an expensive proposition, and field studies have shown that it's probably not worth the expense for quick-service eateries where guests do not linger long in the building. But for full-service dining establishments, it's worth a look. If only fifteen guests a day increase their purchases by $4, staying longer for that extra drink or dessert in the air-conditioned comfort, you'll pay for your dessicant unit in less than a year—while making life a little more comfortable for both customers and employees.

Getting the right mix of temperature and relative humidity is especially difficult in foodservice, when the seated, relaxing customers have different needs—and perceive temperature and humidity differently—than the frantically busy staff members serving them. The American Society of Heating, Refrigeration and Air-Conditioning Engineers (ASHRAE), long regarded as the experts in this field, suggests a restaurant "comfort zone" of 76 degrees Fahrenheit, with 45 percent relative humidity. How you accomplish this is best left up to a qualified HVAC contractor.

Ventilation and Air Quality

The most important reason for ongoing maintenance of your HVAC system is its impact on air quality. Imagine, for a moment, a poorly designed ventilation system. In this restaurant, the guests are treated to each and every smell emanating from the kitchen—even the food that was cooked *yesterday*! Every time the front door opens (which is tough, because incoming customers must fight the pull of a badly balanced exhaust system), a rush of outside air pushes in, along with dust, insects, and uncomfortable drafts. The cleanup crew spends much of its time wiping off a greasy film that seems to settle on everything. The utility bills are higher than they need to be. The nonsmokers complain constantly about the smokers, although the two dining sections look as if they are spaced far enough apart.

Get the idea? Ventilation is part of your overall ambience. At best, it's hard to do perfectly. At worst, it may be jeopardizing the health of your workers and your customers. Indoor air quality ("IAQ") has become a huge concern, prompting lawsuits and legislation on the state level. The U.S. Environmental Protection Agency cites poor indoor air quality as a top public health risk. Many of today's buildings are more tightly sealed. Windows don't even open, meaning the people inside must depend on adequate ventilation from the HVAC system. All too often, they are not getting it.

Over the years, we have all heard the news stories about so-called "sick" buildings, with medically identifiable diseases and symptoms that affect people. Toxic molds and Legionnaires' disease are examples of ***building-related illness (BRI).*** Doctors also now recognize *sick building syndrome (SBS)* as a medical condition, when 20 percent or more of a building's occupants exhibit symptoms of discomfort that disappear when they leave the building. What could possibly cause BRI or SBS? *Legionella,* for example, grows in stagnant water with a source of iron, in a building's air-conditioning system. Air quality can also be seriously compromised by mold spores and dust mites. Chemical contamination may be a culprit. A class of chemicals known as ***volatile organic compounds*** is often to blame, including cleaning solutions, paints, pesticide products, and some construction materials. Because most HVAC systems take in air from the outdoors, consider the locations of any air intake vents. Keep them clear of:

- Automotive exhaust from streets or parking areas
- Air being exhausted from other nearby buildings
- Your loading dock, where delivery truck fumes could be a factor
- Garbage dumpsters and/or unsanitary debris
- Standing water

The primary IAQ issues for most restaurants are kitchen-related odor and smoke, and whether to allow customers to smoke cigarettes. The latter concern, increasingly, is being dealt with by states and cities passing Clean Indoor Air ("CIA") laws that ban smoking in most buildings. In 2003, more than 1,600 U.S. cities passed CIA ordinances, with many more considering them. (Sometimes bars and bowling alleys are exempted, but not always. The state of California went completely smoke-free indoors in 1998.)

The ASHRAE standards for proper ventilation are shown in Table 7-3. The ventilation rate is the amount of *cubic feet of air per minute* that must be replaced per person, and it varies somewhat with the function of the room itself. The three overall principles that can improve indoor air quality are:

- Bring in a sufficient amount of outdoor air
- Manage the flow of air within your interior spaces
- Use high-quality filtration systems
- Keep the HVAC equipment clean and in good working order

In working with HVAC professionals, you may hear a few catchy terms. *Energy Recovery Ventilation (ERV)* is a system in which energy is used to change both temperature and humidity of incoming air. The dessicant HVAC system is a type of ERV. *Demand Control Ventilation (DCV)* is a technology option that allows you to adjust the amount of outdoor air that comes into your system, much like a thermostat allows you to control temperature. DCV systems are popular in foodservice, because the number of guests varies greatly at different times of the day. Computerized sensors vary the amounts of air supplied and exhausted, depending on the occupancy of the space. The sensors are stationed in each "zone" of the

TABLE 7–3

AIR QUALITY STANDARDS

Area	Ventilation Rate	Occupancy Level
Dining rooms	20 cfm/person	70 persons/1000 square feet
Cafeteria, QSR	20 cfm/person	100 persons/1000 square feet
Bars, cocktail lounges	30 cfm/person	100 persons/1000 square feet
Public restrooms	50 cfm/water closet or urinal	
Kitchens (cooking)	15 cfm/person	20 persons/1000 square feet

Source: American Society of Heating, Refrigeration and Air Conditioning Engineers (ASHRAE), Atlanta, Georgia.

building, and they "sense" occupancy by detecting the carbon dioxide levels in the air. Other sensors monitor the air pressure of the system. Installing this type of ventilation system is expensive, and the sensors and controls must be calibrated on a regular basis, but it will save energy and money in the long run by freshening and exhausting air automatically, only as needed.

The "Smoking Section"

The debate about cigarette smoking has been a long and emotionally charged one in the foodservice industry, with no end in sight. Nonsmokers complain about the odors and hazards of breathing secondhand smoke and have plenty of medical evidence to back them up, while smokers claim they're being discriminated against and should be able to enjoy a politely puffed cigarette after dinner. Today, however, fewer people are smoking than ever before. According to the Centers for Disease Control (CDC), only about 21 percent of women and 26 percent of men in the U.S. are smokers, down from 45 percent in the 1960s.

Where there is not a Clean Indoor Air law, many restaurants have tried to walk the fine line between accommodating and infuriating patrons by separating dining areas into smoking and nonsmoking sections. Not surprisingly, the restaurant industry has received considerable advice from the tobacco industry about how to manage the smoking dilemma—plans with soothing names like the "Accommodation Program," "Courtesy of Choice," and "Peaceful Coexistence." What it doesn't change is the fact that most typical commercial HVAC systems are able to dissipate odors associated with cigarette smoke—but *not* the dangerous chemicals *in* the smoke.

A study by the University of California at Berkeley, done before that state's indoor smoking ban, found that even folks who sit in a nonsmoking section inhale the equivalent of one-and-a-half cigarettes' smoke if they stay for two hours.

In 2001, forty Boston area restaurants agreed to participate in a study of their ventilation system performance to measure airborne nicotine and carbon monoxide—only two of the 40-plus chemicals in smoke. From major chains to mom-and-pop eateries, cigarette-smoking machines were placed in the smoking sections of each. The restaurant managers were asked to turn on all air-handling systems, from HVAC to kitchen exhaust, to any specific electronic air cleaners or "smoke eaters" they may have installed. The results? In 30 of the 40 locations, air drifted from the smoking to nonsmoking sections—probably *not* the restaurants' intention!

The study authors' (all members of ASHRAE) conclusions: Lower nicotine concentration in the smoking area did not necessarily reduce it in the nonsmoking area. In other words, simply diluting the smokers' air or venting it out of the area wasn't an effective solution. The pollutants just blew elsewhere. The physical separation of the two areas also didn't seem to make much difference. Instead, the direction of the airflow is the key to keeping smoke out of smoke-free areas, and this study suggested separate HVAC systems for smoking and nonsmoking sections.

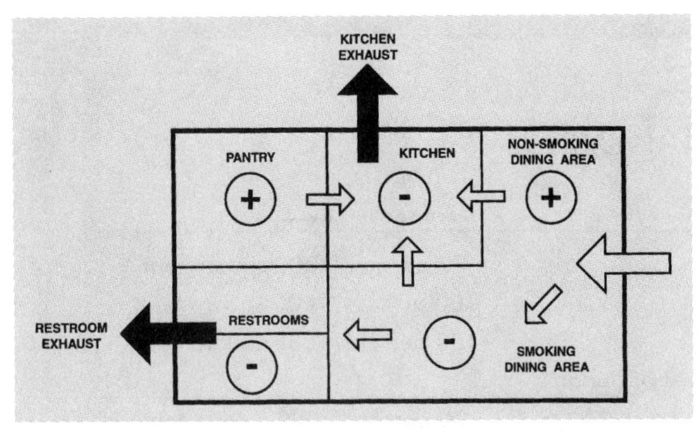

ILLUSTRATION 7-8 A pressure footprint that indicates the airflow of a room and its HVAC system.

Of course, it would be smart before you start making expensive decisions based on "customer preference" to find out just what that preference actually is. Identify the number of smokers and nonsmokers who frequent your business. The other critical consideration is the health of your employees. It is estimated that workers in restaurants that allow smoking are exposed to three to five times more tobacco smoke than in other types of jobs; in bars, four to six times more tobacco smoke. After California's Clean Indoor Air legislation passed, the Journal of the American Medical Association published a survey of bartenders, taken both before and after the law took effect. Before, 74 percent reported respiratory problems and 77 percent complained of eye, nose, and throat irritation on the job. After the smoking ban, the figures dropped to 32 percent and 19 percent, respectively. With health insurance costs not getting any cheaper, it's something to consider.

The next step is to become familiar enough with your HVAC system that you can correctly place the smoking area to limit the drift of tobacco smoke. We've noticed an annoying tendency to seat smokers directly beneath briskly whirring ceiling fans, which only seems to freeze the smokers, blow ashes and paper napkins off the tabletops, and push the smoke toward everyone else.

Your HVAC system is designed to provide ***ventilation***—that is, to blow "fresh" or properly treated, temperature-controlled air into a room, and to send a corresponding amount of "used" air out of the room. All airflow has a direction, and all HVAC systems have areas of positive pressure (to force air out) and negative pressure (to invite air in). Any system's positive and negative pressure and airflow can be drawn as a simple diagram called a ***pressure footprint,*** as shown in Illustration 7-8. Plus signs on the chart indicate positive pressure. A ***positive air pressure zone*** means more air is supplied to an area than is removed from it. Minus signs indicate negative pressure. A ***negative air pressure zone*** is created when more air is removed from an area than is supplied to it. The arrows on the drawing indicate airflow direction.

Briefly, you must locate nonsmoking areas in places where air enters the room and naturally "pushes" smoke away, and place smoking areas near return or exhaust grilles, where the smoke will be drawn out of the room. Remember, air naturally flows from areas of positive pressure to the areas of negative pressure. Slight pressure differences are fine, but large pressure differences are almost like miniature indoor wind gusts. They can cause problems.

Let's look more closely at this footprint, for a small restaurant with a single air supply system that is located in the dining area. Exhaust fans are used in the kitchen to eliminate cooking smoke, fumes, and grease, and in the bathrooms to eliminate odors and stale air. The smoking and nonsmoking sections could be in either "corner" of the dining room, because both corners are close to the incoming air supply and to exhaust areas.

For restaurants of this size, the outdoor air supply is sometimes eliminated, based on the assumption that negative pressure on the kitchen exhaust system will end up drawing in outdoor air more passively, as doors and windows are opened. A small "unit" air conditioner or heater may be installed to recirculate interior air in the dining room, but this would be a mistake. The kitchen needs outside air to keep venting its grease and smoke away from the building, and the dining room needs outside air as a fresh air source.

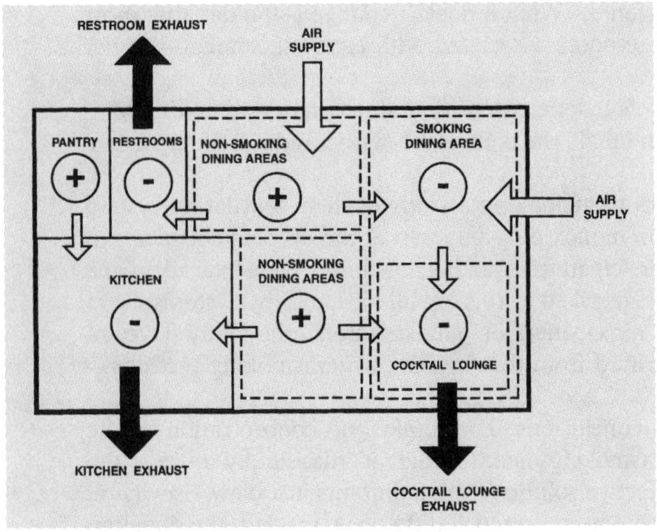

Illustration 7-9 is a more complex footprint of a restaurant with a bar, allowing a sizable smoking area shown in the diagram as negative pressure space. There are two air supply fans. One blows fresh air into

ILLUSTRATION 7-9 A more complex pressure footprint.

the nonsmoking dining room, which forces the airflow into the smoking areas and out the far side of the bar as exhaust. A second fan adds more fresh air into the smoking section, which is also exhausted through the cocktail lounge.

Controlling this system may look easy on paper, and, while it should be relatively simple, adjustments must be made for how crowded the restaurant gets at different times of day. The lounge area's exhaust fan must be timed to operate in tandem with the air intakes in the dining areas; the outdoor dampers in the smoking section's air supply should open automatically, to add an amount of makeup air equivalent to what goes out in the lounge. It's a balancing act that never ends.

HVAC System Maintenance

No matter what type of HVAC system you choose to install, there is one more important design consideration. That is, think about how difficult it will be to get to the components when they need to be inspected, cleaned, or serviced. Preventive maintenance is an essential part of keeping your system working at its best, saving you the most money in the long run and prolonging the life of the system by as much as ten years! After a good cleaning and "tune-up," your system will also automatically be up to 7 percent more energy efficient.

It's best for a professional HVAC service company to put your business on a regular inspection and maintenance schedule. Experts say the most common problems with HVAC systems are:

- Loose belts
- Dirty air filters
- Poorly lubricated bearings

Be certain that the inspections address these items. Air filters must be cleaned or replaced; motors are lubricated; heat transfer surfaces and drain lines are cleaned; switches and thermostats are checked for accuracy. These service calls should be scheduled in the month immediately before your area's cooling season, and again just before the peak heating season.

You might also ask the service technician about the possibility of computerizing your HVAC controls if they aren't already. This allows you to remotely control the system, and provides valuable diagnostic tools that track power usage, temperature limits, refrigerant levels, air filter conditions, maintenance records, and more. After spending more than a million dollars a year in HVAC-related service agreements, the casual dining chain Applebee's installed a company-wide computerized system. With input from the individual location managers, Applebee's headquarters can now remotely control the climate functions in each restaurant over the Internet, without having to put lockboxes on the thermostats.

Basic control options for simpler HVAC systems are on-off switches, which look and work like light switches, and timelocks, which are programmable timing devices to turn fans on and off at certain times of day. Unless you truly understand how to operate a timelock and how to change it when conditions change, you may become so frustrated you'll just disconnect it. Be sure the installer fully familiarizes you with whatever controls you decide on.

The foodservice HVAC system must not only circulate air, it must remove dirt and chemicals as well. This creates the need for air filtration. Air filters remove airborne dust particles. The most widely used standard for filter effectiveness is the ASHRAE *dust spot method,* a rating that labels the filter with the average percentage of dust it removes. There are many bargain filters rated as low as 25 percent, but the suggested minimum rating is from 65 to 85 percent and you can even get 95 percent ones. These higher-rated filters are known as *High-Efficiency Particulate Air (HEPA)* filters. For serious odor removal situations, an activated carbon filter adds an extra amount of effectiveness. Superior filters cost more, but can pay for themselves within a few months with the cost savings from a cleaner system that ends up using less energy. No matter what filter you select, if it is not changed regularly, it won't do its job.

The smoking versus nonsmoking issue has also created a need for products such as the *electronic air cleaner,* which rids the air of gases and odors along with dust particles. This ceiling-mounted unit draws room air into it, sucking the air through an electrically charged field (called an ***ionizer***) and into a series of metal plates. The plates attract the newly charged airborne particles and collect them, ridding

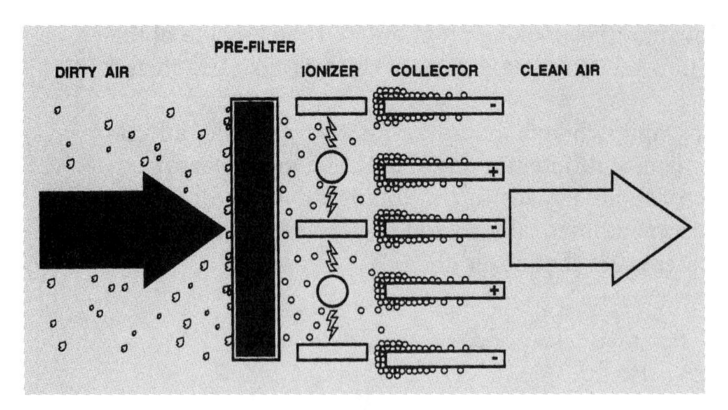

ILLUSTRATION 7-10 **How electronic air cleaners work.**

the air of 95 percent of its impurities (see Illustration 7-10). An optional charcoal filter at the end of the "cleaning" process helps remove odors and some gases; some units require regular filter changes, others do not. A quiet fan returns the cleaned air back into the room.

As a restaurateur, you have the following basic responsibilities in setting your air quality standards:

- **To provide sufficient ventilation for basic indoor air quality, including control of odors, grease, and smoke**
- **To provide a comfortable environment, for both smokers and nonsmokers**
- **To minimize energy use and operating costs**

To accomplish this, we offer a number of suggestions, from design to installation.

Some have been adapted from information in the technical bulletin *Ventilation for a Quality Dining Experience.*

Before Construction. On a floor plan, define how each area of the building will be used, as well as how it may be converted to some other use over time. (If you'll be leasing existing space, ask if the lease includes regular maintenance by a professional HVAC technician.)

During Design. Review the ventilation zones and airflow patterns of the pressure footprint to see that they make sense and comply with standards for that type of use. Evaluate the ventilation specifications of ASHRAE (Standard 62-1999), as well as your local codes and ordinances. Check the locations of outdoor air intakes to be sure they are not near pollution sources. Check the control sequence of intake and outgo for its ability to stay within your standards under different operating conditions. Finally, check the design of your controls, to see that they are simple enough to be easily operated by your future employees.

During Construction. Select construction materials and finishes for low toxicity and easy maintenance. Choose filters that will help you achieve your air quality standards. Evaluate plans for operation, maintenance, and cleaning of your system. Consider painting or rustproofing outdoor units if your climate warrants it.

During Start-Up. Inspect the new construction to ensure it complies with design specifications. Bring in a certified contractor to test, balance, and review your airflow patterns. Review the operation and maintenance manuals for instructions and service information. (Put this on your maintenance cards or charts.) Follow the formal commissioning (startup) procedure for the system, to be sure it operates correctly under a variety of conditions.

Daily Maintenance. Make sure there's enough outdoor air ventilation during and after cleaning and/or pesticide applications. Avoid using strong deodorant, disinfectant, and pesticide products. Use high-efficiency, filtered vacuums to clean all carpets and upholstery. Regularly check and clean exhaust filters and grease traps.

Monthly Maintenance. Check controls and ventilation equipment once a month, just to be sure they are working well and in compliance with your area's regulations. This means checking and changing air supply filters and cleaning condensate pans and humidifiers, as necessary. Always use the correct size of filter and keep a written record of filter replacement. Keep the area around the unit itself free of debris, stacked boxes, anything that would block airflow.

Quarterly Maintenance. Check and clean the fans, air intakes and return grilles, vents and exhaust ducts, and the blades of ceiling fans. Check the dampers, tightening or lubricating them as needed. Check inside the air conditioner for leakage, which should be repaired promptly.

Annual Maintenance. Look closely at all the routine cleaning and pest control products you use to be sure they are the least toxic to do the job. How many do you really need? Are they safe? Keep a notebook of information about them. Check and clean cooling coils and condenser coils in the spring; check and clean heat exchangers and ducts and determine overall furnace efficiency in the fall. Make sure any service person you hire to check your system is looking at belts, fans, filters, thermostats, and the calibration of the system.

Kitchen Ventilation

In the kitchen, professional advice is needed to create a ventilation system. In fact, many cities require that licensed mechanical engineers design these systems. Kitchen ventilation requires changing the ambient air to remove heat, odors, grease, and moisture from the work spaces, primarily the hot line and dishwashing areas. Without proper ventilation, heat and humidity become unbearable, grease slowly but surely builds up on walls and other surfaces, and odors waft into the dining area nearby. Because the doors between kitchen and dining room swing constantly, an improperly cooled kitchen also means more work for the dining room air conditioner, which must compensate for the blasts of hot air. Pilot lights and flame settings are affected by negative pressure. And, perhaps most important, grease buildup in hoods and ducts is the most common cause of kitchen fires. So, as you can see, there are lots of reasons to get it right the first time.

Early kitchens were equipped with propeller-type exhaust fans that were mounted on walls. As kitchen equipment became more complex, designers developed the ***hood*** or *canopy* mounted directly over the cooking equipment, to draw smoke, moisture, heat, and fumes up and away from the kitchen. This led to installation of ducts, completely separate from other HVAC system ducts, connected only to the kitchen exhaust fans. For many years, the national building codes mandated a fixed exhaust volume, determined by the size and type of hood and the cooking equipment beneath it, no matter if cooking was taking place.

Today's exhaust system can respond automatically to the amount of fumes and/or heat being generated below it. Exhaust quantities can be adjusted from 33 to 50 percent with variable-speed fans. The ducts, hoods, and fans do double-duty, by bringing in fresh air and removing contaminated air. Known as a ***balanced system,*** it requires expert design and upkeep to prevent wasting costly, treated air and also to prevent drafts. The system regulates the amount of outside air introduced into a kitchen area, based on a predetermined minimum level of carbon dioxide. If that level is exceeded, the system increases incoming air velocity. Illustration 7-11 is a diagram of a variable-speed hood.

Hoods and fire protection systems have the most stringent legal requirements of any foodservice equipment. The health department is concerned about sanitation; the fire inspector wants to know fire hazards have been abated; environmental agencies want assurances that you're not spewing pollution into the air. In most states, there are also insurance regulations that govern the design and operation of ventilation systems. A commonly accepted procedure nowadays is that a third party (not you, in other words) must come in every six months to inspect your system and make sure it's clean and working properly—or you may be denied fire insurance.

The National Fire Protection Association (NFPA) sets the national norms, and Standard 96 applies to the construction and installation of hoods, canopies, exhaust fans, and their fire protection systems. The NFPA does not make exhaust volume recommendations, but Underwriters' Laboratories (UL) and the Building Officials and Code Administrators (BOCA) do.

There are two basic types of exhaust hoods:

- *Type I* is used for collecting and removing flue gases, smoke, and some of the grease generated by the cooking process. There are two subtypes in this category: "Listed" and "unlisted." Listed hoods meet the requirements of the Underwriters' Laboratory 710 Standard.
- *Type II* is used where steam, odors, and heat are generated, such as over dishwashers and steam tables. It is not intended to handle removal of smoke or grease-laden air.

UL and NFPA standards require that all exhaust hoods be constructed of minimum-gauge stainless or galvanized steel, with a liquid-tight, continuous external weld; and that they include an approved filter or removal device to get rid of grease. Hoods, their filters, and ductwork must all be separated from any combustible material by at least 18 inches. (Smaller clearances are allowed, but

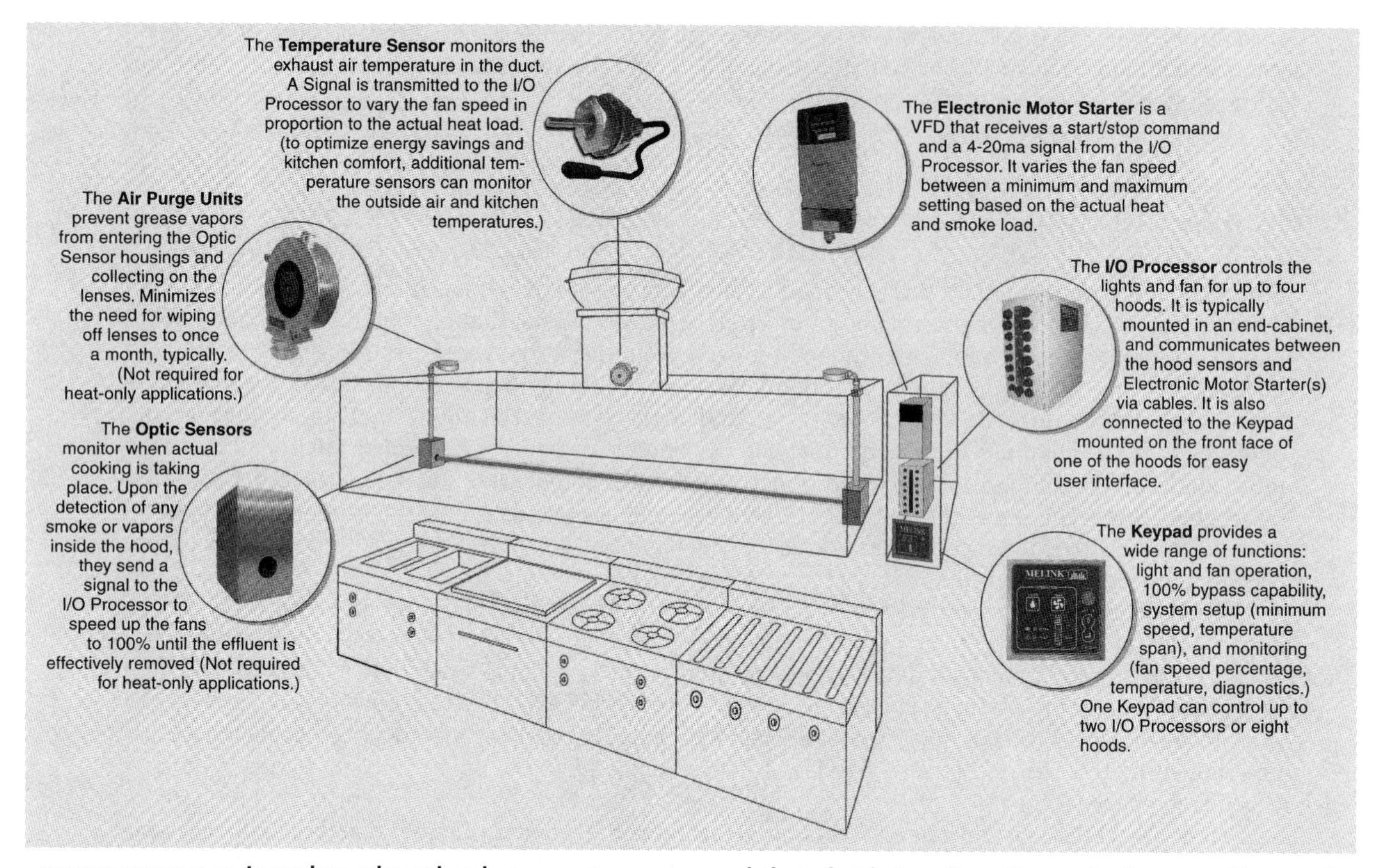

ILLUSTRATION 7–11 In modern exhaust hoods, temperature sensors and photoelectric "eyes" monitor smoke, fumes, and heat and adjust air levels automatically. The kitchen staff can manually override the system if necessary.

SOURCE: MELINK CORPORATION, CINCINNATI, OHIO.

only if an approved fire barrier is installed.) The exhaust system's point of discharge at the top of the building must be at least 40 inches above the roof. The standards also require a fire suppression system, which includes automatic shutdown of cooking equipment and extinguisher nozzles above each piece of surface cooking equipment (broilers, fryers, ranges, etc.) that can be activated manually or automatically as needed.

The amount of exhaust air removed from a kitchen, and the rate of its removal, depends on many variables. These include the size of the hood over the cooking equipment; the design and location of the hood; and the amount of steam, smoke, or vapor generated by the appliances as they are used. For instance, a frying kettle gives off a significant amount of moisture; a convection oven does not. A hood mounted on a wall is not affected by cross-drafts as a hood mounted above an island in the center of the kitchen would be. There are three basic types of hood installations: wall, corner, and island. There are also several different ways the "stale" air is vented away from the hot line, and each type of hood design has a different name.

An expert will use one of two basic ways to determine exhaust requirements. The first common method is to design a ventilation system to provide a certain, specific number of changes of air in a kitchen space each hour the system is in operation. Industry experts recommend that a ventilation system provide 20 to 30 total air changes per hour; or that it exhausts a total of 4 cubic feet per minute (cfm) per square foot of floor space. To calculate the volume of air in the kitchen, use the total dimensions of the room, without deducting any square footage for the space occupied by equipment.

The second method is to maintain a fixed velocity of air across all of the hood openings. In other words, the air has to move a certain number of feet per minute across the face of the hood. Most hood manufacturers indicate in their specifications the air volume necessary for proper operation, which may vary a bit if the hood is installed in a corner or against a wall. The BOCA recommendation is that you'll need an airflow of at least 100 cfm per square foot of cooking space beneath your hood for a wall-mounted unit; 150 cfm for an island unit. Underwriters' Laboratories suggests 150 cfm for wall units and 300 cfm for island units, but makes its calculations using *linear* feet, not square feet.

So, for example, let's say we have an island canopy that is 20 feet long and 4 feet wide for a total of 80 square feet. Using the BOCA standards, we'd need 12,000 cfm: 80 square feet × 150 cfm. Using the UL standards, however, we'd only need half that capability: 20 feet × 300 cfm = 6000 cfm. Again, state or local codes may govern your minimum requirements.

Earlier in this chapter, we talked about positive and negative air pressure zones and the need to create air balance. In the kitchen, because exhaust needs are so great, a system to add makeup air is required. The critical design consideration for kitchen exhaust and makeup air is to build up enough air velocity to capture vapors, grease, and heat at the cooking surfaces, without creating "wind gusts" that blow out pilot lights and make doors difficult to open or close. To achieve the slight difference that keeps air moving without "gusting," makeup air generally replaces about 85 percent of the exhaust air. It is normally introduced into the kitchen from supply grilles (also called ***diffusers***) located as far across the room as possible from the main exhaust, as close as possible to the ceiling. A combination of diffusers that introduce and vent air away from a space can be used to provide specific, directional airflow within that space—regardless of its air pressure relative to other, surrounding spaces. When deciding on a location for the diffusers, make sure you have first decided where to put your pass window and/or pickup counters. Cross-ventilation is usually a good thing, but not if the air blows directly across the food and cools it off before serving!

When selecting a makeup air system, you may hear the terms "tempered" and "untempered." A ***tempered*** system is one that has the capability to heat or cool incoming air, depending on the season; an ***untempered*** system does not. Untempered systems are touted for their energy-saving features, but, while they work well in moderate climates, they may have drawbacks in areas with seasonal temperature extremes.

Two types of exhaust fans may be used to blow air out of the system: propeller-type fans and blower-type fans (also called centrifugal fans).

You've certainly seen ceiling fans, which operate on the same principle as the propeller-type fan, with a group of blades connected to a motor. They don't need much horsepower to move large volumes of air, but they can be noisy and accumulate grease and dirt on their blades, which reduces their effectiveness.

The blower-type or centrifugal fan consists of a motor inside a specially formed housing and blades that incline forward or backward. Blowers cost more than propeller fans to install, but they work better for kitchen exhaust systems because they stand up better to the rigors of heat and static pressure.

Both types of fans have the same job: to remove smoke from the cooking surface. The amount of air they exhaust must be equal to or greater than the amount of air generated naturally by the cooking equipment. Without sufficient ***capture velocity*** to pull the stale air up into the exhaust system, it will roll off the cooking line into the rest of the kitchen. This ***rollout*** condition is what you're trying to avoid by having exhaust fans in the first place.

Kitchen ducts are also subject to a whole host of special requirements so that the hot grease and particles they carry cannot start fires. Ducts must be made of a minimum-gauge steel, welded into a single sheet (not screwed together), and enclosed in a fire-resistant shaft throughout the building. More about fire safety in Chapter 8.

When exhaust fans are located on roofs, grease buildup over time can damage the roof itself. Several manufacturers offer ***grease guards*** to prevent this, installed at the base of the fan to catch grease before it can hit the roof.

Inside the canopy itself, grease buildup can also be a problem. So a grease filter is an integral part of any hood installation, trapping the grease before it enters the exhaust duct. The filter is a set of interlocking metal baffles that remove grease by centrifugal force, collecting it and using gravity to drain it downward into a trough and then into pans attached to the hood. These pans, and the filter system itself, must be easily accessible and removable for cleaning. Some codes specify the minimum distance allowed between the lowest edge of the filter and the cooking equipment below. Frequent cleaning is absolutely necessary to maintain a filter's effectiveness. If grease is left too long, it can harden, "bake" on in the heat of cooking, and hamper the airflow in the exhaust system.

Another method of grease removal, known as ***extraction,*** takes place as the exhaust air moves through a series of specially designed baffles. The air moves at high speed and must change direction in order to go through the baffles, throwing out the grease particles by centrifugal force along the way. The particles collect into a grease cup, which holds more than 6 pounds of grease and can be cleaned and replaced daily, or the particles are collected in grease gutters, which are also cleaned daily.

Large extraction systems for high-volume cooking operations often include a wash cycle that can be activated by pushing a button on the system control panel. Hot water and detergent are sprayed inside the system to wash the day's grease accumulation from it. Afterward, the wash cycle shuts off automatically. This type of cleaning is handy if the chore of cleaning large numbers of filters, fans, and ducts is too expensive and/or time consuming.

Overall, the kitchen exhaust system can be the most expensive single purchase in a kitchen, so be sure that there is one person or company that can be held accountable for its performance. In too many cases, a budget-minded entrepreneur pieces together a hood from one company, ductwork from another, a fan from another, and has somebody's brother-in-law install it all. When the system does not work properly, each supplier blames the other, and the facility is stuck with a system that is a continual problem.

In some situations, space limitations, building configuration, or prohibition of structural changes prevents the use of a standard exhaust system. In these cases, a ventless exhaust system is an interesting new option. The Skydome in Toronto, Canada, has a ventless system, and its owners estimate it cost one-half million dollars less to install it than to try to design and install ductwork for this moving, dome-shaped structure. Ventless systems are also an option if you are leasing space, because you can take the whole package with you when you leave—an impossibility with standard ductwork!

The ventless exhaust system is self-enclosed and uses every possible type of filtering system instead of the usual air intake and outgo: a baffle filter for grease particles, an electronic air cleaner, a charcoal filter, and fans to discharge the newly cleaned air back into the cooking area.

You can purchase individual pieces of equipment with their own, self-contained ventless systems, or you can buy a ventless system that comes already assembled from the factory, to install above existing equipment. They include fire protection systems and automatic shutoffs in case of malfunction. One important note: Most cities will only allow electrical, not gas-powered, equipment under a ventless hood.

7-5 AIR POLLUTION CONTROL

A southern California study found that restaurants, particularly those that do a lot of frying, produce approximately nine times more air pollutants than a city bus! Griddles and deep-fat fryers in the four-county Los Angeles area were found to emit a total of 13.7 tons of particulate matter and 19 tons of volatile compounds per day—as much as an oil refinery! Most of these harmful compounds come from the fats that drip onto open flames during cooking.

This research was done in conjunction with Los Angeles' efforts to reduce air pollution by passing some strict new ordinances that would include, among other things, mandatory pollution control for any eatery that cooks more than 50 pounds of meat and/or 25 pounds of other foods per day. Other large U.S. cities are also interested in this issue, so you might as well be, too.

What exactly is air pollution? Smoke and gases, grease, airborne dirt, and germs begin as individual particles, too small to see until thousands of them are grouped together. One particle is measured in microns—one micron is 1/25,400 of an inch long—and that's what air pollution control equipment must remove from the air system.

Let's use smoke as an example. A single particle of restaurant kitchen smoke is 0.3 to 0.8 micron in size, so it is much easier to measure it in terms of its density or ***opacity***—the degree to which the smoke blocks the flow of light in a room. An opacity level of 100 percent would be solid black; 0 percent would be completely smoke free. The cities that have adopted smoke pollution ordinances usually require an opacity level of no more than 20 percent.

That sounds reasonable, until you consider that the average charbroiler emits enough smoke for a 60 to 70 percent opacity level. Opacity levels can actually be measured by instruments (called opacity meters and cascade impactors) but, more often, the inspector's own eyeballs and experience are used to determine what's in compliance and what's not.

Grease particles are also measured in microns, and are generally larger than smoke particles—approximately 10 microns and up. That's why your grease extractor has to be top of the line, capable of removing 95 percent of the grease it processes.

Cooking odors are molecules generated by the combustion of animal or vegetable products during the cooking process to make an extremely complex mixture of ***reactive organic gases*** (referred to by experts as *ROGs*). Grease absorbs a small percentage of odor ROGs, but most of them float around aimlessly in the air, much too small to be removed by any type of filter. One popular method to remove odor ROGs is to push them through a layer of activated charcoal, which absorbs and retains them. Unfortunately, charcoal is highly flammable, and, when it becomes fully saturated, it begins to release odors—sometimes worse than the original odors! A second common method of odor removal is to use potassium permanganate, which oxidizes and actually solidifies the molecules, then retains them. Either charcoal or oxidation can remove 85 to 90 percent of odor molecules from the air.

Much like electronic air cleaners, there are also air pollution control units, which work using electrostatic precipitation (ESP). Electrostatic cells inside the unit are made of aluminum plates, some grounded and others charged with 5000 volts of electric power. As air enters the unit, it moves through a series of thin electric wires, which charge it with 10,000 volts of power. Smoke and grease particles in the air receive a positive charge from the wires. They're repelled by the positive plates, but attracted by the negative plates, so they collect on the negative plates. Then, on a regular basis, the system is washed with hot water and detergent to remove the accumulation of grease and smoke particles. Often, the fire protection system is wired to the air pollution control unit, so both can go to work quickly in case of a smoky fire. The least sophisticated models will only abate smoke; others will also control odor.

SUMMARY

Lighting is the single most important environmental consideration in foodservice. If your budget allows, hire a lighting design consultant to help you plan the location and light levels you will need, both natural and artificial. Coordinate the lighting design with the color scheme you choose, both for safety and for aesthetics. Take advantage of new lighting technology like LEDs, which may be more spendy at first but will save money in the long run because of their longevity. The colors you use for walls and furnishings will depend on the lighting sources you choose.

Noise control is another consideration. Pay attention to both the frequency and the intensity of the noise level—not just what the guests hear, but the kitchen clamor. Architectural modifications, placement of equipment, furnishings, and draperies can all help modify noise levels. The judicious use of music in your dining area is usually an asset, but only with a good sound system. And remember to pay the licensing fees to use the music, or you'll face fines from the "performing rights" organizations.

Plan your heating, ventilation, and air-conditioning (HVAC) system so it can provide a consistent, comfortable atmosphere during inevitable temperature changes. Your kitchen exhaust system is part of the HVAC, and it should be kept clean and well maintained. Decide whether you want to accommodate smokers, and how to separate them from nonsmokers. Building codes, health codes, and insurance regulations will all have major impacts on your HVAC and kitchen ventilation decisions, which are best left to experts for both installation and maintenance.

The best thing about today's technology is that practically any environment-related system can be automated to save energy by sensing occupancy or activity and adjusting accordingly. Lights can be placed on automatic timers or dimmers; music levels will adjust themselves automatically to fit the crowd noise. Even your kitchen exhaust system can sense when, and how much, cooking is being done and increase (or decrease) its performance. These automated functions are not inexpensive, but they will pay for themselves promptly with time and energy savings.

STUDY QUESTIONS

1. Explain three ways to save on the cost of lighting.
2. In lighting terminology, what is the difference between *efficiency* and *efficacy*? Which would you say is more important in choosing a lighting system?

3. Why do fluorescent lamps work better now than they used to?
4. What are some advantages of using LEDs? Is there a situation in which you would not use them?
5. List two things you can do to reduce reverberant sound in a dining area, and explain why they work.
6. List three ways you can reduce noise in a busy kitchen.
7. If your restaurant's air-conditioning system didn't seem to be working well, what is the first thing you'd check?
8. Would you choose to have a smoking section in your restaurant? Why, or why not?
9. What is a *dessicant* HVAC system, and why would you need one?
10. What are the negative effects of a poorly ventilated kitchen? (Name three.)
11. What is a balanced ventilation system, and why is it important?
12. What's the difference between regular ducts and kitchen exhaust ducts?

CHAPTER 8

SAFETY AND SANITATION

INTRODUCTION AND LEARNING OBJECTIVES

Improving safety in and around your foodservice business can positively impact your staff's morale and productivity, decrease your insurance costs and legal liability, and even attract more customers. The challenge is that "safety" is an intangible—difficult to quantify, and unique to every operation. One constant aspect of food-related safety is sanitation. This entails everything from proper storage to keep rodents and insects out of your pantries; to washing dishes and utensils in the correct temperatures for sterilizing them; to choosing and placing kitchen appliances so they are easier to clean around and behind. Keeping food at safe temperatures is another huge part of working in a professional kitchen. So in this chapter, the wide-ranging topics include:

- Fire safety
- Ergonomics
- Employee comfort and safety
- Flooring and carpeting
- Sanitation
- Food safety
- Waste management

Safety is a state of mind. It requires making a conscious decision to conduct yourself, and your business, in a certain manner. It means identifying any potential hazard in your work processes or in the facility itself, developing safety measures to minimize these hazards, and training employees with an effective, ongoing safety program.

8-1 FIRE PROTECTION

Since we ended the previous chapter with a discussion on kitchen hoods and exhaust systems, let's continue in that vein. About one-third of all restaurant fires originate in the kitchen, and they are generally flash fires on cooking equipment. If the kitchen staff has had the proper training and the right safety equipment is available, these types of range top fires can be extinguished within moments. If not, they can quickly expand into the ductwork, reaching 2000 degrees Fahrenheit as they come into contact with highly flammable grease and lint particles. Therefore, an *automatic fire protection system* is a necessity. In fact, most state insurance departments require a fire safety inspection from an exhaust hood expert before insurance companies can issue a commercial fire insurance policy. As we've mentioned, the site must usually be reinspected every six months to keep the insurance in force. Even if the six-month rule doesn't apply in your area, it is a good idea to have your system professionally cleaned and checked twice a year anyway.

The National Fire Protection Association is the authority on this topic and sets the stringent regulations for commercial kitchen installations. Most canopy manufacturers offer fire protection systems as part of their package, including installation, but you can also hire an independent installer.

An automatic fire protection system, such as the one shown in Illustration 8-1, consists of spray nozzles located above every piece of external (not ovens) cooking equipment on the hot line. There are very specific rules about the numbers of nozzles and their locations:

- Range tops require one nozzle for every 48 linear inches.
- Griddles require one nozzle for every 6 feet of linear space.
- Open broilers (gas, electric, or charcoal) require one nozzle for every 48 inches of broiler surface.
- Tilting frying pans require one nozzle for a surface 48 inches in width.
- Fryers require one nozzle apiece or one nozzle for every 20 inches of fryer surface.

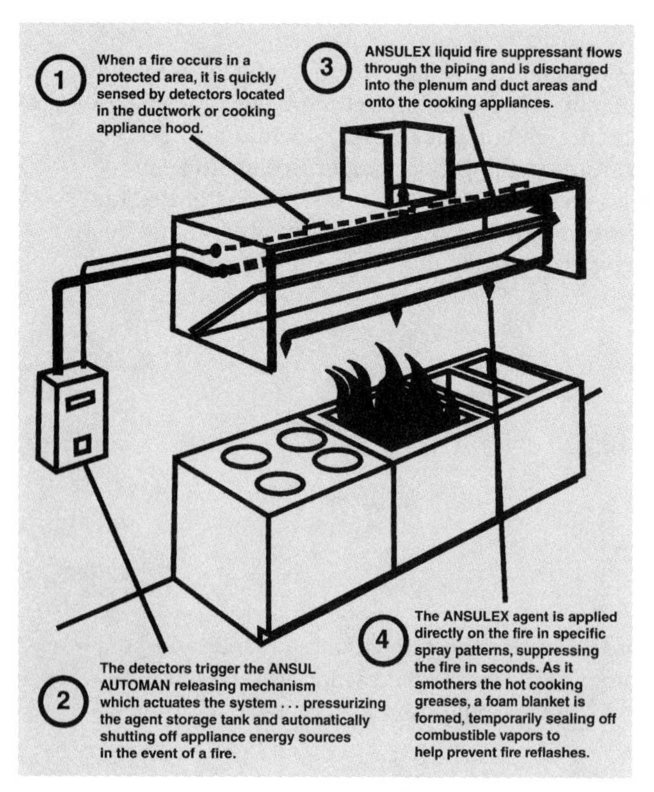

ILLUSTRATION 8-1 **How an automatic fire protection system works with the exhaust hood.**

SOURCE: ANSUL FIRE PROTECTION COMPANY, MORINETTE, WISCONSIN.

Nozzles are placed between 24 and 42 inches above the top of the equipment. (This varies depending on the type of appliance.) The nozzles activate automatically to shoot water or a liquid fire retardant at the cooking surface when the temperature reaches 280 to 325 degrees Fahrenheit. The heat detector may be located in the ductwork or in the hood.

Inside the ductwork, there is also an internal fire protection system—a fuse link or a separate thermostat is wired to automatically close a fire damper at the ends of each section of ductwork. The exhaust fan shuts off, and a spray of water or liquid fire retardant is released into the interior. There are other, similar systems that can be operated by hand instead of automatically. Some keep the exhaust fan running, to help remove smoke during a fire.

In addition to the exhaust system's fire protection, several hand-held fire extinguishers should be mounted on the kitchen walls, and employees should know how to use them. The automated system, when it is triggered, is so thorough that you must close the kitchen and begin a major cleanup, so often a hand-held extinguisher is sufficient for minor flare-ups, and a lot less messy.

Finally, as with any other public building, ceiling-mounted sprinkler systems are also worth investigating, because their installation may significantly reduce your insurance costs. There is a common misperception that, if it

BUILDING AND GROUNDS

TEN COMMANDMENTS OF FIRE SAFETY

1. Make certain that portable fire extinguishers are properly located and mounted; of the correct type, size, and properly identified; and fully charged and maintained.
2. Train all employees in the proper use of portable fire extinguishers.
3. Train employees in conducting a patron evacuation of your restaurants.
4. Post emergency telephone numbers in appropriate locations.
5. Paths of exit must be unobstructed, well lighted, and equipped with emergency lighting and exit signs.
6. Do not permit the accumulation of refuse or rubbish. Follow recognized housekeeping procedures.
7. Clean your kitchen hood, filters, and duct of grease accumulation.
8. Have your kitchen hood's automatic fire protection system inspected and serviced by qualified persons every 6 months.
9. Have your automatic sprinkler system inspected annually.
10. Inspect your electrical system for frayed wiring, prohibited use of extension cords, overloading of circuits, clearance around motors to prevent overheating, and ready accessibility to all electrical panels.

Source: **Reprinted with permission from *Restaurant Business.***

detects even one wayward flame, the entire sprinkler system will douse the whole building, but this is generally not the case. In fact, most restaurant sprinkler systems have heads that activate only when a fire is detected directly beneath them.

Ask your local fire department for suggestions and fire safety training tips for employees.

8-2 ERGONOMICS

The idea of *environmental design* is derived from the field of human engineering, also known as ***ergonomics.*** Ergonomics is an applied science that involves studying the characteristics of people and designing or arranging their activities (sports, on-the-job duties, and so forth) so that they are done in the safest and most efficient manner. This includes designing machines and work methods with safety, comfort, and productivity in mind.

In foodservice, here are just a few ways that ergonomics impact design:

- Planning easy entrance to and exit from the facility
- Placement of public areas to make service easier
- Design of specific work areas to facilitate safer or more sanitary work
- Finding alternatives to repetitive tasks or heavy lifting

To illustrate the many options, let's walk through a well-designed, "human-engineered" restaurant. First, notice that the flow of guests into the lobby area is not impeded by any unnecessary barriers. Comfortable, well-arranged seating eliminates cross-traffic, creates intimate dining areas, and minimizes the clutter of too much movement by wandering guests or staff. The guests' feelings of comfort and security have been taken into account in the arrangement of tables and the control of noise, temperature, and humidity. From the use of lighting and color, it is clear that the space planners paid attention to psychological and physiological comfort factors, and it is also evident that this can be accomplished without a commitment to expensive remodeling projects.

As we proceed to the back of the house, we notice that kitchen equipment has been chosen for easy cleaning and maintenance as well as durability. There are convenient access panels for plumbing and electrical connections. Whenever possible, equipment is mobile and can roll between locations. In some cases, even ovens and refrigerators can be rolled on wheels. Aisles are wide enough to accommodate people and equipment. There are carts to roll heavy supplies around, so people don't have to lift and carry them.

When food is delivered, handling time is minimal, because the storage areas are close and organized efficiently. The big walk-ins are nearest the receiving areas, with smaller, ancillary refrigeration units nearer the hot and cold production areas.

The dish room is arranged to vent heat away and to eliminate unnecessary employee movements. Noise from the dish room doesn't spill into the dining area, because it is well insulated.

The entire back of the house has been designed for ease of cleaning. Employees can work comfortably, because their work areas are well lit, and free of grease and fumes. Work surfaces are poised at the correct height for minimal back strain and to avoid reflecting glare from lights.

In the late 1990s, ergonomics became more than an interesting idea. The U.S. Occupational Safety and Health Administration (OSHA) began requiring some employers to adopt full-scale ergonomics programs to change—or at least, limit—working conditions that contribute to ***musculoskeletal disorders,*** or "MSDs," such as carpal tunnel syndrome, back pain, and tendonitis. OSHA lists the risk factors for MSDs as:

- Repetitive work (typing, chopping, etc.)
- Using power tools
- Being seated for long time periods in the same position
- Maintaining awkward posture while working
- Reaching (either below the knees or above the shoulders) to grasp objects
- Bending or twisting the torso while lifting heavy objects

You can see how quite of few of these activities are part of life in a busy foodservice operation. OSHA-approved plans to minimize MSD risks include several specific steps:

- Have a set procedure for employees to report MSD symptoms or injuries.
- Inform employees of the workplace hazards that may cause MSDs.
- Provide training for all new employees to avoid MSD risks, and repeat it for existing employees at least once every three years.
- Analyze each job and task to identify the MSD risk factors, and reduce them to the extent that it is feasible.
- Ask employees for their input in the job/task analysis.
- Provide workers with medical care (at no cost to the worker) for their job-related injuries.

Most of these steps seem like common sense for a conscientious employer and, according to OSHA, the average annual cost of modifying an otherwise risky job so that it will be safer is only $150. However, at this writing, about half of U.S. workers are still not covered by any type of company ergonomics program. Our suggestion is that you undertake such a program voluntarily. In the long run, it will save you money on Workers' Compensation insurance, employee medical claims, and time lost due to work-related injuries.

8-3 EMPLOYEE COMFORT AND SAFETY

The manager of a foodservice facility should constantly be thinking about the productivity of his or her workforce. Since the early 1900s, when Frederick Taylor began studying workers on American factory assembly lines, productivity and human engineering have fascinated scientists. The concept is not so complex. Anyone who's spent time in a kitchen appreciates the importance of having the right tools and equipment, placed in convenient locations in a comfortable environment.

Exactly what are we avoiding by identifying potential hazards and developing safety measures? Here are just a few of the things your staff members are exposed to on a daily basis:

- Slips and falls, on a wet floor, icy sidewalk, or climbing ladders or steps (Just this category of mishaps results in more than one million hospital emergency room visits annually in the U.S., with 300,000 disabilities and 1,400 deaths.)
- Cuts, from knives, electric slicers, and other sharp kitchen tools
- Burns, from hot oil, hot water, the entire cooking process, and dishwashing
- Back injuries, from moving and handling heavy equipment and bulk supplies
- Bumps and bruises, from dropping things, bumping into things, mid-aisle collisions on a busy evening, and the like

With all that to think about, when employees are first hired, safety should be taught as a way of life and a part of your business culture. This means training them about how to use equipment correctly, explaining why these safety procedures are in place, and posting them in prominent locations—perhaps as "Safety Checklists" for various tasks or appliances. The training, any follow-up discussions, and especially any violations, should be noted in writing as part of each employee's master file.

So what makes for a safe kitchen work environment? We've already discussed the necessity for noise control, adequate lighting, temperature control, and removal of stale air. Now, let's talk about basic equipment needs. Foodservice workers spend about 90 percent of their time on their feet. It's not the standing that's the real problem, but the back and muscle strain that results from having to bend at uncomfortable angles to do repetitive work in a standing position. In an ideal situation, the work surface could be raised or lowered to a height where the person could work without having to bend. More realistically, you can adjust the worktables to the correct height for the persons who use them most often. However, there are some general height recommendations, which are listed in Table 8-1.

Another industry rule of thumb is actually a "rule of elbow"—the height of the worktable should be 4 inches below the worker's elbow. At this height, the person should not have to raise or lower the arms to uncomfortable angles to accomplish tasks. The exposed edges of the table should be smooth and/or rounded to prevent cuts, bruises, and snagged clothing.

And what about the surface of the worktable? A person works best when tools or materials are within 24 to 36 inches of the center of his or her waistline. A good combination of elbow room without much reaching can be achieved if each worker at a table has no more than 5 linear feet of work space; 3 feet if space is tight and supplies can be situated directly in front of the worker. The surface should be brushed or *matte* (not glossy or shiny) stainless steel, and lighting should be designed to minimize glare off both worktables and walls.

Storage of hand tools, small equipment, and supplies is best done on two or three shelves, located in front of the people who will use them, or no farther away than 28 to 38 inches from the standing

TABLE 8–1

HEIGHT RECOMMENDATIONS FOR KITCHEN TABLES

Equipment Use	Suggested Height
Work table, for light work done standing	37–39 inches (for women) 39–41 inches (for men)
Worktable, for heavy work (kneading, chopping)	35–36 inches
Equipment table (for mixers, etc.)	26–28 inches

worker. Items that are more frequently used should be stored at eye level; some under-tabletop shelving could be added for flat pans, cookie sheets, and so on.

In areas such as walk-in pantries and freezers, store the most frequently used items between eye level and waist level to minimize the risk of back injury. Overall, try to avoid the use of ramps, stairs, and ladders in your kitchen, as these are chronic causes of accidents. If you must use them, equip stairways and ramps with handrails, and mark them with yellow "caution" tape or paint at top and bottom or on each stair step. Individual steps should be between 5 and 8 inches tall (the "height" of a step is called its ***riser***) and the step should be from 9 to 11 inches deep (from back to front). Some experts suggest the "safest" steps have depths and risers that are both 8 inches. They also suggest a larger, "landing" area be provided every ten to twelve steps.

If lifting is required, back braces or lifting belts should be available near storerooms and in the receiving area. Wheeled carts or hand trucks should be used to move heavy or bulky supplies.

When using kitchen equipment, the rules are mostly a combination of common sense and attention to detail. Keep it clean, and keep it well maintained. Other good rules:

- Don't leave equipment where someone might trip on it.
- Don't try to use a piece of equipment for something it was not designed to do.
- Read the instruction manual or owner's manual, and pay attention to any warning stickers placed by the manufacturer on the equipment itself.
- Keep electrical cords out of the way and in good condition.
- Check the batteries and/or chargers of battery-powered equipment.
- Pay attention when the equipment is in use. Keep both hands on it when it is operating. Use safety guards if they are part of the equipment.
- Always unplug it before you adjust or repair it.
- Have a "red tag" system, or another clear way to mark equipment that needs to be repaired, so others will know not to use it.

Some restaurants have a policy of "clean it after each use." This is partly to get equipment ready for the next use, but also because cleaning it offers a good chance to notice and document any repair that may be needed.

Hot pads and mitts should be stored at all cooking stations, to discourage employees from using a good old kitchen towel to lift hot plates, pans, or utensils. The best ones are designed with a liquid-and-vapor barrier to prevent burns. These items should be washed regularly, and discarded when they show signs of excessive wear.

Do not ignore the receiving area when it comes to safety concerns. Your dumpsters should be placed inside a locked enclosure, especially if you have mechanical equipment (like trash compactors) as part of the set-up. Sloped-front dumpsters should be equipped with safety legs to keep them from tipping forward accidentally. Floodlights at the receiving door should be equipped with motion sensors that activate whenever employees are working there and minimize the number of insects attracted to a light that's always on. Another way to discourage flying bugs is with an air screen or plastic strip curtains on any open loading dock. And again, lifting belts and/or back braces, hand trucks, or dollies should be stored in this area.

Flooring and Floor Mats

The least you can do for your kitchen staff is provide a bit of cushioning for the long hours they spend on their feet. Kitchen mats, when properly selected and used, provide traction for employees, minimize accidental breakage, and keep floors cleaner as well.

Rubber mats are arguably the most comfortable, but the longest-wearing mats contain nylon cords melded to the rubber. You can also buy mats made of neoprene rubber (the same material wetsuits are made of), solid vinyl, and sponge vinyl. Beveled or tapered edges provide an extra measure of safety against tripping, and are useful in heavy-traffic areas or if you're rolling carts. The mats should not be hard, but textured, which forces people to change posture when standing on them for long time periods.

Before you buy mats, decide where you will put them. Kitchen mats are usually 3-foot squares that can interlock, and they are between ½ and ⅝ inch thick. Thicker mats provide more cushioning but thinner ones are easier to clean. There are special grease-proof mats for areas near fryers, grills, and griddles; slightly raised bar mats, which drain spilled liquid beneath them to prevent slipping; and vinyl loop pile mats for kitchen entrances and exits, which can be specially treated to make them germ-, mold-, and mildew-resistant. For public entrances, mats are only ¼ inch thick (so people won't trip on them) and should be able to trap incoming snow, ice, moisture, or mud. Your logo or a welcoming message can be emblazoned on the mat. If it's going to be placed over a carpeted surface, molded nubs on the bottom side of the mat will grip and adhere to the carpet.

Slippery floors are a safety issue in every kitchen, but mats will improve employee comfort and, in dish rooms, they'll minimize breakage of dishes and glassware. Where wet floors are a persistent problem, as in dish rooms, you also can keep a fan blowing directly onto the floor, and require that workers in these areas wear shoes with nonskid soles.

Even with protective mats, it is the manager's responsibility to see that floors are cleaned often. Today's mats are light, weighing 3 to 5 pounds apiece, for easy lifting and cleaning. They can be taken outdoors to a loading dock or back parking lot and hosed down with high-pressure water and cleaning fluid that melts the grease buildup. Or you can use a motorized floor scrubber/vacuum (sometimes called a "wet-vac"), which has rotary brushes and uses hot water and detergent to loosen grime from mats and floors and then whisks away most of the moisture with an absorbent "squeegee" located at the back of the machine.

The floor covering you select will go a long way toward ensuring a safe working environment. You will hear the term ***aggregate*** to describe nonskid floor surfaces. This means something extra has been added, usually to a top or second layer of the floor material, to give it more friction. It might be fine sand, clay, tiny bits of gravel or metal, Carborundum™ chips or silicone carbide.

There are lots of flooring choices, as shown in Table 8-2, but few of them are truly capable of withstanding the rigors of the foodservice kitchen. Among veterans, unglazed quarry tile is the undisputed favorite. Quarry tile contains bits of ceramic (usually Carborundum™ chips), clay, and/or silicon carbide for slip resistance. Even when it wears down in a high-traffic area, it maintains its skid-proof quality. Quarry tile should be at least ⅜ inch thickness; in areas with heavy foot traffic or where appliances are frequently rolled, a ½ inch or ¾ inch thickness is preferred.

The tile is only part of the puzzle, however. It must be grouted into place, and that grout must be able to withstand significant differences in pressure, temperature, and moisture, as well as exposure to grease and chemicals. The recommendation is a so-called "thick set" mud bed, using either an epoxy or a special cement with latex added for elasticity. Epoxy is generally considered more durable, but either type of grout is susceptible to deterioration if it's constantly wet. The tile is laid onto a "setting bed" of cement, which should be sloped toward the floor drains. This facilitates steam or pressure cleaning, and keeps water from standing beneath the tiles and damaging the floor over time.

Another option is a one-piece floor, poured all at once and made of epoxy, polyester, urethane, or magnesite cement. This is also called a ***monolithic floor***. An epoxy composite floor consists of several layers—the epoxy or resin, then an intermediate "sealer" layer, an aggregate layer to make the floor slip-resistant, and a clear finish (also epoxy) to seal the floor. The resulting floor is 3⁄16 to ¼ inches thick and can withstand extreme temperatures around fryers and other cooking equipment. An epoxy composite floor is roughly half the strength of a quarry tile floor, but its chief advantage is that it is seamless—no grout to clean or maintain. It is skid- and scuff-resistant, resilient, and unharmed by moisture, food spills, acidic substances, and the like. Epoxy is also easy to repair—just grind down the worn or damaged part and pour a new layer there.

Composite sheet vinyl is another type of seamless flooring, laid in sheets on a bare floor with a layer of adhesive. For foodservice applications, an extra top layer of aggregate is mixed with the vinyl for slip resistance. Heat welding is used to seal the perimeter of the sheets onto the floor. Sheet vinyl is attractive, durable, and has excellent resistance to moisture and food acids. It is superior for use in wet areas, although this type of floor is susceptible to scratches, dents, and some stains. Today's sheet vinyl comes in all kinds of funky colors and patterns that mimic everything from wood to slate to jungle prints, so it's suitable for front-of-house areas as well as kitchens.

Hubbelite is a compound mixture of cement, copper, limestone, magnesium, and a few other things. It's half an inch thick, strong, slip-resistant, and comfortable to walk on, and gaining popularity

TABLE 8–2

TYPES OF FLOORING

	Advantages	Disadvantages
Asphalt tile	Low cost Large color selection Moderate maintenance	Least resilient Affected by oils, kerosene Possible cracking or deterioration from temperature extremes
Vinyl tile	Wide color and pattern range Resistant to oils and solvents High resiliency	Prone to scratching and scuffing
Carpeting	Sound absorbent Decorative Comfortable	Greater than average maintenance needs Moderate to high initial cost Affected by spills
Linoleum	Available in sheets or tiles Wide color range Low cost Good resilience	Lacks resistance to moisture and alkaline solutions
Rubber tile	High resilience Wide color selection	Lacks resistance to oils and solvents
Quarry tile	Resistant to moisture Low maintenance Long life	High initial cost Grout subject to deterioration
Terrazzo	Decorative Durable Low maintenance	High initial cost Unsuitable for areas subject to spillage
Wood	Decorative Comfortable	Unsuitable for areas subject to excessive moisture and oil spillage Moderate care and maintenance required
Concrete	Durable Low maintenance Low to moderate initial cost	Dusting and blooming

Source: Reproduced with the permission of the National Restaurant Association, Washington, DC.

in remodels because it can be installed over any existing floor type, including quarry tile. Hubbelite has excellent stain- and moisture-resistance, and the addition of copper and magnesium makes it somewhat insect-repellent and prevents bacteria and mold growth. Cleaning it is also simple—just mop it. The downsides are that Hubbelite comes in only one color, brick red, and that you must choose an installer who is familiar with the product to do a good job with it.

Ceramic tile can be used for both floors and walls, in kitchens or dining areas. It is durable and decorative, and its use can immediately bestow charm and character, or add artsy, ethnic touches to any space. Commercial-grade tile is usually baked clay, covered with a glaze to make it sturdier. New technology has given new life to this ancient material, with many more varieties, colors, styles, and patterns of tile readily available. It is simple to clean, too, with soap and water. The more stylish and decorative the tile, the more expensive it will be, so budget-minded restaurateurs may opt for tile only on half-walls, some countertops, or as accents in display kitchens.

The lowest-cost types of flooring are vinyl tile and sealed concrete. Vinyl tile is considered fairly high maintenance and does not offer maximum traction or heat resistance, although it is easy to install and clean. We'd recommend it only for use in dry storage areas, never in food prep or dishwashing areas. In some jurisdictions, health regulations do not allow its use for commercial foodservice.

A concrete floor sealed with a water-based epoxy coating is a good choice. Be sure the dry thickness of the coating is 1.5 to 2.5 mils. Whether you choose vinyl, tile, or a one-piece poured floor, it is critical that it be properly bonded to the concrete slab beneath it. This is called "curing" the floor, and if it's not done correctly, the floor will eventually crack. No artificial curing agents or sealers should be used. They'll destroy the adhesives used to set the floor and keep it in place.

For public spaces, terrazzo flooring is an attention-getting option. Terrazzo begins with a highly durable cement or epoxy base, into which chips of granite, marble, tile, or even seashells may be mixed. It can be pricey and it takes a long time to install, but it is durable and very attractive.

Carpeting is not an option in the kitchen, but in dining areas, beautiful carpet can make a lasting impression. Worn and soiled carpets, however, can make an equally strong impression: that the place is dirty and the staff is inattentive at best, oblivious at worst. In fact, dust mites, fungi, and bacteria can live and thrive in carpets that are not well cared for. In busy commercial settings, it is probably best to install carpet by the roll rather than squares—fewer edges to lift up and cause people to trip. Carpet has a definite life cycle and will need to be replaced periodically, but it is often preferred in dining areas for its ability to suppress noise.

Floor Cleaning and Maintenance

For floors that are carpeted, you'll need an action plan to keep them clean, safe, healthy, and looking terrific. Here are five practical tips for carpet maintenance:

1. **Make sure the carpet is properly installed. If it's not done right, it will buckle and wear out more quickly. Choose patterns and blended colors over a single, solid color, which shows more dirt.**
2. **Vacuum frequently so dirt will not accumulate. We'll talk more about vacuums in a moment. When vacuuming is not an option, keep dirt at a minimum by using a sweeper.**
3. **Place mats at all entrances; about 70 percent of the soil on carpets is brought in by foot traffic from outdoors and from the kitchen. Entrance mats should be 12 to 15 feet long; 30 percent of soil is trapped on the first 3 feet of mat. Of course, keep the mats clean or they'll only contribute to the problem.**
4. **Implement a spot-removal plan. Ask the installer and/or manufacturer for the correct method (scrape, spray, blot, and so on) and products to clean up those inevitable spills.**
5. **Clean by extraction. Contract with a professional carpet cleaner to deep-clean and remove any harmful bacteria buildup. Most dining rooms could use this service every other month.**

To select the vacuum cleaners to care for your carpet, ask yourself the following questions: Is the area to be cleaned a large, open area or a small, enclosed space? Will the vacuum be carried, wheeled, or rolled on a cart to various points of use? Should the vacuum be able to dust and do edge cleaning as well as carpet cleaning? Are there special requirements for indoor air quality, noise levels, or power requirements?

The most popular commercial-use vacuum cleaners are the *traditional upright* with one motor and *direct airflow* and the *two-motor upright* with *indirect airflow.* The first refers to a vacuum that draws air and dirt directly from the carpet through the motor fan and into a soft cloth dust collection bag that is emptied or replaced when full. The single-motor vacuum is called traditional because it's been around for years and has one basic use, to vacuum carpets. Its nozzle comes in standard widths of 12 to 16 inches. The two-motor label identifies those upright vacuums that have one motor to drive the brush

PLANNING THE DINING ATMOSPHERE

CARPET FIBERS

Here's a look at three carpet fibers and how they can be cleaned.

NYLON. Features high crush resistance, long life, high resiliency, and wide variety of colors. Nylon also has a high melting point, which prevents marks caused by objects dragged across the carpet. Can be cleaned with hot water or dry extraction machines.

POLYPROPYLENE OR OLEFIN. Doesn't absorb moisture. Has ability to resist stains. Polypropylene can be cleaned with hot water or dry extraction methods.

WOOL. Features absorbent fibers, which reduces static build-up. Can last for generations. According to Castex Industries, Inc., when cleaning wool carpet use extreme care and a minimal amount of water.

Source: *Lodging*, the magazine of the American Hotel and Lodging Association, Yardley, Pennsylvania.

rollers and a second motor to provide suction. These units are called indirect because dirt and air must be pulled from the carpet, through the dust bag, and into the suction motor. These vacuums are sometimes touted as ***clean-air vacuums*** because the incoming dust is filtered to the dust bag before entering the motor. Two-motor models are usually larger than direct-airflow vacuums, have a hard-shell case in which the dust bag fits, can vacuum bare floors as well as carpeted areas, and have standard nozzle widths of 14 to 18 inches.

A special version of the two-motor model, called the ***wide-area vacuum,*** has a nozzle width of up to 36 inches, a longer dust bag, and no hose for attachments. You may see them in use in airports or ballrooms.

Don't bother using ampere or horsepower ratings to determine how much suction or power your vacuum is capable of producing. Instead, ask about its *cfm rating*. This is the number of cubic feet per minute pulled through the 1¼-inch orifice that is the typical size of a vacuum's nozzle opening. The greater the cfm number, the greater the airflow and, therefore, the better its cleaning ability. Units that have a higher cfm rating measured at the nozzle opening and a revolving brush roller will have the best overall ability to clean a carpet.

Like any appliance, vacuums require regular maintenance to keep them working at their peak. They have a distinctive motor noise that you'll become familiar with, and your first sign of trouble is if the vacuum "sounds different." This may mean a belt is wearing thin, the brush roller is slipping, the dust bag is full, or the suction motor's fan should be replaced.

A good rule to teach your cleaning crew is, "If it is not commonly found on the floor, don't vacuum it up, pick it up." This certainly prevents many vacuum cleaner breakdowns. Of course, some manufacturers offer magnetic bars in their vacuums to snag paper clips, thumb tacks, and the like, but this lulls the user of the machine into a complacency that usually results in the malfunction of the machine.

When vacuuming is not possible (usually because guests are present), it's good to have a couple of carpet sweepers. They are nonmotorized, with rotary blades that spin with the simple action of being pulled or pushed across a floor. Sweepers work on carpet or bare floors, picking up everything from dust to glass particles to wet or dry food bits and depositing them all in an internal dustpan. They are easy to use and easy to clean.

For noncarpeted areas, the appliance of choice is the *dry steam vapor cleaner.* It's a portable unit that includes a water tank and assortment of hoses, brushes, and wandlike attachments for reaching tough-to-access spots. The hot steam (230 to 330 degrees Fahrenheit) cuts through just about any type of kitchen grime and grease. This hard-working machine can clean floors, restrooms, ceramic tile walls, tabletops, vinyl or plastic chairs, and the grout between tiles. It handily tackles the greasy challenge of the exhaust hood and stainless steel back panel "walls" of the hot line, steaming the grease and melting it enough to be wiped away with a cloth or squeegee.

Modern commercial floors are cleaned with soap and warm water. The cleaning products are formulated to attack oily, fatty residues and soften them for easier removal. You should choose products that work on both concrete and quarry tile, and use them with the correct types of cleaning tools. There are floor-scrubbing brushes designed to clean grout lines between tiles—without having to get on your hands and knees! Brushes with epoxy-set nylon bristles and aluminum brackets and handles are more sanitary than cotton mop heads and wooden handles that absorb dirt and foster bacteria growth.

Several different sets of brushes, mops, and buckets are necessary to prevent cross-contamination between different parts of the restaurant, where different cleaning chemicals may be used. Some operations even designate separate "wash" and "rinse" buckets, which is actually a more efficient way to remove dirt, grease, and oil from floors than using the same bucket for both washing and rinsing. Mop buckets are offered in sizes from 26 quarts to 44 quarts, with ringers to accommodate mop heads from 8 ounces to 36 ounces. Mop sizes are based on the weight of the cotton used to make the mop strands. What we call a "broom" for home use is known as a *floor brush* or *deck brush* in the commercial kitchen. These are 10 to 12 inches wide and come in several different angles for cleaning flat surfaces, reaching beneath appliances, and cleaning along baseboards; wider (36-inch) brushes are called *lobby brushes.* Whenever floors are damp-mopped or scrubbed, you must also have an adequate supply of "Danger, Slippery Floor" cones or signage to put out in the wet area.

For dry cleanup, good brooms are essential. Some manufacturers offer them with boar's hair bristles, nicknamed *Chinese pig brushes,* which are useful for cleaning up fine particles like flour and sugar. Synthetic bristles are good for all-around use. Whatever your choice, an angled broom will be most useful to reach beneath tables and into tight corners.

Cleaning Stainless Steel

As for the cleaning products themselves, select them only after you understand your restaurant's needs: What type of surface are you cleaning? Preserving it is as important as cleaning it. What types of substances will you be cleaning up? Is the water hard or soft? What is the cleaning process, and how often will you use this particular product?

Since so many of the surfaces and appliances in a commercial kitchen are made of stainless steel, your employees will need to know how to keep it looking its best. Most people assume stainless steel is rustproof, but this is not so. Some background information: metals that corrode in a natural environment are known in the scientific world as "active"; those that do not are called "passive" metals. Stainless steel is an alloy containing chromium and nickel, two "passive" metals that shield it against corrosion by acting as a sort-of protective coating. Over time, however, three types of wear can break down this protective film:

- Mechanical abrasion: The use of steel wool pads, wire brushes, and metal scrapers.
- Water: Hot, hard water can leave behind mineral deposits, often called *scales, scaling,* and/or *lime buildup.* On steel, these look like a white film.
- Chloride: This is a type of chlorine found naturally in water, salt, and many foods. If you boil water in a stainless steel pot, the discoloration you may see is most likely caused by chlorides. Many cleaners and soap products also contain them.

Rust buildup makes it impossible to sanitize a surface, so it's important to prevent it by using nonabrasive cloth and/or plastic scouring pads, with alkaline or nonchloride cleaning products at their recommended strengths. After washing, the stainless steel should be rinsed thoroughly, wiped to dampness, and allowed to air-dry completely. (Oxygen helps maintain the "passivity" of the coating.) Some equipment may need to be descaled regularly with a deliming agent. If scaling seems to be a persistent problem, soften the hard water with a filtration system.

Before you wipe down a stainless steel surface, look for the *grain*—microscopic lines that show the direction in which it was first polished by the manufacturer. Wipe parallel to the grain, in the direction of the lines.

BUILDING AND GROUNDS

PURCHASING CLEANING SUPPLIES—INTELLIGENTLY

Read the labels, especially the warnings, and ask these questions before you buy.

1. Is it biodegradable? Whenever possible, biodegradable products are preferred for their minimal environmental impact.
2. Is it nontoxic? Choose products that can't harm workers, guests, or food products.
3. Does it contain phosphates? These encourage the growth of algae, which depletes oxygen and kills aquatic life in waterways.
4. Does it contain chlorine-based bleach? In wastewater, chlorine may combine with other chemicals to form toxic compounds.
5. Is it concentrated? Products that can be diluted in cold water will save money and energy.
6. Does it contain a high concentration of volatile organic compounds (VOCs)? In sunlight, VOCs react with nitrous oxides to produce ozone and photoelectric smog.
7. Is it made from a petroleum derivative? Look instead for a purely oil-based product, with a pH level as close to 7 as possible to avoid irritation if the product comes into skin contact. (A pH level of 1 is the most acidic; 14 is the most alkaline; 7 is neutral.)

Source: *Lodging*, a publication of the American Hotel and Lodging Association, January 1998.

8-4 SANITATION

In foodservice, cleaning is just not clean enough. *Sanitizing* is the goal, which means the additional treatment of equipment and utensils *after* basic cleaning to kill germs and bacteria.

NSF International (the initials stand for the group's original name, "National Sanitation Foundation") sets voluntary standards for the hospitality industry and acts as a national clearinghouse and arbiter of sanitation-related design ideas for equipment and facilities. As its mission statement reads, NSFI is "dedicated to the prevention of illness, the promotion of health and the enrichment of the quality of American living."

Although your goals don't have to be quite that lofty, following the NSFI guidelines will ensure that you will meet a thorough and rigorous list of standards and be able to pass your health inspections. Also, the NSFI allows its insignia (like a seal of approval) on equipment that meets its standards. You may see NSFI Standard 2 on foodservice equipment, NSFI Standard 3 on spray-type dishwashing machines, and NSFI Standard 4 on commercial cooking and hot-food storage equipment.

From NSFI's publication *Food Service Equipment Standards,* items that meet these standards "shall be designed and constructed in a way to exclude vermin, dust, dirt, splash or spillage from the food zone, and be easily cleaned, maintained and serviced." The standards also specify that the materials used in the construction of equipment withstand normal wear and tear; rodent penetration; and any corrosive action of food, beverage, or cleaning compounds. This includes the food contact surfaces, as well as safely rounded corners and exposed parts or edges. NSFI also makes many suggestions for equipment installation, with ease of maintenance and cleaning in mind. Among the terms defined in the NSFI standards are a few you should probably be familiar with, listed here alphabetically:

ACCESSIBLE. Readily exposed for proper and thorough cleaning and inspection.

CLEANING. The physical removal of residues of dirt, dust, foreign matter, or other soiling ingredients or materials.

CLOSED. Spaces that have no opening large enough for the entrance of insects or rodents; an opening $\frac{1}{32}$ of an inch or less shall be considered "closed."

CORROSION RESISTANT. Materials that maintain their original surface characteristics under prolonged contact with foods and normal exposure to cleaning agents and sanitizing solutions.

FOOD ZONE. Also called *food contact surface,* this includes the surfaces of equipment with which food or beverage normally comes into contact; and those surfaces with which the food or beverage is likely, in normal operation, to come into contact with and drain back onto surfaces normally in contact with food or into food. (Yes, read that one twice!)

NONFOOD ZONE. Also called *nonfood contact surface,* this includes all exposed surfaces not in the food or splash zones.

REMOVABLE. Capable of being taken from the main unit with the use of only simple tools, such as pliers, a screwdriver, or a wrench.

SANITIZING. Antibacterial treatment of clean surfaces of equipment and utensils by a process that has been proven effective.

SEALED. Spaces that have no openings that would permit the entry of insects, rodents, dust, or moisture seepage.

SMOOTH. Used to describe a surface free of pits or occlusions and having a cleanable (for food zones) Number 3 finish on stainless steel, or (for splash and nonfood zones) commercial grade hot-rolled steel, free of visible scale.

SPLASH ZONE. Also called *splash contact surface,* the surfaces other than food zones that are subject to routine splash, spillage, and contamination during normal use.

TOXIC. A food-related condition that has adverse physiological effects on humans.

The NSFI standards also require that all utility service lines and openings through floors and walls be properly sealed and fitted with protective shields or guards. When horizontal pipes and lines are required, they must be kept at least 6 inches above the floor and 1 inch away from walls and/or other pipes.

Sink interiors are also considered food contact zones and should meet those minimum standards; wooden bakers' tables, synthetic cutting boards, and shelves all have their own standards.

Interestingly, a great deal of thought is put into how the legs and feet of equipment are built. Here, you'll encounter terms such as *bullet leg* and *gusset,* which you probably never thought you'd need to know when you decided on a foodservice career. So it's very handy to have an organization like the NSFI paying attention to this type of detail and sharing its expertise.

In the following chapters, we'll talk more about how to buy and install equipment—from the legs up!

Food Safety

Most foodservice businesses are spot-checked by health inspectors for cleanliness and correct food temperatures, and cited for code violations. Illustration 8-2 is a sample inspector's report form. Too many restaurants try to please the inspector, when what they should be doing is striving for clean, safe conditions because it's the right thing to do, day in and day out. The food safety system that's been in use since the 1960s is *HACCP—Hazard Analysis of Critical Control Points,* which was developed by NASA to evaluate its methods of assuring that all foods produced for U.S. astronauts were free of bacterial pathogens. Nothing worse than having gastrointestinal problems in space!

Even today, decades later, HACCP is considered the absolute standard for food safety, far more effective than simply spot-checking for violations. It combines up-to-date technical information with

FOOD ESTABLISHMENT INSPECTION REPORT

Business name: ______________________

Date: ____________ **Operator:** ______________________

Address: ______________________ **County:** ______________

City: ______________________ **State:** ____________ **Zip:** __________

Inspection Type: ____________ **Inspector:** ______________________

(Regular, Enforcement, Epidemiology, Follow-Up, HACCP, Investigation, Pre-Opening)

CRITICAL ITEMS – These are items that relate directly to factors which lead to foodborne illness. They require IMMEDIATE attention!

Violation Code	DESCRIPTION	Correction Date

NONCRITICAL ITEMS – These violations relate to overall cleanliness and/or maintenance of food operations.

Violation Code	DESCRIPTION	Correction Date

Inspection Time: __________
(minutes)

HACCP Time: __________
(minutes)

Risk Type L M H

On-site Follow-Up Date:
____

Follow-Up Report Date:
____

Enforcement Inspection Required? YES NO

Enforcement Inspection Date:
____

SCORE

(Number of Critical Violations)

COMMENTS: ______________

(Continue on reverse if needed.)

The items recorded above must be corrected according to the specified dates on this form. Failure to comply will result in an enforcement inspection, the cost of which will be charged to the business operator; or in denial, suspension, revocation, or failure to renew your license. A hearing on any disputed charges may be scheduled at your request.

______________________ ______________________

Business Owner, Operator, or Mgr. on Duty **County Health Inspector**

ILLUSTRATION 8-2 **A typical report form that a health inspector might use when he or she tours a foodservice operation. Code violations would be noted at the bottom of the form.**

step-by-step procedures to help operators evaluate and monitor the flow of food through their facilities. The core objective of the process is to identify and control the three types of food safety threats in any commercial kitchen:

1. **Biological Contaminants or Microorganisms.** These include bacteria, viruses, and/or parasites, which already exist in and on many raw food products and can be passed on by unknowing employees or customers.
2. **Chemical Contaminants.** These can come from improper storage or handling of cleaning products or pesticides, from cross-contamination, or from substitutions of certain recipe ingredients.
3. **Physical Contaminants.** These are the most common cause of food contamination—foreign objects in the food, including hair, bits of plastic or glass, metal slivers, and so on—which can be deadly if choked on.

There are seven basic HACCP steps:

1. Identify hazards and assess their severity and risks.
2. Determine critical control points (CCPs) in food preparation.
3. Determine critical control limits (CCLs) for each CCP identified.
4. Monitor critical control points and record data.
5. Take corrective action whenever monitoring indicates a critical limit is exceeded.
6. Establish an effective record-keeping system to document HACCP system.
7. Establish procedures to verify that HACCP system is working.

The first step is to decide what hazards exist at each stage of a food's journey through your kitchen, and decide how serious each is in terms of your overall safety priorities. On your own checklist, this may include the following items:

- Reviewing recipes, paying careful attention to times for thawing, cooking, cooling, reheating, and handling of leftovers.
- Giving employees thermometers and/or temperature probes and teaching them how to use them. Correctly calibrating these devices.
- Inspecting all fresh and frozen produce upon delivery.
- Requiring hand washing at certain points in the food preparation process and showing employees the correct way to wash for maximum sanitation.
- Adding quick-chill capability to cool foods more quickly in amounts over 1 gallon or 4 pounds, and so on.

There are as many of these possibilities as there are restaurants!

The second step is to identify ***critical control points (CCPs).*** This means any point or procedure in your system where loss of control may result in a health risk. If workers use the same cutting boards to dice vegetables and debone chickens without washing them between uses, *that* is a CCP in need of improvement. Vendor delivery vehicles should be inspected for cleanliness; product temperatures must be kept within 5 degrees of optimum; expiration dates on food items must be clearly marked; utensils must be sanitized; and the list goes on and on.

The third step is to determine the standards and limits for what is acceptable and what is not, in each of the CCP areas, for your kitchen.

The fourth step in the HACCP system is to monitor all the steps you pinpointed in Step 2 for a specific period of time, to be sure each area of concern is taken care of correctly. Some CCPs may remain on the list indefinitely, for constant monitoring; others, once you get the procedure correct, may be removed from the list after several months. Still others may be added to the monitoring list as needed.

Step 5 kicks in whenever you see that one of your "critical limits" (set in Step 3) has been exceeded, and corrective action must be taken.

Step 6 requires that you document this whole process. Without documentation, it is difficult at best to chart whatever progress your facility might be making. If there is a problem that impacts customer health or safety, having written records is also very important.

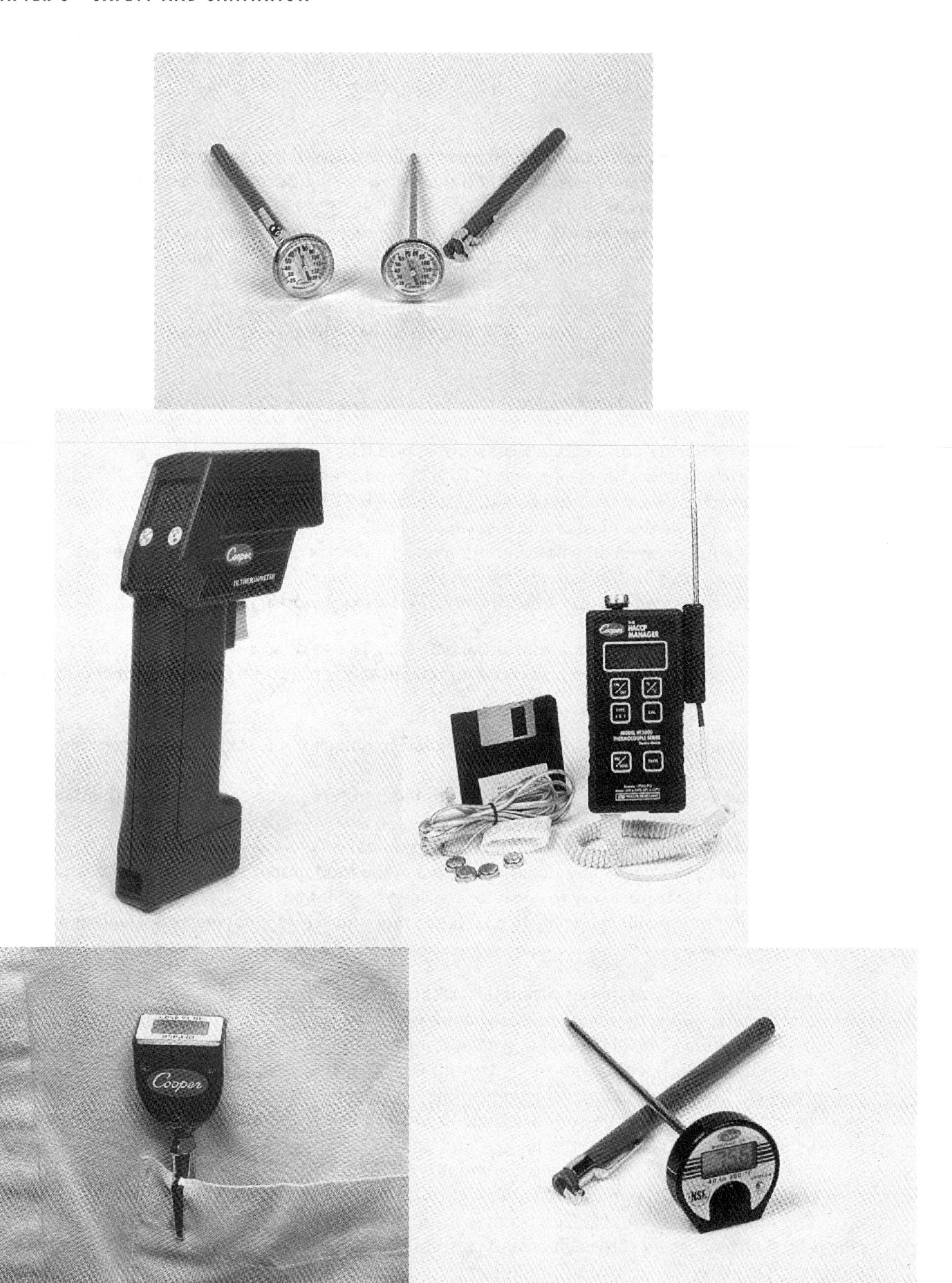

ILLUSTRATION 8-3 There are temperature probes and thermometers to monitor just about any appliance or food preparation process and ensure HACCP compliance.

COURTESY OF COOPER INSTRUMENTS CORPORATION, MIDDLEFIELD, CONNECTICUT.

IN THE KITCHEN

CRITICAL CONTROL POINTS FOR PROTECTING FOOD

1. Raw food is cooked to correct minimum internal temperature.
2. Hot food must be maintained at a minimum temperature of 140°F (60°C) or above while holding and serving.
3. Leftover hot food is cooled quickly and safely to 70°F (21°C) within 2 hours and from 70°F (21°C) to 40°F (4°C) or below within an additional 4 hours.
4. Leftover hot food is reheated quickly (within 2 hours) to 165°F (74°C).
5. Cold food must be maintained at 40°F (4°C) or below while holding and serving.
6. Refrigeration equipment should be maintained at 36°F to 38°F (2°C to 3°C).

Source: International Food Safety Council, Chicago, Illinois.

Finally, Step 7 requires that you establish a procedure to verify whether the HACCP system is working for you. This may mean a committee that meets regularly to discuss health and safety issues and to go over the documentation required in Step 6.

Safe Food Handling

Now that you know some of the temperature and cleanliness requirements for food safety, exactly how do go about achieving them? Part of this critical task is selecting the best equipment for the job. The foodservice industry has been transformed to meet the challenges of HACCP. Perhaps the most stringent requirements have been noted in the area of refrigeration, since HACCP standards emphasize prompt refrigeration of cooked foods, and keeping cold foods at a constant temperature. In the past decade, blast chillers have become standard in most large kitchens. (You'll learn more about this quick-chilling technique and its many applications in Chapter 15.) Display cases and salad bars have improved to maintain constant 41-degree Fahrenheit temperatures without freezing foods.

In other areas of the kitchen, warewashing equipment has been upgraded. Pot sinks feature more powerful sprayers and separate water heaters, helping to ensure that pots and pans are sanitized after use. We discussed mechanized hand-washing systems in Chapter 6. Temperature and humidity monitoring equipment has also greatly improved. It is more affordable, more portable, and easier to use. A single temperature probe doesn't fit all situations, so there are thermometers and probes designed expressly for fryers, freezers, walk-in coolers, storage areas, and so on. Many of them are color-coded so it's easy to see when a HACCP step is being met or violated. Others have remote recording devices to track temperature fluctuations and pinpoint potential problems.

Antimicrobials are making their mark in foodservice operations because of their ability to retard the growth of dangerous microorganisms. The health-care industry has used them since the early 1990s, and it just makes sense that they would also have food-related applications. An antimicrobial is a substance that naturally retards bacteria growth. The companies that make them are somewhat secretive about the contents, but a few of the substances are silver ions, ozone injected into water, chlorine dioxide, and a compound called nano-antibiotic mother granule, or NAMG for short. These are used to do everything from sanitize fresh produce to keep ice machines cleaner. Some antimicrobials can be mixed with plastic, rubber, and other materials in the manufacturing process to form a long-lasting safety barrier on cutting boards, knives and slicing blades, shelf units, appliance legs, and protective gloves. They become part of the molecular structure of the item, so they are effective for its useful life.

Antimicrobials cannot prevent cross-contamination, and there's no claim by manufacturers that they will "kill" microbes. They simply retard their growth by creating a surface that is unfriendly to microorganisms. The U.S. Environmental Protection Agency regulates the use of antimicrobials (and their safety claims), as well as the Food and Drug Administration. NSF International has a certification process for them if they come into contact with food during their common use.

We mentioned protective gloves and, as in so many other industries, foodservice workers are being encouraged to wear them to prevent the spread of germs. There are now "cut-resistant" gloves developed for workers who use knives and other sharp-edged tools. People with latex allergies can choose nonlatex gloves.

There is growing concern that cold foods are most likely to contain food-borne pathogens, since they are not exposed to the heat of the cooking process. Let's take a fresh green salad as an example, and use some commonsense, disease-controlling steps. First, use vegetable brushes on nonleafy vegetables and wash them under running (not stagnant) water. A quick rinse removes most fertilizer and pesticide residue, but it's no longer considered sufficient for most produce. Instead, there are fruit and vegetable cleaning solutions to do a more thorough job. All too often, people use their prep sinks to wash greens—but how well has it been cleaned? Do you also use it to defrost chicken, tempting cross-contamination? Some foodservice kitchens purchase a semi-automatic greens washer. This machine agitates leafy greens (and most can do other types of produce, too) to remove embedded dirt, then tumbles them into a mechanized salad-spinner for drying. We've seen machines that inject ozone into wash water for sanitizing; or you can add a produce cleaning solution to the wash.

After washing nonleafy produce, you can steam most other fruits and vegetables for 15 seconds, then dunk them into ice water or place them in a blast chiller to quick-chill them. The wash-and-steam process helps sanitize them. If you use a blast chiller frequently for chilling instead of freezing foods, you should order one with a "gentle chill" or "soft chill" cycle. Too many people assume refrigerators are designed to cool foods down when, in fact, they are intended to keep cold foods cold! A blast chiller is a faster-working, and therefore safer, alternative.

As salad ingredients are cleaned and chopped, they may be sitting in the prep area at temperatures of 50 degrees Fahrenheit or more. This is not a problem if they are promptly refrigerated to a cooler temperature, 34 to 38 degrees Fahrenheit. (This temperature seems to be optimum for commercial refrigerators, which are opened and closed frequently.) Avoid the tendency to store freshly made salads in large, deep containers—they don't allow quick enough chilling because of the depth and the density of the salad. Instead, spread the salad into containers no more than two or three inches deep to allow faster cooling, and don't put more than ten pounds of product at a time in each shallow container. Cover it with film or foil before chilling, making sure there's no layer of "dead air" between the top of the salad and the film or foil.

When kept at the 34 to 38 degree range, most uncooked foods have a safe-to-use shelf life of seven days. Once they reach temperatures of 41 to 45 degrees Fahrenheit, that drops to only four days.

Hot foods also present risks for food-borne illness. As with cold foods, time and temperature are the keys to safety. Cooking (for the correct time and at the correct temperature) ***pasteurizes*** a food or liquid, sterilizing it by killing harmful microorganisms without chemically altering the food or liquid itself. This is why it's so important to track temperatures through the cooking process, and to cook foods thoroughly.

Once a product is cooked and pasteurized, its shelf life begins again. At least one-third of all food-borne illness stems from improper cooling of foods placed into refrigeration. The accepted standards are:

- **If a food comes off the range or out of the oven at 140 degrees Fahrenheit or hotter, it must be cooled to 70 degrees Fahrenheit or below in less than two hours.**
- **If food is received at a temperature of 41 degrees Fahrenheit, it cannot sit at room temperature for very long. It must be completely chilled (to below 41 degrees Fahrenheit) within four hours.**

In either of these situations, a blast chiller or ice bath is a practical necessity to get the cooling job done quickly enough to keep the food safe. Putting hot food in a freezer is possibly the worst decision you could make. Instead of cooling the entire product quickly and uniformly, a thin layer of ice begins to form almost immediately on the surface of the food. The center or core of the food remains warm and is, in fact, insulated by the ice. This *igloo effect* is dangerous.

8-5 WASTE MANAGEMENT

Foodservice businesses of every size and type contribute to the growing solid waste problems in the United States, but most are doing at least some things to minimize what is sent to the landfill and down the sewer system. Becoming ecologically minded isn't always easy or inexpensive, but the alternatives are worse—a planet cluttered with junk, and higher fees and taxes to try to make headway in cleaning it up. It is estimated that Americans throw away almost 3 pounds of waste per day, per person.

For the restaurateur, there are several good reasons to reduce waste:

- It saves money. Decreasing your waste output decreases the size of the dumpsters you need, and/or lengthens the time between trash pick-ups. You can also use composted materials on your outdoor landscaping, which saves fertilizing costs.
- It protects the environment. Food waste that ends up in garbage disposals and/or landfills increases levels of odor and dangerous methane gas. Oil and grease from cooking and frying must be dealt with by your community's wastewater treatment plant.
- It complies with state laws. Increasingly, more stringent legislation is being passed to manage solid waste. Waste-intensive businesses, including restaurants, are being watched more closely with these new laws in mind.
- It has public relations benefits. Taking responsibility for "doing your part" to preserve the environment and be a so-called "green" business is impressive to your customers, many of whom are taking similar steps at home by recycling and reusing items.

Both the U.S. Environmental Protection Agency and the National Restaurant Association have worked hard to come up with an integrated waste management system that can be adapted for different communities and different types of foodservice facilities. By "integrated," we mean there are several components to the system, and the mix will vary depending on the area, the type of business, and so on. The components are:

- Source reduction
- Reuse and recycling
- Composting
- Combustion or incineration
- Landfill use

We covered maintenance of the grease trap and management of oily waste in Chapter 6. Let's discuss each of the other waste management topics here.

Source Reduction, Reuse, and Recycling

The catchphrase in this area is "P2"—Pollution Prevention. ***Source reduction*** means making less waste in the first place, with a secondary goal of reducing the toxicity of the waste that still exists. In today's "throwaway" society, it's a long-term solution, but more and more products are being designed and used in ways that reduce their chances of ending up in the trash, and market research indicates that consumers think positively about companies making these extra efforts. Some examples of source reduction include:

- Decreased use of mercury in dry-cell batteries, and the introduction of many types of rechargeable batteries
- Advanced resin technology, which means milk cartons and plastic bags made with fewer materials
- Advances in tire design, which have increased the useful life of automotive tires by 45 percent since the 1980s

But food is a different story. The most recent the U.S. Department of Agriculture figures estimate Americans throw away 27 percent of all the edible food available in the nation. If just 5 percent of this discarded food (not even the rest of the trash) in the U.S was recovered, it would provide one day's meals for 4 million people, and save $50 million per year in solid waste disposal costs. In the same report, foodservice waste was blamed mostly on overpreparation, expanded menu choices, plate waste, and sales fluctuations that were beyond the operator's control—things like sudden weather changes that prompted fewer customers to show up on a particular day.

One method of source reduction is to buy less food, using up what you've got before you replace it. Computerizing your inventory will assist greatly in this type of planning. Labeling inventory and using the "first-in, first-out" method will ensure that nothing sits too long on storeroom shelves and goes bad. Ask your suppliers about products that meet your specifications, but are minimally packaged. Ask them to take back and reuse their shipping boxes and pallets. Use and wash linens, kitchen towels, dishes, and silverware instead of disposable paper products.

These are just a few factors that should be considered when beginning a waste reduction program. Yours will be far more effective if you know exactly what it is you classify as "waste." Environmental agencies, and possibly your own trash pickup company, can supply you with a ***waste audit form*** so you can start keeping track of what (and how much) you throw away. You might start by, literally, taking one day's trash and separating it into categories. How much is food waste? How much is cardboard? How much is recyclable?

Using the waste audit information, you can work with your trash pickup service to find out about the resources in your area for recycling. ***Recycling*** is the collection and separation of specific refuse materials that can be processed and marketed as raw materials to manufacture new products. Even as far back as World War II, there was a strong push in the United States to recover paper, steel cans, and other items, but when the war ended, the movement—billed as a form of patriotism—lost its steam. By 1960, Americans were only recycling 7 percent of their solid waste; by 1986, the number had crept up to 11 percent. But during the 1990s, the national recycling rate climbed from 12 to 27 percent, according to the Institute for Local Self-Reliance, a nonprofit group with offices in Washington, D.C., and Minneapolis, Minnesota. At this writing, dozens of individual cities and counties—including Ann Arbor, Michigan; Bellevue, Washington; Crockett, Texas; and Visalia, California—have reduced their municipal solid waste (MSW) to record-setting levels of 40 to 65 percent.

You'd assume just about anyone would think recycling was a wise thing to do, but the entire waste industry butts heads with the federal EPA and some other industries. The EPA believes the waste industry's estimates of how much solid waste can be recovered, collected, and processed is unrealistic, while industries that create new products claim jobs are threatened when people reuse instead of buying new. If there is a happy medium, we haven't yet reached it. The Institute for Local Self-Reliance website (www.ilsr.org) does a good job of keeping up on the ongoing debates.

The National Restaurant Association says that about seven out of ten restaurants have recycling programs. Of these, 84 percent recycle paper and cardboard, 79 percent recycle glass, 74 percent recycle aluminum and tin, and 57 percent recycle plastic. Restaurants do make money from recycling—partly because they pay a little less for trash pickup and/or landfill fees and partly because waste-processing companies pay (usually by the pound) for the materials they receive. It's not a major source of revenue, however, so most restaurateurs view recycling as a break-even situation and a good, community-minded thing to do. Some even donate recycling proceeds to charity.

Another type of charitable effort is to participate in a food reuse program by donating unused produce and/or leftovers to the needy. There are several ways to accomplish this:

- Most states and sizable cities have a food bank, which distributes large volumes of nonperishable goods (dried, canned, and prepackaged foods) to other groups that help low-income families. One of the best-known national networks of food banks is *America's Second Harvest*. Visit www.secondharvest.org to learn more.
- Prepared and Perishable Food Programs (PPFPs) redistribute surplus prepared foods and perishables, usually for use at local homeless shelters. They are sometimes called "food rescue" programs. Most of them offer free pickup of these items. Each has its own guidelines for what it will accept and how to store the food before it is collected. The nonprofit group *Foodchain* has a list of all organizations that accept prepared, perishable food in most areas. Visit

www.foodchain.org to learn more. In some communities, there are also specific produce distribution programs for fresh vegetables and fruit.

- You can also contact homeless shelters, battered women's shelters, and similar organizations in your area and offer to cook for them periodically. Many charitable organizations depend on regular participation from restaurants, church groups, and the like for their mealtime needs. Of course, you are donating your time as well as the ingredients in these cases—but it is well worth the effort.

Concerned about your legal liability in these situations? In the U.S., there is a federal law—the Bill Emerson Good Samaritan Food Donation Act of 1996—as well as individual state laws to address this

BUILDING AND GROUNDS

SOURCE REDUCTION AND RECYCLING IDEAS

- Order condiments in bulk containers instead of single-serving packets.
- Use refillable containers for sugar, salt, pepper, and cream.
- Buy straw and toothpick dispensers instead of individually wrapped items.
- Use permanent coasters instead of cocktail napkins.
- Order items to be shipped in reusable containers such as tubs.
- Use linens for aprons, napkins, and restroom and kitchen towels.
- Use hand dryers in restrooms, or cloth towels instead of paper.
- Design your menu cycles to improve the secondary use of leftover food (yesterday's chicken for today's sandwiches, salads, and soups).
- Use reusable plastic pails or covered plastic containers to store foods.
- Switch to draft beer and fountain soft drinks over bottles and cans.
- Use chalk or marker boards instead of menu inserts to post specials.
- Purchase concentrated, multipurpose cleaning supplies.
- Use dishes and silverware instead of disposables, or use biodegradable plasticware.
- Don't store tomatoes and lettuce in the same container. The tomatoes naturally emit a gas that actually turns lettuce brown.
- Revive wilting fresh vegetables by trimming off the bottom inch or so of their stalks or cores and standing them in warm (not hot) water for 15 to 20 minutes. This works with celery, broccoli, lettuce, and the like.
- Offer discounts to customers who bring their own mugs for beverages.
- Minimize the use of take-out containers; or use ones made from recycled paper.
- Use heat from refrigeration lines to preheat water for the water heater. It takes a load off the condenser, too.
- Check out the EPA's "Green Lights" program for a 30 percent lower electric bill.
- Set up an "Ask for water" policy and put a timer on the disposal to save water.
- Turn organic waste into a rich soil supplement by composting.
- Use a polystyrene densifier, which crushes thousands of foam containers into a dry, dense 40-pound cylinder.
- Make sure you're paying your hauler for full loads—compact trash to maximize bin space.
- Cut trash-hauling fees in half by recycling cardboard. Baling eases labor and storage.
- Make sure bins are well marked and paired with regular trash bins.

Source: *Restaurants and Institutions,* the British Columbia Waste Exchange, and the North Carolina Cooperative Extension Service.

issue. Known as a group as "Good Samaritan laws," they protect food donors from most civil or criminal liability except in cases of gross negligence, recklessness, or intentional misconduct of the donor. Your food bank or PPFP may have an agreement for you to sign, or may provide a "Letter of Indemnification" that spells out the rules and your legal protection.

Composting

In some cases, food scraps, soiled papers, and packaging make up 90 percent of a foodservice operation's waste. They can't all be recycled, but the food scraps can be saved and processed into hog feed; or the whole lot can be composted. ***Composting*** is a bacterial digestion process that naturally breaks down organic materials over time, turning them into soil enhancers that are useful in horticulture and agriculture. One of the most popular types of composting today is to set aside summer's grass clippings and fall's leaves from your lawn to rot over the winter and use as mulch to fertilize that same lawn in the spring. Cities and towns push this effort because up to 20 percent of the waste they deal with is leaves and grass clippings!

Composting can be rather smelly and unsightly, and, until a few years ago, it was unheard of in foodservice. Skyrocketing waste removal costs and vanishing landfill space, however, have revived interest in this back-to-nature option. Some restaurants compost; others allow their employees to take home scraps for their own compost piles.

First, ask about city and/or state composting regulations. Not every area has them, but some restrict composting to preconsumer waste—no leftovers allowed—as a health precaution. Others allow preparation scraps, plate scrapings, napkins, paper towels, and garden, yard, and tree debris.

The program begins with the careful and thorough separation of compostable from noncompostable materials. This prompts most restaurants to compost only back-of-the-house trash, because it is easier to separate and can be monitored by employees. A separation system is set up by placing collection bins in the most logical, convenient locations for the kitchen workstations and traffic patterns. Clearly marked bins are placed near dishwashing areas, prep areas, and so on. You'd be surprised how many items can be composted: eggshells and paper egg cartons, coffee filters and grounds, tea bags, waxed paper and most paper products, and even plasticware that is labeled "biodegradable."

The bins should be lined with plastic bags. Yes, it's true that plastic bags probably aren't the most environmentally conscious option, but they keep the bins clean and prevent odors and rodent infestation, and there is no biodegradable option at this point.

Where to put the slowly rotting compost? Most restaurants do not reuse their own compost, but have it picked up instead for use by their municipality. In the summer months, frequent pickups are a necessity to minimize odor and to keep from attracting insects and rodents. Cleanliness of the waste collection area will help alleviate this problem. It's also important to mix kitchen scraps (high in nitrogen) with leaves and grass (high in carbon dioxide) to help control unpleasant odors. And, if your compost is not bagged, turn and mix the pile regularly to speed up the composting process.

There are some financial benefits to composting, since some cities give composters a break. In East Hampton, New York, for example, businesses are charged $15 a ton to remove compostable materials; $65 a ton for noncompostable materials. Other eateries forge deals with local farmers: our compost for your fresh produce. An added benefit is the goodwill this system generates. Mention it on menus, send out a news release to the local newspaper and/or television station, and you may be the recipient of some good publicity for your environment-friendly ways of doing business.

Combustion

Combustion (or *incineration*) is the burning of trash. Since the 1960s, researchers have experimented with ways to collect and reuse the heat generated by trash burning. By the 1990s, we burned as much as 9 percent of all solid waste; 6 percent of the time, we recovered the energy from this combustion process. However, the incinerators used to do the burning have built-in problems: the ash and smoke they create. Two culprits are of particular concern, dioxins and furans, members of a highly toxic family of gases. Acids and metallic gases are also byproducts of burning, so it is important that materials be carefully sorted before burning to remove anything that creates these chemicals or retards the burning process, including yard waste, glass, and metal.

The good news is that the EPA reports that more than 99 percent of pollutants can be filtered out of combustion emissions before they are released into the atmosphere. Ashes are typically sent to landfills, and the agency has regulations for their safe handling and proper disposal.

The United States is far behind on the international scene when it comes to use of combustion as a waste disposal method, burning only 14 to 20 percent of post-recycled waste. In Japan, 68 percent of waste is incinerated; in Sweden, 56 percent is incinerated.

Landfill Use

The landfill is the destination of most solid waste, but this luxury is most definitely in jeopardy. The number and capacity of landfills dwindle as human health and environmental concerns such as gas emissions and water contamination make headlines. Siting a new landfill—the technical term for starting one—has become more difficult as local, state, and federal governments adopt more stringent policies to help restore public confidence in their waste-disposal decisions. There seems to be a nationwide drive to close municipal solid waste landfills (MSWLFs, in EPA terminology). The number was down from 7,683 in 1992 to 3,581 in 1995. Fewer than half of the landfill sites that close are being replaced. The EPA estimates that by the year 2013 there will be only 1000 landfills in operation.

The remaining landfills are being better managed with a system called *landfill reclamation*. The landfill managers remove whatever recoverable materials they can (and sell them if there's a market), then combust and/or compact much of the remaining waste. This extends the life of the landfill site by using the space more intelligently, but it is also a more costly and equipment-intensive plan for a city or region to undertake. Of course, the fees paid by businesses and homeowners pay for this.

Reality dictates that landfills will always be needed, because there will always be items to discard that cannot be reused or recycled. In managing your business, you must weigh the options carefully to ensure that the landfill isn't just a last stop, but a last resort, for your throwaways.

The "Green" Restaurant

Whether you are building or remodeling a foodservice facility, you can make decisions that impact your tiny spot on the planet and impress your customers at the same time by choosing environment-friendly alternatives. McDonald's, for example, builds most of its new locations from recycled materials. So-called *green building*, also known as ***sustainability***, focuses on conservation of natural, human, and economic resources. "Green" designs incorporate the strengths and limitations of the surrounding environment; the use of native materials; and the addition of ideas that conserve energy, water, and labor. The National Association of Home Builders (NAHB) has spearheaded some of this research, including the use of recycled materials, alternatives to lumber, and insulation materials that keep outgassing (emissions of toxic fumes) to a minimum.

Growing environmental consciousness has an important side benefit—it makes this new technology more and more accessible and less and less expensive. The more willing you are to experiment, the better chance that there is a source of free (or low-interest) funding for your efforts, provided by utility companies or government agencies. The up-front costs may be daunting, but, over time, the long-range benefits of "going green" may be worth it to your bottom line.

SUMMARY

Your restaurant's environment should make guests feel welcome, comfortable, and secure. It should also provide a safe, productive work setting for employees. Safety and cleanliness are top priorities in any

BUILDING AND GROUNDS

DESIGNING A GREEN BUILDING

- Passive solar or radiant heat instead of forced-air heating
- High-efficiency, high-color-rendition fluorescent lights in kitchens and service areas instead of low-rendition fluorescent lights
- Water-based glues instead of glues containing volatile organic compounds
- Water-based, nontoxic, lead-free paints, stains, and finishes instead of oil-based products
- Solid woods instead of formaldehyde-emitting plywood
- Beeswax wood finish instead of polyurethane
- Recycled, reclaimed, or salvaged bricks, lumber, and steel instead of virgin products
- Natural instead of petroleum-based fibers for floor and wall coverings
- Incandescent light fixtures fitted with halogen lamps in dining areas instead of conventional incandescent lights
- Cementitious foam instead of fiberglass insulation
- Flame-retardant, pressed-paper ceilings instead of traditional drywall ceilings composed of gypsum and paper
- Slate, terra-cotta, clay, concrete, recycled metal and copper, cement, and wood-fiber concrete tile for roofing
- Low-E-coated glass, double or triple glazing, recycled aluminum frames, renewable softwood frames for windows
- Water- or milk-based paints void of lead, asbestos, mercury, or volatile organic compounds
- Chlorine-free refrigerants, air quality sensors to control fresh-air supply, passive solar strategies, in-floor radiant heat, hot-water baseboard heating for HVAC system
- Steel, copper, recycled plastic pipes, lead-free solder, nontoxic soft-setting sealant for plumbing
- Recycled aluminum sheets for walk-in coolers and freezers, scrubber system on hood exhausts, efficient hot-water heaters, electric chillers

Source: ***Restaurants and Institutions.***

facility. In the kitchen, sufficient exhaust capacity for hoods (canopies) over cooking equipment is not only necessary for a comfortable work environment; it is also regulated (and monitored) by health and insurance laws and inspectors.

Human engineering or *ergonomics* should be part of the design throughout your facility. This includes everything from nonslip flooring and comfortable heights for work surfaces and stairs, to placement of exits and safe storage of cleaning chemicals. When choosing materials for flooring, select slip-resistant and highly durable material that is resistant to temperature, moisture, and stains and can hold up under heavy equipment and a lot of foot traffic. Carpet is suitable only in dining areas, never kitchens, and it must be kept clean and in good repair.

NSF International sets voluntary standards for equipment sanitation, but everyone in the foodservice industry recognizes these as critical. They cover the design and construction of equipment so that it is easier to clean and sanitize, and readily accessible when maintenance is required.

Most municipalities have adopted the NSFI standards, and also a set of rules for keeping food clean and at safe temperatures known as the HACCP system. The Hazard Analysis of Critical Control Points is just what its name indicates—a system to decide what possible hazards exist in each step of the process, from receiving to storage, and preparation to serving; and doing whatever can be done to minimize these hazards. Documenting and monitoring these steps is a major part of HACCP compliance.

Foodservice businesses also have a civic duty to minimize the waste they generate. Use a waste audit form to keep track of what you are throwing away. You may want to consider recycling, composting, or combusting to lessen your impact on landfills; or you may make arrangements with a local food bank or homeless shelter to donate usable food products you would otherwise discard.

STUDY QUESTIONS

1. What types of fire protection are required in a restaurant kitchen?
2. Why does the height of worktables matter in a kitchen?
3. Where should small tools and hand appliances be stored in a prep area, and why?
4. What is an *aggregate* floor and is it a positive or negative factor in worker safety?
5. What are the main problems with these types of flooring?
   a. quarry tile
   b. composite sheet vinyl
   c. vinyl tile
6. How do you choose a vacuum cleaner for commercial use?
7. When NSF International examines foodservice equipment, what is it looking for?
8. What is HACCP and what is its role in the commercial kitchen?
9. How would you wash fresh fruits or vegetables by hand to sterilize them as much as possible?
10. Why is it not smart to put hot foods immediately into a refrigerator or freezer to cool them?
11. What should your first step be when creating a source reduction or recycling program in an existing restaurant, and why?
12. List two of the recycling tips you think would make the biggest overall difference in the amount of waste output of a foodservice facility, and explain why you chose them.

CHAPTER 9

BUYING AND INSTALLING FOODSERVICE EQUIPMENT

INTRODUCTION AND LEARNING OBJECTIVES

Now, we're ready to find and purchase the equipment that will fit in the foodservice space we have designed. Equipment purchasing is not the world's most glamorous job, but no one can deny its importance. Making the wrong decision—by choosing equipment that is too small or not quite right because it's "such a good deal," or by using a vendor that falls short of your expectations for training, installation, and maintenance—can become a major headache in terms of dollars spent and compromised service.

Equipment purchasing is not just for newly opening businesses, either. Every year, you'll find yourself replacing old or worn-out appliances, partially redesigning spaces, or just becoming aware of new items that do certain things better than what you've already got. Whatever your situation, there's so much to know about equipment that it will take the next half-dozen chapters to go into the specifics.

In this chapter, we will discuss:

- Choosing gas or electric equipment
- Deciding whether to buy, or lease, new or used equipment
- Having equipment custom-built
- New trends and technology in equipment design
- Writing equipment specifications to ensure you'll get exactly what you want
- Installation, service, and maintenance needs

Today, most organizations use a team approach to make equipment-purchase decisions. Often, the end user works with consultants and financial advisers to identify high-cost areas and seek lower-cost alternatives to equip the kitchen properly. The team prioritizes each piece of equipment by comparing costs of annual operation, costs of regular maintenance, and projected costs over the expected life cycle. Together, they develop a plan to achieve the original objectives of the business. Later in this chapter, we'll give specific details on how to analyze a purchase.

The restaurant's concept can be reinforced by some equipment choices—a wood-burning oven for a pizza restaurant, or an open-hearth broiler in a display kitchen, for instance. Remember that the menu has a major impact on equipment selection, as the first part of a much larger picture. What you decide to cook and serve determines equipment needs, which dictates the layout of the kitchen, which establishes your labor needs, which sets your price points, which helps to configure the seating and choose the decor appropriate for those price points. Your goal is to make all the pieces fit together.

9-1 GAS OR ELECTRIC?

In brand-new foodservice facilities, one of the earliest decisions to be made in the planning stages is what energy source will be used to cook the food: gas or electricity? Since it is often a matter of the chef's personal preference, this is one of those questions that will continue to be debated for decades. Without "taking sides," the August 2002 edition of *Foodservice Equipment and Supplies* magazine did a good job of summarizing the advantages of each energy source, which is summarized here:

GAS

1. Overall, natural gas is less expensive than electricity, because it contains a higher cumulative amount of Btus delivered from the point of extraction to the point of use. An example: A supply of 100,000 Btus at the well head, which is then converted to electricity, will have "lost" 73 percent of its original power by the time it is transferred through power lines to the restaurant, delivering only 27,000 Btus for actual use. Take the same 100,000 Btus, keep it in natural gas form, and deliver it through a series of gas pipelines to the same restaurant, and the restaurant receives 91,000 Btus, a net loss of only 7 percent (7,000 Btus).
2. As you have already learned in Chapter 5 of this text, electricity has an additional cost, known as the *demand factor,* which gas bills do not include.
3. Natural gas does not make additional demands on kitchen ventilation systems, which are determined by the cooking process, not the energy source.
4. Technological improvements in gas appliances include infrared fryers with 80 percent fuel use efficiency, and griddles with consistent temperatures on their entire surface. "Boiler-less" combi ovens that use gas have almost eliminated most costly combi oven maintenance problems.
5. Gas-fired bakery ovens produce moister products with longer shelf lives.

ELECTRICITY

1. Electric equipment is more fuel-efficient overall, because more of the energy that it uses goes directly into cooking the food.
2. Electric fryers are more efficient because the heating element (heat source) is located directly in the frying oil, which results in better heat transfer.
3. By design, electric ovens are better insulated, and the way their heating elements are placed gives them more uniform internal temperatures—which results in improved food quality and better product yield.
4. Induction range tops, which use electricity, provide faster heat, instance response, easier cleanup, and they contribute to a much cooler environment. (You'll learn more about how induction cooking works in Chapter 11.)
5. Electric equipment is more energy-efficient, because the way the thermostat controls the temperature, cycling on and off only as needed, means the appliance's actual power use is only a portion of its nameplate rating.
6. Electric utility providers often offer so-called step rate purchasing for commercial customers, meaning a lower cost per kilowatt-hour as consumption increases.

Still not convinced, one way or the other? Here's a bit more information from *Restaurants and Institutions* magazine.

FOODSERVICE EQUIPMENT

SELECTING A RANGE: GAS OR ELECTRIC?

Mustard or mayonnaise? Regular or decaf? Rare or well done? Some foodservice dilemmas are easily answered by examining personal preference. Others require a bit more practical thought and research, especially when deciding on something you'll have to live with and use for a long time to come.

So, what's it going to be: a gas or an electric range? *R&I* asked experts on both sides for the relative benefits.

ELECTRIC RANGES

Kaye Hatch, former executive director for the Electric Cooking Council, lists the benefits of electric-range cooking.

- Greater heat retention that keeps your kitchen cooler.
- No open flame, so fire hazards are lessened
- New technology offers better burners. Halogen, infrared, and traditional coil burners are available now.
- Even heat in the oven chamber makes more consistent products.
- No gas flue to worry about, so the range takes up less space.
- No gas byproducts to vent out of the kitchen.
- Not dependent on a finite fossil fuel. Electric ranges are ideal for places where natural gas is not readily available.

GAS RANGES

Tom Moskitis, director of new product marketing for the American Gas Association, explains the advantages of gas-range cooking.

- Better control of burner temperature. Whether up or down, gas burners respond instantaneously.
- Instant on/instant off burner power. There's no heat-up time and no cool-down time.
- Cheap operating costs. Typically, gas is seven to ten times cheaper than electric power.
- Features that let you control burner power. High-input and low-input burners are available for specific applications.
- Greater moisture content. Gas heat is moist, which is ideal for baking—it keeps foods moist while cooking in the oven chamber.

Source: *Restaurants and Institutions* magazine, a division of Reed Business Information, Oak Brook, Illinois.

As you can tell, the issue won't be decided here and now. But each of these considerations—some subjective, others not—are worth your research time.

Other Basic Decisions

The industry publication *Foodservice Equipment and Supplies* surveys its subscribers every couple of years to see exactly what prompts them to make big-ticket purchases. In the year 2000, the top two reasons stated for buying or replacing kitchen equipment were "food safety concerns" and "labor shortages/

high turnover." Still significant, but not quite as critical as those two, was the operator's "need for greater versatility" from the appliance. What this tells us is that safety, sanitation, and ease of operation are key desires for real-world equipment use. Everything you buy must do its job well, keep workers and customers safe, and not require a lengthy or difficult training process to get new employees up to speed on how to use and clean it.

The same survey indicated almost one-third (29 percent) of the purchases were replacements for existing equipment, while even more (38 percent) were necessary because new locations or new menu items were added, or existing locations were renovated.

Will your equipment be new or used? If you decide to buy new, you can opt to have appliances fabricated to meet the specific needs and dimensions of your facility. This is by far the most expensive option because it requires hiring an equipment consultant to write specifications for each piece of equipment, then paying a premium price to manufacturers to custom-make them.

You can get almost the same custom service for much less money by selecting standard equipment from manufacturers' catalogs, which list and show many color choices and dimensions. To get an idea of what's available, visit manufacturers' showrooms or warehouses or attend a foodservice equipment trade show. At trade shows, special discounts are usually offered; even the display models are sometimes sold at deep discounts so the manufacturers don't have to ship them back to the factory. (However, you will have to figure out how to get the display model from the show to your business!)

Manufacturers' and local equipment dealers' sales representatives have lots of experience and are often willing to help, but they're also trying to sell you their particular wares. If you're truly a beginner, or if you have an unusual kitchen space that you're not quite sure how to cram everything into, consider hiring a foodservice facilities consultant. This person designs kitchens for a living. The term "consultant" may conjure up nightmares of budget-busting fees, but if you are honest with your consultant about what you can spend—and if you ask about fees up front—you can receive objective and invaluable advice. When buying equipment, remember that a mistake will affect your operation for a long time.

So how do you decide the tough questions: To lease or to buy? New or used? Prior to making any large equipment acquisition, ask yourself these questions:

- **Is this piece of equipment needed now?**
- **How will I pay for it?**
- **What capacity or size do I need?**
- **Should I estimate future capacity?**
- **Do I have enough space for this?**
- **Will the kitchen staff use it?**
- **How useful are the options and accessories that are available?**
- **Is the equipment I need available locally?**
- **Is service and maintenance available locally?**
- **Are there local ordinances that affect my use of this (need for increased power supply, ventilation, and the like)?**

What makes a piece of equipment "essential"? If it is the most practical and least expensive means of getting the quality and quantity you need at the right time and place, your kitchen probably can't function without it. If it can be used to accomplish several tasks, it's probably worth having. Think about just how essential it is before you buy it.

9-2 SELECTING ESSENTIAL EQUIPMENT

Several factors will help you decide if a piece of equipment is essential to your operation. This type of calculation has been dubbed ***life-cycle costing*** by equipment expert and author Lendal Kotschevar, Ph.D. Life-cycle costing is a way to analyze what might normally be intangible factors, like durability and quality, as part of the purchase price.

The first is *cost.* Equipment is a long-term investment, so it is not as simple as choosing the most reasonably priced item you can find. Be sure to consider the *useful life* of what you are buying. Ask, "How long can I expect this to last?" A lower-priced choice may not be worth the money if it only lasts one-half or two-thirds as long as a higher-priced model. Ice makers, for example, can be self-contained, or installed with a remote compressor. The installation for the remote compressor is more expensive, but it may extend the useful life of the ice maker and/or reduce maintenance costs, since it is more easily accessible.

The most common method of determining useful life in the United States is to look at the depreciation schedule used by the Internal Revenue Service. Most equipment can't be written off on your income taxes in one lump sum; instead, the depreciation schedule allows you to write off a percentage of equipment costs every year during the standard life of the equipment. The figures are fairly arbitrary, but because so many people must comply with them, they've become the norm.

Illustration 9-1 shows an abbreviated list of some useful life estimates for kitchen equipment and dining room furniture. You'll learn more about depreciation later in this chapter.

In addition to the basic "sticker price," you'll also want to find out about the freight costs to get it to your location, installation costs, standard operating costs, costs of supplies (like filters and chemicals), cost of servicing the equipment (and your downtime when it's not working), and any trade-in or salvage value it may have when you decide to replace it.

Another cost-related consideration is *ease of use.* The more difficult it is to operate and maintain equipment, the longer it takes employees to learn the task and to use it properly. This increases labor costs, which impacts your bottom line. Always think about *labor savings* as an equipment advantage. An example: by using a cook-chill system (described in greater detail in Chapter 15), a chef can prepare his or her specialties, and quick-chill them to near-frozen temperatures where they will keep safely for several days. Then, even if the chef is not there, a less skilled line cook is perfectly capable of reheating it for serving. Self-cleaning appliances may be more expensive initially, but they may also pay for themselves faster by reducing the labor it would take to clean them manually. With these types of options, you can schedule employees' time more productively, at the least cost to the operation.

You'll hear many sales pitches that focus on *energy efficiency*—in fact, you should expect this benefit with any modern appliance. In the first five years of use, you may spend up to three times the initial cost of an appliance on the power required to run it. While almost every manufacturer claims its products are "energy efficient," the true meaning of this phrase to a foodservice operator has more to do with the energy prices in their area. Because it is nearly impossible to predict what power will cost in future years, the best you can hope for is an estimate of annual operating costs and a rundown of whatever features contribute to its energy savings. For some appliances, water use is another necessary calculation.

Projected use is the next major consideration. The tilting skillet, the convection steamer, and the range top with oven are all examples of multiple-use equipment. In a busy kitchen, versatility is key. Combining different functions in a single piece of equipment is another way to increase workers' productivity. Whenever possible, purchase equipment that can do more than one thing for you.

Brand names mean a lot in the restaurant business. Ask any chef who has been around a long time, and you will get some marked preferences for certain types of equipment. While a newcomer to the industry may be swayed by advertising or the recommendations of dealers, the seasoned restaurateur asks kitchen personnel what they like, and why.

Availability of service and repair is a big concern in this cost-conscious era. You'll need a factory-approved service person to install the equipment, to train your workforce to use it, and to repair it periodically. Availability of parts is also crucial, and foodservice operators often find out the hard way that most equipment dealers are not prepared to offer instant repair service. A higher initial cost, if it includes local, dependable service, is usually worth the price.

New pieces of equipment should also come with *warranties,* which cover parts and workmanship for a period of time (usually not more than one year) and then "parts only" for another specified length of time. Generally, a warranty indicates the manufacturer will replace or repair, free of charge, any part that proves not to work properly due to "defects in materials and/or workmanship." Most warranties go on to mention that they're only valid if no one has altered the equipment and if it has been correctly installed and maintained. In fact, most warranty hassles result because the equipment has not been properly installed. You'll read more about the different types of warranties later in this chapter.

Payment terms for the equipment are another important consideration. When you're spending a minimum of $2000 for a commercial mixer, expect some strings attached unless you can afford to pay cash. If money is borrowed from a bank to purchase equipment, the bank technically owns the equipment until the loan is repaid in full. If you're leasing your space, this gets a little sticky. The lease must include information about what to do if you fall behind on your payments, and the landlord must agree to grant the bank a first lien on all financed equipment. This means if the rent is not paid, the bank can repossess the equipment before the landlord can.

If the local bank is reluctant to loan money for equipment, ask the equipment dealer. Many dealers finance or lease entire restaurant installations. An attorney would be helpful to look over the legal aspects of these arrangements.

Analyzing Equipment Purchases

Restaurant owners regularly evaluate purchasing decisions, and their analysis is the same whether the equipment is a replacement piece or an addition, new or used. To justify the purchase price, equipment must generate surplus revenue through savings in utilities, materials, maintenance, or labor. A complete listing of all costs and savings throughout the life of the equipment should be made. Here are a few categories to research:

Costs	*Anticipated Savings*
Purchase price	Labor
Installation	Materials
Maintenance	Utilities
Repairs	Trade-in or salvage value

Keep in mind that the figures you'll come up with will probably be estimates, with a considerable margin for error. The more detailed your research, the better financial decision you can make. You can use the projected years of use set for depreciation by the U.S. Internal Revenue Service, or a median of fourteen years for foodservice equipment, or consult the list in Illustration 9-1.

The National Restaurant Association suggests two other helpful calculations: the simple payback (SB) and the return on investment (ROI).

Simple Payback. This is the amount of time it takes for an appliance to "pay for itself"—not just its cost, but any savings you will realize by using it. Let's assume we are evaluating the purchase of a commercial dishwasher and have received price quotes from two vendors. Vendor A's machine has a price tag of $7,500, with a life expectancy of ten years and "annual savings" (features like lower utility costs and less detergent use) of $1,500, according to the manufacturer. Vendor B's machine costs $9,000, has a life expectancy of ten years, and offers "annual savings" of $2,000.

To calculate simple payback (SB), divide the price of the appliance by the annual savings

EQUIPMENT	PROJECTED YEARS OF USE
Broilers	9
Dish and Tray Dispensers	9
Dishwashers	10
Food Slicers	9
Food Warmers	10
Freezers	9
Deep-Fat Fryers	10
Ice-Making Machines	7
Milk Dispensers	8
Ovens and Ranges	10
Patty-Making Machines	10
Pressure Cookers	12
Range Hoods	15
Scales	9
Scraping and Prewash Machines	9
Serving Carts	9
Service Stands	12
Sinks	14
Steam-Jacketed Kettles	13
Steam Tables	12
Storage Refrigerators	10
Vegetable Peelers	9
Work Tables	13

ILLUSTRATION 9-1 **Useful life of kitchen equipment.**

SOURCE: ARTHUR C. AVERY, *A MODERN GUIDE TO FOODSERVICE EQUIPMENT*, REV. ED., JOHN WILEY & SONS, INC.

Equation 1

Vendor A

$$SB = \frac{\$7{,}500}{\$1{,}500} = 5 \text{ years}$$

Vendor B

$$SB = \frac{\$9{,}000}{\$2{,}000} = 4.5 \text{ years}$$

Equation 2

Vendor A

$$\frac{\$1{,}500 - \$750}{\$7{,}500} = \frac{\$750}{\$7{,}500} = 0.10\ (10\%)$$

Vendor B

$$\frac{\${,}2000 - \$900}{\$9{,}000} = \frac{\$1{,}100}{\$9{,}000} = 0.12\ (12\%)$$

figure. The result is the number of years simple payback requires (see Equation 1). In simple payback terms, it would take six months longer to recover the initial purchase price for Vendor A's dish machine than for Vendor B's. This method, of course, does not take extra features into consideration—only cost.

Return on Investment. You can factor in these extra considerations using the return on investment (ROI) method. To calculate ROI, subtract annual depreciation from annual savings, and then divide that figure by the purchase price. The result is a percentage figure, your return on investment. The equation looks like this:

$$\frac{\text{Annual Savings} - \text{Annual Depreciation}}{\text{Purchase Price}} = \%\ (\text{ROI})$$

Using the same dishwasher quotes, let's determine ROI for the machines from Vendors A and B (see Equation 2).

FOODSERVICE EQUIPMENT

HOW THE "BIG GUNS" DO IT: MAJOR CHAINS' EQUIPMENT PURCHASE REQUIREMENTS

Purchasing in quantity gives quick-service restaurant chains some big advantages when it comes to equipment. Often, a chain can call the shots. But this has the interesting advantage of bringing out the best in their suppliers.

The **Burger King Corporation** expects its equipment suppliers to be experts on fabrication, consolidation, logistics management, and information systems. And, since Burger King has units in other countries, international capabilities are also required. Each year, Burger King opens five hundred to six hundred units domestically. Lead time for new store equipment packages can be from two to eight weeks, depending on how much custom work is involved. The chain expects its equipment suppliers to handle all delivery, installation, and follow-up. The suppliers must also consolidate buyouts, fabricate custom pieces, and replace unusable ones.

Church's Chicken is owned by AFC Enterprises of Atlanta, Georgia. This chain sends out bids and negotiates with the winning suppliers. AFC requires that its equipment dealer accept payment in thirty days, not immediately. AFC also expects the dealer to create an in-store file for each piece of equipment installed at each unit, which saves the operator from having to look up a model number when something needs repairs. Their dealers also take care of all warranty transactions. For this, the dealer receives a fee—1 percent of the cost of the equipment package itself.

Papa John's, a take-away pizza operation based in Louisville, Kentucky, uses three equipment dealers that provide millwork, stainless steel fabrication, equipment buyout, and consolidation services. The suppliers perform the installation, with a deadline of two days from start to finish for each new store. Papa John's allows 28 days from the time a complete equipment package is ordered, until it is placed into a new unit. For "rush jobs," this is cut to 18 days—and somehow, it gets done!

Source: Excerpted from *Foodservice Equipment and Supplies,* September 1999.

The higher the percentage of ROI, the better. Again, Vendor B offers the best buy. Vendor B's machine returns 3 percent more of the original cost throughout its usable life than Vendor A's machine.

An Introduction to Depreciation

United States tax law permits a depreciation deduction for the "exhaust, wear and tear of tangible property" used in the normal course of business. Any property held by the restaurant for the production of income qualifies. More specifically, what does *not* qualify is property used for personal purposes, such as a residence or vehicle not used in the business. Depreciation is also not allowed for food, beverage, or other inventories, land (apart from its improvements such as buildings), or any natural resource.

Usually, the Internal Revenue Service (IRS) allows depreciation on income-producing tangible property that has a useful life of more than one year. Your flatware, glassware, plateware, linens, and uniforms are not depreciated, because they are considered operating expenses, not tangible property. (Their costs can be written off in full in the year they are purchased.)

There are several methods of depreciation. Your accountant or tax adviser can make recommendations, but here are some of the details stipulated in the IRS's yearly publication #534, called *Depreciation:*

- All restaurant equipment is assigned a seven-year useful life. This means that, over a seven-year time period, you can write off the cost of the equipment (one-seventh of the cost each of those years). However, any automobiles, trucks, office equipment, and computers that qualify are assigned a five-year useful life.
- The method of depreciation depends on when the piece of equipment was put into use and how long its life cycle is estimated to be. Again, there are many methods of depreciation, so your accountant is the best source of advice. It is important to note, however, that once you choose a given method you cannot change it without IRS approval.

To see a few examples, let's look at a chart showing how two different methods would treat the same piece of equipment costing $10,000 (see Illustration 9-2).

Do not neglect items other than actual appliances which are also eligible for the seven-year depreciation. These include the gas, electric, and/or plumbing lines and connections necessary to operate all appliances, including computerized point-of-sale systems, exhaust hoods, and fire protection systems. As

PURCHASE PRICE: $10,000
USEFUL LIFE: 7 YEARS

METHOD: STRAIGHT-LINE METHOD			METHOD: 200% DECLINING BALANCE		
Year	7 Years	Dollars	Year	7 Years	Dollars
1	7.14%	714.00	1	14.29%	1429.00
2	14.29%	1429.00	2	24.49%	2449.00
3	14.20%	1429.00	3	17.49%	1749.00
4	14.20%	1429.00	4	12.49%	1249.00
5	14.29%	1429.00	5	8.93%	893.00
6	14.29%	1429.00	6	8.92%	892.00
7	14.29%	1429.00	7	8.93%	893.00
8	7.14%	714.00	8	4.96%	446.00
	100%	10,002.00		100%	10,000.00

ILLUSTRATION 9-2 **The way to depreciate the value of a piece of equipment depends on the method used.**

SOURCE: *UNIFORM SYSTEMS OF ACCOUNTS FOR RESTAURANTS*, NATIONAL RESTAURANT ASSOCIATION, WASHINGTON, D.C.

these items are purchased or replaced, keep and organize all receipts to be able to prove their monetary value for depreciation purposes.

Looking for Top-Quality Equipment

After you know what piece of equipment you want, it's time to research its construction quality, practicality, and ease of use. The best place for observation is probably the equipment dealer's showroom or a food/equipment exhibition, such as the giant trade shows held regularly by the National Association of Foodservice Equipment Manufacturers (NAFEM) or the National Restaurant Association.

Observe carefully when a manufacturer's representative demonstrates how to operate and clean the equipment. Notice where controls are and try the procedures yourself. Controls should be accessible; moving parts should operate easily. Ask about safety features such as guards and shields, and check for hazards: sharp blades, hot surfaces, or open flames workers may have to touch or reach across, protruding or moving parts that may snag clothing or hands.

Think about ergonomic concerns, like whether your staff has the physical ability to operate the equipment; some pieces require significant strength. Are the surfaces too high or too far for a comfortable reach? Repeated bending, for example, can quickly cause fatigue and soreness.

Elsewhere, we've gone into detail about cleaning, but it can't be stressed enough that the cleaning process should be easy and quick to encourage employees to comply with hygiene procedures. Make sure the food contact surfaces can be easily wiped down. Be wary of equipment that requires a multitude of small pieces or fasteners that can be easily lost. On the other hand, be sure that removable pieces are not too large to be properly cleaned and sanitized with your present warewashing equipment. If additional cleaning equipment is part of your plan, make sure it is purchased by the time the rest of your equipment is installed.

The size of the equipment is an important consideration. Most major commercial appliances need a few inches, or as much as 1 or 2 feet, of clearance around them for ventilation, utility hookups, cleaning, or repair. Bring accurate measurements with you when you shop to save yourself some frustrating experiences. Front access is the easiest, but there are several types of easy access to reduce repair and maintenance time and minimize the period of equipment downtime. Another size-related caution: don't buy more than you need. If you're ordering a pot sink and you know your largest piece of equipment will be an 18-by-26-inch sheet pan, why buy a custom-made 30-inch square sink bowl when a standard 28-inch model will do?

On the other hand, don't automatically assume you can't afford options and accessories. Some of them can improve equipment performance, save labor, and add versatility to a new unit. You owe it to yourself to find out what's available before you buy.

There are as many details about equipment construction as there are pieces of equipment. Be aware that the quality and workmanship you choose will help determine the life of your equipment. Before you shop, make a complete list of attributes you are looking for. You will also need this information if you end up ordering custom-fabricated equipment.

First: what is it made of? The substances used to construct most foodservice equipment are stainless steel, galvanized steel, and aluminum.

Stainless Steel. Stainless steel is the costliest and most commonly used material, and for good reason—if cleaned correctly, it is the most resistant to corrosion, pitting, and discoloration. Stainless steel begins as iron, but chromium and nickel are added to form a tough, invisible outer layer that gives it its durability. (You learned how to clean it correctly in Chapter 7.) The most corrosion-resistant is "18/8" stainless steel, meaning it contains 18 percent chromium and 8 percent nickel. An important note: In order for manufacturers to meet NSF International sanitation standards, stainless steel that comes into contact with food must contain at least 16 percent chromium.

The American Iron and Steel Institute ranks stainless steel in five classifications, called *grades,* according to its chemical composition. Each grade is identified with a three-digit number; the ones you'll find most often in foodservice are Grades 304 and 420 and, to a lesser extent, Grade 403. Grade 420 is used for cutlery, cooking utensils, and some cookware.

FOODSERVICE EQUIPMENT

CHOOSING FOOD PREPARATION AND STORAGE EQUIPMENT

Choose only equipment that meets industry and regulatory standards. Check equipment evaluations published by NSF International (formerly the National Sanitation Foundation) and by Underwriters' Laboratories (UL). Never use equipment intended for the home.

NSF standards require:

1. **Food-contact and food-splash surfaces that are:**
   - **easy to reach**
   - **easily cleanable by normal methods**
   - **nontoxic, nonabsorbent, corrosion resistant, nonreactive to food or cleaning products, and that do not leave a color, odor, or taste with food**
   - **smooth and free of pits, crevices, inside threads and shoulders, ledges, and rivet heads**
2. **Nontoxic lubricants**
3. **Rounded, tightly sealed corners and edges**
4. **Solid and liquid waste traps that are easy to remove**

Source: Reprinted with permission from "Foodservice Equipment 2000," a supplement to *Restaurant Business 1995.*

You may be asked what *finish* you desire for your equipment, meaning its degree of polish or shine. The various finishes are given numbers on a scale of 1 to 7—number 1 is very rough; 7 is an almost mirrorlike shine. For most work surfaces, number 3 or 4 (brushed or matte finish) is preferable since it can otherwise reflect glare from lights. The higher the finish number, the more expensive, so even choosing a 3 instead of a 4 can save 10 percent or so on the equipment cost.

Galvanized Steel. *Galvanized* means the iron or steel is coated with zinc. It has the strength of stainless steel, but the galvanized coating or baked-on enamel used to prevent corrosion eventually chips and cracks, thus leaving the underlying steel to rust. Galvanized steel is still a good choice where appearance is not so important, like equipment legs or the bracing that strengthens them. It is not recommended for areas of a kitchen that are usually damp or wet.

Aluminum. Aluminum is a soft, white element found in nature that must be converted to a metal of the same name. It is tempered (mixed with other substances) to improve its density, conductivity, strength, and corrosion resistance, before being used in hundreds of manufacturing applications. Tempered or alloyed aluminum can be almost as strong as stainless steel, but not nearly as heavy. It can be sanitized, is rust-resistant, reflects heat and light, does not ignite or burn, can be polished to an attractive finish, and doesn't get brittle under cold conditions—making it a good choice for refrigeration units. Its thermal (heat) conductivity makes it useful for water heaters, condenser coils, and HVAC system parts. One of the environmental advantages of using aluminum to make appliances is that it is fully recyclable.

Wood. Everyone loves the look of wood, but few people realize the challenges it must survive in a busy foodservice setting. Wood countertops or wall paneling should never be used around service stands, coffeemakers, anywhere there is a lot of traffic or moisture. Never agree to use particleboard in foodservice fabrication, since it loses its shape and consistency when it gets wet. For countertops, plywood is acceptable if it is covered with plastic laminate or wood-look veneer, which should be glued on with an exterior-rated glue typically meant for outdoor use. Again, moisture is the issue, and you want your countertops as moisture-resistant as possible. The best plywood is graded with a three-letter code; if the last letter is "X," that means the glue is exterior-rated. If money isn't an object, request "marine-grade" plywood, which is heavier (and more expensive) than regular.

Solid Surface Materials. In recent years, very attractive countertop options have been formulated from granite, marble, concrete, and synthetic materials like Corian® and Formica®. For long-term quality, many recommend granite because it is not impacted by intense heat, as some other materials are. High-grade granites are quite expensive, but there are lower grades available that are durable and won't break the budget. An interesting website that contains directions (and recommends products) for cleaning, sealing, and polishing a variety of these countertop materials is stonecare.com.

Other Construction Details. The ***gauge*** of a metal (abbreviated GA) means its thickness. The lower the number, the thicker the metal. Pots and pans are usually 18 to 20 gauge steel, since they need to be light enough to conduct heat well. Low-impact surfaces such as counter aprons or exhaust hoods are usually 18 or 20 gauge. But heavy-use and load-bearing surfaces, like worktables and counters in food prep and delivery areas, should be 14 gauge. For surfaces in serving areas, 16 gauge is sufficient.

In terms of cost, the thicker the metal (or lower the GA number) the more expensive it is. This is why you use it sparingly, only in the areas where it is truly needed for safety and sturdiness.

It's important to reinforce equipment that holds a lot of weight or might be impacted by heavy objects. An unreinforced countertop can noticeably bow from the weight of equipment; a storage shelf can even crease or buckle if overloaded. Sturdy leg structure requires horizontal support to prevent wobbling or buckling. Tie rods can accomplish this on mobile racks, cross rails, or worktables.

The least effective reinforcement method is simply to "hem" (turn under) the edges of the enclosing metal sheet frame, which "doubles" the edges and makes them somewhat stronger. This kind of reinforcement may be sufficient for cabinets, but if they are located in heavy-traffic aisles where mobile carts can hit them, more substantial framework is needed.

Equipment is most often held together by welds. ***Welding,*** the joining together of two pieces of metal by heating them, is by far the sturdiest and most permanent method, but also the most expensive. A fully welded piece of equipment will outlast one that has been fastened by other means. It will also cost more to be shipped, if it needs to come from elsewhere, because it will already be fully assembled.

Manufacturers like pop rivets because they're quicker and less expensive, but each pop rivet has a hole in the middle where debris can collect. They cannot be replaced if they snap off. Screws are also less desirable; they tend to vibrate loose from the metal when equipment is in use. The screw can fall out entirely, or the screw hole can become stripped beyond repair. Choose these less expensive options only for light-duty equipment.

9-3 BUYING USED EQUIPMENT

There always seems to be a glut of used restaurant equipment for sale on the market, and for budget-minded entrepreneurs, the lure can be tempting. However, you must think of it like any other type of used-goods sale. The seller usually accepts cash only and will not finance the purchase. The buyer typically accepts the merchandise "as is," with no warranty or possibility of a refund if it breaks down two weeks after purchase.

There are several sources of used equipment. It may be part of an existing business, sold as part of the overall ownership change. An owner may be closing a business and selling off individual pieces. A foodservice equipment dealer may buy the whole lot from such an owner, refurbish each appliance, and resell them piecemeal.

There are some great deals to be had—savings of up to 80 percent over purchase of new equipment—but buying used is only a true bargain if you're buying what you really need *and* it's in good condition. It may be dented or scratched, but will that eventually mean leaks or rust problems? If the current owner is getting rid of it because they've replaced it with something more energy efficient, will it become *your* energy drain next?

To determine how good a deal you're getting, you must first do your homework about new equipment. Find out exactly what it would cost brand new, fully installed, and ready to use. If you are purchasing a similar piece from a used-equipment dealer, pay no more than 50 percent of this "brand new" price and get at least a thirty-day warranty.

Restaurants USA, a publication of the National Restaurant Association, offers these additional guidelines for used-equipment purchase:

- Anything that needs repairs is a risk to the buyer. The availability of service for used merchandise is sporadic and should be determined first.
- Unlike new equipment that can be ordered to exact specifications, used pieces may not fit correctly into a kitchen. "One-stop shopping" to outfit an entire kitchen with used merchandise is almost impossible.
- In addition to its age, you have no idea how much it was used, how "hard" it worked. This can cause unforeseen problems.
- Service warranties, if any, are usually short: 30 to 90 days.
- Used equipment may be up for sale simply because it has been replaced by newer, more energy- and labor-saving models.
- Large equipment can be jarred, and possibly damaged, in moving. Does the sale price include professional delivery to your location?

Equipment resellers may do nothing more than clean a used item before putting it up for sale; or, if they've already refurbished and serviced it, plan to pay more for it. Do business with a reputable dealer who will allow you to have the equipment inspected before the sale by a repair person you know and trust. If you can, ask the piece's former owner, or the local supplier of that brand of equipment, for the written service record on each piece you are thinking about buying. Ask other restaurateurs about their experience with this brand. Is it reliable, or a headache? You can also check the age of the piece by jotting down its serial number, then contacting the manufacturer or local supplier; look for evidence of oil leaks. Some types of older equipment fail to meet current health codes, which is no bargain if they cause you problems with the local health inspector. You might shop only for used equipment that does not have moving parts or electrical components—sinks, tables, shelves, and stainless steel pans.

Finally, here are a few hints from the experts on what *not* to buy used: ice machines, dishwashing machines, and refrigerators. These items are most likely to have problems that crop up after being relocated. Also, gas ranges and ovens are a better used buy than are electrical ones.

Sometimes equipment is sold at auction. In this case, plan to pay no more than 20 percent of "brand new" value; be prepared to pay cash; and bring your truck and enough people to carry whatever you buy, because you will be expected to take it with you.

E-Commerce

A source of both new and used equipment, the Internet auction giant eBay is fast becoming a virtual superstore for foodservice equipment and supplies. eBay reports it sells more than 3,300 pieces of restaurant-related equipment each week, from listings of more than 10,000 items! The site's "Smart Search" feature allows prospective buyers to narrow the wares based on several criteria, including price range, geographic location, item category, and others.

There are business courses held in many communities to teach people how to buy and sell on eBay, and it's a good skill to learn. The process is fairly simple. A seller registers and places goods for sale in an "auction," with detailed digital photographs and a thorough description of each item, along with a minimum acceptable bid and sometimes a "Buy It Now" price—the latter, for someone serious enough to bypass the bidding process and pay top dollar. Potential buyers place bids on the item they want, and it is sold to the highest bidder when the auction closes, in three to ten days. Would-be buyers "watch" their auctions daily, and even hourly, to see if they've been outbid. There are several payment options, and the buyer and seller work out delivery terms if the seller has not already made them clear on the auction site. Both buyer and seller can give written feedback to eBay about the other, which helps keep the transactions hassle-free. However, they are not always, especially when buyer

FOODSERVICE EQUIPMENT

ADVICE FROM THE EXPERTS ON BUYING USED EQUIPMENT

For restaurant operators who do opt for used equipment, here are some tips for making the wisest selections possible.

MICHAEL FORCIER, OWNER OF PASTA MAKERS: "Check out the serial number. Every piece of equipment has a serial number. Call the manufacturers, and they can tell you how old it is. The seller could be selling you a 1965 model, and you don't want to pay much for something like that." Forcier also recommends checking the number of times a piece of equipment has been sold, and to whom.

ED DeMORE, CHAIRMAN OF INDUSTRIAL KITCHEN SERVICE IN BOSTON: "Ice machines are the most problematic machines in the kitchen, but dishwashing machines are the ones that give people the most trouble, because they're abused by the people who operate them. They're also the pieces that, when they go down, cause all hell to break loose. So you're liable to end up with no ice and no clean dishes.

"We're often called in to rebuild a dish machine on site. We do it because people think it's a way of getting by for a few more years, rather than spending $70,000 for a new one. But you don't end up with anything equivalent to a new machine."

DAVID MARCH, MANAGER OF SCHWEPPE'S USED FOODSERVICE EQUIPMENT CO.: "There are different things to look for in different pieces of equipment. See if there are service tags all over the thing. It will tell you if it's a problem piece of equipment. If the service and maintenance record shows 'repair, repair, repair,' on it, look out.

"For refrigeration, check that the door gaskets are not all ripped up. That's a costly thing to replace—$40 or $50 per gasket, plus labor."

JOE DUROCHER, ASSOCIATE PROFESSOR AT THE UNIVERSITY OF NEW HAMPSHIRE: "One of the things I'd look for on a mechanical piece is oil leaks. They all have oil in them somewhere. A leak tells me the unit might have a seal broken or something wrong inside.

"On any kind of chopping equipment, I'd look for blade alignment. On a buffalo chopper, for example, there are three 'cleaner fingers' that stick out and clean the blades. I'd look to see whether they've been nicked.

"The best thing is just to plug it in. Try the thing, although that's difficult with a large-scale piece of equipment. Buy a quality piece of equipment from a reputable manufacturer and you'll probably end up with something that is going to outlast your restaurant."

Source: Reprinted from *Restaurants USA* with permission of the National Restaurant Association.

and seller are individuals (not companies), located far from each other and/or when the on-line description of the item sold was, shall we say, not quite accurate.

A slightly less risky way to buy on the Internet is from the website COMMkitchen.com, which began in 1999 as a partnership of eight top national foodservice equipment vendors. Its virtual catalog, and a series of limited access on-line catalogs, offer more than 20,000 types of equipment, from huge refrigerators to smallware items. Here, instead of purchasing from an individual, you're buying from a well-known company.

Since E-commerce is crucial to businesses of the future, manufacturers are considering an automatic identification system for their products, a sort-of "bar code" as is now commonly used in supermarkets, to ensure exact identification of what is being ordered, and to prevent misbranding or mislabeling of lesser-quality merchandise as "brand name" when it's not. The industry trade group NAFEM—the North American Foodservice Equipment Manufacturers—has a Foodservice E-Commerce Group (FEG) working to develop ethics, standards, and a workable infrastructure for this fast-moving part of the industry.

9-4 LEASING EQUIPMENT

Leasing equipment is an attractive option for first-timers in foodservice. However, like many other types of leases, you end up paying more over the three- to five-year lease than the equipment would have cost to purchase outright. Roughly, the equipment is paid for by the twenty-fourth or thirtieth payment, but you've got 36 to 60 payments total. When you look at it that way, you're paying for a mighty expensive maintenance policy. And if your business doesn't make it, you may still be stuck with the remaining lease payments.

However, there are also advantages to leasing. Generally, it's a way to get the equipment financed 100 percent, whereas a bank will loan an average of 85 to 90 percent for equipment purchases and require you to come up with the rest as a down payment. Leasing leaves you with some cash on hand. And, since the payments are spread over several years or tailored to your cash flow, they may be less expensive month to month than a bank loan. If advancing technology makes a machine obsolete, it's all yours if you bought it. If you leased it, the leasing company absorbs the loss and, next time around, you lease the newest model instead.

The following example shows how to determine the cost breakdowns between leasing and buying the catering equipment to enable you to serve sit-down dinners and/or cocktail receptions for up to 100 guests:

	Rental Costs	*Purchase Costs*
Sit-down dinner	$3,349.75	$31,607.90
Cocktail reception	933.82	6,012.82
Total	$4,283.57	$37,620.72

The $37,620 figure seems excessive if you seldom do this type of catering. However, if you catered at least nine dinners and nine receptions a year, you would be smarter to purchase than rent the equipment.

The most prudent course of action is to do a cost analysis on a couple of specific pieces of equipment.

- To buy: Total price, interest rate, deposit required, monthly payments, depreciation, estimated maintenance costs, and what the equipment will be worth at the time you make your final payment.
- To lease: Total price, interest rate, deposit required, monthly payments, and estimated maintenance costs not covered by the leasing company.

As you read the lease, check that the document contains the following: a specific length of time the lease will be in effect; the dollar amount being financed, as well as the total dollar amount of all the payments and interest (two very different totals, as you will see); and the amount of deposit required.

You must also be clear about the difference between a maintenance contract and a service agreement. The leasing company is technically not responsible for making sure the equipment functions properly. The manufacturer should provide the same warranty or service agreement as you would have if you had purchased the equipment. If the leasing company offers a maintenance contract, it is an agreement over and above the manufacturer's service agreement; it guarantees that maintenance will be provided as needed.

FOODSERVICE EQUIPMENT

LEVELHEADED LEASING

Consider the following suggestions before signing on the dotted line:

DETERMINE WHAT YOU HOPE TO ACCOMPLISH BY LEASING. "Leasing should be a well-thought-out, well-planned financial business decision rather than something you have to fall back on as your only way to get funds for equipment," says Ron Sciortino, president of American Specialty Coffee and Culinary in Atlanta. "You want to know your leasing plan up front."

Sciortino cautions operators against leasing equipment if they're undercapitalized. "I've seen restaurants lease $10,000 to $20,000 dollars worth of equipment because they wanted to go all out and do it right," says Sciortino. "Then six months to a year later, they're out of business. Then we have legal problems and court battles."

- The biggest danger in embarking on a leasing program in an attempt to compensate for inadequate funds occurs when restaurateurs sign personal guarantees on the equipment they lease. "If you sign a personal guarantee, you're personally liable," notes Sciortino. "You could lose your house and a lot more."

EXPLORE ALL YOUR OPTIONS. This detail includes approaching the leasing department at your bank, as well as considering a conventional bank loan so that you can purchase the equipment instead. "Always go to your bank first," says Kay Stephens, owner of Capital Financial Corporation in Cordova, Tennessee. "Depending upon your locality and relationship with your bank, they may give you a lease through their own leasing department, and the rates might be lower. And you might even be preapproved."

PROTECT YOUR CREDIT BY MAKING AN INFORMED BUT FAIRLY RAPID DECISION. "Don't shop a leasing agreement around," warns Stephens. "Each time you shop around, a credit report is pulled on you and your restaurant. To the leasing companies, it either looks like you're shopping around or you've been turned down. If you've been turned down, why would they want to take a chance on you?"

DEAL WITH REPUTABLE SUPPLIERS. "In the lease atmosphere, you're at the mercy of the supplier," remarks Sciortino. "If you deal with a shady individual, it's impossible for you to be safe, so look for those who have a reputation for honesty and for treating their customers with respect."

READ THE LEASE THOROUGHLY BEFORE SIGNING. "You'd be amazed at how many people don't read the lease," says Stephens. "Leases are legally binding contracts drawn up by attorneys."

Source: Reprinted from *Restaurants USA* with permission of the National Restaurant Association.

9-5 NEW TRENDS AND TECHNOLOGY

If it's chosen wisely and used correctly, technology makes business faster, better, and more efficient. Technological trends are driven by several things:

- Ever-changing needs of customers
- Growth strategies of different types of foodservice facilities
- Environmental, ergonomic, and food safety concerns and legislation
- Availability and skill levels of workers in each segment of the industry
- Globalization of commerce

A survey of foodservice operators done in 2003 by *Restaurants and Institutions* magazine showed just how important new technology is to this business segment. Nearly half of the respondents planned to invest in computerized point-of-sale systems to replace their old cash registers and handwritten ordering systems. Thirty-five percent and 26 percent said they are interested in advances that improve food production and labor management, respectively. About half of the fine dining and full-service restaurant owners said they're developing Internet websites. Hospital and school foodservice managers were the most likely to invest in better systems for procurement and inventory management. The common thread? Everyone is anxious to make meaningful improvements.

Although the twenty-first century brought a slowdown in the U.S. economy, equipment manufacturers pressed on to introduce new items that address labor, food safety, and cost control needs of the industry—some in ways that only science fiction fans could have imagined a couple of decades ago!

NAFEM, an organization we've mentioned in this text, has introduced a concept known as the NAFEM Data Protocol, a standard for automating kitchens. It's a computer system that allows bidirectional communication of data between a central PC and various pieces of foodservice equipment. It can be used to automate temperatures of appliances, remind employees about maintenance duties, and even call a manager at home when the walk-in refrigerator quits running after hours, or a vendor when service is required. An "on-line kitchen" has several obvious benefits. It can collect data about whatever you want to track—consumption rates of certain dishes, minimum inventory levels (to determine when to reorder), HACCP-safe holding times and temperatures of foods. You can see which work areas of the kitchen carry what power loads, at what times of day. You can program appliances to heat up or shut off on cue, and track their energy consumption. You can keep maintenance records for appliances and be automatically updated when they need servicing.

At this writing, the idea is new. NAFEM is selling equipment manufacturers on the benefits of signing up for the program and making their products NAFEM Data Protocol compatible, and compliant with data communication capabilities. Learn more about the idea at nafem.org/resources/DataProtocol.

There are plenty more miscellaneous inventions and trends.

Ultraviolet Lamps. The use of ultraviolet light can remove the tiniest grease particles not normally extracted by the baffles in an exhaust hood. High-intensity UV lamps break grease molecules into the more manageable components of water, carbon dioxide and mineral acids—a gray powder that can be easily wiped away. The result is a cleaner duct system and exhaust fan, and additional odor reduction.

Induction Cooking. This type of burner, which uses an electromagnetic field to "excite" the molecules of metal cooking surfaces, is now widely available. Cooks are pleased with induction heat broilers that cook magnetically, without open flames or thermostat controls. It is being used in a wider variety of applications, from portable tableside units to Japanese "shabu-shabu" restaurants. Induction's primary attraction is safety—there are no flames, and it cools off immediately after the pan is removed from the cook-top. Induction burners are also more energy efficient than radiant hotplates or gas burners.

High-Speed Ovens. These units pack a lot of production into a small footprint. They use a combination of air impingement and microwave energy to cook foods at much faster speeds than conventional ovens. In many applications, they don't require a ventilation hood—another bonus.

Cooking Suites. Instead of the standard oven-below-the-range-top configuration, insulating materials have become so effective that small refrigerators or freezer drawers can be placed there instead, making the cooking area more self-contained. Where space is limited, and in display kitchens, these are really popular.

Cook-Chill Technology. Smaller, more streamlined cook-chill systems are now available. They're not just for high-volume production anymore; new systems have been refined for smaller noninstitutional operations. Today's kettle mixers can mix, cook, and chill food in one vessel. We will talk more about the importance of cook-chill systems in Chapter 15.

Mobile Holding Equipment. New cabinets to hold and transport cold and hot food items at safe temperatures use energy-efficient thermoelectric technology, avoiding the use of heating elements and/or refrigeration compressors. Their thick layers of polyurethane insulation allow them to hold food at safe temperatures even when they are not plugged in—up to two hours for cold foods, and eight hours for hot foods.

Other types of new equipment fall into the "labor-saving" category, with the added benefit of protecting workers from repetitive stress injuries. A super-automatic espresso-making machine instantly grinds the correct amount of beans, brews the espresso, steams the milk, and adds it to the coffee. No need to train a barista! With new food-portioning machines for large-scale production, all the workers do is insert the food into the machine. It can be programmed to expel an exact portion into waiting containers. No more scooping, spooning, and ladling. Ergomatix, the Las Vegas–based manufacturer, has sizes from 7-gallon tabletop units to 35-gallon floor models.

"Power Mixers," made by RoboCoupe of Jackson, Mississippi, simplify the task of mixing large batches of food, with a rotating wandlike blade that spins up to 10,000 revolutions per minute. "Power Soak" pot-washers were created by Metcraft, a manufacturer in Grandview, Missouri, to replace having to scrub pots by hand. Instead, the worker places dirty pots into the Power Soak wash tank, where water-jets blast off grease and baked-on food at the rate of 300 gallons per minute. The closed system means the worker can tend to other tasks, then come back later and remove the clean pot, ready to be rinsed and sanitized normally. Metcraft claims this saves 55 percent of the labor in a typical dish room.

Microprocessor controls can be programmed by the manufacturer or the operator to carefully control cooking times and sequences. Systems already exist that can vary the temperature in different parts of a single oven. This technology is making greatest headway in the baking field. Digital readouts indicating critical points such as elapsed cooking time, interior temperature, and ingredient temperatures are being used to simplify operations. "Doneness" can now be measured with moisture readings, vapor content analyzers, and optical sensors that detect color. At specific preset levels, the oven will turn off automatically. Much of today's new equipment features touch screens instead of dials or buttons; their plastic shield prevents grease and dust from entering machinery. Operating instructions are increasingly using symbols rather than words to aid employees who are not fluent in English.

Another feature worth watching is the automatic flame regulator (AFR). It's a spring-loaded valve topped with a mushroom-shaped cap that sits above the gas burner of a range top. When pressed down by a pot or pan, the gas flows and the burner lights. Removal of the pan shuts the stove off completely, but the pilot stays on. In nonpilot units, it maintains the gas at a safe but extremely low setting when the pan is removed.

Display cooking, menu merchandising, and cook-to-order kitchens have spurred numerous equipment innovations, such as salad bars and open grills. Pizza is a strong case in point. Its popularity is driving equipment design, as manufacturers combine microwave and convection methods so that pizza dough can be cooked quickly without becoming soggy or using tray liners. Pizza conveyor ovens have been redesigned with wider conveyor belts, capable of baking a more consistent product in less time. Wood-burning ovens are being replaced by high-temperature stone deck and hearth ovens to avoid the ventilation problems associated with wood burning.

A good way to keep up with the trends is to read the two top equipment trade publications: *Foodservice Equipment and Supplies* (fesmag.com) and *Foodservice Equipment Reports* (fermag.com).

A few equipment innovations are actually the result of regulatory trends. Some items are now height-adjustable, to meet work requirements of the Americans with Disabilities Act. Sinks and storage shelves and cutting boards are made with Microban®, Agion®, or other chemical components to retard bacteria growth and meet HACCP sanitation standards. In the 1990s, the U.S. Food and Drug Administration decided to lower the holding temperatures of cold foods from 45 to 41 degrees Fahrenheit, prompting refrigeration manufacturers to redesign their products to meet the more stringent new requirement. In short, the rules of the game can be changed at will by your federal, state, county, or city government. Often, they are changed by people who are unfamiliar with the restaurant business.

Metric Labeling

We'll discuss CE Marking, a quality control certification in Europe, later in this chapter. If the European Union (EU) has its way, equipment sold into its member countries will also be required to bear metric labeling. This impacts spec sheets, technical manuals, product literature, and advertising—as well as the nameplates on the equipment! At this writing, the "metric only" EU regulation is set to go into effect on January 1, 2010.

It is yet another sign of the globalization of our economy and, if the foodservice industry expects to remain competitive, it must learn to live with the needs of its second largest group of customers in the world (at least, currently)—in Europe. So, for those of you who can't recall the junior high school math classes in which this was surely covered, here is the basic conversion system from U.S. standard to metric.

Area. One inch equals 2.54 centimeters (cm). So, to convert from inches to centimeters, multiply by 2.54. A 6-inch measurement would be:

$$6 \times 2.54 = 15.25 \text{ cm}$$

One foot equals 30.54 centimeters (cm). So, to convert from feet to centimeters, multiply by 30.54. A 3-foot measurement would be:

$$3 \times 30.48 = 91.35 \text{ cm}$$

Volume. One cup (8 ounces) equals 29.57 milliliters (ml). So, to convert from ounces to milliliters, multiply by 29.57. An 8-ounce measurement would be:

$$8 \times 29.57 = 236.56 \text{ ml}$$

Weight. One ounce (of weight, not volume) equals 28.34 grams (gm). So, to convert from ounces to grams, multiply by 28.34. An 8-ounce measurement would be:

$$8 \times 28.34 \times 226.72 \text{ gm}$$

Temperature. From Fahrenheit to Centigrade (C):

From the present Fahrenheit temperature, subtract 32, then divide the result by 1.8.

98.6 degrees F would be:

$$98.6 - 32 = 66.6 \text{ then} \ldots \frac{66.6}{1.8} = 37.0 \text{ degrees C}$$

And from Centigrade to Fahrenheit:

Multiply the present Centigrade temperature by 1.8, then add 32 to the result.

100 degrees C would be:

$$100 \times 1.8 = 180 \quad \text{then} \quad 180 + 32 = 212 \text{ degrees F}$$

9-6 WRITING EQUIPMENT SPECIFICATIONS

Equipment specifications are concise statements about a piece of equipment, written to explain exactly what is needed so that potential sellers can supply exactly what you want. Commonly known as *specs* or *spec sheets,* they can consist of a few sentences, photographs or drawings with accompanying text, or several very detailed pages. It all depends on who writes the specs, and why.

Restaurant owner/operators usually write specs using their own past experience or from reading and reviewing manufacturers' literature. A foodservice consultant can often improve the basic description, adding notes from his or her own experience. Manufacturers write equipment specs in their product catalogs, listing details on capacity, dimensions, utility requirements, and more. The longest and most detailed specs are written by people who must purchase equipment for tax-supported facilities, such as schools and prisons, because they are likely to shop several different manufacturers and must make sure they can justify their final decision.

If you've never written equipment specifications, you can get ideas from manufacturers' catalogs, from trade journals, by consulting with local equipment dealers, or by attending a foodservice equipment trade show. However, in this chapter we'll provide all the information you need.

Legal Challenges. Of course, whenever one manufacturer's product is specified, other products are left out. Some unhappy manufacturers have taken their complaints to court, arguing that this amounts to an unfair trade practice or illegal restraint of trade, under federal antitrust laws. These laws have been broadly written and, so far, no court has declared that writing or using brand-specific spec sheets is illegal or unethical.

There have been some interesting cases in which architects have been taken to court for requesting certain brands of building materials in their construction and design specifications. The rulings to date indicate that these do not violate the antitrust laws as long as the manufacturers compete with each other freely and fairly to influence the architects' opinions prior to the specs being written. In one case, the judge stated that architects' general knowledge of the construction industry permits them to make "informed judgments that are in the client's best interest," and that architects are presumed to react to a healthy competitive environment and to specify what is most appropriate for a project or client.

In much the same way, foodservice consultants make hundreds of decisions on a project. They shop around, test products, learn from past experience and recommendations of others, and meet with manufacturers and equipment suppliers. The whole idea behind specs is to select the brand or model that meets all, or most, of the client's requirements.

Standards for Specifications

A consultant, architect, dealer, manufacturer, or end-user can be a *specifier*—that is, write specs. If you write your own, you can pretty much determine your own format. The ideas on these pages can be adapted to fit your needs.

The Foodservice Consultants Society International (FCSI) and the North American Association of Food Equipment Manufacturers (NAFEM) have developed recommendations for how specification sheets should be written and illustrated. Manufacturers' spec sheets have a standard format. Most recently, NAFEM and FCSI have created the Specifier Identification System (SIS). It assigns an identification code to any new piece of equipment. The code is a series of letters and numbers, which appear immediately following the model number, with an asterisk (*) in between. The SIS code stays with a model from the time the specs are written until the time it reaches the manufacturer's warehouse or showroom floor. Along the way, if a manufacturer has questions for the specifier, they can look on the NAFEM website for that SIS code, find the specifier, and contact them personally. There is no cost for being listed on the site as a specifier.

The terminology in spec sheets is very precise, and for good reason. We've excerpted a few examples and suggestions written by Justin H. Canfield, author of *A Glossary of Equipment Terminology* for the SECO Company, Inc.

Installation. Sometimes referred to as *erection.* It is most important that your specifications, whether they are for a complete installation or a requisition for a single piece of heavy-duty equipment, clearly specify the work that is to be done by the supplier when the equipment is delivered to you. Unless you state in your specifications: "Set in position designated on plan and anchor to floor," the supplier will probably dump it on your shipping platform and *you* will have to "worry it out" from there. Your specifications should also state: "After proper installation has been made as called for and mechanical connections have been completed by others, the supplier will start up and adjust this equipment, including the initial oiling and greasing if necessary, and demonstrate the use of it to any person or persons the owner might designate." Unless you "spell it out" in this or some similar manner, the cheap low bidder will not do it because he doesn't know how! At best, he will say, "Next time a factory man is in town, I will send him around."

It is also important on a complete installation that you stipulate that a competent foreman be provided for the erection and placement of the equipment by an equipment contractor, a foreman who is able to counsel with other contractors in regard to connections required at the time of the installation of the mechanical connections.

In short, spell out the work that you expect the supplier to do; do not leave anything to chance!

Cleanup. Everyone knows what this means, but it is a provision frequently omitted through oversight when specifications are written. Specifications should clearly state: "Equipment contractor will clean up all debris made by his workmen immediately upon completion of installation and remove same from the premises." If this provision is not included in the specifications, the purchaser of the equipment will be left with quite an unsightly mess to clean up—at his or her own expense.

Detail Drawings. These describe the drawings required to be submitted by the equipment contractor to the consultant, owner, and/or architect for approval. The drawings, usually done on computer nowadays, relate to specially built equipment to be supplied by the equipment contractor. These are submitted to the consultant, architect, and/or owner for approval before work is started so corrections, if required, can be made in advance. Whereas floor plans are usually submitted at a scale of one-quarter inch per foot, these detail drawings show the plan elevation and certain cross sections of special fixtures prepared at a scale of no less than three-quarter inch per foot so that all of the details of construction can be clearly indicated.

Guaranty or Guarantee. Everyone knows what this word means, but it is so commonplace it is frequently overlooked. All kitchen equipment specifications written by reliable consultants contain a "guaranty clause." However, people who are requisitioning individual pieces of equipment are cautioned to read the manufacturer's warranty or guaranty and, if it is not suitable for their requirements, they should stipulate what type of guaranty they want. If this exceeds the manufacturer's normal guaranty there will, obviously, be an extra charge. This matter should be carefully weighed to determine if it is worth the extra expense.

Qualifications. It is important, in obtaining quotations for large quantities of equipment, that you state in your specifications the qualifications you require of a bidder: financial ability; the fact that all of the equipment will be manufactured in one shop; that the bidder has the personnel and the engineering facilities to properly design, manufacture, and install the equipment; and that it be of uniform design and finish. Your consultant can help you in the proper wording of this all-important clause.

Installation Instructions. As Canfield also points out, you must be specific in your descriptions of how each item is to be installed. Here's a good example:

> **SHELVES.** It is important in writing specifications that the shelves within a cabinet-type fixture be specified as either "fixed," "removable," or "removable and perforated," whichever is desired. Insofar as overhead shelves are concerned, it should be clearly stated how they are to be mounted. If they are to be placed along a wall, specify the type of bracket (i.e., stainless steel, band iron, etc.). If the shelves are over a fixture in the center of the room, your specifications should state whether they are to be mounted on tubular uprights on all corners and in the center, if necessary, and of what material the tubular uprights are to be made, or if they are to be cantilever-type shelves.

Beginning to Write Specs

Now that you've seen the standard format, the job may seem even more challenging. When jotting down the first notes for your specs, just try to be as practical as possible. Think about who will use the piece of equipment, and what they'll be expected to do with it. Are there particular menu items it will prepare? What capacity do you need? What type of power source will be used? Where will it sit in the kitchen? Does it need to be mobile to serve more than one area? As Canfield hinted, delivery, setup, and installation costs should also be included in the specs. These are often overlooked, resulting in unexpected additional costs.

Arthur Avery, in his *Modern Guide to Foodservice Equipment,* offers a good, thorough outline of general requirements to be included in written specs:

1. The common, easily recognized name of the piece of equipment; for example: Reach-in refrigerator, one-door.
2. A general statement of what the buyer wants: A one-door reach-in refrigerator to be used by the hot-line cooks to store products prior to cooking.
3. Specific classification information; this includes type, size, style or model, grade, type of mounting required, and so on. In some instances, drawings or diagrams will be helpful.
4. Proof of quality assurance: inspection reports or results of performance tests on the equipment.
5. Delivery and installation: who will do it, and when; how much are you willing to pay for it? Put your request in writing here.
6. Any specific requirements about construction. This might include materials used to construct the equipment; utility details; performance parameters; certification by an agency, such as

Underwriters' Laboratories or the American Gas Association; warranty and/or maintenance requirements; and the need to be supplied with instructional materials about installation, use, or maintenance. (You'll learn more about certification agencies later in this section.)

Another suggested list of very practical specifications comes from the Foodservice Information Library's *SPEC-RITE for Kitchen Equipment.* It includes:

1. Who is the purchaser? (Who's paying the bill?)
2. Where should the equipment be shipped?
3. How is the equipment to be shipped, and who pays the freight costs?
4. What specific services are included in delivery: unloading, uncrating, setting in position, leveling, mechanical connections, start-up, use demonstrations?
5. If permits are required, who will secure them? If inspection is needed, who does it, and who pays for it?
6. List any appropriate standards of national agencies (electrical, mechanical) for this type of equipment.
7. Provide mechanical details, such as types of utility hookups, dimensions, and so on.
8. List the interior and exterior colors and finishes you want the piece of equipment to have.
9. Include any other options you would like.
10. Decide to include (or exclude) the "or equal" clause, which states that you will accept something of equal value if your exact needs cannot be met.
11. Require in writing that all custom-fabricated equipment be of uniform design and finish.
12. Provide a deadline for delivery, which includes adequate time for production and shipping.
13. Outline warranty needs, including who will service the warranty, how long it will be, and what it should cover.
14. List installation responsibilities. Who pays for it? Is the cost included in delivery? Should installers be union or nonunion, and at what rate of pay? Are there specific times of day the installation can (or cannot) be performed?
15. If equipment arrives early, before a new facility is completed, where will it be stored, and who pays for storage costs?
16. Taxes: What are they, and who pays them?
17. When and how will the equipment be paid for?
18. If changes are required, who pays for them, and how much?
19. If there are delays, who is responsible for any additional costs ("rush" delivery and so forth) that may be incurred?
20. If the order is cancelled, what would be an acceptable reason? Will there be cancellation penalties, and if so, how much? Who pays?

If you will be doing a lot of spec writing, we recommend the SPEC-RITE publication. It's a workbook to help equipment purchasers buy and sell intelligently. The book first guides the prospective buyer through a series of questions, offers some generic guidelines for each equipment category, and then generates a specification template. The activity is structured to improve communications between the operator and the dealer or consultant.

As you can tell, this is much more complicated than a department store purchase! A few of the topics we've mentioned deserve further discussion. They are certification agencies, warranties, equipment start-up, installation, and maintenance.

Certification Agencies

Certification means that equipment meets a set of minimum standards for safety and sanitation. In the U.S., the Occupational Safety and Health Administration (OSHA) requires that foodservice equipment be certified. It is important, for obvious reasons, that this testing is done by independent third parties—not the manufacturers themselves.

Equipment is tested by several agencies, in the U.S. and internationally. What are they looking for?

- **Materials used to make the equipment must be able to withstand normal wear, corrosive action of food, cleaning products, and even insect or rodent penetration. And nothing that comes into contact with food can impart any odor, color, taste, or harmful substance to the food.**
- **The equipment must be able to be installed, maintained, cleaned, and sanitized properly with reasonable effort.**
- **The equipment must perform as expected, according to its purpose and the manufacturer's promises. A holding unit that has a temperature range of a certain number of degrees must, indeed, be able to hold food in good condition within that temperature range, and so on.**

Some groups not only do the testing—they develop the standards based on their research. All of them also use the standards of the American National Standards Institute (ANSI), to bring some consistency to the process. ANSI does not certify equipment. Its job is to compile all the standards. Another organization, as its name implies, takes a more global view. The International Organization for Standardization (ISO) compiled the standards of 140 nations to develop a set of quality control documents, collectively called "ISO 9000," to set guidelines for manufacturing, production, and management practices. ISO 9000 certification is difficult to attain and considered a real benchmark for doing international business. As our economy becomes more global, expect more companies to achieve it—and more major buyers to expect it. The McDonald's quick-service chain already does.

Certification standards are reviewed by the groups that create them every few years, and may be reaffirmed, revised, or withdrawn. Following is an overview of the major certifying agencies.

NSF International. For our discussion, NSF International (formerly known as the National Sanitation Foundation) is the primary agency that ensures foodservice equipment is manufactured and installed in a safe manner for all concerned—guests as well as employees and managers. Founded in 1948, the organization acts as an authoritative and independent clearinghouse for users, manufacturers, and health authorities to solve sanitation problems together. The name was changed to NSF International in 1995 to reflect its worldwide influence.

The NSF International logo (see Illustration 9-3) is widely recognized and indicates that a particular piece of equipment complies with applicable food safety and sanitation standards. The standards are developed through research, testing, and equipment evaluation. Committees within the foundation made up of government, user, and manufacturer representatives review and determine the standards for equipment manufacturers, fabricators, and installers. You'll find out more about NSF International in the interview with its food equipment program manager, Joseph L. Phillips, following this chapter.

The NSF rule most applicable to foodservice facilities is Standard 2, but there are more than twenty detailed NSF standards dealing with everything from commercial dishwashers to food carts.

ILLUSTRATION 9-3
NSF International logo.
COURTESY OF NSF INTERNATIONAL.

Underwriters Laboratories. Underwriters Laboratories (UL) is the leading third-party product certification organization in the United States. Founded in 1894 to evaluate products for safe use at home and work, the UL tests more than 17,000 types of products, including all electric equipment and appliances used in foodservice settings. The UL standards are compatible with the National Electrical Code and other nationally recognized installation and safety codes.

A UL Mark of approval (see Illustration 9-4) means representative samples of the product have met nationally recognized safety standards for fire, electric shock, and related safety hazards. To the end user, the UL Mark is an accepted symbol of safety certification.

The UL Classification Program evaluates and classifies industrial, commercial, and other products for more specific properties and hazards such as sanitation. Currently, the UL publishes more than 700 standards for the benefit of the entire safety community.

In 1998, the UL started a sanitation certification service as a complement to its safety testing services. Both electric and gas-fired appliance manufacturers can obtain a separate UL sanitation classification, which includes compliance with the appropriate ANSI/NSFI

ILLUSTRATION 9-4
Underwriters Laboratories logo.
REPRODUCED WITH THE PERMISSION OF UNDERWRITERS LABORATORIES, INC.

ILLUSTRATION 9-5 **The CSA International logo.**
COURTESY OF CSA INTERNATIONAL.

standards. This distinct marking, in addition to other UL Marks, is widely recognized by regulatory authorities.

CSA International. Because Canada is the largest trading partner of the United States, it follows that American manufacturers, and others selling equipment in Canada, should comply with Canadian standards. CSA International is a provider of product testing and certification standards for electrical, mechanical, plumbing, gas, and a variety of other products. CSA International (CSA stands for "Canadian Standards Association") marks (see Illustration 9-5) are recognized in the United States, Canada, and around the world. The organization's primary goal is to ensure that products bearing its marks meet the requirements for performance, design, and safety established in the standard(s) to which the product has been tested. If you're planning to do business in or import from Canada, find out about the applicable rules and regulations for any particular product.

ILLUSTRATION 9-6 **Electric Testing Laboratories logo.**
COURTESY OF INTERTEK ETL SEMKO.

ETL SEMKO. ETL stands for Electric Testing Laboratories, a competitor and/or "alternative" to UL. The organization's SEMKO Division certifies both gas and electric equipment, and works with NSF International to allow ETL clients to meet international certification requirements. Its logo is shown in Illustration 9-6.

Conformité Européenne. Not a French major? That means "European Conformity," the health, safety, and environmental protection standard for all products made and/or marketed in Europe. The ***CE Marking,*** as it is called , is required for equipment, toys, and medical devices sold in the 15-nation European Union, and the three members of the European Free Trade Association (Iceland, Liechtenstein, and Norway.) If you purchase equipment in Europe, look for this marking. If you sell equipment to a European customer, you'll need to know more. CEmarking.net is a good Internet website for basic background; or type in "CE Certification" on any Internet search site and you'll find plenty of companies ready to set you up for import/export business.

There are about half a dozen private laboratories, known as third-party testing facilities, that assist the certification agencies. For instance, the Food Service Technology Center is operated by a foodservice engineering and consulting firm, Fisher-Nickel, Inc., to research manufacturers' competing claims. With the blessings of both natural gas and electric power industry groups, Underwriters' Laboratories, and other professional organizations, the FSTC does extensive refrigeration and ventilation research, as well as comparisons of equipment performance. FSTC is located in San Ramon, California, and has been operating since 1987.

If specifications seem somewhat overwhelming by now, here's a hint: For standard equipment such as a range, a manufacturer's catalog is the first point of reference and contains all kinds of information in a digestible format. Illustration 9-7 shows a picture of a range as well as its spec sheet (on facing page), which contains technical information in great detail: dimensions, diagrams, utilities, and a detailed outline showing how to specify this particular information. Most manufacturers have similar sheets for each piece of kitchen equipment they manufacture.

ILLUSTRATION 9-7 **(a) A restaurant range and oven, and (b) one page of the corresponding spec sheet.**
COURTESY OF SOUTHBEND CORPORATION, FUQUAY-VARINA, NORTH CAROLINA.

Warranty Specifics

Whether it is called a warranty or guarantee, the major questions this document should answer are: Exactly what will (and will not) be covered? And for how long? Although there are some standard provisions, each manufacturing company fashions its own warranty.

Models: ☐ **436D-2G, 436D-2T** ☐ **436C-2G, 436C-2T**

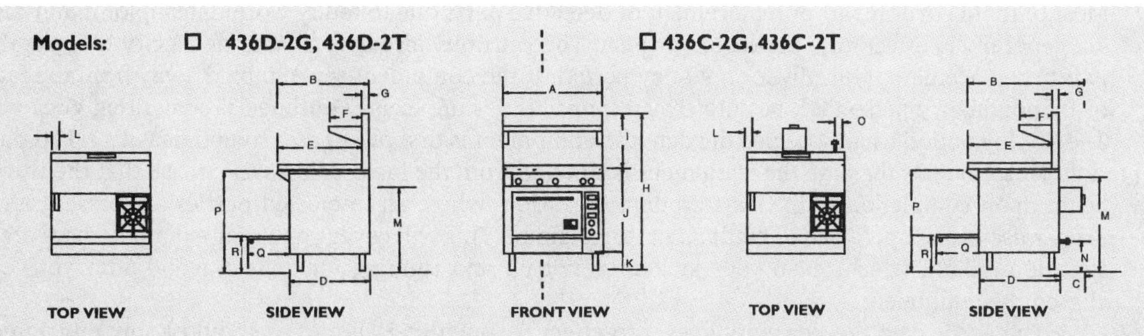

DIMENSIONS ( ) Millimeters

MODEL	EXTERIOR											¾" GAS CONN.		ELECTRIC		Cook Top	Door Opening	Oven Bottom
	A Width	B Depth	C[1]	D	E	F	G	H Height	I	J	K	L	M	N	O	P	Q	R
436D	36.5" (927)	32.5" (826)	–	29.75" (756)	29.625" (752)	12" (305)	2.75" (70)	56.5" (1435)	20.5" (521)	30" (41)	6" (152)	1.625" (41)	30.25" (768)	–	–	36" (914)	15.5" (394)	12.5" (318)
436C	36.5" (927)	40.5" (1029)	8" (203)	29.75" (756)	29.625" (752)	12" (305)	2.75" (70)	56.5" (1435)	20.5" (521)	30" (41)	6" (152)	1.625" (41)	30.25" (768)	10.25" (260)	3.75" (95)	36" (914)	15.5" (394)	12.5" (318)

MODEL	OVEN INTERIOR			BURNERS (BTU each)[2].			CRATE SIZE[3]			Cubic Volume	Crated Weight
	Width	Depth	Height	Open	Oven	Griddle	Width	Depth	Height		
436D	26" (660)	26.5" (673)	14" (356)	2 (26,000)	1 (32,000)	3 (16,000)	45" (1143)	39" (991)	58" (1473)	58.9 cu. ft. 1.67 cu. m.	535 lbs. 241 kg.
436C	26.125" (664)	21.75" (552)	14.25" (362)	2 (26,000)	1 (25,000)	3 (16,000)	55" (1397)	45.5" (1156)	81.5" (2070)	118 cu. ft. 3.33 cu. m	705 lbs 317kg.

NOTES:
1. Dimension **C** on Model 436C-2GL, 436C-2TL includes 2" minimum clearance between motor and wall.
2. OPTIONAL - Hot To Plate in lieu of 2 Open Burners at 12,000 BTU/Burner (24,000 BTU) Total
3. Units shipped with Splasher mounted, unless knockdown requested. Knockdown height 45" (1143mm).

UTILITY INFORMATION:

Gas– 436D-2GL, 436D-2TL total 132,000 BTU,
436C-2GL, 436C-2TL total 125,000 BTU
One 3/4" male connection (for location, see drawing above).
☐ NATURAL
☐ PROPANE
Required operating pressure: Natural Gas 4" W.C., Propane Gas 10" W.C.

Electric–436C-2GL, 436C-2TL only
STANDARD: 115/60/1 Furnished with 6 ft. cord w/3-prong plug. Total max. amps 6.2.
OPTIONAL – 208/60/1 or 3 phase (190 to 219 volts) – supply must be wired to junction box with terminal block located at rear . Total max. amps 4.0.
OPTIONAL – 236/60/1 or 3 phase (220 to 240 volts) – supply must be wired to junction box with terminal block located at rear. Total max. amps 4.0.

CONSTRUCTION (BIDDING) SPECIFICATIONS:

☐ **ALL UNITS –Commercial Gas Convection Oven -** 36½" W x 36 "D (Including 6" high legs.)
Exterior Finish: Stainless Steel front and shelf standard. Stainless Steel sides.
Range Top: Equipped with griddle and 2 each cast iron burners of round, non-clogging design. Center-to-center measurements between burners not less than 12" side-to-side, or front to back. Removable, one piece tray provided under burners to catch grease drippings.
Smooth Polished Griddle — ½" thick, hot rolled, Blanchard ground steel plate with high raised sides, 36½" wide X 29¾" deep.
Rigid Single Deck Stainless Steel Back Shelf. Sturdy Slip-On Design.

☐ **OPTION** –Thermostat griddle control with 2 throttling type thermostats. Temperature range of 100°F to 450°F (add prefix T-).

☐ **436D (STANDARD OVEN) – 32½" Deep**
Interior: Cavity sides top and back aluminized steel. Oven bottom and door lining porcelain enamel finish. Four sides and top of oven insulated with heavy, self-supporting, block type rock wool with oven baffle assembly constructed of Aluma-Ti steel.
Rack and Rack Guides: 2 position rack guides with one removable rack.
Door: Heavy Duty construction with extra heavy duty hinges and unbreakable quadrants. With cool tubular handle.
Controls: Oven thermostat low temperature type adjustable for 150°F to 500°F temperature range. Automatic safety pilot is hydraulic type.
Pressure Regulator: Supplied loose

☐ **436C (CONVECTION OVEN) – 40½" Deep**
Interior: Cavity, oven top, sides, bottom and door porcelain enamel finish.Back aluminized steel. Four sides and top of oven insulated with heavy, self-supporting, block type rock wool.

Rack and Rack Guides: Heavy-duty. 5 sets of rack guides on 2 3/8" centers with 3 removable plated racks (For best results, no more than 3 racks should be used.).
Door: Heavy duty construction with extra heavy duty hinges. Revolutionary single spring and chain device for easy adjustment, with positive door catch, and stainless steel mesh silicone impregnated door seal.
Blower Fan and Motor: 1/4 hp., 1725 rpm, 60 cycle, 115V AC, high-efficiency, permanent split phase motor with permanent lubricated ball bearings, overload protection and Class "B" insulation. Motor mounted to rear of oven. Motor serviceable from front of oven through oven cavity.
Electrical System: Wired for single phase, 115V AC with fused control circuit. 6 ft. cord and 3-prong plug supplied with each deck.
Gas Control System: Includes solenoids, pressure regulator, flame switch safety, pilot filter, pilot adjustment, and manual oven service shutoff valve.
Electronic Ignition: High voltage, spark-type igniter with flame switch safety device. Electronic ignition is activated by Oven Power On Switch, thus igniting standing pilot. With Oven Power switch in off position, the gas supply to the oven is completely shut-off. Operates on 115V system (for 208/236 volt, system incorporates a reducing transformer.)
Oven Heating: Aluminized steel bar burner. Dual flow fan recirculates heat directly from combustion area and within oven cavity.
Controls: Located in slide out drawer, away from heat zone.
Controls include:
Power Switch: Controls power to oven.
Fan Switch: 2-position; on for normal operation, on for forced cool down.
Thermostat: Adjustable for 150°F to 500°F temperature range.
60-Minute Mechanical Timer.
Oven Ready Light: Cycles with oven burner.

MISCELLANEOUS INFORMATION

If using flex hose connector, the I.D. should **not** be smaller than ¾" and must comply with ANSI Z 21.69

If casters are used with flex hose, a restraining device should be used to eliminate undue strain on the flex hose.

For installation on combustible floors (with 6"-high legs) and adjacent to combustible walls, allow 6" (152mm) clearance.

Recommended – install under vented hood.

Check local codes for fire and sanitary regulations.

If the unit is connected directly to the outside flue, an A.G.A. approved down draft diverter must be installed at the flue outlet of the oven.

Two speed motors are not available on Restaurant Range Convection Ovens

Oven **cannot** be operated without fan in operation.

INTENDED FOR COMMERCIAL USE ONLY.

NOT FOR HOUSEHOLD USE.

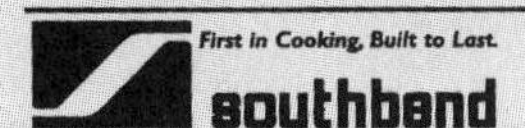

A MIDDLEBY COMPANY

1100 Old Honeycutt Rd. • Fuquay-Varina • NC • 27526
☎ (800) 348-2558 • (919) 552-9161
Fax (800) 625-6143 • (919) 552-9798

RR-04A FORM RR03A/04A PRINTED 07/96

ILLUSTRATION 9–7 (b).

Most of them cover repair or replacement of defective parts due to faulty workmanship or materials and are generally in effect for a period of one year. The warranty also covers labor necessary to make the repair or replacement, typically for a 90-day period. If the equipment is assembled away from the factory, or by nonfactory personnel, be sure that responsibilities are clearly outlined if something goes wrong. The 90-day period usually begins the date the equipment is first put to use; sometimes, it's a 120-day period, which begins the date the equipment is shipped from the factory. It is wise to ask that the warranty begin upon completion of the start-up demonstration, where all interested parties are present and the proper use of the appliance or equipment is explained. This will cover you if, for example, there are construction delays, which mean your kitchen is not up and running for quite a while after you've purchased the equipment.

The exact date the warranty goes into effect is something all parties should know and agree on: the owner/user, foodservice consultant, installer, equipment dealer, and manufacturer.

Warranties also have standard exemptions. Exemptions state that the manufacturer is not responsible for equipment problems that are the result of abuse or improper use. This means the piece must be correctly installed, not altered in any way, and it should have been maintained and operated according to the instructions in its service manual. You would be amazed at how often this crops up during warranty claims. So, to cover yourself, always make sure:

- The equipment is correctly leveled.
- It is connected to the correct power voltage.
- The motor is running in the right direction. (This may sound silly, but sometimes, if electric service has been hooked up wrong, the motor will run backwards!)
- Gas, water, or steam pressure is at the manufacturer's suggested settings (not too high or too low).
- Adequate ventilation is provided.

Manufacturers also usually insist they are not liable for the cost of any lost product or workers' wages to produce that product if the equipment fails.

On heat-producing appliances, warranties typically do not cover basic adjustments of thermostats, Bunsen burners, and pilot lights; nor do they cover replacement of timers, light bulbs, indicator lights, or valve handles. Also, water-related problems (a major source of service calls) are not covered.

Of course, the warranty or guarantee won't do you any good unless you understand how to file a warranty claim if you need to. So make sure the procedure is clearly spelled out for you.

In addition to the basic warranty, other limited warranties of up to three years may be available from the manufacturer at an additional cost. Under this type of warranty, the manufacturer agrees to pay for parts, labor, and "portal-to-portal" (roundtrip) transportation to make repairs during the first 12 months; for the remaining 24 months, the owner/user is responsible only for the transportation costs. Generally, there are mileage and/or drive time limitations specified for transport.

A warranty is not the same thing as a service agreement, in which the buyer pays an annual fee for regular maintenance of the equipment. There is no additional charge for parts, labor, or travel. If you can afford service agreements for your most valuable appliances, and you trust the company that does the maintenance, it does provide a certain amount of insurance against costly breakdowns and downtime.

Parts obsolescence is a real problem in the restaurant equipment industry. The best warranty will allow a five-year minimum time period in which replacement parts and interchangeable assemblies will be stocked and readily available from the manufacturer. Also, look for the manufacturer's ability to provide instructions and service manuals in more than one language (Spanish, in particular) and duplicate manuals if the originals are lost or destroyed.

In looking through warranty materials, you may notice that there are several different kinds. Here is a list of the prevalent ones found in the hospitality industry:

PARTS WARRANTY. Covers repair and/or replacement of defective parts. May or may not cover the labor required to do the repair or replacement or any freight charges involved if parts must be ordered.

LABOR WARRANTY. Covers the labor costs involved in repair or replacement of defective parts. There may be limits on total cost, and the repair person's travel time may not be included.

REFRIGERATION WARRANTY. An extension of the standard parts-and-labor warranty for refrigeration units. Typically, covers the compressor or other parts that might be damaged by compressor failure.

SERVICE CONTRACT. Another name for extended warranty. Many manufacturers offer service contracts to cover repair and replacement costs for time limits beyond the standard warranty. The price of the contract may depend on its length (from one to five years), the type of equipment, and the standard policies of the manufacturer.

CARRY-IN WARRANTY. Covers parts replacement, but only covers labor charges if the piece of equipment is brought to the repair facility for servicing. Labor costs are not covered if repairs are made at the owner/user's place of business.

9-7 START-UP, SERVICE, AND SAFETY

How well you actually "get along" with your new appliance will depend a lot on whether you're happy with the seller, so part of your prepurchase research should focus on the reputation and services of the dealer or manufacturer. An age-old gripe of restaurateurs who purchase new equipment is that dealers and manufacturers "sell 'em and forget 'em." However, the owner/user who makes an effort to keep in touch with these merchants—by sending in warranty cards, calling with questions, or dropping a note of thanks after a successful installation—gets more attention simply by realizing that communication is a two-way street.

Take the time to visit one or two sites where a particular piece of equipment is in use—and, if possible, take the service provider with you on these visits. No amount of sales literature can replace the experience of seeing equipment in operation and talking to the people who use it the most. Your dealer should be willing to provide names of other customers. Be wary if they won't.

Ask about the types of in-house training the seller will provide for your employees. In response to industry growth and customer needs, many manufacturers have implemented standardized, well-organized start-up procedures. These demonstrations familiarize the owner/user and employees with how to use and maintain the equipment and any of its accessories or attachments. It is important to have anyone who will use or clean the equipment present at the demonstration (not just the owner), and that it be conducted by a person who is fully qualified and authorized by the manufacturer to do so. Allow enough time for hands-on experience with the machine, and encourage your employees to ask questions: What kinds of simple, on-site service procedures can we perform ourselves? How often does this need to be cleaned (oiled, adjusted, and so on)? Are there any common problems you've noticed with this model in other kitchens? Where should we store the service manual? Who do we call in case of problems, if our boss isn't around?

A basic thumb-through of the manual should be part of the start-up demonstration, along with warranty information and instructions on how to file a warranty claim. Schedule the demonstration at least two weeks in advance. Have more than one session if you need to, so that all employees who must use the equipment may attend. Ask for videos and/or website addresses in case you need to obtain more information after the demo. Web-based customer service, in today's Internet-driven business world, is especially important. Can you get questions answered, and order parts, on-line?

Finally, it would certainly be an oversight (perhaps quite an amusing one) if you got everyone together for the demonstration and discovered that the equipment was not ready to, shall we say, participate? So, prior to its big debut, make absolutely sure it is correctly installed, hooked up to all necessary utilities, equipped with all its accessories, and ready to do the job. Again, the manufacturer should be able to assist you with this.

Equipment Installation

Depending on the type of equipment, the people who install it may be electricians, plumbers, utility company representatives, or equipment dealers' service personnel. No matter who handles the job, however,

FOODSERVICE EQUIPMENT

SERVICE AND SUPPORT: HOW ONE MANUFACTURER DOES IT

Groen is a Dover Industries company that specializes in steam equipment—steamers, combi ovens, steam jacketed kettles, and so on. Groen President Thomas Phillips, Jr., says its approach to service and support involves a number of different components:

1. Start-Up and Application Training. As training programs evolved and grew more extensive, Groen saw the need for more user-friendly installation and operation manuals. They've met that need by incorporating more graphics—photos, diagrams—and nontechnical language into the resource materials, to make them easier for employees to follow.
2. Warranty and Technical Support. One of Groen's most important long-term commitments is to provide customers with a quality warranty, a parts-stocking service network, and knowledgeable service technicians. With its network of authorized service agents already in place, Groen took the next step and implemented an aggressive "train the technicians" program. Phillips says service personnel are only as good as the training they receive. The training is supplemented by detailed diagnostic service manuals, which are updated regularly.
3. Quick-Ship Inventory. Groen recognizes that not all users can afford to wait on special orders for equipment, and created an "In-Stock" program designed to fill emergency quick-ship requests. It involved a multimillion-dollar inventory investment, to have the most popular models and styles of all its equipment in stock so they'll always be available for shipment within 48 hours.
4. Product Training. The "Groen Training Academy" is a factory-based school that includes a three-day program of hands-on experience with all major equipment categories. It's most often attended by equipment dealers, but others are welcome. While it requires significant staff involvement and resources, Phillips says the response from dealers has been very positive.
5. Accessibility. Groen is committed to being a company that's "easy to find," and has extended its customer service function to 24 hours for emergency technical support. The company website (groen.difoodservice.com) allows customers to access complete operations manuals as PDF files. Spec sheets, CAD drawings, price lists, and performance data are also available on-line.
6. Layout and Design Services. Foodservice consultants and dealers expect up-to-date equipment specs and resource material. When there is a unique situation that requires special assistance, a team of Groen engineers is available to make layout and equipment recommendations.

Source: Adapted from *Equipment Solutions* magazine, November 2000.

both you and your local health department will want it installed so that the equipment and surrounding area are easy to clean.

Not every piece of equipment can safely or conveniently be put on rollers or casters. Other pieces are too tall or too heavy to be wall mounted. So careful consideration should be given to alternate installation methods. Next, we illustrate and describe some of the most common options.

FLOOR MOUNTING. Some equipment (a revolving-tray mechanical oven, for instance) is designed and built to be mounted directly on the floor or on a pedestal. It should be sealed to the floor around the entire base of the equipment.

MASONRY BASE MOUNTING. Reach-in refrigerators, heavy-duty range tops, ovens, and broilers are usually mounted on concrete. The bases should be built at least 2 inches high and coved (rounded) where the platform meets the floor (see Illustration 9-8). The equipment should overhang the base by at least 1 inch, but not more than 4 inches. The equipment must be sealed to the base around the entire perimeter, and all utility connections or service openings through the floor must be adequately sealed for sanitation and to prevent vermin from nesting beneath the equipment.

WALL MOUNTING. Mounting equipment on a wall is the most expensive installation option, but it is very practical for sinks because it allows storage space beneath them. Wall mounting requires, of course, that the wall be reinforced well enough to hold the additional weight without damage to the building. A clearance of at least 6 inches should exist between the lowest horizontal part of the equipment and the floor, to facilitate cleaning. The installer must also make sure that liquid waste, dust, or debris cannot collect between the equipment and the wall itself.

In any installation procedure, remember that all utility service lines and openings through walls or floors must be properly sealed to discourage insect and rodent infestation, an ugly reality when working around food.

When horizontal lines or pipes are required, they should be kept at least 6 inches above the floor and 1 inch away from the wall and/or other pipes. This makes both walls and pipes easier to clean. If the pipes pass through the floor, they should be housed in protective sleeves or guards.

Drain connections should conform to the National Uniform Plumbing Code or to state or local plumbing ordinances. The diameter of the drain lines should be equal to, or greater than, the diameter of the equipment connection provided by the manufacturer. The National Sanitation Foundation provides basic standards for all of these items.

Equipment Maintenance

You've ordered it, installed it, and learned how to work it. Now you've got to keep that piece of equipment in top shape, or your investment won't pay off over the long haul. Most restaurants are filled with so much equipment, made by so many manufacturers, that maintenance requirements are easy to neglect. Almost every seasoned foodservice operator can cite instances in which a well-paid technician was called out for nothing. The "big problem" was a circuit breaker that was tripped and not reset; a pilot light that had blown out; an appliance that didn't work because it got too wet or too dirty; or (perhaps most embarrassing) an appliance that wasn't plugged in. How frustrating—and how expensive.

Let's introduce the fine art of preventive maintenance, or anticipating trouble. It is the best way to control service costs and to assure each piece of equipment a full, useful life. Think of it like servicing your car. If you spend all that money on the initial purchase but never change the oil or rotate the tires, you're asking for trouble.

The tasks will vary according to the age, size, and type of equipment, but all facilities should have a preventive maintenance schedule. This includes a system for regular inspection and cleaning. In Chapter 5, we discussed preparing a master checklist of every piece of electrical equipment in your building, including its location and model numbers or identification numbers. Expand the items on this list to include gas-powered, steam-powered, or battery-powered equipment, and you'll have a complete service list.

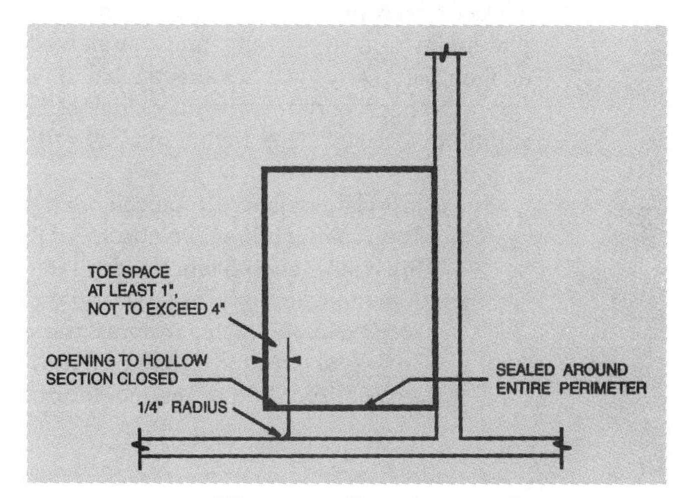

ILLUSTRATION 9-8 When an appliance is mounted on concrete, the special slab is called a masonry base and requires some special installation.

SOURCE: *A MANUAL OF SANITATION ASPECTS OF INSTALLATION OF FOODSERVICE EQUIPMENT*, NSF INTERNATIONAL, ANN ARBOR, MICHIGAN.

				January				February					March			
No. of Units	Time Alloted	Description of Work	Week of Year	1	2	3	4	5	6	7	8	9	10	11	12	13
		Check when completed														

				October				November					December			
No. of Units	Time Alloted	Description of Work	Week of Year	40	41	42	43	44	45	46	47	48	49	50	51	52
		Check when completed														

LEGEND: M - Minutes H - Hours V - Variable SC - Service Contract

Markings X - for completed / - for work scheduled

ILLUSTRATION 9-9 Preventive maintenance schedule, left blank for purposes of illustration.

Make the list into a master maintenance schedule by looking through the service manuals and jotting down the maintenance requirements of each item. You'll find most can be scheduled on a calendar, by day or week. Make sure to allot a certain number of hours for maintenance and specify who will do it (see Illustration 9-9).

Today's computer systems make it easy to keep on-line records, but it is helpful to have hard-copy backups. Some owner/operators keep simple index card files, with one card for each piece of equipment. This equipment information card (see Illustration 9-10) contains all the technical details about the piece, from serial number to date of purchase to manufacturer's website, phone and fax numbers. A separate card file, the equipment repair record or inspection card, can list the itemized history of each piece of equipment: when it was serviced, by whom, for what problems or symptoms, what was done, and what it cost (see Illustration 9-11). Once your maintenance records are organized, keeping them current will save you much time and hassle if anything goes wrong with equipment.

These records will also be invaluable for the other critical type of equipment maintenance, which is troubleshooting. With most types of equipment, there is a prescribed series of steps to follow to diagnose a problem. The dealer, manufacturer's representative, or service technician will often be willing to show someone on your staff what to look for if there is an equipment malfunction. Troubleshooting will help the repair person arrive with the correct parts the first time out, saving you labor costs or the price of an additional service call.

Even the best-maintained kitchen equipment will sometimes require service, simply because parts don't last forever and eventually wear out. If you have the correct replacement parts on hand, you may be able to do the job yourself. One idea is to purchase replacement parts at the same time you purchase the piece of equipment. You're more likely to get a good deal on them, since you are already making the "big ticket" purchase at the same time.

Another theory is to use only Original Equipment Manufacturers' (OEM) parts—and never a generic substitute to save money. Before you buy generic, consider these factors:

- Installing a generic replacement part will void the equipment warranty.
- The use of a generic replacement may modify the original design just enough to increase its fire risk when appliances have thermostats or electric elements.
- A generic part may not perform the same way as an OEM part. For critical tasks like holding foods at certain temperatures, this could be a problem.
- The durability of the appliance may be compromised with the addition of a component that was not specified by the manufacturer, causing other parts to fail sooner.

Safety and Training

The fast-growing Golden Corral cafeteria chain, based in Raleigh, North Carolina, requires its unit managers to complete a one-week intensive course on equipment use and maintenance. It's the best idea we've heard in a long time. They learn how to operate the equipment, and study the instructional manuals, warranty documents, and service contracts. Corporate headquarters spot-checks units around the nation to ensure compliance with its maintenance policies.

Maintenance is being treated as more than an afterthought in today's restaurants, especially the big chains, and that's a good thing. It impacts a variety of other functions: accounting (through life-cycle cost calculations); store operations (through training programs and scheduling of service); and capital spending. Maintenance also has a unique marketing function—the well-maintained restaurant is seen by all as a cleaner, safer place to eat and work.

It is critical to get staff members involved in your company's efforts to care for the equipment. Here are just a few ideas:

- **Reduce the damage caused by carelessness, abusive behavior, and vandalism by holding the staff accountable for the condition of the equipment when they complete a work shift.**
- **Eliminate dents and gashes caused by carts and mobile equipment by providing adequate clearance around equipment. It's not smart design to have people transporting items through what seems like a maze.**
- **Protect equipment with rails, guards, and bumpers, which are offered as accessories, both for fixed and mobile appliances.**
- **Catch little problems before they turn into big ones, with a weekly or monthly check of all kitchen work stations. Look for missing screws, damaged or worn wires and cords, bent panels or hinges. Get them corrected promptly.**
- **Make your staff aware of what maintenance costs the restaurant. Make maintenance the topic of some staff meetings, in addition to training sessions. Solicit opinions from the staff about improvements that could be made.**
- **Make an effort to get "clean" utilities—that is, do everything you can to protect equipment from power spikes with surge suppressors; treat or filter incoming water and air.**

Performing your own maintenance and simple repairs may make you apprehensive, especially if you don't consider yourself handy. However, if you follow these safety lists and read any precautions in your appliance owner's manual or handbook, you can handle many minor items without paying for a service call.

In general:

- **Be sure equipment has cooled down prior to attempting any repairs to it. The repair job will be done quicker and there's no need to blister your hands.**

TYPE OF EQUIPMENT

Manufacturer	Serial No.	Date Acquired	1st year Guarantee Expires	5 year Guarantee Expires

Service Contract	Cost	Weeks PM Service to be performed

Critical Components	Part Number	Nearest Service	Service Company	Remarks

Recommended Spare Parts Inventory

Part Name	Quantity

Miscellaneous Data

ILLUSTRATION 9-10 An equipment information card should be kept on file for every piece of kitchen equipment.

DATE	WORK PERFORMED	PARTS	TOTAL	ACCUMULATED COST / DATE

ILLUSTRATION 9-11 An equipment repair record keeps track of service and repair calls on each appliance. It can be kept on paper or on computer.

- Water and hot oil do not mix. Don't use water near fryers.
- Keep equipment at least 6 inches away from walls.
- Equipment should be cleaned prior to starting any repairs or maintenance. This is critical but very difficult to achieve in a busy kitchen.
- Know where the nearest fire extinguisher is and how to use it.
- Know where the nearest fire alarm is and how to shut it off.
- Know where the circuit breaker is for the appliance you're working on.
- Wear safety glasses.

Now for some specifics that could save your life.

Getting Good Service

No matter how handy you are or how well you've maintained your equipment, the need will eventually arise for a friendly visit by service technicians. The best time to select them is before you need them. In fact, the dealer or manufacturer's representative should be able to supply you with the names and phone numbers of reliable repair shops at the time of the start-up demonstration. You should list these on your equipment information records. Another suggestion is to contact your local restaurant association or even your competitors, to ask for their recommendations. A last resort is to flip open the telephone book and start calling.

IN THE KITCHEN

SAFETY RULES FOR ELECTRICITY

1. When you've turned off a circuit breaker to work on a piece of equipment, always put a piece of tape across it so someone else doesn't accidentally turn it on.
2. After the circuit breaker is turned off, always test the equipment with a volt meter to make sure you turned off the correct breaker and there aren't other circuit breakers that need to be turned off.
3. Always check the voltmeter to make sure it works by testing it in a live outlet that you know works.
4. When doing a "jump" test: Turn off the power, place the jumpers across the switch to be jumped, turn the power back on to observe the results, turn the power off, and remove the jumper.
5. Never leave a switch jumped longer than the few seconds it takes to see the results of the test.
6. Never work on electrical equipment if the floor is wet.
7. When leaving a piece of equipment, tape a big note to the front explaining the situation (e.g., "Out of Order, Will be back with parts this afternoon").
8. When a piece of equipment is "temporarily" fixed, never let anyone use it until it is safe.
9. Never call a job "done" until you have thoroughly tested it and are 100 percent sure it is safe and fully operational.
10. Always know where you can turn off the power quickly in case of a problem, whether it's at the plug or circuit breaker!

Source: Don Walker, *Manual of Gas/Electric Equipment*, Walker Publications.

IN THE KITCHEN

SAFETY RULES FOR GAS

1. Know the location of the gas shutoff valve on the piece of equipment you're working on. Gas valves can jam easily so turn it off before starting repairs.
2. Always know where the emergency gas shutoff is to the entire kitchen, and how to turn it off and on.
3. If you smell gas, turn the equipment off, wait for the gas to dissipate, then look and listen for the site of the leak.
4. If you opened up a gas line in any way, always check for leaks.
5. To check for leaks, use a soap solution—never a flame.
6. Fix all gas leaks promptly, however small.
7. Always cap unused gas lines. Don't just turn them off.
8. When changing gas controls, always check to make sure the control is for the type of gas that you are using (natural or LP). Some controls are interchangeable, but others are specific to one type.
9. Light pilots with a tightly rolled piece of paper at least 12 inches long. This gives you some protection if the gas pops.
10. If the pilot blows out, wait four to five minutes for any gas to dissipate before you try to light the pilot again.

Source: Don Walker, *Manual of Gas/Electric Equipment,* Walker Publications.

How do you decide which service/repair firms are reliable? From Nolan Marks's *On the Spot Repair Manual for Commercial Foodservice Equipment,* here are some pointers:

- How long does it take them to respond to a service call? Then, when you've actually called them, clock them. Ask specifically about service, and prices, on nights and weekends.
- Once they arrive, how long does it take them to diagnose the problem? If it takes more than two trips for one piece of equipment, that's too long.
- Are you charged for more than one trip, even though it's their fault that they can't figure out what's wrong? A bad sign.
- After a visit, does the equipment work fine . . . for a day or two? This indicates they've fixed the symptom, but not the overall problem.
- Are two people sent on the job, when it only takes one? If so, are you charged for both of them?
- Does the repair person show up with few (or no) tools, diagnostic equipment, or replacement parts? This is a sure sign you'll need a return visit, which you'll probably be charged for.
- Does the repair person spend a long time on the phone describing the problem to the "office" instead of dealing with it on site?
- Is the repair person a "part-changer," coming out several times to replace one part at a time, hopeful each time that "This should do it!"?
- Ask about the availability, and possible cost savings, of rebuilt parts instead of new. Ask for the defective parts and inspect them for signs of wear or damage. Of course, if it's a warranty repair, the old part must be returned to the manufacturer.

To summarize what it takes to be a smart equipment owner, Marks also has a dozen handy tips he calls Marks' Maxims. Follow them, and you'll minimize your facility's need for minor but costly equipment servicing and repairs.

Marks' Maxims

1. Check your power source.
2. Read the owner's manual.
3. Clean the equipment.
4. Educate your employees.
5. Use the equipment properly.
6. Know your warranty coverage.
7. Don't let someone do for you what you can do for yourself.
8. Always disconnect the power source before working on equipment.
9. Choose a good service/repair business.
10. Check the obvious.
11. Ask for discounts.
12. Don't be a "part-changer."

SUMMARY

Most foodservice operations use a team approach and do a lot of research to determine which pieces of equipment are best for their business. You can buy new or used equipment, lease equipment, or even have it custom-built. No matter what you decide, you must first decide on the equipment specifications, or "specs," that explain exactly what you want. You can write your own specs, or select a specific manufacturer's "spec sheet" and use it as a guideline to shop around.

Cost, energy efficiency, projected use and useful life, size, brand name, type and length of warranty, availability of service and repairs in your area, and on-line support from the manufacturer are all considerations as you shop for equipment and write your specs. This chapter includes a detailed list of items that can be part of written specifications, including who pays for what if the piece of equipment is late or if the order is cancelled altogether. There are several certification agencies, such as Underwriters' Laboratories, NSF International, and the Canadian Standards Association, which create safety standards for equipment and test appliances to hold them to those standards. It's also important to keep up with trends and technology, by reading the industry equipment publications and attending an occasional foodservice equipment trade show.

In addition to what an appliance actually costs, think about how much it might save you—in labor, materials, utility costs, or even its eventual trade-in value. When you decide on a model, there are three more major steps to getting your money's worth: correct installation that meets your building codes, training of your employees to use the equipment correctly, and regular maintenance of the equipment in your kitchen. Employees' knowledge and attitudes about keeping the equipment safe and sanitary, from day to day, are keys to its longer useful life.

STUDY QUESTIONS

1. List three things to consider in deciding whether you need a particular piece of equipment for your kitchen.
2. What questions should you ask about the equipment seller's ability to service what you buy?
3. Generally, how much less should a piece of used equipment cost than it did when it was new? What should you think about (besides cost) in choosing used equipment?
4. What are equipment specifications, and why do you need them?
5. What kinds of information should be contained in equipment specifications?

6. What kinds of precautions do you have to take to make sure that an appliance is adequately covered by its warranty?
7. What are the five different types of warranties? Which ones would you say you definitely need, and which are optional?
8. What determines if a piece of kitchen equipment should be mounted on the floor, on a base, or on the wall?
9. What information should be listed on an equipment information card?
10. How would you "check up" on an equipment service/repair company? Name at least three things you'd consider in deciding if it was a reliable firm.

A Conversation with . . .

Joseph L. Phillips

TECHNICAL MANAGER, FOOD EQUIPMENT PROGRAM, NSF INTERNATIONAL
ANN ARBOR, MICHIGAN

NSF International (formerly known as the National Sanitation Foundation) celebrated its 50th anniversary in 1994. At this writing, Joe Phillips had been in the Food Equipment Program for 16 of his 20 years at NSF as Program Manager, and more recently as Technical Manager.

Joe began his career in 1976 as a regional representative conducting audits for the company right out of college, with a B.S. degree in environmental health from Ferris State College in Michigan. Except for a two-year stint at Taco Bell, he's been at NSF ever since. Other areas of responsibility have included everything from testing swimming pool equipment to specialty items for the mobile home industry. Joe says his goal is to have the NSF Mark recognized worldwide.

Q: NSFI has come a long way. How did it get to be so big internationally?
A: Many customers were attracted to NSFI because they were sending products to the United States and needed the NSFI Certification Mark to sell them. Today, many countries still don't have their own sanitation requirements for equipment, so they're looking to potentially adopt our standards. Spain and Portugal recently adopted the NSFI Standards.

Q: Why should people use NSFI-certified equipment?
A: It gives them some assurance that the equipment meets minimum requirements for public health. We make sure the materials used are nontoxic, corrosion resistant, easy to clean, and that the design and construction of the equipment doesn't make it easy to harbor vermin. Many products also get performance-tested.

Q: How are your tests and services funded?
A: Primarily through fees that our customers pay for our services. We also conduct special studies for our customers, and for governmental agencies like the Food and Drug Administration and Environmental Protection Agency. And we publish an NSFI Listing book that identifies all the manufacturers and products we certify. It's sent to thousands of people and agencies all over the world, and they use it to purchase equipment or verify that it's been certified. The costs to obtain certification or listing vary, depending on how complex the equipment is and what kinds of testing it requires.

Q: How badly do people need NSFI certification to do business?
A: Getting certified is a voluntary decision the customer makes, but many regulatory agencies will not allow a product to be used in their jurisdiction if it is not NSFI-certified. Our most recent survey showed that 98 percent of the state and local public health agencies surveyed were looking for NSFI listed products, or using NSF International Standards.

Q: How do you update the Standards, or allow for changes in the industry?
A: Our Standards are made by consensus of a committee made up of manufacturers, users, and

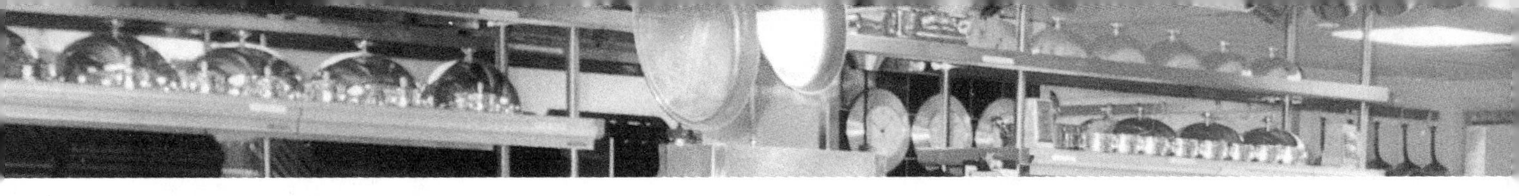

regulatory authorities. The committee is equally balanced between these groups, so no one has an overriding say in the end result. A Standard can be revised at any time if the need arises because of new technology, but must be opened for revision at least every five years.

Q: How does the process work?
A: It's a unique process. A change may begin at the Task Committee level, which is the nuts-and-bolts committee made up of primarily industry people because they are the experts on equipment. Regulatory and user representatives are also on the committee. This group develops a proposal for the Joint Committee, which is equally balanced among all three groups. The document is then balloted, with all negatives being adjudicated or determined to be nonpersuasive. The proposal then goes to the Council of Public Health Consultants for another ballot. This group is made up of public health experts, and once again, any negatives they may see about the proposal are discussed at this stage. From there, the proposal goes to a Board of Directors for adoption. So there are many reviews before a Standard can be changed.

Q: How big is NSF International; how many employees?
A: We have about 300 employees worldwide. There are laboratories at our main office in Ann Arbor, Michigan, and in Sacramento, California. There are also offices in Brussels, Belgium; Nairobi, South Africa; and in Australia. Our field auditors work from their homes.

Q: What exactly do you have in your facilities?
A: We have administrative offices, physical and chemical testing laboratories, toxicologists, microbiological laboratories, and a clean room to test biological safety cabinets.

Q: What is the pass/fail rate for equipment you inspect?
A: The initial evaluation of equipment usually has a high rate of noncompliance with its Standard, but once a customer becomes familiar with the requirements, that rate goes down considerably.

Q: If a student wanted to work for NSF International, how would they go about it? Are there entry-level jobs?
A: There are entry-level positions here, and other levels that require more education or experience. I would recommend contacting our human resource department, or taking a look at our home page on the Internet.

Q: What is NSFI's challenge for the next decade or so?
A: To become as well known internationally as we are in the United States. NSF International is a World Health Organization "Collaborating Centre" for food safety and for drinking water safety and treatment.
I hope we'll be able to have the NSF International Mark and Standards recognized throughout the world.

Q: Have you ever found counterfeit NSFI Marks?
A: We have heard that they exist but have not been able to confirm the allegations. If there is a question about the authenticity of a Mark in the field, it's always best to contact us directly, and we tell regulatory agencies the same thing. NSF International will use every level means to prevent the unauthorized use of our registered Mark. The credibility of the NSFI Mark is all we have! It's important, and we will protect it.

CHAPTER 10

STORAGE EQUIPMENT: DRY AND REFRIGERATED

INTRODUCTION AND LEARNING OBJECTIVES

Any type of foodservice facility begins its food preparation process at the back door, by receiving and storing the raw materials. Food, beverages, and supplies must all be accounted for and properly stored until they're needed. There are two basic types of storage: dry and refrigerated. ***Dry storage*** is for canned goods, paper products, and anything else that doesn't need to be kept cold. ***Refrigerated storage*** is for items that must be kept chilled or frozen until used.

In this chapter, you'll learn more about them both, and about the types of equipment needed to outfit your receiving and storage areas for efficiency, safety, and conservation of space. The list includes:

- Scales
- Pallets
- Carts
- Shelves

We'll also explain how refrigeration systems work in refrigerators, coolers, freezers, ice makers, and specialty systems like beer kegs and soft-serve machines, and discuss how to select them for your operation.

10-1 RECEIVING AND DRY STORAGE

You'd be amazed at how many deliveries arrive at the back door or loading dock of the average restaurant in a week. And they all have one thing in common: They must all be checked for accuracy by someone on your staff. Many owners and chefs go a step further, personally inspecting the quality of fresh items such as produce and seafood and rejecting on the spot those that do not meet their standards or expectations.

Having a well-organized receiving area, setting certain hours for deliveries, and designating employees who are responsible for accepting and storing incoming stock will save you time and money. Illustration 10-1 shows some of the smartest outdoor dock area features.

Merchandise goes from the dock area into the receiving area, shown in Illustration 10-2. The well-equipped receiving area will contain the following basic items.

Scales

Scales are used to weigh items received in bulk, to make sure you're getting the amount you pay for. As food costs increase, checking the weights of items in the receiving area is considered a very strong

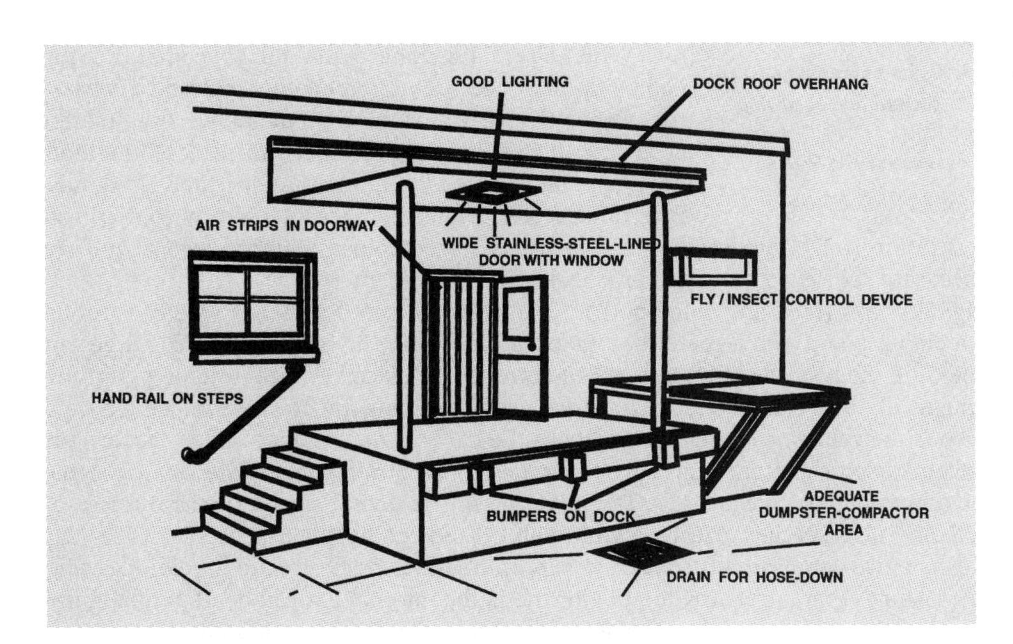

ILLUSTRATION 10-1 Smart features to include in a dock area.

SOURCE: CARL R. SCRIVEN AND JAMES W. STEVENS, *MANUAL OF EQUIPMENT AND DESIGN FOR THE FOODSERVICE INDUSTRY,* THOMSON LEARNING, 1989.

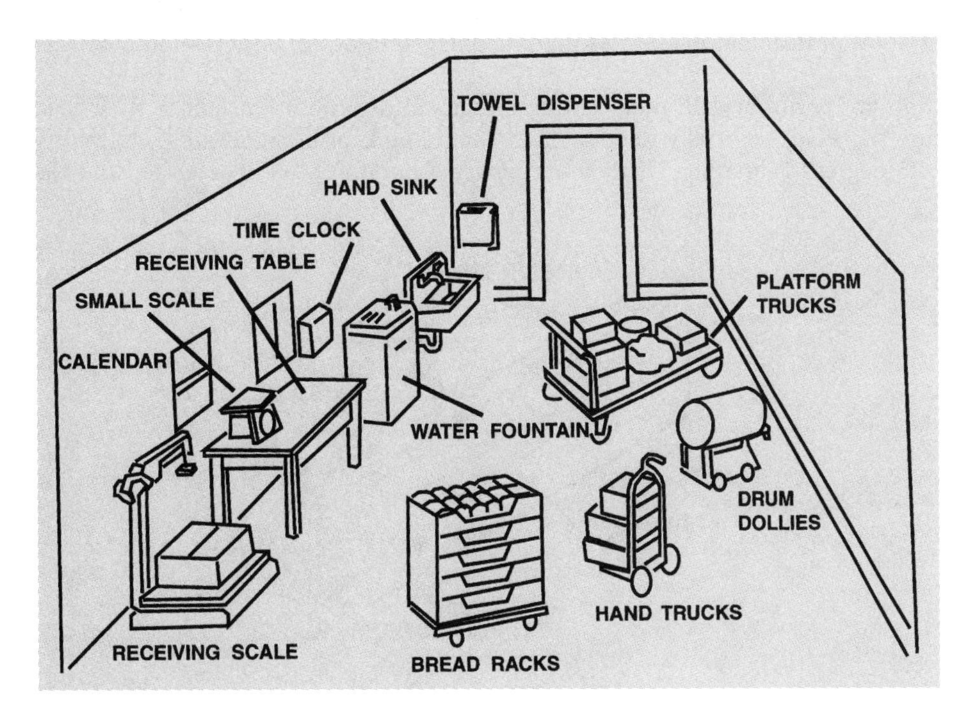

ILLUSTRATION 10-2 A receiving area that works well.

SOURCE: CARL R. SCRIVEN AND JAMES W. STEVENS, *MANUAL OF EQUIPMENT AND DESIGN FOR THE FOOD-SERVICE INDUSTRY,* THOMSON LEARNING, 1989.

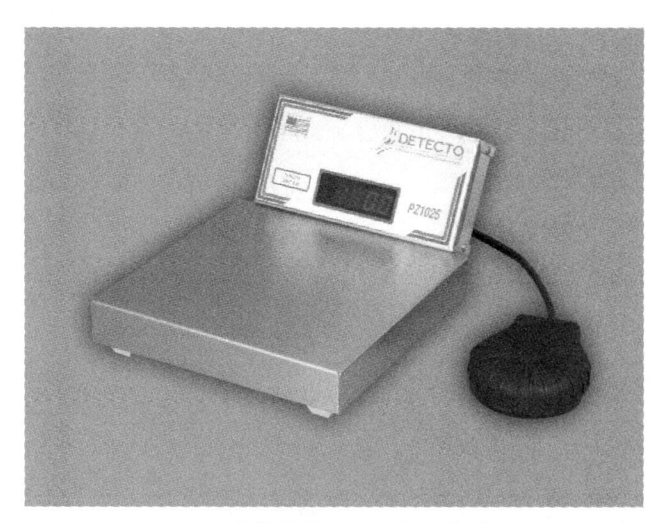

ILLUSTRATION 10-3 A digital electronic scale is compact and easy to read. There are models for both kitchen and receiving use.

COURTESY OF DETECTO SCALE DIVISION OF CARDINAL SCALE MANUFACTURING COMPANY, WEBB CITY, MISSOURI.

"control point" for a smart operation. If the scales aren't accurate, you're wasting time and money. Of course, scales continue to operate even when they're inaccurate, so maintenance is often neglected to keep them in proper calibration. Dust, dirt, and food scraps can easily clog the sensitive inner workings of a scale and build up friction, which causes inaccuracy.

Scales really get a workout in most foodservice operations, so you want a good, sturdy model instead of the inexpensive, lightweight spring-operated ones, which would quickly become inaccurate with heavy use. Scales can be mounted on a receiving table or they can be floor-standing models. Make sure to provide enough space for easy access to the scales in your receiving area. Some of the things you'll be weighing are big and bulky.

The *digital electronic scale* has become the top choice in many foodservice receiving areas and kitchens, too. Capacities vary, but these models have the greatest accuracy as well as being foolproof to read. Some units have the ability to read out in both metric and U.S. units. Electronic scales also have automatic tare weight adjustment buttons (see Illustration 10-3). *Tare weight* is the weight of the container, which is used so you can subtract it to get an accurate *net weight* of whatever is inside the container.

Electronic or digital scales (they are called both) are all manufactured similarly (see Illustration 10-4). They contain a circuit board and a metal bar called a ***load cell.*** The bar has ***strain gauges*** attached to it, which measure the way the bar bends when a product is placed on the weighing platform or tray. The strain gauges literally change electrical resistance when they are stressed by weight, and what is being measured is the voltage change that occurs as the bar is bent. The load cell then transfers this voltage change as an analog signal through a wire attached to the strain gauges. The analog signal is processed by an analog-to-digital converter (A/D converter) into a digital signal. A microprocessor reads the digital signal, and displays the correct weight on an LCD screen for the user to see.

We share these exhaustive details to illustrate how important it is to properly care for digital scales. If they are used to weigh things that are clearly too big for them, the internal bar can bend permanently, thus damaging the scale. If they are dropped or shaken, they can also be permanently damaged. In addition, many of these scales use batteries. The most frequent cause of inaccurate readings and other digital scale malfunctions is a low or dead battery. Scales that are plugged into wall outlets should be kept plugged in and turned on.

Digital scales also do not perform well under extreme temperature conditions, since their load cells are very temperature-sensitive, especially to cold. They should be kept in moderately warm, dry conditions, from 65 to 85 degrees Fahrenheit. If they are moved from one place to another and the

ILLUSTRATION 10-4 The inner components of a digital scale.

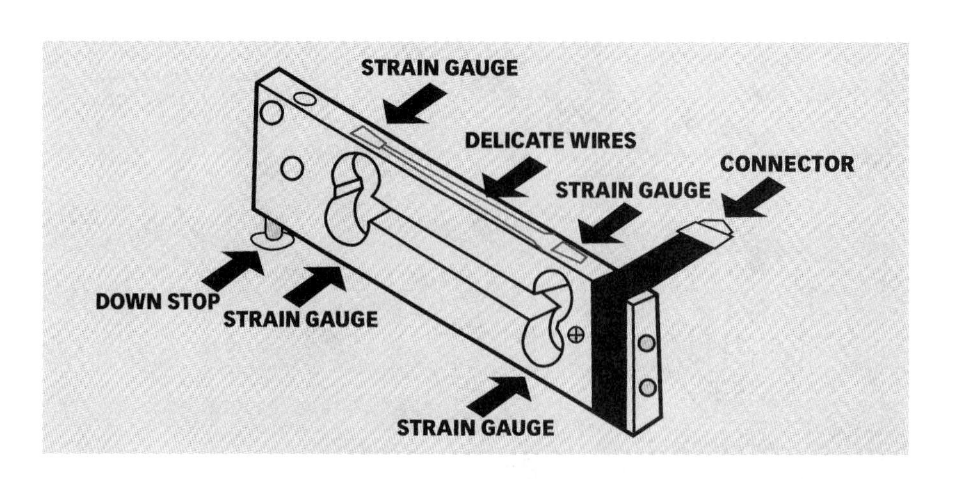

temperatures or humidity levels of the two places are quite different, manufacturers suggest letting the scale "rest," unplugged and unused, for a few hours to adjust to the new surroundings. When you notice ***drift***—the scale industry terminology for a progressive change, up or down, in the numbers shown on the digital readout—it may indicate a temperature problem, or a source of static electricity that is interfering with the scale's balance. Repairs on most models would surpass the cost of a new unit.

There are three other popular types of scales for receiving areas. The dial-type *counter scale* (see Illustration 10-5) rests on a table or bench, with a platform of 12 to 14 inches. It has the capacity to weigh items from 50 to 200 pounds, in graduations of 1 ounce (for a 50-pound scale) to 4 ounces (for a 200-pound scale). The scale has a *tare knob,* which allows you to adjust for the weight of the container. Today's manufacturers often supply weight information in both pounds and metric measurements. Often, when you buy a scale, manufacturers will include the table, bench, or stand to hold it. A very desirable characteristic is that the scale stand be on wheels. Most dial-type counter scales are capable of being leveled for accuracy.

ILLUSTRATION 10-5 **A counter scale is a necessity in a receiving area.**
COURTESY OF HOBART CORPORATION, TROY, OHIO.

Another popular piece of equipment is the *beam scale.* In most instances, it's portable, rolling on four platform casters, but there are also fixed, countertop models that take up less space. Beam scales come in capacities of 100 to 1000 pounds. The upright pillar on which the scale rests is offset to the left of the platform, which allows for a narrower scale. The *counterweights* (sometimes called *beam poises*) are open on one side to allow for easier reading. Counterweights are usually stored on a side bracket located at the top of the pillar; the biggest problem with counterweights is that they can be lost or misplaced if your staff is not careful. There are also double-beam scales; the second beam allows for tare weight adjustment of up to 100 pounds. And some manufacturers have added a balance indicator at the top of the pillar.

A less popular model scale is the *hanging dial scale,* such as the ones you see in supermarket produce sections. It's handy because its double-faced dial allows people on both sides to see the reading, but it usually only has a 30-pound capacity, graduated in 1-ounce or ½-pound increments. The pan, which is often removable, has an 18½-inch diameter and is 4¾ inches deep.

The type of scale you choose for your operation is largely a matter of personal preference. Some say dial scales seem to wear out faster, but are easy to use; others insist beam scales are more accurate, although they take longer to use. Electronic scales are the most expensive. No matter what your choice, remember that the scales will be an integral, hard-working part of your receiving area.

Pallets

A ***pallet*** is a low, raised platform on which boxes of products can be stacked and stored. They come in square or rectangular shapes, usually 36 by 48 inches or 48 by 48 inches, and can be made of steel or wood. In many cities, wood pallets are not acceptable for foodservice purposes, because they are harder to sanitize and could harbor insects or germs.

The typical pallet is 4 or 5 inches high, and inspectors (who usually insist that everything in a restaurant be stored at least 6 inches off the floor) will exempt pallets from this rule. The advantage of storing things on pallets is that they can be lifted and moved by a small hydraulic lift or platform truck, which enables a single person to stack or transport several cases of product at a time. Pallets allow for storage flexibility because they can be moved around easily or up-ended and leaned against walls when not in use.

Carts and Trucks

The terms "cart" and "truck" are sometimes used interchangeably in foodservice, but they are actually different items used for different purposes. A cart is a ***utility cart,*** a rolling shelf unit most often used in restaurants to transport dishes and cleaning supplies. Utility carts can be made of light-gauge metal, usually stainless steel, or molded plastic. The right cart can make life a lot easier on employees by allowing them to organize and move things without wasted steps or mishaps. Recent innovations have produced carts that are ergonomically designed, as well as carts that can fold up for more compact storage.

Depending on its size and construction, a utility cart can hold up to 1000 pounds of supplies or equipment. The shelves are usually solid, but can also be made of wire. Space between shelves should be 18 to 20 inches high. Most carts have three shelves, with a middle shelf that can be adjusted at 1-inch increments. The bottom shelf is typically about 20 inches from the ground. Rolled or raised edges on the shelves prevent spillage of liquids onto the floor.

The correct name for a truck is ***hand truck.*** Also known as a *dolly,* it is a small, sturdy platform on wheels that is designed to support and move heavy items such as cases of canned goods.

There are also ***layer platforms*** or ***flat trucks,*** which have four wheels and are strong enough to hold 10 to 15 cases. Flat trucks should have swivel casters to make them easier to steer. In special situations, hand trucks or flat trucks can be equipped with drum and barrel ***cradles*** to hold beer kegs or other drum-shaped objects steady as they roll.

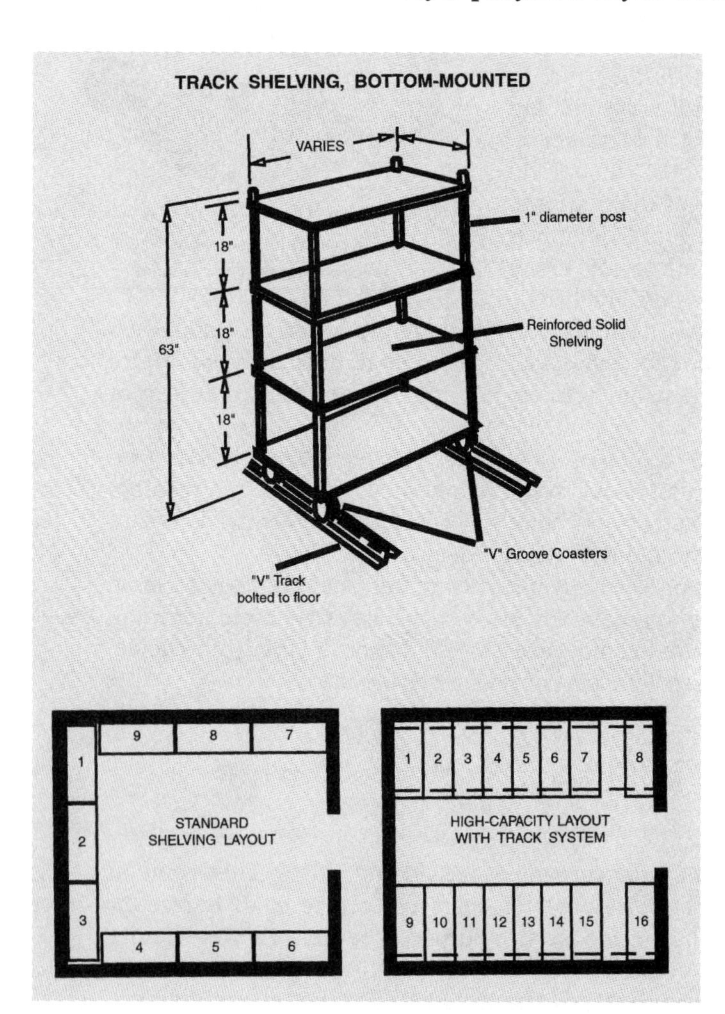

ILLUSTRATION 10-6 High-density storage units consist of shelves (shown in Illustration 4-12) that attach to a track. The track may be installed on either floor or ceiling. This allows the shelves to be moved as needed.

SOURCE: CARL R. SCRIVEN AND JAMES W. STEVENS, *MANUAL OF EQUIPMENT AND DESIGN FOR THE FOODSERVICE INDUSTRY,* THOMSON LEARNING, 1989.

Shelves

Perhaps the top consideration for dry storage shelving is adjustability, to accommodate the heights of various jars, cans, bins, and boxes. The shelves you choose should be adjustable in 1-inch increments.

The type of shelving you'll choose depends on two things: first, where it is to be used, and second, what you plan to store on it. Some shelf types are best suited for refrigerated storage; others for ambient (room temperature) storage. Here are a few of the most popular choices:

- Open-grid wire or mesh shelves allow for air circulation and visibility, but they are difficult and time-consuming to clean because of the many crevices where wires are joined.
- Flat, solid shelves are easy to clean and extremely sturdy. The solid shelf also eliminates the possibility of food products spilling from one shelf to others and contaminating what is stored below. If you use these, order them with a lip around the edges to help contain spills. The downsides? Solid shelving is expensive and it does restrict air movement around food.
- Embossed shelves are alternatives to flat, solid ones. This is a solid shelf, but with ridges or slots on its surface to allow air circulation.

Shelves can be made of polymer (heavy-duty plastic); stainless steel; vinyl-coated steel; zinc-plated, chrome-plated, or galvanized steel; and anodized aluminum. Wire-style shelves can be ordered with hard, synthetic coatings (not the same as vinyl) designed for heavy use; some incorporate the antimicrobials we learned about it Chapter 8.

Shelves are coated not only to prevent rust and discoloration when exposed to humidity and food acids, but also to provide some friction so containers don't slide around easily. Again, your choice depends on their prospective location and function. Most manufacturers will quote a weight load per shelf and, when comparing the same finish and style of shelving, the cost increases as the load limit increases. Stainless steel is the most expensive material, costing at least twice as much as chrome-plated steel shelves.

One shelf coating that definitely won't please the health inspector is paint. Special food-grade paint is required on any painted surface that comes into contact with food, to ensure it is not exposed to lead.

Most shelves attach to upright posts with snap-on plastic clips. Manufacturers have made these very easy to use and adjust, at 1-inch or 2-inch intervals. Polymer shelves sit inside a heavy-duty frame that attaches to the posts. The whole shelf can be removed from the frame and run through a dishwasher to sanitize it.

The width of a shelf should be 12 to 24 inches; any wider and it becomes more difficult to reach items at the backs of the shelves. And, depending on what is being stored, the height requirements between shelves will vary from 12 inches (for gallon containers) to 18 inches (for #10 cans). They can be mounted on solid upright posts, on rolling racks, or on walls. Shelves come in standard lengths of 24 to 72 inches. Shorter lengths tend to be less efficient and more costly per foot of usable space; longer lengths have less weight-bearing capacity because of the increased span to be supported. Most storage areas utilize a combination, but you can't go wrong with a standard 4-foot shelf unit.

You must also match the upright posts to the width and height of the shelves you want. You may decide to trade the leveling foot at the bottom of each post for a caster, to be able to move whole shelf units as needed. Again, you're looking for maximum flexibility as your storage needs change. If you use casters, select the widest, thickest caster that the post will accommodate. As you can imagine, a full shelf unit carries a lot of weight. Remember also that putting casters on anything raises its height by as much as 3 inches.

Height is a consideration because intelligent storage requires that you maximize available space. You can't store food items on the floor (or usually within 6 inches of it) for health and safety reasons, and you can't store it anywhere there is a risk of cross-contamination—near plumbing or sewage lines, under open stairs, etc. Overhead storage is often an overlooked option. Aisles should be 30 to 36 inches wide if one person at a time will be working in the storage space; add 6 more inches if more than one person will be using the aisle at a time. If carts will be rolled in and out of the storage areas, there should be a 10-inch clearance on either side of the cart.

Manufacturers of storage shelving are constantly coming up with new ideas to add space and efficiency. Shelf units with casters

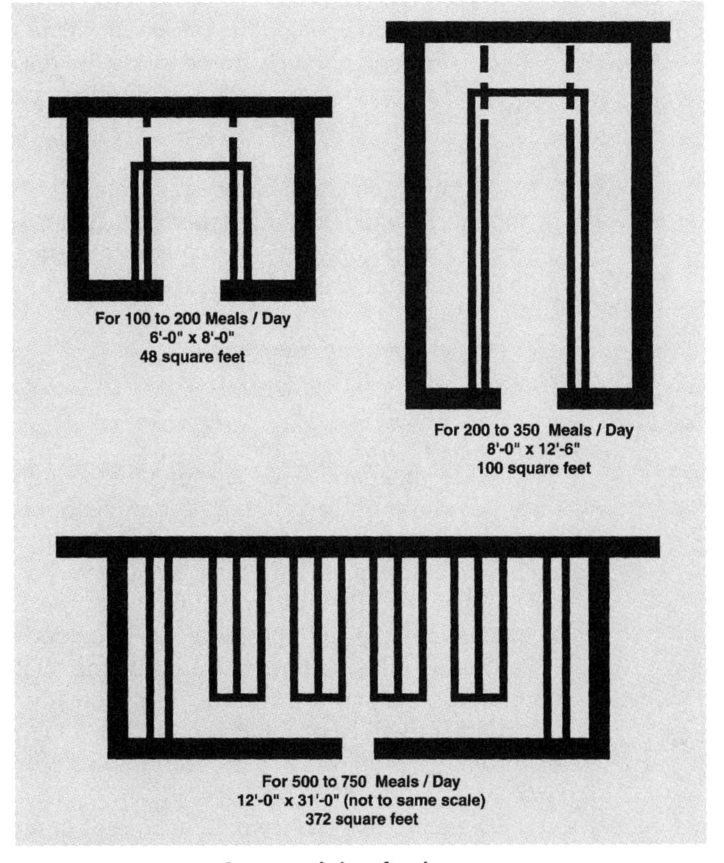

ILLUSTRATION 10-7 **Suggested sizes for dry storage areas.**

SOURCE: CARL R. SCRIVEN AND JAMES W. STEVENS, *MANUAL OF EQUIPMENT AND DESIGN FOR THE FOODSERVICE INDUSTRY*, THOMSON LEARNING, 1989.

can be put on tracks so the user can roll individual shelf units as needed. The tracks can be installed on the floor or overhead; we recommend an overhead configuration, because tracks on floors get clogged with debris of all kinds. The layout of this type of storage area requires only one aisle, since everything can be moved aside as needed, to get to a particular shelf or product. See Illustrations 10-6 and 10-7 for a look at this high-density storage option and how it works.

Finally, ***dunnage racks*** (dunnage is an old-fashioned word for baggage) are recommended for storage of bulky, unopened items such as bags and cases, which would otherwise take up valuable shelf space. Dunnage racks stand about a foot off the floor and come in dimensions from 18 by 24 inches to 24 by 48 inches. The newest models also have removable shelves for easy cleaning. Some establishments routinely designate the bottom shelves of their storage systems as "dunnage racks" for heavy or bulky items.

No matter what type of storage system you choose, as you shop for it, be sure it is quick and easy to adjust, to clean, and to customize as your menus and needs change.

10-2 REFRIGERATED STORAGE

Most people assume that a refrigerator or freezer is designed to make food cold and keep it cold. However, this is a common misconception. In fact, the basic principle of refrigeration is *the transfer of heat out of an enclosed space.* Therefore, cold is the absence of heat. The goal of any refrigerator is to remove heat. Removing heat extends the useful life of the refrigerated food, by protecting it against decay and deterioration.

Now, let's take it one more step. Heat and heat energy are part of any food product, whether it is raw or cooked. Heat enters the refrigerator in three simple ways:

- Through opening the refrigerator door (this is the major cause of heat transfer)
- Through products stored inside the refrigerator
- Through the door edges and rubber grommets (tiny leaks, indeed, but significant because they are continuous)

The more heat that can be removed from a product (without the side effects that can be caused by too much cold), the longer the product can be held in usable condition. To achieve this ideal, there are three things going on simultaneously in the enclosed, conditioned environment that until now *you* probably thought was "just a fridge":

- Temperature reduction
- Air circulation
- Humidity

Wide fluctuations in any of these conditions cause faster deterioration of the stored food; once deterioration begins, the process cannot be reversed. Cold can inhibit, but not completely prevent, the growth of most microorganisms associated with food poisoning. The temperature needed to accomplish this varies with the type of food, but remember the HACCP guidelines introduced in Chapter 8? They require refrigeration temperatures of at least 40 degrees Fahrenheit (or for the food itself to have an "internal temperature" of 41 degrees Fahrenheit).

Overall, the best refrigerator is one that has the ability to hold constant temperature and humidity levels and to circulate air at a steady rate. The other thing to note is that no "conditioned" environment will be absolutely perfect. No matter what it's made of or how well it works, no freezer or refrigerator can fully halt deterioration and certainly can't improve the quality of the food over time.

There are two terms you'll hear a lot in refrigeration: the ***refrigeration cycle*** and the ***refrigeration circuit.*** The *cycle* is the process of removing heat from the refrigerated space. The *circuit* is the physical machinery that makes the cycle possible. It's important to keep them straight as we delve into the inner workings of your refrigerator.

Temperature Reduction

Heat transfer, from inside the refrigerator to the outside environment, cannot take place without a temperature difference between the two. Different materials transfer heat at different rates, which also has an impact. There are four basic components of a refrigeration system. Their names and functions are not unlike a few of those you already learned about in Chapter 7, in our discussion of HVAC systems. They are:

- Evaporator
- Refrigerant
- Compressor
- Condenser

These make up the major physical parts of the refrigeration circuit. The circuit is a closed system. See Illustration 10-8; notice how similar it is to the cooling cycle of an air-conditioning system as shown in Chapter 7.

When you open the refrigerator door, warm air is introduced into the cooled, enclosed space. The warm air rises and is drawn into an ***evaporator,*** a series of copper *coils* surrounded by metal plates called ***fins.*** The fins conduct heat to and from the coils.

The evaporator coils hold liquid ***refrigerant,*** which becomes vapor (gas) as it winds through them. The vaporized gas is pumped by the ***compressor*** into the ***condenser*** (another series of coils surrounded by fins), where it turns back into liquid. As the gas is compressed, its temperature and pressure increase. The *expansion valve* is the small opening between the "low pressure" (evaporator) side and the "high pressure" (condenser) side of the system. This valve allows a little refrigerant or a lot to flow, depending on how much cooling is needed.

An alternative to the expansion valve is the *capillary tube,* or "cap tube" for short. Here's the difference: expansion valves respond more quickly to temperature changes, and are best suited to environments where doors are constantly being opened and closed. They are more expensive than cap tube systems. The capillary tubes work well for storage-type refrigeration, when doors are not opened often.

In either system, the last step is the job of the refrigerator's ***thermostat.*** When it indicates that colder air is needed, the compressor turns on. Air circulates past the two sets of fins, removing heat from the circuit as the refrigerant flows.

There are top-mounted and bottom-mounted refrigeration systems—that is, the condenser coils can be located near the floor, or at the top of the cabinet, above where the food is stored. In foodservice,

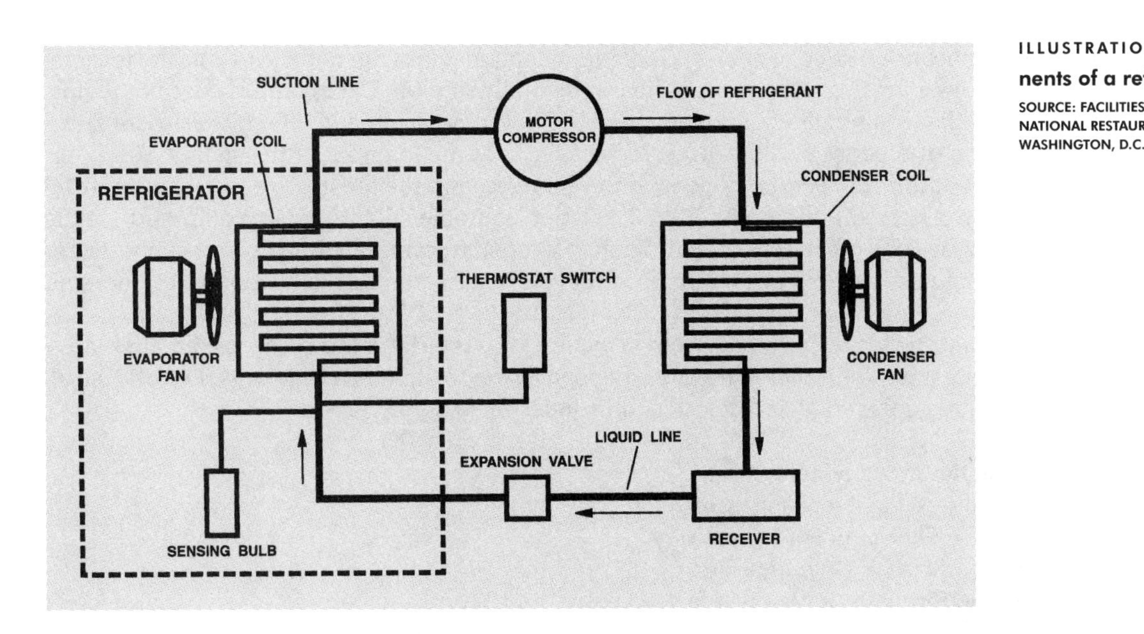

ILLUSTRATION 10-8 **The components of a refrigeration system.**

SOURCE: FACILITIES OPERATIONS MANUAL, NATIONAL RESTAURANT ASSOCIATION, WASHINGTON, D.C.

top-mounted units are worth considering, because the coils don't accumulate grease and dust as quickly as they do closer to the floor.

The goal in refrigeration, as with HVAC units, is a balanced system. The compressor, evaporator, and condenser functions must be sized to work well together.

The Chemistry of Refrigeration

The last decade has seen enormous changes in the HVAC and refrigeration service industries. Back in 1996, the U.S. Environmental Protection Agency's Clean Air Act banned the manufacture of the most common refrigerants, because they contributed to the depletion of the ozone layer in our upper atmosphere that protects the earth from the sun's ultraviolet radiation. These refrigerants were known by the trademark name of their manufacturer, DuPont, as Freon 12 and Freon 22, although we will refer to them here by their generic names, CFC-12 and HCFC-22.

CFC-12 is a chlorofluorocarbon (CFC), a chemical combination of carbon, chlorine, and fluorine. Chlorine is the culprit that zaps ozone. Under the EPA guidelines, CFC-12 was supposed to be phased completely out of use by the year 2000. In its place, related chemical compounds include hydrofluorocarbons (HFCs) and hydrochlorofluorocarbons (HCFCs). However, HCFCs (HCFC-22) are also slated for extinction by the EPA somewhere between 2015 and 2030. HFCs, without chlorine, have skirted the ban so far; in fact, the most widely used alternative refrigerant is now HFC-134a.

All refrigerants are hazardous when exposed to an open flame. Some of them contain butane or propane mixes blended into their formulas. If large quantities of refrigerant are released in a confined area, suffocation is a danger because refrigerant actually displaces oxygen. Breathing refrigerant can cause nausea, dizziness, shortness of breath, and can even be fatal. So any type of refrigerant gas should be handled by a professionally trained technician.

Controversy and health warnings aside, the cooling power of the modern refrigerator comes from the repeated compression and expansion of a gas. As the gas expands, it cools and is cycled around an insulated compartment, chilling the contents inside. Ammonia, new chemical blends, and even space-age technology using sound waves to cool foods are other options being tested. The latter technology is called ***thermoacoustics.*** A loudspeaker is used to create 180 decibels of sound in a tube that contains compressed gases (helium and xenon), which are environmentally safer than CFCs. The sound causes the gas molecules to vibrate, expand, and contract. When they contract, they heat up; when they expand, they cool down. Scientists at NASA believe there is commercial refrigeration potential for thermoacoustic technology.

Another promising new technology is *electromagnetism.* A "magnetic" refrigerator can cool by repeatedly switching a magnetic field on and off. The current prototype is made with gadolinium, a metal used in the recording heads of video recorders. Gadolinium and magnets are not cheap, but the technology shows great potential for two reasons—it is environmentally safe (no CFCs), and it does not require a compressor (no mechanical humming noise as the refrigerator cycles on and off). At this writing, Astronautics Corporation of America in Milwaukee, Wisconsin, is at the forefront of this field.

How does the foodservice operator cope with the changes, and the prospect of expensive new replacements for old workhorse refrigerators? Well, if your equipment is in good repair, you should probably do nothing as long as it lasts except keep it properly maintained. This especially means cleaning the unit's condensing coil once a month to prevent grease and dirt collection that blocks air circulation.

If your refrigerator needs repair, you have two choices: Voluntarily retrofit it to use an alternative refrigerant, or purchase a new unit that is already equipped to use the newer refrigerants. Retrofitting almost always requires more than one service call and includes these steps:

- Recovering the outdated refrigerant
- Changing the coil in the compressor
- Replacing the filter or dryer, if necessary
- Recharging with the new refrigerant
- Checking performance for the first few weeks

The EPA now has a sophisticated set of rules for refrigerator repair. The EPA certifies repair technicians and their equipment, and requires that they recycle or safely dispose of refrigerant by sending it to a licensed reclaimer. The rules also state that "substantial leaks . . . in equipment with a charge greater than 50 pounds" be repaired. This means if the unit leaks 35 percent or more of its pressure per year, it needs fixing. As the owner of commercial refrigeration equipment, you are also required to keep records of the quantity of refrigerant added during any servicing or maintenance procedure.

Your nearest EPA office can provide background and summaries of the rules, as well as lists of acceptable alternative refrigerants that don't contain the ozone-depleting chlorofluorocarbons (CFCs) and hydrochlorofluorocarbons (HCFCs). They're identified with abbreviations and numbers, such as MP-39, HP-80, FX-70, and HFC-134a or R-134a, which probably don't mean much to you as a foodservice employee. However, the important points to remember are:

- **Use an EPA-certified technician, with certified equipment, to do your refrigeration repair work.**
- **Underwriters' Laboratories is now authorized to test and approve alternative refrigerants, so look for the UL label on products.**
- **Keep your maintenance records updated. Violations of the Clean Air Act can result in fines of up to $25,000 per day.**
- **If you have more than one piece of older equipment, plan a gradual phase-out or retrofit program. Don't break your budget by trying to do it all at once.**

This information should also serve as a caution when you are looking at purchasing used equipment. Is the owner getting rid of it because it no longer meets the environmental rules?

With that, we've discussed the first major process going on inside the refrigerator—temperature reduction. So now let's examine the other two processes.

Air Circulation

The refrigerator relies on forced air to transfer heat. Fans inside the appliance move air around. The faster the air flows, the more quickly the heat is removed. So you don't want to do anything to block the airflow.

There are three basic types of forced-air systems in refrigerators. In ***ceiling-type refrigeration,*** a single fan is mounted on the ceiling of the appliance. This is adequate for small-volume interiors but is not used in larger refrigerators. Because it only has a single location, it might allow for "hot spots" in the corners of the interior cabinet.

In *back-wall* or ***mullion-type refrigeration,*** the airflow system takes in air above the top shelf and discharges it below the bottom shelf. The ***duct-type refrigeration*** system is a combination of the first two types. Here, the forced-air unit is located at or above ceiling level, and the air is circulated through a series of small air ducts vented to various spots on the back wall of the cabinet.

Just how important is air circulation? Well, the difference between safe and unsafe raw foods can be as little as 5 to 7 degrees Fahrenheit. Seafood, poultry, or red meats will spoil within 18 to 24 hours if their refrigerated temperature rises above 42 to 45 degrees Fahrenheit, and you already know the HACCP guideline of temperatures no higher than 40 degrees Fahrenheit. Would you rather risk a lawsuit and the resulting negative publicity from food-poisoning allegations, or keep your refrigerator air circulating properly?

Humidity

Humidity is the amount of moisture (or water vapor) in the air. At different temperatures, air can hold different amounts of water. In refrigeration, the type of humidity we are interested in is the ***relative humidity,*** or how much of its maximum water-holding capacity the air contains at any given time, expressed as a percentage. For example, 85 percent humidity indicates that the air is holding 85 percent as much water as it could hold at that temperature. Relative humidity greatly affects the appearance and rate of deterioration of many foods. If the air surrounding the stored foods has a very low relative humidity, for instance, the air naturally picks up moisture from the foods themselves, causing surface discoloration, cracking, and drying. On the other hand, if the air has a high relative humidity, some of the moisture will condense on food that is supposed to be kept dry, causing it to soften or grow mold or bacteria.

Fortunately, most foods do well in a relative humidity of 80 to 85 percent. To achieve this optimum level, manufacturers are concerned that the refrigerator's evaporator coils be large enough to operate at a few degrees' lower temperature than the desired temperature of the appliance. This differential reduces the amount of moisture that accumulates on the evaporator coils, and keeps the moisture in the cabinet of the refrigerator instead. If the coils' temperature becomes too low, however, the moisture will turn to ice crystals and get stuck on the coils. In this case, airflow through the system is blocked and the moisture in the refrigerated space is depleted. As you can see, getting all the factors right is a delicate balance, with your food costs and food quality at stake.

In short, it is difficult to keep frost off the coils, but necessary to keep them frost-free so they will operate properly. Adding heat to the area, to defrost the coils, can compromise the temperature of the food inside. A fairly new concept from Hussman Modular Defrost of Bridgeton, Missouri, does just what its name indicates—defrosts the coil in sections. The automated system defrosts coils at no more than nine minutes per section, and never defrosts adjacent sections at the same time, all programmed by an electronic controller capable of running up to six walk-ins. The idea works for walk-in and reach-in refrigerators, but not freezers. It maintains food quality, and saves energy by keeping the compressors from working overtime to compensate for frozen coils.

Another humidity-control suggestion for inside refrigerated space was pioneered by Humitech International Group, Inc., of Dallas, Texas. Humitech uses a mineral product called ***sorbite*** to absorb moisture and odors.

We mentioned that most foods do well at 80 to 85 percent relative humidity, but fresh fruits and vegetables are exceptions. They require more humidity, up to 95 percent. To increase moisture content, you can slow down the air circulation. This explains why there are separate, closed produce bins in most refrigerators—to hold in natural moisture from the vegetables and to restrict airflow.

Freezers maintain an average relative humidity of only 30 to 35 percent. Any more moisture would automatically raise the temperature because it would hit the coils, freeze in place, and block the airflow, causing the freezer air to become warmer. The low humidity of freezers requires special food storage precautions. Use moisture and vapor-proof wrapping to prevent the surface damage we know as "freezer burn" from occurring if any moisture condenses on the food.

10-3 SELECTING A REFRIGERATOR

Now that you know how a refrigerator works, how do you pick the ones that will work best for your operation? Manufacturers all have printed specifications to look over, and your dealer or equipment consultant will have recommendations. Following are some of the criteria you'll choose from.

FINISHES. The surface of your refrigerator should be as sturdy as its interior components. Popular finishes include stainless steel, vinyl-coated steel, fiberglass, and coated aluminum; the latter comes in rolled, stucco, or anodized styles. Unlike home models, there are not as many porcelain or baked-enamel finishes in the commercial world. In fact, some health departments do not allow these finishes in commercial or institutional installations.

CONSTRUCTION. You can't kick the tires, but certain quality features will be evident, such as overall sturdiness, door alignment, and how securely the handle is attached to the appliance. All-metal, welded construction is a plus, and having a seamless interior compartment is an NSF International requirement. In fact, the NSFI has a whole section on refrigeration, accepted by most cities as minimum standards. Also, look for ease of cleaning and self-defrost features.

INSULATION. The most commonly used type of insulation is polyurethane, in sheets or foam, which has superior insulation qualities and even makes the cabinet a bit sturdier. Make sure it is non-CFC polyurethane foam, with at least an R-15 rating. Fiberglas is also acceptable, although it requires greater thickness to achieve the same results as the polyurethane.

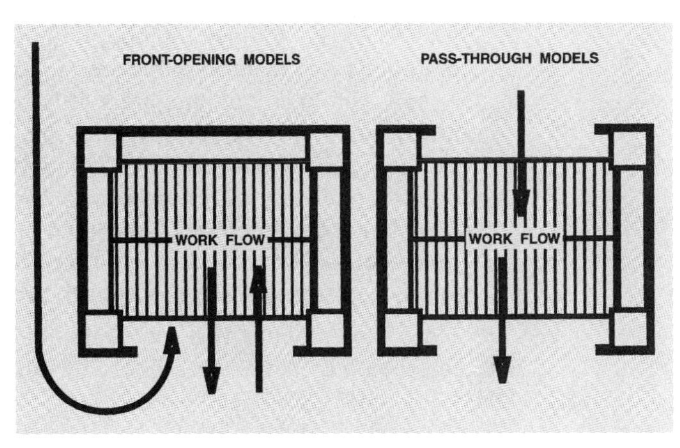

ILLUSTRATION 10-9 **The way that refrigerator doors are opened impacts work flow and traffic patterns in the kitchen.**

DOORS. A small but critical detail is whether you want the door to open from the left or from the right side. There are also half doors (you conserve cooling power by only opening half the refrigerator at a time) or full-length doors; the doors can be solid or made of shatterproof glass; they can have hinges or slide open and shut on a track. The way doors are opened can impact traffic patterns in the kitchen (see Illustration 10-9). Doors can also be self-closing, with magnetic hardware, to prevent being left ajar accidentally. The hinges should be stainless steel, or at least chrome. Look for door gaskets that are easy to snap in place, not the old screw-in kind, since you will probably be replacing them during the life of the unit.

HANDLES. Stainless steel or nickel-plated handles are best. You can select vertical or horizontal handles. They can protrude or be recessed. Be sure the handle is included in the warranty, since handles take a lot of abuse and may have to be replaced periodically.

REFRIGERATION SYSTEM. It may be self-contained or, in the case of very large appliances, a separate unit. As we've mentioned, it may also be top-mounted or bottom-mounted. The accurate electrical current and capacity of the facility must be known so the manufacturer can supply the correct voltage and phase to meet the needs of the space. In some cases, additional expense may be involved to upgrade the electrical system. At any rate, look for the UL seal of approval, a sign that the unit meets basic electrical safety standards. The system may be water cooled or air cooled. The most common in foodservice is the self-contained, air-cooled unit.

Remember, the capillary tube system is for refrigerators used for storage—not much door-opening action. The expansion valve system has quicker pull-down capacity—that is, it can pull the temperature down faster after the unit is opened. It is ideal for busy hot line situations where the refrigerator is constantly in use.

DRAIN REQUIREMENTS. Most new refrigerators provide an automatic defrost system and automatic condensate disposal, which eliminates the need for a separate plumbing connection. Ask about it, however. A reminder: The NSF sanitation standards prohibit drains inside the refrigerator.

ACCESSORY AVAILABILITY. You'll get shelves as standard equipment with a refrigerator purchase. Make sure they are adjustable. For foodservice, there are lots of additional items that might improve efficiency: adjustable tray slides, drawers, special racks for serving pans (called ***pan glides*** or *pan slides*), and dollies or carts designed to convert a reach-in cabinet to a roll-in one. Think about these accessories when selecting the door, too. Certain doors seem to work better with some types of add-ons.

WARRANTY. Most manufacturers provide a one-year warranty on parts in case of defective workmanship or materials; look for a separate five-year warranty on the motor and compressor unit. Some manufacturers also offer extended service warranties.

CABINET CAPACITY. A properly designed refrigerator should provide the maximum amount of usable refrigerated space per square foot of floor area, and must be able to accommodate the sizes of pans you'll be using. There are plenty of complex guidelines for calculating capacity and needs, which will be covered elsewhere in this chapter.

ADAPTABILITY. Because today's foodservice operations have changing needs, manufacturers are building in features to maximize flexibility. One such offering is the *convertible temperature* option. With a flick of a toggle switch, a freezer can be converted to a refrigerator. It can be a pricey addition at the time of purchase, but if food storage requirements change, the option will instantly pay for itself. Another variation is the combination medium-and-high-temperature cabinet, designed to thaw frozen products quickly and safely by introducing warmer air into the cabinet as needed. Additional fans and a temperature-sensing device bring the unit back to its normal refrigeration level when the food is sufficiently thawed. And there are hybrids: cabinets separated into two or three sections, each with different cooling capacities.

More "how-to-buy" tips can be found in the following "Foodservice Equipment" box, reprinted with permission of *Restaurant Hospitality* magazine.

Reach-Ins and Roll-Ins

Most kitchens have too much or too little refrigerated space *at the proper locations to meet their needs.* In other words, the physical capacity may be adequate, but either it's not the correct type of refrigerator space or it's not flexible enough to be used to maximum efficiency. Having the right kinds of refrigeration can actually mean using fewer refrigerators and freezers, an idea that will save energy and money.

FOODSERVICE EQUIPMENT

BUYING REFRIGERATORS

Consider these factors when purchasing a refrigerator.

- If you use a lot of sheet pans or steam table pans, consider pan slides in lieu of wire shelves. Universal-style slides will allow you to use either sheet pans or steam table pans. A sheet pan on slides can also serve as a shelf when both are needed.
- Many reach-in refrigerator manufacturers offer different lengths and widths for the same door configuration. For example, a manufacturer may make a two-door refrigerator in a 48-inch, 52-inch, and 58-inch width. Costs are all very close. If you go with pan slides, then you want the narrowest unit that will fit the slides since anything wider is wasting area in your kitchen. If you will be storing larger items, such as case goods, a larger width may be the best buy.
- To make the most of your worktable, consider undercounter refrigerators. Most can be bought with a worktable top as part of the unit. Garnish pans can be cut into the top for an added means of refrigeration.
- Finish materials can often add to or reduce the cost of a refrigerator, but they also affect durability. An all stainless steel cabinet is the top-of-the-line for most manufacturers. But if you forgo stainless steel inside the box you can save about 10 to 15 percent of the overall cost. If you can accept an aluminum finish on the refrigerator exterior with the exception of the doors, an additional 20 percent or more savings is possible. The tradeoffs are that aluminum is a soft metal and may be dented more easily than stainless.
- Cool down, rapid chill, and blast chillers are becoming popular because health departments are more sensitive to proper food handling procedures and storage temperatures. Any of these rapid-cooling refrigerators will help to cool foods quicker than a standard refrigerator and reduce chances of food-borne illness.

Source: Reprinted with permission of *Restaurant Hospitality* magazine.

So how do you accomplish this ideal? First, you must decide how much capacity you need. As we mentioned in our discussion of space allocation in Chapter 4, the norm in casual restaurants is to allow 1 to 1.5 cubic feet of refrigerated storage space per meal served. In fine dining, this increases to 2 to 5 cubic feet of space per meal served. You will use a refrigerator not only for storage, but to slowly, safely thaw frozen foods 24 to 48 hours before you will need them.

Remember, roughly half of a refrigerator's total cubic footage is usable space. The rest is taken up by the unit's insulation and refrigeration system. (Walk-in coolers also contain aisles, which take up room. More about that later in the chapter.)

Another handy rule is that, in a reach-in refrigerator, 1 cubic foot of space will hold 25 to 30 pounds of food. Divide the total weight of food you'll need to store by 25 or 30, and you'll have a good idea how much cubic footage you will need.

Only after you've determined your capacity needs can you take the next step: deciding how much floor space you have for refrigeration, and what size of unit will fit there. This way, you can calculate how many different units you will need.

The third major step is to look at reach-in, roll-in, and walk-in options. A ***reach-in refrigerator*** is similar to the one you have at home; you pull open the door, reach in, and get what you want. In a commercial kitchen, the problem is that the refrigerator door is opened and closed constantly, in heat that's a lot more intense than a home kitchen. A duct-type system, with louvered air ducts promoting airflow throughout the cabinet, seems to work best to counteract the inevitable blasts of warm air.

Inside the refrigerator cabinet, the wise use of space can increase your capacity by 30 to 35 percent. A simple, heavy-duty pull-out shelf system can allow full use of the bottom part of the unit without making employees stoop to retrieve things there.

Typical reach-in units range from a one-section, single-door unit with 22.7 cubic feet to a three-section, three-door unit with *more than* three times the capacity, at 74.7 cubic feet. Total storage capacity depends somewhat on the number of shelves in the unit; and, of course, the number of shelves will depend on the heights of the products you'll store on those shelves (see illustration 10-10). They can be custom-sized to fit under counters or in small spaces.

There are also convertible reach-ins, basically freezers that can be converted to a refrigerator with the flip of a switch located on the cabinet. Manufacturers offer these in one-, two-, and three-section units, so you can adjust for more refrigerated space or more freezer space, as needed. Adaptability is the key.

The bigger the foodservice operation, the greater the need for a *roll-in refrigerator.* If your operation does a lot of batch cooking, for instance, you will want to have the capability to move large numbers of meals on rolling carts, in and out of refrigerated space. Carts mean less handling, which means less spillage, less heavy lifting, and so forth.

A reach-in unit can be converted to a roll-in by using a dolly on which a half or full rack of product is resting. The rack has swivel casters and is latched on to the dolly. If the height of the dolly platform is compatible with the bottom of the refrigerator cabinet, the person holding the dolly can just position it correctly, tip it forward, and slide the rack of product into the refrigerator (see Illustration 10-11). This seems to work best when the reach-in refrigerator is equipped with 6-inch legs.

The roll-in cabinet is similar to the reach-in except that, instead of 6-inch legs, it has a ramp at floor level so that entire carts can be rolled inside. The floor of the roll-in is stainless steel. Capacities

ILLUSTRATION 10-10 **A commercial reach-in refrigerator allows its user a lot of storage flexibility.**
COURTESY OF TRUE FOOD SERVICE EQUIPMENT, O'FALLON, MISSOURI.

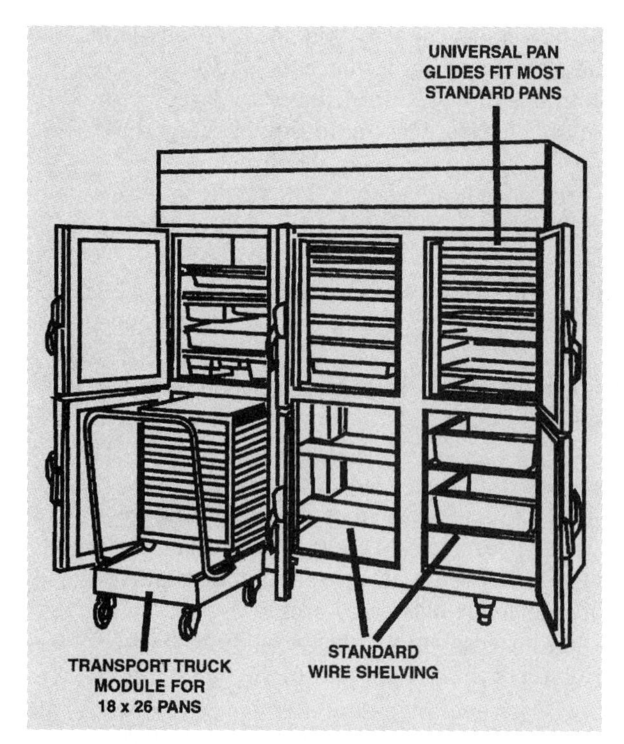

ILLUSTRATION 10-11 **A roll-in refrigerator is sized to accommodate rolling racks of food.**

SOURCE: CARL R. SCRIVEN AND JAMES W. STEVENS, *MANUAL OF EQUIPMENT AND DESIGN FOR THE FOODSERVICE INDUSTRY*, THOMSON LEARNING, 1989.

of roll-ins range from 35.3 cubic feet for one-section units to 113.2 cubic feet for three-section units.

A ***pass-through refrigerator*** is a variation of the standard reach-in. This cabinet has two sets of doors, located opposite each other. Cafeterias and garde manger areas make good use of this special refrigerator, when the kitchen staff places food in one side for servers to pick up at the other. On the service side, which gets opened more frequently, half doors are recommended to help with temperature control. Another recommendation is to check the temperature of this unit more often for food safety reasons, because it is likely to be warmer with all the activity. Choose glazed glass upper doors for easier product visibility, and consider using shelves in the upper portion of the cabinet and half-height carts in the lower section. This would allow a fully loaded tray of prepared salads, for example, to be transferred as needed from its storage in the lower section into the upper section to await serving (refer to Illustration 10-10). The workflow in your kitchen will help you determine the suitability of a pass-through refrigerator.

Other types of specialty refrigerators are made to fit under bars or counters. These are handy in confined areas, particularly around the hot or cold preparation lines. They range in storage capacity from 5.7 to 15.4 cubic feet, with one to three doors. There are also refrigerated prep tabletops, where 8 to 24 pans can sit on a chilled surface instead of using ice. Lately, these prep tops have come under the scrutiny of health departments for exceeding the maximum 40-degree-Fahrenheit temperature to keep food safely chilled. One solution to this is to install the undercounter refrigerator and the prep top above with separate controls and separate refrigeration units, even though they're both run by the same compressor.

Reach-ins and roll-ins have lots of different doors to choose from. The basic things you'll have to decide on are whether you want the door to be full height or half height; on hinges or a sliding track; if it's hinged, what side it opens from; and if the door itself is solid or see-through glass.

Most equipment experts frankly prefer hinged doors over sliding ones. The door seal isn't as tight on sliding doors, and it can be hard to clean the sliding tracks. However, if aisle space is tight, a sliding door can be useful. It takes only half as much "standing room" to open as a hinged door. In a pass-through refrigerator, use sliding doors only on the servers' side; having them on both sides lets out too much cold air too fast.

The question of a full door or half door is also one of temperature loss. The half door is more energy efficient, especially if the unit is located close to the hot line, but it has the disadvantage of only allowing staff to see half of the contents at a time. Inside the refrigerator, you've got to organize the contents well to make half doors work efficiently for you. When it comes to how the doors will open, give a lot of thought to the flow pattern of the workstations in the area. Open doors cannot help but obstruct the flow, and people should be able to see around them and maneuver around them safely. One manufacturer even boasts the ability to change door hinges on the spot within 30 minutes, a feature that might come in handy.

For door hardware, you'll have a choice of magnetic gasketing with self-closing hinges or positive-action fasteners with standard hinges. Eventually both will have to be adjusted, although positive-action fasteners seem to require more frequent tweaking. Both are acceptable if the design and installation of the cabinet is sound.

A final word about reach-ins and roll-ins, and we'll move on. The refrigeration systems can be either air cooled or water cooled, with a heavy-duty (often referred to as a *high-torque*) condenser, which must be kept clean for the system to work properly. In most foodservice settings, the air-cooled, self-contained system is used, because a good kitchen ventilation system will be able to remove the heated air given off by the refrigerator. The reason refrigeration experts don't recommend the water-cooled system for most kitchens is that it requires a water-cooling tower. Although a tower is usually part of the building's regular air-conditioning system, it must be bigger (and take up more space) to accommodate the added demand that refrigerators would place on the cooling system.

No matter which system you choose, correct voltage must be considered before you order or install your refrigerators.

10-4 WALK-IN COOLERS AND FREEZERS

A walk-in cooler is just what its name implies: a cooler big enough to walk into. It can be as small as a closet or as large as a good-sized room, but its primary purpose is to provide refrigerated storage for large quantities of food in a central area. Experts suggest that your operation needs a walk-in when its refrigeration needs exceed 80 cubic feet, or if you serve more than 250 meals per day. Once again, you'll need to determine how much you need to store, what sizes of containers the storage space must accommodate, and the maximum quantity of goods you'll want to have on hand.

The only way to use walk-in space wisely is to equip it with shelves, organized in sections. Exactly how much square footage do you need? The easiest formula is to calculate 1 to 1.5 cubic feet of walk-in storage for every meal you serve per day.

Another basic calculation: Take the total number of linear feet of shelving you've decided you will need (*A*), and divide it by the number of shelves (*B*) you can put in each section. This will give you the number of linear feet per section (*C*). To this number (*C*), add 40 to 50 percent (× 1.40 or 1.50) to cover "overflow"—volume increases, wasted space, and bulky items or loose product. This will give you an estimate of the total linear footage (*D*) needed.

However, linear footage is not enough. Because shelves are three-dimensional, you must calculate square footage. So multiply (*D*) by the depth of each shelf (*E*) to obtain the total square footage amount (*F*).

Finally, double the (*F*) figure, to compensate for aisle space. Roughly half of walk-in cooler space is aisle space. On paper, the calculations look like this:

$$\frac{\text{Linear feet of shelving } (A)}{\text{Number of shelves } (B)} = \text{Linear feet per shelf section } (C)$$

$$(C) \times 1.40 \text{ (or } 1.50) = \text{Total linear footage needed } (D)$$

$$(D) \times \text{Depth of each shelf } (E) = \text{Total square footage } (F)$$

$$(F) \times 2 = \text{Estimated square footage for total walk-in storage needs}$$

Another popular formula is to calculate that, for every 28 to 30 pounds of food you'll store, you will need 1 cubic foot of space. When you get that figure, multiply it by 2.5. (The factor 2.5 means that only 40 percent of your walk-in will be used as storage space; the other 60 percent is aisles and space between products.) The result is the size of the refrigerated storage area you will need. For a walk-in freezer, simply divide your walk-in refrigerator space by two.

Larger kitchens, which serve more than 400 meals a day, may need as many as three walk-in refrigerators for different temperature needs: one for produce (41 degrees Fahrenheit), one for meats and fish (33 to 38 degrees Fahrenheit), and one for dairy products (32 to 41 degrees Fahrenheit).

The walk-in is most often used to store bulk foods. Because this often means wheeling carts or dollies in and out, the floor should be level with the kitchen floor. This leveling is achieved by the use of strips (called ***screeds***) that are applied to the floor.

Coolers don't come as a single unit; they are constructed on site. The walls, ceilings, and floors are made of individual panels. Wall panels should be insulated to a rating of R-30, which means a 4-inch thickness. They come in various lengths and widths, with 12-by-12-inch corner panels at 90-degree angles. They can be as short as 7½ feet or as tall as 13½ feet. The most common type of insulation inside the panels is polyurethane, and the outside walls of the panels can be made of stainless steel, vinyl, or aluminum. Stainless steel is the most expensive, and aluminum—because it's the least expensive—is the most popular choice. If the walk-in is an outdoor installation, aluminum is the most weather resistant. Illustration 10-12 shows how walk-in panels fit together to form the structure. The installer will be sure the unit has interior lighting.

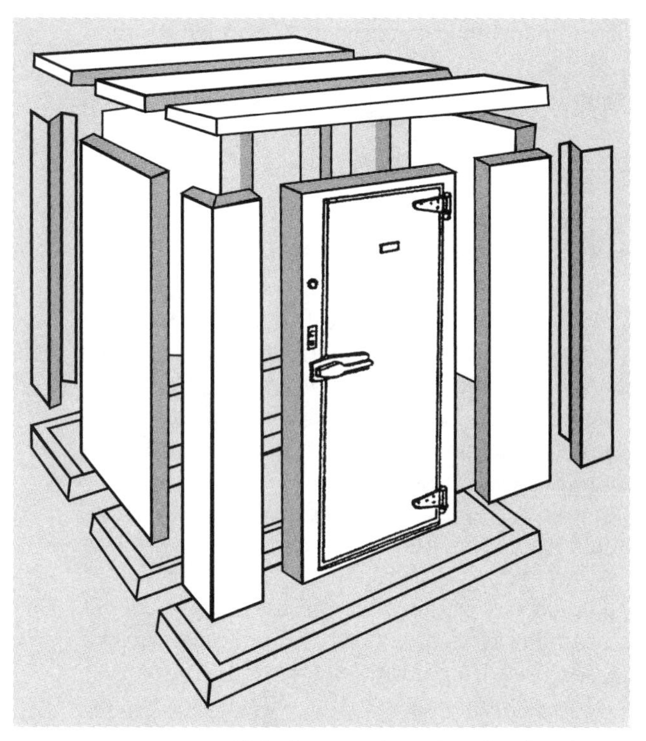

ILLUSTRATION 10-12 This diagram shows how insulated panels fit together to form a walk-in cooler.
COURTESY OF ARCTIC INDUSTRIES, MIAMI, FLORIDA.

The floor panels for walk-ins are similar to the wall panels. Load capacities of 600 pounds per square foot are the norm, but if you plan to store very heavy items (like beer kegs), a reinforced floor can be purchased with a load capacity of up to 1000 pounds per square foot.

The refrigeration system of a walk-in is a more complex installation than a standard refrigerator, primarily because it's so much bigger. Matching the system (and its power requirements) with the dimensions of the walk-in and its projected use is best left to professionals, but it's important to note that a walk-in accessed frequently throughout the day will require a compressor with greater horsepower to maintain its interior temperature than one that is accessed seldom. A 9-foot square walk-in would need at least a 2-horsepower compressor. The condenser unit is located either on top of the walk-in (directly above the evaporator) or up to 25 feet away, with lines connecting it to the walk-in. The latter, for obvious reasons, is known as a *remote* system, and is necessary for larger-than-normal condensing units with capacities of up to 7.5 horsepower.

In a remote system, the refrigerant must be added at the time of installation. For smaller walk-ins, there's also a plumbing configuration called a *quick-couple* system, which is shipped from the factory fully charged with refrigerant. This definitely simplifies installation. However, you may need the added power of a remote system if your kitchen has any of the following drains on the walk-in's cooling ability: frequent door opening, glass display doors, multiple doors per compartment, or an ambient kitchen temperature that's near 90 degrees Fahrenheit.

Modern walk-ins sometimes offer a frozen-food section in addition to the regular cooler space. There are pros and cons to this concept. It may ease the load on the freezer, because it's already located inside a chilled airspace; but it also can't help but reduce overall usable space, because it requires a separate door.

You can also order your walk-in with a separate, reach-in section that has its own door and shelves. Although this may save the cost of purchasing a separate reach-in, some critics claim that a walk-in is not designed to do a reach-in job such as storing uncovered desserts. Do you really want them in the same environment as cartons of lettuce and other bulk storage items? There may be cleanliness or food-quality factors to consider.

The doors should open out, not into the cooler itself. The standard door opening is 34 by 78 inches. Several door features are important for proper walk-in operation. These include:

- A heavy-duty door closer.
- Self-closing, cam-lift door hinges. If the door can be opened past a 90-degree angle, the cam will hold it open.
- A heavy-duty stainless steel threshold. This is installed over the galvanized channel of the door frame.
- A pull-type door handle, with both a cylinder door lock and room to use a separate padlock if necessary.
- Pressure-sensitive vents, which prevent vacuum buildup when opening and closing the door.
- An interior safety release so no one can be locked inside the cooler (accidentally or otherwise).

Other smart features that can be ordered for walk-ins are:

- A thermometer (designed for outdoor use, but mounted inside the cooler) with a range of –40 to +60 degrees Fahrenheit.
- A monitoring and recording system that keeps a printout of refrigeration temperature or downloads to a computer.

- Glass, full-length door panels (like those in supermarkets and convenience stores), sometimes called ***merchandising doors,*** either hinged or sliding.
- Heavy-duty plastic strip curtains inside the door. (One manufacturer claims a 40 percent energy savings with this feature.)
- A ***foot treadle,*** which enables you to open the door by pressing on a pedal or lever with your foot when both hands are full.
- Three-way interior lighting, which can be turned on from outside or inside the cooler, with a light-on indicator light outside. Inside, the light itself should be a vapor-proof bulb with an unbreakable globe and shield.

When space is at a premium, think about whether it is practical to install an outdoor walk-in unit. This is an economical way to add space without increasing the size of your kitchen, and you can purchase ready-to-use, stand-alone structures with electricity and refrigeration systems in place. They come in standard sizes from 8 to 12 feet wide and up to 50 feet in length, in 1-foot increments. They range in height from 7.5 to 9.5 feet. Look for a unit with a slanted, weatherproof roof, a weather hood, and a fully insulated floor. Outdoor walk-ins cost about half of the price of installing an indoor kitchen walk-in, so this is a money-saving idea if it works in your location.

If your demands for walk-in space are seasonal, consider leasing a refrigerated trailer, available in most metropolitan areas on a weekly or monthly basis. They can provide an instant 2000 cubic feet of additional storage space, which can be kept at any temperature from –20 to +80 degrees Fahrenheit. They use basic 60-amp, 230-volt, three-phase electricity. Ask if the lease agreement includes hookup at your site and service if anything goes wrong.

Refrigeration Maintenance

Most refrigerators and walk-ins seem virtually indestructible and problem free, but you'll get longer life out of yours by following these safety and maintenance tips, and the ones on the chart in Illustration 10-13.

AREA	TASK	FREQUENCY
EVAPORATOR	Check for proper defrosting	MONTHLY
	Clean the coil and drain pan	EVERY 6 MONTHS
	Check for proper drainage	
CONDENSER	Inspect/clean the coil if the air supply is near polluting sources (like cooking appliances)	MONTHLY
	Clean the coil surface	EVERY 3 MONTHS
GENERAL	Check/tighten all electrical connections	EVERY 6 MONTHS
	Check all wiring and insulators	
	Check contractor for proper operation and contact point deterioration	
	Check all fan motors	
	Tighten fan set screws and motor mount nuts and bolts	
	For semi-hermetics, check the oil level in the system	
	Make certain all safety controls are operating properly	
	Check operation of the drain line heater and examine for cuts and abrasions	

ILLUSTRATION 10–13 The basic preventive maintenance chart for a walk-in cooler also applies to other types of refrigeration systems.
COURTESY OF KOLPAK, FRANKLIN, TENNESSEE.

Clean the door gaskets and hinges regularly. The door gaskets, made of rubber, can rot more easily if they are caked with food or grime, weakening their sealing properties. They can be safely cleaned with a solution of baking soda and warm water. Hinges can be rubbed with a bit of petroleum jelly to keep them working well.

Dirty coils force the refrigerator to run hotter, which shortens the life of the compressor motor. They should be cleaned every 90 days, preferably with an industrial-strength vacuum cleaner.

Walk-in floors can be damp-mopped, but should never be hosed out. Too much water can get into the seals between the floor panels and damage the insulation.

A refrigerator only works as well as the air that's allowed to circulate around its contents. Cramming food containers together so there's not a spare inch of space around them doesn't help. Also try to keep containers (especially cardboard ones) from touching the walls of the cabinet. They may freeze and stick to the walls, damaging both product and wall.

Use a good rotation system: First in, first out ("FIFO") is preferable. Or put colored dots on food packages, a different color for each day of the week, so everyone in your kitchen knows how long each item has been in the fridge.

10-5 SPECIALTY REFRIGERATION UNITS

Carbonated beverages and ice cream products often require their own storage in a busy restaurant. Because both must be refrigerated, we include them in this chapter.

Carbonated beverages are created on-site at most restaurants by mixing together syrup and carbonated water and propelling it through a system of chilled tubes (called lines) to the dispenser. There are two types of carbonated beverage machines, postmix and premix; and two types of dispensers, fountain and cobra gun.

The ***postmix system*** begins with the syrup concentrate moving through the chilled line, and the carbonated water moving through another line to join at the dispensing head, where the newly mixed soft drink is poured into a cup or glass. A valve for each drink may be electric or manual. The syrup is purchased in 3- or 5-gallon tanks or boxes that sit under the dispenser. Valves on the syrup containers enable easy hookup to the system. Most systems use carbon dioxide as the gas that propels the liquids through the lines.

The ***premix system*** works much like the postmix, except the syrup and carbonated water are purchased already mixed, and propelled up a single line into the dispensing head. Both premix and postmix dispensers usually contain room for ice storage and chutes to dispense ice into cups. Self-contained fountain units come in large stand-alone sizes with a dozen valves or more, and countertop sizes as small as 3 feet in width. They can be purchased from the same company that supplies the syrup.

There are advantages and disadvantages to each type of beverage system. The concentrated postmix syrup is less expensive than the premixed product, but postmix requires more lines and is more complicated to maintain and hook up. In large operations where more than one dispensing site is needed (say, in a cafeteria), you can set up one central propulsion and cooling system and connect it to more than one row of dispensing valves.

In a bar, the dispensing unit is likely to be used only by the bartenders. It is called a ***gun,*** and it's attached to a multiline hose called a ***cobra.*** The top of the dispensing head features up to eight buttons you can push, each one linked to a different line. The *cobra gun* can dispense as many as seven soft drinks and water from a single head (see Illustration 10-14).

Soft-drink dispensers cool beverages one of two ways. One is a small electric refrigeration system that runs on a ¼- or ½-horsepower compressor and can chill up to 700 12-ounce drinks per hour. The other option is a ***cold plate,*** which is a metal sheet chilled by direct contact with ice in a bin. The beverage lines all run through the metal plate, cooling the syrup and water. The larger the plate and the more of its surface is ice covered, the colder the final product.

The temperature of your beverage dispenser is critical, because carbonation tends to fizzle above 40 degrees Fahrenheit, resulting in flat beverages.

Draft Beer Service

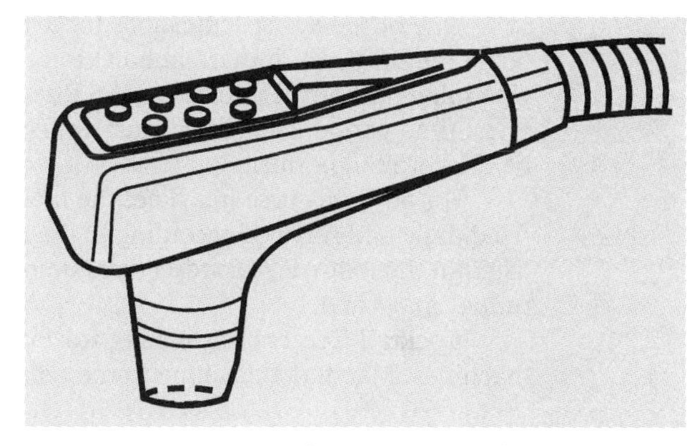

ILLUSTRATION 10-14 **A cobra gun or "speed gun" dispenses multiple liquids.**

The popularity (and profitability) of draft beer requires another type of refrigeration—a system for storage of beer kegs and dispensing of beer. The following paragraph comes from our *Bar and Beverage Book* (third edition, published in 2003 by John Wiley and Sons):

"A draft-beer system consists of a keg or half-keg of beer, the ***beer box*** where the keg is stored, the ***standard*** or tap (faucet), the line between the keg and the standard, and a CO_2 tank connected to the keg with another line. The beer box, also called a *tap box,* is a refrigerator designed especially to hold a keg or half-keg of beer at the proper serving temperature of 36 to 38 degrees Fahrenheit. Generally, it is located right below the standard, which is mounted on the bar top, so that the line between keg and standard is as short as possible. If more than one brand of draft beer is served, each brand has its own system—keg, line, and standard—either in its own beer box, or sharing a box with another brand."

In busy places, where it is inconvenient to tap a new keg every time one is needed, the beer box can be a walk-in cooler, and kegs may be connected in series. Short lines are simply not feasible in this arrangement, so it is imperative that they be kept clean to avoid build-up. Beer lines are made of heavy-duty plastic or stainless steel. Some manufacturers run coolant lines alongside the beer lines to ensure the beer remains cold. A higher-pressure combination of carbon dioxide and nitrogen may be required to "push" the beer through a larger system.

Soft-Serve Machines

Another specialty refrigeration unit is the *soft-serve machine,* which produces frozen desserts from a liquid base mix. There are mixes available to produce ice cream, ice milk, frozen yogurt, gelato, and so on, in many different flavors. Large machines are made for banquet-sized crowds; smaller ones for individual customer servings are placed near the front counters of quick-service restaurants. When shopping for a machine, you'll want to know how much mix it will hold at a time. Capacities range from 8.5 quarts to 72 quarts.

The soft-serve machine pumps air into the base mix, which gives it the soft consistency and increases the volume of the product. The percentage of air forced into the product is referred to as its ***overrun***; for instance, 16 ounces of mix with a 100 percent overrun will produce 32 ounces of frozen product. A 50 percent overrun will produce 24 ounces of product. Overrun accuracy is critical. Too much air will cause a product that is thin and grainy; too little air will cause it to freeze too hard and become difficult to dispense.

There are two types of soft-serve machines. In a ***gravity feed machine,*** the mix is loaded into a hopper. It flows as needed into a cylinder below, where it is frozen, scraped out of the cylinder, and dispensed. Gravity feed machines are simple to operate and clean and are the least expensive.

The *pressurized* soft-serve machine uses an air pump to drive the mix into the freezer chamber, then forces it out through a spigot. Pressurized machines are more expensive, but they control the overrun air better than gravity feed machines.

The refrigeration system is at the heart of the soft-serve machine. The simplest dispensers operate on 110-volt electricity, but machines with more features require 208-volt lines. Most units are air cooled, and there are also very efficient water-cooled systems. Air-cooled machines can be installed just about anywhere, but remember that they vent hot air from the compressor into the environment—not efficient for an already hot kitchen area. Water-cooled machines dump the heat down the drain. There's no problem with ambient heat, but water use and sewage costs may increase slightly.

One of the key specifications for a soft-serve machine is its Btu rating. You may recall from Chapter 5, the Btu (British thermal unit) is a measurement of how much energy a system or appliance uses. One kilowatt-hour is equal to 3413 Btus. The higher the Btu rating, the more mix can be frozen in a given time period. Look for higher ratings in high-volume situations, or where you want the product to be thicker than normal.

For any soft-serve machine, the most important consideration is keeping it clean. It must be sanitized daily and serviced according to the manufacturers' maintenance schedules. Common problems are listed in the following "Foodservice Equipment" box, reprinted with permission of *Restaurants and Institutions* magazine.

Cocktail freezers for making frozen drinks like margaritas and daquiris are kin to the soft-serve machines. Match the machine to the type of cocktail being made. As the amount of sugar and alcohol

FOODSERVICE EQUIPMENT

TIPS FOR TROUBLESHOOTING THE SOFT-SERVE MACHINE

SYMPTOM: FREEZE-UPS

POSSIBLE CAUSE	SOLUTION
Starving the barrel of mix	Don't let mix get too low

SYMPTOM: RUNNY PRODUCT

POSSIBLE CAUSE	SOLUTION
Dull scraper blades	Change blades every 4 to 6 weeks
Outdrawing draw capacity	Calibrate draw handle according to thickness of mix; don't rush dispensing
Kink in water hose (water-cooled machine)	Move machine carefully
Not enough airflow space (air-cooled machine)	Follow manufacturer's recommendations for space around machinearound machine

SYMPTOM: MILKSTONE IN RUBBER GASKETS AND O-RINGS

POSSIBLE CAUSE	SOLUTION
Not using tuneup kit regularly	Replace rings and gaskets every 90 days
Improper maintenance	

SYMPTOM: OFF TASTE

POSSIBLE CAUSE	SOLUTION
Improper cleaning	Clean and sanitize regularly

Source: *Restaurants and Institutions* magazine, a division of Reed Business Information, Oak Brook, Illinois.

in a drink increases, so does the time it takes to freeze the drink. A larger-capacity refrigeration system is needed for high-volume use, such as when you plan to serve frozen drinks by the pitcher, or for drinks that include high alcohol and sugar content.

Soft-serve desserts, shakes, and frozen drinks can be profitable, but the initial cost of the equipment is sometimes prohibitive. Most manufacturers offer a lease, or a lease-to-own option. Calculate how many desserts and/or drinks you must sell per day in order to cover the combined costs of the equipment payment and ingredients.

Ice Makers and Dispensers

All foodservice operations need ice, and the simplest way to meet that need is to have an *ice-making machine* that freezes, "harvests," and stores ice automatically. There are large, stand-alone machines that produce up to 3000 pounds of ice per day; medium-sized, undercounter models that make up to 200 pounds per day; or small, countertop ice makers that deliver as little as 1 pound of ice per hour. You will sometimes see ice makers referred to as *ice cubers.*

Before you buy an ice machine, you should determine not only how much ice you'll need, but *where* you will need it. If there are several sites for ice consumption—garde manger area, bar, wait stations—you might be better off with several smaller machines in various locations instead of everyone hauling ice from a single, large unit. We'll talk in a moment about how to determine your ice needs.

Ice-making machines are refrigeration units. The ice is made when a pump circulates water from a tank. The water runs through tubing to a freezing assembly, which freezes it into a single sheet. The frozen sheet is then crushed or forced through a screen to produce ice cubes. Different types of screens produce different sizes and shapes of cubes. After the ice is crushed or cubed, it is automatically dumped into a storage bin. When the bin fills to capacity, a sensor inside the machine shuts it down until there is room to make and store more ice. Because most of the ice maker's parts come into direct contact with water, it is important that components be made of rustproof materials.

Ice maker capacity is determined by how many pounds of ice the unit can produce in a 24-hour period. However, any machine's output (and the quality of the ice itself) will be affected by several factors:

INCOMING WATER TEMPERATURE. The ideal is 50 degrees Fahrenheit; warmer water makes the machine work harder.

ROOM TEMPERATURE. The ideal is about 70 degrees Fahrenheit. If installed in an environment that has an ambient temperature of 80 degrees Fahrenheit or higher, consider getting a unit with a water-cooled condenser to compensate for hot, humid, or grease-laden air.

INCOMING WATER PRESSURE. The minimum water pressure should be 20 pounds per square inch (psi); recommended pressure is between 45 and 55 psi. Anything higher than 80 psi will cause malfunctions.

WATER QUALITY. Hard water will cause the machine to work more slowly and almost always necessitates some kind of pretreatment before the water enters the machine. The fewer minerals and chemicals in the water, the more quickly and harder it will freeze, and the more slowly it will melt. Filtration is almost always a good idea.

Read manufacturers' output claims carefully and you'll find they are often based on ideal conditions: incoming water temperature of 50 degrees Fahrenheit and ambient air temperature of 70 degrees Fahrenheit. Generally, a 10-degree increase in air temperature means daily ice output decreases by 10 percent. Also examine the water and energy usage figures provided by the manufacturer. You'll note that there is a wide range: from 15 to 27 gallons of water to produce 100 pounds of ice, using from 5 to 10 kilowatts of electricity.

An additional source to check is the Air-Conditioning and Refrigeration Institute (ARI), the national trade association that represents about 90 percent of manufacturers. Ironically, ARI data rates ice machine production capacities using more realistic conditions than does the manufacturers' sales literature—with incoming water temperature at 70 degrees Fahrenheit and ambient air at 90 degrees Fahrenheit. ARI also rates machines by how many kilowatt-hours, and how much water, they need to produce 100 pounds of ice. The group's "CoolNet" website can be found at www.ari.org.

No matter where the ice maker is located, it needs a source of cold water and drainage. Particularly critical is a 1-inch air gap between the ice maker's drain line and the nearest floor drain. This is a necessary precaution to prevent a backflow of soiled water into the ice bin.

Wherever you install the ice maker, proper plumbing will be mandated by your local health department. A recessed floor beneath the unit is also recommended. Along with nearby drainage, this ensures that spilled ice does not melt on the floor and cause accidents. One smart option is to install an ***inlet chiller*** along with your ice machine. About the size of a household fire extinguisher, it collects the water that would normally be discharged from the ice maker into the drain. Instead, the water recirculates first through a series of copper coils in a chamber that contains fresh water on its way into the ice maker. The cold outgoing water chills the coils, which chill the fresh incoming water and allow it to freeze more quickly for faster ice production. The inlet chiller, which has no moving parts and uses no electricity, can save up to 30 percent on the electricity used to run the ice maker and boost its capacity by 50 percent.

There is also a need for air circulation around the unit. An ice maker gives off warm air, like any refrigerator, and should be placed at least 4 inches from the wall to allow for ventilation.

Those are just a few of the factors to consider in your "life-cycle" calculation. Others are on the checklist shown in Illustration 10-15.

Just as there are different machine capacities, there are also various sizes of storage bins. Most operators choose a combination ice maker and storage bin; by adding an extra 20 percent to the total capacity of each, you'll (theoretically) never run out. Ergonomics experts add that bins with a depth of more than 16 or 18 inches are hard to reach for employees who must scoop from the bottom. Look for storage compartments with volume sensors, so production cuts off automatically when the bin is full. When ice tumbles out every time you open the bin, you're just wasting it.

A final important consideration is the length of time it takes the machine to complete one ice-making cycle. Under normal conditions, the whole freezing-harvesting-ejection period should take no longer than 15 to 20 minutes for the finished cubes to hit the storage bin.

Now that we've discussed the machine itself, what kinds of ice do you want it to make? There are many different cubed-ice options.

- Large cubes (*full cubes*) melt more slowly, so they're good for banquet situations where glasses must be set out early. They may be awkward in some glasses and may give the appearance of a less-than-full drink.
- Smaller, *half cubes* stock better into most glasses and are preferred in bar settings.

ICEMAKER LIFECYCLE FACTORS

- Purchase price, tax, freight, start-up
- Installation (remote vs. self-contained)
- Energy (at projected use rates, e.g., 300 lbs./day actual use from a 400 lb. machine)
- Water
- Sewer cost (most machines use more than 12 gals. of water to produce 100 lbs. of ice)
- Preventive maintenance cost
- Supplies (filters, cleaning agents, de-liming chemicals)
- Labor (cleaning, changing filters, etc.)
- Service/repairs cost
- Footprint cost (e.g., by using remote, can you save space to devote to other equipment?)
- Annualized costs—all expenses amortized across projected life expectancy

ILLUSTRATION 10-15 **Factors to consider in determining the useful life and actual costs associated with an ice machine purchase.**

- Even smaller *cubelets* are suggested for soft drinks, because they fill the glass or cup so well (and therefore make your soft drink supply go further).
- *Nugget ice* is flaked ice that has been compacted into high-density, random sized chunks that melt slowly and are easier on blender blades than cubed ice. Nugget ice is a good choice for soft drinks and smoothies.
- Round shapes fit a glass better at its edges.
- Rectangular cubes stack better than round ones, leaving fewer voids in the glass.

Your clientele, your glassware, and your type of service will determine the right shape for you.

In addition, you may have uses for crushed ice or flaked ice. You can buy an ice maker that produces crushed ice or one that makes ice cubes, then runs them through a special canister to crush them when needed. Crushed ice cools faster than cubed ice and is perfect for salad bars or fresh-seafood displays. A flaked ice machine, or ***flaker,*** produces soft, snowlike beads of ice, which are used mostly for keeping things cold, such as fresh fish, bottled wine, or salad bar foods.

How much ice does your operation require? *Foodservice Equipment and Supplies* magazine advises to estimate depending on the type of business:

- 0.9 pound of ice per customer for quick-service restaurants
- 1.7 pounds of ice per customer for full-service restaurants
- 3.0 pounds of ice per customer for cocktail lounges

For most businesses, however, the figure will change daily, and seasonally. You may need more ice during the summer than winter, more during certain peak mealtimes, or more on weekends than on weekdays. So the general rule is to estimate your ice needs, then size your unit and storage bin to accommodate 20 to 25 percent more than your estimate. Running out of ice on a busy night is a restaurateur's nightmare, so most of them avoid it by buying the largest possible ice machine for the space they've got. However, it's probably smarter to calculate how much ice you'll need, and purchase the right machine for your needs. Here's how to do the calculation:

1. Estimate ice usage for one full week. You can figure that every meal you serve will require 1 to 1.5 pounds of ice.
2. Divide the figure by 7.
3. Multiply the result by 1.2.

The final total will be your average daily usage. Also, remember it is two to three times more expensive to *make* ice than to *store* it. With this in mind, never scrimp on the size of your ice-holding bin. Get an ice machine with a bin that stores twice as much as the machine can produce in a 24-hour period. You may only need this large bin capacity two or three days a week, but it's more cost-effective than having your ice machine working all the time.

The other way to store ice is to stockpile it on low-volume days. Keep the machine running, bag the "extra" ice in your freezer, and bring it out as needed on busier days. It's a temporary solution, but if your operation is fairly small, it can work until you decide to upgrade to a larger-capacity machine.

We are noticing a trend toward decentralized ice production for larger-volume businesses. This means that a large machine with its compressor and condenser in a single location (usually outdoors) makes all the ice for a business instead of having several smaller, indoor ice makers in different spots. In the most sophisticated systems, ice is transported by pneumatic tube (think bank drive-up lanes!) to ice bins elsewhere on the property. The bins are automatically kept full without having to transport the ice manually.

From Ser-Vend International, a well-known ice maker manufacturer, come these handy Ice Service Guidelines.

In many foodservice settings, ice is dispensed directly to the guest: quick-service restaurants, convenience stores, cafeterias, and institutional operations, to name a few. So, in addition to ice makers, ***ice dispensers*** are essential. The ice dispenser is also a refrigeration unit that freezes ice in a cylinder or *evaporator chamber.* As water enters the cylinder, it freezes against the wall and is chipped off as needed by a tool (called an *augur*) that looks like a big corkscrew. The ice falls out of the cylinder into

BUDGETING AND PLANNING

ICE SERVICE GUIDELINES

TYPE OF ESTABLISHMENT OR USE	ICE NEEDS PER DAY
Restaurant	1½ pounds per person
Bar/cocktail lounge	3 pounds per person (or per seat)
Water glass service	4 ounces ice per 10-ounce glass
Salad bar	30 pounds per cubic foot (multiply by the times you restock the ice during the day)
Quick-service restaurant	5 ounces per 7–10-ounce drink
	8 ounces per 12–16-ounce drink
	12 ounces per 18–24-ounce drink
Hotel—guest ice	5 pounds per room
Hotel—catering	1 pound per person
Health care—patient ice	10 pounds per bed
Health care—cafeteria	1 pound per person
Convenience store—self-serve drinks	6 ounces per 12-ounce drink
	10 ounces per 20-ounce drink
	16 ounces per 32-ounce drink
Convenience store—cold plate dispenser	50 percent more ice per day than listed above

its storage bin as flakes or chunks. At the push of a button or lever, the ice is dispensed into a container: a paper cup or a hotel ice bucket.

Ice dispensers range in size from huge, freestanding floor models that hold 200 to 600 pounds, often found in hotels and health-care settings, to countertop units that take up as little as 13 inches of counter space, produce 280 pounds of ice per day, and store up to 30 pounds. The largest countertop ice dispensers make 850 pounds of ice per day and store up to 200 pounds.

Manufacturers have come up with an ice dispenser to fit just about any foodservice application: commercial machines that can be programmed for hotels so that only guests' room keys open them; combination ice and beverage-dispensing fountain units, the current standard in fast-food operations; ice dispensers with water-dispensing valves, popular in cafeteria settings. One of the biggest advantages of the dispenser is that it is more sanitary than reaching into an ice bin, even with scoops or tongs.

Finally, remember that an ice dispenser is close kin to the ice maker, with all the same potential considerations. Ambient temperature, water temperature and pressure, plumbing needs, water quality, cleanliness, and proper maintenance all figure prominently in ensuring its long, productive life.

Ice Machine Maintenance and Sanitation

You can order ice makers and dispensers with automatic cleaning features, such as air filters and self-rinsing ice contact surfaces, or you can do it the old-fashioned way. Either way, ice that is dirty, melts too quickly, or lacks uniformity usually signals a dirty or malfunctioning machine. Here are the most common problems and ways to prevent them.

DIRTY ICE. Clean and sanitize the machine regularly to remove algae, slime, and mineral deposits from the water. Bin walls can be washed with a neutral cleanser as long as you are sure to rinse them thoroughly. Always remove ice with clean hands and a sanitized plastic scoop, and never use the bin as a convenient place to stash foods or beverages. Check the air filter on air-cooled models at least twice a month and wash it whenever it is dirty.

SMALL, CLOUDY, OR BROKEN CUBES. This usually signals a problem with the water filtration system. Perhaps the water is not entering the machine at sufficient pounds per square inch because it's being partially blocked. Try replacing the water filter by turning off the water first, then slowly opening the filtration cartridge or canister.

BLOCKS OF ICE STUCK TO BOTTOM OF BIN. Keep the ice machine level and the drain unclogged to prevent melting ice from puddling at the bottom of the bin and refreezing to form blocks.

LACK OF ICE. First, confirm that the electric circuit breakers (for both the machine and the condenser) are on. If it's a self-cleaning machine, be sure it is set to make ice instead of clean; the switch may have been tripped to the wrong position. Check the sensor inside the bin; tighten it if it seems loose. Finally, inspect the water supply. A partially closed valve or insufficient water pressure can interrupt or reduce the ice-making cycle.

MELTING CUBES. The fins on the condenser become clogged when they are dirty, which can interrupt the freezing process or make partial cubes instead of full ones. The fins should be clean enough to see through to allow the refrigerant to reach the right temperature to finish the freezing process. Fins should be inspected and cleaned at least every three months. If your machine has a condenser, it should also be cleaned regularly with a brush or vacuum cleaner.

These few problems are easily handled by your staff, but there are times it's probably best to call the factory-authorized service person: when the ice-making cycle takes far too long, or when ice is not made even though the water supply seems fine.

By health code standards, ice is a food. A mandatory part of employee training should include handling ice so that it remains safe and sanitary for human consumption. This means:

- Use dedicated containers for ice. Don't use containers that have also been used to store food.
- Store ice containers by hanging them upside down, far from the floor. Don't "nest" them in stacks, which is unsanitary.
- Clean and sanitize every ice scoop regularly. Have plenty of scoops, and store them by hanging them outside the machine—not sitting in the ice.
- Employees must wash their hands before they scoop or bag the ice, and technically they're not supposed to touch it at all without using clean disposable gloves.
- No one can eat, drink, or smoke around the ice machines.

Filtering water before it enters the machine will improve ice quality, as well as protecting your machine from lime scaling, chlorine buildup, and slime. (Slime, incidentally, has a scientific name—***biofilm.***) One popular manufacturer, Manitowoc Ice of Manitowoc, Wisconsin, now uses an antimicrobial chemical to make all plastic parts that come into contact with water or ice. This retards, but does not completely eliminate, biofilm growth. Today's ice machines can be programmed to clean themselves automatically during the hours they are not in use, a process that takes about half an hour. Cleaning is recommended weekly.

SUMMARY

Having a well-organized receiving area for incoming goods involves selecting and maintaining accurate scales for weighing products, rolling carts and hand trucks for moving it from place to place, and pallets and shelves for storing it safely above ground. Your options for each are discussed in this chapter.

You'll need two types of storage: dry and refrigerated. Refrigerated storage requires air circulation, humidity (or lack of it) in the refrigerated space, and heavy-duty construction to withstand the rigors of the busy kitchen environment. Because most kitchens have too much or too little refrigerated space at the proper location to meet their needs, you must first decide how much capacity you need, and then determine where to locate it. Often, several smaller units, carefully placed, are preferable to one giant refrigerator.

There are many handy options to consider when ordering a refrigerator, including sturdy thresholds, which prevent tripping; pressure-sensitive vents, which prevent vacuum buildup when opening and closing doors; safety door handles, which prevent accidental lock-ins or lock-outs; glass doors, which allow you to see inside; and temperature monitoring systems, which promote food safety.

In addition to the standard models (reach-in, roll-in, and pass-through), there are a number of popular specialty refrigeration units: underbar refrigerators, carbonated beverage machines, and soft-serve ice cream machines, to name a few. Serving draft beer from kegs requires a separate system with its own refrigeration capability for the kegs.

Walk-in coolers and freezers are excellent storage options. These are best used when equipped with shelves that promote air circulation and are configured for the types of containers you'll be using to store foods. The refrigeration systems for these large units can be placed on their ceilings, or even outside the building.

And no foodservice business is complete without ice-making machines to dispense and store crushed or cubed ice. Your ice needs will depend on how you use the ice. This chapter tells you how to determine the average daily ice usage in your facility. Because it is two to three times more expensive to make ice than it is to store ice, don't scrimp on the size of your ice holding bins. You can also make more ice on slow days and stockpile it for busier ones.

Ice is considered a food, and all related safety and sanitation codes apply to its use in foodservice. It must be kept clean, and water coming into the ice machine should be filtered to keep out impurities that cause scaling and slime and impact the ice quality.

STUDY QUESTIONS

1. Why must incoming deliveries be checked as soon as they arrive at a foodservice business?
2. How would you decide which types of scales to have in your restaurant's receiving area?
3. What is the difference between a cart and a truck?
4. What are the three things going on inside a refrigerator that all work together to cool its contents?
5. What makes the air circulate inside a refrigeration system? Why is air circulation important?
6. What factors would you consider in selecting a door for your refrigerator, and why?
7. Explain the formula for deciding how much refrigeration space you will need for a walk-in cooler or freezer.
8. Name three features you think are important for a walk-in, and explain why you think they are worth having.
9. How do you determine the capacity of an ice-making machine?
10. Explain the advantages of these different types of ice: full, cubelets, and nuggets.
11. What is an inlet chiller?
12. What's the difference between a premix and a postmix soft-drink machine? What are the advantages of each?

C H A P T E R 11

PREPARATION EQUIPMENT: RANGES AND OVENS

INTRODUCTION AND LEARNING OBJECTIVES

The piece of equipment found most often in commercial and institutional kitchens is also a familiar workhorse at home: the ***range,*** commonly referred to in home use as a stove. There are actually two parts to most ranges: the *range top,* for surface cooking needs, and the *range oven,* for baking and other types of preparation that require the food be placed inside a heated cavity. Home models almost always locate both range top and oven on the same appliance; however, in foodservice, you can purchase separate, huge ovens without range tops for volume baking, or range top units without ovens. You can mix and match the pieces to fit your space and cooking needs. In fact, there are almost as many combinations of ranges and ovens as there are types of food!

In this area of equipment technology, professional cooks have reason to rejoice. Never before have science and industry collaborated to offer such a broad array of heat sources and equipment types, particularly ovens. The time when one range/oven combination was sufficient has passed. Today's foodservice kitchen is equipped with innovative, multifunctional "integrated solutions" for greater productivity and a higher food yield.

In this chapter, we will discuss:

- Different types of ranges and ovens, including custom models and options for certain types of cooking
- Sizes and utility requirements of ranges and ovens
- Selection and purchasing basics
- Cleaning and maintenance tips

In this chapter, and the other equipment-related chapters that follow, you will see that the growing use of multifunctional equipment has revolutionized working conditions. Most new appliances are easier to clean, and their designs minimize the need for heavy lifting. They are generally smaller, taking up about one-fourth less space than their traditional counterparts. They work more efficiently, which means shorter cooking times and minimum "weight loss" of the foods being cooked. Cooking cycles and times can be electronically monitored, freeing up employees' time without a loss in food quality.

Before we take a look at some of these modern marvels, here are some guidelines for selecting the right ones for your operation:

Power Requirements. What types of power are available, and what kind of load can your kitchen safely handle without shorts, brownouts, or other malfunctions? What are the installation costs associated with this energy source? After you have your spec sheets in hand, ask an installer to visit your site and talk with you about how your existing power source fits your prospective purchase.

- Gas: Is the existing gas line large enough for another piece of equipment? Only so much gas will flow through a pipe at a given pressure. If existing equipment already draws most of the gas pipe's capacity, the addition of a new appliance could prompt all the gas equipment to function on less gas and, therefore, less power. Gas ranges that function in high-altitude locations (2,000 feet or more above sea level) require adjustments to gas-metering devices. If bottled gas is used, cook times will not be as fast, since bottled gas generates about 25 percent less heat than natural gas supplied directly from a gas line.
- Electricity: If your business will be operating in a high-rise building without a gas line to your floor, electric power is your only option. Should the electrical fuse box be upgraded? Do you have sufficient wiring to service a 208–240 three-phase or single-phase piece of equipment? Bringing in a new line can be costly.

Type of Menu. What are you cooking? How is it most efficiently prepared? Can some items be cooked in advance and reheated, or does the menu focus on individual, made-to-order foods?

Quantity. How much do you need to meet peak demands? How much can be prepared in advance and held hot? Or, if the menu is mostly ***à la carte*** (cooked to order), how much range top space do you need at a time?

Speed. How fast must you be? Cooking at higher temperatures, using microwave ovens, setting the impinger/conveyor oven at its maximum speed—there are many different ways to pick up the pace. Even with slow-cook methods, the smart operator will come in early and get the rotisserie or smoker-cooker going so that, by mealtime, the food is perfectly cooked and ready to serve.

Space Available. How much room do you have? Should you select a few multipurpose workhorses, or is there room for specialized cooking appliances? Can you stack more than one appliance in a space? If so, will people actually use it? Or will they avoid it because reaching and/or bending to get to it is a hassle?

Answering these questions may be tough, but don't let them discourage you. Choosing ranges and ovens, like so many of your appliance decisions, will be a combination of logistics, common sense, and budget considerations.

11-1 BASIC PRINCIPLES OF HEAT

You will notice that often the physical dimensions or power requirements are listed for certain types of appliances. You certainly don't have to memorize these figures; rather, the idea is to acquaint you with the many options available from manufacturers and to prompt you to consider size and utility usage when outfitting your kitchen. Refer back to Chapters 5 and 6 for more information about the different types of utilities.

Before you learn about the types of ranges and ovens available, it would be helpful to understand a few cooking-related terms. This is information you may have already learned in your culinary training; however, it is important enough to summarize briefly the cooking processes that are central to every kitchen.

Conduction is the simplest form of heat transfer. Heat moves directly from one item to another when they are brought into direct contact with each other. Poaching, boiling, and simmering are all examples of conduction cooking.

Induction cooking creates heat when an electromagnetic field is created between stovetop and pan, creating currents that are converted to heat by the natural resistance of the metal pan. This electromagnetic field can only be achieved if the pan is made of ferrous metal, such as cast iron or stainless steel; nonmagnetic copper and aluminum pans will not work. Induction cooking surfaces do not become hot, because the heat is transferred directly into the pan. When the pan is removed from the heat source, the induction process ends automatically, with no residual heat. Illustration 11-1 summarizes the process.

Convection occurs when heat is distributed by moving air, steam, or liquid. The simplest form of convection occurs when you stir a thick sauce, for instance, to circulate it and heat it thoroughly within the pan. Convection cooking was discovered when U.S. Navy researchers placed fans in standard oven compartments to see if food could be cooked more quickly when warm air was blown around it. They found that, indeed, the convection process cooked food faster and more evenly, no matter where in the oven the food was placed. Why? Because the hot air reduces the cool "boundary layer" at the surface of the food, allowing the heat to penetrate the food faster. More about convection ovens later in this chapter.

Radiation occurs when waves of energy hit the food and cause a heating reaction. There are two types of radiation used in the kitchen: infrared and microwave. In a broiler, the heating element becomes hot enough to release ***infrared*** radiation, which cooks food very quickly.

The other type, the ***microwave*** oven, used to be called *magnetron cooking* and was developed during World War II. U.S. airmen noticed they could keep their coffee hot by setting it near the plane's radar equipment box. The short, electromagnetic waves emitted at the speed of light react to moisture, which is found in almost all foods. The moisture molecules in the coffee cup create friction as they try to align themselves with the microwaves, and friction creates heat. Today, this simple principle has found its way into almost every American home. More about microwave ovens later in this chapter.

How Induction Works

Although Induction seems magical in how it works, there is a scientific explanation.

1. An alternating current in an induction coil produces an alternating magnetic field.
2. This magnetic field is instantly transferred and concentrated to the cooking vessel.
3. This concentrated magnetic energy in the cooking vessel causes it to heat up and start cooking.
4. When the vessel is removed from the heat source, the induction unit automatically shuts off.

Note: Induction cooking requires magnetic pots and pans to work effectively.

ILLUSTRATION 11-1 **The induction process.**
COURTESY OF GARLAND GROUP, A WELBILT COMPANY, FREELAND, PENNSYLVANIA.

11-2 THE RANGE TOP

The most heavily used piece of equipment in any kitchen is the range top. It allows us to boil, sauté, simmer, braise, deep-fry, and hold food hot. Today, although new types of specialty equipment have been designed to perform these individual tasks, perhaps with greater efficiency, the range top is still the multipurpose king of the kitchen and the "command post" of most operations.

At first glance, the casual observer would say this workhorse has changed very little over the past 20 years or so. Yes the basic operating functions are still intact. But there are special features that set various manufacturers' models apart and may make a difference in your purchase decision: a narrower profile (24 inches instead of 36 inches) to fit smaller spaces, more energy-efficient burners, additional insulation, and so on.

In determining range top requirements, your selection begins with the menu and your full understanding of how the range will work in relation to other equipment. Consider the answers to these questions:

- How much of the menu is ***à la carte*** (also called "to-order") work? This means items cooked individually instead of in batches. How much will the range be used? Perhaps it will be required only for selected meal periods; or for different functions for breakfast, say, than for dinner.
- Are there other appliances as alternatives to the range top for certain cooking duties: steam-jacketed kettles, steamers, griddles, broilers, fryers?
- Where will you locate the range? In a display kitchen, enameled finishes and brass fixtures are more appropriate, and would be worth the extra money, compared to a typical kitchen hot line that is not seen by the public.
- How will the range fit into the layout, either existing or planned? The layout, of course, should be driven by how the hot line can work most efficiently. In most instances, the hot line will run parallel to a plating and pick-up area, but in some cases, the two "lines" are perpendicular.
- How much space is needed during peak demand periods to hold food hot in pots and pans?

Total daily meal output is often used as a guideline for range top space. The simplest calculation is one restaurant range for every 50 seats, but there are other formulas, too. The requirements listed in Table 11-1 are a bit simplistic, because they assume the presence of griddles and fryers but not steam equipment, which can be used instead of the range top to make stocks, soups, and sauces.

Another set of requirements, from J. Wilkensen's *The Complete Book of Cooking Equipment,* is shown in Table 11-2. As you can see, determining how many ranges you'll need is not an exact science.

Commercial ranges offer a number of options not found in home models. For instance, they may be installed on steel or chrome legs of various heights over 6 inches. They may be equipped with casters for mobility. Stainless steel shelves may be mounted above the back of the range for convenient storage space, and stainless steel *spreader plates* can be placed between ranges and the appliances next to them to increase countertop work space (see Illustration 11-2).

When you look at ranges, always consider how you will clean them. Most of them have a drip pan under the burners to catch spills and grease. How easy is it to remove and clean this pan? Check the location of the controls, to be sure they are easily accessible and protected from spills. Inspect the construction of each burner and order heavy-duty burners if you'll be using large, heavy pots on them. The legs of the range should be attached to heavy steel plates, which are welded to the body.

There are three basic types of ranges, each one suited to different types of kitchen operations. They are the medium-duty restaurant or café range, the heavy-duty range, and the specialty range. The latter has a number of different configurations, several of which will be mentioned in this chapter. Let's look at each type individually, as well as the burners that can be ordered for them.

TABLE 11–1

RESTAURANT RANGE NEEDS

Number of Ranges/Size of Space	Meals per Day
2 ranges: 6 feet in length	300
3 ranges: 9 feet in length	500
4 ranges: 12 feet in length	1000

TABLE 11-2

RANGE REQUIREMENTS BY TYPE OF COOK-SERVE OPERATION

Type of Operation	Meals per Week	Range Area (square feet)	Steam Kettles or Cookers
Restaurant	1,000	24	No
Cafeteria	2,000	10	Yes
Hospital	2,100	24	Yes
Cafeteria	4,000	10	Yes
Cafeteria	8,500	14	Yes
Student union	14,000	36	Yes
Restaurant	14,000	24	Yes

Source: J. Wilkensen, *The Complete Book of Cooking Equipment, Restaurants and Institutions.*

The Medium-Duty Range

This multipurpose appliance is also known as a *restaurant range* or *café range.* It generally measures from 36 to 60 inches in width. Its range top contains six to ten open burners, which are 12 inches square and arranged two deep across the top of the range. As its name suggests, this type of range is suggested for smaller establishments with short-order menus or in settings where there is no need for constant, continuous use, such as church or nursing home kitchens (see Illustration 11-3).

Most often, though, the chef organizes the range top so that certain burners are "reserved" for certain duties. The front row may be used for sautéing, for instance, while finished products are held and kept warm on the back row (hence the popular saying that when something is delayed, it's been put "on the back burner"). There are dozens of different handy range-top configurations (see Illustration 11-4). A six-burner range can be used for quick sauté work in several small pans; a four-burner unit is well suited for large simmering stockpots. If the facility makes a lot of stock as a base for soups or gravies, a ***uniform heat top*** (usually called a *hot top* or an *even-heat plate*) should be considered. This flat surface distributes the heat over a larger area than the burner itself and makes more efficient use of the range top. A *step range* has its back burners raised above the front ones, making access to the back pans easier. You can also order three burners up front and two even-heat plates in the rear for keeping foods warm.

Ranges can be ordered with no oven, one oven, two ovens, or even refrigerated space, underneath. Increased insulation provides sufficient temperature separation for units with refrigerated bases, a popular choice in smaller kitchens. Oven sizes are based on the dimensions of a typical baking sheet—18 by 26 inches. If a facility has minimal baking or roasting needs, one oven will be sufficient. Many cooks like the convenience of having an oven below their range top, but feel the performance of the

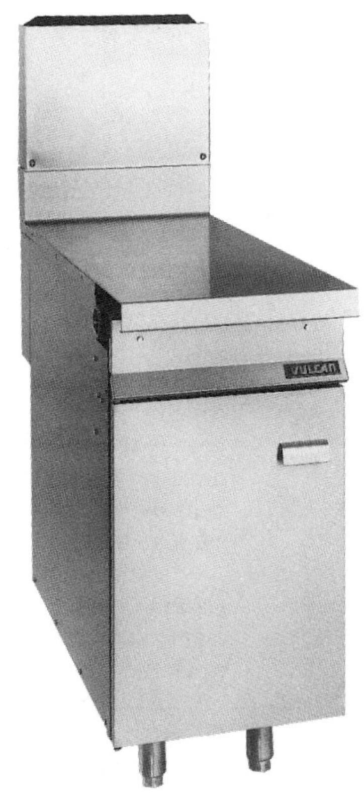

ILLUSTRATION 11-2 **Spreader plates fit between appliances to give you more room to work. Many are also cabinets that provide extra storage space.**

COURTESY OF VULCAN-HART, DIVISION OF ITW FOOD EQUIPMENT GROUP, LLC, LOUISVILLE, KENTUCKY.

ILLUSTRATION 11-3 **A medium-duty or café range.**

COURTESY OF VULCAN-HART, DIVISION OF ITW FOOD EQUIPMENT GROUP, LLC, LOUISVILLE, KENTUCKY.

oven is not always equal to that of a stand-alone oven. And placement of an oven below the range top also limits its size, when stand-alone ovens can be ordered in larger dimensions.

Most manufacturers offer three sizes of range tops: 24-, 48-, or 60-inch widths, which may be added to in 12-inch modular dimensions to increase capacity. The back panel of the range may be either a small shelf located above the cooking surface or a simple 5-inch "stub" back that helps prevent food from spattering onto the wall. Many ranges also have 6-inch adjustable legs so that the appliance can be leveled. As mentioned earlier, a smart option is to replace the standard legs with 6-inch swivel casters, at least 1½ inches thick, so that the range can be moved for cleaning and servicing. Two of the casters should be lockable so the appliance won't move when in use.

Types of Burners

There are also several different types of burners. You will hear some referred to as ***drop-in burners,*** which means they are ordered in modules of one or two burners that "drop in" to the range top and are connected there. This makes it easy to mix and match different types of burners depending on what you plan to cook. There are freestanding burner units, too.

The most common *open burner* is covered by a one-piece, heavy-duty metal grate on which to set pots and pans. Typically, the grate is 12 inches wide and 25 inches long. An open gas burner is sometimes called a ***grate top*** or *graduated heat top.* Some grate tops are made in concentric circles called *rings,* which can be removed individually, enabling the chef to set a pan closer to the flame. The heat may emerge from the burner through *jets* or rings. Jets concentrate the gas flame and aim it directly at a spot, allowing food to cook quickly in highly concentrated heat; rings disperse the heat over a wider area and are better for melting, sautéing, or cooking duties that require steady, even heat.

We've already mentioned hot tops, which can be installed instead of open burners. And there are ***griddles,*** which are flat like hot tops, but made of a thicker steel plate. A griddle has the major advantage of a trough for convenient removal of grease from its surface, as well as a removable drip pan to catch the grease and short "walls" on three sides that help prevent spattering.

A combination ***broiler/griddle*** may also come in handy. This hybrid includes tubular burners for the griddle and infrared radiant bulbs for the broiler.

Perhaps the biggest breakthrough in recent years is in induction cooking, with the ability to better control heat. The latest technology includes a current sensor on the coil, which detects whether a pan is on the cook top and precisely controls the power applied (and, therefore, the heat delivered) to the pan. Most induction cook tops offer a temperature range from 90 to 440 degrees Fahrenheit, in 5-degree increments. If a pan's material or dimensions are not suitable for the range—too small, or not magnetic—the burner will not heat.

In general, the larger the range, the greater the number of burner options and combinations that are available.

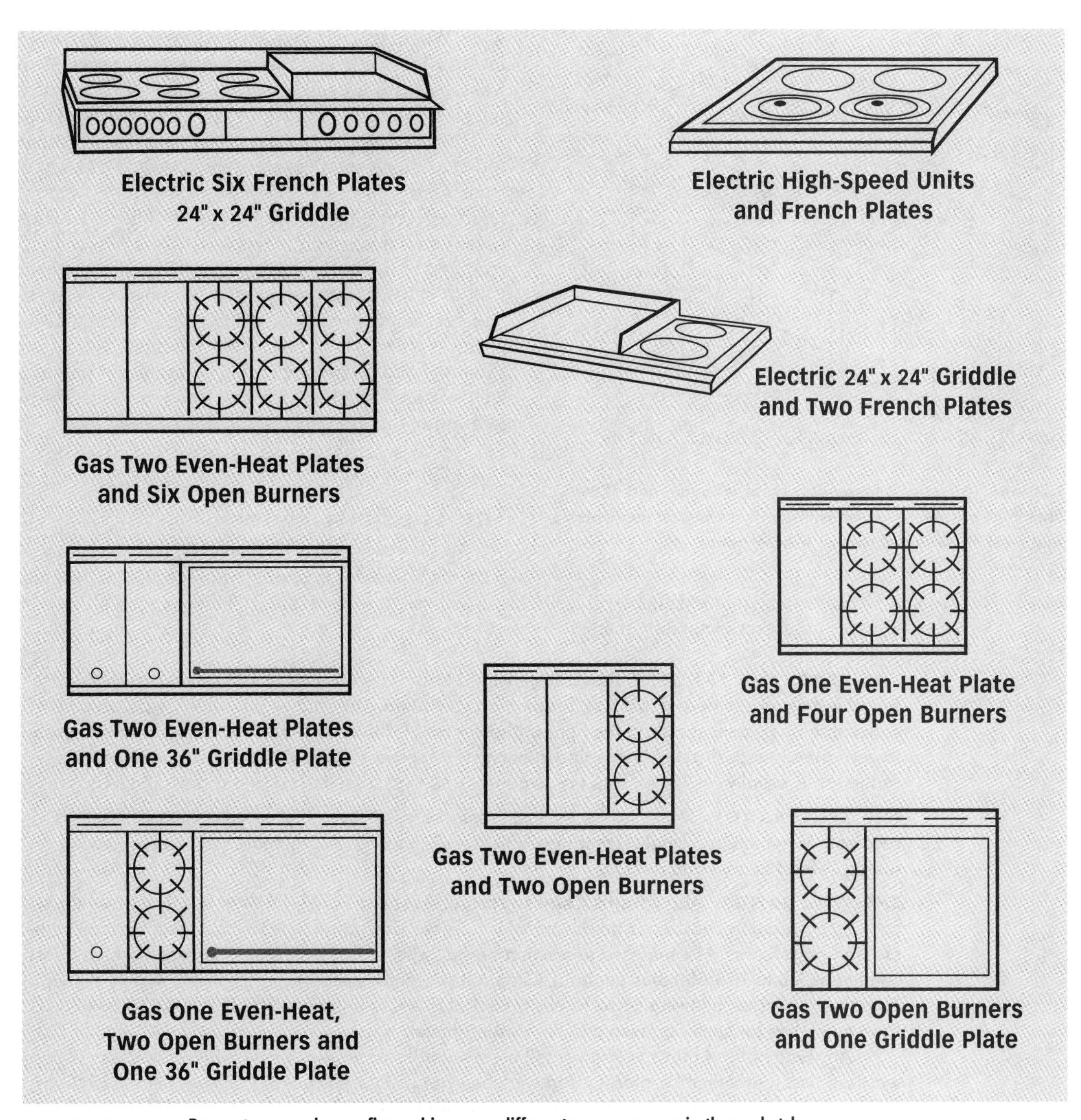

ILLUSTRATION 11-4 **Range tops can be configured in many different ways, as seen in these sketches.**

The Heavy-Duty Range

This model is similar to the medium-duty range, but is made of heavier materials to withstand the rigors of high demand and large, heavy pots and pans (see Illustration 11-5). The ***heavy-duty range*** is best suited for long hours and high-volume cooking. Its four open burners are rated up to 30,000 Btus per hour, per burner—able to cook hotter and faster than the 20,000-Btu output of the café range burners. Heavy-duty range tops fit together to form a solid line, or ***battery,*** in almost any combination of grate tops, hot tops, or griddles. Its modular design makes it easy to customize. Burners may be 12 or 18 inches

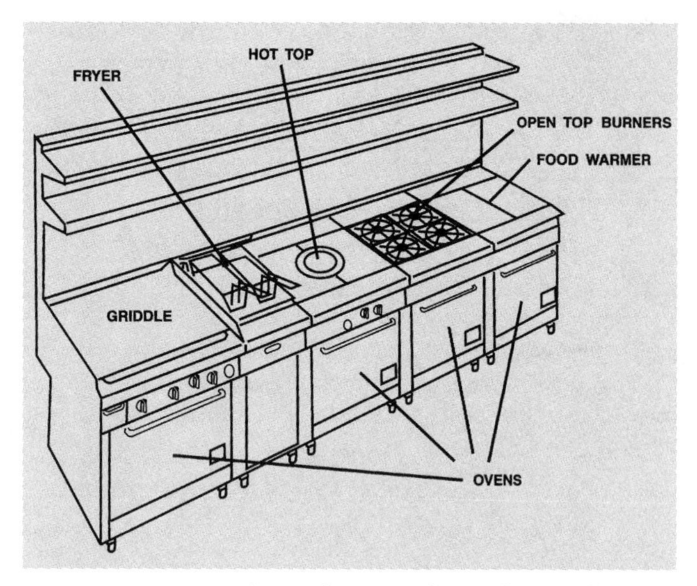

ILLUSTRATION 11-5 **A heavy-duty range is usually part of the "hot line" or cooking center, which includes several appliances that must all be vented with an exhaust hood.**

wide. Where space is tight, half-width ranges (12 to 17 inches wide) with only two burners may be specified. A front rail with a ***belly bar*** keeps cooks from accidentally leaning on the hot surface when they're working.

An additional feature of the heavy-duty range top is a 7-inch stainless steel shelf located directly in front of the burners. The heavy-duty range has its manifold (the pipe in which gas is delivered to the range) on the front of the range. The café range has its manifold at the back, so it can't sit as close to the back wall as a heavy-duty range. For gas-powered ranges, another handy option is a high flue riser made of stainless steel, which helps vent the flue gases to the exhaust hood. It minimizes spills inside the range as well as protecting the back wall from spatters. One or two storage shelves may be mounted on the flue.

The Specialty Range

If there's a popular type of food or method of cooking, you can be sure some manufacturer will create a custom range to facilitate it. Following are a few current options in this ever-expanding market.

THE STOCKPOT RANGE. A short range with a large open burner used to heat stockpots; also found in bakeries, where it is handy for melting chocolate. The range-top burner is a series of concentric rings, concentrating the heat at the center of the burner with a gradual heat decrease toward the outer perimeter. Its heating capacity is impressive—55,000 Btus per hour—and the range top is usually an 18-inch square. Double- and triple-sized units are also available.

THE TACO RANGE. Designed for Mexican restaurants. Pans and pressure cookers fit into its recessed burners. On a single gas range, you can accomplish the multiple duties of preparing meats, refried beans, and rice.

THE WOK RANGE. Also called a ***Chinese range.*** As you probably know, a *wok* is a bowl-shaped cooking pot used to cook foods quickly in Asian cuisine. This range features recessed, circular burners with rings that can be adjusted to accommodate large woks for stir-frying under very high heat conditions, up to 106,000 Btus per hour. "Step range" models set the back burners slightly higher than the front ones, allowing cooks to easily control six woks at a time. There may be removable trays or shelves for spices or even a built-in water faucet.

Any type of food that's cut into small pieces can be cooked in a wok with a minimum of fat, which makes it practical for glazing and sautéing, not just Asian cooking. In commercial kitchens, woks usually have two handles, but you can find them with a single handle, more like a big frying pan, that take up less cooktop space. Either type of wok fits on this unique burner.

THE TABLETOP RANGE. A partial range top that consists of two burners. It is used where space is at a premium. The burners can be situated side by side, for a depth of 16 inches, or front and back, for a depth of 28 inches. The 12-inch burners have a normal heating capacity of 20,000 Btus per hour.

Electric Range Tops

Thus far, we've discussed gas-powered range tops, because gas is by far the most widely used heat source in the commercial kitchen. However, there are sites where the only available source of energy is electricity, and there are plenty of ranges to fit this situation. Instead of Btus per hour, electric burner heat input is described in kilowatts (kW).

The voltage needed for an electric range is 208 to 240 volts, either one phase or three phase. This is an important detail because compatibility of voltage and phase must be determined when the appliance is ordered. Otherwise, you may find yourself making expensive modifications—or returning your shiny new range to the manufacturer.

Hot tops and griddles are also available for use on electric ranges. Like their gas counterparts, the range-top components are modular and easy to mix and match.

The proper installation of an electric range will determine its efficiency. It must be carefully unpacked and assembled by an experienced professional. First, the nameplate should be checked to be sure that the voltage, ampere rating, and number of phases are all correct and that they match the available power supply. The electrician should install a ground wire and connect the equipment for safe use. The appliance should be plugged in to an outlet that is convenient for unplugging so it can be disconnected to clean behind and beneath it.

Before the unit is placed in service, anyone who will be using the range should read the instruction manual. A quick run-through of its switches, thermostats, and other parts will save a lot of time and trouble in the future.

An oily protective coating is put on some appliances at the factory to prevent them from getting scratched in transit. This film can be softened by rubbing the appliance exterior with a soft cloth saturated with cooking oil. After the coating has been rubbed off in this way, the range can be washed with a clean cloth and mild soapy water and rinsed clean with another damp cloth. Harsh detergents or steel wool will damage the surface. Electric appliances should not be hosed down or steam cleaned, because the excess moisture may damage their electrical components.

After the range body has been cleaned, it's time to turn on the range tops and ovens, with a procedure known as ***burn-in.*** Turning on the burners and/or ovens and letting them heat up before you need them will burn off any protective coating, smoke, or odors and ready them for regular use. Tubular high-speed heating elements don't need this process, but other heating surfaces do. The hot top can be set at its highest heat for 30 minutes or so, the oven at 400 degrees Fahrenheit for 2 to 3 hours.

Trying out the ovens and burners is a good idea anyway, because you can also check the performance of the appliance against its specification sheet. The manufacturer claims the range top will preheat from room temperature to 350 degrees Fahrenheit in 4½ minutes. Does it? Check it! When all burners are turned on to the same temperature at once, do they all heat evenly, or is one sluggish? It may be a loose wire or a defective control knob or heating element. Whatever the case, find out early—before you're in the middle of a rush and discover it the hard way.

Heavy-duty electric ranges are 36 inches square and are found in high-volume foodservice operations. Medium-duty ranges measure 30 inches square and are usually seen in smaller establishments. Each size of range has several combinations of surface units available.

Rectangular Hotplates. These are 12 by 24 inches wide and are capable of temperatures ranging from 250 to 850 degrees Fahrenheit. Their heat controls can be either so-called ***infinite-heat knobs,*** which allow for small adjustments, or standard low-medium-high knobs. Because they're made of solid, 1-inch-thick cast iron, rectangular hotplates respond slowly to temperature changes. They only use 5 kilowatts of electricity per hour, but they take as long as 15 minutes to heat and longer to cool down. If a kitchen requires a lot of griddling, a griddle is still a better way to go than these hotplates.

French Hotplates. These are smaller (10 inches in diameter) and lighter in weight than rectangular hotplates, but the "burners" are solid, not coils. Made of iron, they are typically used for medium-volume cooking and à la carte-style cooking. Their power use is not even a whole kilowatt per hour, so they provide an economical method of cooking small quantities of food. However, they can require up to 15 minutes to heat to a maximum temperature of 500 degrees Fahrenheit. When cool, the round French hotplates can be scoured with a damp cloth and mild abrasive. Illustration 11-6 shows the difference between a standard, coil-type hotplate and a French one.

High-Speed Surface Units. These are a type of French hotplate with a kick, so to speak. They're designed for practically instant heat (within 2 minutes) and quick response to temperature adjustments, and yet their power usage is still a modest 2100 to 2600 watts per hour. High-speed surface units are tubular in shape, 8 inches in diameter, and supported by a rugged 10-inch ring. The tradeoff when either type of

(a)

(b)

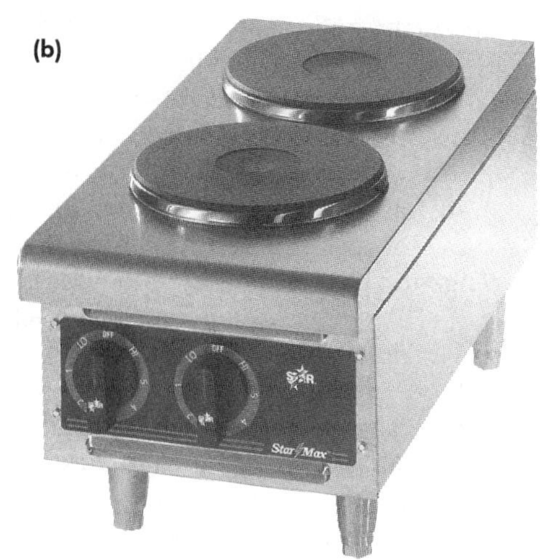

ILLUSTRATION 11-6 These two drop-in, two-burner hot plates show the difference between a coil-type (a) and a French hot plate (b).

STARMAX™ ELECTRIC HOT PLATES COURTESY OF STAR MANUFACTURING INTERNATIONAL, INC., ST. LOUIS, MISSOURI.

French plate is used is that manufacturers suggest pots and pans be chosen correctly to fit the smaller diameter and output of the burner. As an example, one manufacturer specifies that the bottom surface of the pot or pan should not exceed 10 inches and its overall capacity should not be larger than 16 quarts.

Electric Induction Range Tops. The induction cooktop has become a symbol, of sorts, of the modern kitchen. Trends in meal consumption away from home continue to move toward fast, light, tasty cooking that uses a minimal amount of fats and oils. As you read earlier in this chapter, induction cooking does not require an external heat source. The smooth, solid surface of the induction range top is usually made of tempered ceramic or glass, with circular markings to indicate where to place the pans (see Illustration 11-7). There are even concave induction burners for woks nowadays. The evenly distributed heat makes induction cooking ideal for even the most fragile ingredients.

This type of range is also highly efficient, using less energy than either its gas or traditional electric counterparts. (Its burner units are also called ***hobs.***) Some hobs automatically revert to a power-saving "standby" mode when not in use. However, they only work when the right kinds of cookware are used. In a word, it must be magnetic. Most steel and cast-iron pans are fine, but those made of aluminum, copper, and some types of stainless steel are not magnetic and therefore will not work on an induction range top. Pans that have a nonstick surface are able to be heated, but the nonstick coating usually cannot withstand the intense heat of induction and will become unusable in a matter of days. The best pan for induction cooking is completely flat on the bottom, maximizing its surface contact with the range top. An empty pan left on an induction surface too long or too often will transfer heat back into the hob, which can damage its interior circuits.

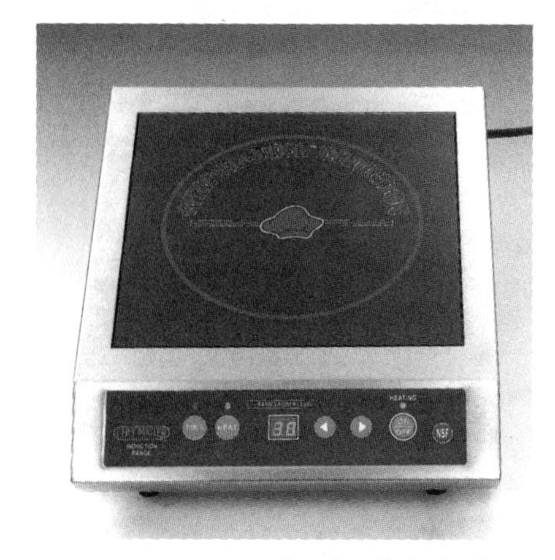

ILLUSTRATION 11-7 An electric induction burner.

COURTESY OF CHEF*MASTER, OLD BETHPAGE, NY.

Since it is not porous, an induction appliance is ready to cook very different foods after a quick clean-up without the risk of taste transfer. It works quickly, does not heat up the kitchen, and requires less ventilation (exhaust and supply air) than more traditional range tops. Notice we said *less* ventilation . . . not *no* ventilation! The earliest induction cooktops were not required to be placed under an exhaust system, but this is no longer the case in some jurisdictions. Ventilation is most definitely needed, to discharge the intense internal heat these units generate and prevent the circuits in the unit from overloading. Induction ranges should not be used near other heat-generating equipment—fryers, griddles, broilers—where circulating hot air might be picked up by the range's intake fan and the range will not be able to cool itself down.

The heaviest-duty induction range top requires a 208–240 volt electrical outlet. It generates from 3500 to 5000 watts of power, and has a hefty price tag ($3,500 or so). There are also portable, one- or two-burner hobs for off-site catering. They can be plugged into standard 120-volt electrical outlets, use from 1200 to 1500 watts of power per hour, and can be heated from 160 to 440 degrees Fahrenheit.

Factors to consider when purchasing induction cooking equipment include the power requirements, approval of reliable safety and sanitation authorities (like UL and NSFI), overall costs, availability of qualified repair service, and whether the cookware you plan to use is suitable.

11-3 THE RANGE OVEN

The range oven, located beneath the range top, is the principal method of large-volume, dry-heat cooking in most restaurant and commercial settings. Like range tops, range ovens can be powered by gas or electricity. Just remember, even if the appliance cooks with gas, the oven still has electrical needs: for the timer, lights, and, in the case of convection ovens, to run the fans inside the oven.

These ovens are as hard-working as any commercial kitchen appliance, and they're usually on all day long.

The basic range has a single oven below the range top. It may be used for roasting, baking, braising, smoking; for finishing sautéed and grilled items; or simply for storing hot food until needed. Even its swing-down doors are useful, as a holding platform to support heavy pans. Most commercial gas range ovens have a 40,000-Btu capacity, with heavy-gauge, double-wall construction (known as a ***flame spreader***), which distributes heat evenly throughout the cavity. Electric range ovens require from 1250 to 5000 watts of power per hour.

The oven cavity is large enough to hold a standard sheet pan, which is 18 by 26 inches. Like home kitchen ovens, they come with at least one chrome-plated rack on which to place pans, and the position of the rack can be adjusted up or down inside the oven cavity.

In commercial use, there are quite a few smart options to consider when shopping for a range oven. Many of them evolved to help busy chefs use both oven and range top in tandem with minimal hassle. They include:

- Oven controls mounted on the side, instead of at the front of the appliance. The knobs are temperature sensitive and can be damaged by repeated blasts of heat. Also, side-mounted controls minimize the chance that a cook standing at the range top will lean on the oven control and accidentally "adjust" it.
- Infinite-heat controls that allow the most precise adjustments are preferable to knobs with only a few heat settings (the old low-medium-high controls).
- The oven door should be hinged on the bottom and should open flush with the deck, so that food can be slid in and out easily on sheet pans.
- The handle should be smooth and well insulated, so that it will stay cool and can be opened safely without using a potholder.
- Fully open, the door should be able to support full pans of food, as much as 200 to 250 pounds. The door should also be counterbalanced, to allow it to stay partially or totally open without being held in place manually.
- Heat-treated glass oven doors are preferable, because you can see the food inside without having to open the door. However, in our experience, most operators fail to keep the glass clean. Instead of being useful, the result is an eyesore.
- The ***deck*** (the "floor" or bottom of the oven inside the oven cavity) should be made of at least 14-gauge steel, with raised sides and back to help catch spills. Try to choose a deck that is removable for cleaning.
- Ask about insulation, which is crucial to the oven's ability to hold heat. A minimum of 2 inches of rock wool insulation is recommended.
- Easy-to-clean surfaces and a self-cleaning cycle for the oven cavity are recommended.
- Ovens must be leveled for some products to be baked correctly; cakes and cheesecakes are among the most sensitive items. Leveling requires not only that the floor be perfectly flat, but that the oven legs themselves be height adjustable. We've already mentioned the wisdom of mounting the oven on adjustable casters, to permit rolling when necessary.

11-4 CONVECTION OVENS

The first convection ovens were installed in restaurant kitchens in the 1960s, and today's units boast up to 40 percent more energy efficiency than their predecessors. Convection, as you'll recall, uses fans to circulate heated air around the oven cavity, reducing cooking times by 25 to 35 percent. Because the heat transfer is so much more efficient, foods can be baked or roasted at lower temperatures, which minimizes shrinkage and maximizes yield per pound. Standard recipes may have to be altered for best results in a convection oven.

Manufacturers have engineered the airflow in these ovens for better uniformity, giving operators an even quicker finished product with better results. So, in recent years, convection ovens have all but replaced conventional ovens everywhere—except under the range top.

Convection ovens come in three basic sizes:

- Full size, which accommodates standard 18-by-26-inch sheet pans.
- Bakery depth, which accommodates standard sheet pans placed either lengthwise or widthwise in the oven.
- Half size, which holds the smaller, 18-by-13-inch half-sheet pan.

Because of their very precise airflow patterns, convection ovens don't do as well with a variety of pan sizes. Therefore, it makes sense to determine the pan sizes you plan to use before you buy the oven. The key to successful convection oven use is proper air circulation around the food. It is therefore critical that the oven not be overloaded or improperly loaded. The food is placed on pans, which are loaded onto shelves (racks) inside the oven cavity. The number of racks is determined by the height of the food being cooked. Like the cook-and-hold oven, many convection ovens automatically hold food hot after it's been cooked.

The type and sophistication of controls are other critical decisions. Top-of-the-line models can run preset programs with variables of time, temperature, and internal fan speeds; but some operators feel that when the controls are too complex, it limits who is able to use the oven correctly. Think about the skill levels of the employees who will be doing the cooking rather than selecting the most highly technical option.

One option that can be useful is a two-speed fan. Convection ovens with slow-roasting capabilities use a lower fan speed for low-temperature cooking. A lower speed or pulse option is also handy for delicate products, such as muffins and cakes, when a higher-velocity fan might botch the results.

TABLE 11–3

HOW EVEN IS EVEN HEAT?

Cooking Uniformity Test:	A1 Half (Electric)	A2 Full (Electric)	B1 Half (Gas)	B2 Full (Gas)
Rack #1 (top) temp. (°F)	150	158	142	130
Rack #2 temp. (°F)	135	129	131	124
Rack #3 temp. (°F)	137	123	124	116
Rack #4 temp. (°F)	136	131	126	116
Rack #5 temp. (°F)	160	133	160	159
Max. temp. difference (°F)	25	35	36	43

Ice water melt test—Temps recorded from pans of water when the pan reaches 160°.
***Source: Food Service Equipment and Supplies* magazine, July 2001.**

Since convection ovens' biggest selling point is their heating uniformity, *Foodservice Equipment and Supplies* magazine put some of them to the test in July 2001. Table 11-3 shows the interesting results. Convection cooking, as explained earlier in this chapter, uses fans to circulate heated air around the oven cavity, reducing cooking times by 25 to 35 percent. Because the heat transfer is so much more efficient, foods can be baked or roasted at lower temperatures, which minimizes shrinkage and maximizes yield per pound. Standard recipes may have to be altered for best results in a convection oven.

The typical oven is about 6 feet tall, 3 feet wide, and 3 to 4 feet deep (see Illustration 11-8). There are also half-size ovens, some made to fit on countertops. Special models can be ordered for baking, wide enough to accommodate baking sheets by length or width. The airflow design is critical in these ovens to achieve balanced heating and browning.

ILLUSTRATION 11-8 Fans inside the convection oven blow hot air through the oven cavity, reducing cooking times.
COURTESY OF LANG MANUFACTURING COMPANY, EVERETT, WASHINGTON.

Solid-state controls can maximize the convection oven's cooking possibilities. They can be programmed for different temperature settings for the most common items cooked in them. Electronic sensors inside the oven will slightly lengthen cooking time to compensate for temperature drops caused by the door being open too long; other models feature electric meat probes that allow three different products to be cooked at different times and temperatures.

When ordering a convection oven, pay special attention to its doors. Full-sized ovens have double doors, which open simultaneously when one or the other is pulled open. They can open from each side or from top and bottom. A single, counterbalanced door—more like a traditional range oven—is also available, hinged at the bottom or on either side. Another option is a single or double pane of glass on the door.

Convection ovens can be powered by gas or electricity. There is a general feeling among chefs who do a lot of baking that the electric ovens provide a moister product. Gas models are required to have a venting system; check your local ordinance about electric models.

Their popularity and versatility have prompted many manufacturers to create interesting new hybrid forms of convection ovens. There are now double-oven models, with two separate cavities and control panels. They don't necessarily take up more space or use more energy, either. Bakeries may choose the combination proofer and convection oven, with separate cavities for letting bread rise and then baking it. The proofer has special humidity controls; the oven has a built-in steam generator. There are also dual-compartment steamers and convection ovens, which work independently of each other and allow the same floor space to do double-duty.

11-5 OTHER OVEN TYPES

Deck or Stack Ovens

An oven manufactured with more than one cavity and set of controls is called a ***deck oven*** or *stack oven,* because the heated, insulated boxes are "stacked" on top of each other in double-deck or even triple-deck configurations. Deck ovens are needed when production is high and space is limited. Depending on the needs of the kitchen, different types of ovens can be stacked in any configuration—one regular oven and one convection oven, for instance (see Illustration 11-9).

The term "deck oven" comes from the way the oven is used: Food is set directly onto the deck, or bottom of the oven cavity, to cook (although some deck ovens also have interior racks and/or multiple decks to pack more product into the compact space). The deck itself is made of either stainless steel or ceramic; bakers prefer ceramic decks or "stone hearth" decks for more even distribution of heat. At least one manufacturer offers an "Air Deck Oven," using impinged hot air to eliminate hot and cold spots in the traditional deck.

ILLUSTRATION 11–9 **Deck ovens are nicknamed "stack ovens" because multiple compartments can be stacked atop one another.**
COURTESY OF VULCAN-HART, DIVISION OF ITW FOOD EQUIPMENT GROUP, LLC, LOUISVILLE, KENTUCKY.

Contemporary gas deck ovens are generally classified in four broad categories:

- The traditional style deck oven: each individual oven is either 8 inches high (for baking) or 12 inches high (for roasting) and, as we mentioned, can be stacked. The smallest ones hold two half-sheet pans (each 13 by 18 inches); the largest hold eight full-size sheet pans (each 18 by 26 inches).
- The motorized, convective deck oven: a single baking cavity equipped with three separate, horizontal baking hearths, made of perforated, nickel-plated steel. This oven has a reversing fan system that circulates air evenly and enhances its heat transfer capabilities.
- The vaulted deck oven: a single baking cavity with a larger, arched opening that provides easy access. Some have a secondary burner located under the oven cavity to increase baking speed.
- The turntable deck oven: the largest of the "family" stands more than 6 feet tall, with 3 or 4 horizontal, rotating, circular baking decks with diameters of 48 to 56 inches, perhaps made of ceramic ("stone hearth"). Multiple access doors maximize its efficiency.

The same basic guidelines for purchasing a single range oven also apply to deck ovens: doors that open flush with the deck, insulated handles, and so on. Insulation requirements are greater for multiple ovens: 4 inches of rock wool or fiberglass are recommended.

Another important recommendation is to order individual control panels for each deck, enabling them to be used simultaneously for a variety of duties. Temperatures for each oven range from 175 to 550 degrees Fahrenheit. Deck ovens may be ordered with or without steam capability.

Control compartments for gas-fired ovens are located below each deck; electric ovens may have controls either directly below or at the side of each deck. Electric ovens also have two sets of heating elements, just like a home oven—one on the top (for broiling) and one on the bottom (for baking). There are separate, three-position heat switches for each element, also located in the control compartment. An observation: Side controls are easier to access than bottom controls, especially when the oven doors are opened frequently.

Both gas and electric deck ovens have flue vents at the back of the appliance, which may be controlled by a hand-operated lever, also found in the control compartment.

Each deck holds two 18-by-26-inch sheet pans. Electric and gas requirements for each model are listed in manufacturers' catalogs and, as you've already learned, should be checked and rechecked to ensure compatibility with the kitchen before purchase.

At least one manufacturer has introduced a blower-and-duct system within each deck, so that a regular oven has the ability to function as a convection oven with the flick of a switch.

Impinger/Conveyor Ovens

One meaning of the word *impingement* is "to strike sharply," and that is the basic premise of the air impingement process in cooking: to blast high-velocity, heated air into the oven cavity. The air is aimed to be concentrated on food that travels horizontally on a moving conveyor belt made of stainless steel or wire mesh (see Illustration 11-10). The belt can be adjusted to move at different speeds for different lengths of cooking time. The air moves with enough force to displace the natural layer of colder air that directly surrounds the piece of food being cooked, resulting in a shorter cooking time than traditional ovens.

Impinger/conveyor ovens are standard equipment in many busy food-service facilities. Because they're automatic, quality output is consistent with minimal staff supervision, and very little training is required to operate the ovens. The hot-air jets are located above and below the conveyor. Some ovens even have separate controls for different zones within the oven, allowing the air temperature and pressure to be set independently. This so-called ***zoned cooking*** is another feature of the most technically advanced ovens.

The air is forced through ***finger panels,*** which look almost like screens, laced with small holes. Different types of panels are used to cook different types of food, but the basic premise is that items that cook quickly (a pizza, for instance) need panels with fewer holes to restrict airflow, while thicker items requiring longer cooking time (such as lasagna) need more holes in the panels to allow more hot air to hit them. Depending on the type of food, it can be placed in pans on the conveyor, from thin aluminum for fish or frozen French fries, to thicker stainless steel for pork chops. Porcelain cookware is also acceptable. If you'll be baking protein products—burgers, sausage, chicken, and so on—look for ovens with built-in grease and smoke controls.

There are four different heat sources for impinger/conveyor ovens: Infrared and quartz models are electric; while natural-convection and forced-convection models may be gas powered or electric. No matter what the heat source, most models need a small, electric motor to move the conveyor belt. The gas-fired ovens operate from 39,000 to 180,000 Btus; most are in the 70,000- to 75,000-Btu range. The electric models require 10 to 27 kilowatts per hour and three-phase power. All of them require exhaust canopies. Temperatures in impinger/conveyor ovens range from 300 to 600 degrees Fahrenheit, and the conveyor speeds are adjustable. The conveyor belt is usually about 7 feet long, although only about 3 feet of it is located within the baking chamber. There are no doors on either side of the oven, although there is usually a small viewing door near the middle so you can check the progress of the cooking line or place items halfway down the belt if they don't require the full conveyor length to cook. There are also much smaller countertop models with heated tunnels of only 3 or 4 feet (see Illustration 11-11).

The most enthusiastic users of this oven are carry-out pizza restaurants, because of the speed and ease with which pizza can be cooked, but their manufacturers say conveyor ovens are versatile enough to be used a hundred different ways. Impinger/conveyor ovens can be double stacked and/or placed on

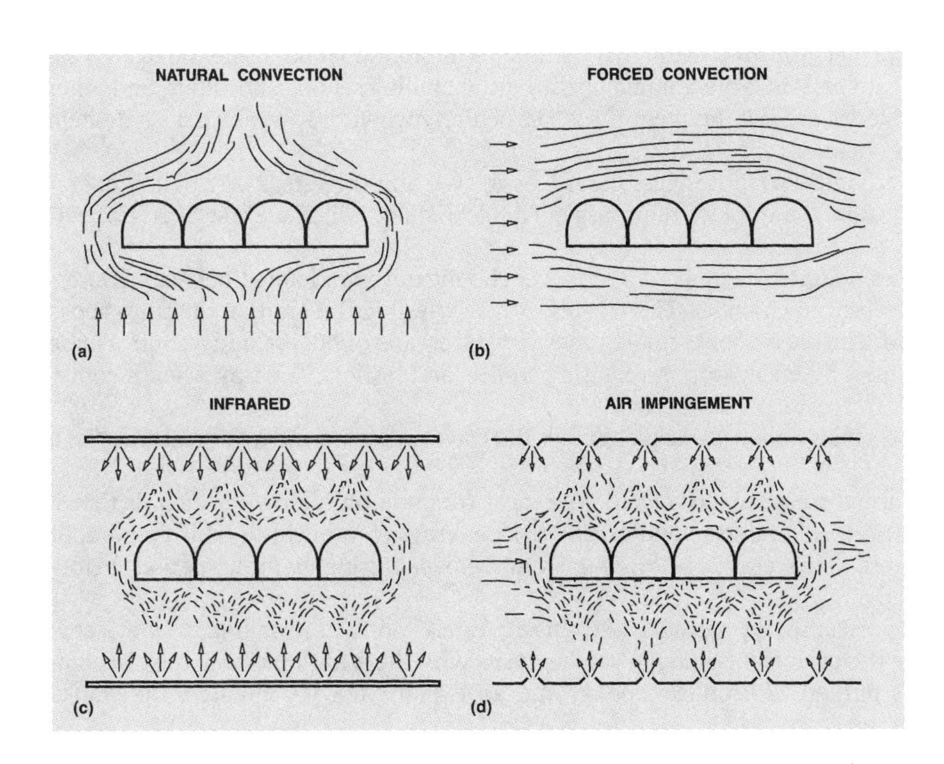

ILLUSTRATION 11-10 These diagrams show how different types of heat penetrate food differently. (a) Natural convection is the least efficient method of penetrating the coldest air layer, directly around the food. (b) Forced convection is a little more efficient than natural convection, because heated air is blown horizontally across the food, displacing the cold air around it. (c) Infrared, radiant heat cooks the surface of the food, but does not impact the cold layer of air around it, which can affect cooking time. (d) Air impingement aims air at the food from different directions, forcing aside the cold boundary layer and prompting fast, uniform surface cooking.

SOURCE: BASED ON ILLUSTRATIONS FROM LINCOLN FOODSERVICE PRODUCTS, INC., A WELBILT COMPANY, FORT WAYNE, INDIANA.

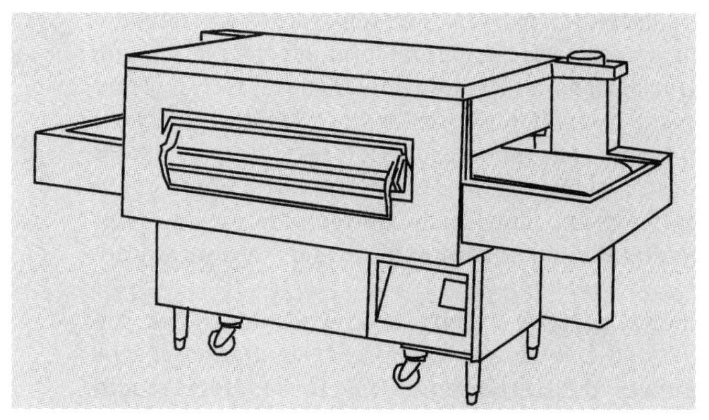

ILLUSTRATION 11–11 Using a conveyor oven, you can control cooking speed two ways: by the temperature inside the oven and by the speed of the conveyor belt.

stands with casters. Options include twin belts, which run side by side at different speeds to cook different foods simultaneously.

Pizza Ovens

The pizza oven is a special high-temperature application of the deck or stack oven, which can be set from 300 to 700 degrees Fahrenheit. The deck (also called a *hearth* in this case) is made of two pieces of 1-inch-thick ceramic or a single piece of steel. And, since pizzas are so flat, the height of the oven cavity can be as low as 7 or 8 inches. There are multi-pizza tabletop models for places where space is tight.

The power requirements for pizza ovens are 82,000 Btus per hour per deck for gas-fired models and 7.2 kilowatts for electric ovens. One manufacturer has produced a model with a removable center divider; each half of the oven has its own burner and thermostat. Other manufacturers have introduced ovens with an ***air door*** or *air deck*. These pizza ovens have no solid door; instead, a blower fan circulates a curtain of air across the opening to prevent heat from escaping. To further increase energy efficiency, at least one manufacturer has doubled the thickness of insulation in its pizza oven walls, claiming a 49 percent energy savings over conventional ovens. Another twist is a gas-powered pizza oven with a visible flame inside that simulates the look of "wood-burning," for display purposes. Its exterior can be ordered with an attractive brick or tile face. All of the atmosphere, and none of the smoke!

As you can see, there is no shortage of creativity when it comes to getting something cooked. Attend any restaurant equipment show to catch up on the latest trends.

Mechanical Ovens

The main characteristic of the ***mechanical oven*** is that the food is in motion while inside the oven compartment. Mechanical ovens are used in large-volume operations, including schools, hospitals, and other group-feeding situations. Bakeries are also frequent users, particularly when they produce a single item in volume, such as loaves of bread or rolls.

There are two basic mechanical oven designs: the *revolving oven* and the *rotary oven*. Their operation is easier to understand when you think of their respective nicknames, which are the "Ferris wheel" and the "merry-go-round"!

The revolving oven does operate much like a Ferris wheel. Flat trays are loaded between two rotating wheels inside the heated oven chamber. The wheels turn slowly, and the food on the trays cooks as it rides around and around. The oven door is small, to restrict the escape of hot air, and a control knob is located near the door that lets a person stop the rotating wheels and retrieve any tray when it comes parallel to the door.

The rotary oven operates on a similar principle, except the trays rotate around a vertical axis like a merry-go-round.

Mechanical ovens are usually assembled on site. The most common problem with them is that it can be difficult to level the trays, either end to end or side to side. Trays with burnt or built-up food, or trays that have warped because they were not allowed to sit in the oven during warm-up periods before baking, are also hard to level.

Because they require so much space, mechanical ovens are rarely found in restaurants. One exception is the Southwest, where the oven is modified to smoke meats with the addition of a firebox (a compartment in which wood is burned to produce smoke) and an exhaust fan (to circulate the smoke through the oven).

ILLUSTRATION 11-12 A rack oven is tall and just wide enough for rolling racks of product inside it.

COURTESY OF LANG MANUFACTURING COMPANY, EVERETT, WASHINGTON.

Rack Ovens

The ***rack oven*** has gained popularity in restaurants because of its efficient use of space. When floor space is at a minimum, you can purchase a tall, thin oven chamber into which racks of product can be rolled (primarily for baking or roasting). The smallest rack oven is 4 feet square and 6 feet tall. It may take up a bit more space than a full-sized convection oven, but it can accommodate as many as 20 sheet pans simultaneously; double rack ovens hold 40 sheet pans. Some racks lock into their oven slots and then rotate during cooking for more even heat distribution.

Add the options of a self-contained steam system and built-in hood, and you have a baking and roasting workhorse! Of course, you'll also need the proper-sized bakers' racks for this oven (see Illustration 11-12). The racks are loaded with food, then wheeled into the baking chamber on heat-resistant casters. Purchase racks that allow you to space the individual pans from 2 to 6 inches apart.

The best rack ovens are super-insulated so they require minimal clearance around their sides and back, and they're flush with the floor so the racks can be wheeled in and out easily.

Remember, if your rack oven makes its own steam for baking, you must be able to provide water connections and drainage.

Cook-and-Hold Ovens

The ***cook-and-hold oven,*** or *retherm oven* as it is sometimes nicknamed, is an excellent choice for preparation of convenience foods (see Illustration 11-13). Schools and hospitals use it to reheat large

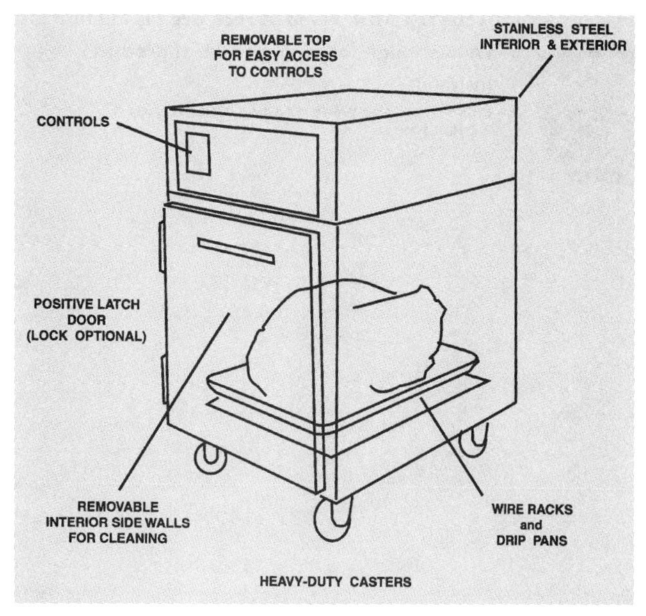

ILLUSTRATION 11-13 Cook-and-hold ovens allow you to prepare foods in advance and hold them up to 24 hours.

quantities of frozen, prepared dishes—ravioli, macaroni and cheese, dishes made with sauces or gravies—and keep them hot until they are served. The cook-and-hold advantage is its low, steady temperature, which heats a product without drying it out or burning it.

Where large quantities of roasted meats are consumed, cook-and-hold capability is also a must. Radiant and convection heat combine to cook foods slowly while keeping shrinkage to a minimum. When the cooking cycle is finished, the oven's preset timer switches it to a holding mode that keeps the food warm. Holding is actually recommended—for beef roasts, for instance—because enzymes within the meat provide their own natural tenderizing process as the finished roast sits for at least 90 minutes after cooking.

There are three basic types of cook-and-hold food production:

1. One type of oven uses natural convection (no air movement), cooks at a slightly higher temperature, and maintains a humidity level of 90 to 95 percent.
2. Another roasts at slightly lower temperatures with a slow-moving air current and humidity levels of 30 to 60 percent.
3. A third option is the "smoke-and-hold" oven, which can cold-smoke (that is, smoke at low temperatures), then slowly roast anything from salmon to tomatoes to hazelnuts. A regular cook-and-hold can be converted to a smoke-and-hold with a firebox attachment (that allows the oven to burn wood and charcoal), and an ash collection tray. Smoke-and-hold ovens have slightly higher electric power requirements. They also need exhaust hoods in this case, which are not required otherwise.

In addition to being well suited to high-volume cooking, they are energy efficient because, although cooking times are longer, temperatures are lower and require less total energy than conventional ovens. Some foodservice businesses do their slow-cooking overnight, since they are safe and low-temperature enough to be left unattended.

The temperature range for a cook-and-hold oven is from 140 to 245 degrees Fahrenheit; 140 degrees Fahrenheit is the lowest safe holding temperature that prevents harmful bacteria from multiplying. Some ovens have timers (up to 100 hours of roasting time can be programmed) or they use electric meat probes to determine when products are done. Either way, the oven automatically returns to its holding temperature when prompted. If a programmed oven experiences a large temperature change, it will trigger a warning beeper and/or caution light. Tabletop ovens can be ordered with a carving station on top, for use in buffet lines.

Because they cook so slowly, meats in a cook-and-hold oven should be placed on wire racks so they'll brown uniformly; a sheet pan can be placed on the oven deck to catch any drippings.

The power requirements of cook-and-hold ovens range from 120-volt single phase to 380-volt three phase. No special ventilation is required, so the ovens can be rolled on 5-inch casters. The capacity for a single cook-and-hold oven is about 90 pounds of product; stacked double ovens can hold up to 180 pounds.

Smoker-Cookers

Commercial meat smokers are often called ***smoker-cookers.*** The upsurge in popularity of barbecued meats, particularly pork ribs, has created a cottage industry of sorts as barbecue restaurants and civic clubs sponsor cooking teams that enter outdoor cook-off competitions with their "secret recipes" and traveling meat smokers in tow on wheeled trailers.

The smoker-cooker is a type of oven, made to create smoke. There are two types of smokers:

1. **One popular style of smoker looks like a cylindrical metal container, with grills suspended inside on which food is placed. In fact, this smoker-cooker doubles as a grill if it's kept open while the cooking is done. Attached to one end of the cylinder is a firebox, in which charcoal or wood chips are placed to make a fire. When wood is used, it is often dampened with water or beer to make it smoke more as it burns. The use of liquid also helps the meat stay moist as it cooks. Portable models are mounted on trailers, or at least equipped with casters or wheels (see Illustration 11-14).**
2. **The cabinet-style *oven-smoker* uses gas or electricity to start the fire instead of matches. Wood or charcoal is added to the firebox. A draft door on smaller, home-use models—or damper on the smokestack of larger units—controls the even draw of smoke and heat through the cavity. The food inside cooks slowly, taking on a subtle, smoky flavor.**

A smoker-cooker can be as simple as a stove-top unit that smokes a few portions of fish at a time, or as complex as a conveyor-driven machine that can be loaded with hundreds of pounds of pork ribs. Temperature control is a major consideration when selecting a smoker. Look for a unit that can produce the correct heat and smoke levels for the foods you intend to smoke; seafood, poultry, and beef each have very different requirements. Foods are either cold-smoked (at temperatures below 100 degrees Fahrenheit) or hot-smoked (at temperatures from 165 to 185 degrees Fahrenheit). Chefs will tell you a side of salmon, intended to be smoked at a low temperature, can be ruined within minutes if the smoker temperature is too high.

Barbecue aficionados also seem to have strong feelings about which types of wood are best, but the choices include oak, hickory, and mesquite. Some people also add nutshells, stalks of herbs, or grapevines to the fire to produce flavor variations.

Another factor is the heat source. For long-term smoking (more than just a few hours), an electric heating element is best for consistent temperature control, and far less labor-intensive than a fire that must be constantly tended. Shelves and/or hanging hooks for meats should be adjustable in larger models. Ventilation is an issue in urban settings. To address environmental concerns, an oven-smoker may be equipped with a precipitator unit to filter and cleanse discharged smoke before it can be exhausted to the outdoors. Cleanup is also a challenging chore unless you have a unit that can be hosed out with a high-pressure spray, a drip pan for easier grease cleanup, and removable grills or racks. Thermometers should be well sealed so smoke doesn't condense inside and make them hard to read. Temperatures should be checked near the top of the smoker, where they are generally the highest.

The dimensions of full-sized smoker-cookers vary from 9 to 12 feet in length and from 5 to 6 feet in height. More important is the grill space, which varies on commercial models from 18 to 30 square feet. For smoking, about 10 pounds of charcoal and up to four pieces of wood are needed; for use as a grill, as much as 40 pounds of charcoal may be needed for a full-sized smoker-cooker.

An alternative to the standard smoker-cooker is the ***pit smoker,*** a combination meat smoker and pressure cooker. Food is placed into a cooking chamber and the doors are shut. As the temperature inside rises, pressure (15 psi) builds up along with the smoke, forcing a faster cooking process as the food is smoked with a handful of wood chips. None of the smoke escapes from the cabinet, so no exhaust hood is required. Perhaps purists don't consider this "real" smoked barbecue—but it turns out 45 pounds of ribs in 90 minutes, or 40 pounds of beef brisket in two and a half hours.

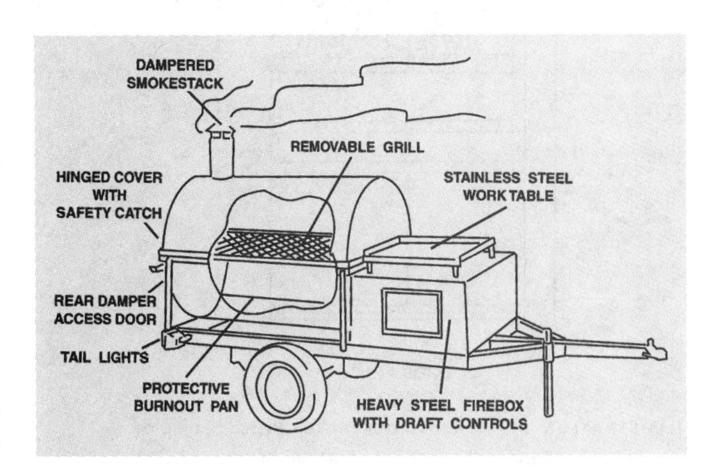

ILLUSTRATION 11-14 The smoker-cooker uses wet wood or charcoal to create a moist, smoky, slow-cooking environment.

Rotisserie Ovens

Food trends come and go, but the universal appeal of rotisserie cooking endures. ***Rotisserie*** comes from an old French word meaning "to roast." The sensory appeal of skewered meat turning over and over as it cooks over dancing flames evokes mouthwatering sights and smells in most of us.

The modern-day rotisserie oven contains rows of metal spikes, called spits, on which meat is placed. One (or more) small electric motors rotate the spits as warm, moist air circulates through the oven cavity, slowly roasting the meat while the moisture minimizes shrinkage (see Illustration 11-15). Other rotisserie models are equipped with either hanging baskets (for placing food that won't stand up to skewering) that rotate carousel-style; or a vertical, moving ladder with product on each "rung." The vertical rotisserie is often called a ***continuous cooker.*** An advantage of a vertical machine is that it allows the operator to add more product during the day, always at the "bottom" position on the ladder. Programmable models allow presetting of cook times and temperatures for different types of food.

Rotisserie ovens are usually used for baking whole chickens, and capacities range from six whole (3-pound) chickens to 70 of them. The small, countertop models make an attractive display, with their glass doors and halogen lights. Occasionally, rotisseries are used for barbecued items such as sausage or ribs. Accessories include heavy-duty, angled spits that are large enough to hold a slab of prime rib or a whole turkey, and baskets for fish or vegetables. Multiple products can be cooked simultaneously in a rotisserie oven, as long as care is taken not to cross-contaminate. Raw chicken juices shouldn't drip onto vegetables, and so forth. Better yet, there are models with separate cooking chambers, side by side or stackable units.

Cooking power can be provided by gas or electricity. In gas models, an electric power source is also needed to run the motor that turns the spits. A single motor can turn all the spits or, in larger units, multiple small motors turn different sets of spits. The heat source is located at the top and/or bottom of the oven cavity; some have a second heating element in the center of the rotating shaft. A few rotisseries burn wood or charcoal to impart a smoky flavor. (Not all jurisdictions allow wood-burning rotisseries, citing air-quality problems or insect problems associated with wood storage.) As an alternative, gas models may be equipped with ceramic "logs" at the base of the unit to simulate the open fire look.

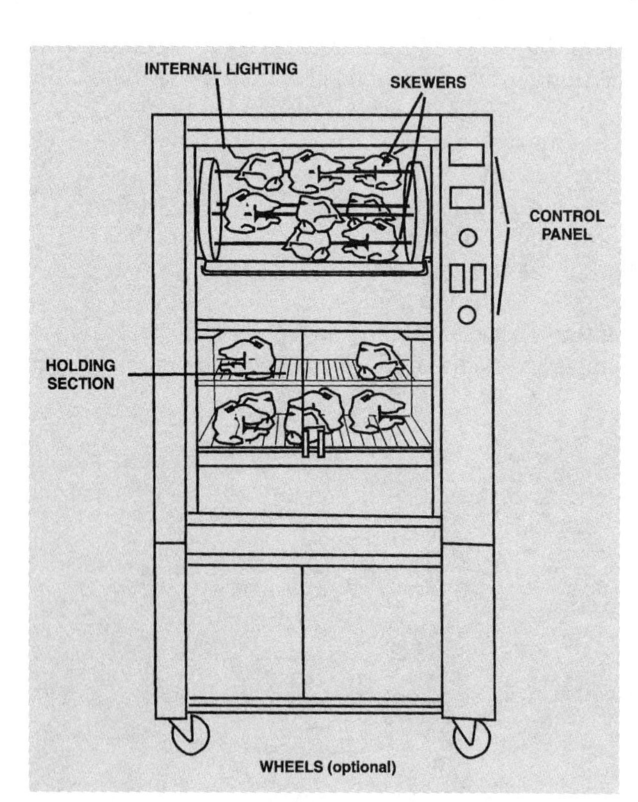

ILLUSTRATION 11-15 Rotisseries make attractive display ovens, as well as cooking slowly and minimizing meat shrinkage.

COURTESY FOODSERVICE EQUIPMENT AND SUPPLIES SPECIALIST.

Some chefs prefer gas-fired infrared burners, which produce fast, high-intensity heat that melts away the layer of fat just beneath the skin of the chicken. Inside the oven, there is a water pan to provide moisture, and a drip pan for grease removal.

Restaurants with outdoor seating or extensive off-premise catering business sometimes buy portable rotisserie ovens. Although the spits are still rotated electrically, the cooking is done with LP gas or a meat smoker–style combination of wood and charcoal. The spits are often adjustable from 13 to 24 inches from the charcoal surface. Before buying, always find out how much weight the spit can support so you'll know not to overload it.

A clean, well-stocked rotisserie oven can be a wonderful sales tool. Customers can see the food as it cooks, and, because they can be made without added fat, rotisserie meats are considered healthful. Take-home food chains like Boston Market have capitalized on this trend. One result of good rotisserie cooking is a moist, tender chicken that has been basted by juices from surrounding chickens as they revolve in the oven.

As attractive as they are, one complaint about the rotisserie is that it generates a lot of heat and requires good ventilation. Care should be taken for at least 36 inches of clear space around it, so employees can work nearby without discomfort. Heatproof gloves should be used to avoid burns while removing and reloading the spits. The visual appeal of this oven also makes regular cleaning absolutely necessary. All removable parts should be cleaned daily. Spattered grease is a major problem, so the drip pans, spits, and drains should be easy to remove and replace without tools. Overcooking and undercooking are concerns, and both are often the result of poor maintenance. The fans that circulate the air, for instance, must be cleaned regularly.

There is one more rotisserie choice to make: whether you need your meat in one batch at a time or in a continuous flow. Batch production is ideal for institutional settings that require a high volume of product all at the same time. Continuous production works for restaurants that need smaller amounts of cooked product at staggered times. Rotisseries can also be ordered with a warming cabinet where finished product can sit while a new batch is being cooked.

Wood-Fired Ovens

A ***wood-fired oven*** contains a well-insulated cavity in which wood is burned. The heat generated by the burning wood is stored and then released slowly and evenly for a flavorful method of roasting and baking. The heat is retained within 4-inch-thick stone or brick blocks, which store enough to cook for long periods of time without having to replenish the wood. The wood-fired oven cooks like a traditional oven, with heat generated at the bottom of the cavity that rises to the top, but it cooks much more quickly. It can retain temperatures from 350 to 620 degrees Fahrenheit, even if the fire is tended only once an hour.

The accepted name for this type of oven is evolving. Most manufacturers call it a "wood-fired" or "wood-burning" oven, but others favor ***brick oven***—from the days when ovens were usually made of brick—or even ***stone hearth oven.*** No matter what you call it, the intense heat of this oven produces benefits found in no other type of baking or roasting equipment. Its fiercely hot, fast-cooking method seals in meat juices, caramelizes sugars, and produces full flavor profiles for customers, for a wide variety of menu items.

Most operators decide to use gas-powered models that offer efficiency, fresh-roasted flavor, and visual appeal. This is partly because in so many urban areas and commercial buildings, it is difficult to get permission to use a purely wood-burning oven without an existing fireplace flue or a special variance. In terms of ambience, it probably doesn't matter to the customer whether the open flames are created by wood or by gas jets. Wood-gas combinations (sometimes called ***gas-assist ovens***), or all-gas ovens, are also easier for the staff to use. Infrared burners at the oven floor maintain a constant temperature, with natural wood or adjustable gas flames boosting it as needed. These ovens vent through a flue collar located above the door, and there are very specific standards for proper venting. A gas-powered oven can share hood space with other equipment under a standard exhaust hood. But if yours is an all-wood, or wood-gas combination, the situation changes rather drastically. The National Fire Protection Association (NFPA) Standard 96 says the by-products of solid and nonsolid fuels cannot be mixed in the same ventilation system, meaning if the oven does burn wood, it must have a separate chimney that vents to the outdoors, and a hood with specs to suit wood fuel.

Most manufacturers custom-make these heavy (5000 pounds or more) ovens. The domed interior walls are made of alumina (unrefined aluminum), high-temperature ceramic, or refractory cement. Typical dimensions of the oven exterior range from 4 to 8 feet, with an interior cavity size of 9 to 31 square feet. It is important to note the ***thermal headspace*** of the oven. The more headspace it has, the more heat stays inside the oven cavity instead of being vented away. The deck of this oven is called its floor, which can be made of individual bricks or tiles, or cast as a solid piece of alumina, ceramic, cordite, clay-and-aluminum, etc. Those who favor bricks or tiles say they are easier to replace when one cracks. One-piece-floor proponents believe there is better heat retention in a single piece without seams or gaps.

Insulation is another point of differentiation. Some manufacturers put it between the oven's steel outer shell and its domed roof, in the form of spun ceramic fiber or a cast mixture of perlite and cement. Others insist it is smarter to wrap the whole oven in a blanket of insulation before putting on its outer cover.

There is no argument that a wood-fired oven is a hefty acquisition. The weight of the oven makes installation a tricky and technical job. Some floors must first be reinforced with structural steel; venting, ductwork, and air pollution prevention measures may be expensive.

An optional *mantle* is a sturdy ledge that extends the work space near the oven door; a wood storage box (located below the door) is also optional but recommended. A metal ***ash dolly*** on wheels is a safe, efficient means of storing and transporting wood ashes, which should be removed daily from the oven when it is cool. All ashes and coals should be doused with water to be sure they are fully extinguished before final disposal. After use, the oven cools—a relative term—to a temperature of about 400 degrees Fahrenheit. It takes at least an hour to bring it back to its peak operating temperature (550 to 625 degrees Fahrenheit).

Firewood is chosen for these ovens based on the smoky flavor it may impart, how easy it is to ignite, and how much moisture it contains. Properly dried firewood should not contain more than 20 percent moisture, or it produces too much smoke and not enough heat. You can buy a meter that will tell you the moisture content of wood, which can be used to check incoming loads before you pay for them! Pressed-wood products are never appropriate for a wood-burning oven. They contain chemicals that may damage its interior. It's a good idea to nurture a relationship with a reliable wood supplier, because wood quality and delivery schedules are so important. You must also have a separate storage area for curing newly delivered wood, so as not to introduce termites and other wood-borne pests into the kitchen.

11-6 MICROWAVE OVENS

Some people in the hospitality industry just don't consider the microwave oven to be a real an oven! This debate seems to have raged ever since this handy little box made its first appearance in foodservice in the 1950s. The fact remains, however, that the ability to microwave food has had a profound effect on most households. Demographers claim the microwave oven has had the most influence on our eating patterns since the development of the freezer. Most American households have at least one microwave in service. At home, they're used mostly for reheating.

In foodservice, today's professionals have been forced to rethink their views of microwave cooking. Once frowned upon as a "necessary evil" for kitchens with very limited space, they have come to be indispensable in the well-equipped kitchen—and not just for "retherm" (reheating). Commercial microwaves have more power and durability, and better warranties, than ever before. They're easy to install and operate. They don't take up much space. They work quickly without affecting taste or nutrient content, and some microwaved foods can be cooked and served in the same dish, minimizing cleanup. Another asset: They help keep the kitchen cool and comfortable.

When you're shopping for microwaves, remember the home-use models are not at all appropriate for commercial use. A home model is not meant for continuous use and might burn out within a few hours under the rigors of restaurant cooking. Commercial doors are also much more sturdy than home ones, made to be opened and closed hundreds of times a day. In fact, most health inspectors and fire insurance companies require commercial-grade microwaves; and the manufacturer will not honor a home microwave's warranty if it is being used in a commercial setting.

Here's how the cooking process works. An electronic magnetron tube located in the back ceiling of the oven converts electricity into microwaves. There are between one and four tubes, depending on the size of the oven. A fan (sometimes called a ***wave guide***) located in the top of the oven pushes the waves from the tube into the oven cavity, where ***wave stirrers*** distribute them evenly to prevent hot spots in the appliance. The waves themselves are not hot, which keeps the inside of the oven cool (see Illustration 11-16).

Many people believe that an item in a microwave heats "from the inside out," which is not exactly true. In fact, the waves heat only the molecules of moisture (water or fat) in the food. Both the microwaves and the molecules have positive and negative electrical poles, so the molecules try to align themselves with the microwaves. This creates intense friction—2.5 million times per second—which makes the heat that cooks the food. Exactly where the waves heat depends on where the moisture is located within the food. For instance, because there is more moisture in the center of a baked potato, the highest temperature is attained there. Anyone who has baked a potato in a microwave may have noticed that if you take the potato out immediately and cut it open, you're more likely to find a rather hard, lumpy consistency. After microwaving, the potato should be wrapped in a clean kitchen towel and allowed to sit for five to ten minutes, to fully distribute all the heat that has built up in the center. The result will be a much better, more evenly cooked potato.

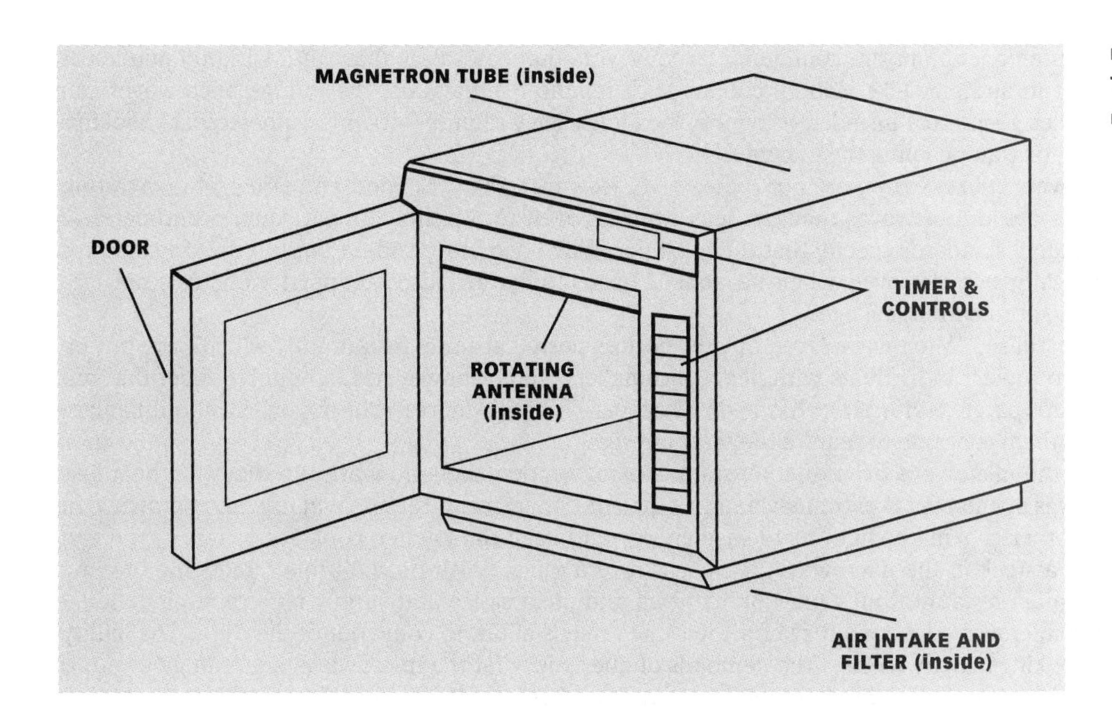

ILLUSTRATION 11-16
The major parts of a microwave oven.

Microwaves can be used to safely defrost frozen foods, too. Vegetables and seafood can be steamed or poached in the microwave with excellent results. High-powered countertop models cost the same or less than most tabletop steamers, and require less power, no water supply or drain lines (and therefore, no deliming process), and no preheating or "standby" mode.

You've surely heard that metal dishes or aluminum foil should never be used in a microwave oven. This is because the metal deflects the waves away from the food, preventing cooking and possibly aiming them back at the magnetron tube, overheating it. (Yes, you've probably noticed the inside walls of the oven are metal, but they deflect the rays back toward the food. A big difference!) Paper, plastic, glass, and ceramics are all good alternative materials for containers. If the containers are large, their contents will probably have to be stirred or rearranged partway through the heating process for best results.

The microwave oven should never be operated empty. Something is needed inside to absorb the energy, even if it's only a cup of water. Most microwaves are equipped with an automatic shutoff if they are left on, to prevent them from shorting out. If your oven is spattered with food and not cleaned regularly, it puts a strain on the heating element, which can cause a breakdown or, at the very least, shorten the life of the appliance.

So much for cooking tips (and with microwaves, there are hundreds of them). In the present generation of commercial-grade microwave ovens, power is more important than size. The more power you have, the faster foods will cook. Power is measured in watts. For low-volume operators such as snack bars and service stations that only use them for warming, a smaller, 700-watt microwave oven is sufficient. Coffee shops, diners, and fast-food eateries will need at least a 1000-watt microwave to heat precooked foods to serving temperature. High-volume restaurants, hospitals, and so on can get 1400- to 2700-watt microwave ovens, capable of defrosting and retherming in bulk quantities. Cavity size and cooking speed are the major differences between models. Large cavities accommodate two dinner plates or 10-by-12-inch pans.

You'll notice the lighted digital display and control panels are very similar to home-use microwave ovens, but commercial models have programmable cooking times and some can do a computerized inventory of different types of dishes that are cooked in them. A buzzer sounds at the end of each cooking cycle. Smaller units operate on 120 volts of electric power, while larger, heavy-duty ovens require 208 to 240 volts.

The cabinet sizes of microwaves range from 13 to 25 inches wide, 16½ to 25 inches deep, and 13 to 19 inches high. It is generally a real space saver for a busy kitchen.

We just said size wasn't important for microwaves, but there are a few gigantic microwaves that blast up to 40,000 watts! A model that's popular on cruise ships has a cooking cavity 8 feet by 4 feet, and is used primarily for speedy defrosting. It can thaw 85 pounds of chicken in 90 seconds. Even larger ones in food-processing operations are configured as conveyors that can safely thaw up to 12,000 pounds of frozen product in an hour. Like walk-in coolers with remote compressors, these huge microwaves can have their power generators installed in remote locations. They require 440–480 volt electrical hookups and plumbing for water-cooling the generator.

Microwaves, unlike X rays, are not radioactive. However, the U.S. Food and Drug Administration has limited the amount of waves that can leak from an oven to 5 milliwatts per square centimeter of oven cavity. Safety standards specify that the oven door have two independent but interlocking systems that automatically stop the oven when its door is open. Doors are also equipped with seals and absorbers to prevent radiation leakage.

You may still see "Microwave Oven in Use" notices posted at some restaurants, which were put up in past years to protect individuals with heart pacemakers. Today, however, the popular belief that microwaves disrupt pacemaker activity has been disproven. It's still smart to check your local ordinances, since many of them continue to require the warning signs.

Because the microwave oven offers instant cooking on demand—no warm-up times, no heat loss when the door is opened—it is extremely energy efficient. Smart restaurateurs will use the microwave in tandem with other cooking appliances to save energy and keep the kitchen cool. For instance, you can partially cook a steak in the microwave, then sear it to a quick finish on the broiler, reducing broiling time. Or an Italian restaurant may precook its pasta and meat sauce and simply retherm an individual serving in the microwave just as a pizza for the same table is about to come out of the oven. The ability to schedule cooking activity to meet the demands of guests is what the microwave oven can give you. A couple of manufacturers offer combination microwave and convection ovens. This single unit can heat and brown pastries, poultry, and other foods that require both heating and careful control of their exterior color. These combination ovens heat and brown quickly, with a 1000-watt microwave and convection cooking of up to 475 degrees Fahrenheit. Half a chicken in this combination oven for just four minutes looks like it has been roasting on a spit for 30 minutes! It's a practical, versatile combination you may want to consider.

As you make your purchase decisions, be honest with yourself about exactly what you will use the microwave for. Melting, cooking, and defrosting are distinctly different tasks, requiring different wattages and speeds. You may need several microwave ovens, each for different tasks.

11-7 OVENS FOR BAKERIES

The quality of bread served by a restaurant is, to some customers, a measure of the overall quality of the establishment. Chains and independent outlets now offer an ever-growing variety of artisan-style breads, rolls, croissants, muffins, and pastries. Adding freshly baked goods to a menu adds profit and value in the eyes of the diner, and the sights and scents of baking are enticing enough to create a market niche. But what types of appliances do you need to achieve this? Whether they are produced on-site or prepared in commissaries and finished on location, production of bakery items relies on durable and (in most cases) high-capacity ovens and proofers, as well as holding equipment that presents the products attractively to customers.

Perhaps the most basic question for a bakery operation is whether to use rack ovens or deck ovens, both of which have already been described in this chapter. The rack oven offers high volume for items such as bagels, producing them quickly—but the deck oven, which "hearth-bakes" breads directly on a ceramic deck, produces a better crust and is more appropriate for artisan breads. Deck ovens are also adaptable enough to use for entrées, casseroles, pizzas, soft pretzels, and cookies. Not everything must be placed directly on the deck; items can be placed in pans for baking too. The disadvantage is that a deck oven usually requires more space than a rack oven to produce the same quantity of product. Deck ovens use more power than rack ovens, and also require more skilled labor to use them correctly. They are trickier to use with frozen or partly baked dough. So, in smaller spaces and/or for use by unskilled employees, a computerized, programmable rack oven is ideal.

Specialty ovens used for point-of-sale bakeries now allow such options as baking flat breads (those that don't have to rise) in two to three minutes, so customers can have fresh bread to order. Wood Stone, a manufacturer of wood-fired ovens, has also introduced a line of gas-powered and gas-assist hearth ovens designed for high-volume bread production.

An alternative to the wood-fired oven for baking is the ***steam tube oven,*** which looks like a wood-fired oven with its high, domed cooking cavity and larger opening. This very heavy oven is made of concrete, with hollow tubes placed above and below the cooking cavity, and a gas-powered "firing chamber" at the bottom of the oven made of brick. The tubes are filled with water, which turns to steam in the intensity of the 580-degree Fahrenheit heat. The steam circulates through the tubes, then flows back to the fire chamber as water, to be reheated and recirculated. The high temperature and moisture combination is perfect for baking.

The oven tubes may also be filled with thermal oil, an excellent heat conductor, but this increases the cost and complexity of the system, requiring pumps, additional controls, and a very knowledgeable staff member to maintain the oven. Using water in the tubes requires no maintenance and no moving parts.

Proofers and Retarders

Proofing is the process of warming dough to allow the yeast to activate and cause the dough to rise before baking. There are specialty appliances that both start and stop the proofing process, called *proofing cabinets* (***proofers***) and ***retarders.*** It's true that yeast products can proof just sitting on a shelf, but if you want them to reach their optimal size and texture consistently, you'll use a proofer. It maintains the correct temperature and humidity for the type of baked goods you are making. Incorrect humidity is a concern because it may cause dough to dry out, crust too soon, split open, or form irregularly.

After the dough has risen, you could probably just put it in the refrigerator to stop the yeast growth. But, again, there's an appliance to do the job more precisely. The retarder offers humidity control. For small spaces, you can order a combination proofer/retarder, which can be programmed to retard at night, then proof in the early morning hours so you can bake fresh bread for the breakfast rush. For larger kitchens, there are proofing cabinets that fit beneath an existing oven, as well as roll-in proofers and even mobile proofers. Many of them do double duty as warmers for food cooked in other appliances.

A combination oven/proofer can turn any foodservice operation into a full-service bakery. Small enough to fit behind a counter, but powerful enough to yield reasonable quantities, they consist of two chambers—the proofer (usually on the bottom) with temperature and humidity controls, and the oven on top, both with clear glass doors for easily checking on both processes. The oven is most often a convection oven, and can be used for almost any cooking application; the proofer does double-duty as a holding cabinet.

11-8 NEW OVEN TECHNOLOGY

You've probably learned by now that the science of cooking is constantly being toyed with in appliance manufacturers' laboratories, and the results have been impressive. The most common new arrivals recently have been combination appliances that can be used for multiple functions; and computerized, highly programmable ones. However, new cooking systems have also been developed that mostly tout incredible speed. They are generally referred to as ***super-cookers,*** multitask ovens that cook a wide variety of foods faster than anything else on the market—or so their manufacturers claim.

A combination of microwave and convection cooking, sometimes called ***combi-wave,*** adds the speed of microwaving to the browning power and efficiency of convection cooking. Traditional microwaves have never been very good at browning, crisping, and broiling, but add the forced air of convection fans at temperatures of up to 475 degrees Fahrenheit and the problem is solved. The downside is that the combi-wave unit is small, so its output is limited, and employees may have to do some experimenting to hit on the right combination of ingredients, time, and temperature for the best results. The units are programmable, though, and easy to use.

Rapid-cook ovens combine radiation with heat impingement, a technology pioneered by TurboChef, Inc. Instead of the heated air hitting the food and bouncing off, as in typical impingement cooking, the rapid-cook unit contains a fan that sucks air all the way around the food and out the bottom of the oven cavity (see Illustration 11-17) at a rate of up to 60 miles an hour! This means the heat transfer between the air and food is constant, resulting in a very uniform method of browning and crisping. At the same

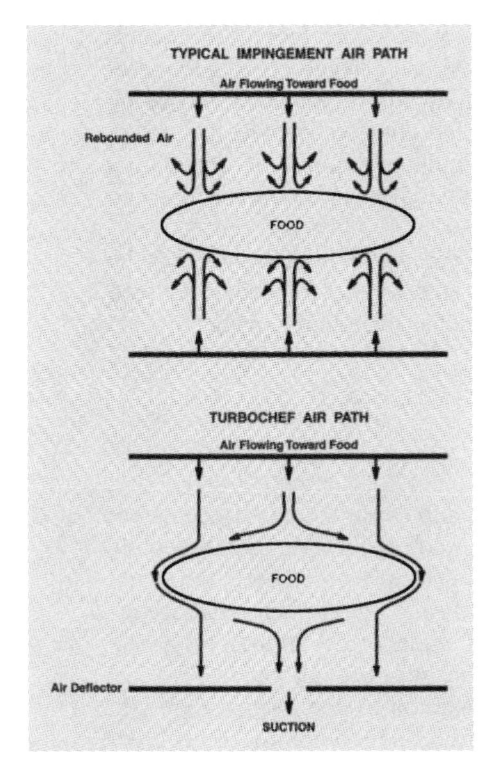

ILLUSTRATION 11-17 TurboChef, Inc., of Dallas, Texas, combined microwave and impingement cooking for its patented oven. Air hits the top of the food being cooked, then is sucked around the food and out the bottom of the oven cavity by a powerful fan.

time, microwaves penetrate the food. Dishes cook 10 to 15 times faster than on a conventional griddle or broiler, or in a conventional oven. A pizza in 70 seconds? A rack of lamb in 55 seconds? Cinnamon rolls in 8 seconds? No problem, says the manufacturer. And no exhaust hood is required, making these ideal for confined spaces.

TurboChef's rapid-cook model is a countertop unit, no more than 3 feet square, which uses 208- to 240-volt, three-phase power, programmable for multiple types of menu items. Note that whenever microwaving is a component of the cooking process, the types of pans that can be used are limited to nonmetal, like plastic or glass.

Another innovation is the FlashBake oven, which uses a combination of intense light and infrared energy. The light waves (not microwaves) penetrate the food, cooking it quickly, while the infrared waves brown the food surface. The idea here is to control and balance the two energy sources by computer, to cook both interior and exterior simultaneously in the shortest possible time.

You can program up to 50 cooking cycles on the FlashBake oven control panel. Its footprint is less than 6 feet square, and it weighs only 121 pounds. It requires 208 to 240 voltage and one- or three-phase electricity, about a 60-amp circuit at maximum power. The door locks when it's cooking and unlocks automatically when the cycle is finished.

Speed is, once again, the driving force behind this type of oven technology. FlashBake manufacturer's literature boasts a 2-minute cooking time for a boneless chicken breast that would otherwise take about 7 minutes to broil; and 1 minute for French fries that would require 10 to 13 minutes in a fryer.

Super-cookers may not be miracle-working replacements for all other appliances, but they are worth a second look. Work with manufacturers, ask a lot of questions, and witness a couple of demonstrations to ensure any particular super-cooking appliance really meets your needs before you spend the money.

11-9 CLEANING AND MAINTENANCE

In Chapter 6, we discussed basic troubleshooting for burners on gas range tops. Here are some additional tips for oven components and types.

The very purpose of an oven is to put out a lot of heat, so it is important that the vents to dissipate that heat are always kept clean and free of debris. If food is overdone, or underdone, this often signals that airflow is somehow being blocked. It may be as simple as moving the appliance out farther from the wall so that airflow around it is not restricted.

Pilot lights and gas connections should be checked on gas-powered models, and wiring on electric models. Periodically check the doors, to make sure their seals are tight and that they are closing correctly. Otherwise, you are probably wasting heat that escapes through a misaligned door.

Cleaning your commercial appliances is as easy as cleaning your dishes. Most equipment surfaces are made of stainless steel, which resists corrosion and is practically unharmed by moisture, detergents, food acids, salts, or anything else corrosive. These are solid sheets, not just a coating or surface that can be chipped off.

Appliance Surfaces

The secret of maintaining stainless steel lies in cleaning it frequently to prevent buildup of surface deposits, which may be harmful over time. Ordinary food or grime can be removed with soap and water, applied with cloth, sponge, or fiber brush. Be especially careful with abrasive cleansers, though—use them gently and rub in the same direction as the polish marks on the steel surface so any small scratches cannot be seen.

Baked-on food on range tops and ovens requires a bit more effort. You can make a paste out of water and any of the following substances:

- Ammonia
- Magnesium oxide
- Powdered pumice
- French chalk

Rub as gently as possible in the direction of the polish marks or use a scouring sponge or stainless steel wool for more resistant stains. Stainless steel wool is different from plain steel wool, which is too abrasive and should be avoided. Also, avoid using steel scrapers, wire brushes, or files. All of them contain iron particles, which may become embedded in and rust your stainless steel surface.

The exterior appliance surfaces can be further enhanced by the use of lemon oil or a good grade of furniture polish, wherever it is not in contact with food and when the polish odor is not objectionable. Polish removes grease, fingerprints, and smudges from stainless steel finishes. Equipment manufacturers will have suggestions for cleaning products and methods, so be sure to ask when they demonstrate for your staff. You can also contact the American Iron and Steel Institute's Committee of Stainless Steel Producers in New York City for its recommendations.

Microwave Care

Microwave interior cooking cavities should be wiped out daily with a soft sponge or cloth, mild detergent, and warm water—do not use oven cleaner! If foods are cooked on and appear to be solidly stuck on walls or ceiling, boil a cup of water in the oven first, which will help loosen them. There's a spatter shield on the oven ceiling, which should be removed and cleaned weekly. Also check the air intake and discharge areas of the microwave. Keep them free of dust and debris, and make sure they're not blocked or the oven will overheat. Some people say if an oven's controls are malfunctioning, unplug it, wait about one minute, plug it back into the wall and try again.

Microwave ovens appear to be very sturdy, but they are subjected to rigorous use in most restaurants. Don't slam their doors, and place dishes inside carefully. And of course, never put foil or metal objects in a microwave.

Conveyor Oven Care

Conveyor ovens put out a lot of heat, so it is especially important that their cooling fans and filters be checked and cleaned, at least weekly. The oven exterior, interior chamber, conveyor belt, crumb pans, and the inspection window should be wiped off daily, but only with whatever cleaning products are recommended by the manufacturer—and only when the oven is fully cooled. Some electric-powered ovens have their own computerized cleaning cycle, but should still be wiped clean to remove debris.

Deck Oven Care

Exteriors and interiors of deck ovens can be cleaned daily with mild detergent and water on a cloth. It is not advisable to spray water directly into the oven cavity. In terms of routine maintenance, when deck ovens malfunction, it is often because they are not completely level or that the vents that remove heat from the oven cavity have been blocked. Manufacturers also recommend a "break-in" process for "curing" a new deck or hearth, especially a ceramic one. There are heat shields under the deck that should be checked periodically. If you notice the deck doesn't seem to be heating evenly, these shields may need to be replaced.

Rotisserie Care

Check burners and fans for debris. Any heat transfer surface should only be cleaned with products that are recommended by the oven manufacturer. Some cleaners, while effective, are corrosive enough to damage ceramic or metal parts.

On gas units, the manifold valve should be open all the way, and the quick-disconnect on the gas line should be securely closed. Carbon and grease build up on the hub assemblies and gears of the motor that turns the spits so, in additional to external cleaning, the drive system should be partially disassembled for a major cleaning at least twice a year. If the rotisserie is consistently overloaded, or loaded improperly, it may automatically shut itself off to protect the motor. Correct loading and cleaning will prevent this.

Wood-Fired Oven Care

Like deck ovens, wood-fired ovens must be properly leveled or the cooking results are uneven. A new oven should be "fired up" as hot as possible for an hour a day, for about a week, before you cook anything in it. This curing process tempers the floor and walls of the new oven. There are damper controls that must be adjusted manually. If the oven uses a gas pilot light, it should be kept clean.

A good way to wipe out the surface of the (completely cooled) oven is with a damp mop. After a full day's work, there is no need to wait around to empty out the ashes from the wood fire. Just close the oven door, and do the cleaning in the morning before starting the fire again.

SUMMARY

The most heavily used piece of kitchen equipment is the range top. Unlike your stove at home, in foodservice you can mix and match a range top with either an oven cavity or a refrigeration cabinet beneath it, to create the appliance that's right for your commercial kitchen. On the range surface itself, you then select the types of burners and/or cooking surfaces you want. These decisions are based on the types of cooking you'll be doing and whether there are other appliances in the kitchen that could perform some of the tasks you would otherwise do on a range. Also consider how easy it will be to clean your range, because you will be cleaning it often.

The three basic types of ranges are the medium-duty or café range; the heavy-duty range, which is made for higher volume and large, heavy pots and pans; and specialty ranges for certain types of cuisine. Whether to choose a gas or electric range is largely a personal preference.

With ovens, there are even more choices. You can order a single oven, or a deck oven with more than one cavity and separate temperature controls for each. Large-volume operations may want to consider a mechanical oven, which moves the food through the cooking cycle on a rotating wheel or conveyor belt.

The old-fashioned flavor and ambience of cooking with wood is available with the use of smoker-cookers and wood-fired ovens. In many jurisdictions, however, wood burning is regulated for air-quality reasons. Wood/gas combinations or gas-fired ovens with simulated fireplace "looks" are two ways to avoid these problems.

Microwave ovens, once frowned upon by chefs who didn't consider them "real ovens," have become kitchen staples for safe but speedy thawing of frozen products as well as retherming precooked dishes. Microwave technology has also been combined with convection and/or impingement to create so-called super-cookers that cook foods in a fraction of the time of conventional appliances.

This chapter has introduced and explained more than a dozen different oven types—how they work and why they're used. Now, the choices are yours!

STUDY QUESTIONS

1. What is a café range? How does it differ from a heavy-duty range?
2. What is the difference between conduction and induction?
3. Would there be any advantages to using a French hotplate instead of a rectangular hotplate on your electric range top?
4. Name three important considerations when you are ordering a door for your new oven.
5. What's the difference between a range oven and a deck oven?
6. Why would you choose a mechanical oven for your foodservice business?
7. Explain how an impinger/conveyor oven works.
8. What should you take into account when buying a smoker-cooker? List three important considerations.
9. Why should you choose a commercial instead of a home-use model of microwave oven for restaurant use?
10. What are three considerations you should make to avoid complications when installing a wood-fired oven?

A Conversation with . . .
Kathy Carpenter

HOSPITAL FOODSERVICE CONSULTANT
GRAND JUNCTION, COLORADO

After 30 years of working in hospitals, Kathy Carpenter left her job as director of food services at St. Mary's Hospital in Grand Junction, Colorado, to start a new phase of her career as a consultant and teacher. She has a bachelor's degree in nutrition and food sciences from Colorado State University, and began her hospital career with a college internship at Highland General Hospital in Oakland, California.

Carpenter spent 16 years as a clinical dietician, and worked her way up to the director of food services position at St. Mary's. At the time of this interview, she was helping friends open a bagel bakery and restaurant in addition to her health-care consulting business.

Q: What made you want to become a dietician?
A: It was purely accidental. I wasn't doing very well as a microbiology major in college, and I figured I'd better get out of it quick! I thought of nutrition because I really like to cook; my father had a cardiac condition and my mom had cancer. The more I got into nutrition, the more I saw how it related to my family background. I had a real empathy for patients.

Q: Do you have to be a dietician to work in hospital foodservice?
A: No, but it is a definite asset. I've had district managers from so many fields—one was an English major, one was an archaeology major, another was a human resource person—and they all drifted into foodservice! In fact, I can probably name five food-service directors who have no college-level health-care background at all.

Q: What do you enjoy about hospital work?
A: I like the interaction between the disciplines I'm interested in. For example, I like seeing what food can do for a patient's recovery, or how food influences a staff member who is stressed out. In a hospital setting, you can honestly do a little bit of everything, from fancy party catering to selecting product for the vending machines.

Q: How do you stay up to date on the latest nutrition and medical trends?
A: A registered dietician is required to have seventy-five hours of continuing education within a five-year period, so I've managed to keep up with that my entire career. And you can call any other hospital foodservice director and ask for any information you need.

Q: In the hospital, you worked for a large foodservice corporation (Marriott). What are the pluses and minuses of working for such a big organization?
A: What's good is that you literally always have a job. If you're in good standing, you can transfer within the system to a different department, or from hospital to hospital. The disadvantage is that it's really big, and it loses some of its personal touch and caring for its people. They get so caught up in the business and legalities and contracts that they forget there are people behind all those contracts.

Q: The corporation you worked for also has other divisions in

addition to hospitals: business and industry and education. What would help someone decide which division they would enjoy?
A: It depends on the person's background and education, and also about what they like about other people in general. In the hospital division, you need a keen awareness of and empathy for the patient's needs, since there are special diets and preparation methods for the food.

In the education division, the ones who do well really like caring for kids, or dealing with students in dorms. If you like the flair and showmanship of demonstration cooking, you might end up in the business or industry division, because those clients enjoy the marketing part of it.

Q: If you think you might be interested in hospital foodservice, are there courses you can take in school to give you a taste of it?
A: I'd say either general nutrition or physiology because it really is true, you are what you eat. My one-year internship was very valuable; I spent six months working with patients individually, and another six months on administrative work where we got into everything from kitchen design to sanitation, ordering and receiving, and personnel management. You do have to know quite a bit about what types and amounts of nutrition to deliver in a particular meal. There are low-fat, low-sodium, cardiac, and diabetic requirements, just to name a few.

Q: Do you ever have very difficult patients who refuse to eat?
A: A lot of times they won't eat. It's not always that they're being difficult; sometimes they just don't feel well enough to eat. That's where a foodservice director's interaction with the dieticians is invaluable. An understanding of their suggestions and requirements allows the director to communicate with the necessary people to get whatever a person wants as fast as they want it. If somebody says they want rice cakes and I can understand the rationale, we can have rice cakes there within an hour.

Q: So, do you have a lot of contact with patients?
A: You can if you want to. It helps to develop your menus and specials with the patients in mind. At St. Mary's we consistently were at the top of the scale in our quarterly ratings by patients as well as cafeteria customers, and I think our interaction with patients and staffers on the floor had a lot to do with our success.

Q: If somebody is not going to be happy in hospital foodservice, what should be their first clue?
A: Hospitals are open 15 to 18 hours a day, 365 days a year. Don't do it if you don't like long hours, 7 days a week, adjusting your schedule at a moment's notice, and being dedicated to your staff and the type of clients you work for. It's not easy by any means, but it's very rewarding.

Q: Does the schedule affect your personal life quite a bit?
A: I adapted to it, and it can actually be quite flexible. Most hospital kitchens don't do a full evening meal service—usually the latest they're open is 8:30 or 9:00 P.M. The vending machine operation can be stocked in the daytime. Some hospitals choose to keep their cafeterias open all the time, but they can often do this with a skeleton crew of three or four people. As a dietician, when my children were small I was able to work part-time, or to have set hours when I needed them.

CHAPTER 12

PREPARATION EQUIPMENT: FRYERS AND FRY STATIONS

INTRODUCTION AND LEARNING OBJECTIVES

Despite its reputation for being greasy and unhealthful, fried food is a part of almost every nation's cuisine. When done properly, frying produces food that is light, not greasy in texture, with an attractive, crisply browned surface and a delicious flavor.

To fry means to cook in hot fat or oil, and there are several different frying methods. They include sautéing, stir-frying, panfrying, and deep-fat frying. While the basic frying techniques remain the same, the equipment needed to accomplish them in a commercial kitchen has evolved to be safer, more energy efficient, and easier to clean. In past decades, it was generally accepted that you had to sacrifice either performance or energy efficiency. Today, that is no longer the case.

The fryers we will discuss in this chapter fall into several basic categories. The most common is the deep fryer (which used to be known as a deep-fat fryer), in which food is immersed in hot oil. There are also pressure fryers, conveyor fryers, and air fryers. A host of options, including oil-less fryers, are now available to foodservice outlets.

In this chapter, we will discuss:

- How fryers work
- How to maximize the life of the fryer and the frying oil
- How to choose the type of fryer needed for your foodservice operation
- The components of a fry station
- The latest developments in fryer technology

12-1 DISSECTING THE FRYER

In foodservice, the fryer is a piece of equipment as standard as a range top. There are only isolated exceptions—perhaps nursing homes, hospitals, or health food restaurants—in which fryers are not an integral part of the kitchen. They can operate on electricity, natural gas, or even propane gas.

The receptacle in which the oil is placed is known as a ***frying kettle*** or *frying bin*. It is usually made of 16-gauge stainless steel and should have rounded (coved) corners for easy cleaning. Kettles come in a variety of widths, from 11 to 34 inches; their depth, from front to back, should not exceed 24 inches. This is primarily a safety feature, so that the person using the fryer will not have to reach very far over the heated oil. The kettle will have fill lines on its interior wall to indicate the proper oil level. The newest fryers feature insulated kettles, which manufacturers claim increases energy efficiency by about 10 percent.

Frying kettles are usually cube shaped, but some are Y shaped, widest at the top and tapering to a thin cone at the bottom. Heat is applied to the upper part of the Y, and the narrow bottom part functions as the cold zone, a term we'll explain in just a moment. Smart foodservice operators will match the shape of the kettle or bin to the type of food they'll be frying. For instance, to fry a large number of doughnuts simultaneously, you need a long, wide area. (A 34-by-24-inch frying bin can make up to six dozen doughnuts at once.) Doughnut frying bins also have a swing-up drain board to allow the doughnuts to "dry" after the frying process.

The size of a kettle is a measurement of its capacity to hold cooking oil. An 11-by-11-inch kettle, for instance, is referred to as a "15-pound fryer" because it can hold 15 pounds of oil. The largest kettle, 34 by 24 inches, can hold 210 pounds of oil. Some manufacturers rate the sizes of their fryers by the amount of french fries they will produce in an hour. More about how to measure output later in this chapter.

The food is lowered into the hot oil in a ***fryer basket***. These are also made of wire mesh stainless steel or chrome-plated steel. The mesh allows oil to flow easily through the basket and surround the food. Baskets come in two depths: 4 inches and 6 inches. A fryer can have a single basket, double basket, or

FOODSERVICE EQUIPMENT

SAFETY TIPS FOR FRYERS

- There is an open flame inside the fryer. The unit may get hot enough to set nearby materials on fire. Keep the area around the fryer free from combustibles.
- *Do not* supply the fryer with a gas that is not indicated on the data plate. If you need to convert the fryer to another type of fuel, contact your dealer.
- *Do not* use an open flame to check for gas leaks!
- Wait five minutes before attempting to relight the pilot to allow for any gas in the fryer to dissipate.
- Never melt blocks of shortening on top of the burner tubes. This will cause a fire, and will void your warranty.
- Water and shortening *do not* mix. Keep liquids away from hot shortening. Dropping liquid frozen food into the hot shortening will cause violent boiling.
- At operating temperature, the shortening temperature will be greater than 300 degrees Fahrenheit. Extreme care should be used when filtering operating-temperature shortening to avoid personnel injury.
- Ensure that the fryer can get enough air to keep the flame burning correctly. If the flame is starved for air, it can give off a dangerous carbon monoxide gas. Carbon monoxide is a clear odorless gas that can cause suffocation.

Source: Pitco Frialater, a division of the Meddleby Corporation, Elgin, Illinois.

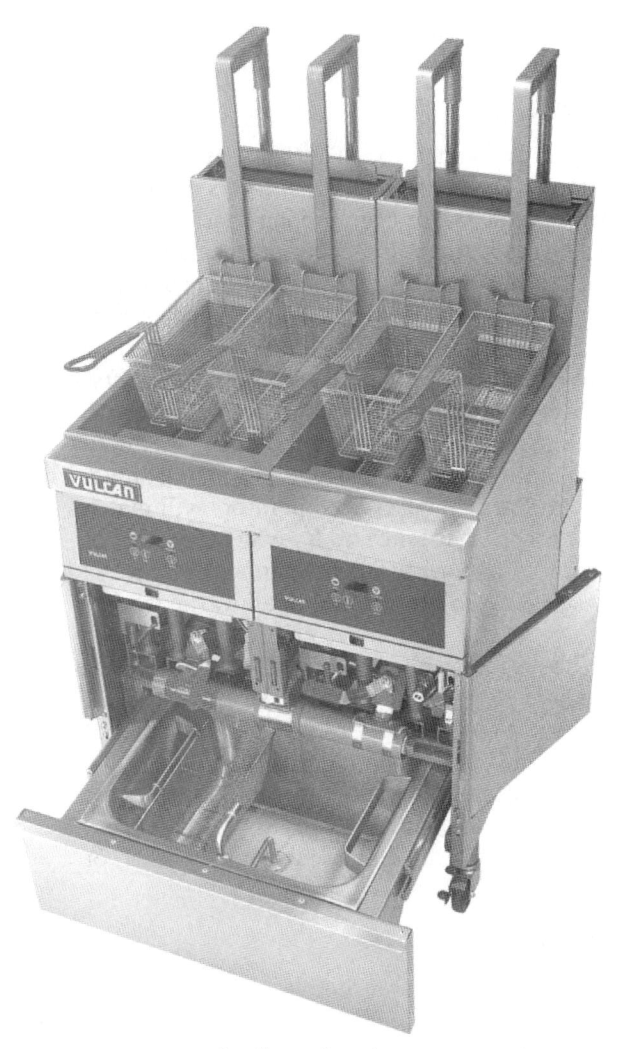

ILLUSTRATION 12-1 **A split-vat fryer has two separate kettles and sets of controls.**

COURTESY OF VULCAN-HART, DIVISION OF ITW FOOD EQUIPMENT GROUP, LLC, LOUISVILLE, KENTUCKY.

more, depending on how many will fit side by side in the kettle. The basket, which often looks like a wire mesh saucepan, has a long handle on one side and a metal hook directly opposite the handle so that it can be hooked in place on the kettle or on a ***basket rack*** for storage. Make sure the handle is insulated for comfortable use. Baskets are often made of copper or brass and coated with stainless steel. Be sure none of the metal is exposed, because frying oil breaks down faster if exposed to it.

As you might imagine, a steel kettle full of hot oil can be a burn hazard if not enclosed somehow. So the kettle fits into a separate, fabricated metal ***fryer cabinet,*** which is also made of 16-gauge steel. The kettle, at least in smaller models, should be easily removable for cleaning. Other safety tips are provided here, from fryer manufacturer Pitco Frialater.

The smallest fryers are called *drop-in fryers.* The kettle "drops in" to its metal cabinet, where it fits snugly. The controls for the fryer are located on the front vertical surface of the cabinet. The bottom of the cabinet is removable for cleaning. The drop-in fryer is mounted on a countertop. Most manufacturers provide sealing gaskets for mounting, which fit on the bottom of the kettle. Drop-in unit capacity ranges from 15 to 30 pounds.

The *countertop fryer* is a drop-in fryer that stands on 4-inch legs. The legs are either stainless steel or plastic.

The *freestanding* or *floor model fryer* is the workhorse of any operation that mass-produces fried foods. It has a capacity of 28 pounds or more, and its cabinet stands on four adjustable 6-inch legs or rolling casters. If it's on casters, the front wheels should lock and the rear wheels should be able to swivel. A freestanding fryer with two kettles and two separate sets of controls is called a *split-vat* floor model (see Illustration 12-1).

Freestanding models often have spreader plates like ranges or collapsible shelves to provide extra work space. If the bottom part of the cabinet is empty, it can be used for storage. However, in some large units, that's where the ***filtration system*** is located, which keeps the frying oil clean by filtering particles out of it.

Other handy options are automatic touch-time basket lifts (controlled by a timer, they lift the full basket out of the oil automatically) and a ***crumb tray*** at the bottom of the kettle, to remove and clean.

12-2 HOW FRYING WORKS

What happens in the fryer kettle? Let's take a closer look. Oil can be poured in as a liquid or as a solid (lard) that melts in the fryer. How the oil is heated depends on the type of fryer you purchase—we'll introduce the different heating options, and explain how they work, in a moment. The oil reaches an optimum frying temperature, from 325 to 375 degrees Fahrenheit. Hotter temperatures than this—say, above 400 degrees Fahrenheit—are not good for frying, because the oil begins to decompose and burn. The oil also begins to break down if it is exposed to the following items, all of which are found routinely in the frying process:

- Water (foods with a high natural moisture content)
- Sediment (crumbs, flour, food particles)

- Salt
- Oxygen

This is why filtering and regularly changing fryer oil is so important. Over the years, equipment manufacturers have improved fryer heating techniques to help prolong the life of each batch of oil. Today's fryers have heating elements located about 2 inches from the bottom of the kettle. They divide the kettle into two "zones": the ***cooking zone,*** which is the hot oil above the heating elements; and the ***cold zone,*** which is the 2-inch space between the heating elements and the bottom of the kettle (see Illustration 12-2). This cold zone is critical to the frying process and to the life of the oil. This is where all the crumbs and debris fall during frying, preventing them from mixing with the oil and damaging the fried foods. Because it is cooler than the rest of the oil, the cold zone prevents these crumbs from cooking or burning, which would deteriorate the cooking oil. More about preserving the life of your oil later in this chapter.

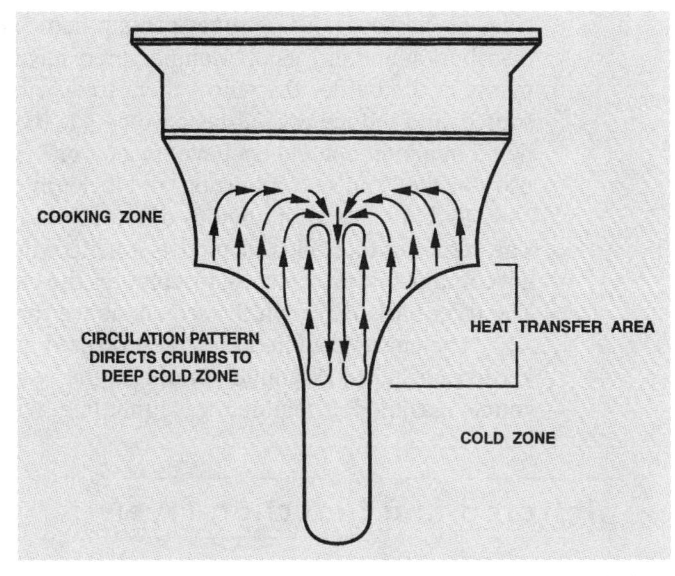

ILLUSTRATION 12-2 **A deep, narrow cold zone in the kettle helps ensure that crumbs and debris don't mix with the hot oil above.**

An important term you'll hear in comparisons of fryers is ***heat recovery time.*** This is the time it takes for the oil to return to optimum cooking temperature after cold (or frozen) food is dropped in to be cooked. For the most part, manufacturers have shaved heat recovery time to two minutes or less in commercial fryers. Why? If the food is placed in oil heated to less than 325 degrees Fahrenheit for more than two minutes, it begins to absorb the oil. Instead of frying crisply, it becomes soggy and greasy and takes on the taste or smell of any other food the oil may have come into contact with. A speedy heat recovery time prevents this.

So a fryer overloaded with thawing french fries has to work harder, cook longer, and compromise oil quality as the moisture on the fries drips into the hot oil. Water in oil also causes dangerous spattering.

Most fryers have control panels at the front of the unit, with an on/off switch and a melt/fry switch (the latter to distinguish melting oil from frying product). The melt cycle is used only to liquefy a solid block of new shortening that has been placed in the kettle. This is not supposed to be done when the fryer is already at its optimum cooking temperature—the resulting ***temperature shock*** can scorch the shortening and overwork the heating elements of the fryer.

A red light signals that the fryer is "on." On manually controlled fryers, a dial can be set to the desired oil temperature, but many are now computerized and can be preset for cooking different products. Most modern fryers have two thermostats. The ***cycling thermostat*** regulates frying temperatures up to 400 degrees Fahrenheit. The ***high-limit thermostat*** is a safety feature to detect overheating. The high-limit thermostat will turn the fryer off automatically if the oil temperature reaches 435 degrees Fahrenheit. Some electric models, if attached to a vent hood, will also shut off automatically when the hood's fire extinguishing system is activated.

When automatic basket lifts are used, there are additional push-button timers that can be adjusted from 0 to 15 minutes' frying time, and automatically reset. Each basket is governed by a separate timer. Other, more recent innovations include electronic sensors, which measure oil absorption, heat, and doneness, signaling when the food is ready to eat.

Gas Fryers

Gas-powered fryers have made major quality strides in recent years. Today, they require less gas to operate and their energy efficiency ratings have almost doubled from only a few years ago. Gas-fired fryers are heated in one of two ways: with an atmospheric burner (which mixes air with gas to ignite a flame) located under the frying kettle; or by injecting the gas flames through tubes located along the bottom or sides of the kettle. These tubes contain flame slots, or baffles, which aim the flames for maximum heat distribution and efficiency. Manufacturers have added larger tubes for faster heating, and made improvements in the baffles that allow them to distribute heat more evenly, extract more heat from the energy source, and reduce wasted heat. Some gas models have infrared burners, which we'll discuss in greater detail in a moment. Others have "instant on" electric igniters instead of relying on gas pilot lights, or they use so-called "pulse combustion" to efficiently fire up when turned on.

Perhaps the most obvious difference between electric and gas fryers is the way they are cleaned. The kettles of most electric models are lifted out for cleaning, while the gas models are not. Also, gas fryers contain a ***fryer screen*** that separates the cold zone at the bottom of the kettle from the rest of the oil. The fryer baskets rest on the screen, above the burner.

The energy required for a gas-powered fryer ranges from 25,000 Btu for a standard, 11-inch square kettle that holds 15 pounds of oil, all the way up to 260,000 Btu for the largest 210-pound kettle. Of course, gas models require a gas input line, which should be at least ¾-inch in diameter.

Infrared and Induction Fryers

One type of gas-powered fryer uses infrared burners to cook the food. Remember, infrared heat—like the sun's rays—transfers to objects only on direct contact. Because the heat is directed to the food itself, not to the surrounding area, it penetrates the food more quickly than other types of heat transfer (see Illustration 12-3). The end result is faster cooking time with less energy use, and frying is no exception. Infrared fryers are 20 to 70 percent more energy efficient than their electric and standard gas-fired counterparts. It would take a conventional fryer 120,000 Btus of natural gas to do the job of an 80,000-Btu infrared fryer.

The infrared burner consists of ceramic plates or metal screens full of tiny holes (about 200 per square surface inch). A mixture of gas and air is forced through these holes by a blower or fan, burning at a surface temperature of 1600 degrees Fahrenheit. The oil is heated so quickly that the frying process takes only a fraction of the time conventional fryers require, making infrared a good choice for high-volume operations. Infrared fryers have fast heat recovery times between batches and allow lower cooking temperatures, which means the oil lasts longer.

The latest electric model is an induction fryer, which cooks with electromagnetics—when the metal container comes into contact with the induction burner, heat is transferred immediately to the container (and thus, to the food). In this case, no heating elements are immersed in the frying medium, and frying can be accomplished at 600 degrees Fahrenheit instead of the nearly 800 degrees required for conventional fryers. This results in a cooler "cold zone" for induction fryers, and cooler flue temperatures—all making for a more comfortable kitchen.

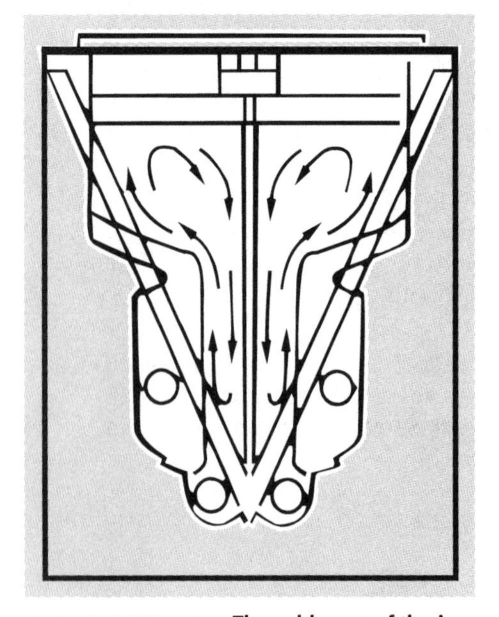

ILLUSTRATION 12-3 **The cold zone of the infrared gas fryer is at the lower sides of its kettle. Circulation of the hot oil forces debris down and away from foods being fried.**

Electric Fryers

In electric fryers, the heating elements are contained in stainless steel tubes immersed directly in the oil. This makes them very energy efficient. The amount of the element that comes into contact with the oil is important; more surface area contact heats faster and minimizes heat recovery time. As the hot oil rises around the elements, it creates a rolling action that quickly heats the oil at the top part of the fryer, leaving a cool, quiet zone at the bottom of the kettle where sediment and food particles settle.

The elements swing up from near the bottom of the kettle for cleaning. Generally, you lift them out of the kettle, turn them on, and they'll burn themselves clean. If this isn't done regularly, burnt food particles stick to the heating elements, compromising their effectiveness and using more energy.

From room temperature, it takes about six minutes for most electric fryers to preheat to 350 degrees Fahrenheit using liquid oil. Solid shortening takes longer, because it must first melt. The proper procedure is to set the thermostat to 250 degrees Fahrenheit and pack the shortening tightly around the elements. The lower temperature permits the shortening to melt slowly and cover the elements without smoking. Many electric fryers have a "melt" setting that pulsates the temperature gently during the melting process. After melting, the thermostat can be readjusted to the desired frying temperature.

Electric fryers typically use less oil per unit of product than their gas counterparts. Their cold zone is not as deep, and the heat they generate throughout the kettle is uniform. Power requirements for electric fryers vary with their size and capacity, ranging from 5.7 to 36 kilowatts per hour and from 208 to 240 volts.

Several manufacturers have introduced electric induction fryers to the market. In these fryers, heat is created inside the tubes by induction coils. The metal tubes create a magnetic field inside that generates heat, even though the tube itself is not connected to any power source. An induction fryer can heat oil efficiently when its tubes are about 600 degrees Fahrenheit, instead of the 750 to 800 degrees Fahrenheit needed for a conventional electric fryer. It's important to note that the *oil* doesn't get that hot—the *tubes* get that hot, in order to transfer sufficient heat to the oil. The more moderate temperature prolongs the life of the frying oil, and the fryer itself gives off less heat. Induction fryers have large cold zones at the bottom of the kettle that are about 100 degrees lower in temperature than the cooking zone.

Electric induction fryers have many of the same advantages as infrared gas fryers. They're well suited to large-volume cooking, they extend oil life (as much as 35 percent), they have quick heat recovery times, and their lower temperatures reduce oil spattering. No elements must be moved for cleaning. The most common induction kettle size is 15½ by 13¾ inches; it uses 14 kilowatts per hour to heat 50 pounds of frying oil. Most have internal filtration systems, with a dump station located to the right or left for handy disposal of crumbs and/or used oil.

Computerized Fryers

Computer-controlled fryers can easily be programmed to turn out consistent product every time, regardless of the size of the batch. They take a lot of judgment out of the frying process, which can mean better product consistency, less waste, and less time spent having to train employees to use them.

A computerized fryer can, for instance, fry 16 chicken quarters in 16 minutes at 325 degrees Fahrenheit; a load of chicken tenders at 360 degrees is finished in four minutes; and full-capacity loads of french fries or small shrimp take only 90 seconds. Timers show how much longer the cooking cycle will last, and a bell or alarm signals the end of the cycle. The computerized fryer will automatically raise and lower baskets and shake them for a preprogrammed time period to allow grease to drip off. Kettle temperatures can be read instantly by sensors inside the kettle, and, if overheating occurs, the unit can shut off, sound an alarm, flash a warning light—or all three! And, in a dual-kettle model, the computer keeps track of both cooking cycles at once. The advantages are obvious: Expert estimates indicate you can cut the labor costs associated with frying foods by up to 30 percent by automating the process as much as possible.

Cashing in on the computerization trend, a couple of manufacturers sell remote "computer systems" for regular frying kettles. These are basically electronic probes that can be installed inside the kettle, attached to the heating elements, or dropped into the kettle as the product cooks, to keep track of oil temperature and signal doneness. They may work off the same power source as the fryer or plug into a standard wall outlet.

12-3 FRYER CAPACITY AND INSTALLATION

We've already mentioned that fryers come in many sizes, and that they're often identified by the number of pounds of oil they will hold. One way to gauge capacity is to know that, for every 5 pounds of oil the kettle can hold, you can load and fry one pound of food. So, a fryer that holds 50 pounds of oil should be able to fry up to 10 pounds of food per batch.

Another way manufacturers rate a fryer is by how many pounds of french fries it can fry in one hour. The general rule is that a fryer will produce product equal to one and a half to two times the weight of oil it will hold. For example, a countertop fryer with an oil capacity of 15 pounds can produce about 25 pounds of french fries per hour. Remember, this rule does not necessarily apply to *all* fried foods, but it's helpful in determining basic fryer capacity.

Finally, consult the capacity-per-hour guidelines put out by the manufacturers, usually in chart form. They'll give you examples of product, frying temperature, and hourly output. Table 12-1 is an example.

TABLE 12–1

FRYER COOKING GUIDE

Food	Temperature° (degrees Fahrenheit)	Time° (minutes)
Potatoes, French fries (3/8-inch cut)		
Raw to done	350	6
Blanched, only	350	3
Browned, only	350	3
Commercially treated	350	6
Frozen, fat blanched	350	2
Potato chips	350	3–4
Potato puffs	360	1½
Seafoods		
Frozen breaded shrimp	350	4
Fresh breaded shrimp	350	3
Frozen fish fillets	350	4
Fresh fish fillets	350	3
Fresh breaded scallops	350	4
Breaded fried clams	350	1
Breaded fried oysters	350	5
Frozen fish sticks	350	4
Chicken		
Raw to done	325	12–15
Croquettes	350	3–4
Turnovers	350	5-7
Precooked, breaded	350	3–4
Miscellaneous		
Breaded veal cutlets	350	3–4
Breaded onion rings	375	1½–2
Precooked broccoli	350	3
Precooked cauliflower	350	3
Precooked eggplant	360	3–4
Breaded tamale sticks	360	3
Fritters	375	4–5
French toasted sandwiches	375	1

TABLE 12–1

FRYER COOKING GUIDE (CONT.)

Food	Temperature[a] (degrees Fahrenheit)	Time[a] (minutes)
Yeast raised doughnuts	375	1
Hand-cut cake doughnuts	375	1½
Doughnuts	375	2–3
Glazed cinnamon apples	300	3–5
Corn on the cob	300	3
Turnovers	375	4–5

[a]Allow for minor variations from these suggested times and temperatures according to the weight, texture, density, and other characteristics of the foods you use.

Source: Texas Utilities Electric Company, Dallas, Texas.

Also ask about the fryer's Btu input. The higher its Btu input, the better chance it will maintain an even temperature, and the faster its recovery time. For gas fryers, start with the basic premise that it requires 600 Btus to fry 1 pound of french fries. Because most gas fryers operate at about 50 percent efficiency, you can estimate their production capacity by dividing the gas input (total number of Btus) by 1200 (600 Btu times 2). Example: a 60,000-Btu fryer, divided by 1200, equals a 50-pound capacity of french fries.

The key is to buy a fryer with both the desired capacity and the shortest possible recovery time. Many professionals have found two smaller-sized fryers do the job better than a single, large one. Dual fryers give you the ability to cook two different products at the same time or to turn off one fryer during slow times. Another note: Some foods (french fries, for instance) can be blanched prior to final frying, which decreases frying time and therefore increases overall fryer capacity.

Hot oil is a dangerous commodity, not only because it can cause serious burns, but because it is highly flammable. The National Fire Protection Association requires that a fryer be located at least 16 inches away from any piece of equipment that uses an open flame, such as broilers or range tops. Unless it is oil-free, the fryer must also be located under a vent hood, because it gives off grease-laden moisture (as well as gas fumes, in gas-powered models).

The sides and back of the fryer must be at least 6 inches from the walls to be ventilated properly, allowing an unobstructed flow of combustion air. If a gas fryer is installed in a spot where incoming airflow (to its burners and blower motor) is restricted, it will build up abnormally high temperatures and eventually short-circuit the electrical part of the motor—its controls.

Fryers must always be mounted on sturdy legs or heavy-gauge metal stands. Installing them flat would also restrict air circulation. The fryer should be correctly leveled, from front to rear and from right to left. Before use, it must be calibrated.

For all of these reasons, fryer installation is not something that most general electricians or plumbers are qualified to do. By all means, hire a professional installer. The manufacturer or equipment dealer can recommend one.

For both safety and convenience, it's smart to flank the fryer with counter space instead of other equipment. After all, you'll need a place to put incoming and outgoing food. This may not be possible, because so much equipment must fit under the (usually limited) vent hood space, but it's worth considering. As an alternative, place the fryer at the end of the hot line and set a work table or rolling cart beside it.

Also, remember to clean the fryer thoroughly before its first use. The manufacturer will provide cleaning instructions, but it never hurts to break in a new kettle by boiling a solution of one part vinegar to ten parts water in it, to remove the manufacturer's grease that is often used to shine and protect a new kettle before delivery. The kettle should be drained and cleaned regularly and always rinsed with clear water until all vinegar odor is gone.

12-4 CARE AND CONSERVATION OF FRYING OIL

The flavor, aroma, and texture of fried food depend largely on the quality and condition of the frying oil. There are a number of ways foodservice professionals prolong the life of their oil, and there are plenty of reasons to do it. Other than the food itself, the oil to fry it in is your most expensive food-related cost. We've already mentioned that, in cooking, oil is affected greatly by temperature, moisture, food particles, salt, and more. In fact, as it is heated, oil changes its chemical properties even without food in it! So far, science has been unable to provide us an antioxidant or antifoaming agent that will stop oil from naturally deteriorating with use. However, there are ways to minimize this deterioration.

First, some background about frying fats and oils. The oil used for frying is different from oil used for baking or making salad dressings. Butter and most animal fats are not suitable for frying, because they contain high percentages of free fatty acids, which break down quickly. Frying fats and oils fall under the broad classification of *lipids,* which contain two other fat-related compounds: cholesterol and lecithin. The latter, in food, is an emulsifier, meaning it keeps something in suspension in a liquid. All of these products contain both "good" fats (polyunsaturated and monounsaturated), which prompt a human body to produce so-called "good" (HDL) cholesterol and remove so-called "bad" (LDL) cholesterol from the bloodstream back to the liver for reuse or excretion; and "bad" fats (saturated), which can cause a build-up of "bad" (LDL) cholesterol in the bloodstream.

The apparent difference between a "fat" and an "oil" is that fats are solid at room temperature, while oils are liquid. In the past, semisolid fats were used for frying. They had such high melting points that some restaurants found it easier to melt them on the range top first, then pour them into the frying kettle for use. Fortunately, technology has given us newer, liquid frying fats with longer life expectancies and minimum decomposition. Most of them are hydrogenated oils made from nuts or vegetables: soybean, corn, peanut, cottonseed, canola, or sunflower. The term ***hydrogenated*** means the oil has been chemically combined with hydrogen molecules, which makes it more solid at room temperature and increases its stability. Soybean oil is the most widely used for french fries. Peanut oil is most popular in Chinese cooking.

Interestingly, all fats and oils are identical when it comes to calorie content. Whether it's lard, butter, chicken fat, canola oil, olive oil, or whatever, they all have 252 calories per ounce, or 9 calories per gram.

Because frying oils are so widely used and such a big part of most foodservice budgets, there is quite a bit of ongoing research in this area. Keep your eyes open for the latest developments. In many instances, you'll choose an oil based on what you'll be frying. Chicken may require oil with a distinctive flavor, while french fry oil should have a bland, neutral taste.

The best frying oil is one that has enough stability to withstand the rigors of high frying temperatures, which is called "having a high ***smoke point.***" If the oil produces thin, bluish-white smoke as it heats, it is a sign of overheating. Soon after, you'll notice a sharp, pungent odor reminiscent of a carnival midway (where food aplenty is being badly fried!). Turn down the kettle temperature immediately.

Another sign of oil deterioration is foaming, which happens as the fat breaks down into its oily components, becomes thick or gummy, and takes on a rancid odor. This sometimes occurs when the heating elements are not completely covered by the oil, causing them to overheat, smoke, and deposit a layer of black carbon onto the heating elements. Particularly with electric fryers and their movable, swing-up elements, be sure they are not exposed to the air during frying.

When the kettle is not in use, one good oil-saving technique is to turn it down to a lower setting, between 190 and 275 degrees Fahrenheit, then cover it. Certain manufacturers have added an automatic turn-down feature as an option. A kettle not in use should also be covered, but only if you have first let the oil cool to its lower temperature. Covering a hot kettle will accelerate the breakdown of the oil, because moisture condenses on the hot lid.

Another way to increase the usefulness of your frying oil is to control the rate of ***fat turnover.*** This is the amount of fat used each day compared to the total capacity of the kettle. Experts recommend a total daily turnover of oil—that is, if the kettle's capacity is 30 pounds, a total of 30 pounds of fresh oil should be added in small amounts throughout the day. This won't cause the kettle to overflow because, as food is fried, it retains from 10 to 40 percent of the frying oil. The small amounts of new oil, added to maintain the kettle at its fill line, keep the overall batch of oil fresh. Another similar rule is that, at all times, fresh oil should account for one-third of the total oil in the kettle. One important caveat: Don't add fresh oil to oil that is already brown or foamy. You're just wasting it.

Fat turnover is determined in large part by how much oil is absorbed into the products you're frying. This is known as ***oil take-up.*** The oil take-up of a french fry, for instance, is 20 to 40 percent, which means the finished fry has that amount of frying fat in it. Undesirable oil take-up rates occur if the oil temperature is too low or if the food is allowed to stay in the oil too long.

Filtration

Research done by the PG&E Food Service Technology Center in San Ramon, California, has shown no relationship to "oil longevity" between gas or electric fryers, but the same tests showed that oil with a typical useful life of 10 days can extend that time to 15 days or more if it is filtered properly. In fact, the single most important way to protect your oil is to filter it regularly to remove crumbs. A little debris is a by-product of any fried food, and your job is to strain the debris out of the oil so it doesn't stay suspended in the oil and burn. Removable crumb trays and *skimmers,* to skim floating particles off the oil, are helpful. But nothing works quite as well as filtering. Most fryers today come with automatic filtering systems. However, for small kettles, you'll probably have to filter by hand.

What you'll need is a stockpot *larger* than the capacity of the fryer kettle, a metal filter holder, and a filter that fits the holder. Filters can be made of paper, cloth, or fine-mesh wire screen; the disposable paper ones are the handiest, but filter papers are messy. Some say the wire screen is the most eco-friendly, but it is also the most expensive. It is *very* important that the oil temperature is *below* 100 degrees Fahrenheit before filtering, or you risk getting burned.

Most electric kettles can be lifted out and their contents poured through the filter into the stockpot. In gas fryers, there's usually a piece of pipe screwed into the bottom of the fryer than can be loosened, causing the oil to drain out through the filter and into the stockpot. This is the case with most countertop, drop-in, and some of the smaller floor models. After filtering, the oil must be returned to the kettle. Some systems require a hose and nozzle, which is aimed manually into the kettle as oil is pumped back in. Others pump the oil directly back into the kettle, through a hole in the kettle wall or a ring that encircles the kettle.

There are two types of automatic filters: The built-in system is stored beneath or beside the fryer; the portable system sits in its own cabinet and can be rolled and attached to fryers anywhere in the kitchen.

Portable Filtration System. For large, floor model fryers, a portable filter system is recommended. It is typically a four-wheeled cart, mounted on rolling casters, which can often be stored underneath the fryer kettle. An electric pump pushes the oil through a filter at the rate of 5 to 7 gallons per minute. The pump has a ⅓-horsepower motor and operates when plugged into a standard 120-volt electrical outlet. Portable systems are also handy because they allow you to wheel the used oil away for disposal (see Illustration 12-4). Most have a safety feature—that the filter will not turn on unless the fryer is turned off. This is critical. A fryer can be ruined if it is turned on when the vat is empty.

You will probably pay as much for a portable filtration system as you do for the fryer itself, but if you use a single filter for several fryers, it's money well spent. Before you buy, find out:

1. How efficient is the system at sucking all the oil from the filter pan? Some pans are constructed such that it is impossible to extract all the oil.
2. Can solid shortening be used with this system? Typically, the answer is no, because it can solidify and clog both filter and lines.

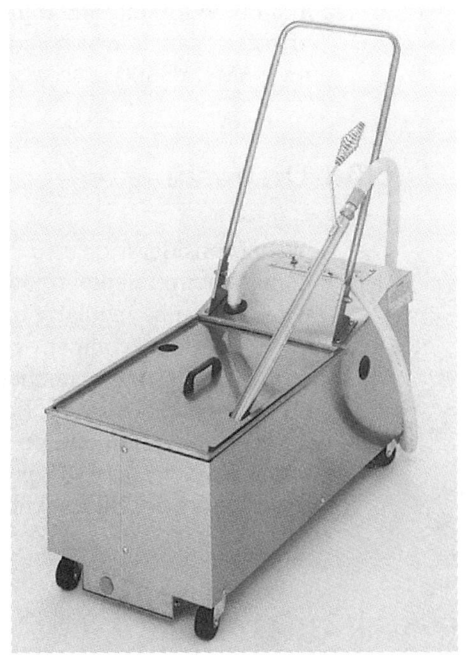

ILLUSTRATION 12–4 **A portable filtering system can roll on casters to serve more than one fryer.**

COURTESY OF FRYMASTER CORPORATION, A WELBILT COMPANY, SHREVEPORT, LOUISIANA.

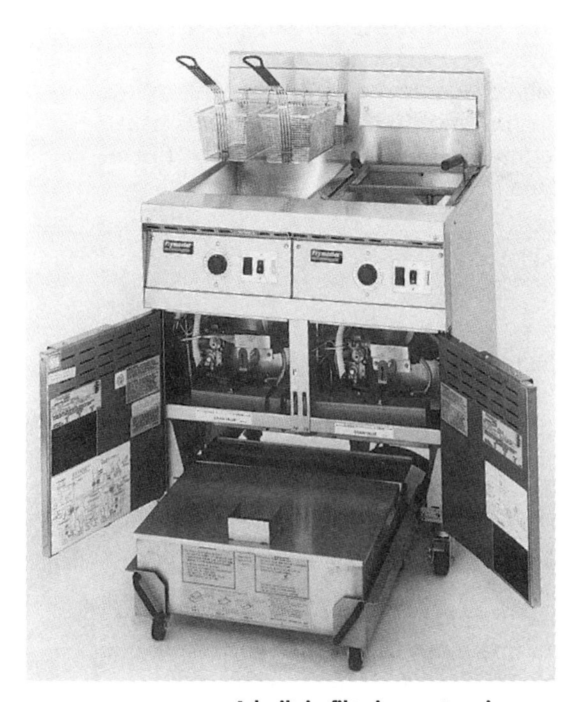

ILLUSTRATION 12-5 **A built-in filtering system is contained in the cabinet below the fryer.**
COURTESY OF FRYMASTER CORPORATION, A WELBILT COMPANY, SHREVEPORT, LOUISIANA.

3. **How is the filtered oil returned to the kettle? Each manufacturer seems to have a different type of line for accomplishing this.**

Built-in Filtration System. This is contained in the cabinet below the fryer. A built-in system automatically transfers the oil from the kettle to a tank below the kettle, where filtering occurs. Fryers with built-in filtering systems are more expensive but, where space is at a premium, they are especially handy. A single built-in system can also be set up to filter the oil from several side-by-side fryers (see Illustration 12-5).

The built-in filter system automatically shuts off the fryer and opens the kettle drain, then turns on the pump and opens the return line (so the filtered oil can be returned to the kettle), in a sequence of steps. These steps require a person to pull a couple of handles on the machine, and perhaps to manually turn on the pump.

Most manufacturers offer a product called *fry powder* to "condition" frying oil. They claim that adding the recommended amount of powder to the oil during the filtering process has an antioxidant effect, absorbing particulates and improving its life between 25 and 40 percent. But these add-ins are controversial, and probably not needed if sufficient care and cleaning routines are followed.

Filter manufacturers continue to make safety strides, with the goal of keeping hot oil away from kitchen workers. No matter what type of system you select, the filtration schedule is critical. Frying oil should be filtered a minimum of once a day; where consumption is high or the menu includes a lot of breaded items, filter after each eight-hour work shift. Use the filtering time to continue maintenance of the fryer and filter, keeping them free of residue. The filtration system should also be checked periodically to make sure that all the lines are clean and snugly connected, with O-rings properly seated and not worn out; and that elements, temperature controls, and so on are clean and secure. These types of maintenance, which should be outlined in your operator's manual from the manufacturer, will save from $60 to $500 in service call costs per visit.

Oil Disposal

A final consideration is to dispense with filtration altogether, and let an outside contractor handle it. Mahoney Environmental of Joliet, Illinois, sells a system to pump used oil through pipes to a remote tank where it can be regularly hauled away.

A national company called Restaurant Technologies, Inc. (headquartered in Eagan, Minnesota), offers an interesting agreement—buy oil from them, and they install two tanks at your site, one for fresh oil and one for waste oil. Both tanks hook up to a port on an exterior wall, where MVE tanker trucks regularly tap in, fill up the fresh oil tank, and drain the waste oil tank. The company says a large chain customer may pay 10 to 15 percent more overall for its "cooking oil management system," but will save in other ways with the convenience of it.

12-5 PRESSURE FRYERS

Pressure fryers provide some of the most versatile and profitable cooking applications available for use in small spaces. Today's pressure fryer can turn out 14 pounds of chicken in less than ten minutes.

This specialty appliance cooks food with a combination of hot oil and steam. The steam may be generated from moisture in the food, or it may be added during the cooking process. Manufacturers claim that their pressure fryers cook food more quickly than regular fryers, and that the food is tenderized by

FOODSERVICE EQUIPMENT

HINTS FOR SUCCESSFUL FRYING

- Do not bring the fryer to full frying temperature without having food in the kettle.
- Never salt food over the kettle.
- Never fry bacon in the kettle.
- Never load the fryer basket more than two-thirds full. Overloading results in soggy food that is not uniformly cooked.
- Bite-sized foods cook more uniformly and don't stick together if the basket is lifted and shaken now and then during the frying process.
- Try to fry pieces of food that are similar in size, so they'll all be done at about the same time.
- Soap residue is an enemy of the frying oil. Always use *only* the cleaning solutions recommended by the fryer manufacturer.
- Clean the kettle by boiling water in it—not once, but twice. After the second boil, rinse the kettle and wipe it out thoroughly.
- Keep the exhaust hoods above the fryer clean. The last thing you want is grease from the vent hood dripping into your kettle.

the use of steam. Because we know moisture breaks down cooking oil, the system is designed to prevent condensation on the lid from falling back into the oil.

The pressure fryer is basically a fryer with a tightly sealed lid, usually a rubber gasket (see Illustration 12-6). The kettle is filled with cooking oil. When food is added to the hot oil, the lid is shut and the steam that naturally builds up inside raises the pressure in the kettle. The pressure causes the food inside the kettle to tumble and roll, like clothing in a washing machine. The frying time is shorter, and the oil temperature can be lower, which means less oil is absorbed into the food. This frying method saves energy and cooking oil in addition to being fast, making pressure frying a popular choice for quick-service restaurants. When done correctly, the fried food emerges crisp, golden, and moist.

Pressure fryers can be powered by gas or electricity. Gas models, depending on capacity, require between 40,000 and 80,000 Btus per hour. Electric models require from 120 to 208/240 volts and operate (again, depending on capacity) on 5.2 to 13 kilowatts per hour.

Like other types of fryers, pressure fryers come in countertop and floor models and should stand on legs or casters to facilitate ventilation. The smallest countertop models hold 10 pounds of oil and produce 6 pounds of food per nine-minute frying cycle; the largest floor models hold 43 pounds of oil and fry 14 pounds of food per frying cycle. Control panels and thermostats are similar to those on regular fryers, except that pressure fryers also have a pressure gauge. When shopping for this appliance, look for programmable, computerized models that adjust themselves for the load size, can be used with or without the steam pressure function, and have a built-in oil filtering system.

A cousin to the pressure fryer is the ***controlled evaporation cooker,*** also known as a *CVAP* or *Collectromatic Fryer System.* The CVAP technology involves holding food at peak quality in a combination of heat and steam. A heater and thermostat control the amount and temperature of the steam. CVAP fryers can be used with their lids open or closed, and have built-in continuous filtration capability.

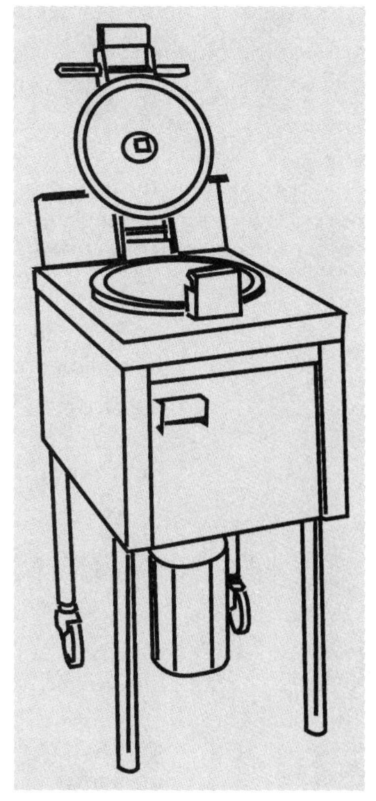

ILLUSTRATION 12-6 **A pressure fryer.**

12-6 THE FRY STATION

The frying process requires an assembly line setup when large volumes of food are being produced. The complexity of the operation will dictate the need for a ***fry station*** and whatever storage it requires (see Illustration 12-7).

Fry stations include the fryers themselves, landing space for food as it moves into and out of kettles, food warmers, basket racks, a filtering system, and a product storage area. In the typical fry station, you'll see:

ILLUSTRATION 12-7 **The fry station contains everything needed to prepare and hold fried foods before serving.**
COURTESY OF FRYMASTER CORPORATION, A WELBILT COMPANY, SHREVEPORT, LOUISIANA.

FOOD WARMERS. These are fixtures that hold 250-watt infrared bulbs with heavy-duty chrome shades, in sizes ranging from two to eight bulbs. There are many styles available, including warmers that can be mounted on walls, counters, shelves, or on their own rolling carts. A two-bulb unit installed under a shelf can warm an area the size of a standard hotel pan—about 12 by 20 inches. Most food warmers plug into standard electrical outlets (see Illustration 12-8).

LANDING SPACE. Counter space is absolutely necessary to a successful fry station, where the food can "land" as it goes in and out of the fryer. Collapsible shelving, either beside or in front of the fryer, is preferable. Rolling carts or tables, the same height as the fryer cabinet, are the next best option.

FILTER SYSTEM. We've already discussed the importance of filtration in the frying process. In a fry station where space is tight, the top of the filtering unit is often designed to work as counter space or as a place to mount food warmers. Portable filtering systems can be used where space is at a premium, so they can be stored elsewhere.

DUMP PANS. As their name suggests, ***dump pans*** are used to "dump" food as it moves into or out of the fryer. They're made of stainless steel. Dump pans for finished food are usually perforated to allow some grease drainage.

BAGGING AREA. This is where employees stand to put the finished product (usually french fries) into bags or other containers for sale. Doing this directly in front of the fryer only hinders the frying process. Think about the flow concept, and you'll see that the critical decisions in laying out a fry station are where to put pans and where to bag product.

BASKET RACK. This metal rack is a necessary storage tool for busy fry stations. Full baskets of food can hang suspended on it, ready to be submerged into the kettle. Basket racks come in different sizes; the largest can hold up to 32 baskets.

12-7 BUYING A FRYER

Whatever type of fryer you select, examine it for four crucial characteristics, over and above whether it fits your budget:

CAPACITY. Can it meet the volume you require, based on your menu and peak demand amounts?

SIMPLICITY. Is it easy to use and clean? Can you train new employees to use it (or, if it's computerized, to program it) correctly and safely? Does it have a full, clear set of instructions?

Nowadays, most operators choose fryers with built-in oil filtration systems, just because they are easier to clean.

RELIABILITY. Does it require minimal maintenance? Is a local service person available for repairs or questions after the initial installation?

SPACE ALLOCATION. Does it fit in the space you have in mind? Will it complement, or conflict with, other kitchen operations? If it must be vented, can the exhaust hood handle the additional load?

ILLUSTRATION 12-8 **Food warmers use heat lamps to hold food at safe temperatures for short periods of time until it is served.**

COURTESY OF ALTO-SHAAM, INC., MENOMONEE FALLS, WISCONSIN.

There are a number of additional considerations, primarily because fryer manufacturers now offer so many options:

- *Atmospheric or infrared burners?* Gas-powered fryers with traditional atmospheric burners (that mix gas with air from the surrounding atmosphere) are simpler overall than infrared burners, which also contain a blower or fan to force air through the system.
- *Nonbreaded or breaded products?* For nonbreaded products, a traditional kettle—wide at the top, with a narrow cold zone at the bottom—is preferable. For breaded products, an open fryer and wider cold zone at the bottom works well. Only a fryer with tubes (either gas or electric) works for products like fried chicken—partly because it is impossible to heat such a large amount of oil except with tubes immersed in the oil, and partly because the cold zone must be located below the heating tubes to catch the debris from breaded chicken.
- *Solid-state or computerized controls?* If you fry only a few items that work well in a similar temperature range, you'll do just fine with a fryer with basic, solid-state controls that can be adjusted by hand, and a timer and beeper to alert you when the frying cycle is done. However, if you will be frying many different types of products that require a variety of times, temperatures, and cooking cycles, computerized controls are preferable.
- *Gas pilot or electric ignition?* The electric ignition is preferred today, since it eliminates the need for a pilot light.
- *Automatic or manual basket lifts?* This requirement seems to depend on the menu type and the experience level of the staff. Quick-service restaurants with limited menu offerings and workers who specifically "man" the fryers don't seem to mind the manual lifting of baskets in and out of the oil. Operations with more varied menus and multiple uses for the same fryers find auto-lift models a good solution.

12-8 FRYERS OF THE FUTURE

As we've mentioned, frying is big business in most restaurants, so new types of fryers are always on the drawing board. Here, we list a few of the latest options to consider when conventional fryers simply don't work in a particular space.

Oil-Free (Greaseless) Fryers

And you thought this was a contradiction in terms! Known by a variety of names—including hot air fryer, greaseless fryer, no-oil fryer, and forced air food system—this type of machine is well suited for small jobs where speed is still important in the frying process.

The greaseless fryer is often a countertop model, but there are higher-volume units up to 3 feet tall. Food is loaded into some type of basket and inserted into a heated chamber, where it is blasted with hot, fast-moving air resulting in a product that resembles "fried" food, without the frying. There are two general types of oil-free fryers: ones that use infrared or radiant heat to speed the cooking process; and ones

FOODSERVICE EQUIPMENT

FRYER LIFECYCLE FACTORS

- Purchase price, tax, freight, installation, start-up
- Energy
- Hours of use per day, calculating percentage of use at full load and at idle
- Average energy use per hour (Btus or kW)
- Labor (for example, dedicated employee vs. using basket lifts)
- Preventive maintenance (costs for labor and supplies to clean, filter, oil, and the like)
- Shortening usage and disposal
- Service repair costs
- Some types of units may have higher component repair/replace rates
- High-volume usage may "age" units faster
- Level of staff training may affect how much abuse equipment receives
- Additional energy costs (added load of HVAC)
- Annualized costs: all expenses amortized across projected life expectancy

Source: ***Foodservice Equipment Reports*****, a Gill Ashton publication, Skokie, Illinois, May 2003.**

that flash-heat the oil found naturally in the food. Either way, the process results in a crispy, "just-fried" flavor. Frozen foods can be placed in these fryers, and a microprocessor gauges how large the load is, and just how frozen it is, then heats accordingly. See Illustration 12-9 for a closer look at the interior workings of the ventless fryer.

As the hot air turns the frozen moisture into steam, the steam is exhausted away. However, because no oily vapors are released, the machine does not need to fit under an exhaust hood. Instead, it contains a built-in grease trap and air filtering system that must be cleaned regularly. A greaseless fryer might be appropriate for a low-volume operation, such as a convenience store, where foodservice is not the major thrust of the business. In delis and bistros, bars with appetizer menus, and mall and airport quick-service eateries, it can be used to prepare onion rings and french fries, pizza, battered fish, some types of sandwiches, steamed vegetables—all oil-free.

Another type of greaseless fryer uses electricity to steam frozen or blanched french fries and quickly air-dry them. The countertop model cooks about 3½ pounds of fries in 4 minutes.

Current greaseless fryers on the market need from 2100 to 6000 watts of power to operate. They run on 110 or 220 volt electricity, and can heat up to 550 degrees Fahrenheit, with cooking times of up to six minutes.

The obvious advantage of no-oil frying is that you don't have to continually purchase and store oil or shortening. They are easy to use and clean, and eliminate costly hood and venting requirements. They may also lower insurance costs. An interesting disadvantage is that so many of these fryers look like microwave ovens—your employees must remember that they can't reach into the oven cavity, or use plastic in it. Most no-oil fryers cook smaller-volume batches of food, not sufficient for a very busy operation. And finally, the use of cooking oil imparts a certain distinctive flavor to foods, part of the reason people love them so much! Air-fried foods do taste slightly different than their oil-fried counterparts.

Ventless Fryers

When you can't make (or can't afford) expensive roof modifications, there are self-contained fryers, often called ***ventless fryers***. "Ventless" means the fryer has an internal air filtration system to capture grease and return cleaned air to the kitchen without venting it to the outside.

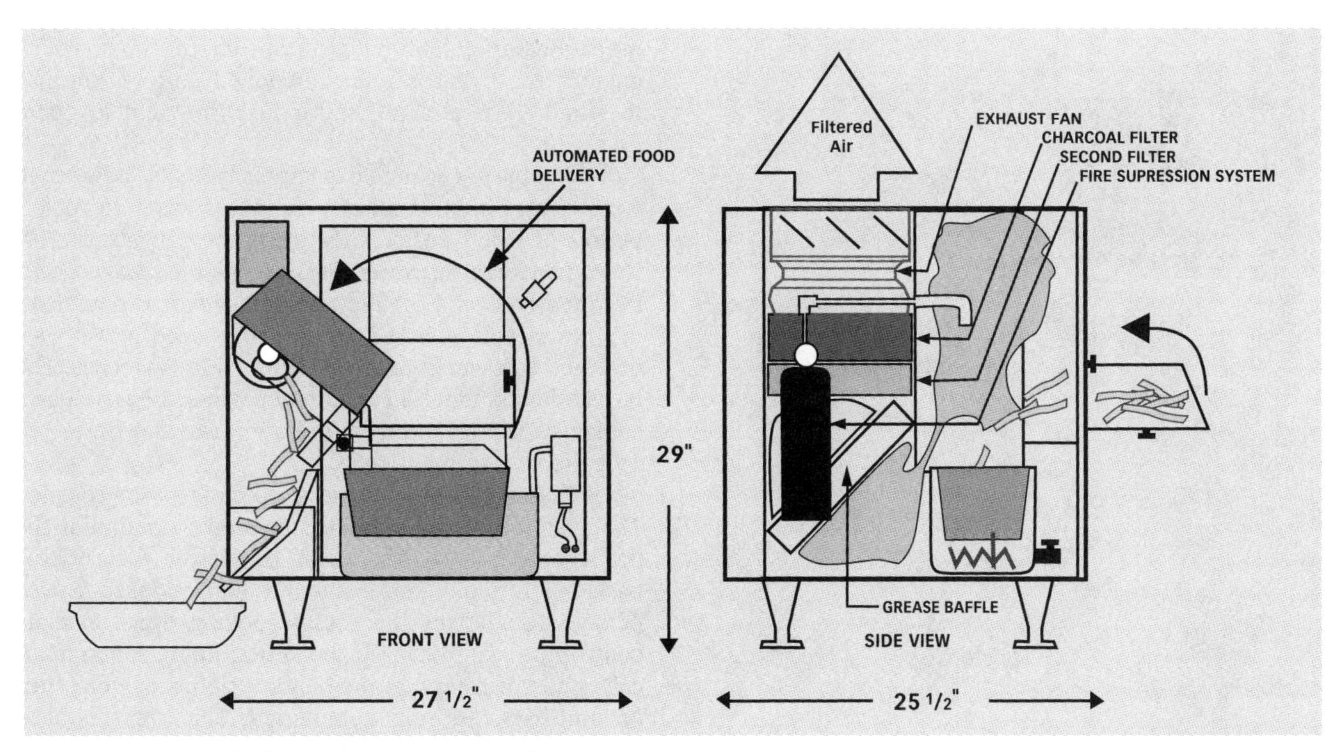

ILLUSTRATION 12-9 **An interior view of a ventless fryer.**
THIS AUTOFRY™ MODEL DIAGRAM IS COURTESY OF MOTION TECHNOLOGY, INC., NATICK, MASSACHUSETTS.

For mall food courts, sports arenas, sites with very limited space, and/or in historic buildings where modifications are not permitted, ventless hood frying is perhaps the only choice. However, it is important to check first with local fire officials to be sure your building and fire codes permit their use.

These fryers began as countertop models, about 34 inches high, on 4-inch adjustable legs. Place the food in a front hatch, push a button, and the fryer automatically slides it into a basket and sends it into the hot oil. When the preset cooking cycle is finished, the basket automatically tilts itself for ten seconds to drain, then dumps the food into a chute, where it rolls into a catch pan located outside the unit. The self-contained fryer takes six minutes to heat its 27 pounds of oil, and can cook for up to ten minutes per cycle. There are now much larger, floor-standing models.

One attractive feature of the self-contained fryer is its built-in systems: filtering, draining, and air purifying. The drain setup eliminates the danger of handling hot oil by draining the oil directly from the fryer into a sealed safety canister. And, because the smoke and grease-laden moisture from the frying process pass through an air purifier, there is no need to vent this fryer under an exhaust hood. When the oil temperature reaches 413 degrees Fahrenheit, or when the unit has operated for more than 16 consecutive minutes, it shuts itself off.

Another option is a stand-alone ventless hood unit with a built-in fire suppression system that can be purchased separately and used with a traditional fryer.

Pasta Cookers

The nearest cousin to the fryer is the pasta cooker, growing in popularity for small-batch cooking. It's a kettle seated in a cabinet that is similar in construction to the fryer kettle. It comes in both electric and gas-powered models, and is designed to cook, rinse, chill, and reheat—a system tailor-made for perfect pasta. The two-tank cooker looks like a two-kettle fryer. On one side, pasta is cooked in bulk in stainless steel baskets; on the other, it is reheated in convenient portion-sized plastic baskets (see Illustration 12-10).

There are two advantages to using a pasta cooker. One is speed—there are gas models with more than 150,000 Btus of heat input, which will bring a full tank of water to boil within ten minutes. The

ILLUSTRATION 12-10 **Pasta cooker.**

COURTESY OF VULCAN-HART, DIVISION OF ITW FOOD EQUIPMENT GROUP, LLC, LOUISVILLE, KENTUCKY.

other advantage is having an automatic timer, which ensures the pasta is not overcooked and no one has to stand there and test it for just the right al dente doneness.

The pasta cooker's capacity is rated by how much water it can hold. It takes 1 gallon of water to cook 1 pound of dried pasta, and the pasta will absorb $1\frac{1}{2}$ times its weight in water. This means the smallest gas-powered cooker, a 6-gallon pot that holds 6 pounds of dry pasta, will yield 15 pounds of cooked pasta. Gas-powered kettles range in size from 14 to 24 inches. The largest holds about 17 gallons of water. Electric pasta cooker capacities are slightly larger, ranging from 7 to 19 gallons of water.

There are manual and automatic pasta cookers. The manual models require someone to maintain the proper water level and raise and lower the cooking baskets into the water. In the automatic cooker, a push of a button begins the preset cooking cycle. Sensors control the water level in the kettle, filling it automatically and ensuring that the heating elements won't turn on until they are fully submerged. This prevents accidental damage to the kettle from elements that might overheat and scorch if the water level is too low. Automated arms lower and raise the baskets full of pasta, which must still be moved and rinsed by hand after cooking.

Because cooking pasta produces excess starch, a pasta cooker also has an overflow valve and drain mechanism at the front of the unit, allowing starch and foam to be drained off regularly. The largest, floor units may have an optional pasta-rinsing sink beside the cooking and reheating kettles. The baskets that fit into the pasta cooker can be full size or half size (sometimes called a ***split basket***). For individual orders of different types of pasta, a carousel is a convenient option. It holds up to nine plastic perforated cups, which can be cooked simultaneously.

A sauce warmer is another option on some models, capable of warming a standard-size steam table pan.

Thermostats on the front control panel gauge water temperatures from 60 to 250 degrees Fahrenheit. A series of indicator lights alert you when the kettle is being filled, heated, or simply when the power is on.

Pasta cookers have some basic plumbing requirements, because they must be hooked up to a water source and have the capability to be filled and drained automatically. A 1-inch (or $1\frac{1}{4}$-inch) drain pipe is located at the front of the cabinet and extends 6 inches up from the floor. Your local plumbing code will specify whether this pipe should be open and whether it should contain a grease trap.

SUMMARY

Frying is the process of cooking food in hot fat or oil, but with today's "greaseless" frying options, it can also be done with very hot, fast-moving air that cooks the natural oil already found in some foods.

Done properly, this method produces food that is as tasty as it is attractive. Traditional fryers are made up of several basic parts: the kettle or bin, where the oil is placed and heated; the basket, where food is placed to be lowered into the kettle; and the cabinet, which houses the whole process, with room for the kettle and equipment to clean and filter the system. The kettle contains an all-important cold zone, located below the heating elements. It is a deep "well" that is colder than the rest of the kettle, where crumbs and sediment can fall.

To prolong the useful life of your oil, it is absolutely necessary to filter and change the frying oil regularly. Frying oil can break down when exposed to water, sediment, salt, and oxygen, all of which are common in the frying process. It can also burn if heated too much, especially if it already contains debris that hasn't been filtered out. Signs of oil deterioration include foaming, smoking, and a pungent odor. Some fryers are drained and cleaned by hand; others contain their own internal filtering system. Portable filtration systems can also be purchased and hooked up to the fryers as needed.

When shopping for a fryer, ask about its heat recovery time. This is the time it takes for hot oil to return to its optimum cooking temperature after you drop in a batch of cold food. The heat recovery time should be no more than two minutes, to prevent the food from absorbing too much oil and becoming soggy.

Fryers are powered with electricity or gas, and there are also gas infrared fryers that work more quickly than traditional fryers, as well as pressure fryers that use a combination of heat and low-temperature steam to cook food. A fryer is chosen based on its output capacity, usually listed in pounds per hour by manufacturers.

Many restaurants serve enough french fries to warrant the setup of a fry station, an area that includes one or more fryers and filtering systems, landing space for the food, warmers, racks to hang fryer baskets, a bagging area just before serving, and storage space for products.

Fryers should always be installed by a professional who is familiar with this very specialized type of equipment. Traditional fryers turn out enough heat that they must be installed under hood ventilation; greaseless fryers are the only options in spaces where hood ventilation is either impractical or impossible.

A "cousin" to the fryer is the pasta cooker. Like a fryer, it contains a kettle seated in a cabinet, and is designed to cook pasta, then reheat it as needed in individual portions. Since the kettles hold water, they require a plumbing source in addition to gas or electricity.

STUDY QUESTIONS

1. What factors combine to break down frying oil over time? Can any of them be prevented to lengthen the life of the oil?
2. What is the *heat recovery time* of a fryer, and why is it important?
3. In determining a fryer's output, what is more important: how much oil it holds at one time or how much product it fries in a certain time period?
4. What kinds of precautions are taken to make sure a fryer is safely installed?
5. Why would an *automatic turn-down* feature be beneficial on a fryer?
6. What is a *pressure fryer,* and why would you use one?
7. In addition to the fryer itself, what else do you find in a *fry station*? List at least four of the six components.
8. What are the signs that your frying oil is deteriorating and should be changed?
9. How does a *pasta cooker* work?
10. What are the four things you should find out about a fryer, once you've decided it is in your price range?

CHAPTER 13

PREPARATION EQUIPMENT: BROILERS, GRIDDLES, AND TILTING BRAISING PANS

INTRODUCTION AND LEARNING OBJECTIVES

In foodservice, the terms *broiling, grilling,* and *griddling* seem to be used interchangeably since they are all types of dry, radiant heat cooking. In many cookbooks, the words "broiler," "grill," and "griddle" are often used as though they refer to the same piece of equipment. So, for the purposes of this text, let's clarify these terms.

A ***broiler*** is a piece of surface cooking equipment with its heat source located above the food (an ***overbroiler***) or below the food (an ***underbroiler***). The food rests on a *grate,* and marks are left on the food from its contact with the hot grate. The act of creating these "burn marks" is called ***scoring*** or *branding.*

A ***griddle*** is also a piece of surface cooking equipment. Its heat source is located below the food, and the griddle is usually a smooth, solid surface. However, a griddle can be ordered with grooves in it, for other types of cooking.

Technically, a ***grill*** is not a cooking appliance. Rather, it is a grid of metal bars set over a heat source, on which food is placed to cook. However, outdoor barbecues for home use are often called grills, and the word "grill" is also used commonly in restaurant names, as in "Clancy's Bar and Grill."

You will see the words "broiling" and "grilling" used interchangeably, but "griddling" refers specifically to cooking something on a solid-top griddle. The point of all this wordplay is to help you notice all three terms, and make it a point to clarify in your own mind exactly what is meant when they appear in recipes, sales literature, or any other text.

To further confuse the novice, each of the two appliances has several common names. In this chapter, you will learn to recognize those, as well as the following topics:

- How broilers and griddles operate
- What different types of broilers and griddles are on the market, and their uses
- Maintenance tips
- Installation and utility requirements

And, finally, we'll introduce you to the ***tilting braising pan,*** another versatile piece of equipment that complements (and even replaces) the griddle in some kitchens.

13-1 BROILERS

At its simplest, broiling is an encounter between food and fire, with intense, direct heat creating a beautifully browned exterior and moist, flavorful interior. Professional chefs say the pure, true flavor of food is best showcased when it is broiled. But this preparation method has become more precise and complex with the assimilation of "world flavors" into U.S. kitchens, creating the need for bolder flavor profiles using spices, herbs, chiles, and other gourmet ingredients. Today's chef soaks bundles of herbs, grapevines, or mesquite chips in water and throws them onto the fire to add aroma and flavor, then finishes the broiled food with a dollop of herbed butter.

Broiling is a dry-heat method of cooking, which means little or no liquid is used in the process. Instead, the food is cooked by its proximity to direct radiant heat. Broiling gives foods an appetizing appearance, complemented by an inviting aroma that is the result of juices dripping down on the radiant heat source to produce the unique, smoky flavor. It's a popular cooking method because it is convenient and quick, but most consumers don't realize what an art form it is, or the value of the skilled and experienced "broiler man" in a kitchen—who, nowadays, can just as easily be a "broiler woman."

The ideal in broiling is to cook the surface of the food perfectly, while keeping the interior of the product tender and juicy. Because direct heat will perform the former function much more quickly than the latter, broiling requires close attention and a true sense of timing. To get it right, there is no substitute for experience.

The chef must also take care to select only top-quality meats and vegetables, since broiling will not mask the imperfections of inferior produce or tough cuts of meat. Broiling should not be restricted only to steaks, because this cooking method is easy and popular among health-conscious diners who want fresh, simply prepared foods.

To achieve these ideals, each operation requires something different from its broiler(s). Some need large volume; others want a broiler that is attractive as well as functional, for front-of-the-house exhibition cooking. Broilers are generally lumped into three broad categories:

- The *overhead broiler* (also known as *overfired,* or an ***overbroiler,*** or a *hotel broiler*)
- The *charbroiler* (also known as *underfired,* or an *open-hearth broiler*)
- The *specialty broiler* (their names vary; you'll "meet" them later in this chapter)

Let's take a closer look at each of these categories.

The Overhead Broiler

The term "overhead broiler" refers to the fact that its heating element is located above the food being broiled. It is a heavy-duty piece of equipment designed for high-volume output. In the typical battery of commercial kitchen appliances, it is installed beside the range. It measures the same width and height as the range.

A big advantage of this type of broiler over other types is the separation of the radiant heat source from the drip pan at the bottom of the broiler. This eliminates most of the smoking and danger of grease fires.

Modern broilers usually preheat in less than two minutes. The broiler compartment is heated by burners, either ceramic or infrared. The burners may be located at the top and center of the broiler, spaced evenly across the top width of the broiler, or placed on the sides of the unit, aimed inward at the food. Gas burners use between 65,000 and 100,000 Btus per hour; electric burners use from 12 to 16 kilowatts per hour. Gas infrared broilers can reach full operating temperatures within 90 seconds, with a blower that mixes air and gas and can cut conventional broiling time in half (see Illustration 13-1).

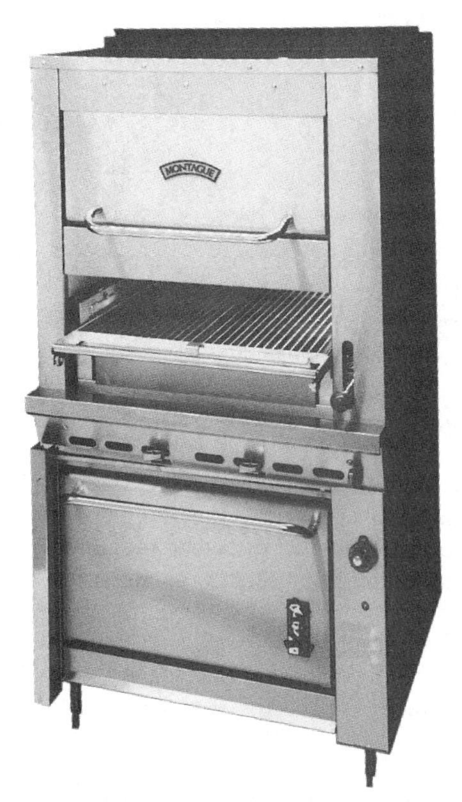

ILLUSTRATION 13-1 An overfired gas broiler with a conventional oven compartment below.

COURTESY OF MONTAGUE COMPANY, HAYWARD, CALIFORNIA.

There's no wasted space on a commercial broiler. You'll find a high shelf directly above the broiling area, an overhead oven instead of the shelf (heated by the same burners as the broiler, used primarily as a warming or holding space), or a standard range-type oven underneath the broiler. When an oven is part of the appliance, it has its own separate control panel.

If there's no need for an extra oven, you can install a second broiler unit. The ***double broiler*** needs a space at least 76 inches tall, including room for its 6-inch legs or casters. One criticism of the double broiler is that it can be uncomfortable to use. When the top broiler is working, the chef's face is parallel to the hot broiler; when the bottom broiler is working, the chef's body gets the heat. So experts suggest using a double-broiler arrangement only when space is at a premium and production needs are high. In any case, to reduce energy consumption and give the chef a break, make sure the broiler cavity is well insulated, and turn it off whenever it's not in use.

Every broiler has a *flue,* which vents smoke and fumes out the rear of the unit. Most installation guidelines suggest at least an 8-inch clearance at the top of the flue. Local ordinances also usually require broilers to be located under the exhaust hood and their exhaust requirements are substantial, from 900 to 1200 cubic feet per minute (CFM).

The ***grid assembly*** is the metal grill on which food is placed to be broiled. It should roll in and out smoothly on ball bearings, and should be easily adjustable to place it closer to (or farther from) the heat source, depending on what's being broiled. Today's manufacturers offer up to a dozen stop-lock grid adjustments, between 1½ and 8 inches away from the burners. To score or brand steaks for a charcoal-fired look, the broiler operator preheats the grid by raising it closest to the burners for five to ten minutes before loading it. Sometimes—in a high-volume banquet situation, for instance—the meat is scored hours, or even days, ahead of time, then placed in sheet pans, refrigerated, and finished in convection ovens prior to serving. Cooking speed is determined by the distance between the food and the heating elements, plus the size and/or thickness of the product being broiled. A chart of suggested broiling times for meats is reproduced in Table 13-1, courtesy of Texas Utilities, an electric company.

At its topmost position, the grid is capable of reaching temperatures near 600 degrees Fahrenheit. So it's obvious the grid should have a pistol-grip-style handle, which stays cool to the touch and is easy to hold. A sloping ***grease tray*** located under the grid reflects heat up and aims grease runoff into a ***drip pan*** at the bottom of the broiler cavity. The drip pan can be removed for cleaning.

Use of an overhead broiler requires practice. A series of switches controls the heat. There is a separate switch for specific zones (areas) of the broiler, and each switch can be adjusted to low, medium, or high heat. The three heat settings can be misleading to a newcomer. "Low" produces a heat setting one-fourth that of "High" over the entire section of the broiler controlled by that switch. "Low" is not for broiling, only for holding foods. "Medium" provides broiling heat, and "High" produces cherry-red radiant heat. The broiler should be preheated for five to ten minutes before use. During idle periods, the broiler can be used for browning casseroles, making croutons, or keeping food warm.

In overhead broiler installation, there are a couple of important considerations. If the broiler is located beside a fryer, open-hearth broiler, or griddle, consider ordering it with stainless steel sides. Cleanup of the inevitable spatters will be much easier. Install it well away from side and back walls for adequate ventilation space.

Avoid locating a refrigerator next to the broiler. You'll waste a lot of energy as the two run constantly to compensate for each others' cold and heat. However, you do need to have a refrigerator near the broiler area, so the broiler operator has easy access to it.

Finally, consider casters and so-called "quick disconnects" to move the broiler more conveniently for cleaning. Ask the manufacturer or installer about this, so it can be done without disrupting the gas or electric supply or the fire suppression system.

TABLE 13–1

OVERHEAD BROILER COOKING GUIDE

Meat	Thickness (inches)	Weight Range (pounds)	Total Time[a] (minutes) (One-half time on each side) Rare	Medium	Well Done
Beef					
Club and rib steak	1	1–1½	6–8	8–10	10–12
Porterhouse	1	1½–2	6–8	8–10	10–12
Porterhouse	1½	2½–3	8–10	11–13	14–16
Porterhouse	2	3–3½	10–12	13–15	16–18
Sirloin	1	2½–3½	6–8	8–10	10–12
Sirloin	1½	3½–4½	8–10	11–13	14–16
Sirloin	2	5–5½	10–12	13–15	16–18
Ground beef patties	¾	6–8 ounces each	3–4	4–6	6–8
Tenderloin	1		6–8	8–10	10–12
Lamb					
Rib or loin chops (1 rib)	¾	2–3 ounces each	8–10	10–12	
Double rib	1½	4–5 ounces each	11–13	14–16	
Lamb shoulder chops	¾	3–4 ounces each	8–10	10–12	
Lamb shoulder chops	1½	5–6 ounces each	11–13	14–16	
Lamb patties	½	6–8 ounces each	4–6	6–8	
Ham and sausage					
Ham slices	½	9–12 ounces each	6–8		
Ham slices	¾	1–1¼	8–10		
Ham slices	1	1¼–1¾	10–12		
Pork sausage links		12–16 to the pound	4–5		
Broiling chickens (drawn) halves	1–1½		18–22		

[a]It is recognized that broiling times may be varied by chefs to suit individual tastes and preferences.

Note: The suggested broiling time is based on a thoroughly preheated broiler with units switched on high heat. The distance of the grid from the heating units will vary from 1 to 2 inches for thin and rare steaks to 3 to 4 inches for thick, medium to well-done steaks.

Source: Texas Utilities Electric Company Dallas, Texas.

The Charbroiler

Charbroiling is the term for broiling food with flames, smoke, and radiant heat. It is popular in display kitchens, because it's showy and fun to watch, but charbroiling also imparts a nice, charcoal flavor to the food. The *underfired* broiler has its heating elements located beneath the food. The flames and smoke result when the cooking juices drip onto the heating elements. The open flame

ILLUSTRATION 13-2 **A gas charbroiler.**
COURTESY OF WOLF RANGE COMPANY, COMPTON, CALIFORNIA.

look has also prompted the name *open-hearth broiler* (see Illustration 13-2).

Unlike the overfired broiler with its movable grate and grid assembly, the underfired broiler cooks food at a fixed distance from the heat source. Its grate may have only two or three positions. Instead, it is the cooking temperature that varies. The chef can broil at anywhere from 300 to 700 degrees Fahrenheit, but charbroilers are at their best when they are between 550 and 625 degrees Fahrenheit. Heat is transferred in three ways: conduction (from the heat of the broiler grate); convection (as heat transfers from grease and vapors produced by burner combustion); and radiation (as heat reflects upward from the radiants).

The underfired broiler is much more compact than the overfired. It preheats faster, and its infinite-heat controls allow precise temperature control that takes much of the guesswork out of producing uniformly high-quality broiled foods. For comparison with overhead broilers, we've reproduced a chart of suggested cooking times for charbroiled foods in Table 13-2.

The four basic fuels used to create heat for charbroilers are charcoal, wood, gas, and electricity. They vary in their efficiency, ease of use, and ability to impart flavor to food. Charbroilers are the modern, high-volume answer to dad's backyard charcoal barbecue, and the ones

TABLE 13–2

CHARBROILER COOKING GUIDE

Wells Power Charbroiler

Steaks	Doneness Desired	Infinite-Heat Control Setting[a]	Suggested Cooking Time (minutes)
Frozen			
½ inch thick	Rare to medium	8	4–7
¾ inch thick	Rare to medium	7	7–8
1 inch thick	Rare to medium	7	12–14
1½ inch thick	Rare to medium	6	23–25
Refrigerated			
¾ inch thick	Rare to medium	High	4–5
1 inch thick	Rare to medium	High	7–8
1¼ inch thick	Rare to medium	High	12–13
1½ inch thick	Rare to medium	High	14–18
Canadian bacon	Medium	400	2–3
Minute steaks and boiled ham	Medium	500	2–3
Hamburgers	Medium to well	550	2–4
Sausage patties	Medium to well	550	3–4
Ham steaks and beef tenderloin	Medium	600	5–7
Club steaks	Rare to medium	700	4–7
New York strip steaks	Rare to medium	700	5–10

[a]Infinite-heat controls are adjustable from 1 to 8 and to "High." Thermostat is adjustable from 300 to 700 degrees Fahrenheit.

Source: Texas Utilities Electric Company, Dallas, Texas.

that use charcoal are simply larger versions of the backyard model. Gas charbroilers use ceramic, stainless steel, cast iron, or igneous rock ("lava rock") to form a radiant bed above the gas burners that heat it, and so these items are known as *radiants*. When highly porous lava rock or ceramic is used, it must be replaced periodically to avoid grease or carbon buildup. Electric models may have long, cylindrical heating rods, a cast iron grate located above electric burners, or heating units embedded in the underside of the broiler grates (see Illustration 13-3).

Burners underneath the radiants may be made of steel alloy or cast iron. Steel alloy is the least expensive. One thing to look for in selecting a broiler is how the burners are shaped and spaced. There are straight tubes, U-shaped, H-shaped, and S-shaped configurations. A burner produces a row of flames for every 6 inches or so of its width, although some models "stretch" that to 9 inches. Just remember, the closer the rows of flames are together, the more even the bed of heat will be. If the rows of flames are far apart, they create hot and cold spots on the broiler surface. This is not necessarily a bad thing if you're trying to cook two steaks simultaneously, one medium and one well done, and know where to put them on the broiler to achieve these different results. But if the temperature variance is too extreme, or if you are not familiar enough with the appliance, you may have problems.

The burners beneath some radiant beds include ***reflectors***—also made of stainless steel or cast iron—that absorb heat and aim it toward the grate in a more concentrated fashion than a simple open flame. Stainless steel radiants are less expensive, preheat quickly, and are easier to clean than their cast-iron counterparts. Cast iron may be slow to heat, but it holds its heat longer and is therefore handy for long shifts of high-volume production. Both steel and iron radiants are popular among high-volume operations because they provide even heat and require minimal handling—no turning stones, replacing lava rock, and so on. However, rocks and briquettes look great for display cooking, and some "broiler men" like moving them around to get exactly the results they want. (Incidentally, there are kits to convert a radiant broiler to a briquette one.)

Manufacturers also offer a variety of grates, and spacing of the rods on the grate is important. The optimum is probably $5/16$ of an inch, and no more than one-half inch spacing; otherwise you will drop pieces of foods with light or flaky surfaces. Here are your "really grate" choices:

FLOATING ROD. Usually chrome plated, ***floating rods*** are easy to clean and prevent sticking. They are pretty much standard on most charbroilers today. They're also called *free-floating grates,* because they allow for repeated expansion and contraction without warping.

CAST-IRON FLAT. These grates are excellent for heat retention, but awful for foods that tend to stick. They are also more prone to warping.

CAST-IRON WAVY. These also retain heat and are made with the score-markings that have come to identify broiled foods.

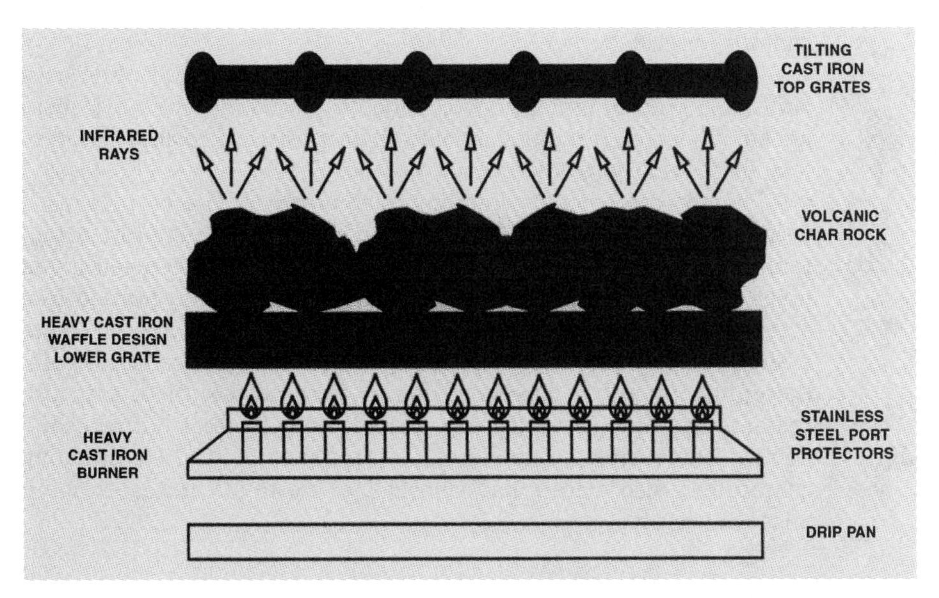

ILLUSTRATION 13-3
A side view of the inner workings of a gas charbroiler that uses lava rock.

MEAT GRATES. These contain fine marks to make the broiled meats look like they've been scored. They are designed to deliver maximum heat for thick cuts of meat.

No matter what type of charbroiler you choose, the combination of heat, smoke, and flame raises safety concerns. Grease drips onto the heat source during cooking, although most grates are tilted slightly to allow at least some of it to flow away. An accumulation of grease in the drip pan can catch fire. Some manufacturers have added baffles to separate the drip pan from the heat source, but the simplest precaution is to keep an adequate water level in the drip pan, check it, and clean it frequently.

Charbroiling produces a lot of smoke, as well as grease. Most city ordinances require the broiler to be under an exhaust canopy and that an air filter be located no more than 48 inches from the surface of the broiler. A separate flue and/or exhaust system may be required for wood or charcoal broiling. In short, investigate local fire and health codes before you purchase anything that broils using live flames.

There are three types of charbroilers that require special mention: the drop-in, countertop, and freestanding models. The drop-in charbroiler is electric, installed in a stainless steel counter with a gasket separating the broiler from the countertop (see Illustration 13-4). The heating element doubles as the grid in this compact unit, which measures from 16 inches square to 16 by 32 inches and uses only 10 kilowatts of power per hour. The drip pan may be a drawer, equipped with a splash guard and a funnel-like receptacle, all removable for cleaning.

A simple front control panel includes a switch (with "Off," "Broil," and "Clean" settings), a dial (with "Off," "Low," and "High" settings), and a signal light indicating whether the unit is on.

Despite their small size, there are very specific installation requirements for drop-in charbroilers. Their exhaust requirements are higher, ranging from 900 to 1200 cubic feet per minute. The unit must be located at least 1 foot from any side wall, at least 5 inches from a back wall, and about 4 inches from any other piece of countertop equipment.

Drop-in charbroilers are meant for restaurants or other foodservice facilities with low production needs. However, they may be a valuable addition to any small operation strapped for space. They come in widths that range from 15 to 42 inches, with a standard depth of 24 inches.

The countertop charbroiler is a stand-alone unit. It may run on gas or electricity. The electric countertop model looks similar to the drop-in charbroiler. It has 4-inch plastic or stainless steel legs, which are both removable and adjustable, and stands about 16 inches high. This broiler can fit snugly in a space no more than 2 feet square, and uses 10 kilowatts of power per hour.

The gas-powered units seem to be more popular than the electric ones, perhaps because the grates come in a variety of widths, from 24 to 72 inches. Gas countertop charbroilers are taller than electric ones, from 25 to 36 inches, and they generally hold more food. They can be mounted directly onto a counter or on a stainless steel stand. The burners are arranged so that there's one for every 6 inches (or 12 inches) of broiler width. Each burner uses between 15,000 and 45,000 Btus per hour, and there are usually individual heat controls for each burner.

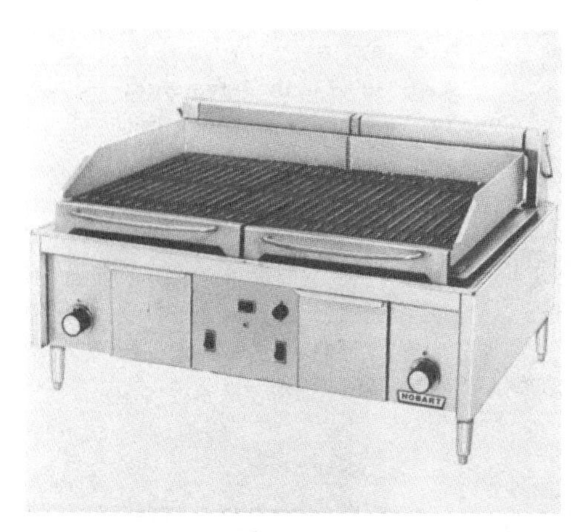

ILLUSTRATION 13-4 A drop-in charbroiler for countertop use.
COURTESY OF HOBART CORPORATION, TROY, OHIO.

Like their full-sized counterparts, these smaller units often have adjustable grids or grates on which food is placed for broiling. If these are not adjustable, they at least slant slightly forward, to allow juices to flow into a built-in grease trough.

Finally, we have freestanding or floor model charbroilers, used in large facilities as part of a production line-type output or in restaurants with a brisk à la carte business. The broiler space itself is not much bigger than the largest countertop unit, but it is housed in a stainless steel cabinet for storage beneath the broiling area. It's generally placed on a floor stand, on casters to facilitate cleaning. Again, the grates are either positioned flat or they can be titled; cast iron radiants or ceramic "coals" are heated by individual burners; and there is one burner for each 6 inches of broiler width. Freestanding charbroilers also feature backsplashes, of up to 10 inches, to minimize spattering.

13-2 SPECIALTY BROILERS

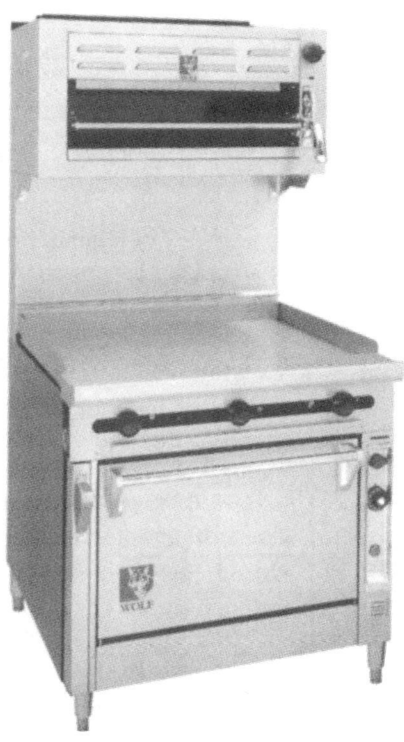

ILLUSTRATION 13-5 **The salamander is often installed above the hot line, so it can be used as a food warmer as well as a broiler. This model also has a fry-top range (griddle) and oven below.**

COURTESY OF WOLF RANGE COMPANY, COMPTON, CALIFORNIA.

Salamander. Perhaps its unusual name derives from the fact that the ***salamander*** is so adaptable, fitting into its surroundings much like the lizard that is its namesake. A miniature version of the heavy-duty broilers, it measures from 10 to 13 inches deep and from 23 to 28 inches wide. It's small enough to rest on a shelf or counter, or it can be mounted directly above a range or spreader plate. During slow times when the main broiler is turned off, it can be used as an energy-efficient backup broiler. Or, when things are really hectic, the salamander makes a convenient plate warmer (see Illustration 13-5).

Gas-powered salamanders may have anywhere from one to six infrared burners. Depending on the number of burners, its gas input requirements are from 30,000 to 66,000 Btus per hour. Electric salamanders vary from 5.2 to 7 kilowatts per hour, and require outlets of either 208 or 240 volts. Typically, control panels include a switch to turn the broiler on to "High," "Medium," or "Low," as well as controls that allow the operator to turn on the left- or right-side elements separately or all together.

Infrared salamanders offer several advantages. They preheat in about 90 seconds instead of the 12 to 15 minutes a radiant-heat model takes. When not in use, radiant-heat units are kept in "standby" mode, which is not necessary with infrared units. And infrared units keep the kitchen cooler, because they heat only the food, not the surrounding airspace.

Grids or grates are removable for cleaning and can be positioned and locked in place between 1⅓ and 4 inches away from the heating units. Grease deflectors are attached to the underside of the grids, channeling hot drippings to a drain pan or drawer that is also removable.

If the salamander will be installed on a high shelf, or above any combination of hot range equipment, it should sit at least 18 inches above the lower workspace, and the shelf should be specially reinforced to prevent an unexpected tumble onto the hot top below. Some manufacturers charge extra for heavy-duty wall mounting brackets, but they are worth the cost. If it's going to sit on a counter, you might also consider more expensive stainless steel legs, which are sturdier than plastic.

Cheesemelter. The ***cheesemelter*** is another type of specialty broiler. Its name explains its most common use: to melt cheese, primarily on Italian, Mexican, and Tex-Mex dishes. While it accomplishes that task quite nicely, it can also be used to brown, poach, and boil. However, its primary function is to "finish" food, not cook it, so cheesemelters often have lower overall heat output than salamanders. Some units are constructed so that they only turn on when food is placed on the grid or rack, thus saving energy (see Illustration 13-6).

Installation methods are similar to the salamander—as a countertop or wall-mounted unit—but there are a few models with a pass-through feature similar to the pass-through refrigerator with doors on two sides.

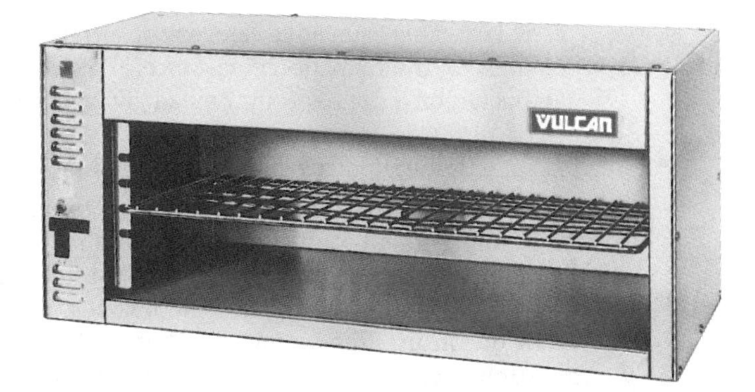

ILLUSTRATION 13-6 **The cheesemelter is used most often for what its name implies: melting cheese atop foods just before serving.**

COURTESY OF VULCAN-HART, DIVISION OF ITW FOOD EQUIPMENT GROUP, LLC, LOUISVILLE, KENTUCKY.

FOODSERVICE EQUIPMENT

BROILING TECHNIQUE TIPS

1. Tender cuts of meat are best suited for the intense heat of grilling. Cuts should be thick to avoid overcooking before properly browned. Pat food dry and season with salt and pepper before grilling or brush with marinade toward the end of cooking. Skewer small pieces of food, such as shrimp, for easy handling and consider firmer, fattier fish species that are forgiving if overcooked. Less fatty or thinner fillets of fish can be grilled if brushed with oil.
2. Just as the food must be correctly prepared, a properly primed grill is essential. To prevent food from sticking, use a cloth moistened with a high smoking point oil to grease grill rods. Also oil metal skewers and soak wooden ones in water. Some chefs brush on or dip ingredients in oil (allowing excess to drip off) before placing them on a preheated grill. Scrape off any burnt particles throughout service.
3. Place food on preheated grill presentation-side down first and resist the temptation to move food before a good sear is established. Turn food over only once during cooking; constant flipping will interfere with carmelization and impede juices from moving to the center of the food where moisture is retained. To create crosshatch grill marks, rotate the meat, fish, or chicken a quarter turn and continue cooking before flipping it.
4. Section the grill into different temperature zones, with the hottest for searing, a moderate temperature for cooking, and lower heat for holding food. Dividing the grill by types of food can also can prevent the transfer of undesirable flavors and add speed to the line. Because food continues to cook after it leaves the heat, advise novice cooks to remove the food just before the desired temperature is reached, particularly leaner foods that are less forgiving.

Source: ***Restaurants and Institutions*** **magazine, a division of Reed Business information, Oak Brook, Illinois, March 15, 2002.**

Most cheesemelters, from the 2-foot widths to the 10-foot widths, have only a single, adjustable (three-position) rack on which to place food. However, additional racks can often be installed.

Cheesemelters can be powered by gas or electricity. The infrared units seem to be the most popular, because they preheat within one or two minutes and eliminate the need to leave them warming indefinitely on a standby mode. As with other broiler types, the burners are individually controlled, and there is a removable drip pan.

Food Finisher. This innovative category of surface cooking equipment has emerged in response to operators' demands for multifunctional, resourceful pieces of equipment. "Finishing" typically means putting the final touches on a dish. This may mean giving it a final sear to toast or brown, melt or crisp something. The term is also used to indicate reheating previously or partially cooked foods. The food finisher does all of this, and quickly, requiring about half the time as products heated in a conventional oven.

Whether it's a food kiosk that needs fast cheese melting over nachos, or a hotel that provides limited-menu 24-hour food service, the simplicity of the machine facilitates easy operation. It looks like a cross between a microwave oven and a toaster oven (see Illustration 13-7). Unlike a salamander, which stays at full power throughout the business day, the food finisher can switch itself to standby mode when not in use, remaining "on" but at 20 percent of its full power until it is needed again. Some restaurants use these instead of cheesemelters and/or salamanders.

Conveyor Broiler. There are three important advantages to conveyor broilers: speed, reduced labor, and consistent results. These broiler types cook faster because foods are exposed to both top and bottom heat. A burger that would normally take four or five minutes to broil can emerge from a conveyor broiler in less than two minutes. If you must broil quickly to keep up with volume, or if you simply need a broiler than can react quickly to a surge in demand, this is it.

The downside of a conveyor broiler is that the food items must be uniform in size and thickness to cook evenly. For hamburgers or frozen waffles, this is no problem; for steaks or chicken breasts, it might be. The largest conveyor broilers have as many as three separately controlled, programmable conveyor belts and heat chambers that can all work at once, on different settings. Conveyor broilers can be all-gas, all-electric, or a combination of gas burners on the bottom and electric on top. The smallest countertop unit measures 15 by 23 inches, requires 208 220–240 volts of power, and can cook at least 100 hamburgers per hour.

ILLUSTRATION 13-7 The food finisher is countertop-sized and can do a number of last-minute duties quickly, including reheating, melting, and toasting.
COURTESY OF HATCO CORPORATION, MILWAUKEE, WISCONSIN.

Wood-Burning Broiler. Folks who prefer the "comfort foods" of yesterday to the diners of today may select the old-fashioned, *wood-burning broiler.* It's making a comeback, although few manufacturers produce broilers that burn real wood or charcoal anymore (see Illustration 13-8).

Of course, the commercial model has to be larger than grandma's old wood stove and certainly more efficient. The wood or coal is contained in a firebox inside the broiler. A damper controls the flow of incoming air, which, in turn, controls the intensity of the fire. The fire heats the cooking grates, on which food is placed.

The unit is made of double-wall stainless steel; between these two steel layers are fireproof insulation and a firebrick lining that holds the heat inside. Ashes are removed through two drawers at the bottom of the unit.

Wood-burning broilers come in sizes that give you cooking surfaces as small as 4 square feet or as large as 10 square feet. At least one manufacturer has added an optional rotisserie unit to its wood burner, which may be removable or installed as a permanent fixture.

Rotisserie. This brings up the interesting point that some people consider the rotisserie a type of broiler. After all, it also cooks food over a flame or heat source. You already read about rotisseries in Chapter 11, but there are a couple of specialty appliances that fall somewhere in between these three definitions. One is the *hot dog roller grill* so popular at movie theaters and in convenience stores. The other is the *gyros machine,* a countertop oven with a vertical rotating spit in a glass-front cabinet to permit public display. Gas jets or electric heating elements line the sides of its semicircular wall. A traditional Greek food, the gyro (pronounced "year-oh") is made of a chopped lamb or lamb-beef mixture and formed into a cone-shaped slab specifically to fit onto the spit. The broiled layers are thickly sliced off the cylinder and made into delicious sandwiches.

The third and most recent import to the U.S. is the *churrasco,* an open-flame broiler from Brazil that cooks meats on long, thick spits made to resemble swords. It is the showpiece of South American *churrascarias,* barbecue restaurants that specialize in this type of cooking and casual, all-you-can-eat menus, called *rodizio* service. They've become popular in many U.S. metropolitan areas for their huge portions and reasonable prices.

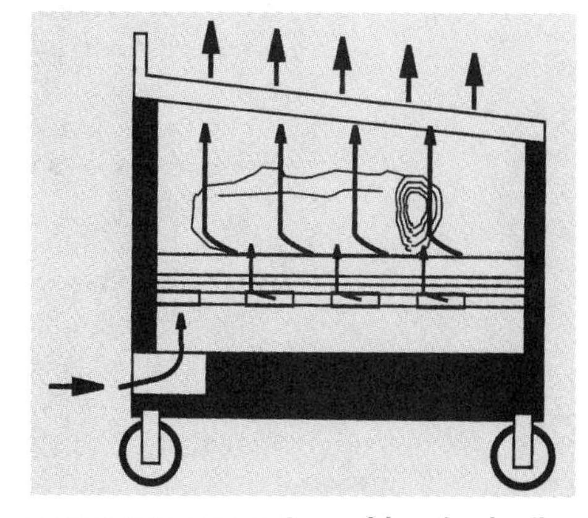

ILLUSTRATION 13-8 A wood-burning broiler creates heat in a deep firebox below the food.

Portable Broiler. We just mentioned the Greek gyros machine, with its side-mounted broiler elements and display case. Another popular option you'll see in mall food courts, at street festivals, corporate picnics, and so on is the *portable broiler,* which offers up to 10 square feet of cooking area. Its heat source can be liquefied-petroleum gas, wood, or charcoal. The wood- and charcoal-burning models have grates that can be adjusted to four positions above the flames. It's wise to order an optional hood for your portable broiler; although it weighs more than 70 pounds, the hood can be removed if not needed, and will keep the appliance in much better shape when not in use. LP-gas models use a 40-pound fuel cylinder that is mounted beneath the firebox area for easy portability. Liquefied petroleum gas provides the advantage of lighting all four burners with the push of a button. Each burner has a separate control knob. One fuel cylinder produces a total of 130,000 Btus.

Comal. The *comal* is a specialty broiler used in Mexican restaurants for display cooking of fajitas. The meat is positioned on a grill over gas-heated briquettes.

13-3 BUYING AND MAINTAINING BROILERS

As you ponder your broiler choices, there are a few points that must be considered:

YOUR MENU. If delicate items will be broiled—say, gratin-type casseroles—you'll need a broiler that heats from the top. For just about every other item, either top or bottom heat will work equally well. Infrared broilers are capable of the highest temperatures, but they work best with foods that are of uniform thickness (like a T-bone steak). Radiant heat works better with foods that have "highs and lows" (like a lobster tail, thicker in the middle than it is on the end). The few other exceptions are specialty items, like Greek gyros, for which a specialty (vertical) broiler is appropriate.

THE HEAT SOURCE. Again, there are multiple choices here: electricity, natural gas, liquid propane (LP) gas, charcoal, hardwood. Electric broilers offer a trouble-free, clean cooking source. The benefits of gas use are high-heat cooking and glowing flames that are attractive to customers in exhibition cooking. Remember, the heat output of a gas broiler drops as much as 25 percent with LP gas compared to natural gas, so be sure to calculate this when determining the size of cooking surface you'll need.

Charcoal briquettes offer quick, long-lasting heat. An alternative, particularly for exhibition broiling, is the lump "charcoal" made from chunks of hickory, oak, or other hardwood. These have none of the petrochemicals found in traditional charcoal, and the random sizes of these pieces make them especially handy—use the smallest pieces for a quick-starting fire, and larger ones for a longer-lasting fire. Their use also means finding a reliable supplier and a place to store them. Fresh wood chips can be used, even in gas and electric models, to impart a smoky flavor.

THE AMOUNT OF FOOD. This includes how much you must cook at one time, as well as how much will be cooked during a set time period. If you're broiling chicken halves for an outdoor catered event and plan to serve them directly from the broiler, you'll need a bigger cooking surface than if you plan to batch-cook them early and hold them in a warmer. When broiling steaks to order, there should be enough broiler surface to hold the maximum number of steaks that may be ordered at any point in time. Size requirements will be determined by a few things:

- The temperature of the food to be broiled
- The size of the pieces to be broiled (both thickness and surface area)
- The temperature to which most of the food will be cooked (medium, well done, etc.)
- Distance between the food surface and the heat source

Once you have determined those basics, you'll find differences between manufacturers in these areas:

GRATE MATERIALS. Cast iron has long been the standard on underfired broilers, but a growing number of them are available with stainless steel grates that minimize food sticking to them. In either case, the grates must be examined closely to ensure they can stand up to the rigors of daily use.

GRATE POSITION. On some broilers, the grates lie flat, while on others, they are slightly sloped. The sloped grates help drain grease away, which is helpful in reducing flare-ups. This minimizes the amount of smoke and potential fire hazards, and makes it a little safer to lean against the front of the appliance, which is the natural tendency as you broil. Some grates are adjustable—they lie flat to be used like a range top, or slope for broiling. Grates should also be easily adjustable "up and down," to move them closer or farther from the heat source depending on what you're broiling.

TEMPERATURE ZONES. Distance from the heat source is key, but some broilers also offer separately controlled temperature zones. The overhead or hotel broiler has two burners with separate control knobs. Some charbroilers offer the highest heat at the back of the grate, medium heat in the middle, and lowest heat at the front. Portable underfired broilers can have up to six sections, each with its own controls.

GREASE TRAY. The location and size of the tray or pan that collects drippings are important. It must be easy to remove and large enough to catch grease from the entire grate area. The pan must be able to hold water, since this minimizes smoking and is helpful in providing moisture as food is cooked. The pan should have baffles in it, to help prevent spilling as it is removed and cleaned.

PREHEAT FUNCTIONS. Some broilers are ready to use within a few minutes; others require as much as 15 minutes to get to optimum broiling temperature. The preheating times are usually longer with charcoal and wood-fired models.

RADIANTS. Depending on whether you want an overfired or underfired broiler, the gas and electric models' heating elements may be made of stainless steel or ceramic. Stainless steel radiants heat up very quickly, but the ceramic "briquettes" or lava rock can be attractive. Infrared models generate the highest heat, up to 1500 degrees Fahrenheit and far hotter than ceramic radiants can reach. The number and type of radiants in these broilers will affect the consistency of the cooking. Around the edges of the grate, you may find hot spots or cooler zones.

PORTABILITY AND FLEXIBILITY. You can purchase a griddle top for use with some broiler models, which will allow you to cook a wider variety of foods. An optional rotisserie or clamshell (discussed elsewhere in this chapter) is available with some models. And, if you plan to use your broiler for off-premise catering, order it with heavy legs and large-diameter casters, bigger than the standard 4-inch casters.

Before we move on to our discussion of griddles, there are a few troubleshooting and maintenance hints that should make life with your broiler a lot easier.

First, the grates or grids must be kept clean for best performance. Food and the inevitable carbon buildup that is part of the broiling process must be scraped from the grates with a spatula or, better yet, a wire *broiler brush* designed for this purpose. Allow this gunk to accumulate and you risk contaminating the food. Daily, or as often as necessary, lift the grates off the broiler and take them to the pot sink for a good, thorough scrub. The grid assembly can also be lifted from the broiler and put in the pot sink for cleaning with a soft wire brush. If your broiler is electric, never use water to clean the heating elements! Wipe surfaces of the unit with a damp cloth, but clean the tubular elements by turning them on high and allowing them to burn off any food residue.

Like many metal surfaces, grids and grates should be lightly greased before use (or after washing) to prevent food from sticking to them. This "seasoning" also helps prevent rusting.

Drip pans should be checked often and emptied and cleaned daily. Just because your unit has a big grease pan doesn't mean you should wait until it's full to dump and scour it. Drip pans can hold up to 1 quart of water, which prevents accumulated grease from catching fire. Check the water level frequently.

Consider replacing lava rocks or ceramic briquettes every six months. The intense heat of broiling will cause the radiants, and even the lower grates, to warp or crack. Check them periodically and change them as needed.

Cooking tips: Preheat the grates before using them to mark or brand foods. Don't press the natural juices out of meat as it broils. Turn it only once on the grate, and don't salt it until it has been cooked and removed from the heat. Salt can damage heating elements.

Inspect and clean the ventilation system above your broiler often, both for safety and for health reasons. On gas-powered broilers, check the burner ports to make sure they're clear of debris. If they are blocked, you can clear them with a wire brush or a drill, fitted with a properly sized bit. If the gas pilot light won't stay on, it may be due to clogged orifices or an air draft that is blowing them out. Scout the kitchen area for the source of the draft.

If an electric-powered broiler has an element that does not seem to be getting hot enough, there's a possibility it's not getting enough voltage. If your unit needs a 240-volt outlet but is plugged into a 208-volt outlet, it may work—just not quite well enough.

Finally, you may live in a region with strict air pollution guidelines that require you to control smoke or grease output from your broiler (especially underfired units) by installing expensive rooftop turbines and electronic precipitators. If this is the case, read on. The grooved griddle, which we'll discuss later in this chapter, may be an alternative for you.

13-4 GRIDDLES

As mentioned earlier in this chapter, the griddle is a piece of surface cooking equipment, a flat-top appliance on which food is heated from below by gas or electric heating elements located directly under the flat surface. In most instances, the food is cooked on one side at a time, and the metal griddle top is coated with oil to prevent it from sticking (see Illustration 13-9).

No matter what the size of restaurant or type of cuisine, the griddle is practically indispensable. It is usually placed in the center of the hot line, or à la carte cooking section of the kitchen, right in the middle of the action along with range tops and fryers.

The benefits to cooking on a griddle are many. They include:

- **The coagulation of proteins, such as those found in eggs, fish, and meat**
- **Flavor enhancements, such as those that occur during browning meats or toasting buns**
- **Tenderizing certain items, like steaks and chicken breasts**
- **Development of a crispness or crust for foods like potatoes or bacon**

However, a griddle's chief disadvantage is that it is not energy efficient. Unless its surface is entirely covered with burgers or pancakes to be flipped, the heat simply escapes into the exhaust hood under which the griddle must sit. Even locating a griddle too close to an outside door can impact its efficiency when cold drafts of air hit it in the winter.

The griddle's cooking surface is also called the ***plate.*** Surrounding the plate, you'll find the ***trough*** and ***splash guard.*** The trough (also called a *gutter*) is a recessed area that surrounds the griddle top on three sides. Grease buildup and food particles are scraped into the trough with a spatula to keep the surface clean and smooth. At the back or far side of the griddle, the splash guard (also called a *fence*) is a 3- to 6-inch-high piece of metal that prevents food from falling behind the griddle, as well as controlling grease spatters.

Whatever the size of griddle surface, you want one that heats evenly, without so-called "hot spots" or "cold spots." Modern research has enabled griddle manufacturers to even out these spots with strategic placement of heating elements and more sensitive thermostats. Most griddle plates are divided into sections with separate heat controls, so that different foods can be cooked simultaneously. These sections are called ***heat zones.*** As you face the griddle, its heat zones run horizontally—from left to right, not from front to back. Each heat zone is between 12 and 18 inches wide, so a large griddle may have up to six of them. Electric griddles seem to have more precise heat zone control than gas griddles; for the latter, a separate thermostat is recommended for each burner to gauge the temperature of each 12-inch zone.

ILLUSTRATION 13-9 **A countertop griddle.**

COURTESY OF VULCAN-HART, DIVISION OF ITW FOOD EQUIPMENT GROUP, LLC, LOUISVILLE, KENTUCKY.

Zoning is a fundamental concept in griddle cooking. Theoretically, you can cook different foods at very different temperatures side by side on two separate zones, a huge advantage in fast-paced kitchens. However, extreme temperature differences may affect each zone to a certain extent, so it's best to cook foods that require widely differing temperatures in nonadjacent zones. Each zone can be set for cooking temperatures from 150 to 450 degrees Fahrenheit.

Beneath the plate, the gas griddle burners are either atmospheric (conventional, individual gas flames) or infrared (combustion within a ceramic housing that heats up and glows instead of producing a flame). Infrared technology eliminates one maintenance problem—the build-up of carbon soot on the underside of the griddle plate from flames hitting it for hours at a time.

Griddle tops range in length from 18-inch countertop models to 72-inch floor models, with depths (from front to back) of between 20 and 32½ inches. If aisle space permits, you might consider adding additional depth, which is available in 6-inch increments. That doesn't sound like much, but on a 24-inch griddle, it adds another square foot of cooking space, or another 3 square feet on a 5-foot-long griddle. A 7-inch-wide *work board* can also be ordered to provide landing space at the front of the plate. However, before you start adding depth, think about the person who has to lean over the griddle to do the cooking. Only add as many "extras" as make sense ergonomically.

Griddles can be retrofitted with a heated top plate, and this plate can be "contact" (the top plate actually touches the food) or "noncontact" (the top plate heats from above, but without touching the food). This also increases the power requirements for the griddle, so be sure your system can handle it. The top plate is typically ordered to fit on just one heat zone, either the far left or far right side, and it may be smooth or grooved.

Yes, most folks think of a griddle as a perfectly smooth surface, but there are also griddles with grooved plates, to create the seared look for burgers or steaks. Grooved plates can cover the entire surface of a griddle or just a section of it. Generally, the grooves are ⅜ of an inch wide, spaced $^{3}/_{16}$ of an inch apart. Some appliances can be fitted with detachable grooves and griddles so they can work like griddles or broilers.

As with broilers, griddles also come in countertop and floor models; the countertop ones may be drop-in or freestanding units. Each can be ordered as a gas or electric appliance.

A drop-in countertop griddle will vary in width from 2 to 6 feet, and in depth (front to back) from 2 to 2½ feet, which gives you a cooking space of between 400 and 1663 square inches. (Or think of it this way: You can cook 32 3-inch burgers on the smallest griddle, up to 120 at a time on the largest.) Like other griddles, countertop models have thermostatically controlled heat zones for every 12 inches of surface width. They're on 4-inch adjustable legs made of plastic or stainless steel. The grease troughs run the full width of the griddle plate, at the front and rear. Incidentally, it's a handy feature to have a trough at both front and rear, because some people prefer to scrape forward, while others prefer the grease and food particles be scraped to the rear of the griddle and out of sight.

The power requirements of electric countertop units vary from 8 to 32.4 kilowatts per hour, and they're available in single-phase or three-phase installations, from 208 to 480 volts. Gas griddles usually have a rating of 20,000 Btus per hour per burner, and there's a minimum of two burners on the smallest unit. Some manufacturers boast 30,000-Btu-per-hour burners. Again, the higher the Btu capacity, the faster the heat recovery time, and that's important when you're loading your griddle with frozen foods.

Countertop griddles may also be ordered with work boards on the front, which give the operator a 7-inch surface for resting plates and product. And you can choose two grease drawers, one at each side of the unit, instead of a single drawer beneath the plate. Griddles can be ordered in combination with open burners; or a special bar can be installed on one side of the griddle onto which a pan can be hooked. The pan can be filled with buns, for instance, to load with burgers or hot dogs as they come off the griddle.

You can make a countertop griddle into a floor-standing model by simply mounting it on a stand that places the griddle surface no more than 36 to 38 inches from the floor. (That's the most comfortable work height for most people.) Or you can buy a floor-standing unit. Either way, the smartest choices for legs are 3½-inch swivel-type casters, with front casters that lock in place to prevent rolling when the equipment is in use, then roll for cleaning or repositioning. In gas models, a quick-disconnect feature also enables easy movement. However, the basic location of the griddle is usually determined by city ordinance: under the exhaust canopy.

Floor models are often sectioned so that different areas can be used for different foods. A griddle in a breakfast operation, for example, will cook hotcakes, bacon, and hash-browned potatoes simultaneously. Before you buy, decide on the different combinations you'll need.

13-5 SPECIALTY GRIDDLES

Heavy-Duty Griddle. A variation of the floor model griddle is the *heavy-duty griddle* with an oven below. It's considered heavy duty because its gas burners are rated at 30,000 Btu per hour. Sometimes, it's referred to as a ***fry-top range*** (see Illustration 13-10).

You can also order griddles with a broiler below, but this is only recommended for low-volume operations (less than 100 meals in one seating) because of the obvious discomfort that kind of heat would cause the operator. Most recently, several manufacturers have introduced infrared griddles, touting energy efficiency and cooking temperatures that remain constant no matter how much food is loaded on them.

Steam Griddle. In recent years, at least two manufacturers have introduced steam technology to griddles. When water is heated under pressure—in this case, in a sealed chamber beneath the griddle plate—it creates superheated steam with a temperature of up to 400 degrees Fahrenheit. Since steam produces amazingly even heat, the temperature does not vary more than 2 degrees anywhere on the plate. This eliminates hot and cold spots, and allows foods to be cooked closer together. Steam griddles also come with top lids (instead of plates). When the lids are closed, steam is injected into the cooking area for even faster results.

Teppanyaki Griddle. Another variation of the floor model griddle is the ***teppanyaki griddle,*** specifically designed for the exhibition cooking so popular at Japanese steakhouses. The oversized teppanyaki is 8 feet long and 6 feet wide, with a countertop around three sides that allows seating of up to eight guests around a 12-square-foot cooking area. The chef stands on the fourth side—slicing, dicing, and searing the customers' food to order, in a fast-paced presentation that's as tasty as it is fun. The three-inch trough is closest to the chef, with a large drain hole and grease drawer beneath. The teppanyaki griddle plate is ¾ inch thick; gas burners are rated at 30,000 Btus per hour each; electric burners have a 208- to 240-volt rating.

Clamshell Griddle. The ***clamshell griddle*** is a unique piece of equipment that cooks food on both sides simultaneously, therefore reducing cooking time by half. Imagine a big waffle iron (without its patterned surfaces) with a 2-foot hood that is lowered over food on the griddle below. The hood automatically ignites when lowered, reaching a temperature of 165 degrees Fahrenheit within seconds. The hood is spring loaded, and can be raised to add or remove food without substantially affecting the cooking speed. Lift it up, and it turns off automatically. The hood of a clamshell griddle does not close all the way, allowing a 5-inch clearance for the food when it is in its lowered position. Its cooking speed allows a 36-inch clamshell to boast the same food output as a 60-inch griddle (see Table 13-3).

The gas-powered clamshell has a stainless steel frame and 1-inch-thick griddle plate. The plate can be flat or sloping. A ***sloping plate*** is tilted slightly to allow grease to drain into a large, removable pan. The hood contains a single infrared burner; each foot of griddle surface is heated by a specially designed burner (27,000 Btus per hour) and controlled by a thermostat. Clamshells come in widths (lengths) of 3 feet and 4 feet, and can be ordered with grooved griddle plates in 1-foot widths.

ILLUSTRATION 13-10 **A fry-top range is a griddle on top, with an oven below.**
COURTESY OF LANG MANUFACTURING COMPANY, EVERETT, WASHINGTON.

Contact Grill. The electric answer to the gas-powered clamshell is called a *contact grill*—and there's that troublesome term "grill," popping up again. At any rate, that's what it's called. The contact grill is a square griddle plate, smaller than a clamshell but also able to envelop the food and cook it on both sides at once. Models are available on legs or casters, with shelves or small steam tables mounted beneath. This appliance requires 12 kilowatts of electric power per hour.

There are two more incarnations of the two-sided cooker worth mentioning. One is Vulcan-Hart Corporation's Duplex Cooker. Every 12-inch griddle section has a top layer (called a *platen*) that can be lowered onto the

TABLE 13-3

GRIDDLE COOKING GUIDE (IN MINUTES)

	The Clamshell	Griddle	Charbroiler
4-ounce boneless chicken breast	3½	7	6
Hamburger (4-inch patty)	1½	3½	3½
Steak (¾ inch thick)	3½	7	6½
Grilled sandwich	1½	4	n/a
Halibut (¾ inch thick)	3½	7	6
Snapper	1½	4	n/a
Cheese melting	½	n/a	n/a
Hash browns	4	8	n/a
Sausage	2	4	n/a

Source: Lang Manufacturing Company, Everett, Washington.

griddle. The plate height can be readjusted to accommodate different types of foods, and each griddle/platen zone has its own controls (see Illustration 13-11). The Instant Burger (made by Smokaroma of Boley, Oklahoma) is another two-sided appliance. This one is unique because it uses direct energy transfer to cook hamburgers within half a minute. No oil is necessary for the Instant Burger griddle, and it doesn't require preheating or an exhaust hood, either. Like a microwave, it uses energy only when there's a product being cooked. This specialty griddle does well in small operations such as convenience stores and delis. Hot dogs, sausages, and chicken breasts can also be cooked in it.

Panini Grill. These toasted, grill-marked sandwiches stuffed with fresh ingredients and melted cheese are very popular, and easy to make using this appliance, a type of contact grill which cooks both sides of a sandwich at once. Sometimes called a *sandwich press,* it is versatile enough to use with breakfast items, flat breads, Reubens, and other sandwich

ILLUSTRATION 13-11 **The duplex cooker is a griddle divided into sections. Each section has its own "hood" that lowers, like a clamshell, onto the cooking surface.**
COURTESY OF LANG MANUFACTURING COMPANY, EVERETT, WASHINGTON.

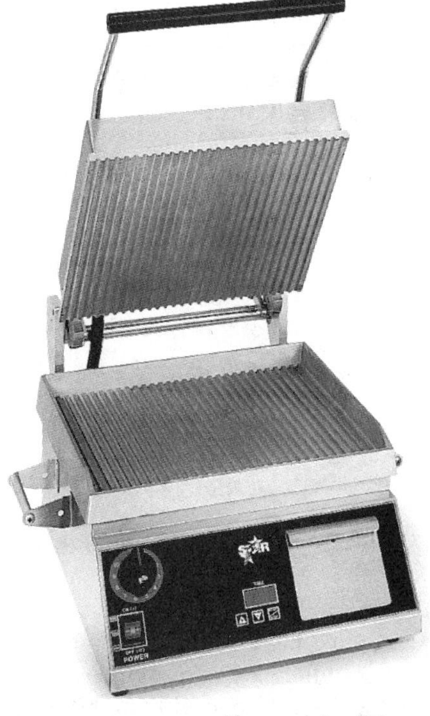

ILLUSTRATION 13-12 **The panini grill is a type of clamshell griddle that grills or toasts both sides of a sandwich at once.**
COURTESY OF STAR MANUFACTURING INTERNATIONAL, INC., ST. LOUIS, MISSOURI.

types as well. Most panini grills are countertop models that do not require special ventilation. There are multiple combinations of flat or grooved surfaces for top and bottom platens that can heat up to 570 degrees Fahrenheit. Look for controls that allow you to adjust top and bottom cooking surfaces separately (see Illustration 13-12).

The biggest difference between models is the material the surfaces are made of. Aluminum is nonstick and heats quickly, but also loses heat quickly. Cast iron retains heat well, but takes longer to heat. Stainless steel heats quickly, is known for its even heat distribution, and is the most durable of the three.

Mongolian Barbecue. The ***Mongolian barbecue*** is a flat-surface griddle, 4 feet in diameter, which may run on either electricity or gas. At its namesake restaurant, the customer fills a bowl with thinly sliced vegetables and chicken, beef, or fish. The bowl is handed to an attendant, who tosses the ingredients on the griddle and sears them in less than one minute. The story behind the name "Mongolian barbecue" is that the Mongols, who invaded China centuries ago, placed their shields over fires and used them to cook on—the first griddles!

13-6 BUYING AND MAINTAINING GRIDDLES

One of the most important considerations when shopping for a griddle is its heat recovery time. This becomes critical if you'll be using foods straight from the freezer instead of from the refrigerator, because the colder the food is when you place it on the griddle, the lower the surface temperature drops. In tests conducted by the American Gas Association, it took 77 Btus of gas energy to bring 1 pound of beef from 40 to 140 degrees Fahrenheit. However, when the beef was frozen to a temperature of 0 degrees Fahrenheit, it took 196 Btus of gas energy to bring it to 140 degrees Fahrenheit—more than twice the energy output for frozen food as for chilled food.

So, when a manufacturer recommends your griddle burners should each have an average gas input of 30,000 Btus per hour, don't decide it's "way too much." You may only use that kind of energy when the griddle is being preheated or when it is totally covered by frozen foods, but you do want to have that capability, and more Btu power helps cut your heat recovery time.

Another way to decide if the gas input is sufficient is to take the total Btu input figure you get from the manufacturer and divide it by the total number of square inches of surface cooking space. The figure you come up with should be at least 100 Btus per square inch. If it's not, the griddle might not meet your needs, especially if you'll be cooking a lot of frozen product.

To control the heat on the griddle surface, there are several options. Basic manual controls are ideal for users that require immediate control over the flame and careful monitoring of the food by the cook. For consistent heat that does not require monitoring, there are three types of thermostats:

1. *Hydraulic modulating.* The user sets the temperature level, and a thermostat monitors the surface temperature, automatically adjusting the burner flames below as needed. Accuracy of this type of thermostat is plus or minus 20 to 30 degrees Fahrenheit.
2. *Snap action.* This is an electric analog thermostat. Again, the user sets the temperature and the thermostat adjusts the burner flames—but since, as its name implies, it "snaps" into action quickly, the temperature variation is not so wide, only 7 to 10 degrees Fahrenheit.
3. *Solid state.* This is the most sophisticated of the thermostat types, which adjusts temperature within 2 to 5 degrees Fahrenheit.

There should be one thermostat for each burner on your griddle.

The griddle plate can be ordered to specifications. Griddle surfaces are made of several different materials and thicknesses, so you'll need to give some thought to exactly what you'll be cooking. If you'll need to preheat quickly and make frequent temperature adjustments, a thinner (3/8- to 1/2-inch) griddle plate is recommended. If you'll be using mostly frozen products, a thicker (3/4- to 1-inch) plate will hold heat without extreme temperature fluctuations. In *Food Equipment Facts,* Carl Scriven and James Stevens describe these griddle materials from which to choose:

CAST-IRON PLATES. These hold heat extremely well, but are porous and do not contract and expand. Frozen foods tend to stick to cast iron, and the surface requires constant cleaning.

POLISHED STEEL PLATES. These eliminate the food sticking problem, but the shiny surface causes some product shrinkage.

COLD-ROLLED STEEL PLATES. These have excellent heat transfer properties; the product adheres well without sticking or shrinking; and they're easy to clean.

CHROME FINISH PLATES. Again, these have excellent heat-retention and transfer properties; they're easy to clean, using a blade or scraper, cold water, and a special dry chemical powder.

No matter what type of griddle plate you use, you must season it, both before first use and after each cleaning. By now you're probably familiar with ***seasoning,*** which means to lightly coat a cooking surface with oil. This prevents food from sticking, while protecting the surface from rust. Failure to season a griddle consistently will result in poorly cooked products and, eventually, permanent damage to the griddle plate.

When first installed, preheating the griddle to 200 degrees Fahrenheit will help remove any of the heavy rust-preventive coating applied at the factory. Use a cloth dampened with cooking oil or a grease solvent to remove the coating, which has been softened by heating.

Then, one manufacturer suggests this five-step seasoning method:

1. Set the griddle temperature at 375 to 400 degrees Fahrenheit.
2. Pour unsalted cooking oil onto the griddle surface.
3. Allow the oil to heat until it begins to smoke.
4. Wipe the griddle surface thoroughly with a porous cloth.
5. Put on a second coat of seasoning oil and repeat steps 3 and 4.

Seasoning should only be done after cleaning, which in this case means to remove all bits of burnt and carbonized food from the surface. This is usually done using a metal spatula or griddle scraper, with long, firm strokes. Never strike the griddle surface with the edge of the spatula, as this can easily nick the surface and eventually affect its cooking ability.

Griddles should be cleaned when they're warm but not too hot; manufacturers suggest between 150 and 175 degrees Fahrenheit. Some manufacturers sell specially made griddle cleaner, which is spread across the plate, allowed to stand for five or ten minutes, then scraped or wiped into the trough. Afterwards, the plate must be sponged thoroughly with water to remove all cleaning residue before seasoning. Some chefs prefer to use water or ice to clean the griddle; the ice provides some gentle friction to help loosen baked-on particles, although it is critical that the griddle be absolutely cold when ice is used or the temperature shock may warp the plate. Cleaning may also include lemon juice or vinegar, to cut grease and leave the griddle plate looking shiny. Finally, it is always easier to maintain a "clean as you go" policy, instead of trying to scrape accumulations of burnt food and carbon off the griddle.

A couple of other notes about griddle care: Electric griddles can be scraped clean with a ***griddle stone,*** also called a *brick* or *pumice,* which is pulled over the plate in the same direction as the grain of the metal. As tempting as it may be to use steel wool on some of the tough, baked-on burnt food, don't do it. It will scratch the surface. (The other thing that will warp a griddle is using it as a range top. Don't set full pans or kettles on it.)

In addition to the griddle surface, the controls and connecting cables should also be wiped clean with a nonabrasive cleanser, and grease troughs should be emptied at least once a day.

13-7 TILTING BRAISING PANS

We include this piece of equipment in this chapter because it is actually a griddle with raised sides, which gives it the appearance of a big, flat-bottomed kettle. The ***tilting braising pan*** is often called a *tilting skillet* or *tilting frypan.* Whatever the name, it should definitely be awarded a medal for outstanding achievement in the line of duty! As you'll see, it is perhaps the most versatile piece of equipment in a foodservice kitchen (see Illustration 13-13). When professionals have to choose between a tilting steam kettle and a tilting braising pan, they consistently choose the tilting braising pan.

ILLUSTRATION 13-13 **The tilting braising pan is actually an appliance, not a "pan," that can fry, roast, steam, braise, poach, and proof. It can be square or round, and its large size and ability to tilt make it handy for large-volume cooking.**
COURTESY OF VULCAN-HART, DIVISION OF ITW FOOD EQUIPMENT GROUP, LLC, LOUISVILLE, KENTUCKY.

Why? Well first, it is designed for easy operation. It consists of four parts: a large griddle surface for heat transfer, vertical sides to contain liquids and spatters, a tilting pan body to make loading and unloading food easier, and an attached cover to retain heat and control the cooking process. The braising pan is heated from the bottom (beneath the griddle area) by gas burners or electric elements. This, and its hinged cover, allows the pan to perform the duties of a range top, a griddle, a kettle, a steamer, a cook-and hold oven, a bain marie, and more. Its sturdy construction conserves energy, and its ability to handle so many cooking tasks mean less labor, because it saves staff the chore of lifting and transferring foods from one pot or pan to another. Floor models can be put on casters for wheeling product out to serving areas, although they must be able to anchor to a wall to prevent them from tipping over when tilted.

A restaurant kitchen could use its tilting braising pan at breakfast as a griddle, to cook scrambled eggs, potatoes, and bacon; for lunch, to make large batches of chili, soups, or stews; and, at dinner, to poach salmon, steam vegetables, slow-roast a pot roast, or mix a chicken casserole.

Let's walk through the typical functions of this appliance to show you exactly how this is possible. Note that the temperatures we suggest are estimated ranges—specific recipes may call for some experimentation.

1. *Griddle Cooking.* Its raised sides may get in the way a bit, but the griddle area of a tilting braising pan comes in sizes ranging from 9 square feet to almost 24 square feet. Keep the cover open, set the thermostat at 300 to 400 degrees Fahrenheit, and the griddle area will fry pounds of bacon, sausage, and burgers, or grill sandwiches and French toast.
2. *Pan Frying.* Heavy loads of chicken or fish, or doughnuts to be deep-fried, are no problem here. Pour the oil into the pan, heat it from 350 to 400 degrees Fahrenheit, and keep the cover open for frying. A thermostat regulates the temperature to minimize oil breakdown.
3. *Kettle Cooking.* Tilting braising pans can do all types of liquid batch cooking, with capacities that vary from 10 to 40 gallons. Wide temperature ranges provide for anything from simmering to rapid boiling. Having a griddle at the bottom enables the operator to brown meat, then pour everything else in and switch its use from griddle to kettle. Close the cover during kettle cooking and set the thermostat at 200 to 225 degrees Fahrenheit.
4. *Steam Cooking.* Add 2 inches of water to the bottom of the pan and it becomes a steamer. Racks can be purchased to hold as many as three full-sized pans above the water. Make sure the pans are perforated, to allow the steam to surround the food in them. The lightweight racks are easily removed for storage. During steaming, the cover is closed and the temperature is set from 300 to 400 degrees Fahrenheit.
5. *Oven Roasting.* The same wire racks can be loaded with meats to be roasted at very low temperatures (175 to 275 degrees Fahrenheit) to minimize shrinkage and hold in natural juices. Of course, the cover is closed during roasting.
6. *Braising.* A speedy way of preparing meat is to braise it. This means quickly brown it with an open cover, then lower the temperature, add a small amount of liquid to the pan, and simmer

until fully cooked and tender. The recommended temperature for simmering is 225 to 295 degrees Fahrenheit.

7. *Poaching.* Add 2 inches of water to the bottom of the pan and set the thermostat at 200 to 225 degrees Fahrenheit, closing the cover for cooking. If it's eggs you're poaching, be sure to acidify and salt the water first.
8. *Proofing.* Add 2 inches of water to the bottom of the pan, place the pan of dough on a wire rack so it doesn't touch the water, turn the thermostat to a gentle 80 to 100 degrees Fahrenheit, and wait until the dough rises.
9. *Holding, Thawing, Serving.* Fill the tilting braising pan with 4 inches of water and set the thermostat at 175 degrees Fahrenheit. With that, you've transformed it into a bain marie, a steam table that keeps vegetables or sauces warm on a serving line. You could also hold rolls, bread, or pastries in it without using water and with the cover closed. If the unit will be used regularly in this way, rolling casters and a quick-disconnect feature for its gas hookup are recommended.

Now that we've seen the full range of capabilities, how about its construction details? Most tilting braising pans have ¾-inch-thick stainless steel griddle plates at the bottom; even thicker plates are recommended if you'll be preparing a lot of sauces. The sides of the pan are 10-gauge stainless steel, either 7 or 9 inches tall. Covered corners are recommended for easier cleaning. Several manufacturers mark lines on the sides to indicate capacities (10 gallons, 20 gallons, and so on).

The hinged cover is made from a solid sheet of 16-gauge steel, with a one-piece tubular handle welded on. The cover is supported by heavy hinges and a spring-loaded activator. At the rear of the cover, a ***drip shield*** catches condensate and funnels it back into the pan.

Because the pan is too heavy to lift manually, its tilting function is accomplished with a hand crank or an electric motor. Both are self-locking to ensure precise control.

The heating unit, located beneath the griddle plate, runs more than a hundred small heat transfer lines around the cooking surface to ensure even heat distribution. It has two thermostats—one to control cooking temperature and the other to detect overheating and shut the unit off automatically if its temperature exceeds 450 degrees Fahrenheit. When that happens, it takes about 30 minutes to allow the pan to cool down and reset the high-limit thermostat.

Tilting braising pan units stand on stainless steel legs no more than 3 inches tall, making the total height of the appliance 20 to 23 inches. Where space is at a premium, the pan can be mounted in a cabinet to add extra storage space. Tabletop versions of the tilting braising pan require as few as 28 square inches of counter space, with capacities of 10 to 12 gallons. Even a 30-gallon model can fit into a space less than 3 feet square. They can be fueled by gas or electricity or even hooked up to special batteries for portability. The pan is supported by a ***trunnion*** (a term usually used for cannons!), which means a set of sturdy steel pins that allows it to pivot.

The type of culinary operation often signals the size of pan you'll need. The largest high-volume uses require a 30- or 40-gallon pan. The type of product you'll prepare most frequently also helps with the decision. If you will be stewing foods, the formula is the same as for a steam-jacketed kettle: Multiply the number of portions by the portion size, then divide that figure by 128 ounces.

If you'll be using it as griddle space, think about the number of portions that can be cooked on 1 square foot, then determine how many portions you'll have to cook at one time. This will give you the total square footage needed.

Optional accessories and features may not be so "optional" once you see everything the tilting braising pan can do! Steamer conversion kits include wire racks and perforated pans; casters are recommended to allow easy movement of the unit for cleaning and repositioning; fill faucets can be ordered to put water directly into the pan from a faucet mounting bracket so employees will not have to carry pots of water to the pan. Also, get the "gallon marker" feature, which allows you to deliver a measured amount of water into the pan without hand-measuring it. Sealed, water-resistant controls are preferable because liquids are so often part of this type of cooking. Coved (rounded) corners make for easier stirring and cleaning. Work trays and pan carriers can be ordered to fit most pans. Pan carriers make it easier to pour large quantities into the pan no matter how it is tilted.

Finally, there are two special considerations for installing your tilting braising pan. First, because the pans give off steam and other vapors that are natural byproducts of the cooking process, most cities require they be installed under an exhaust hood. Traditionally, you'll find them located next to the steamers or steam-jacketed kettles, because they share the characteristics of large-batch cooking equipment. They must be able to tilt in any direction without obstruction.

Second, consider installing the pan where its pouring lip will be located over a floor drain or, at the very least, over a recessed floor area. Occasional spills can't be avoided, and this will make cleanup much easier.

SUMMARY

Many of the most popular types of surface cooking equipment are discussed in this chapter, including broilers, griddles, and grills. There are three basic broiler types: overhead, charbroiler, and specialty broilers. Overhead broilers have their heating elements over the food; a charbroiler's heat comes from beneath the food, which is why it is sometimes referred to as an underfired broiler. Charbroiling uses flames, smoke, and radiant heat.

The most recent category of specialty broiler is the food finisher, a countertop appliance that quickly reheats, browns, or crisps foods just enough to "finish" them before serving. Its cousins include the versatile salamander, a miniature broiler that can also be used as a plate warmer; and the cheesemelter, which does just what its name indicates and can also brown, poach, and boil small batches of food.

Broilers should be purchased based on the type of menu, the amount of food that must be cooked at one time, and the type of heat source that works best for what you will be broiling.

One thing all broilers have in common is that they must be kept clean. Food and carbon buildup can be scrubbed away with a special broiler brush. Another similarity is the need for an adequate ventilation system for smoke and grease, which are inevitable byproducts of broiling.

The griddle is the flat-top appliance you've seen in fast-food eateries, used to cook hamburgers, pancakes, hash browns, and more. The cooking surface of a griddle is called its plate, and most griddles are divided into sections, called heat zones, with separate temperature controls so you only have to heat as much space as you need. You can order an appliance with an oven or broiler under the griddle; a specialty griddle that heats on two sides simultaneously, known as a clamshell, is also available. There are several styles of clamshells and so-called "contact grills," named for the fact that a heated upper plate, as well as the lower plate, comes into contact with the food.

Griddles must be cleaned regularly, but they must also be seasoned—lightly coated with oil to prevent rust as well as food sticking to them.

And finally, an incredibly versatile appliance—the tilting braising pan—combines the large, flat surface of a griddle with raised sides to contain liquids or spatters, and a tilting body to make it easier to load and unload food in large quantities. The pan is heated from the bottom, either by gas or electricity. You can use the tilting braising pan to griddle, panfry, roast, braise, poach, steam, and more. Operators also like it because its tilting function minimizes the need to lift large amounts of food in heavy pans.

STUDY QUESTIONS

1. What is the difference between an overfired broiler and an underfired broiler? For what would you use each of them?
2. What is the grid (or grid assembly) of a broiler?
3. Many broilers come in three models or styles: drop-in, countertop, and freestanding. What is the difference, and why would you choose each of them?
4. What is a salamander, and where in the kitchen would you put it?
5. List three things you should check if your broiler does not seem to be getting hot enough.

6. How does the temperature of the food you are cooking impact a griddle?
7. How do you clean and season a griddle?
8. What is a clamshell?
9. What makes a tilting braising pan so versatile?
10. How do you decide what size of tilting braising pan you will need for your foodservice operation?

A Conversation with . . .

Christophe Chatron-Michaud

GENERAL MANAGER, AUREOLE
LAS VEGAS, NEVADA

Name any four-star chef or restaurateur in the last two decades, and Christophe Chatron-Michaud has probably worked with them—David Bouley, Daniel Boulud, Robert Meyzen, Michael Mina, Charlie Palmer, and Jean Georges Vongerichten, to name a few. His current position is General Manager of Aureole, the show-stopping restaurant in Las Vegas' opulent Mandalay Bay hotel complex, where he works with an energetic young executive chef, Philippe Rispoli. Christophe has strong feelings about his employees' energy level and work attitude, which are the very attributes that propelled him to top-level restaurant management.

Q: What made you decide to choose this as a career?
A: It wasn't really a choice!
I was studying to become a chiropractor in Paris and at the same time, I was a musician. I had a hard time doing both, making a living and studying. I ended up leaving my studies and went back to my parents, as I was in need of some finances! I started a job in my hometown of Megeve, in the Alps, behind a deli counter, slicing ham and getting out cold cuts for the customers. It was a temporary thing, while I tried to decide what I was doing next.

One day my father came to me and proposed that I visit his friend in southern France who was a captain in a very high-end restaurant, a three-Michelin-star restaurant at the time, Le Moulin de Mougins, owned by a very famous chef, Roger Verger. There was a busboy position available, and I took it not because I had great aspirations to be a busboy, but because I had always wanted to go to the South of France! I just wanted to see if I liked being in the restaurant business.

Q: Obviously, you ended up liking it!
A: I liked the fact that it allowed me to be independent, to make my own money and find a passion for being around people. And the food coming out of the kitchen was, of course, amazing! I admired the food, the quality of service, and the famous people who would come in, especially during the Cannes Film Festival. It was overwhelming for me, and I got hooked into being part of the hospitality industry there.
I decided to work very hard to move up in the business. Little by little, due to hard work, good attitude, and attendance—being always available, whenever they needed me—I moved up pretty quickly to an assistant captain position. I held that for about a year, before I was asked to move to another three-Michelin-star restaurant as a captain! It was owned by Louis Outhier, called L'Oasis, with a little bit more recognition and prestige at the time.

Q: In Europe, service careers have been "careers"—not "jobs," like they often are in the United States. Is this still true?
A: It depends on which level of service. There are casual restaurants in Europe too, where students take jobs to supplement their income, just as they do in the U.S. But I understood right from the beginning that if I was going to pursue a high-end restaurant job, it was going to be a career. Most of the people I was working with had either come out of hotel and restaurant school, or had started in other restaurants and worked their way up.

Q: Why did you decide to come to the United States?
A: I knew my job well and was lucky enough to have worked in some of the best restaurants in Southern France. I thought, "Why don't I utilize that to travel a little bit?" I was single at the time, and I had heard Mr. Verger was in a partnership in Florida with other French restaurateurs at Epcot Center. There weren't any job openings there, but he referred me . . . to the owner of a new restaurant in Westchester County, about 45 minutes outside of New York City. It was called La Panetiere, meaning a "bread box" where you put bread when it comes out of the oven. When I called the owner, he said, "I need somebody right away, and the position is maître'd." It was another step up—which I could not refuse—and, within a few weeks, I had my tourist visa and was on my way. I was twenty-two years old, and I hardly spoke any English! I had a hard time understanding people, but the owner was very understanding. He knew I had had a lot of experience, and had the technical skills to train the staff on tableside service, which was very popular then. Even the guests were understanding. They made a real effort to speak French, to help me out. That's what made me stay in America—I was amazed by how welcoming and helpful everyone was.

Q: Your resume includes a dozen more truly great restaurants, and we don't have time to talk about each of them here. But is it safe to say this is a very mobile business—you move up, you move on, you move around?
A: Yes. At the beginning of your career, if you want to acquire as much experience as possible, you must learn as fast as possible at each restaurant and then move on. That is how you build the resume and increase your knowledge—by always going for a better position and working with another great chef. I think you have to give a job at least two years, as long as you are evolving and learning. And when it's time to move up, you will know it, because the job becomes routine. If you are serious about this business, you will start enjoying food, eating out on your own, and discovering the competition.

Q: Eventually, you transitioned into general manager positions—not just fine dining, but trendy nightclubs. What does a G.M. do?
A: Your responsibility becomes the entire restaurant, front and back of the house. The chef, the hiring, health and safety decisions fall to you, as well as profitability and cost controls. It is a lot of paperwork. You're directly involved with budgets and numbers, and thank God in high school, I had done a lot of math and physics! I had no problem understanding basic accounting.

Q: Is it scary, having to run a restaurant to show the investors it is a successful business?
A: It is scary, because the investors will come directly to you. The chef/owner will focus on the quality of the food. You, on the other hand, have the responsibility of dealing with the food costs! They like to work with the best foie gras, lobster, truffles, caviar, et cetera. And somehow you have to manage them, and control the prices so that you make money at the end. Labor, wages, benefits all factor in. I was very fortunate that all the chefs I have worked for attracted customers because of their reputations. It is easy to be the general manager of a busy restaurant, where people pay a high price for a high-end meal. I'm sure it would not be so easy otherwise.

Q: How did you get to Las Vegas?
A: For a very long time, I wanted to open my own restaurant in New York. I had a chef partner, we had a business plan, and raised the money. But we could not deal with the real estate prices. Some places were charging $30,000 a month for rent! I mean, how do you manage that? I got so discouraged. I was working as a consultant when a friend called and told me [chef] Michael Mina of San Francisco was taking over a restaurant at the MGM Grand Hotel in Las Vegas and needed help. After fifteen years in New York, I was married and had two children in one of those small New York apartments, and I thought, "Why not?" It was a great offer, great weather, great quality of life. So we moved.

Q: How do you stay competitive in a restaurant scene like Las Vegas?
A: It is very difficult, very competitive. But for me, the level of service and food still hasn't reached what I was exposed to in New York City, so I have a little bit of an edge on that and I try to train the staff to get them up to that level. We rethink things all the time, and we eat out at the competitors to see what they are trying. At Aureole, the food concept is "Progressive American Cuisine," and it's a vast subject, so we are able to include a lot of culinary styles. We also have three different dining areas, which offer three different experiences—real fine dining in Swan Court, an area overlooking a beautiful pond with swans with tableside service; the bar, with tapas and interesting wines by the glass; and the central dining room, which we're treating more as a modern, elegant bistro.

Q: How important is atmosphere to what you are trying to accomplish?

A: It is one of the most important aspects of any restaurant, and it's not just the decor. It's the hospitality you feel during your dining experience. People expect a smile, eye contact, warmth—they expect to be made comfortable as soon as they walk in. You have to take them into your hands and show them what you have to offer. There must be a very positive attitude and energy, from the first impression to the time they are on their way out.

Q: And how do you accomplish that when the business itself is getting to be so much more technical—the point-of-sale system, the computerized wine list, and so on?
A: Actually, point-of-sale systems are very beneficial because they allow the wait staff to spend more time with the guests. There's not so much running back and forth to the kitchen to hand in handwritten orders. You can basically communicate from that little computer to the kitchen, letting the chef know about special orders, if the guests want to wait before a course, if they are in a rush, et cetera. You can be more attentive to the customers, as you are on the floor all the time.

Q: If someone walks into a restaurant for a job interview and doesn't have much experience, do they have a chance? What are you looking for?
A: We go through what we call behavioral interviews. What I look for now is not necessarily someone with a lot of experience, but somebody that has energy, warmth, great eye contact—that sends me good, positive vibes. We can train and fine-tune them to tailor them to our standards, if they have the energy and the willingness to learn.

Q: What is your advice for students just learning the foodservice business?
A: You want to be able to evolve and to be happy with this business. It takes a lot of hours and a lot out of your life, so you must be passionate about what you do. You have to have a genuine passion for food, cooking, wines, or hospitality, which ties you into your work. For me, at the beginning, it was the wines. I loved tasting wine, studying it, pairing it with foods, and talking with the chefs about it. Be a food and wine lover, is what I would encourage. Love people, love food, love wine. This is what you have to study!

CHAPTER 14

STEAM COOKING EQUIPMENT

INTRODUCTION AND LEARNING OBJECTIVES

More and more foodservice operators are turning to a cooking medium that predates written history, and yet is remarkably well-suited to our fast-moving culinary culture. Steam cooking makes light and flavorful food quickly, and without added fat or calories. You can steam foods with as little equipment as a saucepan and a metal or bamboo basket that fits inside of it, but for foodservice there are steam appliances that turn out much larger quantities.

Institutional foodservice has led the way in the acceptance and use of steam equipment; except for hotel banquet kitchens, the commercial restaurant sector was slower to embrace steam cooking. The hesitation was understandable—traditional chefs want to regularly inspect food as it cooks, which is not the correct way to use steam. However, new technology, dual-use appliances, and health-conscious dining trends have prompted new interest in steam cooking. The programmable combination oven/steamer (commonly known as a ***combi oven***) offers three powerful cooking methods: hot air, steam, and a combination of the two, which amounts to the work of several separate appliances. It will also hold hot food at its optimal temperature safely and without loss of moisture, flavor, or overall quality.

In this chapter, we'll cover what steam equipment can and cannot do, to date. Whether you want to make baked potatoes or bratwurst, to warm enough Danish for an army or cook just a single serving of fresh vegetables, you'll learn everything you need to know to choose and use steaming equipment. In Chapter 6, we introduced the rudiments of steam energy. So now, we'll explore your equipment choices. They include:

- Steam-jacketed kettles
- Pressure steamers
- High-pressure and low-pressure steamers
- Pressureless steamers
- Combination pressure/pressureless steamers
- Specialty steamers
- Combination oven/steamers

We'll also detail the sizes of equipment on the market, accessories available to use with steamers for all manner of special applications, plus tips for choosing, installing, and maintaining steam equipment.

14-1 WHY STEAM?

The principle behind steam cooking is simple: when water reaches its boiling point, at temperatures of 212 degrees Fahrenheit or higher, steam is its gaseous form. It takes only 180 Btus of energy to make a 1-pound block of ice into boiling water, but to evaporate the same water and make it into steam vapor requires 970 Btus. The energy of these Btus is contained within the steam and is used to cook foods quickly but gently.

The secret to steam cooking is that it is such an efficient heat transfer medium. The energy carried by the steam——about six times more energy than boiling water——transfers immediately upon contact with a cooler surface. This transfer can occur directly (in a steamer, for instance) or indirectly (through the wall of a steam-jacketed kettle). The ***heat transfer rate*** is the measurement in Btus of how much heat goes from one substance to another in a given time and space. (Actually, per square foot per hour per degree of temperature difference. How's *that* for a complex formula?) Steam's heat transfer rate is 300 Btus, compared to a rate of 3 for air, 7 for circulated air in a convection oven, 38 on a griddle, and 85 when boiling in a pot of water.

To illustrate the difference, if you stick your hand into an oven cavity that's been preheated to 400 degrees Fahrenheit, you'll feel the heat but you won't burn your arm; but put your hand over a boiling tea kettle and the 212 degree Fahrenheit steam will scald your hand immediately. The latter is simply more effective heat transfer. In a closed steamer, however, food will not burn. As long as the temperature of the food is less than 212 degrees Fahrenheit, the steam will continue to condense on the food surface and deliver 970 Btus of energy as it cooks.

Steam cooking uses less energy overall than conventional methods, because power is only needed to keep the steam up to pressure while the equipment is operating. But in order to achieve higher temperatures than 212 degrees Fahrenheit, the steam must be held under pressure. A quick physics lesson about steam: Water at its boiling point (at sea level) cannot exceed 212 degrees Fahrenheit, no matter how much additional heat is applied. Instead, we must raise the pressure in the cooking chamber to raise the temperature of the steam. At 5 psi, water and steam temperatures rise to 222 degrees Fahrenheit; at 10 psi, 239 degrees Fahrenheit; and at 15 psi, the temperature hovers at 250 degrees Fahrenheit. Conversely, lowering the pressure decreases the temperature.

This is often referred to as ***superheated steam.*** The name sounds impressive, but superheated steam doesn't offer any big advantages in the cooking process. Its most common use is to "dry out" the steam to cook certain types of foods that don't require as much direct contact with moisture. Steam with a small amount of superheat in it is often called ***dry steam.***

Steam's reduced cooking time allows "closer-to-service" scheduling for many delicate foods that don't keep well under heat lamps or on steam tables. In some commercial situations, a small pressureless steamer gives the option of cooking vegetables to order, which might allow a restaurant to stock smaller quantities of more exotic fresh produce. In some steam appliances, almost any liquid can be used, including broth, a basic stock, or even wine—each of which imparts a distinct flavor.

The reason fresh vegetables are often steamed is that this method preserves their flavor, color, and texture, presenting a plate that is good-looking as well as nutritious. Comparison tests have shown that vegetables prepared in a steamer contain one-third to one-half more of their natural nutrients than the same vegetables prepared in a stockpot (see Table 14-1). Steam cooking also tends to yield more than other methods, because the product doesn't shrink as much. Food handling is minimal, since most items can be prepared, cooked, and plated right out of the same pan. This results in less labor cost, with less stirring and handling of the food.

Interestingly, steam cooking equipment actually reduces the number of pots and pans to be bought, used, and washed. The food simply does not burn on the pan, which reduces the need to scour and scrub. You can use the same unit for initial food preparation, reheating, and reconstitution. The equipment itself eliminates the lifting of heavy stockpots on and off the range.

TABLE 14–1

STEAMING OR BOILING?

This Chart Shows the Dramatic Savings of Nutrients When Steam Is Used in Preference to a Stockpot. There Is a Close Relationship between Nutrients, Color, Flavor, and Texture.

Vegetable	Cooking Method	Loss of Dry Matter %	Loss of Protein %	Loss of Calcium %	Loss of Magnesium %	Loss of Phosphorus %	Loss of Iron %
Asparagus	Boiled	14.0	20.0	16.5	8.8	25.8	34.4
	Steamed	7.09	13.3	15.3	1.4	10.4	20.0
Beans, string	Boiled	24.6	29.1	29.3	31.4	27.6	38.1
	Steamed	14.2	16.6	16.3	21.4	18.8	24.5
Beet greens	Boiled	29.7	22.2	15.9	41.6	44.9	43.1
	Steamed	15.7	6.9	3.8	14.1	14.0	24.5
Cabbage	Boiled	60.7	61.5	72.3	76.1	59.9	66.6
	Steamed	26.4	31.5	40.2	43.4	22.0	34.6
Cauliflower	Boiled	37.6	44.4	24.6	25.0	49.8	36.2
	Steamed	2.1	7.6	3.1	1.7	19.2	8.3
Celery	Boiled	45.4	52.6	36.1	57.1	48.7	—
	Steamed	22.3	22.3	11.6	32.4	15.7	—
Celery cabbage	Boiled	63.2	67.1	49.7	61.6	66.1	67.6
	Steamed	38.3	33.5	16.3	32.6	30.2	44.1
Spinach	Boiled	33.9	29.0	5.5	59.1	48.8	57.1
	Steamed	8.4	5.6	0.0	17.8	10.2	25.7
Beets	Boiled	30.9	22.0	18.7	30.9	33.6	—
	Steamed	21.5	5.4	1.5	29.4	20.1	—
Carrots	Boiled	20.1	26.4	8.9	22.8	19.0	34.1
	Steamed	5.1	14.5	5.1	5.6	1.0	20.7
Kohlrabi	Boiled	33.6	23.2	27.8	40.4	27.7	51.7
	Steamed	7.6	1.0	1.0	14.3	7.7	21.3
Onions	Boiled	21.3	50.2	15.6	27.8	40.2	36.1
	Steamed	11.0	30.7	7.1	15.7	31.5	15.9
Parsnips	Boiled	21.9	13.3	11.4	46.8	23.7	27.6
	Steamed	4.6	20.0	4.2	18.2	5.7	8.1
Potatoes	Boiled	9.4	—	16.8	18.8	18.3	—
	Steamed	4.0	—	9.6	14.0	11.7	—
Sweet potatoes	Boiled	29.0	71.5	38.3	45.3	44.4	31.5
	Steamed	21.1	15.0	22.1	31.5	24.3	25.1
Rutabagas	Boiled	45.8	48.6	37.1	42.7	57.2	50.0
	Steamed	13.2	15.7	13.4	3.4	24.6	14.3
Average for all vegetables	Boiled	39.4	43.0	31.9	44.7	46.4	48.0
	Steamed	14.0	16.0	10.7	18.6	16.7	21.3

Source: **North American Association of Food Equipment Manufacturers, Chicago, Illinois.**

What can't steam cooking do? It cannot bake, and it cannot brown. Braising is possible in a steam-jacketed kettle, but meat can't be browned. For both browning and baking, you'll need a combination of steam and dry heat, which is possible in the "combi" units we'll discuss later in the chapter.

As handy as they are, steam equipment is only effective if it fits your kitchen layout. You'll need to look at the locations of water lines and drains and, for steam-jacketed kettles, ventilation hoods are necessary. Does it make sense to centralize steam equipment in one part of the kitchen, or can you work more efficiently with a large steamer on the hot line and a couple of smaller ones in the à la carte area?

Steam and Water Quality

Before we introduce the different types of steam equipment, it cannot be stressed strongly enough that the steam used in cooking must be clean and produced from potable water. Most experts recommend that water used to make steam be treated, usually by installing a water softener, to eliminate many service problems. The recommended water quality standards are:

- **Total dissolved solids of no more than 80 parts per million (ppm)**
- **Water pH factor of 7.0 to 8.0**
- **Water hardness not to exceed 2.0 grains**
- **Total alkalinity not to exceed 20 ppm**
- **Maximum allowable silica 13 ppm**
- **Maximum allowable chloride 30 ppm**

The amounts of iron, chlorine, and dissolved gases in the water all affect steam generation. The most frequent problems with steam equipment are caused by mineral buildup over time, on the interior sides of the units and on the parts that water comes into contact with, like heating elements and different types of sensors. Regular cleaning can minimize but not completely eliminate these problems. Because water supplies vary from location to location, consult the local water treatment agency (usually the municipality) before you install any steam-generating equipment.

Also, be aware that warranties and/or service agreements on steam appliances do not cover repairs for breakdowns created by water problems. And, like many other appliances, steamers must be level for proper operation. Your warranty won't be honored if your steamer is improperly installed. You'll learn more about water quality and its impact on steam generation in the sections on "Steam Generators" and "Cleaning and Maintenance of Steam Equipment," also in this chapter.

14-2 THE STEAM-JACKETED KETTLE

The steam-jacketed kettle cooks much more quickly than a stockpot on a range top, and uses less energy. It can stir-fry, stew, and more. No kitchen today should be without this versatile piece of equipment, but choosing the best unit for your needs requires some careful calculation—since they come in sizes from one quart to 160 gallons! Each type of kettle has particular qualities and operational requirements. Kettles are used to prepare soups, stews, gravies, puddings, sauces, pasta, eggs, and rice. They can partially precook foods, like fresh vegetables, for finishing later; or they can "finish-cook" foods that are precooked in other equipment. A kitchen that prepares whole cuts of meat, fish and poultry can use a kettle to cook the trimmings for other purposes, rather than waste them. Kettles can simmer all day, combining the leftover meats and julienne of vegetables to make stocks. We'll continue extolling the kettle's virtues after we look at some of its more technical aspects.

A steam-jacketed kettle works like a combination double boiler and bain marie. One round, hemispheric bowl is sealed within another, with about 2 inches of space separating the bowls. Steam is introduced into this space (see Illustration 14-1). The interior bowl is welded along its top to the outside of the exterior bowl (the jacket). Specially constructed baffles within the jacket provide even distribution of heat inside the kettle. The pressure of the steam can be adjusted—for rapid cooking, it is increased; for slower cooking, it is lowered.

Steam does not actually come into contact with the food, but stays in the 2-inch space as described. As it enters and condenses on the inner surface, it transfers its heat to the stainless steel wall, which in turn transfers heat to the food. The steam expands to fill the entire space between the kettle walls, with no hot spots or temperature variations. Because the kettle is almost always covered during the cooking process, it is fast and allows less heat to escape into the kitchen. And, because the kettle transfers heat through its entire jacketed sides and bottom, it offers three to four times more surface area for heat transfer than the same size of stockpot. In fact, the kettle is about 65 percent more efficient than its range top counterpart!

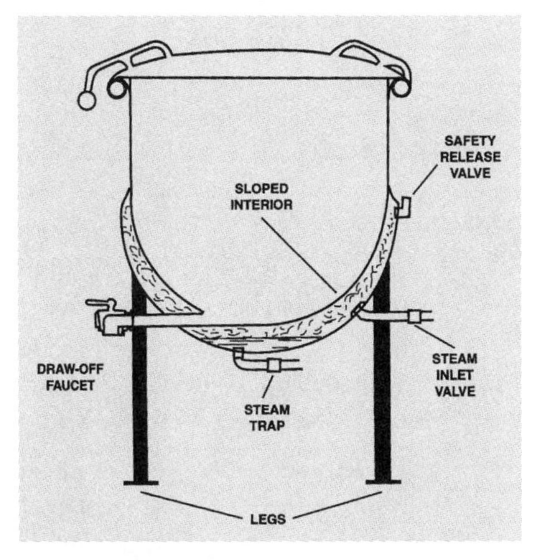

ILLUSTRATION 14-1 **The construction of a steam-jacketed kettle allows it to cook food slowly and evenly. As shown here, the steam does not come into contact with the food.**

Look back at Table 6-1 in Chapter 6 for the variety of steam pressure settings and the corresponding temperatures they generate. Remember, only a few degrees are lost as the steam condenses on the wall of the inner bowl, so most of the heat goes right to the food without scorching or burning it.

To summarize, the usefulness of the steam-jacketed kettle is based on its superior ability to transfer heat rapidly and maintain uniformity of temperature throughout the heated surface of the kettle. In a busy kitchen, this allows it to be a real workhorse. In addition to the uses mentioned earlier, these kettles can boil large quantities of water, reheat food, slow-cook sauces, and even make coffee or tea in mass quantities with efficiency and consistency. For banquet service or other special functions, coffee can be drawn from the fixed kettle directly into large coffee pots or urns for serving. (There's another option for coffee, too, which we'll mention later.) Illustration 14-2 shows a common restaurant kitchen setup for two kettles.

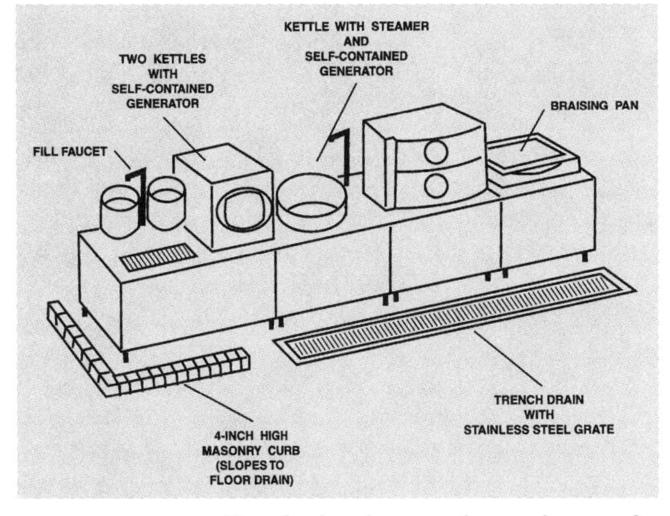

ILLUSTRATION 14-2 **Many foodservice operations make room for more than one steam-jacketed kettle.**

SOURCE: CARL SCRIVEN AND JAMES STEVENS, *MANUAL OF EQUIPMENT AND DESIGN FOR THE FOOD SERVICE INDUSTRY*, THOMSON LEARNING, 1989.

Most steam-jacketed kettles are made of stainless steel. Choose one with an especially high grade of stainless steel lining (called a "316-type" lining) to prevent corrosion if you'll be preparing highly acidic products such as tomato-based soup, sauces, and pizza toppings.

During the cook-chill process, which we will discuss in the following chapter, the steam-jacketed kettle is used to both cook and refrigerate food. If yours is a direct-steam kettle, you can connect it to a source of cold water or ice and use it to cool hot foods or mix cold salads. You'll learn more about direct-steam kettles in a moment. One enterprising school foodservice worker uses the kettle to soften the burned-on crusts in baking pans before washing.

At least one manufacturer has created an insulated kettle—an important achievement, because workers can brush against it without worrying about getting burned. No matter how hot the inside is, the exterior remains cool and safe to the touch. An added bonus is that the kitchen stays cooler. In a recent issue of *Foodservice Equipment and Supplies* magazine, operators were asked what other improvements they would like to see in steam-jacketed kettles. Their responses included:

1. Fill marks on the kettles' interior walls, so they can be refilled without premeasuring.
2. More sensitive temperature ranges; some respondents feel their kettles are either "too hot" or "too cool." The ability to cool off more quickly is needed by some users; others had a "holding" mode on their wish list of features.
3. Kettle manufacturers should provide a better and/or frequent preventive maintenance schedule.

FOODSERVICE EQUIPMENT

STEAM-JACKETED KETTLES AT WORK

At the Summer Shack restaurant in Cambridge, Massachusetts, chef Jasper White has brought the steam-jacketed kettle to the front of the house. Two 80-gallon kettles, dramatically lit with colored spotlights, are located on the "lobster line," the focal point of this fresh seafood eatery. An overhead suspension system of baskets, chains, and hooks efficiently moves live lobsters from their 1200-gallon holding tank to waiting steam kettles. Each kettle holds about 40 lobsters; about 400 are processed nightly.

The kettles are also used for preparing soup, and for specialty clambakes—wire mesh bags are filled with mussels and clams, corn, sausage, and potatoes and placed in the kettles, supported by hooks inside the rim. The food is steamed to perfection! These nontraditional displays of steam kettle cookery help attract an overflow crowd to the Summer Shack.

At Pickle Bill's in Grand River, Ohio, chef Jeff Richards uses small (32- to 72-ounce) kettles, sometimes called ***oyster kettles,*** to create soups and chowders. He feels that the steam kettle helps keep the flavor in the soup. Customers can watch the line cooks shuck oysters and assemble their soup ingredients "to order."

At the Big Bowl Asian Kitchen restaurant chain, large steam-jacketed kettles are used for making stocks from scratch, usually in the morning; then filled with water and used to cook the noodles for the many Asian dishes prepared there.

At Harrah's Casino in Lake Tahoe, Nevada, the line cooks use highly automated kettles to produce foods in very large quantities. Water fill rates, water and steam temperatures, and cooking times can all be preprogrammed, and there is a sensor system that tracks product temperature in each kettle throughout the cooking process. The cooks also appreciate the ability to program the arm speeds in agitator kettles.

Varieties of Steam-Jacketed Kettles

To harness its versatility, manufacturers have developed a wide variety of kettle types and sizes using different mounting devices and sources of steam.

Deep Jacketed. This kettle is more cylindrical than round (hence the name "deep"), so it's a good alternative where floor space is limited but output must be high. It is stationary and can be mounted on the floor or wall. Floor mounting may be on tubular legs or a pedestal (see Illustration 14-3). It usually includes a one-piece hinged cover, steam inlet and outlet valves, a safety valve, a draw-off line, and faucet. Stock or soup can be drained through a removable strainer that fits into the draw-off pipe.

You'll get maximum versatility with two smaller, 30-gallon kettles than with a single, 60-gallon model.

Shallow or Full Jacketed. Similar in construction to the deep-jacketed type, this kettle is wider and shorter. Its width makes it easy to empty, but it is not desirable for certain products. For instance, when cooking a large quantity of chicken, the sheer weight of the meat might mean that the bottom layers of chicken come out looking squashed instead of appetizing.

Two-Thirds Jacketed. A deep cylinder with a hinged cover, two-thirds of its jacketed area is surrounded by steam. Like their deep-jacketed cousins, these kettles are used for foods with high liquid content, are easy to stir, and require little attention. The height of the kettle allows for vigorous stirring without spillover. Again, this kettle requires minimum floor space for the amount of work it can do.

Tilting or Trunnion. Designed with a pouring lip, this kettle is mounted on pivots (called ***trunnions***), which allow the appliance to be easily tilted so that its contents can be emptied completely and comfortably without having to lift it. A self-locking feature is included to tip the kettle at whatever angle you need. Larger tilting kettles use a hand-cranked wheel or motor-driven tilting device; smaller countertop models use a hand lever (see Illustration 14-4).

Tilting kettles are also available in smaller, tabletop sizes. Ranging in capacity from 10 to 40 quarts, tabletop models are installed individually or in groups on counter space or tables. The table can be custom-made, fitted with a small gas or electric steam-generating unit, but the trend is toward self-contained electric kettles, which are simply plugged in, without having to retrofit an existing countertop. They also come with their own stands.

These kettles are ideal for preparing small quantities of vegetables, sauces, and cereals. Among other things, they are useful in bakeries to warm syrups and fruit or chocolate sauces or to make puddings. The self-contained electric unit is factory equipped with rust inhibitors and a water purifier to prevent mineral buildup within the kettle steam chamber. The unit also features a sight glass to check the water level in the chamber.

ILLUSTRATION 14-3 **A deep-jacketed kettle is deeper and narrower than a regular steam-jacketed kettle, and can fit where space is tight.**

COURTESY OF VULCAN-HART, DIVISION OF ITW FOOD EQUIPMENT GROUP, LLC, LOUISVILLE, KENTUCKY.

Sizing and Selecting Kettles

Since Chapter 9 covered the fundamentals of purchasing equipment and writing your own "specs" for it, let's assume you've already considered cost, operating life, availability of repairs, and so forth. Now, your task is to choose which kind, what size, and how many kettles you want to buy. Although it sounds like a simple job, there's a lot to think about. The steps you'll go through to make the decision are:

1. **First, determine the kinds of foods you'll serve and estimated volumes you'll need, especially peak volumes.**
2. **Look carefully at your available counter and floor space. How much room do you have for kettles? You may be able to fit a stationary kettle into a hot line, but a large tilting kettle will have to be freestanding to be able to tilt correctly.**
3. **Consider your workers' ability to handle the food amounts, especially if you'll be buying a large kettle.**

One calculation for determining kettle size is to estimate its output, which is roughly 8 pounds of meat or vegetables (or 4 pounds of poultry) for each gallon of kettle capacity. These numbers take into account shrinkage (which is minimal) and allow some ***head space***—the term for the empty space between the top of the food and the rim of the kettle. Most manufacturers recommend that 15 percent of the total kettle volume be left for head space.

ILLUSTRATION 14-4 **A tilting or trunnion kettle can be locked into place for filling and cooking, or tilted for easier cleaning.**

COURTESY OF GROEN, A DOVER INDUSTRIES COMPANY, ELK GROVE VILLAGE, ILLINOIS.

Many manufacturers also have charts that suggest kettle sizes based on either numbers of meals to be served or portion sizes of those meals. Table 14-2 is a sizing chart, reprinted with permission from the *Handbook of Steam Equipment* by the North American Association of Food Equipment Manufacturers.

Typical kettle sizes in today's market range from a few quarts for tabletop models, up to 160 gallons for floor models. Always consider the space available for installing a steam-jacketed kettle and find out building regulations in advance of purchase. Don't be surprised to discover that kettles must be installed under exhaust hoods! And be flexible as you make tradeoffs. If you're short on space, you may have to use one larger kettle instead of two smaller ones, simply because the big one takes up less room.

Here are some guidelines gathered from a combination of foodservice industry publications and personal experience:

- Use 10- and 20-quart kettles for gravies, sauces, and the like.
- Use a 20-quart tilting tabletop kettle for preparing small quantities of sauces and the like.
- Use a 20-gallon kettle for vegetables.
- Use a 40-gallon kettle for stewing.
- Use a 50-gallon kettle for soup stock.
- Use one 30-gallon kettle for 600 meals.
- Use one 40-gallon kettle for each 800 meals served during peak periods.

TABLE 14–2

KETTLE SIZING GUIDE

Number of Meals Served Daily	Number and Size of Steam-Jacketed Kettles
100–250	(1) 20-gallon kettle[a]
251–350	(1) 30-gallon kettle
351–500	(1) 40-gallon kettle
501–750	(2) 30-gallon kettles
	or
	(1) 60-gallon kettle
751–1000	(2) 40-gallon kettles
1001–1250	(2) 40-gallon kettles and
	(1) 20-gallon kettle
	or
	(1) 60-gallon kettle and
	(1) 40-gallon kettle
1251–1500	(3) 40-gallon kettles
	or
	(2) 60-gallon kettles

[a]Obviously, two 10-gallon kettles could be substituted.

Source: *Handbook of Steam Equipment,* North American Association of Food Equipment Manufacturers, Chicago, Illinois.

Power Sources for Steam-Jacketed Kettles

Either kettles are *self-contained* and generate their own steam, or they are *direct-steam* models and obtain their steam from a remote source—although nowadays, few modern facilities have access to direct steam sources. Self-contained kettles recycle water condensate and operate without water or drain plumbing connections; their temperatures range from 150 to 298 degrees Fahrenheit, plenty of heat for warming, simmering, boiling, and braising duties. All such kettles have a safety valve that releases jacket pressure at preset levels and provides a way to add distilled water to the system. The factory-installed distilled water, rust inhibitor, pressure gauge, temperature controls, water sight gauge, and low-water cutoff make this a very desirable piece of equipment because it's almost foolproof.

Electric (Self-Contained) Models. Steam is generated by immersing electric heating elements into a water reservoir located below the kettle jacket. Typical electrical ratings for floor-mounted models will vary from 10.8 to 16.4 kilowatts. The voltage varies from 208 to 480 volts, and three-phase installation is the norm, although some manufacturers provide single-phase installation for the 20- or 40-gallon kettles.

Tabletop models require as little as 30 kilowatts; they are available in a single- or three-phase configuration and use 208 to 480 volts.

Gas (Self-Contained) Models. In these kettles, gas or propane burners boil the water to produce steam. Most gas models are stationary but several manufacturers offer tilting models. The energy usage varies quite a bit with these models, depending on their size. Floor models range all the way up to 129,000 Btus per hour; tabletop models require as few as 31,000 Btus per hour for a 20-quart model.

Direct-Steam Models. The remote energy these kettles receive can come from a main boiler designed for a whole facility, a boiler designated solely for kitchen use, or a boiler located near the kettles solely to provide steam for their needs. Boiler units are built to comply with the American Society of Mechanical Engineers (ASME) Code for unfired pressure vessels. The steam is piped into the jacket and flows to the inner wall where it releases its energy load, condenses, and drops to the bottom of the jacket. The piping is usually fitted with a valve to regulate steam flow; it may also have a pressure-reducing valve. Once the condensate (now hot water) collects, the steam trap functions as an automatic valve and discharges the water. This valve action prevents any loss or variation in the steam pressure during the cooking process. The condensate can flow into a floor drain or be returned to the steam boiler, depending on the design. A safety valve is sensitive enough to correct for as little as 2 to 5 pounds past the desired pressure.

A chief advantage of the direct-steam kettle is that cold water can also be pumped into the jacket instead of steam. This is useful for chilling foods or cooling them after cooking.

Mounting Steam-Jacketed Kettles

Mounting a kettle means securing it to a surface so it is safely anchored and doesn't tip unless you want it to. Don't be confused by the fact that you can attach a floor model to a wall! In fact, there are four ways steam-jacketed kettles can be mounted:

1. *Pedestal Mount.* A stainless steel base is factory-welded to the kettle; the outer circumference of the pedestal is flanged down vertically to seal it to the floor. Being on a pedestal provides open space, making it easy to clean under the base. Because you need a way to get water out of the chamber, a 2-inch draw-off pipe is part of the kettle. Another option is a stainless steel yoke stand or table, with or without a tilting mechanism.
2. *Tubular Leg Mount.* Stainless steel adjustable floor flanges support tubular legs, making a tripod that is welded around the kettle. Again, you can install trunnions to enable the kettle to tilt.
3. *Wall Mount.* This installation provides the highest degree of sanitation and is often used in conjunction with modern buildings' power distribution systems, which are known as ***raceways.*** In other words, the kettle can be installed on a cooking line and attached directly to

ILLUSTRATION 14-5 A cabinet-mounted kettle is, as its name implies, a kettle that can be fitted into a cabinet or hot line.
COURTESY OF CLEVELAND RANGE, A WELBILT COMPANY, CLEVELAND, OHIO.

the building's major power (and water) sources. Many manufacturers produce wall-mounted units especially designed to attach to raceway systems. It is critical to coordinate this installation with the kettle manufacturer as well as the raceway system designer. With or without a raceway, the wall must be strong enough and constructed correctly to support a full, heavy kettle, with or without a tilting mechanism.

4. *Cabinet Mount.* These kettles can also be tilting or stationary, but they are usually part of an entire cooking battery. The complete assembly is available with legs as a standalone unit; or, the entire cabinet can be wall-mounted, which is often referred to as a console unit. Keep in mind that this mounting usually features a built-in drain to attach directly to a permanent floor drain, so your placement will be affected by the plumbing (see Illustration 14-5).

Accessories and Special Uses

Here are the options, both standard and nontraditional, that are available for most kettles.

1. *Cooking Baskets.* Not unlike fryer baskets, these stainless steel inserts are used to load, cook, and remove products prepared in boiling water. They're especially useful when cooking vegetables, pasta, rice, and potatoes. Normally, single baskets are used but floor-mounted models can accommodate triple-basket inserts.
2. *Covers.* Some stationary models come with covers and most manufacturers offer them as an option on all their models. As mentioned earlier, covering the kettle retains more moisture and nutrients, speeds cooking time, saves energy, and keeps the kitchen cooler.
3. *Tables and Stands.* A manufacturer will mount all self-contained tabletop kettles on tables if you request it, and the same goes for direct-steam models. Usually, two or more tabletop kettles can be mounted on one table. When this happens, a pouring sink is usually included and positioned in the "pouring path" of both kettles. It's worth shopping around for this handy option; some manufacturers also include a hot- and cold-water faucet with a swing spout to service the two kettles. A kettle on a stand is shown in Illustration 14-6.
4. *Agitator Mixer.* This is a mechanical mixing unit that virtually eliminates the need for hand stirring. Mixers make it easy to prepare large quantities of delicate cream sauces and other food products that need constant stirring. Heat is evenly distributed by twin-shaft agitators and scrapers.
5. *Steam-Jacketed Coffee Urn.* You can make from 20 to 150 gallons of coffee without a hot-water source. The fast-heating action of the kettle quickly brings cold water to the proper temperature for coffee brewing.
6. *Steam-Jacketed Oyster Cooker.* You shuck 'em, and there's a special kettle to cook 'em! This is a direct-steam piece of equipment; one manufacturer sells them in 32-, 64-, and 72-ounce capacities.

Cleaning Steam-Jacketed Kettles

Steam-jacketed kettles are relatively easy to clean and maintain. We'll talk about cleaning in detail later in the chapter, but here are the basics. It's best to turn the steam off and let the kettle cool down before emptying it for cleaning. Scrub the walls with a plastic brush and open the drain to let the soapy water drain out as you work. Be sure the kettle is fully rinsed, inside and out. Clearly, the kettle should be installed where it can drain into a floor drain. For tilting kettles, the drain placement should align with the pouring path. Tabletop units should have a sink to drain into. If you don't have time to clean your kettle immediately after use, at least fill it with water and turn on the steam to heat the water and start the cleaning process by loosening any food particles.

ILLUSTRATION 14-6 **Stands and tables, like this one, can be ordered to fit gas and electric kettles.**

COURTESY OF VULCAN-HART, DIVISION OF ITW FOOD EQUIPMENT GROUP, LLC, LOUISVILLE, KENTUCKY.

14-3 PRESSURE STEAMERS

As you consider a steamer purchase, think first about the menu you'll be working with. Delicate items, like fish fillets, benefit from pressureless convection steamers because the cooking process keeps them whole and preserves their texture. "Sturdier" items, like pork loins, can be started in a high-pressure steamer for a terrific tenderizing effect, and finished later on a grill. Also think about the numbers of portions: Single? Small batch? Hundreds at a time?

Pressure steamers, also known as pressure cookers or ***compartment steamers,*** are ideal appliances for cooking fresh, defrosted, or loosely packed foods. They cook at up to 228 degrees Fahrenheit, with steam pressures that range from 2 psi to 15 psi. Where volume is required and all meals are served at once, the compartment steamer is the appliance of choice. Generally, it is a floor model, available with two, three, or four compartments. Its downside is that you must be careful not to overcook, and, if not properly cleaned between uses, there is the potential for residual flavor transfer—in other words, you'll taste the last dish along with the one you're now cooking. Some popular applications of pressure steamers are:

- **Stews, pot roasts, ham, and roast beef. Shrinkage is minimal, and steamed meats are tender, moist, and flavorful.**
- **Potatoes and other root vegetables can be cooked very quickly in high quantities: 100 pounds per compartment in thirty-five minutes. Most other vegetables are cooked in five to ten minutes. Frozen vegetables take even less time.**
- **Pasta and rice can be boiled faster than in pressureless steamers.**

We already know that steamers cook by transferring heat from steam onto a cold food product. In pressure steamers, however, the cold air that comes off of the product is vented away from the cooking chamber as steam is introduced. As temperature inside the chamber increases and cold air is removed, a steam trap slowly closes and steam pressure builds up inside the chamber. This pressurized steam (also called ***lazy steam***) works its way through the food and cooks it from the outside. Frozen block products don't cook well in this situation because the outside becomes overcooked before the inside is cooked properly.

Pressure steamers come in high-pressure (15 psi) and low-pressure (2–5 psi) models. Low-pressure compartment steamers provide higher productivity at lower operating cost than pressureless models when cooking single items in volume. However, high-pressure units are usually faster and more economical. They cook at about 250 degrees Fahrenheit and are regularly used for such basic kitchen duties as blanching and reheating. With most steamer models, you can put the food on sheet pans or steam table pans.

Pressure steamers may run on electricity (24 to 48 kilowatts per hour) or gas (170,000 to 300,000 Btus per hour). A typical unit will be 36 inches wide and 33 inches deep; four-compartment units will be as tall as 67 inches. When shopping for pressure steamers, be sure they are equipped with heavy-duty doors and gaskets and pan slide racks that are all easy to clean; compartments with seamless fabrication to prevent leaks; an adequate steam condenser and drainage system; and pressure gauges and safety devices. A separate timer for each compartment is also recommended.

Low-Pressure Steamers

The power requirements for low-pressure steamers are the same as for pressure steamers. They differ from the standard pressure steamers because they contain their own boilers, built into a stainless steel cabinet underneath the steamer. You can also purchase low-pressure steamers that use direct steam from a central source, or have steam coil heat exchangers. The steam coil heat exchangers can take incoming steam from other building systems at 35 to 50 psi and boil it into clean steam for use by the low-pressure steamer. If you're using direct steam, the experts recommend 40 to 50 psi of incoming steam with a flow of 34.5 psi per hour, per compartment. A pressure-reducing valve can reduce compartment pressure to as low as 2 psi.

The low-pressure compartment steamers have a combination pan-and-shelf slide unit that is really handy. Pans may be pulled out two-thirds or more of their length without tipping to check the progress of cooking food. Simply remove the center assembly or partition of the slide unit. Each compartment can hold standard-sized sheet pans.

High-Pressure Steamers

High-pressure steamers do not offer the high-volume cooking capacity of their compartment steamer or low-pressure counterparts. A high-pressure steamer can only hold three standard-sized hotel pans, compared to eight for the compartment steamer. However, they are in heavy use in places where small batches are in continuous demand, such as large table service restaurants. The high-pressure steamer operates with a total pressure of about 15 psi, which raises the cooking temperature of the food and therefore reduces the cooking time, as shown in Table 14-3.

The high-pressure steamer comes in countertop and cabinet-mounted models, both gas and electric. The gas models use 40,000 Btus per hour, and the electric ones run on 12 kilowatts per hour.

High-pressure direct-steam countertop steamers are small and quick, recommended for small batches of frozen vegetables. Despite their size (18 by 22 by 26 inches), they can produce up to 900 (2½ ounce) servings of vegetables within an hour. Be sure to use a steam coil model, which is the only one capable of purifying the steam for culinary use.

14-4 PRESSURELESS STEAMERS

Pressureless steamers, also called *convection steamers,* cook food at 212 degrees Fahrenheit. There are several differences between pressurized and pressureless steam cooking. First, during the pressureless process, the steam comes into direct contact with the food. Second, the doors may be safely opened at any time during the cooking cycle to check, rotate, or season the food—unlike other types of steamers, where this is not recommended.

Pressureless steamers are smaller and don't cook as quickly as their pressurized counterparts, but they are so easy to use that they account for more than half of all steamers sold in the United States. They provide higher-quality end results for frozen foods, fresh vegetables, and seafood. The combinations of moisture and relatively low temperature provide a better taste and texture. And, like other steamers, there's no problem with food burning, sticking, scorching, or drying out. Most professionals suggest that small-quantity, batch-cooking needs are best met by a pressureless steam cooker.

TABLE 14–3

APPROXIMATE STEAMER COOKING TIMES

Food	Pounds per 100 Portions	Compartment Steamer (5 psi)	High-Pressure Steamer (15 psi)
Apples	30	15 minutes	10 minutes
Asparagus	36	15	8
Broccoli	40	12	8
Brussels sprouts	24	15	8
Cabbage	16	20	10
Carrots	16	25	12
Lima beans	20	20	10
Peas	30	8	5
Rice	26	30	15
Spinach	36	7	4
Potatoes	30	30	20

Source: *Foodservice Equipment and Supplies Specialist.*

In the convection steamer, the steam is injected into numerous inlets arranged to create a *spray* that circulates inside the cooking chamber. The heat transfers from steam to the food by the turbulence this creates—a kind of "forced convection."

Some units have a fan to help circulate the steam around the chamber, and a vent system is used so that air is continually eliminated from the cooking chamber. This technique provides fast, moderate cooking at a steady temperature of about 212 degrees Fahrenheit. In addition, venting prevents the buildup of gases, odors, or any other by-products of the cooking process that could compromise food quality. Some people compare pressureless steaming to microwaving, only without the microwave's drawback of unevenly cooking or drying out some foods.

The preferred pan is perforated, 12 by 20 inches in size or smaller. However, solid pans may be used for "messier" foods such as pasta, rice, meat loaf, casseroles, or stews. The pressureless pure steam environment is capable of 300 percent more efficiency than a combination of air and pressurized steam. And, because the steam is not pressurized, the door of the unit can be opened any time during the cooking process and the temperature will remain at 212 degrees Fahrenheit.

Here are a few more practical kitchen realities for which pressureless steaming can be helpful:

1. Untrained personnel will be less likely to have accidents with the pressureless steamer.
2. If food must be seasoned or handled during cooking, you can open the door.
3. You can cook large quantities faster than, say, in a stockpot or oven—although it's still slower than pressure steaming.

In recent years, manufacturers have created higher-power and higher-capacity units to enable the pressureless steamer to compete with the low-pressure steamer. New models, with Btu ratings of 150,000 to 300,000, have greatly increased volume capacity and accelerated cooking times. You can also purchase a large unit with more than one boiler for more than one compartment, two powerful blowers in each compartment, and two large steam entry ports for better steam distribution.

You'll find the basic pressureless steamer is usually less expensive than pressure steamers, and it saves energy as well. The typical features of a pressureless steamer include:

- Multiple steam generators (one for each cooking cavity) do not require pressure reheat valves or gauges, because they are not pressurized, so there are fewer components to fail. Energy can be saved by using only those compartments that are needed.
- A powerful blower is installed in each cavity to increase the velocity of steam within that cavity.
- A simple, dependable 60-minute timer is necessary. You'll also want a warming mode that keeps the boiler simmering between uses and the cooking cavity warm and ready to power up within seconds.
- "Easy-clean" features include an automatic deliming process. There's usually a removable side panel where you pour a deliming solution into the appliance, as well as a warning light to tell you when to do it to prevent overheating or other malfunctions.
- A specially hinged door has a heat-resistant gasket and an easy-open latch. Always look carefully at appliance doors, because they'll be opened and closed thousands of times every year. Also, look for a door that is easy to change (to open from either left or right) in case you want to move the steamer. A magnetic switch should cut power to the blower whenever the door is opened.
- Simple utility connections mean faster, inexpensive installation and easy disconnection if necessary for cleaning or servicing. A wide sink with a drain under the door of the cooking compartment reduces the hazard of wet or slippery floors near the steamer, and provides a convenient landing spot for hot pans.

Connectionless Steamers

The newest type of convection steamer touts itself as being connectionless and/or "boilerless!" Yes, that means it requires no traditional plumbing hookups. Water is added manually into a tank that contains one or more heating elements. A source of water and an electrical outlet, and they're good to go. There are a couple of advantages to this: there is no boiler to clean, and this type of steamer is easy to move as needed. A fan circulates steam within the cooking cavity.

The connectionless steamer is not a substitute for higher-volume, fully plumbed steamers, but can be useful in low- to medium-volume operations. It can hold three to six sheet pans. This type of unit uses less water—perhaps 8 gallons *a day,* compared to 30 gallons *an hour* for a traditional steamer—and no water filtration system is necessary. It is electric, with some manufacturers submerging the heating elements in the water tank, and others not. Some models have low-temperature vacuum steaming capability. The boiler-less, connectionless steamers are quite new—their reliability and popularity are still unproven. But they are yet another example of the evolution of foodservice equipment to meet a wide variety of needs.

Steam and Power Requirements for Pressureless Steamers

Most facilities, particularly small and medium-sized restaurants, will opt for a pressureless steamer with a self-contained generator that makes its own steam. However, pressureless steamers can also use direct steam from a central supply or an external boiler, in buildings that already supply steam for water heating or temperature control.

Unless your unit is self-contained and creates its own steam, you must determine whether the available steam supply is sufficient to run your pressureless steamer. For example, a steam boiler rated at 2.2 BHP (boiler horsepower) can generate enough steam for most two-compartment (six-pan) steamers, but *not* enough to run *two* steam appliances in the kitchen at the same time. The manufacturers will be clear about each unit's steam requirements, but it takes roughly 0.75 BHP to run one steamer compartment; or 1 BHP for every 20 gallons of steam-jacketed kettle capacity. Water pressure should be no more than 60 psi and no less than 30 psi. Keep in mind the distance steam must travel—it loses some pressure on its journey through pipes—as well as the sizes of piping and fittings used.

Countertop models will have their own self-contained steam boilers, which might be powered by gas or electricity. Either way, they need a clean, direct water supply and appropriate drain lines. Countertop units go through a preheat phase before the cooking process begins. The ones that preheat faster will use more power.

Electrically powered steamers require from 5 to 48 kilowatts; a 48-kilowatt model is capable of delivering 4.7 BHP. Gas-operated units use anywhere from 45,000 to 250,000 Btus. Under normal conditions, it takes six minutes or less to preheat.

Sizing and Selecting Pressureless Steamers

Because we've been discussing countertop models, let's begin with their typical dimensions. On the outside, a countertop model is 26 inches wide, 19 inches high, and 29 inches deep, including its utility connections. The stand on which it can rest measures 24 inches wide, 25 inches high, and 29 inches deep. These figures will increase proportionately with a stacked system, where two units are placed atop each other, or the quad system, where two units are stacked on top of two more, for a total of four steamers side by side (see Illustration 14-7).

Floor models usually have two compartments and their overall size is 24 inches wide, 33 inches deep, and 58 inches high, although some are 36 inches wide. This includes the base.

Generally, the cabinet base of floor models houses the boiler and is made of stainless steel with a full-perimeter angle frame and reinforced counter mountings. Most have hinged doors and adjustable stainless steel legs. The desirable height for a steamer is 32 inches.

Although overall dimensions are important for planning your floor space, you must also think carefully about the interior capacity of the steamer. Table 14-4 provides estimates for pressureless steamers based on the number of meals served per hour. We assume that each compartment can hold three to four 12-by-20-by-2½-inch pans or two to three 12-by-20-by-4-inch pans. As with steam-jacketed kettles, you'll find that most manufacturers offer these types of guidelines with their steamer models, too.

Manufacturers will also provide suggested cooking times for various foods, although these are all relative. As with ovens, ranges, or any other cooking appliances, a good chef will make allowances for temperature fluctuations; food freshness, size, and shape; the depth and type of pan used; and the desired degree of doneness.

A common complaint among new steamer users is that they are so fast, it's easy to overcook items. Use the manufacturer's manual until you are comfortable deciding on cooking times yourself, and remember to set the timer. It automatically cuts off the steam at the appropriate moment and alerts you with its loud buzzer. With a little practice, you'll find steamers are great for blanching items such as chicken before breading, cooking thinly sliced meats that don't need browning, freshening stale breads, and reheating almost any type of food.

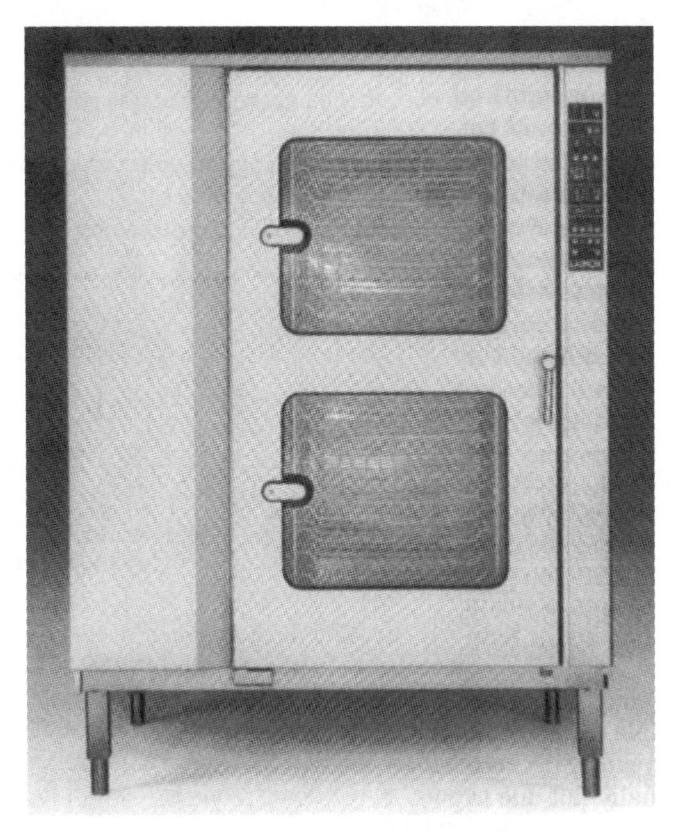

ILLUSTRATION 14-7 **Pressureless steamers can be stacked, one compartment atop another, or located side by side.**

COURTESY OF CLEVELAND RANGE, A WELBILT COMPANY, CLEVELAND, OHIO.

TABLE 14–4

FOODSERVICE STEAMER NEEDS

Meals per Hour	Compartments
0–200	One single-compartment steamer
200–400	One double-compartment steamer
400–600	One triple-compartment steamer
600–800+	One quadruple- or two double-compartment steamers

FOODSERVICE EQUIPMENT

STEAMER LIFECYCLE FACTORS

- Purchase price, tax, freight, start-up
- Installation cost (standard steamers require water hook-up, drainage, and so on; connectionless don't)
- Energy cost (hours of use per day at average percentage of peak energy use)
- Water cost (much higher with standard boiler steamers)
- Sewer cost (ditto)
- Preventive maintenance
- Labor (cleaning, deliming)
- Supplies (deliming chemicals, and so on)
- Service/repairs (more likely with standard units)
- "Real estate" cost (connectionless steamers more mobile, allowing greater flexibility in kitchen layout and/or menu changes)
- Annualized costs: all expenses amortized across projected life expectancy

Source: *Foodservice Equipment Reports,* a Gill Ashton publication, Skokie, Illinois, May 2003.

14-5 SPECIALTY STEAMERS

This category includes a few highly specialized steam appliances designed for very limited but common purposes.

Flash Steamer. The ***flash steamer,*** also called a *steam food cooker,* is a countertop model used to heat sliced meat sandwiches, melt cheese, and warm rolls and other pastry products without making them soggy. There are "closed" and "open" flash steamers. In both cases, distilled water is heated and converted into steam inside a sealed cavity. In the closed steamer, the steam escapes through tiny holes and enters the food, which creates some pressure and velocity in the system. In the open system, a steam generator sprays water over the bottom heated surface. This creates steam, which naturally rises into the food area.

There are two sizes of flash steamers. The full-sized model accommodates a 4-inch-deep steam table pan and the half-sized model takes a 2½-inch-deep pan. There are two operating styles as well.

You can either push a button to energize the water pump or solenoid valve, which allows water to enter the steam generator, or press down on a handle or arm lever to start the process. They require 120-volt electric power (15 to 20 amps), and have simple controls for rapid reheating of precooked foods. The steam food cooker can also hold cooked foods until needed.

Cleaning on a regular basis is a must for long and trouble-free use of flash steamers. Only distilled water should be used, because mineral deposits will find their way onto the heating surfaces.

Spatula Steamer. The spatula steamer—so named because of the insulated handle on its perforated basket, making it look like a spatula—is for reheating individual portions. It's great for things like late-night hotel room service and à la carte restaurant orders. It can be used either to cook or to reheat, shooting steam at the food as many as 14 times in a 2½-minute cycle. It contains a built-in boiler that uses plain tap water to generate steam when plugged into a 110- or 208-volt electrical source. A release valve at the back of the unit allows excess steam to be safely released.

Needle Injection Steamer. Finally, there's the right steamer for a perfectly warmed Danish! The ***needle injection steamer*** offers a fast, simple, and consistent way to reheat or warm porous food products. In this appliance, needles are arranged in a specific pattern to heat one type of product. When food is pressed down onto the needles, steam is released directly through the needles into the food. Although this direct internal steam penetration is efficient from a heat transfer point of view, it only makes sense to have this steamer if the type of food you'll be heating with it makes enough money for you to justify the cost. Pastries do well in this environment, but dense items such as baked potatoes or bratwursts would do poorly in this type of steamer.

Pasta/Rice Steamer. Other steamers can cook pasta and rice perfectly, but there are *pasta and rice steamers* especially designed to cook an 8-ounce portion of refrigerated pasta in 30 seconds or the same serving of frozen pasta in 90 seconds. This piece of equipment releases steam at high velocity into a small, enclosed cavity. According to its manufacturer, it's also ideal for quick reheating of other precooked items.

Steamer/Fryer. The public's interest in more healthful cooking has led to the development of a *steamer/fryer.* For foods that already contain enough oil to brown, the steamer/fryer can cook a product with the same taste and texture as if it had been fried. After the food is steamed, the steamer/fryer switches to "Brown" mode to create a crisp, golden-brown exterior. Potatoes, chicken pieces, breaded cheese sticks, onion rings, and some types of seafood turn out well in the steamer/fryer.

14-6 COMBINATION OVEN/STEAMER

The *combi oven* is essentially a union of the steamer and the convection oven (which cooks by circulating hot air) into a single appliance that can provide either of these processes, or both at once. This unit, with its variable humidity controls, consumes up to 60 percent less energy than traditional cooking appliances. Today's combis are technologically sophisticated and highly programmable. They can be set to turn on and off automatically, and programmed to cook dozens of different products with specific settings and steps. However, most of these are a combination of three distinct cooking modes:

1. "Steam" mode emulates the speed of a pressure steamer, without the pressure.
2. "Hot Air" mode allows convection oven heating with the added feature of humidity.
3. "Combi Cooking" uses a combination of superheated steam plus hot air.

The combi oven moves air around mechanically within the moisture-filled cooking cavity. By combining convection heat with the steam's moisture, the result is a moister, fresher cooked product with a longer holding life. The moisture keeps nutrients and flavor in the product, while the air movement speeds up the cooking process. Browning also takes place faster and more evenly in the presence of moisture (see Illustration 14-8).

The chief benefits of combi ovens are their speed and versatility. They are handy in preparing all sorts of foods, from roast chicken to steamed vegetables, poached salmon to pot pies, soups to

ILLUSTRATION 14–8 **The combination oven/steamer, nicknamed "combi oven," combines moisture and convection heat and can take the place of several pieces of cooking equipment.**
COURTESY OF RATIONAL COOKING SYSTEMS, INC., SCHAUMBURG, ILLINOIS.

desserts. This is also the unit to use when preparing meats or baking rolls, breads, or pastries that require perfectly browned crusts. With a combi oven on the hot line, waste is no longer an issue. Vegetables can be prepared just before they're served, instead of sitting and getting soggy. Daily specials can be plated and cooked in advance, then steam-reheated and sauced for service. Pasta can be cooked in advance, and reheated in less than a minute.

The first combi oven was electric, introduced in the 1970s by Rational Cooking Systems, Inc., of Schaumburg, Illinois; in the mid-1980s, the company added a gas-heated model. Rational is still the industry leader in sales and in this technology, with more than 80 patents for its "Combi-Steamer" models and related inventions. (You can find out more on the company's websites, www.rationalusa.com and www.rationalfpp. com.) The original units were enormous and expensive, found only in high-volume foodservice sites like hotels, casinos, and hospitals. Over the years, smaller and simpler combis have been introduced. Most recently, they've become a major tool in the home meal replacement segment, with many take-home and supermarket applications. In high-volume settings, food may be cooked early in the day, rolled into a blast chiller for a quick cool down, then plated, garnished, covered with plastic wrap, and put in a display case. When a customer selects it, the combi can retherm it right in the store.

Combi ovens not only cook many kinds of foods—they cook them fast. Production times are up to 40 percent quicker than conventional ovens. For example, a typical restaurant rotisserie cooks a whole chicken in about 80 minutes. If you're willing to forgo the "look" of the rotisserie, a combi oven will have the same chicken ready to serve in 35 minutes, with the same taste, moist interior, and nicely browned exterior texture. The oven's convection airflow means it can cook at lower temperatures, increasing the yield of meat products up to 30 percent. Many combis use computerized temperature probes to sense when meat is fully cooked and automatically stop when the meat reaches the desired core temperature.

The combi's high-temperature settings rival a pressure steamer when it comes to cooking time. Fresh steam is directed into the cavity from a self-contained steam generator, and rapidly circulated by a power blower. All frozen and most fresh vegetables do well in this mode, if you remember that load size and the type of product will influence its cooking time. Shellfish, seafood, meats, and poultry also cook nicely at high-temperature settings, with no additional liquid added. Dehydrated items, however—rice, pasta, and cereals—must be covered with water. The ratio of water to rice is two to one; water to dry pasta, three to one.

Low-temperature steam enhances the cooking of items such as frozen chopped spinach, fresh leafy vegetables, and fresh broccoli or asparagus. The low-temperature setting simply circulates the steam more gently.

The "hold" mode is not intended for primary cooking, only to keep foods warm or reheat something that's already been cooked and then refrigerated. The hold temperature is preset at 145 degrees Fahrenheit with a light moisture level to preserve food quality, although it's probably not wise to hold most foods for more than an hour or their overall quality and appearance are bound to suffer a bit. Large cuts of meat should be tempered (marinated or tenderized) before putting them in the hold mode, where they can then remain for several hours. For instance, if you know a roast will be sitting for a while, adjust the combi's hold temperature to 10 degrees below the desired final temperature.

So combi ovens save time, increase yield, and are versatile. Did we mention they also save space? Most models can be stacked. They must be installed under an exhaust hood, but they are smaller than many hot line appliances, so not as much hood is needed. (When you consider that restaurant exhaust systems cost between $500 and $800 per linear foot of hood, every inch counts.)

More refined than the removable strainer for steam-jacketed kettles, the air inside a combi oven circulates through a ***fat filter,*** trapping food odors and grease particles before the air reenters the cooking chamber. This filter is mounted on the baffle at the rear of the oven, where it is easily removable after the oven racks are removed. (Make sure the oven is cool before you try to remove the filter.) At least one manufacturer recommends having two fat filters, so you can wash one while the other is being used. They're easy to wash, either by hand with a mild soap or by running them through the dishwasher.

Tabletop combination oven/steamers can hold from 10 to 20 steam table pans. If you prefer using sheet pans, most can accommodate a stainless steel slide rack that fits neatly, is easy to remove for cleaning, and can hold up to 10 sheet pans. Most manufacturers also have stands on which to put tabletop units.

Some floor model combis can be adapted to fit on rolling carts, which is especially handy if your operation will do a lot of cook-chill preparation. The loaded cart can roll handily from oven to refrigerator. The dimensions of these rolling units vary, but they generally come in full and half sizes. When determining the correct size for your kitchen, remember to calculate another 6 inches of height for rolling legs.

Water, Steam, and Power Requirements for Combis

For electric combi ovens, the air is heated by a separate heating coil in the oven interior, which requires from 10 kilowatts for countertop models to 72 kilowatts for floor models. Gas-operated units are floor models only; their Btu requirements range from 90,000 to 225,000 per hour.

In the electric combis, you have a choice of steam production systems: a built-in steam generator, or steam injection. Choosing steam injection does away with a generator and its associated cleaning and deliming—water is simply sprayed into the hot oven cavity, where it turns immediately to steam. It works well for small quantities of product and costs less to operate than a boiler; but the boiler is still the best choice for high-volume cooking. For gas models, a built-in generator is the only choice. It is located either in back of or beneath the cooking cavity.

Most models have a "Steam Generator Standby" mode, which allows the generator to switch pressures without a cool-down period. Steam generators empty from inside the oven, flushing their systems completely to eliminate water buildup, scales, or lime deposits, which will save money you may otherwise spend on costly water treatment. As mentioned earlier, however, a water-softening system is recommended to reduce the risk of contamination. Manufacturers recommend even lower levels of total dissolved solids (less than 60 ppm) for combi ovens than they do for pressureless steamers (less than 80 ppm). In other respects, however, water requirements are similar: an alkaline content of less than 20 ppm, and a pH factor greater than 7.0. Combi units are sensitive enough, and used often enough, to recommend a regular cleaning schedule.

They require a ¾- to 1¾-inch cold-water pipe with 45 to 75 psi of water pressure, as well as a separate, indirect waste pipe with air gap. You must also have a water supply shutoff valve and a backflow preventer. As you might have guessed, it is best to have a professional plumber handle installation.

All units have automatic controls to preset and regulate the oven temperature. The amount of moisture inside the oven is relative to the temperature setting; that is, even when cooking with steam, as the temperature increases, humidity decreases. Obtaining full saturation, or 100 percent relative humidity, is not possible, because steam in the combination mode is pressureless and combi cooking almost always requires hot temperatures (above 212 degrees Fahrenheit). However, to some degree, you can adjust and control the oven's humidity.

Many models offer programmable, computerized controls that do more than implement the various cooking combinations. An employee can choose the amount and temperature of the steam, as well as the fan speed to circulate it. A computerized combi can be programmed for up to 99 cooking modes, each of

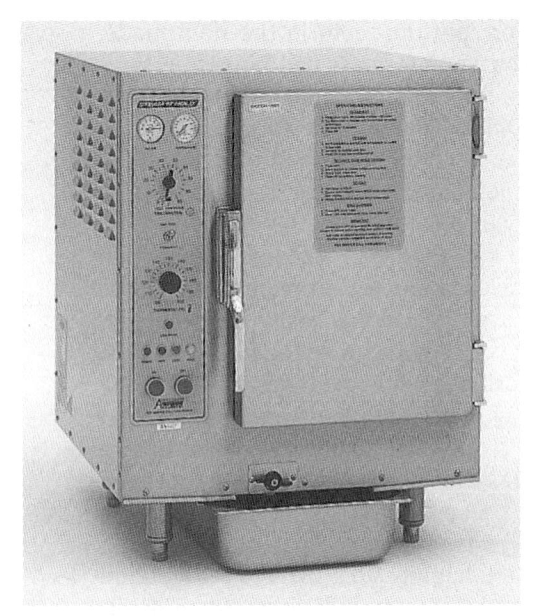

ILLUSTRATION 14-9 **The vapor oven or vacuum steamer is a combination oven and low-pressure steamer.**
COURTESY OF ACCUTEMP PRODUCTS, INC., NEW HAVEN, INDIANA.

which may contain up to five steps. A correctly programmed combi oven allows a cook to put a certain food into the oven chamber and press a button. What a luxury!

Combi controls, whether solid state or computerized, are extremely temperature-sensitive. If the unit is going to be installed next to other hot line equipment (broilers, fryers, ovens, or the like) a *side shield* is a requirement to protect the combi oven from heat damage.

Vacuum Steam Cooking

This variation on the combination oven/steamer is commonly known as a ***vapor oven.*** Preparing food with this low-temperature steam appliance means the food is never exposed to high or uneven temperatures. Vapor ovens are relatively new pieces of equipment, electrically heated (using from 4 to 6 kilowatts per hour) and built into a heavy-duty stainless steel cabinet (see Illustration 14-9).

Here's how vacuum steam cooking works. The cooking chamber holds water, which is preheated to the desired temperature. Inside the chamber, a vacuum pump reduces the air pressure by pulling air out of the cooking chamber. This lowers the boiling temperature of water to a (relatively) cool 140 or 150 degrees Fahrenheit, creating a low-temperature steam with a superior heat transfer rate. This means food can be cooked with a minimum of temperature difference between the food and the steam. Steam has the ability to transfer heat uniformly—that is, it condenses on all exposed surfaces depending on their temperature rather than their proximity. More steam will condense on the colder parts of the food. The steam continues to condense there until the food's surface temperature reaches the same temperature as the steam itself. This process naturally lowers the temperature in the cooking chamber, which causes the thermostat to turn the heater back on. The ongoing cycle means food is cooked evenly and thoroughly without having to turn it, baste it, or watch it constantly. A timer on most units will shut off the steam and put the oven in hold mode when the food is done.

This type of cooking is perfect for delicate items like fresh green beans, broccoli, and fish fillets. Vapor oven temperatures range from about 140 to 200 degrees Fahrenheit. Timer, thermostat, and cooking cycles are all set using an electronic control panel.

Energy savings is one big advantage of vacuum steam cooking. It uses only 10 percent of the power of a conventional steamer, because it makes only as much steam as it needs. There is no excess to be condensed and drained.

14-7 STEAM GENERATORS

There are three sources of steam for steam cooking equipment:

- From a central, outside source, usually a building's existing steam supply, that may be tapped into with piping to run appliances
- From a freestanding ***boiler*** or *steam generator* (the terms are used interchangeably in the industry) that heats water and makes steam for two or three appliances all at once
- From an appliance that is equipped with its own built-in boiler

We've already mentioned the importance of "clean" steam, uncontaminated by chemicals in the water. Centrally supplied steam from an outside source may be economical, but the piping for such a system is expensive, which may mean you'll have to group your steam equipment together to save money. Central steam supplies are becoming a rarity, though, so most foodservice businesses have freestanding steam generators, or appliances with built-in boilers.

Steam production is measured either in pounds per hour (pph) or in boiler horsepower (BHP). A good general guideline is that 1 BHP is needed for every 20 gallons of kettle capacity, and ¾ BHP is needed for each steamer compartment.

Steam generators can be purchased to run on electricity, gas, or steam coils. If you're shopping for a generator, first look at what is available in the market, because models vary quite a bit. Then be sure to match the capacity of the boiler to the total demands of the steam appliances it is designed to service.

If you'll be attaching multiple pieces of equipment to the same boiler, they should be turned on one at a time, never all at once, to give the generator time to provide sufficient steam and pressure. A few more boiler basics follow.

Electric. Constructed according to standards of the American Society of Mechanical Engineers (ASME), when you turn on its heat and water switches, the unit fills with water automatically. It also drains automatically under pressure when the switches are turned off. It uses three-phase power, between 24 and 48 kilowatts, depending on its size. An electric unit can generate steam at 5 to 15 pounds per square inch. If two separate pressures are needed—for example, 5 psi for one unit and 15 psi for another—a pressure-reducing valve to provide the 5 psi is necessary.

Available dimensions range from 24 to 36 inches wide, 33 to 36 inches deep, and 28 to 33 inches high. The stainless steel cabinet is mounted on a 24- or 36-inch-wide base, typically on four 6-inch stainless steel legs. Its two front doors have magnetic latches and all piping is confined within the cabinet.

Gas. Also constructed according to ASME guidelines, the gas-fired model is rated (sometimes called its ***firing rate***) between 140,000 and 300,000 Btu; about 60 percent of that figure actually reaches the food being cooked.

The gas-fired boiler provides 15 psi of pressure and has an optional pressure-reducing valve that lowers pressure to 5 psi if necessary. The dimensions and control systems of the gas model are similar to the electric model.

When writing specs for a gas generator, always specify the elevation where it will be installed. In communities higher than 2000 feet above sea level, special gas orifices must be used.

Steam Coil. With either electric or gas, you can use a coil system. Steam coils circulate steam to heat water, but they need another type of boiler to create the steam. In order to be part of a steam generation system, coils need a minimum incoming water pressure of 20 psi and maximum incoming pressure of 50 psi. They can produce 2.6 BHP and 85 pounds of steam per hour. A 120-volt, single-phase electrical connection is necessary to operate the coil controls. Consider using a reducing valve to facilitate the various steam-operated equipment connected to this generator.

Boilers are generally "blown out" or drained at the end of a busy day, and refilled the next day. To maintain its efficiency, a boiler needs to be rigorously cleaned to prevent buildup of lime, scales, and other mineral deposits on the boiler's walls and around the water-heating element. This buildup also forms acidic conditions inside the boiler, accelerating spot corrosion and causing eventual leaks. The frequency of this cleaning depends on whether your area has hard or soft water, but it must be done even if the boiler has its own water filter. Some boilers are easier to clean than others, with a "port" that allows you to add descaling agents or inspect the interior, or a warning light that comes on when cleaning is due. Metallic corrosion can be controlled by installing a ***cathodic protector,*** a metal device mounted on the inside of the boiler, suspended in the water. You cannot eliminate acidity and scaling, but you can minimize it with a good water treatment system and regular maintenance. Service companies say forgetting to delime is the most common cause of repair-related calls.

Finally, no matter which power source you use, comes another firm reminder to comply with the manufacturer's water-quality requirements. Equipment failure caused by inadequate water quality is not covered under warranty.

14-8 CLEANING AND MAINTENANCE OF STEAM EQUIPMENT

Now that we've cleaned the boilers, let's turn our attention to the rest of the steam equipment, where scale buildup is also a problem. It forms on any surface of the appliance, from interior walls, to water-sensing probes, to heating elements. In fact, each quarter-inch of lime-scale thickness increases the energy required to produce sufficient steam for the appliance by 40 percent! So it pays to descale and delime, and it is a lot less expensive to do so than to replace equipment. Some full-featured appliances, including combi ovens, have automatic deliming programs that sense buildup and will alert the operator to run a deliming cycle.

You'll get greater energy efficiency from steam-operated equipment if you use it at full capacity. That means a couple of smaller steamers, well loaded, are usually better than one large one that is used less or operated half-full. Whenever possible, cook similarly sized pieces of food together; it's faster and the food will be more uniformly prepared.

Another energy-saving practice: Place an external steam generator as close as possible to all the steam-operated equipment to reduce heat loss in transit. And remember, cooking speed can triple with units (e.g., vapor ovens) that introduce steam directly onto food.

Also explore the option of cooking food to near-doneness in steamers and then finishing it on a griddle, fryer, broiler, or even in the oven. You can save energy by steaming first, then using the more conventional cooking appliances to add the color, crust, or flavor your customers crave.

Steamer maintenance means making sure steam is properly contained, and that safety valves work to prevent ruptures of the cooking cavity from too much interior pressure. Check all gaskets regularly for leaks and notice the door mechanisms while cooking to ensure they fit tightly. Once or twice a week, the safety valve should be lifted and checked to make sure it has not corroded to the point of being ineffective under excessive pressure. Check pressure gauges, pilot lights, timers, and thermostats for proper calibration. Always keep the user's manual handy on the job site.

Stainless steel surfaces, both interior and exterior, are key to prolonging the useful life of these appliances. They are usually easy to clean. A degreasing spray can be applied to the interior, then the steam mode is activated to loosen any food particles. A built-in spray hose rinses away the grime. Some experts consider steam equipment a pot washer's best friend, for its ability to steam off caked-on food particles on pots and pans, too. Here are a few more cleaning specifics:

- Allow the unit to cool with the door open. Remove the shelves by unlatching them; scrub them in the pot-washing sink with detergent and hot water; rinse and drain.
- Always scrub both interior and exterior with a soft brush. With every cleaning, check the compartment drain to clean out any clogs.
- Rinse the inside and outside of the steamer with clean, hot water; then close the door and turn on just enough steam to heat the interior. You must wipe down the exterior with a clean dry cloth; the stored heat should be enough to dry the interior.
- Install and latch the shelves.

For the interior oven cavity, the latest technology in combi ovens is a self-cleaning model. Clip a device into a slot on the back of the oven, and enter a number onto the control panel that indicates the level of cleaning desired (from "1" for light cleaning after bread-baking to "7" for heavy, greasy jobs like roasting chicken). The cleaning device has arms that move around, dispensing correct proportions of water and detergent, then a drying agent.

14-9 SITE SELECTION

This is the last section but, in many respects, the most important, because your whole choice of equipment often rests upon the floor space, plumbing connections, and other utility availability in your kitchen. Here is a summary of the key factors to keep in mind when sizing up a potential steam installation site. They are excerpted from the *Handbook of Steam Equipment* published by NAFEM in 1994:

1. Most steamers require a floor drain for removing condensate from the cooking compartment and the boiler.
2. Determine the water quality and provide for a water softener if needed.

3. Floor drains in front of tilting pans and kettles help contain liquids that spill on the kitchen floor.
4. Consult the manufacturer's requirements for sizes of water inlets and drains. An air gap is recommended between steam generator and floor drain.
5. The floor drain should be located outside the perimeter of the equipment. This is to prevent the venting steam from rising and condensing on the unit's electrical components.
6. Look at your city or county ordinances for proper ventilation and fire suppression requirements for steam-operated equipment.
7. Consider the proximity of electrical power to the equipment. Even if you're using a gas-fired boiler, you may need electricity to power the appliance itself.

SUMMARY

Steam cooking is a very efficient way to transfer heat to food. Steaming also preserves the flavor, color, and texture of fresh foods, keeps them moist, and reduces waste and shrinkage. You can use a steamer to prepare food to the point of near-doneness, then finish it on a griddle, broiler, and so on. The faster cooking time is useful when you need to prepare delicate foods quickly, especially those that don't keep well under heat lamps or on steam tables.

There are several different types of steam equipment, all very popular in foodservice. The steam-jacketed kettle is constructed like a bowl within a bowl. In the small space that separates the two bowls, steam is pumped in and cooks the food without coming into direct contact with it. Its large area of surface heat makes the steam-jacketed kettle extremely versatile. Kettles range in size from 1-quart tabletop models to 200-gallon floor models, and they must be installed under a ventilation hood. They can be mounted in cabinets, on walls, on legs, or on pedestals.

A related appliance, the pressure steamer or compartment steamer, can be used to boil rice or pasta and cook meats or root vegetables. Inside this steamer, the cold air that comes off the food product is vented away from the cooking chamber as steam is introduced. The pressurized steam cooks the food from the "outside in." There are low-pressure and high-pressure steamers, as well as pressureless steamers. The designations refer to the pounds per square inch (psi) of steam that may be generated, which determines the maximum temperature the steam can reach. "Pressureless" steaming is actually steaming at very low pressure, which is best for delicate foods.

During pressureless steaming, the steam comes into direct contact with the food, and—unlike other steamers—the door or lid may be opened safely at any time to check or season the food. Pressureless steamers are also called convection steamers.

Combination steamers can be used either with or without pressure; and combination oven/steamers can also cook with or without steam. The major benefits of combi ovens are their speed and versatility, allowing them to take the place of several pieces of equipment. Today's combi is a technological marvel, programmable to cook dozens of different foods even when the cooking processes require multiple steps.

Steam equipment requires steam-generating capacity. There are three ways to do this: by tapping into a building's existing steam system; by purchasing a freestanding steam generator (or "boiler') that can create steam for several appliances; or by purchasing equipment with its own built-in boiler. Freestanding generators should be placed as close as possible to the equipment to reduce the loss of heat and pressure as the steam is piped to its destination. In all types of steam generation, water quality is a critical factor. Mineral solids from water build up over time and can coat the insides of appliances, heating and sensor systems, and so on. That's why filtration and regular cleaning are requirements for proper steam equipment operation.

STUDY QUESTIONS

1. Which qualities make the steam-jacketed kettle a superior heat transfer unit?
2. If space is limited in your facility, which piece of equipment could you use instead of a coffee urn to make large amounts of coffee?

3. Is wall mounting an option if you buy a floor model appliance?
4. Briefly explain the benefits and disadvantages of pressure versus pressureless steaming. Why would you choose one or the other?
5. How important is steam water purity and what can you do to maintain it? If equipment malfunctions as a result of contaminated water, is the restaurant or the municipality responsible?
6. What is the desirable height for a steamer to rest above the floor?
7. How accurate are suggested food cooking guidelines? What are some considerations in determining cooking time?
8. When would you prefer a pasta and rice specialty steamer to any of the other, more versatile units?
9. Of all the steam cooking units described, which has the highest heat transfer ratio?
10. Your budget says you can only afford one type of steamer unit for your new quick-service restaurant. Which one would you choose, and why?

CHAPTER 15

COOK-CHILL TECHNOLOGY

INTRODUCTION AND LEARNING OBJECTIVES

Most foodservice businesses already do some advance food preparation. ***Cook-chill*** is one of these techniques, used to streamline production and improve food safety at the same time. Cook-chill is the process of cooking food in quantity, then rapidly chilling it. The cooked food is not frozen, but cooled so quickly that it does not stay in the "danger zone" (from 41 to 135 degrees Fahrenheit, or 5 to 57 degrees Celsius) long enough to support harmful bacteria growth. Once it's been quickly chilled, holding the food under proper refrigeration (at 34 to 40 degrees Fahrenheit) prolongs its shelf life, allowing it to sit for at least five days and, in some cases, up to 21 days before serving. A product that stays fresher longer means certain menu items can be prepared twice a week instead of daily, saving time and money.

The technique was developed in Germany more than 40 years ago for the government-run hospitals there. It was an attempt to control labor costs, which is still one of its advantages today. Cook-chill technology is not, however, a way to stockpile leftovers. It is a system of quantity food preparation designed to create a stock of safely prepared and refrigerated foods that can be used as needed. That's one reason why it is also sometimes referred to as the ***cook-to-inventory*** method of food preparation.

In this chapter, we will discuss:

- The uses, benefits, and drawbacks of cook-chill technology
- The cook-chill process, including recipe adaptation and food safety
- The equipment used in the cook-chill process

Cook-chill is not to be confused with an old but popular technique called ***ice shock,*** when food is cooked and then immediately plunged into ice water. Instead, it is a complete cooking and cooling system that requires the purchase of several different pieces of equipment. There's a certain mystique about it, because the process separates meal preparation from meal service. However, the advantages are obvious. Let's take a look at them.

15-1 WHY USE COOK-CHILL?

It seems there has always been a debate in commercial food preparation between cooking "from scratch," and using convenience products. You might consider cook-chill a way to satisfy both sides of the argument. The food *is* made fresh, "from scratch." It's just packaged and chilled quickly for later use instead of being served immediately.

Users of cook-chill agree that consistency—the uniformity of product—is its major selling point. At one time, the perception was that this type of system would be useful only in institutional kitchens with constant high-volume needs, like prisons or school systems. But in recent years, cook-chill technology has received the blessing of respected culinary professionals. A restaurant can hire a gourmet chef, for instance, to establish recipes with food quality, flavor, and presentation standards that may then be carried out uniformly by a staff with far less training.

From an investment standpoint, centralized production allows restaurant operators with multiple locations to produce their high-end, signature items with less cash outlay for real estate, conventional kitchen equipment, and expensive ventilation systems. The food is delivered cold to multiple locations. The chilled items are removed from the refrigerator as needed, and brought to serving temperature by heating (commonly referred to as ***rethermalization*** or "retherm") in smaller, more portable appliances. When done correctly, the cook-chill process produces foods with no loss of color, flavor, texture, or nutrients. The advantages are many:

SUITABLE FOR ANY SIZE OF OPERATION. The advantages to high-volume feeding situations are obvious, and as we've mentioned, chain restaurants can use it to centralize production and distribute to smaller units. Even a modest-sized restaurant, however, can make use of cook-chill technology for its banquet and catering needs. It can also be used as a way to keep a good stock of all menu items on hand at all times by precooking and properly storing them.

EFFECTIVE TIME MANAGEMENT. Restaurants can organize their kitchen staff's time for best results, cooking high-volume items when business is slow and having them ready to use when the kitchen is busiest. Labor savings (of 10 to 40 percent) can be realized, because highly skilled workers can produce the core menu items, while relatively unskilled workers can keep up with tasks like portioning and reheating.

OTHER RESOURCE MANAGEMENT. Equipment and space can be used more efficiently, because a central kitchen can turn out product for several sites. Also, ingredients can be purchased in larger quantities, providing some savings.

MENU FLEXIBILITY AND DIVERSITY. Because foods are prepared in advance, some experimentation is possible—and mistakes are likely to be caught in production, instead of on the guest's plate! Chefs can embellish precooked ingredients to offer a greater overall selection of dishes, because each is not cooked individually. Traditional recipes sometimes have to be modified slightly, which we'll discuss elsewhere in this chapter.

ABILITY TO SERVE SPECIAL NEEDS. For places like hospitals, senior centers, and schools, nutritional requirements are easier to meet because foods can be portioned ahead of time and special diet restrictions can be taken into account. Food safety for these higher-risk dining populations is improved.

INCREASED FOOD SAFETY. Time-temperature abuse is one of the biggest factors in food-borne illness outbreaks—that is, preparation and storage methods that keep foods in the danger zone for bacteria growth too long. The whole cook-chill concept is a series of procedures designed to minimize this critical period, and scientific research shows it works very well. Check out Illustrations 15-3 and 15-4, later in this chapter.

SERVICE IMPROVEMENTS. Most cook-chilled foods simply have to be reheated, so the kitchen and the wait staff have more time to garnish, improve presentation, and attend to the needs of the customers. The restaurant or cafeteria can also offer a wider variety of foods, no matter what the time of day.

REDUCED WASTE, IMPROVED PORTION CONTROL. No matter how the number of guests fluctuates, only the meals ordered must be reheated. This means no more partial batches of your uneaten "daily special" hit the trash can at the end of the day. It's easier to keep tighter control over food waste and portion sizes when you can package them individually, or in bulk, as needed.

A MORE RELAXED WORK ATMOSPHERE. The critical time factor between cooking and serving each item is greatly reduced, relieving pressure on production staff and, quite simply, improving productivity. The kitchen isn't on a constant deadline.

INCREASED PROFITABILITY. If food can be ordered and served promptly, your turnover rate improves and so does your profitability. Serving sites can be smaller, and you can keep minimal inventories on hand. There is tighter control over expensive kitchen expansion or hiring extra kitchen employees. In short, it can mean more efficiency, greater choices for guests, and better service.

Little wonder that cook-chill systems can now be found in every foodservice venue, from fine dining to in-flight commissaries, and luxury cruise lines to supermarket home meal replacements.

15-2 HOW COOK-CHILL WORKS

The theory behind cook-chill is simple—hot foods must be cooled through the bacterial danger zone quickly (within 90 minutes) and uniformly. Once chilled, the food can be stored in standard reach-in or walk-in refrigerators and, when ready to serve, it just needs to be rethermed. The cook-chill production system is also simple to operate if it is well managed. Illustration 15-1 shows the basic steps in the system; Illustration 15-2 shows the range of equipment that comprises a cook-chill system.

Quick-chilling requires more sophisticated refrigeration than a walk-in, which can't do the job fast enough. This includes introducing high-velocity cold air all around (above and beneath) each container of food. The U.S. Department of Agriculture (USDA) mandates that cook-chill equipment be capable of bringing food from a cooking temperature of 180 degrees Fahrenheit to its chilled storage temperature of below 40 degrees Fahrenheit in less than two hours. Today's commercial equipment is, in fact, capable of doing the job in less than one hour, depending on the product. (Most manufacturers' estimates of chilling time are based on "incoming" food product that is between 140 and 160 degrees Fahrenheit.) Before we discuss the equipment used to accomplish this super-chilling, let's cover the all-important steps that lead up to that point.

Selection of Raw Materials. As with any cooking endeavor, your final product is only as good as the quality of the raw ingredients. Especially if you are working with fresh meats or seafood, it is vital to check your vendors' handling and distribution methods.

Storage Conditions. After purchasing the best-quality raw products, keep them in safe, sanitary storage so they will be in prime condition when you cook them. This means following basic hygiene principles and using a storage area that is the appropriate temperature and humidity. If some materials arrive frozen, thaw them in the refrigerator, not at room temperature or in the microwave. Either of these exposes food to high enough temperatures that bacteria could grow. It can also thaw them unevenly, leaving cold spots at the core of the food that may result in uneven cooking.

Preparation. Prepare the food for cooking, just as if you were going to cook and serve it that day. Again, the use of sanitary surfaces and separate hand tools for different raw foods is a must. Ideally, this prep work should take place in an area separate from the cook-to-order and plating areas of the kitchen.

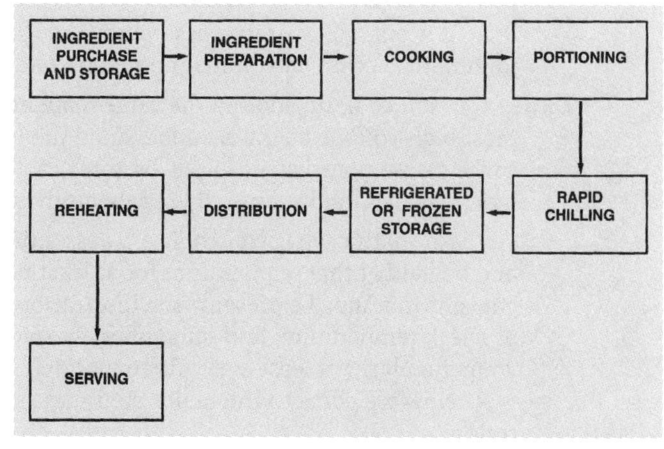

ILLUSTRATION 15-1 **The basic steps in a cook-chill system.**

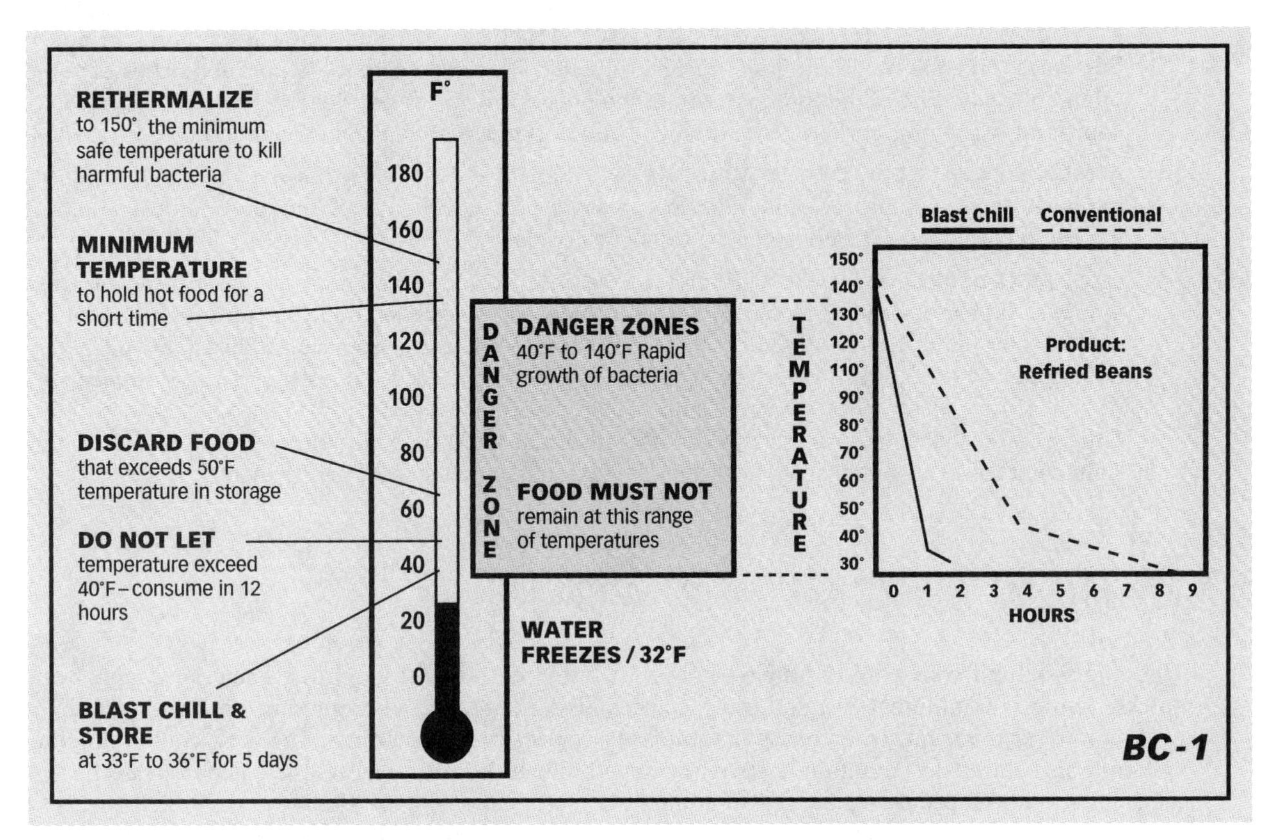

ILLUSTRATION 15-2 **The food danger zones that impact time-temperature abuse. As you can see, a blast chiller does the job far more quickly than conventional refrigeration.**
COURTESY OF THE ELECTRIC FOODSERVICE COUNCIL, FAYETTEVILLE, GEORGIA.

Special care must be taken to make the quick-chill process more efficient for meat. Make sure cuts of meat do not weigh more than 6 pounds or measure more than 2½ inches in thickness.

Recipe Modification. Since the cooked foods will "rest" a while before being served, some minor recipe alterations may be necessary. Flavors continue to develop; spices get spicier; starches change consistency. The first step in cook-chill recipe development is to create the optimum flavor and presentation profile for a single dish, then convert it to batch production. Equipment manufacturers suggest reducing sugars and salt content by 25 percent, reducing other spices and seasonings by 10 percent, and adding 20 to 30 percent fewer leafy herbs—particularly sage and bay leaves, which grow stronger with lengthier food contact. They also recommend not adding all the liquid called for in a recipe until the end of the cooking process, using it to correctly adjust the texture just before cooling.

It is smart to create recipes that call for standard quantities of ingredients—full cans, round numbers—to minimize errors in volume production and eliminate half-empty containers that must be stored.

Cooking. You cook the food in the same manner, and using the same equipment, you would normally use. Most high-volume operators understand the benefits of batch cooking, and the use of cook-chill methods can increase your batch output by three or four times! Favorite "batch"-type items, such as casseroles, meats, sauces, and soups, all lend themselves to this preparation.

No matter what you cook, it is essential that its core temperature reach 165 degrees Fahrenheit, and be held at that temperature for at least two minutes, to destroy any pathogenic bacteria or microorganisms that may be present (see Illustration 15-3). A variety of probes and thermometers may be used to check temperature; and remember, thermometers can be out of calibration, so check the accuracy of yours regularly, at least every three months.

Now we come to the major steps that make cook-chill unique from any other type of food preparation.

Prechill Preparation

Once the food is cooked, the chilling process must begin as soon as possible—that is, within 30 minutes. Load the food into shallow, stainless steel steam table pans, 2½ inches deep, 12 by 10 inches or 12 by 20 inches. Don't load the cooked food any higher than 2 inches deep in the pan. Deeper containers may be used, but only if the chilling unit is efficient enough to lower the food temperature fast enough. Also, some foods may require even smaller portioning to facilitate chilling. It will depend on the size, shape, and density of the food—the denser the product, the longer time required to chill.

You may wonder why we recommend covering the food (with plastic wrap or flat, stainless steel lids) when the pans would cool even faster uncovered. This is done primarily for safety reasons, but keeping pans covered also reduces surface dehydration of the food, eliminates flavor transfer when chilling different foods at the same time, and prevents excessive icing on the coils of the chiller's refrigeration system when too much moisture circulates inside the unit. The moisture content of the food, its ability to retain heat, and its temperature when it reaches the chiller all impact total chilling time.

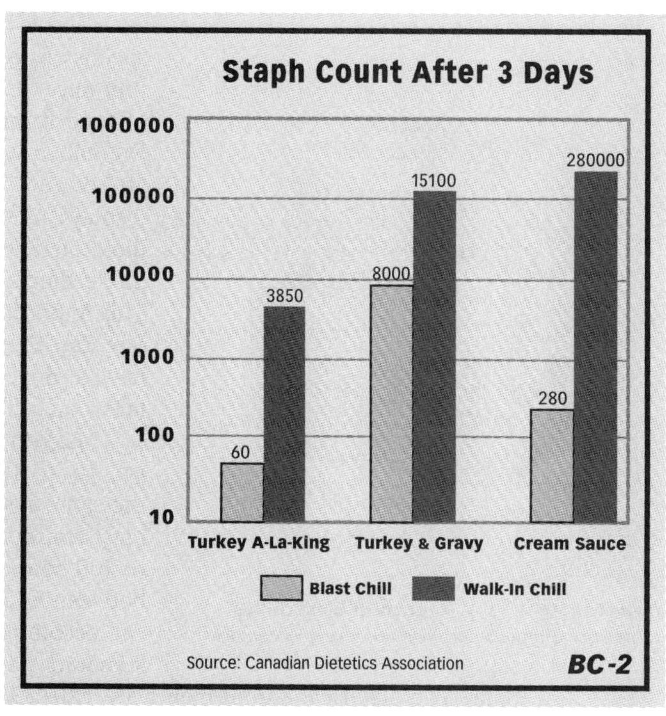

ILLUSTRATION 15-3 **An example of the exponential growth of bacteria on just a few products, using two types of chilling.**
COURTESY OF THE ELECTRIC FOODSERVICE COUNCIL, FAYETTEVILLE, GEORGIA.

If you're using containers made of materials meant to insulate food, this will also affect chilling time. Other containers, made of disposable materials, must have been stored under completely sanitary conditions before use, or you risk introducing bacteria as you fill them with hot food. As you can see, there are many variables, almost all of them health related.

Each cook-chill system user must decide whether to portion foods before or after chilling takes place. There is no clear-cut choice, so you've got to consider how and when you'll be using the food. For a banquet, you might take it from the chilling pan, arrange it in a serving pan, and chill it that way. For room service or on a cruise ship, you'd make individual portions, perhaps on covered plates or in vacuum-pack bags. For a buffet, the food might be put into a chafing dish insert. Whatever the choice, portioning must be done within 30 minutes of cooking so the food can be safely chilled. Regular measurement of food temperature during this time, and throughout the cook-chill process, is recommended.

There are two kinds of quick-chilling. ***Blast chilling*** circulates cold air at high velocity around pans of food to cool it quickly. ***Tumble chilling*** immerses packed food products in cold liquid.

Blast Chilling

We'll start with blast chilling, which is appropriate for solid foods such as chicken parts, beef rounds, whole casseroles, burger patties, and so forth. The food containers are covered, and then either placed on carts to roll inside the ***blast chiller*** or put directly on shelves inside the chilling unit (see Illustrations 15-4 and 15-5).

Today, there are blast chillers with conveyor belt systems for high-volume production, all the way down to undercounter models that hold three pans. Like refrigerators, they are available in reach-in or roll-in models. Some feature rotating racks to circulate food as it chills; some are on casters to be easily rolled between production areas.

ILLUSTRATION 15-4 **A reach-in blast chiller.**
COURTESY OF ALTO-SHAAM, INC., MENOMONEE FALLS, WISCONSIN.

Inside the blast chiller, powerful fans blast the food with cold air at speeds of up to 1300 feet per minute to quickly draw off the heat. The unit must be capable of reducing the temperature of a 2- to 3-inch layer of food from 150 degrees Fahrenheit (or so) to less than 40 degrees Fahrenheit within 90 minutes, when fully loaded. It should also contain an accurate temperature display and built-in food probes with digital displays, as well as a timer that will alert kitchen personnel with an audible buzz or bell when it's done. Most blast chillers have a holding mode that kicks in automatically when the blast chilling is complete. This important feature allows you to chill a batch of food at the end of one day, then come back and remove it the next morning. Some also have a "delicate" or "soft chill" cycle for light loads or low-density products, and a "flash freeze" mode when frozen food is the desired outcome.

As you might imagine, these units require a lot of power. The typical 10- to 12-pan chiller has a 1- to 2-horsepower motor—essentially the same system used in a 240-square-foot walk-in cooler. The largest blast chillers in institutional settings are capable of handling from 90 to 400 pounds of food per chilling cycle, which means a daily output of between 540 and 2400 pounds of food. The 90-pound-capacity chiller can accommodate ten 2-inch-deep hotel pans (the flat, 12-by-20-inch standard pans); the 400-pound chiller can handle forty-four such pans. The electrical requirements are 120/208–240 volts, 60-cycle, single-phase power for the smaller models; 208–230 volts, 60-cycle, three-phase power for the larger models. Larger condenser unit motors are between 3 and 10 horsepower.

ILLUSTRATION 15-5 **A roll-in blast chiller.**
COURTESY OF ALTO-SHAAM, INC., MENOMONEE FALLS, WISCONSIN.

The capacities of the smallest chillers, which hold only a few pans, range from 18 to 30 pounds of food per load. Medium-sized chillers can handle 45 to 100 pounds of food per load. A bonus for the undercounter chiller user is that some feature a self-contained evaporator, making a nearby floor drain unnecessary.

In many kitchens, blast chillers are located close to the cooking areas to help meet that critical thirty-minute deadline between stove and chiller. For the medium-sized units, you'll need an area about 5 feet square and 7 feet tall. For the larger models, it's more like 9 feet tall, 9 feet long, and 5 feet deep. (Of course, the huge industrial models will require more space.) The condensing unit takes up the most room and can be located with the chiller or, with proper installation, elsewhere in your kitchen.

Some chillers have optional printers that can be linked to their temperature probes. You can receive a printout that lists the product's core temperature, performance details about the chilling cycle and the refrigerated cabinet itself, and the compressor's running temperature (to check for overheating or icing up). An audible alarm feature can be set to alert you if any of these components fails. Others contain optional ultraviolet lights that sanitize utensils overnight.

Pay close attention to the design of the chiller. Check for easy access to the evaporator components because, like any refrigerator, these will need to be cleaned regularly. Look for removable racks, shelves, and shelf slides, since they are also easier to clean. Automatic defrost and evaporation features will save time and hassle in operating your blast chiller.

Buying a Blast Chiller

Selecting a blast chiller is not just a matter of size and temperature. Other decisions:

DIRECT VERSUS INDIRECT AIRFLOW. Direct airflow is cold air blown directly onto the pans. It is very effective, but can mean more spattering and/or drying out of foods that are not covered. The velocity of the airflow, and the way it is balanced, are critical considerations. A blast chiller without sufficient air velocity or balanced airflow will be unable to evenly chill all its contents.

SIZE CONSIDERATIONS. There are several things to think about here. One is the footprint of the appliance. The other is the volume of product that can be chilled *per cycle*. The manufacturer's specifications should include two measurements of volume: the number of pounds, and the number of pans the chiller can accommodate, per cycle.

REACH-IN VERSUS ROLL-IN. Reach-ins are designed to hold a certain number of pans. Their shelves may or may not be adjustable (adjustable is, of course, better). Roll-ins are designed as open cavities into which roll-in racks full of product are wheeled.

SELF-CONTAINED VERSUS REMOTE COMPRESSOR. How much available space do you have to install the chiller? Smaller units are self-contained; larger ones have the option of a separate, stand-alone compressor that requires installation of refrigeration lines between compressor and chiller, more space (for the compressor/condenser), and another electrical connection.

CHILLING VERSUS FREEZING. Some cook-chill systems may also be used as cook-freeze systems. You can buy a blast chiller that has a blast freeze mode, and freeze the finished product after quick-chilling. Comparing the two processes, a University of Wisconsin study determined that it takes 18 times more Btus of energy to prepare, chill, store, and reheat meat loaf in a cook-freeze system than it does in a cook-chill system. But for long-term food storage, freezing is the way to go.

COMPATIBILITY WITH OTHER EQUIPMENT AND YOUR PRODUCT FLOW. These are the logistical considerations. What types of equipment are used to cook the food that is being chilled? Make sure the pans or roll-ins will fit into the chiller, or you'll have to spend time dishing hot food into different containers for chilling. And is there adequate reach-in and walk-in space for the amount of product the blast chiller will produce?

Tumble Chilling

Blast chillers are generally considered more versatile than tumble chillers, because they work with just about any type of solid food product and are available in a great variety of sizes. But using them to cool liquids—soups, stews, gravies, and sauces, as well as some types of pasta—presents practical challenges. Putting these foods in shallow pans gets messy, and the process holds potential for contamination. High-velocity cold air can cause them to spatter, but placing them in buckets in a walk-in doesn't chill them fast enough. So for these foods, ***tumble chilling*** is most appropriate. And, when it comes to reducing food temperatures quickly, the tumble chiller is much more efficient than the blast chiller.

The whole tumble chilling process requires several pieces of equipment. First, pumpable foods are cooked in a special steam-jacketed kettle (called an ***agitator kettle*** or *mixer kettle*), which includes an agitator or mixing arm (see Illustration 15-6). The agitator is needed for uniform suspension of solids and continuous mixing action as the food is pumped from cooking vessel into containers. Manual mixing isn't precise enough to accomplish this. The goal of the agitator is to lift and fold the food gently so it isn't damaged in the process. Use of a kettle is also handy because, by adding cold water into the kettle jacket, the food can be precooled to a certain extent when cooking is complete.

The smallest kettles hold 60 gallons; the largest hold up to 200 gallons, but you never want to fill them any fuller than 180 gallons to facilitate mixing. Agitator kettles must have a 3-inch steel valve at

ILLUSTRATION 15-6 An agitator kettle, which can mix large quantities of food.
COURTESY OF GROEN, A DOVER INDUSTRIES COMPANY, ELK GROVE VILLAGE, ILLINOIS.

the bottom that is used to flush foods through the kettle and on to the next step, the ***pump-fill station.*** Here, kettle-cooked foods are transferred (pumped) directly from the kettle into flexible plastic tubes designed to withstand the temperature extremes of both hot filling and rapid chilling (see Illustration 15-7). To flush efficiently, the solids in foods (chunks of vegetables in soup, for instance) should not exceed 1 inch in diameter.

The plastic casings are disposable and come in sizes from 1 quart to 3 gallons. When full, they are sealed, weighed, and labeled in an area some kitchens refer to as their ***metering station.*** They can be refrigerated, frozen, or reheated.

At the pump-fill station, you'll have a pump with a variable-speed motor, a sink with a drain, a rack to hold the plastic casings, a clipper to snip off the ends of full casings, and a label printer. The whole thing can be put on a rolling cart so it can be used for multiple kettles, and will take up a space of about 3 square feet, almost 7 feet tall. The motor is 1½ horsepower and uses 208/230-volt, 60-cycle, single-phase electricity.

By the time the casings leave the pump-fill area, they should be filled, sealed, trimmed, and labeled. From here, the casings are manually transported to the ***tumble chiller*** or, in very large operations, perhaps placed on a conveyor belt. The tumble chiller is a rotating drum full of cold water (34 degrees Fahrenheit) that gently kneads the full casings as they roil around inside (see Illustration 15-8). This speeds the cooling process to less than one hour, and can drop the food temperature from a steaming 180 degrees Fahrenheit to 40 degrees Fahrenheit. The ice water is supplied by an ice machine (called an ***ice builder***), which can be installed either indoors or outdoors with correct plumbing connections.

Tumble chillers come in sizes for 85 to 300 gallons of product per cycle. The smallest units are 6½ by 4 by 8 feet; the largest take up 10 by 6½ by 10½ feet. Some tumble chillers employ a chute that allows loading as the unit is running. Tumble chillers also require a 2-inch water supply pipe and a 3-inch drainpipe.

Many tumble-chilled foods can be stored safely for 35 to 45 days before rethermalization.

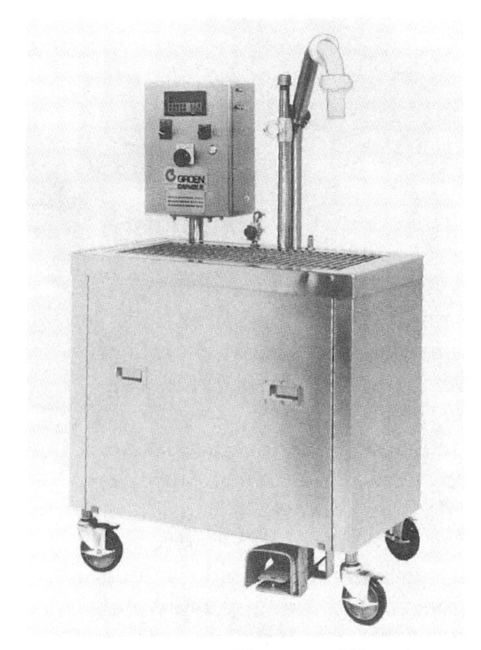

ILLUSTRATION 15-7 The pump-fill station.
COURTESY OF GROEN, A DOVER INDUSTRIES COMPANY, ELK GROVE VILLAGE, ILLINOIS.

Other Chilling Options

There are a few old-fashioned ways to chill food faster. They probably seem inconvenient compared to blast and tumble chillers, but they are used nonetheless, especially for small batches.

Experienced chefs know that you must first reduce the size or quantity of what you're trying to cool. Any food cools more quickly in small or shallow containers. After dividing into smaller quantities, an *ice water bath* may be used. Place the pan full of food into a larger container (another pan or a sink) filled with ice water, still loaded with plenty of ice. This is labor intensive, but effective.

Large batches of food can be precooled (before blast or tumble chilling) by stirring them with a hollow, heavy-duty plastic paddle filled with water and frozen. Of course, this requires workers to fill and freeze the paddles, and do the stirring.

Steam-jacketed kettles can be used to cool food by filling them with icy-cold water instead of steam. And some recipes that require water, like soups, can be prepared using less water—then the remaining water can be added cold, to help cool the finished product.

You can turn your walk-in cooler into a super-chiller with special ***conduction shelves.*** A solution of water and antifreeze circulates through the shelf interiors, cooling food rapidly as the pans sit on them. Its manufacturer claims the refrigerated shelves also decrease the work the walk-in has to do, without affecting other foods stored inside (see Illustration 15-9). Some shelf systems are on rolling casters, equipped to hold pans or individual plates, and come in standard sizes that can roll directly from a blast chiller to a retherm unit.

For large cuts of meat (hams, turkeys, etc.) or other foods not suited to kettle cooking, you can buy a combination ***cook-cool tank,*** also known as a *chiller-cook tank* or a *cook-chill tank*. In this appliance, uncooked food is vacuum packed into its plastic casings, then loaded onto wire racks. The full racks are stacked inside the tank. Hot water circulates inside the tank, slow-cooking meats in their own juices and producing a very high yield. Then, on a pretimed cycle, the hot water is drained away and replaced with icy cold water to begin the rapid-chill process. Cook-cool tanks can be used to prepare most types of meats and vegetables. The refrigerated shelf life of food cooked this way is similar to tumble-chilled items.

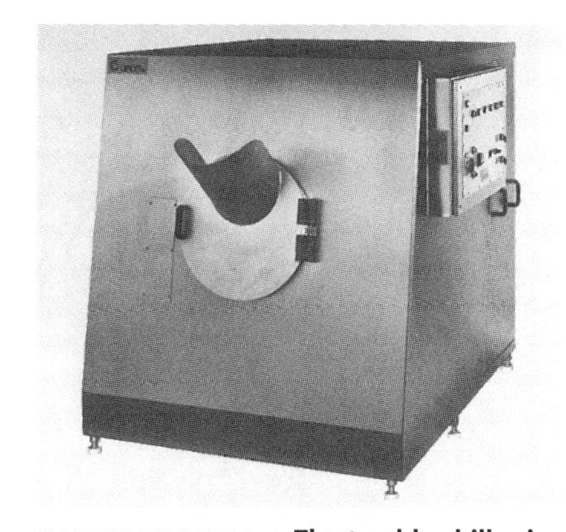

ILLUSTRATION 15-8 **The tumble chiller is a rotating drum full of cold water.**
COURTESY OF GROEN, A DOVER INDUSTRIES COMPANY, ELK GROVE VILLAGE, ILLINOIS.

Cook-cool tanks come in a variety of sizes, ranging from 200- to 2500-pound capacities. The largest models contain multiple rows of wire trays on which to put food. They use a 3-horsepower recirculating water pump, and need floor space that's about 8 feet long and 4 feet wide. The tank should include a meat probe to record internal cooking temperatures. Electrical requirements will depend on the size of the tank, but they'll be 208-, 230-, or 460-volt, three phase.

A final word about ice builders, because they are required accessories for both cook-cool tanks and tumbler chillers to provide sufficient amounts of ice water. Your total daily production figure will determine the capacity of the ice builder. The ice builder makes ice water and pumps it to a heat exchanger located on the tank (or chiller). There, the ice water chills the distilled water that is used to cool the food. As mentioned earlier, ice builders don't have to be installed directly adjacent to the appliances they serve, and they can be located either indoors or outdoors. For added efficiency, they can also be used to produce ice during off-peak hours.

ILLUSTRATION 15-9 **Conduction shelves are refrigerated to cool food faster.**
COURTESY OF THERMODYNE FOODSERVICE PRODUCTS, INC., FORT WAYNE, INDIANA.

15-3 STORAGE AND DISTRIBUTION

Now that you have a batch of precooked, chilled food, where do you put it until you need it? The typical commercial refrigerator in general use in your restaurant kitchen is not the answer, because it is opened and closed constantly. Your goals are to keep the precooked food at a constant temperature and to prevent cross-contamination between the precooked food and other foods in the refrigerator. Your best option is to install a dedicated reach-in or walk-in unit specifically for cook-chilled foods. Most are available with monitoring systems and alarms, to alert you if the temperature strays from that ideal range of 32 to 38 degrees Fahrenheit. If this is not possible, consider some of the conduction shelves mentioned earlier for an all-purpose walk-in.

Employ a first-in, first-out method of stock rotation, so that none of the food exceeds the expiration date on its label. Labeling is the other key to successful storage. The labels on each container must specify the type of food, date of preparation, and destination if the food will be used at an off-site location.

If for any reason the food reaches a temperature over 41 degrees Fahrenheit but not more than 50 degrees Fahrenheit, it should be used within 12 hours; if it is accidentally allowed to reach more than 50 degrees Fahrenheit, it should be destroyed because it has become potentially dangerous to eat. It is a good idea to regularly send out samples of your stored foods for laboratory testing, just to check their safety.

Despite these cautions, cook-chilled food that has been correctly prepared and stored will make you proud to serve it! It will look and taste as good and fresh as the day it was prepared. Most of the time, food travels only a few feet to its point of rethermalization, but this is not always the case. As you've discovered by now, temperature control is key to the success of this system, and it is especially important when transporting food from its preparation site to other locations. Insulated carriers, refrigerated carts, and refrigerated vehicles are all used for off-site catering and satellite distribution. In just a moment, you will also learn about carts made for reheating and serving cook-chill foods. At the very least, prechilled, insulated coolers should be used for short journeys.

When the food reaches its destination, a temperature reading should always be taken to ensure it has not warmed to a temperature above 45 degrees Fahrenheit. Then it must immediately be placed in the appropriate refrigerated storage until use. If the food must be transported hot because there is no retherm capability at its destination, its core temperature is also critical. It must be 140 degrees Fahrenheit or higher.

15-4 RETHERMALIZATION

Rethermalization, or reheating, is the conclusion of the cook-chill process, and also a key control point in the prevention of contamination. The guidelines:

- **If a cook-chill food is supposed to be eaten cold or at room temperature, it should be consumed within 30 minutes of leaving the refrigerator.**
- **If it must be reheated, this should take place promptly. This means taking the food "through" the temperature danger zone once again, as quickly as possible—bringing it to an internal temperature of 165 degrees Fahrenheit for at least 15 seconds, within two hours of its removal from refrigeration. A probe thermometer inserted into the center of the food item is the best way to determine if this core temperature has been reached.**
- **If the rethermed food has been allowed to cool to room temperature, it should never be reheated or returned to chilled storage. For safety reasons, it must be destroyed.**

There are specially built rethermalization units designed to work with cook-chill systems, but you can use just about any cooking appliance: convection ovens, steamers, kettles, combi ovens, and infrared units. If a traditional oven is used, extra care must be taken not to dry the food out. Food should always be covered when it's rethermed, to retain moisture, and your choice of methods and appliances depends mostly on how the product is packaged. Many experts feel the microwave oven is the best way to retherm individual portions that have already been plated. Food that has been bagged and chilled should always be heated in moist conditions—by placing the bag in a bain marie, a steamer, or a combi oven—or you can remove it from the bag and heat on a range top or in an oven. High-volume commissaries can retherm covered dishes with conveyor impingement ovens.

The ***rethermalization cart*** is a recent invention to meet the needs of health care facilities and others that want "roving" capability. These specialty carts are capable of bringing preplated cold food up to proper serving temperature in less than one hour and holding it at optimum temperature for extended time periods (see Illustration 15-10). Sometimes the carts roll; sometimes the trays of food can detach and be rolled.

Retherm carts offer a wide variety of options. Units that can be switched from moist (convection) heat to dry (radiant) heat offer the most flexibility—lasagna, for instance, requires moist heat, while dry heat retains the crispness of fried chicken. Some have a small refrigerated compartment in addition to heating capability; some can be connected to an electrical outlet and programmed to start heating at a certain time; others can be plugged into their own compatible freestanding chiller units. The controls for the carts can be manual or automatic, and most have secured control systems to guard against tampering. HACCP compliance is easy when an on-board computer monitors every push of a button, every internal temperature, and every door opening.

ILLUSTRATION 15-10 A rethermalization cart can warm plates of chilled food as it is wheeled from place to place.
COURTESY OF ALTO-SHAAM, INC., MENOMONEE FALLS, WISCONSIN.

When looking at retherm carts, ask about the dishes that may be used with each type. Some carts accommodate disposable paper and plasticware, while others require the purchase of special, compatible plates because the dishes come into direct contact with a heating element. Some of the carts have mechanisms to "lock" ceramic dishes in place as they roll to prevent breakage; others do not. Other considerations will depend on your intended use for the cart, but you should have a good idea of the following:

- Any size constraints, including loading docks, freight elevators, or regular elevators to maneuver through
- How quickly the food should be heated, and how long it must be held at that temperature
- The type of electrical power available at your remote locations
- The flexibility of your menu

The last point is important because serving from a cart sometimes means changing the way the food is presented to customers. For space, or quicker reheating, you may have to slice baked potatoes in half, for instance, or wrap some foods in aluminum foil. It's hard to get toast to stay crisp and dry in a retherm cart, but soups, stews, and hot cereals adapt perfectly.

15-5 BUYING AND USING A COOK-CHILL SYSTEM

When selecting cook-chill equipment, the first thing to determine is whether its capacity is sufficient to meet your peak production needs. This is important for two reasons: to determine overall output and to facilitate careful timing of the process so that newly cooked food can always begin its rapid chilling within half an hour.

The best way to begin shopping for a system is to visit foodservice facilities that already use cook-chill. If it's a retherm cart you're looking for, ask a hospital employee how convenient it is to push around to patients' rooms. Some foodservice businesses form committees with goals of making sure the entire operation understands and supports the cook-chill process before they purchase the components. Many manufacturers also allow you to test the pieces of equipment on site for a few days before finalizing your purchase.

Let's look now at how cook-chill technology is used in some specific foodservice situations. The first example is provided by Williams Refrigeration Limited, a British company with its U.S. offices in Maywood, New Jersey. The company describes a small resort restaurant, open six days a week, serving approximately 600 meals per week. It seats 50, serves two meals per day, and turns each table once during those meals.

The restaurant's owner realized business was being turned away on the one day (Monday) the restaurant was closed, but saw no other way to give his kitchen staff a day off—until he purchased a blast chiller, capable of chilling 175 meals per day. By using the cook-chill system four days a week, he could increase his sales by 100 meals (by staying open Mondays) and allow the kitchen staff to take two days off per week instead of one! Take a look at the schedule in Table 15-1 of meals produced (and consumed) before—and after—the cook-chill system. See the difference?

In this operation, an upright blast chiller was used, capable of handling 50 pounds of food per cycle in standard hotel pans. Each chill cycle lasts from 68 to 87 minutes, and the blast chiller automatically reverts to storage mode following the completion of the cycle.

A suburban school district in Duncanville, Texas, has used uses cook-chill technology for more than a decade to plan its foodservice production and deliver more than 7000 meals per day to 12 schools from a single kitchen. Six cooks staff the central kitchen, located in a 2000-student high school. The major components of the 5000-square-foot production center include:

- 1250 square feet for ingredient storage
- 532 square feet for preparation
- 345 square feet for test kitchen
- 720 square feet for cooking area

The hot-food area uses a total of three agitator kettles: one 200-gallon, one 100-gallon, and one 60-gallon. Along with the requisite pump-fill station and tumble chiller, this equipment provides about 90 percent of the food items served at the schools. An air pressure system is used in the preparation area and during repacking, to vacuum pack some prepared items (such as cut produce) for longer storage life.

In the bakery area, three full-time bakers turn out all the buns, cookies, doughnuts, and sheet cakes for the district. Here, you'll find two 80-quart mixers, automatic dividers, round and sheet pans, a large, roll-in proofer, a rotary oven, and a doughnut maker.

Finished products await distribution in an 8000-square-foot storage and distribution center, which features about 740 square feet of walk-in refrigerated space. The storage area is divided to hold about 20 percent prepared foods, 40 percent frozen foods, and 40 percent dry goods.

TABLE 15–1

MEAL OUTPUT COMPARISONS: SMALL RESORT RESTAURANT

Meal Output BEFORE Installation of Blast Chiller	M	Tu	W	Th	F	Sa	Su
In storage	0	0	0	0	0	0	0
Production (meals cooked)	0	100	100	100	100	100	100
Sales (meals sold)	0	100	100	100	100	100	100
Inventory (meals on hand)	0	0	0	0	0	0	0
Meal Output AFTER Installation of Blast Chiller							
In storage	0	75	150	50	125	200	100
Production (cook-chill)	175	175	0	175	175	0	0
Sales (meals sold)	100	100	100	100	100	100	100
Inventory (meals on hand)	75	150	50	125	200	100	0

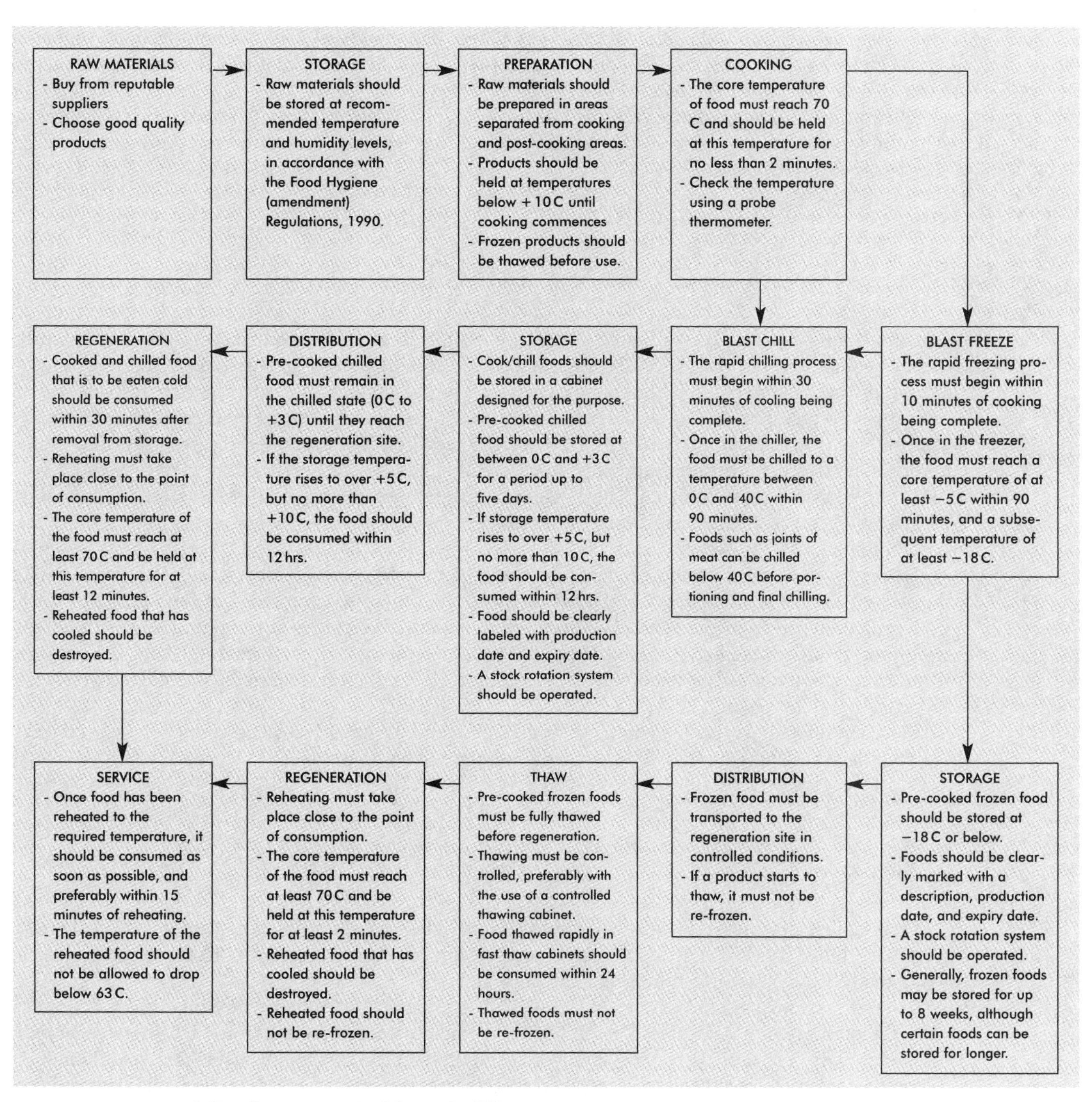

ILLUSTRATION 15-11 A flowchart summary of the cook-chill process.
SOURCE: WILLIAMS REFRIGERATION, INC., MAYWOOD, NEW JERSEY.

In each satellite school, mini-kitchens of about 450 square feet contain combination range top and convection ovens, plus steam table setups. Food items arrive in bags and are boiled or baked and placed in steam table pans for cafeteria service. The satellite kitchens do prepare a few of their own items: french fries, some cookies, and cornbread. For cobblers, the filling is cooked and chilled and the shell is made at the central bakery; at the satellite kitchen, the cook simply pours filling into the shell and bakes it in the convection oven. At the end of the lunch period, satellite kitchens send dirty dishes and silverware back to the central kitchen's warewashing system.

The benefits of this system to the Duncanville School District include the ability to offer students choices of three hot entrées daily, adding variety to the menu. Waste has been reduced, because the cooks at the satellite kitchens open only as many bags of food as they need. The rest remain refrigerated for later use. Labor costs and energy use are minimized, and quality control has improved. A computer program stores and analyzes recipes, helps in meal planning, and tracks sales to determine which items are the biggest "hits" with students.

Illustration 15-11 provides a summary of the cook-chill process. As you can see, the convenience and multiple benefits of cook-chill technology make it an expenditure worth considering, no matter what your foodservice needs.

15-6 CLEANING COOK-CHILL EQUIPMENT

All the food safety precautions you are taking with cook-chill are for naught if you don't scrupulously clean the equipment between uses. The agitator kettle should be sanitized between batches using a bleach-and-water solution of 5 tablespoons bleach for every 20 gallons of kettle capacity, and one cup of bleach for a 60-gallon kettle. This solution should be heated to lukewarm (about 80 degrees Fahrenheit), and then pumped from the kettle through the food product fill hoses and into a drain. Some pumping stations have "constant pumping" functions that can be used with a detergent and warm water solution to clean between batches of product, but only bleach can provide the sanitizing function. Whenever a system is cleaned or sanitized, the process must end with a clean-water rinse to flush out any detergent or bleach residue.

Some components of tumble chillers, like the food pumps, can be removed and disinfected separately, which should be done on a daily basis. Remove the pump, brush it clean of food particles, soak it in the manufacturer's recommended solution, and lubricate both sides of the pump seal with an approved food-grade grease before reinstalling. The agitator arm, the scraper blade(s), and the strainer for the water circulator pump can also be removed and cleaned. Again, lubricate all seals as you reinstall parts.

Most manufacturers include complete disassembly and cleaning instructions with the equipment. Read them, learn them . . . and then keep them on hand for new employees to learn as well.

SUMMARY

Cook-chill is the process of cooking food in quantity, then rapidly chilling it. The cooked food is not frozen, just chilled quickly and thoroughly to prevent harmful bacteria growth. This method creates a stock of ready-made refrigerated foods for the busy kitchen.

Government regulations specify that cook-chill equipment be capable of bringing food from a cooked temperature of 180 degrees Fahrenheit to its chilled storage temperature of below 40 degrees Fahrenheit in less than two hours; most commercial equipment can do the job in less than one hour.

In the cook-chill process, food is prepared just as if it were to be served immediately. The chilling process must begin within 30 minutes after the food is cooked. Some foods should be divided into smaller portions to facilitate quick chilling. There are two kinds of chillers: blast chillers, which circulate cold air at high velocity around the food; and tumble chillers, which immerse packaged food in cold liquid. Tumble chillers require an ice builder, a machine that produces ice water to chill the food.

Foods with high liquid content are packaged and chilled in a pump-fill station, where they pass through flexible plastic tubing and into plastic casing that can be refrigerated, labeled, and frozen or reheated. The reheating process is known as rethermalization—"retherm" for short. Safe retherming means reheating the food to a temperature of 165 degrees Fahrenheit for at least 15 seconds, within two hours of taking it out of refrigeration. If cook-chill food reaches a temperature of more than 50 degrees Fahrenheit before you are ready to serve it, it should be destroyed. There are a variety of refrigerated and insulated containers for safe food storage, but it is always wise to check and recheck temperatures.

Because they are not going to be used right away, it is especially important to label cook-chill foods with their contents, the production date, and often, the weight of the item. A first-in, first-out method of stock rotation is recommended for cook-chilled products, just like all others.

STUDY QUESTIONS

1. Describe two of the advantages of the cook-chill method, and explain why you think they are important.
2. What is the optimum temperature range at which cook-chill food should be stored until it is served?
3. Why would it be especially important to check a potential food vendor's storage methods before buying ingredients to use in the cook-chill process?
4. What are the special considerations for cooking and chilling meats, and why?
5. At least six factors affect the time it takes to chill cooked foods. Name three of them.
6. What is rethermalization? Do you need to set aside space in your kitchen, or use a particular piece of equipment, to carry out this step of the cook-chill process?
7. What information must be provided when you label a cook-chill product?
8. Why should cook-chill food be stored separately from other types of food whenever possible?
9. What is an ice builder?
10. What is the difference between tumble chilling and blast chilling? What factors determine which of these two methods you should use?

A Conversation with . . .

Allan P. King, Jr.

FOODSERVICE DESIGN CONSULTANT
RENO, NEVADA

"Never in his wildest dreams" did Allan King think he would become a designer of restaurants and foodservice facilities when he looked for a summer job in 1962. He was already set to begin a master's program in electrical engineering at the University of California. Instead, he was hired as a draftsman, sketching designs for customers of a restaurant supply house. It launched a whole new career.

More than 30 years later, King owns his own design firm, belongs to the Foodservice Consultants Society International, and is a leader in an industry organization that didn't exist when he started in the business. You can see his work at some of the famous hotels and casinos in Reno and Las Vegas, as well as numerous restaurants, correctional facilities, schools, hospitals, and community centers in the western United States. King also received a National Historic Preservation award in 1994 for his redesign of a vintage opera house in Eureka, Nevada.

Q: What exactly does a foodservice design consultant do?
A: In the simplest terms, you take what a client has in terms of menu, and come up with a functional floor plan for the equipment needed to produce that menu. You try to build in enough flexibility so the client can add to it or deviate slightly from it. But if it starts out as a Mexican restaurant and then somebody wants to turn it into a Chinese place? Well, none of us has a crystal ball!

Q: So in design, do you start with the number of meals you would like to serve?
A: It's all menu driven, what type of restaurant you want, what type of meals you want to serve, what you expect to do in volume. If it's a correctional facility, you know you'll be serving three meals a day; in a school, it's two. So you work backwards: What's your average plate cost going to be? What do you expect your profit to be? We've found in the most common type of restaurant, the coffee shop, you have to have at least 25 seats or you won't generate enough volume to cover your costs. Hopefully, the normal restaurant will have three turns a day.

With a 30-seat restaurant, at today's prices, I don't know how you could ever generate enough volume to be profitable, unless you filled all 30 seats three times a day and made $100 a seat for it! There are so many expenses.

Q: How do professional designers charge for their services? Is there a difference per square foot in design charges for a fast-food restaurant versus a fine-dining establishment?
A: I've never broken it down that way. Some designers may base their fees on a percentage of the cost per square foot of construction, but usually the fee is a percentage (from 4 to 8 percent) of the estimated costs of the foodservice equipment. Some people add on ½ percent more for ancillary charges, like telephone and fax if long distances are involved.

I came out of the contract sales business, so I had a fairly good idea of costs when I started. We've done enough projects over the years to have a pretty good idea of what different types of facilities will cost, so now my company has a formula that's built into the computer, showing how many hours we've put into projects of each type and size. We can make a pretty good estimate of our real charges.

Q: But how would someone who's just starting out make estimates, or know if the estimates they're getting are reasonable for design services?
A: Like the rest of us did, by trial and error. Our fee is 4½ percent, so if you figure a restaurant's equipment costs will be $100,000, we'll charge $4500. There are other out-of-pocket costs sometimes, too, like document production, shipping, phone, and fax charges, which we keep track of and bill the client.

Experts have already come up with figures for the amount of cubic storage per meal needed for dairy products, or frozen foods, or dry storage. With those figures, we can come up with a "guesstimated" square footage of the production space needed for the kitchen, including walk-ins. If you know the kitchen equipment is going to be $120 to $150 per square foot, you can compute costs.

The thing that's so tricky now is that schools, correctional facilities, and even the fast-food places are using more and more pass-through food product—it arrives in a frozen or refrigerated state, and all the operator does is the retherm [reheating] process. This means the freezer space, which is usually half of your total refrigerator space, should be a lot greater to store all the premade product. In those cases, we modify our formula and divide the refrigerator space into thirds; we'll say one-third is refrigerator and two-thirds is freezer, for instance.

Kitchens are also different animals now with the advent of the cook-chill system, which has been a learning process for all of us. For example, in Reno we have a client with five different restaurants preparing clam chowder on Fridays. Why not take the best recipe, have the chef prepare 200 gallons of it, and put it in Cryovac casing, which can hold it from twenty-one to forty-five days depending on temperatures, humidity, and food chemistry? It'll take the chef no longer to prepare two hundred gallons of it than it did to make forty. Now, the chefs at all five restaurants can go to their walk-ins every Friday and retherm, and you'll get the same quality as if it was freshly cooked . . . but your cost of labor has gone down immensely. And when we do taste tests, even the chefs can't tell the difference!

Q: How did you come to be a restaurant designer?
A: I worked in two restaurant supply houses, sold equipment, and also became a partner in a commercial refrigeration firm. Most of us in this business gained our layout and design experience from supply houses, because they provide design service to their customers as a means of selling equipment to them. It's a tool for them. It's only in the last ten or fifteen years that independent restaurant consultants have been widely accepted. The biggest competition we all still face is the restaurant supply house, with its own internal design service. In those cases, the price of the design is really buried in the cost of the equipment.

Q: So how do you convince people to hire you when they could get the service for free?
A: Sometimes it's not so "free," because you may be married to what the supply house buys best, so that their profit margins go up. We don't owe allegiance to any factory. A very big supply house buys such enormous quantities that they can make deals on certain equipment, even if it's not the best choice for their particular need. Usually we've found in the bid process that we can save the cost of our fees in the difference between what the supply house suggests, and what we can help the client put together. Members of FCSI [Foodservice Consultants Society International] are not allowed to sell equipment. Our incomes are based on services only.

Q: If somebody wanted to get into this business, should they go to work for a restaurant supply house?
A: Absolutely, or go work for a designer.

Q: You started with an engineering background, which gave you some technical aptitude. What qualities do you think it takes to be successful as a designer?
A: Engineering really teaches you to think through a problem and keep moving while you're doing it. It would be nice if you could build walls wherever you want! But sometimes there's an elevator in the way, and you have to find a way to accommodate your client's needs and an architect's columns at the same time.

Students really need a restaurant design course, and warewashing should be way up on the list. Also, you simply have to be interested in food and have the "fire in the belly" to design. You have to read through to the bottom line and not be swayed by cheerleaders from the [equipment] factories.

When I stop to wonder if our advice to people to spend millions of dollars one way or the other is the right thing, it's overwhelming to me. You have to believe what you're doing is right.

CHAPTER 16

DISHWASHING AND WASTE DISPOSAL

INTRODUCTION AND LEARNING OBJECTIVES

The dishwasher may well be the most expensive piece of foodservice equipment you will ever buy. A medium-sized restaurant, for example, will spend as much as $20,000 for the machine itself. Outfit the entire dish room—with tables for soiled and clean dishes, a few dish racks, and a booster heater for the final rinse—and you're looking at a total investment of $40,000 or more.

In addition to "money pit," the dish room has another unfortunate designation—in most businesses, it receives the least attention and is typically run by the least trained (and perhaps the least appreciated) employees. The purpose of this chapter is to instill a new appreciation for this critical foodservice sanitation function, so you'll never take your dish machines—or the people who use them—for granted.

In this chapter, we will discuss:

- The different sizes and types of dishwashers
- Booster heaters
- Food disposals and waste pulpers
- Specialty washers

Let's begin by defining ***dishwashing*** as the process by which dishes, glasses, flatware, and so on are cleaned using a combination of three things: hot water, detergent, and motion. In mechanical dishwashing, there are three essential parts:

1. Sufficient amounts of water at hot enough temperatures to sanitize. These temperatures are mandated by NSF International, and dish machine manufacturers comply with the NSF International rules. They specify the amount of water, minimum temperatures, and length of time of the wash and rinse cycles to ensure sanitation (see Table 16-1).
2. Detergents that can soften water, penetrate food particles, and loosen them from dish surfaces. This is known as a detergent's ***wetting action.*** Detergent is not inexpensive, and can

TABLE 16–1

SPECIFICATION REQUIREMENTS FOR HOT-WATER SANITIZING MACHINES

	Minimum Wash Temperature	Minimum Sanitizing Rinse Temperature	Maximum Sanitizing Rinse Temperature	Sanitizing Rinse Pressure (range)
Stationary rack/ single temperature	74°C (165°F)	74°C (165°F)	90°C (195°F)	138 kPa ± 34 kPa (20 psi ± 5 psi)
Stationary rack/ dual temperature	66°C (150°F)	82°C (180°F)	90°C (195°F)	138 kPa ± 34 kPa (20 psi ± 5 psi)
Single-tank conveyor	71°C (160°F)	82°C (180°F)	90°C (195°F)	138 kPa ± 34 kPa (20 psi ± 5 psi)
Multiple-tank conveyor	66°C (150°F)	82°C (180°F)	90°C (195°F)	138 kPa ± 34 kPa (20 psi ± 5 psi)

Source: NSF International, Ann Arbor, Michigan.

become a real budget drain if it is not used carefully. Water temperature works in tandem with the detergent to clean and sanitize.

3. Adequate time for water recirculation. This includes sufficient volume and velocity of the water, as well as a machine designed to spray the dishes for the correct amount of time to rinse them fully.

These factors, and more, will be covered in detail in this chapter.

16-1 CHOOSING A DISHWASHING SYSTEM

There are numerous variables to consider before you set up your dishwashing area. As you write your equipment specs, think about these things:

- Space available in your kitchen
- Proximity to service areas, such as wait stations, and (equally important) distance from eating area
- Type of service; how food is delivered to guests
- Power and plumbing sources and their locations
- Water quality
- Energy-saving and water-saving features
- Local building codes and health ordinances
- Detergent quality and uses
- How dishes are handled (and where they are placed) both before and after washing
- Your proposed budget
- Access to reliable workers

A word about the last item: The cost of labor for the whole dishwashing process is more than 1 cent per dish. This doesn't sound like much until you see just how many dishes are used per guest, per day! The averages can be found in Table 16-2, and these don't even include your pots and pans, or the preparation and cooking utensils. Selecting the proper dishwashing system will have a significant

TABLE 16–2

DISH NEEDS IN FOODSERVICE

Type of Operation	Dish Count per Guest
Fine dining	20
Casual, family	15
Cafeteria	10–12, plus tray
Counter service	6–8

impact on your bottom line. Most foodservice consultants believe the first places to cut costs are the non-income-producing, unskilled labor areas of a restaurant, which includes the dish handlers. However, making the process more automated (so that it uses fewer workers) will cost more initially than a basic system. In short, it's a tradeoff you must decide on. First, you must analyze the whole dishwashing function. This means:

- Determining the volume and types of wares and utensils you'll be washing. How can they be cleaned quickly and at the least cost?
- Studying individual job functions. How many times are dishes handled before they reach the dish machine? Do employees have to lift the full, heavy bus tubs in order to put the dishes onto racks to load into the machine, or are they kept on the same level for unloading and racking? Are there logical places to put the clean dishes?
- Determining your peak activity hours. If you can wash and return dishes to service quickly, you can maintain a smaller inventory, saving money and still taking good care of your guests.
- Deciding how your guest fits into the process. In some foodservice settings, the guests clear their own tables. A conveyor system can be installed to send them directly from dining area to dishwashing area.
- Analyzing the service records of the dishwashing equipment. If your dishwashing staff is faced with frequent breakdowns, dish breakage, or general inefficiency, you should consider purchasing new machines or perhaps redesigning the system.

It helps to know a little bit about how mechanical dishwashing works. If you're a dirty dish, there are only two ways to go: Either you move through a spray pattern on some kind of conveyor, or you hold still in a dish rack and the spray pattern moves around you. The motion is necessary to cover all of the dish surfaces completely. To perform either of these functions properly, as well as to control costs, the foodservice manager should note there are six fundamental rules:

1. Maximize the use of sanitation personnel in the kitchen.
2. Control detergent use by prescraping or prewashing.
3. Use enough water (and the energy to heat it), without wasting either.
4. Control machine use by running only fully loaded dish racks through the washer.
5. Use the proper wetting agent, which promotes faster rinsing and drying and minimizes spotting.
6. Lay out the dish room efficiently, which will help reduce breakage.

Speaking of layouts, there are as many configurations as there are dishwashing systems (see Illustration 16-1). Resist the temptation to choose whatever model seems to fit best in your square footage. Instead, design the system with productive work flow, safety, and sanitation issues in mind.

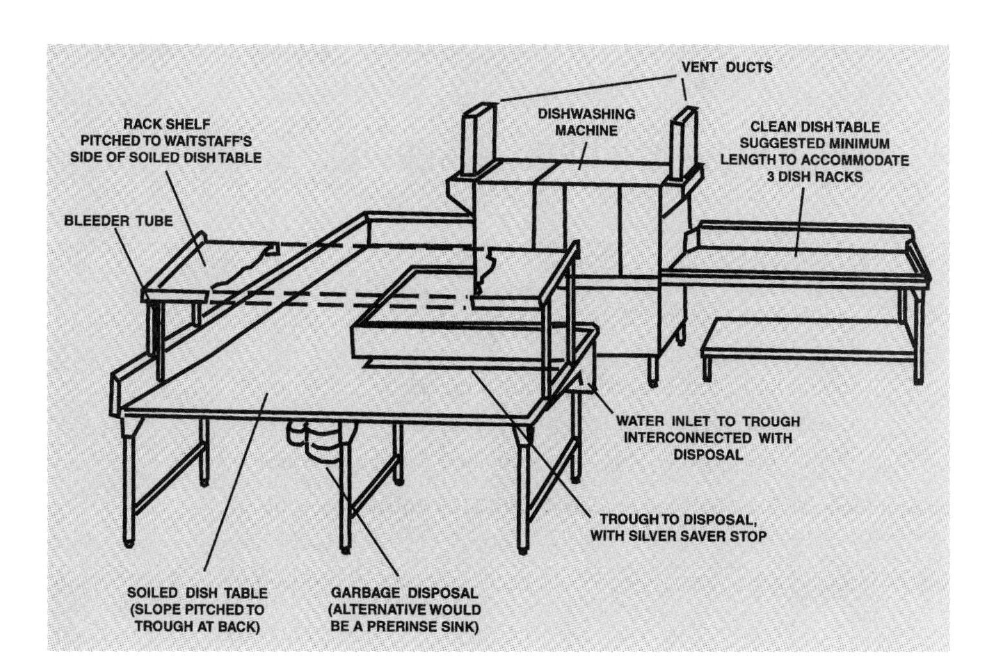

ILLUSTRATION 16–1 A dish room layout.

SOURCE: ADAPTED FROM CARL R. SCRIVEN AND JAMES W. STEVENS, *MANUAL OF EQUIPMENT AND DESIGN FOR THE FOODSERVICE INDUSTRY,* THOMSON LEARNING.

High-Temp or Low-Temp?

The amount of room you will need overall depends in part on the type of dish machine you choose. Dish machines often perform their sanitation functions at a steamy 180 degrees Fahrenheit, so you'll need space for ventilation hoods, which are normally required. The dishwasher that rinses at 180 degrees Fahrenheit is known as a *high-temp* model. A high-temp dish machine requires a ***booster heater.*** Ordinary 140-degree-Fahrenheit dishwater is piped into the booster heater to bring it up to the necessary 180 degrees Fahrenheit for use during the rinse cycle. Most of the newer-model dish machines have their own built-in booster heaters, but others offer them as a separate unit, which requires more space and separate installation.

Where venting is a problem, most cities will allow the use of a *low-temp* dishwasher. This model rinses at the lower, "regular" wash temperature of 140 degrees Fahrenheit, but uses chemicals instead of super-hot water to do the sanitizing during the rinse cycle, usually some type of chlorine, stored in a separate unit that can be wall-mounted or located under a counter. This can be obtained from (and usually installed by) your detergent supplier. It works by injecting a preset amount of chemical solution directly into the rinse water. Some cities also require that a litmus-paper test be used periodically to check the strength of the sanitizing solution. Table 16-3 lists a few guidelines for chemical concentration and rinse water temperature.

Many dish machines are ***field convertible,*** which means they can be changed from hot-water sanitizing to chemical sanitizing or vice versa, by following the manufacturer's instructions.

There are a couple of things to consider in the high-temp versus low-temp debate: Using a chemical sanitizer can save money on energy costs, since water doesn't have to be quite so hot. It also reduces ventilation requirements. But some utensils don't react well to the chemicals, notably aluminum, pewter, and silver-plated items. The sanitizing solutions should be a minimum of 50 parts per million (ppm), as per NSF International guidelines. If the sanitizing chemicals aren't properly balanced to the amount of water, spotting and a "detergent taste" can linger on the dishes.

With a high-temp machine, one benefit is faster drying time. Of course, it takes an adequate power supply to provide that "boost" of heat to the water, and the plumbing lines must be nearby to handle high volumes of water that go from booster to dish machine.

The rule about dishwashing is that dishes soiled by foods that contain protein usually require very hot water to dissolve and "power-off" the food residue; but coffee cups and most glassware don't really need scalding water to get them clean. The only glassware exception may be wineglasses—purists feel

TABLE 16–3

DISHWASHER NEEDS IN FOODSERVICE

Meals per Hour	Style of Dishwasher
Up to 50	Counter or undercounter
50–250	Single tank, door style
250–400	Single tank, conveyor
400–750	Single tank, conveyor with prewash option
750–1500	Double tank, conveyor with prewash option
1500 or more	Flight-type conveyor (or, where space is tight, a carousel)

Source: Carl R. Scriven and James W. Stevens, *Food Equipment Facts,* John Wiley & Sons, Inc., Hoboken, NJ.

they should be cleaned with scalding water and a minimum amount of detergent (or no detergent at all), to prevent residue on the glass that interferes with the taste and clarity of the wine. Some wineglasses are also quite delicate, and it's not unusual for very fine dining establishments to assign someone to hand-wash all the wineglasses.

Both high-temp and low-temp machines will be impacted by your area's water quality. They have the same scaling and chemical build-up problems as any other appliance that uses tap water, in addition to food particles that can plug up lines and solenoids. Water filters and, in some areas, water softeners will be necessary components of successful mechanical dishwashing.

Dishwasher Sizes and Ratings

Here's a step backward for you: How about deciding whether the volume of dishes you'll be using is sufficient to require a mechanical dishwasher in the first place? Most foodservice operators decide almost immediately they couldn't live without one—not only for sanitation reasons, but to save time. Even the least sophisticated machine can complete one of its cycles in a couple of minutes. An undercounter machine may take 90 seconds to four and a half minutes. You may not need a continuous dishwashing operation; you might store the soiled dishes and wash them periodically in a flurry of activity.

Let's talk first about a small breakfast-and-lunch-only operation, which serves no more than 50 meals per day. In this case, a single-tank, undercounter machine will suffice. An "average" commercial operation serving 200 meals per day should have a single-tank, door-style dishwasher. An operation serving 300 to 500 meals per day would probably choose a conveyor-style dishwasher. And a large, high-volume operation (such as a hotel banquet department or airline kitchen preparing meals for passengers) would need a flight-type conveyor-style washer, which comes in lengths from 9½ feet up to 60 feet. Later in this chapter, you'll learn more about each of these options. Meanwhile, Table 16-4 lists the sizing suggestions from the authoritative handbook *Food Equipment Facts* by Carl Scriven and James Stevens.

Dishwashers are often sold with promises that they will clean so many dishes, or so many racks of dishes, per hour. Although these claims are not necessarily untrue, in real-life dish rooms, a machine rarely achieves its full potential.

Carl Scriven and James Stevens have also devised some basic space guidelines for the dish room. They are given in Table 16-5. You may also refer back to Chapter 4 of this textbook for additional dish room space guidelines.

TABLE 16–4

DIMENSIONS OF A DISH ROOM

Meals per Hour	Dishroom Area Square Footage
200	100
400	200–300
800	400–500
1200	600–700
1600	800–900

Source: Carl R. Scriven and James W. Stevens, *Food Equipment Facts,* John Wiley & Sons, Inc., Hoboken, NJ.

TABLE 16–5

DATA PLATE SPECIFICATIONS FOR THE CHEMICAL SANITIZING RINSE

Sanitizing Solution Type	Final Rinse Temperature	Concentration
Chlorine solution	min: 49°C (120°F)[a]	min: 50 ppm (as NaOCl)
Iodine solution	min: 24°C (75°F)	min: 12.5 ppm, max: 25 ppm
Quaternary ammonium solution	min: 24°C (75°F)	min: 150 ppm, max: 400 ppm

[a]For glasswashing machines that use a chlorine sanitizing solution, the minimum final rinse temperature specified by the manufacturer shall be at least 24°C (75°F).

Source: NSF International, Ann Arbor, Michigan.

There are several basic categories by which most commercial dishwashers are rated.

Machine Ratings. The number of full dish racks per hour that a machine can wash is its rating, an indication of its maximum mechanical capacity. A typical dish rack measures 20 by 20 square inches. A good rule of thumb is to estimate the machine capacity at 70 percent and the average rack capacity as follows:

- Sixteen to eighteen 9-inch plates per rack (which will vary, of course, based on dish diameter)
- 25 water glasses per rack (again, this will change based on glass size)
- 100 pieces of flatware per rack

Different types of racks can be purchased to accommodate different types of dishware: peg racks for cups or bowls, mesh cups to contain flatware, and so forth.

Other variables that go into figuring the actual productivity are the type of dish room layout, the length of time the dirty dishes sit before washing, the hardness of the water, the efficiency of the machine operator, and fluctuations in the flow of soiled dishes into the dish room.

Pumps and Motors. Highly efficient motors and pumps ensure the proper volume of water at the required pressure levels to remove food from dishes. Motors range in size from ½ to 5 horsepower. If dishes or glasses are plastic, you might consider the use of ***reducing collars.***

When the dishwasher fills with water, the water is circulated by a pump. Pump efficiency is measured in gallons per minute (gpm) and ranges from 45 for small, undercounter models to 240 for the largest, flight-type machines.

Heating Equipment. The water heater keeps water hot in the tank(s) of the dishwasher. The smallest electric models use only 1.2 kilowatts; the largest, 23 kilowatts. Gas is also used to heat dishwater, resulting in new technology similar to infrared gas burners. Gas is extremely efficient and puts out less radiant heat in the kitchen. The decision to use gas or electricity depends largely on what is available in your area and how much each costs.

Rinse Water. The rinse cycle is critical for sanitation. Allowable water pressure for this cycle is between 15 and 25 pounds. It can be checked at the pressure gauge installed at the inlet side of the final rinse valve, and read while the rinse water is flowing. The *rate of flow* is also important, which means the machine must be replenishing its storage tank even while a rinse cycle is taking place. Finally, the rinse water temperature should be 180 degrees Fahrenheit in most situations. As already mentioned, it may take a booster heater to accomplish this.

Conservation Features. The other important consideration in today's resource-conscious environment is water and power consumption. Some of the newest dish machines are advertised to reduce water consumption by 30 to 60 percent compared to their competitors of similar sizes. The idea is good, but look at the machine's washing capacity! Using less water may require running it more often. Is that really cost-effective?

An alternative feature is the use of sensors in the machine that detect the presence of dish racks and shut down the pumps and exhaust fans when the machine is empty. Also ask about the amount of insulation in dish machine construction, as the better insulated it is, the more comfortable the ambient temperature in the dish room, which also saves on utility costs for cooling the space.

16-2 TYPES OF DISHWASHERS

The dish machine is the major component of your dish room operation. Now let's look at some specific types of dishwashers for different sizes of foodservice operations.

Undercounter Dishwashers

At peak performance, this unit can wash from 18 to 40 racks per hour, one rack at a time, making it ideal for small operations. Some machines have slides on which to rest removable racks; others contain racks that remain in the machine on rollers and can be slid in and out. Removable racks are recommended, because this allows the operator to rack dirty dishes in advance and have them ready to load as soon as others are washed (see Illustration 16-2).

Undercounter dishwashers use from 3 to 5 gallons of water per wash cycle, and only a few kilowatts of power to maintain the water temperature in the tank. The time for each wash cycle varies from 1:30 to 2:20 (minutes/seconds). The ideal pump capacity is 45 gpm.

High-temp models may have a built-in booster heater that provides just over 1 gallon of super-hot rinse water, 40 degrees Fahrenheit hotter than the wash cycle; most models also feature an optional booster with a 70-degree heat increase.

Low-temp undercounter machines are usually smaller, only 1.5 gallons in capacity. They get their water from a standard hot water heater, at 120 degrees Fahrenheit, and do their sanitizing with chemicals.

The best interior design seems to be dual wash arms at both top and bottom of the washing area, sending high-pressure sprays that clean dishes thoroughly. The double arms are more expensive than a single arm, and you may not need them if your dishes are not heavily soiled.

The height of the dishwashing cavity is between 11 and 17 inches. The largest height seems handiest, because it allows the operator to load and unload 16-by-18-inch cafeteria trays as well as bulky but often-used preparation items such as mixing bowls, electric mixer accessories, and so forth.

As its name implies, the undercounter machine is usually installed under a counter with a drop-down front door or beneath a specially designed dish table with a sink. The sink setup may contain a heavy-duty sprayer for prerinsing dishes; a removable, perforated ***scrap basket*** to scrape waste into; an electric food disposal unit; a corrugated drain board area; and/or an overshelf. Some models also have a switch that activates a deliming cycle. Plumbing is located beneath the drain board to save floor space.

There are also freestanding models with stainless steel sides and top, which measure a little less than 3 feet in height and 2 feet square in width and depth. They can be mounted on 6-inch legs or on an 18-inch stand. The latter is handy because it raises the machine to counter level, which reduces bending to load and unload.

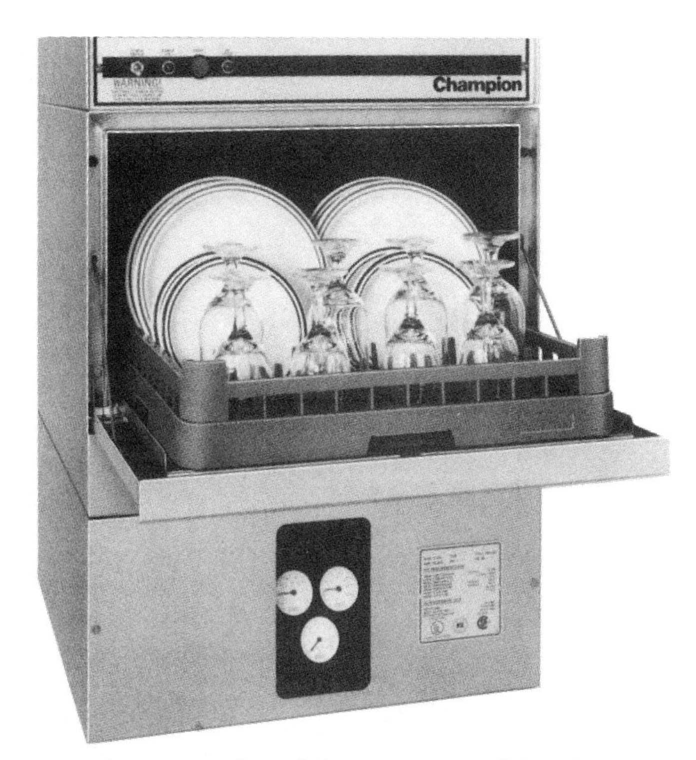

ILLUSTRATION 16-2 **A small, low-temperature dishwasher designed to fit beneath a counter.**

COURTESY OF CHAMPION INDUSTRIES, ALI GROUP, WINSTON-SALEM, NORTH CAROLINA.

Like most mechanical dishwashers, the wash and rinse cycles are fully automatic, controlled by electric timers when the door is closed and latched. If necessary, the cycle can be interrupted by lifting the door handle; it will resume when it is latched again. A light goes out when the cycle is complete. On the front door, gauges indicate wash and rinse temperatures, as well as rinse pressure.

Undercounter models are powered by electricity only, at 120 volts. Models with built-in heaters will require 208/240 volts.

Glasswashers

A variation of the undercounter dishwasher is the glasswasher, often installed in bar settings. Getting glasses clean (without breakage) is apparently such an art form that at least one manufacturer's machine has eight different cycles for different types of glassware! The normal glasswasher washes and rinses with tap water, provides a final rinse with sanitized water, and blow-dries the glasses. A heavy steam cycle ensures removal of food particles; a low-temperature cycle cleans plasticware without melting it, and so on.

Water temperatures in these units range from the 150-degree (Fahrenheit) wash cycle to a superhot 212 degrees Fahrenheit for quick drying. This is a benefit because the faster the glasses dry, the less likely they are to emerge with water spots on them. The full wash-rinse-dry cycle takes about 20 minutes and uses 3 gallons of water for each fill. The required water pressure is 20 psi.

Most glasswashers look like undercounter dish machines, with the pull-down door and a variety of racks designed to hold different types and sizes of glasses. However, you can also get a conveyor-type glasswasher that will turn out nearly 1000 clean glasses per hour (see Illustration 16-3). This machine has three separate pumps, which dispense precise amounts of detergent, sanitizing solution, and rinse aids, and a 3-kilowatt water heater. Glasses enter the machine on a polypropylene conveyor belt, are washed at 140 to 160 degrees Fahrenheit, and rinsed twice. The final rinse is with cool, distilled water so the glasses come out ready to use instead of hot to the touch.

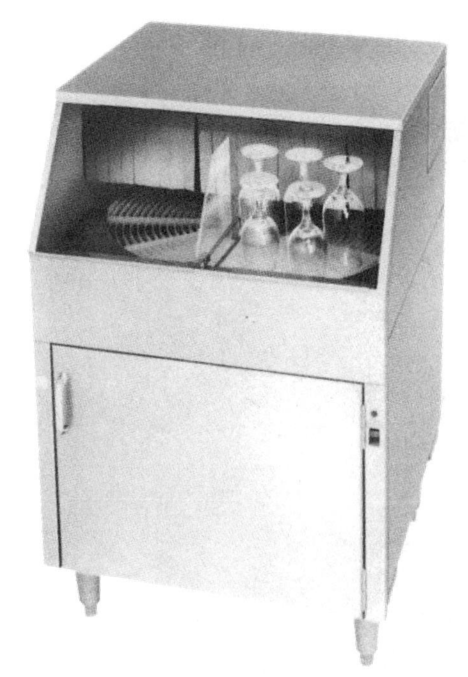

ILLUSTRATION 16-3 A glasswasher is designed specifically to wash and sanitize cups and glasses without water-spotting or chipping them.

COURTESY OF CHAMPION INDUSTRIES, ALI GROUP, WINSTON-SALEM, NORTH CAROLINA.

There's also a pass-through-type fully automatic glasswasher that is capable of washing 2000 glasses per hour. The conveyor belt on this unit can run in either direction, at 14 inches per minute. Installed at countertop height, its width varies from 48 to 72 inches, depending on the amount of space you want for loading and unloading at either end of the machine.

Single-Tank, Door-Style Dishwashers

The common name for this very popular model is the ***stationary-rack dishwasher.*** For a foodservice operation that generates 750 to 1250 dishes and related items per hour, this is the typical recommendation. The single-tank, door-style dishwasher is easy to operate and can handle from 35 to 55 racks per hour. You'll usually find them installed between two dish tables, one for stacking soiled dishes, the other for stacking clean ones. (A helpful guideline is to make sure the soiled-dish table is 50 percent larger than the clean-dish area.)

Most manufacturers offer single-tank machines with a hot-water sanitizing mode or two chemical sanitizing modes. A "normal-duty" chemical sanitizing machine requires a water temperature of 120 degrees Fahrenheit and is able to wash 62 racks per hour; a "light-duty" model requires 130-degree-Fahrenheit water and is able to handle up to 80 racks per hour. A lot of single-tank machines are ***field convertible,*** with the ability to be set up either for high-temp or low-temp sanitizing.

A 1-horsepower motor powers the unit and, depending on its size, will need from 100 to 120 volts to 400 to 460 volts of electricity. The water tank holds 16 gallons, kept hot by a 5-kilowatt electric heating element or, in the case of gas heat, an energy-saving burner with an automatic pilot light igniter (so the pilot light is not on continuously). The pump capacity is 160 gpm. The stationary-rack machine uses about 1.2 gallons of water per rack.

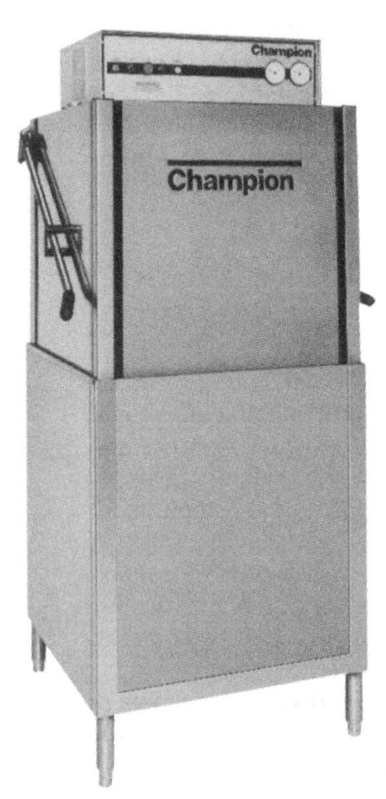

ILLUSTRATION 16-4 A single-tank, door-style dishwasher.

COURTESY OF CHAMPION INDUSTRIES, ALI GROUP, WINSTON-SALEM, NORTH CAROLINA.

The unit is made of stainless steel, including the wash arms inside (see Illustration 16-4). Located above and below the dish racks, the arms are removable and interchangeable. Two nozzles are mounted on the arms to spray and sanitize during the rinse cycles. Always check dishwasher arms once in a while to make sure they rotate easily.

The newest dishwashers feature microcomputer controls that can be programmed to time each cycle. It begins automatically when the door is lowered, stops immediately if it's opened during the process, and resets itself. Draining and overflow prevention is also regulated, with a large, bell-type automatic overflow and drain valve. Perforated strainers and removable scrap baskets submerged beneath the unit make it practically clog-proof.

The dimensions of the average stationary-rack dishwasher are 67½ inches high, 26½ inches wide, and 29¼ inches deep; in short, it'll fit into a 4-square-foot floor space. The door opening can range in width from 15 to 19 inches, which is a consideration if you'll be washing serving trays in it. Because the booster heater is almost always built into the unit, you needn't allow additional space for it. However, your site selection may depend on accessibility to an exhaust canopy, as most local ordinances require this machine be vented at a rate of 100 cubic feet per minute (cfm).

Handy options for your single-tank dishwasher include fabricated steel ***dish tables*** (see Illustration 16-5). You'll want to order them with a 6-inch-high back-splash and perhaps an overshelf to store glass racks. A drain tube on one

ILLUSTRATION 16-5 **Sturdy, stainless steel dish tables give you landing space for both dirty and clean dishes in the dish room.**
COURTESY OF HOBART CORPORATION, TROY, OHIO.

edge of the shelf will allow water to drain onto the table below. Undershelves are also available; they're 6 inches off the ground and handiest below the "clean-dish" side of the table.

The "dirty-dish" side should include a sink for rinsing and scraping duties. Most are 20 inches square and 6 inches deep. The center opening at the bottom of the sink should be at least 3¼ inches wide to accommodate installation of a drain or disposal unit and a removable scrap basket. If no electric disposal is installed, a sink flange with a 2-inch lever drain is recommended. Also ask for a heavy-duty sprayer to use by hand to prerinse dishes.

Moving Dishwashers

Machines that move dishes through their cycles on conveyor belts are known as ***moving*** (or *conveyor*) ***dishwashers***. The most basic of these models is the ***rack conveyor***, designed to transport full racks of dishes through wash and rinse. A single-tank rack conveyor can wash anywhere from 125 to 200 racks per hour; a double-tank machine can do 250 to 300 racks per hour (see Illustrations 16-6 and 16-7).

ILLUSTRATION 16-6 **A diagram of the parts of a conveyor dish machine.**

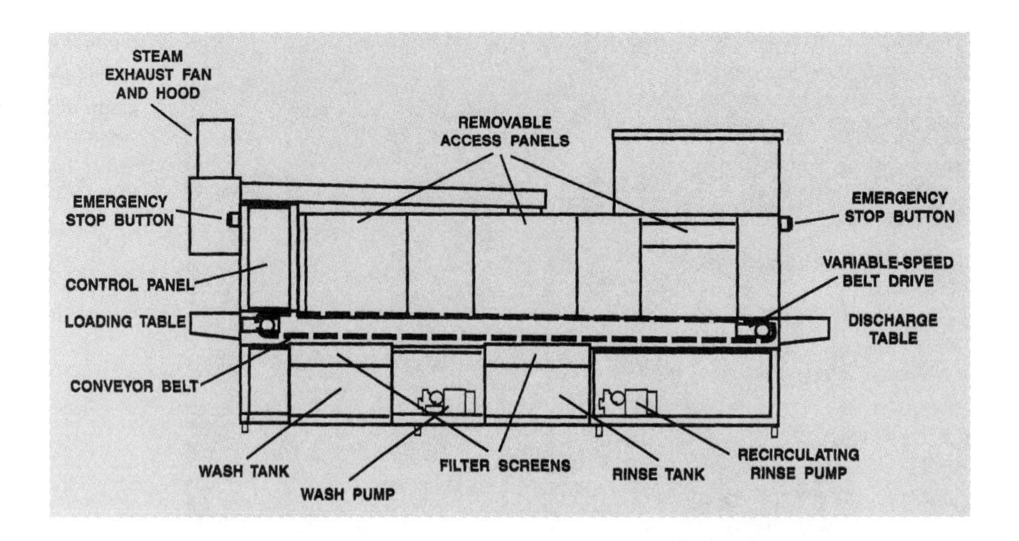

ILLUSTRATION 16-7 **A double-tank conveyor dishwasher.**
COURTESY OF CHAMPION INDUSTRIES, ALI GROUP, WINSTON-SALEM, NORTH CAROLINA.

Rack conveyors generally include more sophisticated options than stationary dishwashers, such as recirculating prewash or power prewash cycles, corner scraper units, and automated activators, which run the machine only when racks are in it. Human labor requirements depend mostly on the number of racks to be handled; when a multiple-tank machine is running at full demand, two people may be needed to load and unload racks. However, you'll soon read about automatic loaders and unloaders, which could cut labor needs to a single employee.

Rack conveyors move the dishes either by chain or pawl. The *chain* works somewhat like a bicycle chain, pulling the conveyor belt along; a *pawl* is like a piston that uses a single rod to pull the racks through the machine.

The dishwasher is constructed of 16-gauge stainless steel. Flexible strip curtains made of plastic hang at both ends of the dishwashing chamber to keep water inside; curtains also are used to separate the wash and rinse functions. A large inspection door located near the front of the machine allows easy access to its interior; if the door is opened midcycle, the pump motor shuts off automatically as a safety feature. Likewise, if the movement of racks is obstructed somehow, an adjustable overload mechanism is triggered, which prevents damage to the machine or the racks. Conveyor speed is set by NSFI standards for each machine. When chemical sanitizing is used, the final rinse uses a liquid bleach solution.

The 2-horsepower motor is controlled by a single on-off switch. Its electrical specifications range from 200 to 230 volts (single phase) to 400 to 460 volts (three phase). Some models use the same motor to drive the conveyor and the pump; others have a separate, ¼-horsepower motor to drive the conveyor. The pump circulates water at 195 to 265 gallons per minute, and self-draining pumps drain water out of the tank between cycles.

Washing is done by two fixed spray assemblies, one at the top of the unit and one below. Both are removable (no tools needed) for cleaning. Perforated strainers cover the wash tank's water surface to prevent debris from clogging the drain. The rinse cycle is conducted by double-action rinse arms, also above and below. The machine is plumbed so that half of the rinse water goes back to the wash tank as makeup water, while the other half goes down the drain. The water tank is filled manually by turning a fill valve (or, in some cases, pushing a button) in front of the machine.

A space-saving option is the ***side loader,*** a cantilevered table that lets the operator load racks from the front of the machine. It is available in 24- and 30-inch lengths to accommodate different sizes of racks. An extended pawl bar connects the side loader firmly to the machine. A side loader is helpful where space is small, because it allows a dishwasher to be located closer to a corner of the room.

Another popular feature is called the ***curved unloader.*** This 38-inch addition to the machine helps push racks off the conveyor at a 90-degree angle to the machine. Each rack clears the machine before the next rack is discharged.

Both single-tank and double-tank conveyor-type dishwashers can be ordered with a ***prewash option.*** This is a separate chamber from the regular washing area, where dirty dishes begin their odyssey. Sometimes, it's referred to as a *power prewash,* and most machines have a separate, 1- or 2-horsepower motor for this function. Inside the chamber, stainless steel arms with water jets on them shoot a high-velocity spray to effectively scrape all kinds of tableware. The food residue is collected on perforated strainer screens (also called ***scrap screens***) and then drops into a strainer basket (or scrap basket) below.

At a length of about 7 feet, double-tank machines are large enough for four separate compartments for their various duties: power prewash, regular wash, power rinse, and, finally, sanitizing rinse. Three large inspection doors allow access to the chambers and, if they're opened during a cycle, the action immediately shuts down until they're lowered again. Each compartment has its own set of spray arms and/or directional water jets. A benefit of these large machines is that they provide better separation of the wash and rinse cycles; also, with two rinse cycles, detergent is removed thoroughly from the dishes. If you've ever eaten at a restaurant and tasted soap on a glass or utensil, you know how important a good rinse cycle is.

Many models recirculate their final rinse water back into the wash tank. The water in the tanks may be heated by electricity, gas, or steam. Inside each tank, a thermistor sensor continuously monitors and adjusts the water temperature, and a float device protects against the water level becoming too low.

More Options for Conveyor Machines

To eliminate the need for an exhaust system, several manufacturers have introduced a ***condenser.*** Mounted over the vent of the dishwasher, air flowing out of the dishwasher passes over water-cooled coils inside the condenser. A ½-horsepower centrifugal blower (1000 cubic feet per minute) helps remove the moisture from the air and returns the dried air to the dish room area. Louvers on the unit can be moved to send the airflow in any direction. The condenser is activated automatically by the pump switch located on the control panel.

Because the coils are water cooled, access to a water source is a prime installation consideration. The required incoming water pressure is 20 psi; it should flow at a rate of at least 6 gpm, at a starting temperature of no more than 55 degrees Fahrenheit. As the air passes over the cool water, the water naturally heats up. It can then be piped back into the prewash tank or discharged to the drain. Another water-saving feature: The rate that cold water flows into the condenser is regulated automatically by the passage of dishes into the machine.

A condenser unit adds about 2 cubic feet to the dish machine, but because it is installed at the top of the machine it fits anywhere there is a standard 7-foot ceiling.

A system to speed the drying of dishware is the ***blower-dryer,*** found in conveyor models as well as flight-type dishwashers. Blower-dryers can be all electric or steam heated. The all-electric blower-dryer is a 5-foot-long stainless steel chamber with forced circulation of high-velocity hot air inside. It's designed to dry china, silver, and even plasticware quickly. A heavy-duty fan draws room air through an electric heater and into ducts at the top of the chamber. The heated air is forced vertically downward onto passing dish racks; then, baffles below the conveyor aim the air upward again. So that all this hot air doesn't blast anyone who happens to be unloading the machine, large fan intakes at the outgoing end keep cooler, room temperature air moving over the unloading area. An exhaust system to the outside of the chamber is necessary to carry off the moist air from the drying process. This is accomplished by a vent stack with a control damper, which can be connected to the dish room's main exhaust system.

Adding a blower-dryer adds 5 feet of length and almost 2 feet of height to your dishwashing area, plus additional exhaust requirements of 1200 to 1400 cubic feet per minute. You'll need more electrical capacity, too, because the unit has a 2-horsepower blower motor and a dozen 3500-watt air heaters that use 42 kilowatts of power per hour. An installer will wire the blower-dryer to your dishwasher motor switch, so that it operates only when the dishwasher is on.

A steam-heated blower-dryer functions just like the all-electric model except for the way heat is delivered to the drying process. Instead of electric heaters, we find double rows of high-efficiency tube-type heat exchangers. Also, the condensation that gathers on each coil must be drained away. Steam requirements to dry dishes properly are 110 pounds per hour at 20 to 25 psi.

Like other types of dishwashers, conveyor machines must be ventilated. The exhaust requirements vary based on the size and capacity of each machine. Overall, however, built-in hoods will increase their efficiency as they make for faster air drying and a more comfortable work environment.

Another option: Stainless steel ventilating ***cowls*** (seen on the machine in Illustration 16-8) are hood-shaped covers that can be fastened to each end of the conveyor, then connected to an exhaust system. Cowls add 16 to 20 inches to the overall height of the unit. Dampers can be installed in the ducts of the exhaust system to ensure exhausting doesn't occur too rapidly, which could chill the final rinse water below acceptable temperatures.

Suggested exhaust requirements are as follows: for single-tank machines, 200 cfm at the entrance, 400 cfm at the clean-dish side; for multiple-tank machines, 200 cfm at the entrance, 500 cfm at the clean-dish side. And, for a machine with a blower-dryer unit, requirements for the clean-dish side jump to 1400 cfm.

Circular Dishwashing Systems

The highest-volume foodservice operations might want to look at circular conveyor systems, which can handle from 8000 to 24,000 dishes per hour. An oft-cited industry survey ranks these carousel-type conveyors

ILLUSTRATION 16-8

The cowls on the ends of this three-tank machine are vents that are connected to an exhaust system.

COURTESY OF CHAMPION INDUSTRIES, ALI GROUP, WINSTON-SALEM, NORTH CAROLINA.

in three categories based on rack capacity per hour: type 1, 180 racks; type 2, 387 racks; and type 3, 480 racks per hour.

The advantage of a circular system over an equally large flight-type dishwasher is the ability to operate it with fewer personnel. Although the dishes go through the inside of the machine in a straight line, the exterior part of the conveyor belt curves to form a loop from the exit to the entrance of the dishwasher (see Illustration 16-9).

A circular dishwasher can have one, two, or three water tanks. Its rack conveyor is driven by a heavy-gauge stainless steel chain assembly, with a motor of either ½ or ¾ horsepower. The power of the motor depends on the length and configuration of the setup, as well as the size of the machine. The conveyor can move clockwise or counterclockwise, and the machine can be purchased with its inspection windows on the outside or the inside of the "circle," so be sure to specify this.

A 2-horsepower motor runs the wash and rinse cycles; an additional 1-horsepower motor can be included for a power prewash cycle. Electrical requirements may be 200 to 230 volts or 400 to 460 volts, three phase. Water temperatures are 150 degrees Fahrenheit for the wash, 160 for the rinse, and 180 for the final, sanitizing rinse. The unit will use between 130 and 348 gallons of water per hour (depending on its size); its venting requirements are 200 cfm at the loading end, 500 cfm at the unloading end. (Add a blower-dryer and this increases to 1400 cfm at the unloading end.)

ILLUSTRATION 16-9 A circular dishwashing system can handle the highest volume of dirty dishes with fewer employees needed to operate it.

COURTESY OF CHAMPION INDUSTRIES, ALI GROUP, WINSTON-SALEM, NORTH CAROLINA.

This curve requires a slightly different setup, often consisting of: a straight table for soiled dishes (with waste disposal and spray nozzle), the dish machine, a curved soiled-dish table, and a curved clean-dish table.

This is a big work space, so have it designed specifically to fit your needs. The experts generally suggest allowing no less than 6½ feet for the curved tables and about 11 feet for the straight table. At least one manufacturer, however, claims to be able to handle 130 racks per hour in a linear space of just over 9 feet. This configuration is usually not a true circle; an oval-, triangle-, or L-shaped configuration might fit your dish room better. In short, the modular nature of the components gives you a lot of options.

As long as you are designing exactly what you want, here are a few add-ons to consider, depending on your budget:

- Built-in electric or stream booster heaters to increase the water temperature 40 to 70 degrees Fahrenheit
- Prewash units
- Automatic tank fill, with a manual bypass
- Food waste disposal unit
- Blower-dryer or condenser
- Longer dish tables on either end
- Hand sinks and/or silverware soaking sinks (these require separate hot/cold water and drain connections)
- Storage shelves above or below
- Vent cowls and locking dampers
- High vertical clearance inside the machine (up to 2 feet, to accommodate large pots and pans)
- Top-mounted control panel (saves space and protects the controls from water contact)
- A hose reel, with a 25-foot hose (for cleanup)

Flight-Type Conveyor Dishwashers

Flight-type dishwashers are also called *rackless conveyors* or *belt conveyors*. Perhaps they are called "flight-type" because the dishes run through in a line as straight as an airport runway. The machine itself is the "aircraft carrier" of the dish room, and the length of the "runway" can be customized from 13 feet up, usually in 1- or 2-foot increments; the width of the machine can also be customized, up to 5 feet across (see Illustrations 16-10 and 16-11). These are real workhorses, found mostly in hospitals, prisons, dormitories, and so on. A flight-type machine is recommended if more than 8000 dish "pieces" are handled per hour, which translates to between 600 and 800 meals an hour.

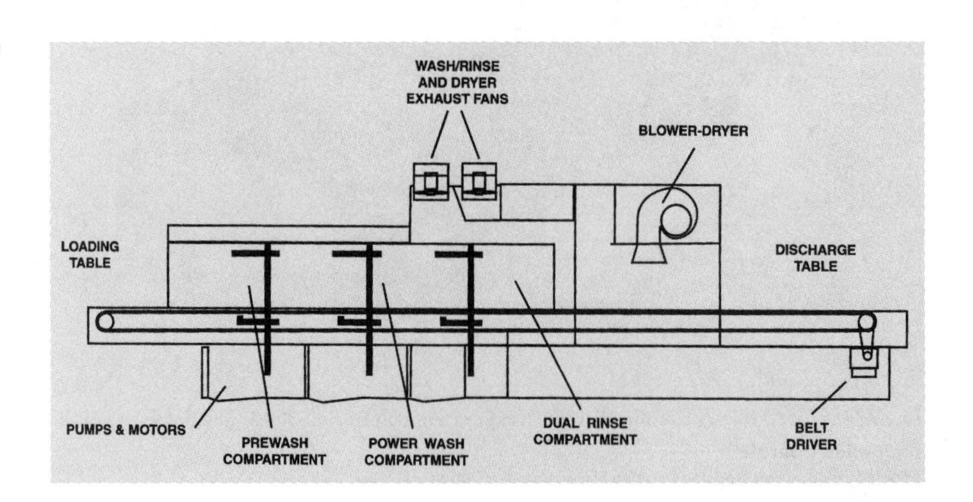

ILLUSTRATION 16-10 **On a flight-type dish machine, dishes are stacked directly onto a conveyor belt that moves them through the wash and dry cycles.**

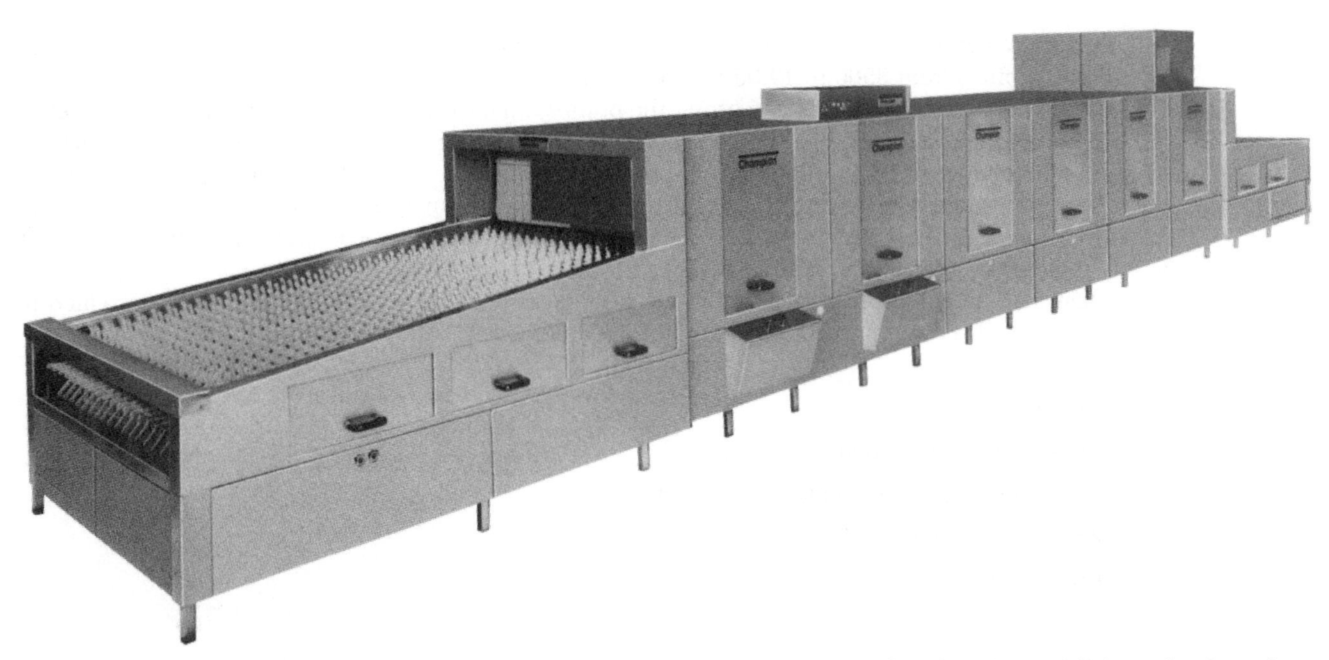

ILLUSTRATION 16-11 **The flight-type dishwasher can be as long as needed to accommodate the volume of dishes to be cleaned.**
COURTESY OF CHAMPION INDUSTRIES, ALI GROUP, WINSTON-SALEM, NORTH CAROLINA.

The flight-type dishwasher operates on the same principle as the rack conveyor, except that the dishes and/or cafeteria trays are placed by hand between rows of plastic-tipped pegs on a continuous conveyor. They travel through the wash-rinse-sanitize cycle in a vertical position. Dishes don't need racks, but someone's got to load them onto and off of the conveyor. Special racks for glasses, cups, and flatware are made to fit right on top of the pegs; flatware may be placed first in a rack and then sorted into perforated cylinders. It usually takes two employees to keep the dishes going through smoothly, one at each end.

The heat and humidity these machines produce are formidable, and ventilation can be a challenge. You may need to vent and replace the dish room air as much as six times an hour.

The most basic machine takes up an absolute minimum of 8 feet, but requires additional space to load and unload. The typical configuration is a 3- to 5-foot loading section, a prewash section, an 8-foot power wash section, a power rinse, a fresh-water final rinse, and a drying/unloading section that will range in length from 5 to 11 feet. A total realistic length to plan on is 20 to 25 feet.

Inside the dishwasher, splash baffles and flexible plastic strip curtains separate the spray systems. The curtains are easily removable for cleaning, as are the scrap screens and baskets in the prewash section. Like other dishwashers, these can sanitize using hotter water or lower temperatures with chemicals.

Here are some basic electrical, water, and steam specifications:

- There are several motors with different horsepower levels depending on their function: prewash (2 to 3 HP), wash (3 HP), rinse (3 HP), conveyor motor (½ HP), blower-dryer (2 HP).
- Electric heating coils, if used to make steam, need 20 kilowatts of power. A steam blower-dryer needs steam pressure of 20 psi to provide the required 75 pounds of steam pressure per hour.
- Each water tank (prewash, wash, and rinse) holds 40 gallons.
- Power requirements for water heating: 20 kilowatts for washing; another 20 for rinsing.
- Pump capacities: prewash (150 gpm), wash (240 gpm), rinse (240 gpm).
- Exhaust requirements: 500 cfm at loading end; 1000 cfm at unloading end.

Remember to match your utility needs to peak demands, not average demand. You will be using a lot of energy! Each 3 horsepower pump requires 2.2 KW each; for water-heating, an electric tank uses roughly 25 KW per cycle. Water consumption will range from 325 to 425 gallons per hour if the dish machine's output is from 8500 to 19,000 pieces per hour.

As an energy-saving and safety feature, the pumps, final rinse cycle, and optional blower-dryer all stop whenever the conveyor stops. There's also an automatic shutoff feature whenever dishes reach the end of the unloading section, unless they are removed by the operator. Whenever inspection doors are open, the pumps and conveyor stop. A lever must be raised to open the drain. If a water tank is accidentally drained, a low-water protection device will turn off the water heater until the tank is sufficiently refilled. In short, the unit almost thinks for itself!

There's a handy formula to determine the dish capacity of flight-type machines. It was created by NSF International and is reprinted here from Scriven and Stevens's *Food Equipment Facts:*

THE VARIABLES

C = Maximum dish capacity per hour

V = Speed of conveyor (in feet per minute)

W = Width of conveyor (in inches)

D = Distance between pegs

THE FORMULA

$$C = \frac{120 \times V \times W}{D}$$

Notice that even with this formula, foodservice authorities suggest dishwashers be rated at only 70 percent of their stated capacity, because not every rack or every peg is loaded perfectly by the hard-working staff in your dish room.

Flight-type dish machines are the largest and most expensive dishwashers on the market, so manufacturers offer quite a few options to make them worth your consideration. You can usually bargain for the length of your loading and unloading areas, extra water- or energy-saving capabilities, noise-reducing insulation packages, variable-speed conveyor controls, dryer attachments, special designs for trays, and good prices on racks and flatware troughs. For correctional facilities, there are even theft-proof and tamper-resistant features.

In addition, computer technology is making these machines more sophisticated all the time. Solid-state microcomputer controls include sensors inside the machine; visual displays or readouts of problems or service reminders; and automatic sorting, loading, and unloading units that can be programmed to gently separate plates and bowls by size. Strong magnets can pick up cutlery from the trays as they pass by. Wash and rinse arms can be automatically preset to the height of the incoming dishes to provide maximum cleaning without breakage. And, at the discharge end of the system, the warm, clean plates or trays can be automatically stacked into spring-loaded lowerators to hold their temperature for new, hot meals.

Sounds amazing? Yes, it is, and amazingly expensive, too, with the largest, hospital-sized system costing a cool $300,000. However, at least one facility claims that its investment in such a state-of-the-art dish system has saved $35,000 in annual energy costs and cut labor costs by 12 full-time employees. Its daily dishwashing functions—some 8000 lunch trays a day—can all be performed within 4 hours.

16-3 BOOSTER HEATERS

As mentioned earlier in this chapter, the booster heater is used to raise the water temperature to 180 degrees Fahrenheit for a dishwasher's final rinse cycle. How hard the booster heater must work depends in part on the incoming water temperature, which is typically 110 or 120 degrees Fahrenheit but could be as high as 140. This means an additional boost of 40 to 70 degrees Fahrenheit. Electrical requirements vary drastically with the size of the booster and the incoming water temperature. For instance, a small unit will use 4 kilowatts of power to heat 40 gallons of 140-degree (Fahrenheit) water per hour, or 23 gallons of 110-degree water per hour. The largest booster can heat 588 gallons per hour if it comes in at 140 degrees, but only 335 gallons if it comes in at 110 degrees.

Most of the new dish machines have built-in booster heaters but, if yours requires a stand-alone model, it should be located as close as possible to the dish machine so the water won't lose heat as it travels through the pipes between the two. Often, it can be installed on an undershelf of the clean-dish table. The typical booster is small, less than 3 feet high and 2 feet wide. Its storage capacity is only 6 gallons, and a low-water cutoff feature eliminates the chance that the heating element will burn out if the tank is only partially full (see Illustration 16-12).

ILLUSTRATION 16-12 A booster heater is a small tank in which hot water is "boosted" to higher temperatures for sanitizing dishes and flatware during the rinse cycle.

COURTESY OF HATCO CORPORATION, MILWAUKEE, WISCONSIN.

Gas-powered booster heaters have not been as popular as electric boosters, but several manufacturers offer them. They have firing inputs ranging from 50,000 to 500,000 Btu, heat very quickly, and can be connected to either a natural gas or a propane gas supply. These units are vented into the same exhaust system as the dish machine. Gas booster heaters cost more initially than electric ones but, in areas of the country where gas rates are substantially cheaper than electric rates, they can be a bargain in the long run. Even the largest gas-powered boosters can pay for themselves in utility cost savings within three to four years.

The newest generation of booster heaters is infrared. They can deliver a lot of hot water efficiently without adding to the ambient heat of the dish room like the gas-powered models.

Steam-heated boosters are often found in large facilities such as hotels and hospitals, where steam is generated by gas-fueled boilers and used for a variety of cooking needs, not just dishwashing. This type of booster needs steam pressure of at least 10 psi to work effectively. In some cases, it also requires electricity to heat the water to produce the steam. The steam-heated booster uses between 34 pounds of steam per hour (for an undercounter dishwasher) and 205 pounds per hour (for a flight-type dishwasher).

16-4 DISHWASHER MAINTENANCE

Dish machines' cycles can often be preprogrammed, but that doesn't mean they should be any shorter than the manufacturer's recommendations. If the water isn't hot enough, the detergent solution is weak, or the machine doesn't wash or rinse long enough, proper cleaning and/or sanitation will not occur.

Special care must be taken in storing and using the chemicals associated with dishwashing. Every item must have a Material Safety Data Sheet (MSDS) on hand, stored where employees can refer to it in case of emergency. This is a federal regulation, enforced by OSHA.

It's important to use the chemicals designed to work with your particular machine, and to follow the manufacturer's instructions about correct quantities. Too much leaves residue; too little doesn't do the job properly. In both dish machines and pot sinks, you can use a test strip that checks the concentration of sanitizing solution. This should be done every time a new "batch" of solution is mixed with water, and perhaps up to 90 minutes later, to ensure it is not too diluted.

Assuming the installation and start-up procedures have been properly performed, some routine preventive maintenance will keep a dish machine running correctly. Here are a few tasks, grouped by how often they should be performed. They first appeared in the January 2000 issue of *Food and Service News,* a publication of the Texas Restaurant Association, and are the advice of Frank Murphy, training director of Ecolab's GCS Service Division in St. Paul, Minnesota.

DAILY MAINTENANCE

- Leave the dish machine door open when the equipment is not in use. If there are curtains, remove and clean them and hang them up to dry at the end of each day.
- Drain, clean, and flush the water tanks.
- Brush and/or spray the scrap screens until they are clean. Do not clean them by banging them against something to loosen the scraps!
- Remove the spray pipes and flush them clean. Take care to reinstall them correctly.
- Look at the final rinse nozzles; brush off any hard water deposits.
- Check water pressure and temperatures.
- Inspect the water pump shafts for leaks.
- Refill the chemicals.

WEEKLY MAINTENANCE

- Check all the water lines and drain/overflow tubes for leaks and make sure they're tight.
- Remove mineral scales that may have accumulated on heating elements.
- Clean any detergent residue off the machine exterior.
- Remove and inspect each spray arm. You can unclog the rinse nozzles with a straightened paper clip.
- Check the pawl bars for signs of wear and/or restricted movement.
- Check the idle pump and final rinse levers for restricted movement.

Dishwashers must also be delimed periodically. Before the deliming process can begin, take the time to flush any sanitizing chemicals out of the lines by running the machine empty for six cycles. Additional troubleshooting tips are in the chart that is Illustration 16-13.

16-5 FOOD WASTE DISPOSERS

Waste and garbage handling is not a glamorous topic, but it's a daily concern in any foodservice operation. How you dispose of waste can impact the public's perception of your establishment, as well as your health inspections—especially when there are odor, insect, and rodent problems. We discussed the options back in Chapter 8 of this text, and the dish room is part of your waste management plan.

Scraping and prerinsing dishes before they enter the dishwasher is a critical part of ensuring cleanliness. It also helps to maintain the dish machine by keeping food scraps and gunk out of the lines and equipment. Before we dissect the waste disposal unit itself, let's discuss the prerinsing process.

The Prerinse Station

A prerinse station is as simple as a pot sink and a spray nozzle or faucet with a flexible hose, and a food waste disposer to get rid of the scraps. One way to save water and still do an adequate prerinse is to install a low-flow spray nozzle. The idea is much like water-saving shower heads in hotels—it restricts water use but still allows a sufficient flow to blast food particles off most dishes. The water doesn't have to be any hotter than regular hot tap water, 110 to 120 degrees Fahrenheit.

A single nozzle in a dish room doesn't sound like it can amount to much in the way of conservation, but tests in California have proven otherwise. The California Urban Water Conservation Council in Los Angeles estimates a medium-sized restaurant that uses its dish room six hours a day saves 300 gallons of water per day, plus the money used to heat the water—total savings of $1018 per year, or about $85 a month.

Symptom	Possible Cause	Self-Help Steps	Savings
Machine Will Not Start	Door not closed Loose door switch Main switch off No rack inserted Overload protector tripped	Secure door Tighten switch in mount Check disconnect Place rack in unit Reset overload in control box	>$60
Low or No Water	Main water supply off Drain overflow tub unseated Machine doors not fully closed Stuck or defective float Clogged "Y" strainer	Turn on water supply Place and seat tube Secure doors Check and clean float Clean or replace	>$60
Continuous Water Filling	Stuck or defective float Drain tube not in place	Check and clean float Look for drain tube in tank	>$60
Wash-Tank Water Temperature is Low	Incoming water temperature too low Defective thermostat Low steam pressure	Raise to 140°F in sanitizer models, 120°F for chemical sanitizer models Check the setting Check and adjust if necessary	
Any Motor Not Running	Tripped due to overheating	Reset overload in control box	>$60
Insufficient Spray Pressure	Clogged pump intake Clogged spray pipe Scrap screen full Low water level in tank	Clean intake screen Clean Must keep clean and in place Check drain and overflow tube	
Insufficient or No Final Rinse	Inproper setting on pressure rinse setting Clogged rinse nozzle and/or pipe Clogged "Y" strainer	Set PSI flow to 20–22 psi on pressure-reducing valve Clean Clean	>$60
Low Final-Rinse Temperature	Low incoming water temperature	Check that booster temperature setting is at 180°F	>$60
Poor Washing Results	Wash arm clogged Improperly scraped dishes Wares improperly placed in racks	Clean Check scraping procedures Use proper racks and don't overload	>$60

ILLUSTRATION 16-13 **Some troubleshooting tips for dish machine maintenance, along with the projected savings for not having to call a service technician!**

In a separate study, the Food Service Technology Center in San Ramon, California, tested different low-flow prerinse spray nozzles and compared the various costs associated with using them. The results for two popular "flow rates" (one regular and one low-flow) are shown in Table 16-6.

A list of food products to be avoided when loading the disposer should be prominently displayed in the prerinse station. This includes eggshells, onion skins, rice, corn husks, celery, banana peels, bones, and the outer leaves of artichokes, to name a few. Common nonfood culprits include flatware, rubber bands, and bits of plastic, wood, or cloth that sometimes find their way to the bottom of the sink in a busy kitchen.

TABLE 16–6

SAVING MONEY WITH LOW-FLOW SPRAY NOZZLES

Nozzles: Gallons Per Minute	1.8 gpm	3.2 gpm
Daily Usage (hours)	4	4
Gallons Per Day	430	430
Annual Heat Cost	$ 750	$ 1310
Annual Water Cost	$ 420	$ 750
Annual Sewer Cost	$ 620	$ 1120
TOTAL ANNUAL COST	$ 1790	$ 3180

Based on $0.60 therm, $2 per unit of water, $3 per unit of sewer.

Source: Food Service Technology Center, San Ramon, California.

The Disposer

The ***food waste disposer*** is easy to ignore until it doesn't work correctly! Hidden beneath the sink, it isn't an attractive or income-producing piece of equipment. On the other hand, it is a convenient, sanitary, and environmentally conscious way of disposing of some types of waste that would otherwise have to be bagged and sent to landfills, attracting rodents and insects in the meantime.

ILLUSTRATION 16–14 **The anatomy of a food waste disposer.**

COURTESY OF GENERAL SLICING/RED GOAT DISPOSERS, A STANDEX COMPANY, MURFREESBORO, TENNESSEE.

The disposer is almost always installed as a component of the dishwashing system. However, in large kitchens where lots of fresh produce is prepared, a disposer is also installed in the prep sink and another in the pot sink. Most of this discussion will focus on the disposer as part of the dishwashing process (see Illustration 16-14).

The electric disposer provides a quick, tidy method of dealing with some of the food waste in our kitchens. It is designed to grind up food scraps, which are similar in chemical content to the human waste already flushed into sewers. The food waste goes into the kitchen's drainage system, through the grease trap and into the sewer. The ground-up and liquefied waste ends up being a better bargain for most towns and cities because it costs less to deal with waste water ($34 to $56 per ton in a wastewater treatment plant in 2003) than the other popular waste disposal options—$60 per ton in a landfill, or $120 per ton in an incinerator.

Today's commercial-quality disposer can handle almost any of the waste food products generated by a restaurant, except perhaps in areas with water shortages or very old sewer systems. If your building uses a septic tank instead of being hooked up to a municipal sewage system, you may choose a pulper or collector instead of a disposer, both of which you'll learn more about in a moment.

It is ironic that some large cities, primarily because of sewage system crises, have banned food waste disposers altogether for commercial uses like restaurants; while others encourage disposers because they cut down on landfill volume. Even where disposers are welcome, there are often strict code requirements for restaurant use. These include rules for connecting them properly to the sewer systems, separation of the disposer and grease trap pipes, types of valves used, and installation of vacuum breakers.

A professional plumber will be able to assess your situation and meet these specifics. But first you'd better be sure you can have a disposer installed in the first place.

How Food Disposers Work

There are two types of commercial disposers: ***continuous feed*** and ***batch feed.*** Continuous feed is the most popular. When switched on, it immediately shreds whatever is in it, and keeps shredding until you turn it off. Batch-feed units can handle only a certain amount of waste at one time.

When waste is put into either type of disposer, it enters a grinding chamber. Cold water is flushed through the chamber to help wash the waste through the grinder. The water flow depends on the size of the grinder motor, from 5 gallons per minute for a ½-horsepower grinder to 10 gallons per minute for a 5-horsepower grinder. Inside the chamber, small protruding bars are attached to a revolving rotor. As the rotor spins, the bars jam the pieces of food into contact with a ring of sharp steel teeth around the inside wall of the chamber, which shreds them. The bits of waste mix with the water to form a pulp and, when the pieces are small enough, they are washed into the drain. Depending on the manufacturer, the rotor may be called a turntable, grinding table, flywheel, or grinding wheel; the steel ring may be called a shredder ring or a grinding sleeve.

Disposers are available in several sizes. They are rated most often by horsepower, but sometimes by the number of meals served (not ground up!) per day. Size guidelines, based on the number of meals served and the location of the disposer, are shown in Table 16-7. A 5-HP unit is usually more than enough to service a large restaurant, and a 2-HP unit is sufficient for a pot-washing or prep area, but there are disposers with up to 10-HP motors! You might think more horsepower means more grinding ability, but that is not the case. Yes, you'll get more power, but you also need a larger rotor or the extra power is wasted. Rotors come in diameters from 5½ to 15 inches. The opening at the top of the disposer (known as its ***throat***) should also be bigger if the motor is bigger. Throat openings range from 3 to 8 inches.

Disposers have different electrical requirements, which are also based on motor size: 115 to 230 volts, single phase, for the lower-horsepower motors; 200 to 460 volts, three phase, for the higher-horsepower motors.

The problem with using an undersized disposer is that food is more likely to jam it, and disposer malfunctions never seem to happen at convenient times. One important lesson for kitchen employees is

TABLE 16–7

RECOMMENDED DISPOSER SIZES (IN HORSEPOWER)

Persons per Meal	At Dish Table	Vegetable Prep	Salad Prep	Pot Sink
Up to 100	¾ hp	½ hp	⅓ hp	⅓ hp
100–150	1	¾	½	½
150–175	1¼	1	¾	½
175–200	1½	1½	¾	¾
200–300	2	1½	¾	¾
300–750	3	1½	¾	¾
750–1500	5	3	1½	1½
1500–2500	5	3	1½	1½

Source: Carl R. Scriven and James W. Stevens, *Food Equipment Facts*, John Wiley & Sons, Inc., Hoboken, NJ.

that if their own teeth won't chew it, neither will the disposer. Of course, some of the problems are plumbing related. When installing a disposer, pay particular attention to the sizes of incoming water lines and drain pipes. For example, the drain pipe for a $\frac{1}{2}$- to $1\frac{1}{4}$-horsepower motor should be no less than $1\frac{1}{2}$ inches in diameter; for the 5-horsepower motor, the drain pipe should be a full 3 inches in diameter.

Look for a disposer that allows automatic reversal of the direction the flywheel rotates—a feature that increases the life and efficiency of the unit and helps to free it when it jams. Most disposers come with a small dejamming wrench that should be stored near the unit. The wrench is used to free the flywheel if reversing its direction doesn't do the job. Remember, the power must always be turned off before you start using the wrench! Most manufacturers will provide detailed instructions on this process. Read them to avoid very serious—and very preventable—accidents. Another option to improve the life of your disposer is a control setting to flush water through the disposer even when it's not in use, to keep the lines clear.

Have your disposer installed in a sink that is at least 20 inches square and 6 inches deep. Make sure there is a "silver saver" vinyl guard for the drain (usually $3\frac{1}{2}$ inches in diameter) to prevent tableware, jewelry, bones, and so on from falling into the disposer. A removable molded scrap basket with rack guides and a spray nozzle for rinsing dishes are also recommended for the sink. The electrical controls for the disposer will be located below the working surface of the soiled-dish table. Most models include, at the very least, an on-off switch, an automatic reversing button, and a reset button.

Waste Pulpers and Collectors

Buildings on septic systems or with large kitchens (more than 1500 meals a day) may want to add a ***waste pulper*** to their disposal repertoire, because it can handle paper products as well as food waste. This machine is capable of reducing much of your trash into 15 percent of its original volume. Paper napkins, styrofoam containers, plastic straws, cardboard boxes, milk cartons, aluminum foil, and more are all potential fodder for this system, which uses as little as 1 or 2 gallons of water per minute, as opposed to 7 or 8 gallons needed to flush waste through a disposer. Pulpers also improve sanitation and reduce rodent and insect problems, by breaking down waste that might otherwise be sitting in plastic bags in your dumpster for days at a time.

Here's how the pulper works: Waste is dumped into a trough that feeds the pulper's input tray (see Illustration 16-15). Inside, the waste is drawn into a whirling vortex of water and through a flatware trap (just in case any utensils get caught in the machine). Then, precision cutters mince the waste into bits. The ***slurry*** (consisting of 95 percent water and 5 percent waste) moves into a water press, or water extraction chamber. Here, an augur spins the waste against a fine-mesh screen to squeeze out most of the water. The water is returned to the pulping tank for reuse, while the clean, semidry slurry (now pulp) is expelled down a chute, usually directly into bags for transfer to the dumpster.

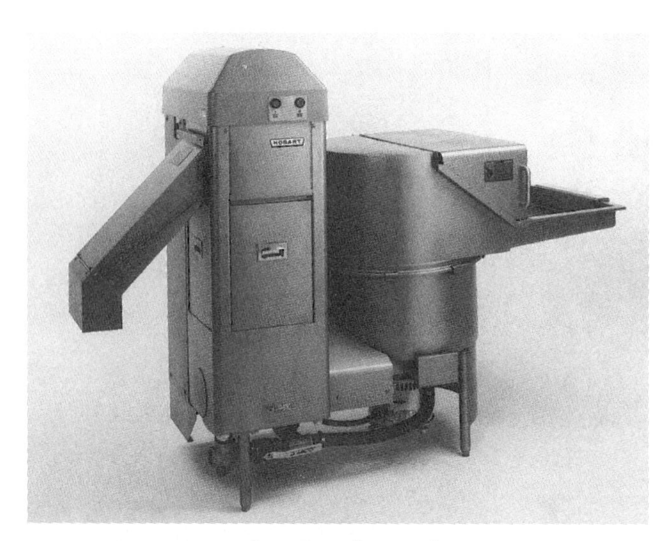

ILLUSTRATION 16-15 A waste pulper system.
COURTESY OF HOBART CORPORATION, TROY, OHIO.

Very large operations have several waste pulpers at work throughout their kitchens, with custom piping to a single, large water press unit located adjacent to the dumpster. That way, no labor is required to haul the bags outside. This is a major consideration because, although pulpers are effective at reducing the volume of trash, the actual weight is the same—or heavier, because of the added water content.

The issue of waste odor has been addressed by some manufacturers, who offer automatic dispensers of deodorant or sanitizing solution, injected as the slurry reaches the water press. Thus, the pulp emerges odor-free as well as semidry. Advances in technology produce a drier, flaked product that accelerates the composting process. But like disposers, there are limits to the types of materials that can be processed by a pulper. Glass, ceramic, metal, and cloth are not suited for this type of machine.

Waste pulpers vary in size from 2 by 4 feet (with a capacity of 350 pounds of waste per hour) to 4 by 7 feet (900-pound-per-hour capacity). They run on three-phase electrical current. Motors for the most common sizes range from 5 to 7½ horsepower, although there are now "mini-pulpers" for smaller operations, with motors from 1 to 3 HP, and huge custom units with motors from 10 to 40 HP. The water press in each unit has its own motor. Pulpers require at least a ½-inch water line and a 3-inch drain pipe.

Perhaps the biggest drawback of waste pulping is the initial cost of the equipment. Smaller units may cost $8,000 to $15,000, but the largest ones can run as high as $125,000. Manufacturers have several methods for determining optimum size. One factor is the amount of waste to be processed per peak hour of business, and the general rule is that one pound of waste is generated for each guest served. There are variables, however—a half-pound of cooked vegetable waste, for example, is not the same as a half-pound of waste that includes chicken bones and packaging material. Some manufacturers suggest a "waste profile" be created for the operation, so you'll know exactly what, and how much, you have to dispose of.

The other concern is noise. Grinding things into bits is not a tidy or quiet process, The best place to install the machine may be where some soundproofing can be used, and you'll want to install it with seismic pads under the legs, which will help with vibration noise.

Some training is necessary to get optimum use from your pulper. It should be cleaned daily, and most have an automated cleaning cycle. The augur screens should also be removed and rinsed, and the internal components sprayed clean with water. When the first pulpers were introduced, there was a widely held belief that every day, some cardboard should be the final item run through the pulper to absorb small food particles and help with odor control. It makes sense, and manufacturers say there's no harm in doing this, but it is no longer necessary.

If you do business in an extremely rural area, both pulpers and disposers may be out of the question. A third option is the ***scrap collector,*** a perforated pot that sits in a sink while water recirculates through it. A collector can be used in a dish room, for instance, where scraps are scraped into it instead of a garbage can. The water "cleans up" the waste as much as necessary, dissolves part of it, and leaves the rest to be periodically dumped into the trash. Collectors are even better water-savers than pulpers, because they use only 2 gallons of water per hour.

16-6 WASHING POTS AND PANS

Like waste disposal, pot washing is an unappealing but essential function in any kitchen. The goal here is to clean and sanitize all the cooking and preparation utensils and equipment, with procedures that are similar to those in the dishwashing area, but on a larger scale to accommodate the largest items in the kitchen. All that, and it must still be suited to the person responsible for the pot washing, so he or she can get it done quickly and effectively. For years, how well these items were cleaned depended entirely on the work ethic of the person doing the scrubbing. As you'll see . . . not anymore!

Mechanical Pot/Pan Washers

That old standby the three-compartment sink has been a staple in commercial kitchens and has also become associated over the years with the drudgery of pot washing. Most people, even if they need a job badly, resist the idea of standing for hours over dirty, lukewarm water, scrubbing baked-on gunk off of large pots and pans, and who can blame them? This, combined with labor costs and high turnover, has prompted the invention of the ***mechanical pot/pan washer.*** It's better for the business owner, too. Depending on the menu and volume of business, mechanizing this part of the kitchen can save up to 75 percent on labor costs, and 50 percent on water and detergents.

The mechanical pot-washer is similar to a dishwasher but customized for the messy, bulky work of cleaning cookware. It takes from 4 to 15 minutes to clean, sanitize, and dry just about any pot or pan, and the machine is designed to accommodate many at a time. Pots and pans can be checked periodically

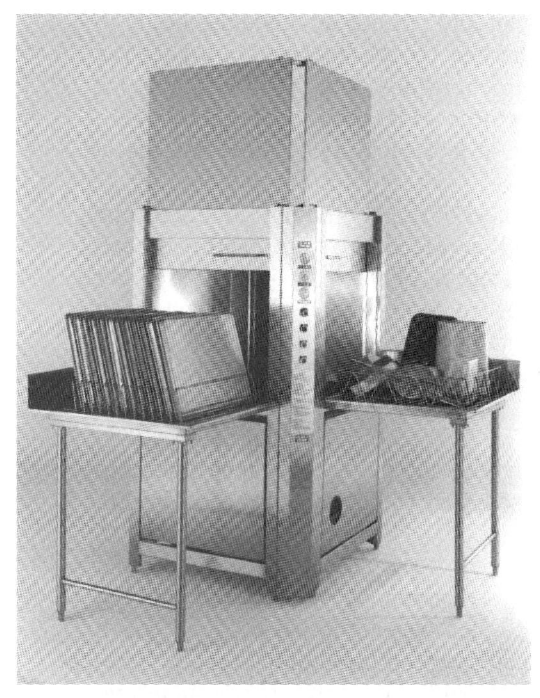

ILLUSTRATION 16-16 The mechanical pot/pan washer is also known as a pan/rack washer. It is a kitchen workhorse built to fit pots, pans, lids, and utensils.

COURTESY OF CHAMPION INDUSTRIES, ALI GROUP, WINSTON-SALEM, NORTH CAROLINA.

and taken in and out of the machine, or left in longer, as needed (see Illustration 16-16). However, NSF International standards require that the machine have a "120-second wash cycle, using at least 20 gallons of 150-degree-Fahrenheit water per square foot of rack, and a minimum of 15 seconds for the final rinse cycle, using at least $\frac{6}{10}$ of a gallon of 180 to 195 degrees Fahrenheit water per square foot of rack."

Detergent for mechanical pot-washers should be a nonfoaming type, designed to penetrate baked-on food and grease. The turbulence of the water as it circulates in the machine allows it to peel away the softened food residue.

Undercounter pot-washers are designed to save space. They can wash and sanitize sheet pans, cake pans, candy molds, machine parts, or mixing bowls of up to 40 quarts in size. An undercounter machine is about 19 inches tall, 28 inches wide at the door, and operates with a 2-horsepower pump motor. A 3-kilowatt booster heater provides the additional water heat for sanitizing. The wash water is heated by an electric immersion heater that uses 6 kilowatts per hour.

The undercounter model will accommodate six sheet pans, while counter-height machines fit ten sheet pans. This slightly larger unit may also have a shorter wash time, as fast as three minutes for a full cycle. Handy pass-through models are also available, with doors on both sides. Remember, in any installation, to allow enough room to open doors fully for loading and unloading.

The ultimate in pot/pan washing is the machine that will fit an entire rack of sheet pans. These are installed in a "pit" or used with a ramp so that multitiered racks can be rolled in for washing.

Mechanical pot/pan washers also save labor costs, because they can go virtually unattended about 75 percent of the time. Of course, some restaurants have an employee whose job is to load and unload the pots and pans. In most cases, this isn't the smartest use of that person's time.

As in any dish room, it is critical to have enough workspace to separate the clean items from the dirty ones. Loaded racks of pots and pans can also be quite heavy, so having adequate space for unloading and drying is important. It takes up to half an hour for a full rack to air-dry. A ***scrapping area,*** where as much food and grease as possible is removed by hand, is necessary as part of the incoming soiled-pan area and, if the same person does the loading and unloading, a hand sink is required for proper hand washing between tasks.

Power Sinks

There is one creative alternative to the mechanical pot/pan washer that requires more human involvement. It is a combination of high pressure and hot water in a motorized three-compartment sink known as a ***power sink*** (see Illustration 16-17) or *continuous motion sink.* A heater in the sink keeps the water hot so it doesn't have to be drained and refilled as often, which saves water. An industrial wash pump with an electric 1½- to 3-horsepower motor, also installed in the wash sink, creates a high-velocity water spray of about 115 degrees Fahrenheit. This is enough water to blast soil from most pots and pans without hand-scrubbing them. Nozzles are installed below the water level, angled to create a rolling, bubbling "hot tub" effect on the submerged pots and pans. A presoak area, a rinsing area, and a sanitizing sink make up the rest of the installation, all with gooseneck, swiveling faucets. Food waste disposer, drain boards, and shelving complete this dish room setup, so you'll need to decide whether you have the space for all of it. Power sinks do come in different configurations, from straight lines to L-shapes and U-shapes.

The difference between competing power sinks is the way water moves around in the sink—a tornado or whirlpool action, or hot-tub-like jets. The locations of the intake nozzles are different, too. Be sure to view the water flow when the sink is full of pots and pans to see if it meets your needs. The tank (sink) sizes range from 61 to 112 gallons in volume; the motor requires either single- or three-phase electricity, 208- to 240-volt and/or 480-volt. Prices range from $6,500 to $8,500 for a basic 9-foot model, up to $12,000 or $15,000 for the largest industrial models. When traditional three-compartment sinks cost between $1,600 and $2,500, it's a big expenditure—but manufacturers say with the savings in labor, water, utilities, and detergent, a power sink can easily pay for itself within a year.

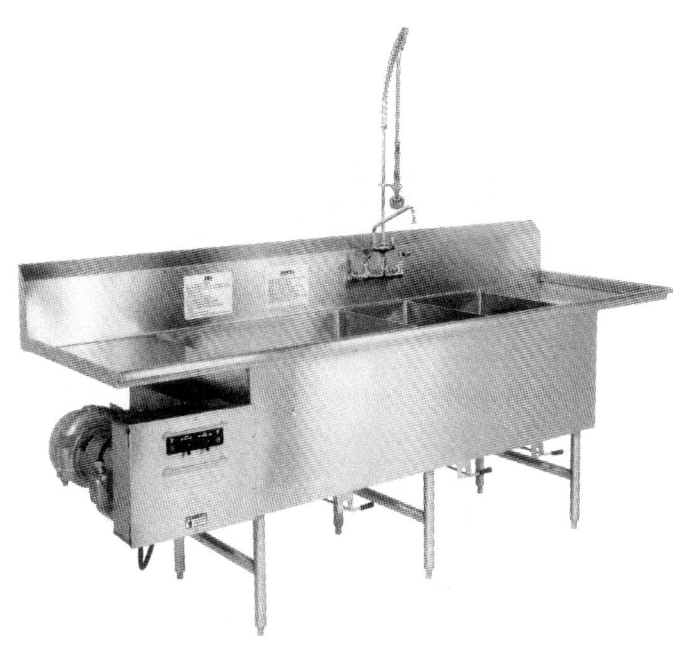

ILLUSTRATION 16-17 **A power sink uses a small motor to circulate hot water in the wash sink and "power" grime off pots and pans.**
COURTESY OF HOBART CORPORATION, TROY, OHIO.

Recent developments include power-wash units that can be retrofitted onto existing pot sinks. A small stand-alone pump mounts to the drain opening of the sink, filtering the water and sending it back churning into the sink compartment. The price of this option is about $2,000.

To summarize this chapter, dishwashing is typically one of the most expensive kitchen functions, in terms of labor and energy costs. As you shop for the right machines for your dish room, be aware of their hot-water requirements, energy use estimates, and energy-saving options.

SUMMARY

Commercial dishwashing requires hot water, detergent, and motion. The water must be hot enough to sanitize the dishes, and there are health regulations about the amount and temperature of the water, as well as the length of the wash and rinse cycles. Motion is necessary to ensure complete coverage of all the dish surfaces.

Dish rooms are designed based on available space and utility connections, proximity to service areas, building codes and health ordinances, how food is served and how dishes are handled, and the quality of both water and detergent. Your peak hours and volume of dishes are also major considerations.

Even the least sophisticated dish machine can complete a wash cycle in less than one minute; the smallest machines, under five minutes. Some dish machines require a separate tank, called a booster heater, that provides additional water heating for sanitizing during the rinse cycle.

Dishwashers are "rated" or sized by the number of full dish racks per hour they are capable of washing, but experts caution that, because most racks are not loaded perfectly, a more realistic capacity is 70 percent of what the machine's manufacturer claims. The rate of flow, the speed at which the machine replenishes and heats its water storage tank, is also an important consideration.

Dishwashers can be freestanding or undercounter, and there are specialty machines for washing glassware.

Commercial dishwashers are either single-tank or double-tank, depending on how much hot water you require. The largest volume foodservice operations have conveyor or flight-type dishwashers. Other specialty machines wash pots and pans. All dish machines contain a scrap screen or basket to collect food particles. A blower-dryer can also be added to dry dishes faster.

Heat and moisture make dish rooms a tough place to work unless your dish machines are installed under a vent and, often, ventilation is required by law. Some models have a built-in condenser, which removes moist air from the machine, dries and cools it, and releases it into the dish room.

The food waste disposer is an integral part of the dishwashing system. Not all cities allow the use of disposers, usually depending on the area's sewage system, but some cities encourage the use of disposers, because they cut down on landfill use. Where the latter is true, there are strict requirements that disposers be connected properly to water and sewer systems and not interfere with grease trap operation.

Commercial disposers can be continuous feed or batch feed. A continuous-feed machine shreds whatever is in it as soon as it is turned on; a batch-feed machine shreds only a certain amount of waste at a time. Their electrical requirements are based on their motor size.

Facilities with large waste output may also want to consider installing a waste pulper, which can mash paper and styrofoam products, aluminum, and cardboard into a watery slurry along with food. There are also smaller models, known as mini-pulpers. The advantage of a pulper is that it turns out "cleaner" waste that is easily composted. But the systems are expensive, and even pulpers can't accept all types of waste.

STUDY QUESTIONS

1. Joe's Cafeteria serves about 800 people during its daily lunch rush. According to the guidelines in this chapter, how many lunch dishes will Joe and his staff need to wash?
2. Based on the previous number of dishes, what is your recommendation for a dishwashing system for Joe's Cafeteria, and why?
3. Why is it important to vent the air released by a dishwasher?
4. When discussing a dishwasher's water tank capabilities, what is the difference between the rate of flow and the pump efficiency?
5. When shopping for a glasswasher, why is it important to find out what temperature it uses to dry glassware?
6. What three things should you consider before deciding to add a blower-dryer to your dishwashing system?
7. Are there any advantages to a circular dishwashing system over a straight-line one?
8. Calculate the dish capacity of a flight-type dishwasher that moves at 5.2 feet per minute, has 2 inches between pegs, and is 45 inches wide.
9. Why is it important for a food disposer to be able to reverse its motor?
10. What is the function of the water press in a waste-pulping machine, and why do you think it is necessary?

A Conversation with . . .

Jim Hungerford

CO-OWNER, BOISE APPLIANCE AND REFRIGERATION COMPANY
BOISE, IDAHO

Jim Hungerford is co-owner of Boise Appliance and Refrigeration Company, along with Roy McMurtry. They operate a full-service maintenance and repair firm for restaurant equipment. Mr. Hungerford has been in the foodservice equipment business since 1982.

Q: When a restaurant first opens, can you tell whether it will be able to stay in business based on the equipment decisions the owners make?

A: With any new business, you have to keep a close watch on your cash flow. What I have experienced with first-time restaurant operators is, they try to save money by purchasing used equipment when they would probably be better off to buy new. Most dealers of used equipment only warranty it for thirty days, but new equipment carries a full one-year parts and labor warranty. That alone keeps their maintenance budget down, at least for the first year.

It's extremely hard to say whether or not a restaurant will make it in today's marketplace, because there is so much competition. In my opinion, it is very hard for the mom-and-pop operation to succeed without a lot of capital and a unique specialty item or concept to interest the customer.

Q: Most chefs seem to have strong opinions about using gas or electric appliances. From a service standpoint, what do you think?

A: The local utility rates would be a factor in choosing the type of equipment, for obvious reasons, but my personal preference would be all gas equipment. It's quicker, it's easier to maintain, and, to me, gas appliances give off a more even heat.

With electrical equipment, you have a lot more components that make up the inner workings of the appliance. From a service standpoint, electrical equipment is generally more expensive to repair than gas equipment. Electrical problems can also be harder to diagnose, due to the more prevalent use of solid-state controls that contain integrated circuitry.

Q: What is the most common service problem you see with ranges and ovens?

A: Most often, it is calibration of equipment and the inexperience of the people who operate it. All new restaurant owners should familiarize themselves fully with the owner's and operator's manuals. We receive a lot of service calls simply because the owner is not familiar with how to use the equipment and since this is not a factory warranty situation, the owner is responsible for paying for the service call.

Q: Is cleanliness of the equipment a problem?

A: It's a really big problem. The french fryers are a prime example. We have had customers bring fryers in for repair that are in such bad shape, since they have never been cleaned, that it is actually cheaper for them to buy a new one than to try and repair the old one. The wiring and electrical controls can become so grease-soaked that they will either stop working or become a fire hazard.

Q: So how often should you wipe them down?

A: I believe all cooking equipment should be wiped down and cleaned on a daily basis. This should be done at the end of every workday, and there should be one person designated to perform these tasks.

I would also suggest purchasing a fryer with a built-in filter system, or at least a portable oil filter. Oil is expensive, and filtering it will save on your oil costs.

Q: What about griddles? They look indestructible, but I know there are some maintenance tips for them. What's the most common problem you see?
A: The extreme heat of the griddle can sometimes cause the side and backsplash to become separated from the cooking surface. Should this happen, you have the possibility of oil leaking into the burner assembly [on gas equipment] and into the controls (on electrical equipment].

It is also important that the correct utensils be used on the griddle surface to prevent grooves and gouges. You can have a griddle buffed, but it's a very expensive procedure.

Q: How about steam-jacketed kettles?
A: Steam-jacketed kettles are virtually maintenance-free. The most important thing to be aware of is to use distilled water only, to prevent contamination within the system.

On convection or pressure steamers, a steam generator is used to produce steam for the cooking compartments. In areas where the water has a high mineral content, it is important that the generator be opened up at least twice a year. There are service agencies that can inspect and descale the generator to remove mineral deposits. Failure to do this regularly can ultimately destroy the steam generator and might even be hazardous to employees working with it.

We suggest a water filtering system for convection steamers. I'm also a big fan of a process called reverse osmosis. It's a filtering system with three or four cartridges, a pump, and a membrane filter that purifies and stores water in a tank for the steam appliance. The appliance uses water out of the storage tank instead of tap water.

Q: And how about refrigeration units?
A: Of course, not cooling is the biggest problem. It is the restaurant's responsibility to clean the condenser coils to keep them free of grease and dirt that blocks airflow. You can keep coils cleaner if you buy a roll of thin, black foam that can be used as a filter. Cut a piece off to fit right over the condenser. Change it every so often.

I would also recommend checking with your local service company to see if there's a quarterly preventive maintenance program they offer for refrigeration equipment.

Q: If you were a restaurateur, would you designate certain people in the kitchen to take responsibility for certain pieces of equipment?
A: That is a good idea, but the biggest problem is high turnover of personnel. If you knew you were going to have the same people, that would be a great idea. I've found the cleanest places are usually the schools, nursing homes, and any government facilities. It's partly because they have more regulations, and also because their people have been there longer than in a restaurant. Some even have their own maintenance people.

Q: Let's talk about common problems you see with dishwashing systems.
A: Dishwashers are pretty simple to use. There are high-temp machines, which use 150-degree wash water and 180-degree rinse water; and low-temp machines, which use standard household-type hot water and clean and sanitize dishes with the use of chemicals.

On the high-temp machines, the most common problem is that the booster heater isn't heating the water hot enough. On low-temp machines, spotting is the most common problem. It's usually due to not adding the right amounts of wash and rinse chemicals.

Q: How about waste disposal systems?
A: The most common problem is jamming the turntable. This is usually caused by something other than food, like silverware, dropping into it. Also very often, they'll mop the floor with those stringy mops and dump the mop water into the sinks. The mop strings get wrapped around the seals, causing the unit to leak water down into the motor housing and destroying the upper and lower bearings.

Q: Now if you're a foodservice operator and you want to get a service contract from a company like yours, what should you expect? What is reasonable to expect?
A: We try to run our business so that a service call is a same-day call. If it's an emergency—if a refrigeration unit is down or something they absolutely can't get along without—we try to get somebody out there within two or three hours. But, normally, within a 24-hour period from the time you call is reasonable. Truthfully, though, the customers don't feel that way. They think you should have somebody sitting here waiting around to answer their call within ten minutes! The fast-food chains are the worst.

Q: When someone calls you for service, what do you need to know over the phone?
A: The first thing I always try to get is the brand name, and what is it doing? Is it running at all? We can fix a lot of things right over the phone. You'd be amazed at the calls we get where the appliance has been accidentally unplugged, or there's a reset button, but nobody knows to push it. If I had any advice, it would be: Just try to keep your people as informed as possible on the use and care of the equipment.

CHAPTER 17

MISCELLANEOUS KITCHEN EQUIPMENT

INTRODUCTION AND LEARNING OBJECTIVES

As the labor supply in the foodservice industry continues to tighten, you can increase the productivity of your kitchen with the food processing equipment you'll learn about in this chapter. Numerous machines have been invented to complete tasks faster or more simply, that once required far more training and skill. But with price tags that may require a year or more to recoup the initial investment, you may wonder which ones are truly worth the money.

There are hundreds of handy kitchen devices so, in this chapter, we'll confine the discussion to those machines that are in most frequent use across a broad scope of the industry. A few of the selections are new, and there are also new technological developments for the venerable workhorses. The list includes:

- Food mixers and attachments
- Food slicers, cutters, and grinders
- Blenders and juicers
- Toasters
- Food warmers
- Coffee brewers
- Espresso/cappuccino machines

You'll learn about which features are important when selecting these items—and which are not—as well as tips on determining the correct sizes for your operation, and caring for your shiny new appliances.

17-1 FOOD MIXERS

The most popular mixer in today's restaurant kitchen is the ***vertical mixer,*** also known as the *planetary mixer.* The latter term refers to its mixing action, which is like the rotation of a planet around the sun: The mixing arm (also called a beater or ***agitator***) revolves around the inside of the mixing bowl, while also rotating on its own axis. The bowl itself remains stationary, except on a few very large models. This motion provides thorough, effective mixing action.

Food mixers are identified by size, by the capacities of their mixing bowls. Typical sizes range from 5 to 80 quarts, but there are huge, 120-quart to 250-quart capacity mixers for special applications such as bagel production. The smaller units can sit on tables or countertops; the larger ones are floor models. The standard for small and medium-sized operations is the 20-quart model; larger establishments may find a 60-quart model necessary, with a back-up 20-quart mixer for occasional, additional capacity. These choices should be made by first establishing:

- What foods you will be mixing
- How much will be mixed in a single batch
- How often the mixer will be used in a day

As an example, if mashed potatoes are on your menu, assess how many batches you will need in a day, and in what time periods they must be made. Your restaurant may require all its mashers in a single, one-hour time window, each and every day; or it may be a high-volume location that uses 8-pound batches of potatoes, as many as 30 times a day. Seen this way, your answers to the questions—what, how much, how often?—are critical. The most frequent cause of mixer malfunction is overwork. Overloading the bowl past its rated capacity, over and over again, shortens the machine's lifespan and squanders your investment.

There is also a second size consideration, the size of the motor. Don't buy a mixer strictly based on its horsepower, but on the kinds of products you will be mixing. For most restaurants, a 20-quart machine with a ½ HP motor is sufficient. But for stiff doughs, like bagels or pizza, higher horsepower is recommended. The thicker the dough (more flour and less water) in the bowl, the greater stress is placed on the mixer motor. This is why it is helpful to understand the ***absorption ratio*** of the flour—that is, the weight of the liquid, divided by the weight of the flour. The dough can only hold a certain amount of water to reach its desired consistency, and this amount is expressed as a percentage of the weight of the flour. (The ideal absorption ratio is 50 percent, which means twice as much flour as water.) Lower absorption ratios mean stiffer dough, which is harder to mix.

The other caveat is to allow enough room for volume increases when mixing yeast-based dough. You can't load what amounts to 20 pounds of ingredients in a 20-pound mixer and call it good! Consider the space that the product will require in the bowl after mixing, as it rises.

The most common mixer sizes in foodservice are the 12-quart and 20-quart models, but the 5-quart mixer is the standard workhorse for any small operation. It can mix dough, make mayonnaise, and mash potatoes with a ⅙-horsepower motor. It weighs about 44 pounds empty, so it's hefty but portable. However, caution should always be taken to set it on a firm platform. This prevents an unattended mixer from "walking," a natural consequence of the mixing action in which the vibration moves it, a little at a time, until it falls off a table or countertop. Special stainless steel tables (sometimes called ***benches***) can be purchased for tabletop appliances like mixers, with locking casters, mounting holes to hold the appliance in place, pegs to store attachments, and a shelf beneath. Because a standard 20-quart mixer weighs at least 200 pounds, these rolling tables are a smart idea.

The 30-quart mixer is a floor model recommended for facilities that make bread or other dough products in large quantity. It is the smallest mixer model with a four-wheeled ***bowl dolly*** (or *bowl truck*), which permits the operator to roll the mixer bowl into position rather than lifting it. This mixer also sits above the floor on four solid legs, permitting thorough floor cleaning beneath it. At this size, motors can range from ¾ to 1½ horsepower for the toughest mixing jobs. The 30-quart mixer weighs about 330 pounds.

The largest mixer models are 40 to 80 quarts, for large kitchens, busy pizzerias, and retail bakeries. Mixers with a capacity of 30 quarts or more that use at least 200 volts of electricity can be ordered with a power bowl lift, a separate switch that raises and lowers the bowl into place. Otherwise, they are equipped with a hand-operated lift. Larger mixers can also be used with smaller bowls—you can put a 30-quart bowl on a 60-quart mixer for a smaller job, for instance—as long as you fit the mixer with a ***bowl adapter*** to hold the smaller bowl firmly in place. These super-size mixers come with motors of up to 2½ horsepower.

How Mixers Work

ILLUSTRATION 17–1 **A vertical mixer.**
COURTESY OF HOBART CORPORATION, TROY, OHIO.

All vertical mixers work basically the same way (see Illustration 17-1). The mixing mechanism is contained in a horizontal housing that is parallel to the table or floor. The power hub for mixing attachments is located on the front side of the housing. The motor is located in or near this upper housing; the housing is supported by a vertical column, mounted on a flat base or heavy feet, which also contains the lift mechanism that raises and lowers the bowl to the agitator. (The smallest mixers may not have a lift mechanism; you just unlock the upper housing and tilt it back to disengage the agitator and bowl.)

Attached to the base of the vertical column is the sliding yoke or arm that holds the mixing bowl. Arm and bowl are made to fit snugly together, and the bowl is stabilized by a flat wheel or lever or by an electric drive in the largest mixers. Some mixers allow a variety of positions for their bowls; others, only "up" or "down." Mixing bowls have flat bottoms to help stabilize them during the mixing process.

The mixing mechanism consists of a vertical shaft of heavy steel connected to the gear drive. The planetary action allows full contact with all the contents of the bowl during every revolution of the shaft and agitators. The agitators fit onto the mixer with a "push up and twist" motion that engages locking pins, the same as the beaters on your mixer at home. They're made of aluminum or (more expensive) stainless steel.

Electrical controls operate the motor and transmission, which drive the agitator at different speeds. Most mixers have three speeds, but some have four. Speed selection is controlled by a simple lever that shifts gears. In simple terms, mixers have transmissions that are either gear-driven or belt-driven:

- **Gear-driven machines come with three or four fixed speeds, preset at the factory. They must be stopped in order to change speeds, which is a key operating point. Generally speaking, the higher the horsepower of the gear-driven motor, the more likely it is to successfully mix heavy ingredients.**
- **Belt-driven machines are also called *variable speed mixers,* because this type of motor allows you to increase the speed while it is operating.**

The gear-driven mixer will usually cost more than a belt-driven mixer, because it provides consistent results with longer life and fewer maintenance problems. The belt-driven machines can, over time, experience belt slippage and require lubrication, especially if they are overworked. For safety reasons, be sure the on-off controls on your mixer are large and plainly labeled. NSF International has additional safety standards for mixer construction.

Technology improvements allow the mixer to do double-duty with the attachment of other tools to a hub on the motor called the Power Take-Off, or PTO. Vegetable slicers and meat grinders are just a couple of handy assemblies that may be attached and removed as needed.

The smaller mixer sizes (5-, 12-, 20-, and 30-quart) can operate on 110/114V electrical current. Models that are 40 quarts and larger will require 112/120V, three-phase 220V, 208/240V and even 230/460V for the largest, industrial-sized models.

In an industry where change is the norm, the mixer is one machine built to last with a little simple maintenance. Wear and tear on a mixer is determined by the number of batches and types of product being mixed. Liquids, of course, are easier on the machine than solids. If dough is what you'll be mixing most, regular cleaning is a must to prevent damaging buildup of ingredients. Also remember these important maintenance tips:

- **Use the attachments for the purposes they were designed. If a paddle (used to mix batter and pudding) is used for something as thick as bread dough, the motor will work harder and the resulting bread will be inadequately kneaded by the wrong attachment.**

- Never overload the mixer. It makes the mixing task messier and more difficult, and could cause the motor to burn out.
- Never bypass the safety interlocks. These are magnetic locks that prevent the machine from running if the bowl is not in its proper position. Running the mixer without these locks can seriously injure the operator.
- Never change gears while the mixer is running. Doing so will strip the gears.

Attachments and Accessories

In addition to mixers, you will need the attachments (agitators) that go with them. There's an attachment for whatever you want to do, from whipping marshmallow to cutting shortening into dough to chopping cabbage. The "standard" accessories include bowls and a variety of aluminum agitators, including beaters (also called *paddles*), whips (wire whisks, also called *balloon whips*), and dough hooks (also called a *dough arm*). These are seen in Illustration 17-2, and they allow you to adapt the mixer for multiple tasks. Different sizes of mixing bowls may be used on most mixers, and these will require attachments that work with the particular size of bowl.

Bowl guards are heavy wire shields that fit over the bowls during mixing to prevent injury to workers' hands. Once an optional accessory, the bowl guard is now a required OSHA safety feature, and mixers are designed so that they cannot operate unless the guard is in place. (Older models can be retrofitted to comply with this requirement.) Most manufacturers offer a chute that can be used, even with the guard in place, to add ingredients during mixing. Lids and splash covers are also available.

To keep food inside the mixing bowl at a certain temperature, use a ***bowl jacket.*** The bowl fits inside it, and the jacket can be filled with ice or hot water as needed.

With its sturdy motor, the mixer is almost like a power tool. As we've mentioned, most manufacturers offer other attachments that use the motor to accomplish different kitchen tasks—such as slicing, dicing, and chopping—to increase its versatility. There are also speed-drive attachments that more than triple the speed of the vegetable-slicing function. These extras can double the utility of the mixer, as long as you remember they also double wear and tear on the motor.

As you can now see, it's a mistake to purchase a more moderately priced mixer designed for home use. They are very similar, but the commercial mixer has greater horsepower and is designed to meet the rigors of foodservice use.

Mixer Variations

The chief advantage of the *hand-held mixer* is its portability. It saves the time and mess of ladling ingredients from a pot into a mixer or food processor to be beaten. The hand-held mixer can be fitted with different agitator attachments for beating, stirring, and blending. Hand-held mixer sizes reflect the capacity of the container they are able to mix, from 1 pint to 20 or 30 quarts. The agitator shaft can be from 10 to 40 inches in length. These versatile tools will last indefinitely if you remember never to put

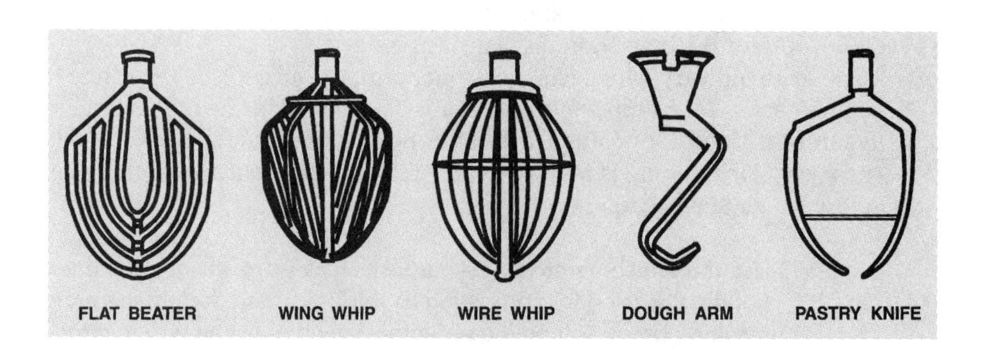

ILLUSTRATION 17-2 **There's a different attachment for any mixing job. Here are a few of the most popular ones.**

them in the pot sink or dishwasher, but to clean them individually by hand, keeping the motor dry.

The ***spiral mixer*** is designed to provide thorough kneading and smooth, evenly mixed dough in less time than the planetary mixer. It works gently and does not significantly increase the temperature of the dough during mixing, which helps to prevent early fermentation of the yeast.

It has a dual timer, which allows the operator to program the machine to shift automatically from one speed to the other during the kneading process. Spiral mixer capacities range from 120 to 250 quarts, but they can mix batches as small as 4 pounds if necessary. Their motors range in size from 7½ to 10 horsepower.

Originally, the ***vertical cutter mixer*** came to the United States from Europe. The "VCM" is almost a combination mixer and blender, a mixing bowl with a motor mounted at the bottom and a shaft that projects up from it. Removable sleeves with cutting or mixing blades can be fitted onto the shaft (see Illustration 17-3). Its advantage is its high mixing speed. Dough or batter can be thoroughly agitated in about 30 seconds; a bowl filled with water and a dozen heads of iceberg lettuce becomes perfectly chopped salad greens in less than five seconds! VCMs can be used to chop meats or vegetables, make slaw, mash potatoes, mix salad dressings, and stir cake batters. A dough-kneading sleeve can also be ordered.

ILLUSTRATION 17-3 **The vertical cutter mixer (VCM) can chop, puree, and knead dough as well as mix.**
COURTESY OF HOBART CORPORATION, TROY, OHIO.

The bowl is attached to the machine, but has a pouring lip and tilts a full 90 degrees to pour out its contents. Clamp-down lids are transparent, or there is a viewing portal in them, to allow the operator to safely check on the contents of the bowl during processing. A rotating lever on top turns a baffle inside the bowl that scoops food from the edges toward the blade. The smallest VCMs stand about 2 feet tall; the largest floor models, 4 to 5 feet tall.

Power requirements for their two-speed motors vary depending on the size of the VCM. The 4- to 6-quart tabletop models have 1-HP motors; 30- to 45-quart models have 5-HP motors; the largest units (up to 130 quarts) are driven by 25-HP motors!

The last two items on the "mixer variations" list are not really mixers, but we're putting them here because they involve mixing action for different purposes. The ***vacuum tumbler*** is a motorized machine that gently kneads pieces of meat with a marinade that contains salt. The salt extracts the proteins in the meat to form a natural coating that, when the meat is cooked, seals in its juices. Because the coating is evenly distributed, the meat surface also cooks with a nice, even color. In addition to producing juicy, uniformly cooked products, vacuum tumblers can increase the yield of most meats.

Food-borne illness outbreaks have created new interest in the *salad greens washer* and *salad spinner*. Greens washers immerse produce in water, with gentle agitation that loosens dirt and grit from hard-to-clean items like lettuce and spinach. Some greens washers inject small amounts of ozone gas into the water, which works as a natural sanitizer. Then the greens are transferred to a salad spinner to remove the excess water. You've probably seen hand-cranked spinners for home use, but motorized spinners are required for volume drying. Look for a salad spinner that has two speeds, since more delicate greens (spinach) dry better at slower speeds, while hardier items (romaine) can be dried at higher speeds. Washed-and-spun greens that are promptly bagged and sealed will last up to five days in the refrigerator.

17-2 FOOD SLICERS

The ***food slicer*** is intended to replace the old chef's knife and the seemingly endless chore of slicing meats and cheeses. Its use in portion control is also universally acknowledged. Because it can cut a variety of foods more accurately than a knife, the results are uniform thickness and maximum output. Illustration 17-4 shows the major components of this important machine.

Choosing a slicer is a lot like choosing a mixer—you have to think about what foods you will be slicing and how much the machine will be in use. A medium-duty slicer with a 9- or 10-inch blade will

ILLUSTRATION 17-4 Use of a slicer cart can help control portion sizes and reduce waste.
COURTESY OF HOBART CORPORATION, TROY, OHIO.

be sufficient for two or three hours a day, but if you'll be slicing for more than four hours a day, or if you'll be slicing a lot of cheese—which is notoriously tough on slicers—a high-volume machine with a 12-inch blade or larger should be considered.

The next decision is whether your machine should be manual (hand-operated, slicing to order), or automatic (capable of slicing large batches without hands-on supervision).

Slicers are usually identified by the diameter of the cutting knife or blade, and the size does impact the machine's efficiency. The most popular are the 10- and 12-inch disk-shaped blades. A typical slicer can hold products that range from $7\frac{1}{2}$ to 12 inches in diameter in its carriage. The food rests in the carriage area, where the operator holds it firmly in place with a handle as the ***feed grip*** gently pushes the food toward the spinning blade for slicing; or the food is loaded into a chute and gravity-fed toward the blade. Automated models have a carriage drive with two speeds, low (36 strokes per second) and high (51 strokes per second). It works without an operator, but should still be checked periodically, because it continues the slicing motions even when it runs out of food.

Slicers can be adjusted for various thicknesses of finished product, from a paper-thin $\frac{1}{32}$ of an inch to $1\frac{3}{4}$ inches thick. When not in use, setting the slicer's gauge plate (or *index knob*) at "zero" is an important safety precaution. Slicer accidents are common in foodservice—so much that some manufacturers have now made it impossible to detach the carriage from the slicer unless the gauge plate is set at "zero." A *center plate interlock* is a feature that prevents the slicer from being operated when the carriage is removed for cleaning. The *blade ring safety guard* is a permanent edge at the back of the blade that protects anyone handling the blade during operation and cleaning. These thin metal guard plates shield the top and back edges of the blade, and the feed grip is positioned so it cannot strike the blade or guards.

Any heavy-duty slicer should have a $\frac{1}{2}$ horsepower motor. You will find gravity-feed slicers with standard $\frac{1}{3}$ HP motors, but you'll get better results with the slightly larger motor. There are manual, two-speed automatic, and variable speed automatic models. A *no-volt release* prevents unintentional start-up of the slicer. After a power outage, or if it is accidentally unplugged, machines with no-volt release protection must be restarted.

Food slicers should be made of anodized aluminum or stainless steel, and designed with maximum sanitation in mind. Look for seamless surfaces, rounded corners, and corrugated feed grips to prevent foods and liquids from lodging in crevices. Slicer blades are made of stainless steel or carbon steel, and there are ongoing debates about the merits of each and whether (like other knives) they should be cast or stamped. The Food Service Technology Center in San Ramon, California—which we've mentioned elsewhere in this text for its state-of-the-art appliance-testing research—has found that both substances and both manufacturing methods are equally well suited for the job. Some manufacturers tout chrome-plated blades, but the sharpening process will soon wear the chrome off the cutting edge anyway.

In addition to the standard models, there are a few specialty slicers. If it is bread you want to slice, consider a specially designed bread slicer. A typical food slicer will do the job, but with too much waste. There are special slicers to core tomatoes, make French fries, and slice and stack large quantities of lunchmeats. Some can even process two foods at once, then layer them alternately for commercial sandwich making.

Everyone seems to dread slicing cheese because, as it reaches room temperature, it sheds moisture and becomes gummy, making the slicing process sloppy and wasteful. Keep cheeses and meats refrigerated until just before slicing to minimize this problem.

Slicer Safety and Maintenance

We've already mentioned the built-in safety features of food slicers, and there are plenty of other slicer-related safety rules. The first is that no one under age 18 is legally allowed to operate a slicer.

Slicer operators should use clean, well-fitting, cut-resistant gloves, especially to clean the machine; they should also avoid wearing loose-fitting clothing or jewelry that could become caught in the slicer.

Slicers should never be used to cut solidly frozen products. That's a job for a meat saw. Plastic or paper wrapping must be removed from foods before slicing, and never attempt to cut through a tin can or container of food using a slicer.

If the food is not being cut satisfactorily, the most common culprit is a dull blade. Built-in blade sharpeners, stored under the slicer base, are standard features and there are several guidelines for correct sharpening:

- The slicer has two types of stones, a sharpening stone and a honing stone. The latter is to smooth out any burrs that develop on the blade during sharpening.
- The blade should always be completely clean and dry before it is sharpened, and should be wiped clean of debris after the sharpening process.
- Avoid over-sharpening—a few seconds is sufficient. Longer sharpening will reduce the life of the blade.
- When you replace the blade, replace the sharpening and honing stones. Old stones can damage new blades.

Inexperience makes slicing dangerous, and well over half of the slicer-related accidents occur when the machines are being cleaned. This is why the manufacturer's safety features and guidelines are so important. The entire slicer must be cleaned and sanitized daily, and the blade should be cleaned frequently during the day. Blades must also be lubricated with food-grade oil, never vegetable oil. The motor must also occasionally be oiled, which is a job for a service technician.

Aluminum parts should be cleaned by hand, never with bleach or in a dish machine. One innovative idea is the BladeRunner, a chunk of material saturated with quaternary ammonia, which cleans and sanitizes the blade as it is sliced like a food product and then discarded. KatchAll® Industries International makes the BladeRunner.

Finally, cross-contamination must be prevented in slicer use by cleaning or switching blades between slicing cooked and raw products; by slicing meats and cheeses while they are cold, and returning unsliced or unused portions to refrigeration immediately.

17-3 FOOD PROCESSING

The food processor was introduced to the European market in the 1950s by the French company Robot Coupe (pronounced "ROH-boh koop") and was marketed in the U.S. by a trio of investors who saw it at work in a New Orleans restaurant and formed Robot Coupe U.S.A. in Jackson, Mississippi, still a dominant player in the market. Cuisinart (known for creating processors for home use) and Waring (famous for its blenders) are other major manufacturers.

Food processing is the term for changing the shape, size, and/or consistency of food. The equipment is versatile, user-friendly, and surprisingly safe considering the inherent hazards of chopping, grating, etc. There are two basic types of food processors—the continuous feed and the bowl style—as well as a third type of machine that can do both.

The *continuous-feed processor* has different-shaped openings for different types of cuts, such as strips, shreds, coins, crinkles, and juliennes. Food is fed through a chute at the top and sliced by a blade. There are as many as 20 interchangeable blades (sometimes called ***plates***) available; a few of them are shown as Illustration 17-5 and there are specialty blades for grating nuts, spices, chocolate, bread crumbs, etc. Be sure your equipment package includes a ***hard-products slicer,*** a blade for processing foods such as carrots, potatoes, and cheese. Also, notice whether the design of the machine allows you to easily catch the outgoing food in standard-sized pans. The continuous-feed processor is designed to expel finished product through a chute and keep working as new food is fed into it. It is the best choice when a consistent cut or high-volume output is needed.

The *bowl-style processor* has a removable, rotating bowl and a two-blade assembly that rotates at the bottom of the bowl to do the cutting (see Illustration 17-6). Different types of assemblies produce the

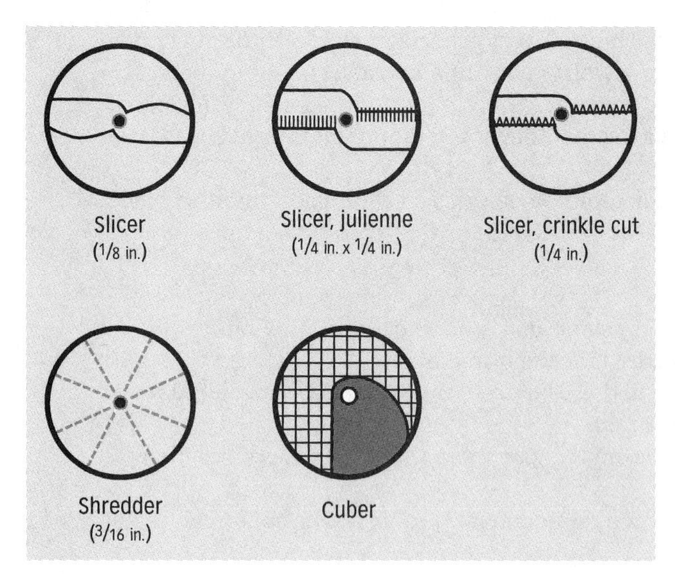

ILLUSTRATION 17-5 Food processor plates, or blades, produce different sizes and shapes of product.

different cuts. Handy accessories include a dicer, a meat chopper, and a scraper that can be operated from outside the bowl as it rotates. With this type of processor, you fill the bowl, process the food, empty the bowl, and refill it. The advantage of the bowl-style processor is that it cuts quickly and without pressure, so delicate items like parsley can be minced without becoming mushy. The rotating action also helps mix the food as it's being cut. As a safety precaution, bowl-style machines cannot be turned on unless a bowl cover is locked in place, and the cover cannot be raised while the machine is turned on.

Combination food processors have a single-motor base, which can be used with a continuous feed attachment or a bowl for cutting and mixing. They work well for kitchens that require maximum versatility to prep a wide variety of foods.

Like mixers, food processors require an assortment of bowls. Some are tall and narrow (to minimize spattering) and others are wide and short (for faster processing). Purchase your bowls, like the processor itself, based on what foods they will be used to process.

Food processors operate with motors of ½ to 1 horsepower and a single on-off switch. Like other small appliances, there are belt-driven and gear-driven motors. Electrical requirements vary from 115-volt, single-phase power to 460-volt, three-phase power. A wise option for the bowl-style processor is a "pulse" function, which prevents overheating when processing large chunks of food.

How quickly do processors work? Their blades whir at more than 1700 revolutions per minute! The output for two popular sizes is shown in Table 17-1. A standard ½ HP combination-style processor is capable of handling 350 pounds of product in an hour, depending on the product and the type of cut. The best way to ensure maximum output is to outfit the machine with large feed attachments and cutting plates with large surface areas.

The most difficult food item for any processor is cheese. The job will be much easier if the cheese is refrigerated until processing, and dusted lightly with cornstarch to prevent sticking. If your processor will be used primarily for grinding cheese, consider buying one with a slightly larger motor.

ILLUSTRATION 17-6 A bowl-style food processor.

COURTESY OF HOBART CORPORATION, TROY, OHIO.

Processor Safety and Maintenance

Employees must be instructed to keep their fingers out of the hoppers and discharge chutes! There are tools to push product into the processor, called ***stompers*** or *pushers,* which should always be used instead of hands! If anything must be cleared from inside the machine during use, it is not enough to simply turn it off and reach in. You must also unplug it, so you don't accidentally lean over and trip the on/off switch with your hand in the bowl.

You'll get maximum useful life from your processor if it is cleaned after each use with a nylon brush that is able to reach into the crevices of the machine without damaging the seals on the motor. Stainless steel, aluminum, and plastic parts can be wiped clean with a damp cloth and the manufacturer's suggested sanitizing solution. The removable parts—cover, discharge chute, bowls, etc.—can be washed in hot, soapy water, rinsed, and air-dried. The processor base that contains the motor must never be submerged in water.

TABLE 17–1

FOOD PROCESSING TIMES

Food Product	4-Quart Processor	6-Quart Processor
Minced vegetables	2 pounds/20 seconds	3 pounds/20 seconds
Soft cheese	2 pounds/20 seconds	3 pounds/20 seconds
Ground meat (raw)	2 pounds/60 seconds	3 pounds/60 seconds
Ground meat (cooked)	3 pounds/60 seconds	4 pounds/60 seconds

Source: Hobart Corporation, Troy, Ohio.

Be sure the plates are easy to change, sharpen, and sanitize. They should never be run through a dish machine, but hand-washed and allowed to air-dry. Drying them by hand only invites an injury. Also never leave a blade (plate) submerged in a sink, as an unsuspecting person could reach in and cut themselves. Manufacturers make special plate storage racks, and these should be used.

Manual Food Processing

A final word about food processors: Non-motorized units with simple hand cranks are also available. These are an effective and inexpensive option.

The names *food chopper* and *food grinder* are used interchangeably, because most of these machines do both, depending on the attachment they are fitted with (see Illustration 17-7). The chopper/grinder is a simple machine with a single-pole toggle switch that controls a push-button on-off function.

ILLUSTRATION 17–7 **A food chopper/grinder is used primarily for meats.**

COURTESY OF HOBART CORPORATION, TROY, OHIO.

A rubber cap typically protects the push-button assembly. Larger machines may have a magnetic starter with an automatic reset button that will shut off if the unit becomes jammed, overheated, or overloaded.

The operator loads meat into the machine by pushing on a hard plastic *stomper* or *pusher*; plastic is a better choice than wood, because it can be sanitized. Stainless steel knives or plates inside the machine perform the chopping functions, and the same pressure that pushes the food into the machine forces the ground meat out through another opening. It can emerge into a pan or a sausage casing attached to the machine. The diameter of the output opening can be adjusted by hand. Electrical requirements for chopper/grinders range from 115-volt, single phase, to 460-volt, three phase.

There are also manually operated *vegetable cutters,* designed for individual processing of items like "blooming onions." Some can core, wedge, and dice; others can perform one simple but otherwise time-consuming task, like slicing cabbage or tomatoes, coring apples, or cutting potatoes into French fries. Most manual cutters are tabletop units that can be solidly clamped to a prep counter or wall-mounted. They are good choices for small operations, or when the budget prohibits purchase of motorized processors.

In caring for these items, remember that their metal parts are susceptible to food acids, so they must be hand-washed frequently to keep their sharpness, and always air-dried.

Blenders

An inventor named Fred Osius patented the first blending machine in the U.S. in 1933, but it was big band leader Fred Waring who made it famous. Osius knew that Waring had been an engineering student who loved gadgets, so he took his blender to a concert, talked the security crew into letting him backstage, and convinced Waring to back further research for the new machine. It was first marketed as the Miracle Mixer, and finally the Waring Blendor® (with an "o").

High-capacity, heavy-duty blenders—the kind that can crush ice cubes in seconds and blend them into drinks—didn't come onto the commercial foodservice scene until the late 1980s. Depending on your menu and theme, you may need heavy-duty blenders and/or bar mixers. The *blender* can perform most light food-processing duties, with multiple settings for speed, power, and so on. It can chop, puree, mix sauces, liquefy, make milkshakes, and crush ice into slush for frozen drinks like margaritas. The ***bar mixer*** is designed to work primarily with ice, liquids, and solids, and is used strictly for making drinks at bars and nightclubs. It is quicker and usually quieter than a blender. Some bar mixers have their own integrated ice bins that dispense exact amounts into the container, eliminating the bartender's need to measure it by hand.

Your equipment choices in this category depend on answers to a number of questions:

- How much blending must be done in a day's work?
- How quickly do you need to produce a drink?
- How much space do you have?
- Where do you want to locate the machine?
- How much noise are your customers and staff members willing to endure?
- And finally, what will you be blending—frozen or cooked items, especially fragile, hard, or chunky ingredients, etc.?

Both commercial blenders and bar mixers are larger than those made for home use, and built to handle about 100 blending cycles per day. They are identified by the capacity of their containers, from 1 quart to 5 gallons. They rest on a heavy, die-cast base with rubber feet for added stability. Motor sizes range from ½ to 3 horsepower. The strongest ones are used for juicing vegetables.

Most commercial blenders have two speeds, but some offer variable-speed control or a pulse switch for intermittent higher-speed uses. Frankly, its power is not as important as your ability to control it. If you can't select the blade speed (measured in RPMs) and the timing for each task, consistent results are not possible. Digital technology has made the modern blender extremely user-friendly—highly programmable, with sensors that monitor and modify blade action and overall designs that are engineered for uniformly blended output. Some manufacturers will even program the electronic chips in their highest-end blenders specifically for certain bars' signature drink recipes.

Inside the blender, a toggle switch activates the motor, which turns a set of blades attached to the clutch. The clutch, which you'll see when the container is removed, takes a lot of abuse—particularly with a lot of start-and-stop action—so it's constructed of steel or flexible neoprene rubber. There are 4- and 6-blade assemblies, easily removable for cleaning. Different manufacturers offer different angles, lengths, and shapes of blades for different purposes.

The container where blending is done is most often made of stainless steel, but glass or polycarbonate materials are also used. (Polycarbonate is the material used to make aircraft windows, tough but translucent.) The only trouble with poly containers is that they can't be cleaned in dish machines. Stainless steel retains heat and is used for hot soups and sauces; glass is rarely used in foodservice because of the risk of breakage. Containers can be ordered with one or two handles, whatever makes lifting easier. The container is usually covered with a lid made of rubber or vinyl, often a two-piece ***filler cap*** so ingredients can be added during the process. Container capacities range from 1 quart to 5 gallons. The 1-quart model is sufficient for as few as a dozen frozen drinks per hour.

If the blender is not already programmable, a timer is a worthwhile option, especially for bar mixers. It frees the operator to handle other tasks and limits wear and tear on the appliance from unnecessary starts and stops.

In smaller restaurants, a ***spindle blender*** performs the mixing functions, from making milkshakes and sauces to whipping cream to adding ingredients to frozen yogurt. Ice cream parlors couldn't function without them! They come in countertop and hand-held models, with a single mixing head or up to five heads to make five drinks simultaneously (see Illustration 17-8). Some have three height settings to accommodate different sizes of stainless steel mixing cups.

Even when you turn the unit on, the blender head does not start turning until an activator at its tip makes contact with the mixing cup. Cup vents, usually made of stainless steel, hold the cups in place during blending.

Spindle blenders with multiple heads have a separate small head for each motor, which means one can be serviced without affecting the others. It's important to check the motors periodically, because they have transmissions that require oil. The manufacturer's manual will provide recommendations.

Counter space with a water source and/or drain nearby is necessary for a multiple-head spindle blender. If space is tight, a single-head, hand-held model will be fine. Longer rods are available for use with deeper pots and kettles.

The body or casing that houses the blender motor has sealed seams to protect it and simplify cleaning. Blenders need to be kept clean and, oddly enough, protected from unnecessary moisture. This is because most units have a cooling fan in the base; water can get sucked into it and corrode internal parts. Some models have recessed cooling vents to help prevent this. Syrups and other sticky ingredients can also clog toggle switches and buttons, so it's smart to spray the machine with an alcohol-based cleaner at the end of the day.

Noise is a significant issue with all the grinding and glugging noises that busy blenders make. Most manufacturers have crafted some type of encasement of polycarbonate or acrylic, to fit around the blender and allow access through a hinged door. The blender may or may not shut off automatically whenever the door is opened, but it's a nice feature. While no one can eliminate blender noise altogether, the goal is to reduce the noise level to below 72 decibels, the standard loudness of a normal two-person conversation.

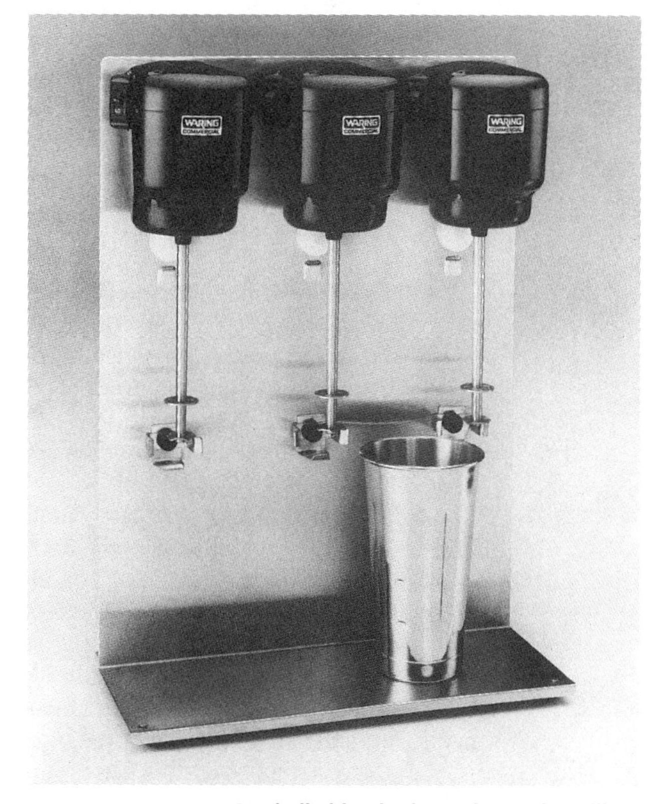

ILLUSTRATION 17-8 **A spindle blender is used to make milkshakes at soda fountains.**

COURTESY OF WARING PRODUCTS DIVISION, NEW HARTFORD, CONNECTICUT.

Juicers

The *juicer* or *juice extractor* is a cousin to both the blender and food processor. Nothing quite compares to the olfactory and visual appeal of offering customers a freshly-squeezed glass of orange juice with breakfast, at the health club, or at the bar as part of a tropical cocktail. You can choose from manual-feed models, or hoppers that can be loaded with oranges that roll automatically into an extractor. The machine, with a motor from ½ to ¾ HP, crushes the fruit, separates the skin and pulp from the juice, and dispenses the juice into an individual glass (smallest models) or a chilled reservoir (largest models). Of course, the fruit must always be washed before being loaded into the machine, since its external skin comes into contact with the juice during the mechanical squeezing process.

These sturdy machines are made of stainless steel and aluminum. The unit itself can't be submerged in water, but most parts are removable for cleaning—which can be a sticky job! Most juicers are appropriate for a variety of fruits and vegetables, although they are most often used to juice citrus fruits. Prices for commercial-model juicers range from $2000 to about $7000 for the largest, standalone machines.

17-4 TOASTERS

The incredible variety of breads now available in foodservice require choosing toasters that provide versatility and ideal conditions for toasting each type of product. Do your customers expect their bagels toasted on one side, or both? How much toasted bread is your kitchen expected to turn out on a busy morning? As with other appliances, the toasters you choose will depend on the types and sizes of food products being toasted, the volume of business, speed of service, amount of labor available to you, and how much space you have available.

This is another appliance that most surely requires a commercial model instead of those designed for home use. Just one difference says it all: the home toaster weighs up to 3 pounds; the commercial four-slot toaster weighs between 18 and 20 pounds. Technology has greatly improved commercial toasters, adding innovations like solid-state controls that allow easy time and temperature adjustments, inch-thick insulation to keep the appliance and kitchen cooler, and power-saving switches with a standby mode. There are both electric and gas-powered toasters, but since the gas-powered models must be vented, most users specify electric ones for ease of installation.

In toasting any food, you have to think about the process and the product. It is a three-step process:

- First, you remove the moisture from the product with heat. (That is why day-old bread makes better toast. It's already drier, doesn't stick in slots or on conveyors, etc.) It does not take a lot of heat to evaporate moisture, but it does take time.
- Second, you cook the product throughout, so it is warm and crisp.
- Third, you add the color. In a continuous (conveyor) toaster, this is done by boosting the wattage at the exit end of the conveyor belt, for a nice final browning.

The two basic toaster types are batch toasters and continuous toasters. *Batch toasters* are often the familiar slot variety, which lowers the bread into slots between heating coils. The finished toast pops out automatically or is manually lifted in and out of the slots. The slot or "pop-up" toaster can be ordered with wider slots to accommodate "Texas Toast," bagels, and other, thicker types of bread products. Its downside is that someone must select the degree of doneness, then stand there and wait for the bread to pop up, butter and serve it while still warm. The pop-up toaster is also available in two- or four-slot models. Some four-slot models have longer, wider (1½-inch) slots to toast burger buns, with heating elements that toast only one side of the bread. These produce buns for more than 90 burgers an hour.

Commercial batch toasters can produce 125 to 190 slices of bread per hour (2-slot models) and up to 250 slices of bread an hour (4-slot models). The manual-lift models have footprints of about 9 by 14 inches. There are also single-slot pop-up toasters. Individually, their output may not seem impressive, but banks of these models are capable of large production and are built for long, heavy use.

There's also a pop-down toaster, where bread is dropped into the slot and comes out toasted, resting against a door at the bottom of the toaster. They require more vertical space than pop-up models, since they're taller, but they can be reloaded more quickly. Both pop-up and pop-down styles are appropriate in situations where the servers, or even the guests in self-serve settings, make the toast.

Heating elements toast the bread, and there are two types of toaster heating elements: ***quartz sheath*** and ***metal sheath.*** Proponents of the quartz elements claim that quartz radiates heat better than other substances and indeed, a good one can go from cold to "full toasting" mode in 30 seconds. The downside: quartz is brittle and can break if the appliance is dropped or handled roughly, or even during cleaning.

Others are metal-sheathed elements, which manufacturers claim can remove excess moisture from the bread as it browns. The metal sheath may take 15 minutes or longer to reach "full toast" mode. However, once it has heated up, it holds its heat longer and is much more sturdy overall.

You can order toasters with different numbers of heating elements, from two to seven of them. No matter what type or number you choose, an internal temperature monitor is an important feature of a good toaster. This is because after several consecutive uses, the toaster is already hot and, therefore, starts the next batch of toast at a higher temperature. Without an internal monitor, a toaster may pop the bread up too soon, only slightly toasted.

The other popular type of batch toaster is a ***drawer-style toaster,*** with a drawer or platen onto which bread is placed to slide into a heated chamber for toasting. In drawer-style toasters, there are two types of heat: contact and radiant. Contact toasters work like griddles. The item to be toasted is placed on the heated surface, and a lid or weight holds it flat. This is a perfect way to toast hamburger and hot dog buns. Radiant-heat toasters look like cheesemelters or home-use toaster ovens. A drawer or tray is loaded with food and slides under a heating element that toasts the top surface of the food. This type of toaster is commonly used for heating open-faced sandwiches, a process that takes from 2½ to 5 minutes. Its footprint is about 10 by 20 inches.

The ***continuous toaster*** (or *conveyor toaster*) is a high-volume appliance (see Illustration 17-9) for toast output of 150 slices or more per hour. Some operators use a regular conveyor oven for this job, since it can bake cookies, pizzas, and so on in addition to making toast. Food is placed on a conveyor belt at the front of the toaster. Color and doneness are controlled by varying the heat, the conveyor speed, or both. In fact, many operators prefer a single control for both, because it's difficult to set separate controls to work together. Most conveyor toasters also feature a switch that allows you to turn off one of the heating elements and toast food on one side instead of two. At the end of the conveyor belt, the food slides from the belt into a chute that empties into a collection tray. The biggest conveyor toasters have multiple belts running side by side, each with its own heating elements and controls.

There are both horizontal conveyors and vertical conveyors. The horizontal conveyor is the best choice for extremely high volume—more than 500 slices per hour. It has a slightly larger footprint than the vertical conveyor, but you can stack more than one horizontal conveyor toaster on top of each other and run them simultaneously.

Vertical conveyor toasters feature longer cooking chambers and are known for producing excellent quality product, but they cannot be stacked. The "vertical" designation means that bread, for example, is loaded into baskets or pockets at the top of the appliance. The conveyor revolves up, over the top, and down the back of the toaster, and the finished toast is dropped onto a tray at the bottom. The available space may determine whether your operation uses a vertical or horizontal machine. The vertical models have a 22-by-15-inch footprint and stand almost 3 feet tall. Horizontal models have an 18-by-12½-inch footprint and are only 14 inches in height. Vertical toasters usually cost more than horizontal ones.

ILLUSTRATION 17-9 A conveyor toaster is a type of conveyor oven, designed for high-volume output.
COURTESY OF TOASTMASTER, A MIDDLEBY COMPANY, ELGIN, ILLINOIS.

Toasters of all types require single-phase electricity, ranging from 120 to 240 volts. Select yours based on the volume you'll need at peak production times, and remember that, no matter what the manufacturer's claims, the output of a toaster depends on several factors: the density and moisture content of the product being toasted and whether an attendant will be present to keep it fully and promptly loaded. Because toasters are used to make everything from toast to frozen waffles, the figures will vary. Some common "toaster bragging rights"—and our own real-life estimates—are as follows:

- **Four-Slot Toaster.** Listed from 200 to 360 slices an hour, an average of 150 to 180 slices is more realistic, and only if an employee attends the machine.
- **Pop-Down Toaster.** Listed from 240 to 420 slices an hour. Again, an average of 150 to 180 slices is more realistic.
- **Batch Toaster.** Fully loaded with 12 burger bun halves per batch, manufacturers suggest figures of 200 to 300 bun halves per hour, enough to make 100 to 150 burgers. In actual operation, these figures are 50 to 75 percent attainable.
- **Conveyor Toaster.** Bun production is rated higher than toast output from the same machine, and the range is from 150 to 1200 slices per hour. Real-life figures depend on the density and moisture of the product, as well as the skill of the operator in loading and unloading the machine.

There are specialty toasters for buns and bagels. The bagel toasters are conveyor models, with a higher clearance (2-inch is standard, so these are wider to allow "taller" bagels to pass through). They run higher wattages and have a different pattern of heating elements than standard machines. The *contact bun toaster* uses a heated platen to sear the bread quickly and evenly on both sides. The panini grill, which we introduced in an earlier chapter, is a type of contact toaster. The idea behind these quick-toast appliances is to seal the bread a bit by toasting it, so that it doesn't get soggy when made into a sandwich.

17-5 FOOD WARMERS

Keeping food hot while it's waiting to be served without compromising its quality is the classic foodservice challenge. Fortunately, there's a piece of warming equipment to fit almost every situation. Most fall into one of three categories: food warmer, soup warmer, or drawer warmer. There are also steam tables and/or hot-food tables.

Warming or *holding* is the process of a controlled cool-down, from hot-out-of-the-oven to a safe and stable serving temperature. The type of food, the length of time it will be held, and the way it will be served all impact the equipment choice, but the absolute first requirement of any food warmer is that it can maintain a minimum holding temperature of 135 degrees Fahrenheit. This minimum temperature refers to the internal temperature of the food, not its surface temperature, and is required by law in all jurisdictions for food-borne illness prevention.

The ideal food-holding appliance for time periods exceeding one hour is a humidified, closed cabinet with convection heat. Why? The closed cabinet protects the food from temperature changes, and the moisture content prevents it from drying out. Any holding cabinet must be equipped with sufficient power to reheat rapidly any time the door is opened.

Where a closed cabinet is not feasible—say, for short holding periods (15 minutes to an hour) in a busy self-serve or quick-service environment—the best choice is an open warming unit that provides conductive heat from the bottom and/or radiant heat from the top. Open warmers are susceptible to air drafts, which can be minimized either by turning up the unit's temperature slightly, or adding extra wattage to the radiant elements above. The disadvantage to short-term warming is that is cannot keep food moist. Some foods may be covered or wrapped for holding this way, but it's not the ideal. And what to wrap it in? Tinfoil holds in moisture but reflects radiant heat rather than absorbing it; paper absorbs the heat better but dries the food out faster; and plastic wrap will melt in the intense heat required to keep the food safely hot.

Here are some of the food-holding choices available today: *Overhead warmers* simply refer to the fact that the warming lights are placed above the food. These can be affixed to a counter, suspended from a wall or ceiling, or they can be freestanding, portable warmers attached to a base suitable for food placement. Because they give off so much heat, it is not advisable to mount warmers any closer than

3 inches from walls or ceilings. They may blister your paint! Most warmers turn on and off with a switch, but the more expensive ones have a knob for adjustable heat control. Overhead warmers use *warmer lamps* to work effectively, and they are different from your basic heat lamp. This type of bulb uses infrared lighting to produce heat, from 250 to 275 watts. Like other types of light bulbs, there are round ones and tube-shaped ones for larger heating spaces.

There are several specialty warmer models, such as the buffet warmer for banquet and cafeteria settings. It has sneeze guards and, in addition to overhead lamps, insulated cables that run beneath the food pans to provide gentle warming from beneath. Thermal shelves, found in pizzerias and bakeries, are made of insulated material with lamps above that can be adjusted with a dial for different heat settings. Pass-through windows can also be outfitted with warmer capability (see Illustration 17-10)

ILLUSTRATION 17-10 Food warmers can keep food hot and appetizing until it is delivered to the customer.

COURTESY OF ALTO-SHAAM, INC., MENOMONEE FALLS, WISCONSIN.

You can also choose a *base-heated warmer,* either electric or gas. These operate by heating a well into which food containers are placed. Most models require that water be in the well, but some are operated without water. A thermostat controls the heating element, which uses from 500 to 1200 watts. Multiunit designs use up to 1600 watts. Base-heated warmers open at the top to accommodate food pans, and there are adapters available to accept various pan sizes.

Like its electric counterpart, the gas-powered food warmer is about 14 inches wide and 24 inches deep. It stands less than a foot in height, but adjustable legs are recommended, which add another 3 to 5 inches. In buying a gas model, consider where the gas fitting will be placed so that it doesn't interfere with the use of the unit. It is also very important that air intakes on the bottom of gas units be kept clean.

ILLUSTRATION 17-11 A drawer warmer can accommodate standard pan sizes, in different drawers with separate controls.

COURTESY OF ALTO-SHAAM, INC., MENOMONEE FALLS, WISCONSIN.

Soup warmers are cousins to base-heated warmers, and they're far more versatile than their name suggests. Use them to heat sauces, chili, melted cheese, hot fudge, and more. They come in many sizes and styles, from round to rectangular, and can be placed on a countertop or dropped into cooks' tables or salad bars. The countertop models may remind you of an electric crockpot, but these are made of stainless steel. The most common countertop models accommodate inserts of 4, 7, or 11 quarts. For larger volume, rectangular 12-by-20-inch pan units are available.

Water is required for most warmers, and it is highly recommended. The electric element at the base of the unit heats a reservoir of water, which, in turn, heats the contents of the immersed pan. The advantage of water is that it transfers the heat evenly to the food container, eliminating scorching and resulting in a steady, even temperature that is controlled by thermostat.

A ***drawer warmer*** is a heated cabinet with one or more drawers that slide open to load or remove food. Many offer humidity control, which is a plus for some types of food, and separate controls for each drawer. Drawer warmers are available as countertop or drop-in appliances (see Illustration 17-11).

NSF International safety standards require these warmers be equipped with a temperature gauge on the outside of the unit to be able to easily monitor interior temperatures without having to open the drawers. Temperatures range from 60 to 200 degrees Fahrenheit. Drawer warmers use 120- or 208/240-volt electricity and from 450 to 1350 watts, depending on the size of the unit and the number of drawers.

Steam and Hot-Food Tables

Another option for keeping food warm between cooking and serving is the hot-food table or steam table. What's the difference?

The typical ***steam table*** contains a shallow tank of water that is heated with electric or gas-fired elements to temperatures of between 180 and 212 degrees Fahrenheit. (See Illustration 17-12). Used

ILLUSTRATION 17-12 Steam tables are designed to keep food at warm, safe temperatures in banquet, buffet, and cafeteria settings. This one has a built-in oven below, for cooking, then serving above!

COURTESY OF THERMODYNE FOODSERVICE PRODUCTS, INC., FORT WAYNE, INDIANA.

mostly for moist foods, shallow steam table pans full of food are placed in the water, keeping them hot until needed. An advantage of many steam tables is that they can be heated with or without water.

The ***hot-food table*** accomplishes the same task by heating multiple compartments of hot air. The operator fills the compartments with standard-sized serving pans of food. Individual compartments are controlled with separate thermostats, so several foods can be held in the same table at different temperatures. These may be called *waterless hot-food tables,* since no water is required and, therefore, no plumbing connections.

Both types of heated tables are available in a variety of lengths and sizes, depending on how many openings are needed for different pans. The openings, or the appliances themselves, are often called *hot wells.* An individual well measures 12 by 20 inches, just enough to fit a steam table pan. Table 17-2 shows the most common table lengths.

A major concern with heated tables is that users keep the food pans covered as much as possible to conserve heat and energy. Food at the bottom of the

TABLE 17–2

COMMON SIZE STEAM TABLES: GAS AND ELECTRIC (DEPTHS VARY WITH ACCESSORIES)

Gas

Number of 12″×20″ Openings	Typical Length
2	2′ 6″
3	3′ 8″
4	4′ 10″
5	6′ 0″
6	7′ 2″

Electric

Number of 12″×20″ Openings	Typical Length
2	2′ 6″
3	3′ 8″
4	4′ 10″
5	6′ 0″
6	7′ 2″
7	8′ 4″

Source: ***Foodservice Equipment and Supplies*** **magazine, a division of Reed Business Information, Oak Brook, Illinois, February 1999.**

pan may be adequately heated, but at the top of an uncovered pan, the much lower temperature may be dangerous. Casseroles, sauces, and stews must also not spend too much time on the hot-food table before being disposed of.

NSF International suggests users of holding equipment play it safe by dividing large batches of food into smaller portions for holding and serving. Smaller portions ensure more thorough and even heating, minimizing food safety risks. It is also a standard rule that, no matter what it's being heated in, food that has been held for more than two hours should be discarded.

Where sauces and soups are served, you may also have use for the ***bain marie,*** a small, uncovered variation of the steam table. The bain marie is a single container used for cooking and also for holding, with a thermostat that controls its temperature. The bain marie was first mentioned in Chapter 6. It comes in sizes that range from a 2-foot square, to 30 by 72 inches. The smallest electric-powered bain maries use 3 kW of power per hour; the largest ones use almost 10 kW.

17-6 COFFEE MAKERS

The best-selling nonalcoholic beverage in restaurants is coffee, and the specialty coffee market has grown to more than $7 billion a year in sales with nearly half of the U.S. population counted among its customers! In the early 1990s, there were fewer than 500 coffee bars in America—today, there are more than 8,500. This represents a major cultural shift in little more than a decade. We probably have Starbucks, the upscale Seattle coffee bar chain, to thank for this. In the same time period, Starbucks has grown from 84 locations to more than 6,000, in many foreign countries.

The best news for foodservice operators? Customers are now accustomed to paying $3 and up for gourmet coffees and specialty drinks, just about any time of day! You'd better learn to make them well, and this starts with top-notch equipment. Incidentally, there are also commercial tea-making machines, which may be useful depending on your menu and customer preferences. But we'll focus here on coffee making.

Your first considerations in purchasing coffee-brewing equipment are the amounts you'll need to produce and the types of coffee drinks your establishment will offer. The requirements are far different for a place that needs 10 gallons of coffee per hour and a single brew versus the place that needs 1 gallon per hour of 10 different brews. The market is cluttered with a wide selection of brewing options and appliances for coffee, tea, and espresso drinks. Before we introduce them, let's list all the factors that influence the quality of brewed coffee. Some, but not all, are directly related to the coffee-making machine:

- **The type, freshness, and quality of the coffee bean**
- **The way the beans are ground (finely ground coffee takes less time to brew, but can be stronger in flavor than coarsely ground coffee)**
- **The overall water quality (you may need to filter or purify water before using it for brewing)**
- **The temperature of the brewing water (optimum 195 to 205 degrees Fahrenheit)**
- **The quality of the coffee filter**
- **The uniform extraction of the coffee from the ground beans (the volume and way the water makes contact with the coffee grounds during brewing)**
- **How well the equipment has been cleaned**
- **The holding temperature of the finished product (optimum 185 degrees Fahrenheit, held no longer than 30 minutes)**
- **Whether the finished coffee is stirred for consistent strength**

It's almost a science, and there are plenty of coffee aficionados who know a good cup from a bad cup and won't hesitate to tell you about it. And there's one more big variable: the ratio of coffee grounds to water. One pound of coffee can be brewed with 1¾ to 3 gallons of water, with very different results. You should learn the strength of brew your customers expect.

Speaking of water, we've mentioned in other chapters the importance of filtering it. With coffee, this is critical. What you're selling is, in fact, mostly water—so you might as well treat it like the valuable resource it is. Minimize its mineral and chemical content, and anything that might create an off-taste. You'll not only brew better coffee; you'll help your brewing equipment last longer.

Coffee drinks can enhance your image as a foodservice provider, so they offer plenty of merchandising opportunity. You can purchase green coffee beans and roast them yourself or whole beans to grind as needed, both adding an aromatic component to your dining area. Gourmet blends are another way to scent the air with hazelnut, peppermint, cinnamon, and other luscious flavors that may entice customers to buy.

For all this hoopla, *coffee brewers* are surprisingly simple machines. Let's dissect the brewing process:

Water is added to the machine, either manually or by a ¼-inch water line that is plumbed into the unit. For the latter, proximity to a water line and a floor drain is required.

Coffee is placed in a brew basket (or filter basket), in a filter that keeps the grounds from falling down into the brewed coffee. The filter may be made of disposable, porous paper or reusable metal mesh. Cone-shaped filters allow slightly more contact between water and grounds (and therefore more flavor extraction) than filters that are flat on the bottom. Since the grounds swell during brewing, it is important that the total depth of your basket be able to handle about 50 percent more coffee grounds than you'll be putting into it. Most brew baskets will be about 3 inches deep.

The water is heated and either allowed to drip straight through the coffee grounds or sprayed uniformly over the grounds through a spray plate full of small holes. The spray plate may be flat or rounded, and is usually made of stainless steel. It should be wiped down regularly to prevent build-up of lime (from the water) and the natural oils from the coffee beans, since the holes can clog easily. The pattern of holes on the spray plate should ensure that the water uniformly hits and wets the whole basket of grounds, not just a spot or two in the center.

The coffee emerges from a spout or faucet into a glass or metal carafe, which sits on a warming burner, or into an ***airpot,*** an insulated thermal container with a pump. It can hold the liquid hot and fresh for up to an hour without flavor loss, and allows customers to serve themselves. The chief advantage of an airpot is that it increases the usable life of the coffee. In a traditional, open-top coffee pot on a standard burner, you can't guarantee the flavor will be satisfactory for more than about half an hour before it starts to taste stale and burnt.

Self-serve coffee is a staple in hotel lobbies, banquet set-ups, convenience stores and more. Large airpots are often used for this purpose, and the newest piece of equipment is the 1.5-gallon ***satellite brewer*** (see Illustration 17-13). Its output exceeds the standard 10- or 12-cup pot, but is limited enough so that the coffee doesn't sit around for long periods of time. The satellite brewer can be programmed to make smaller batches as needed, is compact (18 to 25 inches in width) to fit in small spaces, and can be connected to a plug-in "docking station" that gently heats to maintain correct coffee temperatures and will shut off automatically when optimum temperature is reached. This is a big improvement over single-temperature burners on most coffee brewers that are either "on" or "off."

ILLUSTRATION 17-13 The satellite system is a combination coffee-maker and docking station with sensors to hold the coffee at optimum temperature until it is ready to serve.
COURTESY OF BUNN-O-MATIC CORPORATION, SPRINGFIELD, ILLINOIS.

For individual tableside coffee service, small airpots work well, and so does the ***French press.*** In this stylish glass carafe, ground beans and water are combined and left to steep for a few minutes. Then, the server (or the guest) holds the knob of a plunger that reaches inside the pot and presses the used grounds to the bottom of the pot before pouring out the finished brew.

Coffee makers range in size from the ½-gallon decanter to 80-gallon *coffee urns* for banquets and cafeteria settings. The most common-sized coffee urn can brew 1½ to 3 gallons within 10 minutes. Inside, a metal spray plate sprays hot water across the coffee grounds. Temperatures can be set for different strengths of brews, and most urns give you "Half" and "Full" settings to brew a partial urn or a full one. A control panel at the front of the urn has buttons for "Brew" and "Standby," and some have a light or alarm to indicate when the urn is almost empty.

Coffee urns generally have three serving taps, two for coffee and a third for hot water, as well as a metal drip tray beneath the taps to catch spills. Unlike the 1 square foot needed for a coffee brewer, urns are almost 3 feet tall and at least 20 inches around. They use 240-volt, three-phase electricity to operate efficiently (see Illustration 17-14).

Remember, the natural oil from coffee beans creates a film on your coffee-making equipment. It will work best for you if you wipe it down daily, including spray plates, heads and faucets, and carafes.

Espresso Machines

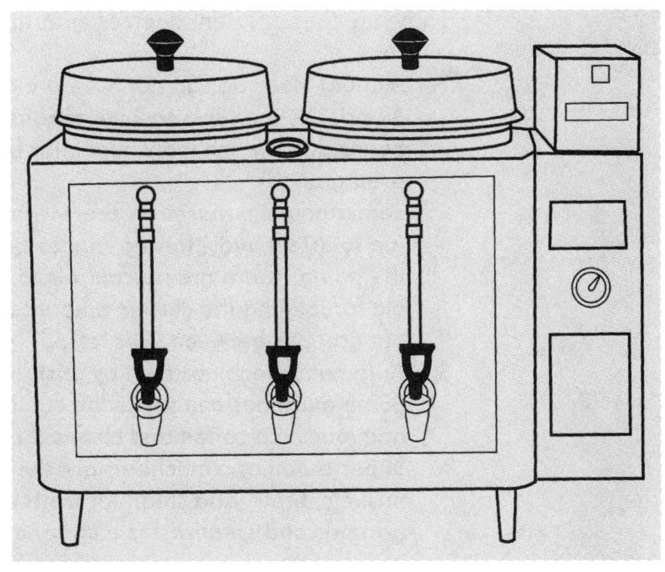

ILLUSTRATION 17-14 **Coffee urns are the largest types of coffee makers, sometimes with several taps and capacities of up to 80 gallons.**

The name ***espresso*** (and no, it's not "*ex*-presso") means "to press" or "to press out" in Italian. It was first coined as a name for that country's strong coffee in small, made-to-order servings. Coffee beans are ground into a fine powder just before use—much finer than regular coffee grounds—and packed into a small filter basket. Hot water is pressed through the powder to create a very strong, aromatic cup of coffee.

The name ***cappuccino*** is a bit more difficult to translate into English. This traditional Italian morning beverage is made simply from espresso coffee topped with milk that has been steamed into a puffy foam. The foam often has a little peak on top, so the drink got its name from the peaked hoods worn by the Capuchin monks. (Use that trivia tidbit to impress someone next time you're in a coffee bar!)

The popularity of these specialty coffees and others has created a profitable market niche in restaurants and bars, a demand for the espresso machines used to make them, and for experienced operators, who are called ***baristas.*** But until recently, the complexity of espresso-making equipment kept many foodservice businesses out of the loop, unwilling to figure out how to learn all the tricks and train the baristas, and where to put the equipment.

The basic espresso maker is actually two small, electric machines on the same base: a coffee grinder and a pump that adds water to the grounds to brew the coffee. Machines have several heads, depending on how many cups or "shots" can be brewed at once. Be sure that the cups you want to use will fit under the heads, because some machines are designed only for the traditional, small demitasse cups. Most also have a stainless steel valve on one side of the machine that uses high-pressure steam to foam the milk when needed for cappuccino and latte drinks (see Illustration 17-15).

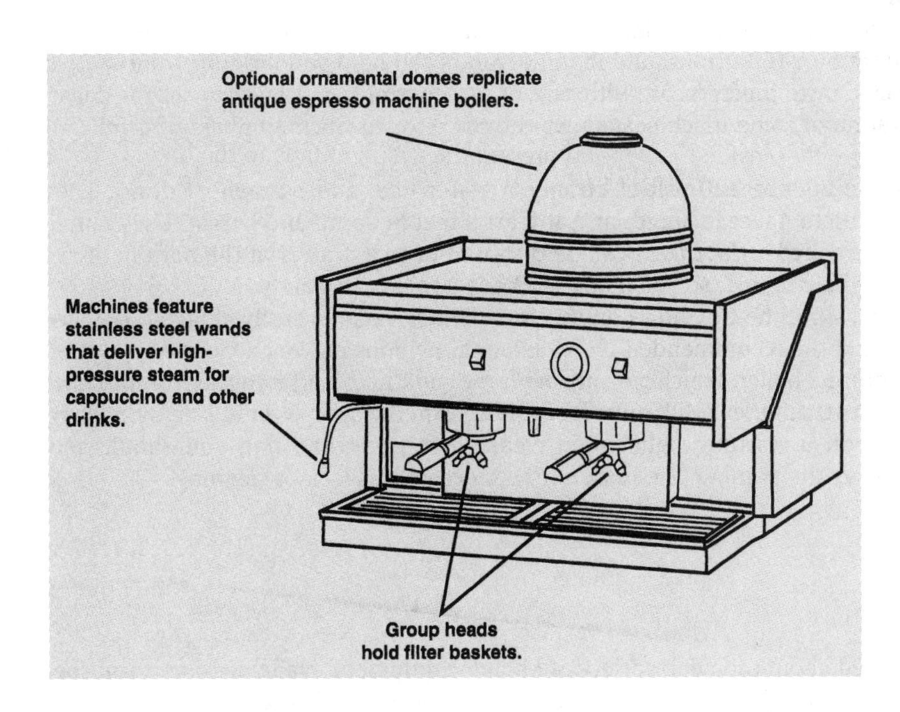

ILLUSTRATION 17-15 **Espresso machines create the water pressure, temperature, and steam necessary to make these strong coffee drinks.**

SOURCE: *FOODSERVICE EQUIPMENT AND SUPPLIES*, A PUBLICATION OF REED BUSINESS INFORMATION, OAK BROOK, ILLINOIS.

There are four different degrees of automation for espresso machines:

1. Manual machines do not have a motor. They operate with a lever that is driven by a hydraulic pump. The manual machine requires the most training to use, because the barista grinds the beans, loads each individual filter basket, watches as the espresso fills each cup, and discards used grounds.
2. Semiautomatic machines brew with the touch of a button, but the barista must stand and wait (up to 30 seconds) for the shot to be completed, then manually stop the action by turning off the pump. There are no controls to adjust the strength of the brew, so the barista is responsible for loading the correct amount of coffee grounds into the basket, then waiting, dumping out grounds between uses, etc.
3. Automatic machines start by push button, and stop themselves when the shot is finished. Some machines can sense the cup size and will stop before it overflows. The barista grinds and loads the coffee and cleans the machine.
4. Super-automatic machines are the only ones that grind the coffee, load it into the filter baskets, brew, and steam or froth milk, all automatically. This model also discards the used grounds and sterilizes its components to stand ready for the next brewing cycle.

Then again . . . how much automation can you afford? The super-automatic machines range in price from $6,000 (a small-volume unit, for up to 50 cups a day) to $20,000 for high-volume coffee bar use. This is a significant investment, but industry figures indicate an 85 percent gross profit return on specialty coffee drinks, and the more automated the machine, the more labor savings you will realize. They can also be leased, which is a good idea if you're not sure how well your employees will adapt to using the machine.

Another major consideration is how quickly you want to be able to make the drinks. Brewing time per shot can range from 20 seconds to 1 minute. An efficient barista can make about 70 cups of espresso per hour with each head on the machine, but latte or cappuccino drinks will take longer. There are machines capable of fewer than 100 servings per hour, and up to 350 servings per hour.

Italian-made machines are known for their all-metal brewing chambers, which use hotter water and produce a darker brew. Swiss-made machines most often use a plastic brewing chamber, and produce a slightly lighter espresso. Look for both Underwriters' Laboratories and NSF International listings, and quiz the seller to be sure local repair and maintenance service is available in your area. To give you an idea of the complex engineering involved in espresso-making, consider that steam is produced inside the unit at 220 degrees Fahrenheit, while milk (inside the same unit) must be held below 39 degrees Fahrenheit to meet NSF International safety standards—all within a space that is two feet wide, or less! The control panel features from 6 to 16 buttons, and manufacturers can help you program your system for custom drinks or to specific taste preferences, with one of the control panel buttons set for decaffeinated coffee. Computerization of some machines allows the operator to track numbers of drinks, or sales information by server.

Espresso machines require 208- to 220-volt electricity, a water line, and permanent drain. They can take up a lot of space, too, up to 4 feet in length and at least 2 feet in depth and height. Used coffee grounds (the spent grounds are called *coffee cakes*) can be disposed of in a drawer at the bottom of the machine that must periodically be emptied, so you'll have to have a trash can handy.

Proper cleaning is critical, since the tiny holes on the spray plates become easily clogged. A water filtration system is essential, and the recommended daily cleaning includes putting a cleaning solution (sometimes in tablet form) into the boiler, which runs through the milk lines and into the frothing nozzle. The process takes from 2 to 10 minutes. Milk must be flushed from the system daily. The more computerized machines will turn on a warning light when cleaning is necessary, and you should pay attention to it. Manufacturers say the number one reason for service calls is lack of cleaning.

SUMMARY

In this chapter, you learned about the wide variety of kitchen equipment available to perform specialized preparation tasks. Perhaps the most popular appliance is the vertical mixer, with an electric motor and agitator arm(s) to take the muscle strain out of mixing and kneading. Mixers are sized based on the capacity of their mixing bowl and the size of their motor; choose yours based on the kinds of products you'll be mixing.

Hand-held mixers are smaller, portable versions of the vertical mixer. Vertical cutter mixers ("VCMs") let you cut, mix, and blend foods. Electric slicers can cut foods more thinly and accurately than a knife, improving portion control.

Food processors have interchangeable blades to slice and dice food into a variety of shapes and sizes. The most difficult food for any processor to handle is cheese, which seems to slice best when it is still cold from the refrigerator. There are also food choppers, grinders, and blenders—their names indicate their role in the kitchen, but today's appliances are capable of performing additional tasks depending on the attachments that can be used with their motors.

Commercial toasters are sturdier and faster than home models. They turn out more than 200 slices of toast per hour! There are pop-up models, pop-down models, and drawer models where food is placed on a drawer that slides into a heated chamber. High-volume operations may use a continuous toaster, placing food onto a conveyor belt with variable speeds for different degrees of doneness.

There are also many types of food warmers. Some are heat-producing lamps; others are wells or tables with heated compartments where food containers are placed. Heating is accomplished with heating elements beneath the food pans, using water and steam or dry heat. Strict health and safety requirements mandate minimum internal food temperatures—not just surface temperatures—for cooked foods being held before serving, and generally, food that has been held for more than two hours should be discarded.

Finally, this chapter introduces coffee-making and espresso-making equipment. What you buy will depend on the amounts you need to produce and the types of coffee drinks your establishment will offer. There are many variables in good coffee-brewing, including the grind (coarse, medium, fine) of the beans, the type of filter, and the design of the spray plate that sprays hot water over the grounds—but the most important factor is water quality. Filtered water is always preferable to minimize off-tastes.

Espresso-makers range from manual to super-automatic, depending on how skilled your operator ("barista") is and how much work you want them to have to do. The espresso-maker uses an internal boiler to heat water for brewing, and to make steam to froth milk.

No matter which kind of machine you use, or which type of container you use to store coffee, these must be kept clean. Coffee beans produce oil that creates a film buildup on equipment.

STUDY QUESTIONS

1. Why would you want to know about the PTO on a commercial mixer?
2. When would you need to use a bowl adapter?
3. Why should you always set the gauge plate of a slicer to zero when it's not in use?
4. What is the best way to grate cheese in a food processor?
5. What is the difference between a blender and a spindle blender?
6. How and why would you use a honing stone?
7. Why would you use a vertical conveyor toaster instead of a horizontal one?
8. What is the minimum temperature at which a food warmer should operate, and why?
9. Why should you keep foods covered while on a steam table?
10. Technically, what is the difference between a coffee maker and an espresso maker?

CHAPTER 18

SMALLWARE FOR KITCHENS

INTRODUCTION AND LEARNING OBJECTIVES

If you're in foodservice, you can't live without all the little necessities it takes to mix, measure, cook, and store food. As a group, these items are commonly known as ***smallware.*** Manufacturers will tempt you with hundreds of styles of cookware, lengths of knives, or shapes of containers, each with its own unique attributes and accompanying sales pitch. So how do you decide what (and how many) to buy? Do you really need a kiwi peeler? A tomato corer? Japanese fish tweezers, for extracting tiny individual bones from fish fillets? Well, maybe . . .

The most effective way to make these decisions is to divide the category of smallware into subcategories and tackle it in more manageable "chunks." To that end, this chapter contains information about:

- **Hand tools and knives**
- **Measuring tools and kitchen scales**
- **Range top cookware**
- **Oven cookware**
- **Serving and holding containers**

Remember, although these purchases are minor compared to the high-dollar equipment we've covered in past chapters, they are just as important to the success of your operation. In the foodservice kitchen, little things *do* mean a lot, and the more professional your workers are, the more likely they are to have some very definite preferences about the tools they use to complete their work.

18-1 HAND TOOLS

Have you ever tried using a spoon when you really needed a ladle, or a ladle when you really needed a spatula? Then you know how important it is to match the proper utensil to the task at hand. The incorrect hand tool may, in some cases, compromise quality, efficiency, cost control, or even sanitation. And yet kitchen utensils are frequently misused. A hurried cook will use a spoon to pry open a stubborn container lid or poke holes in a metal oil canister with a fork. This kind of "double duty" surely shortens the useful life of the tool.

In some cases, it takes longer to complete a task if you choose the wrong utensil. Someone who is hand-whipping cream, for instance, will take twice as long if he or she uses a French whip than if he or she uses a piano or balloon whip, made of more delicate wire. On the other hand, the French whip is a better choice for mixing thicker substances, such as pancake batter.

You may want to consider a mechanized hand tool, depending on the needs of your operation, to improve efficiency. If you serve a lot of hamburgers, dressing most of them with sliced onion and tomato, a mechanical slicer beats doing the work by hand. With a single pull of its handle, the slicer cuts through an entire onion or tomato, creating uniform slices. This uniformity also makes it easier to decide how much product to buy, because you can count on a certain number of slices from similarly sized produce.

Also, because mechanical tools work more quickly, they leave food exposed, and at room temperature, a shorter period of time than doing the job by hand. In some situations, this is more sanitary.

Another sanitation concern: Teach all employees to use different tools for each task to avoid cross-contamination. Keep some spatulas specifically for the grill; others to use elsewhere in the kitchen. Instruct them never to wipe or rinse a spoon from one pot, then use it to stir a completely different dish in another pot.

Serving Spoons

Speaking of spoons, let's examine them more closely. In the commercial or industrial kitchen, spoons are made of stainless steel, 15 or 20 gauge. They vary in length from 15 to 20 inches. The bowl of the spoon is elliptical and may be solid, slotted, or perforated. The handle may also be steel or heavy-duty plastic. If it's plastic, the handle should be heat resistant. (Some are heat resistant to 230 degrees Fahrenheit, others up to 375 degrees Fahrenheit, so consider your needs.) In either case, a round hole is punched into the end of the handle, allowing you to hang the spoon from a hook.

Recent variations in spoon technology include the *triple-edged spoon*. The radius of the spoon is the same shape as the interior of the pan (see Illustration 18-1). Triple-edged spoons generally have plastic handles. And there's also the ***spoodle,*** also known as a *spoonladle, portioner,* or *spooner* (see Illustration 18-2). As you can see, it's a serving utensil that combines the stirring capabilities of a spoon with the portion control capabilities of a ladle. Its bowl is round and may be solid or perforated. The spoodle's plastic handle has a thumb notch on top and a stopper at the bottom. Together, these features are helpful to maintain a good grip on the handle and to prevent the utensil from slipping into the pot when not in use.

You can buy the spoodle in sizes ranging from 2 to 8 ounces, and most manufacturers have handily color-coded the various sizes. Table 18-1 is a sample of colors and sizes from Vollrath, a smallware manufacturer.

One manufacturer has introduced spoodle handles made of polycarbonate, a type of plastic that is reported to be heat resistant to 270 degrees Fahrenheit, as well as dishwasher safe. However, because cooks tend to rest serving spoons on boiling hot range tops, you're better off using spoodles for buffets or serving, not for cooking.

ILLUSTRATION 18-1 **The triple-edged spoons, shown on the right, differ in shape from conventional spoons.**

COURTESY OF VOLLRATH COMPANY, LLC, SHEBOYGAN, WISCONSIN.

Wire Whips

Whips were created for that brisk type of hand-stirring that introduces air into the ingredients to make them fluffy, frothy, or otherwise well blended. They're used for everything from whipping cream to mashed potatoes.

ILLUSTRATION 18-2 **The spoodle can be used for stirring, scooping, and portioning.**
COURTESY OF VOLLRATH COMPANY, LLC, SHEBOYGAN, WISCONSIN.

A whip is made of 18-gauge steel. It consists of a handle, with long, curved wires coming out of it to form big loops. Most of the handles are stainless steel, but there are also whips with plastic or wooden handles.

The type of whip you use depends on the consistency of the food you're working with, and whether you plan to stir it, or aerate it. The standard kitchen whip is often called a ***French whip***; its wires are fairly thick and stiff, and it is best used for batters, mashed potatoes, and other heavy foods. French whips come in lengths ranging from 10 to 24 inches. For lighter or thinner concoctions—whipped cream, consommés, Hollandaise sauce—the finer wires of the ***piano whip*** (or *balloon whip*) are recommended. Compared to the elongated French whip, there are more wires on the piano whip and they form a rounder shape, designed to whip more air into the mixture. Piano whips come in lengths of 10 to 18 inches.

The smaller version of a whip is called a *whisk*. Specialty products in the whip/whisk family include:

- Whisks with flat loops (instead of rounded ones) called *sauce whips*. The flatter coils allow the user to scrape the insides of a pot or pan more thoroughly.
- The *ball whip* has a steel ball, about the size of a marble, caged inside a sheer mesh ball, which is trapped in the loops of a regular whip. The theory here is that the friction from the ball flying around within the whip aerates a mixture more quickly.

On commercial-grade whips, the wires and handle should be welded smooth, so that bits of food can't get caught in tiny crevices. In fact, there are National Sanitation Foundation standards for whip construction, so make sure the style you choose is NSF approved.

Food Turners and Spatulas

The most common utensil used to flip food over is the *offset spatula,* also called a *hamburger turner.* This works equally well on broilers, griddles, sheet pans, and more. Impervious to heat, the wide *offset blade* (the part that comes in contact with the food) is made of stainless steel with a shiny chrome finish, riveted to a durable wooden or plastic handle. The newest generation of spatulas is silicone-coated for high-heat uses.

TABLE 18–1

TYPES OF SPOODLES

Handle Color	Capacity (ounces/milliliters)	
Red	2	59.1
Beige	3	88.7
Green	4	118.3
Black	6	177.4
Blue	8	236.6

Source: **Vollrath Company, Sheboygan, Wisconsin.**

The blade of the spatula may be solid, slotted, or perforated; the end of the blade may be straight or rounded. Although the standard total length of the spatula is 14 inches, there are also 10-, 12-, and 20-inch models available.

A variation of the spatula is the *pancake turner,* which is totally stainless steel. Another specialty tool is the *pie server,* also often used for serving pizza slices. Its blade is triangular and flexible, 5½ inches long with a wooden or plastic handle.

There is one category of hand tool a kitchen can't be without: the spatula/scraper. For scraping food off the sides of pots, mixing bowls, and storage containers, this hand tool should be NSF approved. Look for spatula/scrapers with a flexible, thermoplastic blade and a polypropylene "I-beam" handle (see Illustration 18-3). These range in length from 9½ to 16½ inches and are dishwasher safe. Although thermoplastic means heat-resistant, these are not made for use on hot surfaces such as griddles or range tops. They will melt!

ILLUSTRATION 18-3 **Multipurpose spatula/scrapers.**
COURTESY OF VOLLRATH COMPANY, LLC, SHEBOYGAN, WISCONSIN.

Members of this family include:

- The straight spatula (also called a ***palette knife***) with a 9-inch flexible blade with a rounded end, used to spread icing on cakes or for bowl scraping.
- The sandwich spreader is a shorter version of the palette knife; its blade is only 4 inches long.
- A ***bench scraper*** or *dough knife* is a must for bakeries; its stiff metal blade, with rounded edges and wooden handle, is used for scraping and cutting dough. A plastic version of this tool is available, but it doesn't wear as well as the metal ones.
- There are specialty scrapers in different shapes that aren't so much for cooking, but for scraping the last bits out of prep containers, condiment jars, and the like. They're handy in the dish room too, for scraping baked-on or dried-on residue off of pots and pans.
- A ***chan*** is a wooden spatula preferred by some Asian chefs to use in stir-frying and wok cooking.

Tongs

Used primarily on serving lines and for lifting foods out of pots and other containers, commercial-grade tongs should be made of stainless steel. However, heavy-duty plastic tongs of different colors may be used in some prep areas—like cutting boards, the color system is used to avoid cross-contamination. Some tongs are spring loaded, with either one or two flat springs. They range in length from 10 to 16 inches. ***Pom tongs,*** also made of stainless steel, do not contain springs. Pom tongs are available in lengths from 6½ to 12 inches.

Depending on your needs, there are several different types of specialty tongs, including those with scalloped edges, flattened edges, or plastic handles. Various manufacturers make tongs for tossing salad, serving spaghetti, and picking up pastries or meats on a buffet line. The tiniest tongs are made for picking up individual sugar cubes. Tongs are certainly handy, but before you get carried away purchasing them, ask yourself if there is any other kitchen tool that can do the job as well.

Kitchen Forks

Forks for commercial kitchen use have two tines, made of solid, 18-gauge stainless steel. You can select wooden or metal handles—the all-metal ones have holes at the end of the handle to facilitate hanging. Forks can be ordered in lengths from 11 to 20 inches.

There are a number of specialty forks, including the bayonet, which measures 12 inches in length and has two 6 inch tines; the broiler fork, with two long tines and a long (almost 16-inch) handle; and a carver fork, which is only 9 inches in length.

Strainers and Colanders

Strainers are hand tools made of fine mesh, used to separate solid foods from liquids. They're cup shaped or cone shaped and have handles. ***Colanders*** are perforated containers, generally bowl shaped, used to wash and rinse food items. Both are indispensable in the kitchen, but there are a number of types to suit different needs.

The basic strainer is a round-bottomed, cup-shaped hand tool made of either screen-type mesh or perforated metal. It is used for rinsing and draining pasta, vegetables, and more. The bowl of the strainer may be as shallow as 2 inches or as deep as 6 inches; its interior diameter varies from 4½ to 13½ inches; and its total length, from 12 to 33 inches. You can order strainers with single or double layers of mesh. Variations include the following:

- The ***China cap*** is a metal mesh cone welded to a stainless steel band and handle. It is used to strain stocks, soups, sauces, and other liquids, so the mesh on the cone is fine, from $\frac{1}{16}$ inch to $\frac{3}{32}$ inch. China caps are measured by the size of the cone, from 8 inches deep and 8 inches in diameter to 12 inches deep and 12 inches in diameter. It's handy to get one with a hook at the outer rim of the cone so you can hang it over the stockpot (see Illustration 18-4).
- The *chinois* is a type of China cap, made with finer mesh. It's used when a liquid requires greater clarity or smoothness.
- The ***sieve*** looks like a metal coffee cup with fine mesh screen at the bottom and a side handle. It is used to sift flour and other dry ingredients. Some sieves have wooden frames instead of metal; wooden ones are lighter, but metal ones are much sturdier.
- The ***food mill*** is a hand-cranked type of strainer used primarily to puree foods; some smaller models are useful for grinding spices. Food is placed in the hopper of the mill, and it's ground by being forced through a perforated disk by a flat blade, cranked around and around by hand. Most commercial food mills have interchangeable disks with various-sized holes.

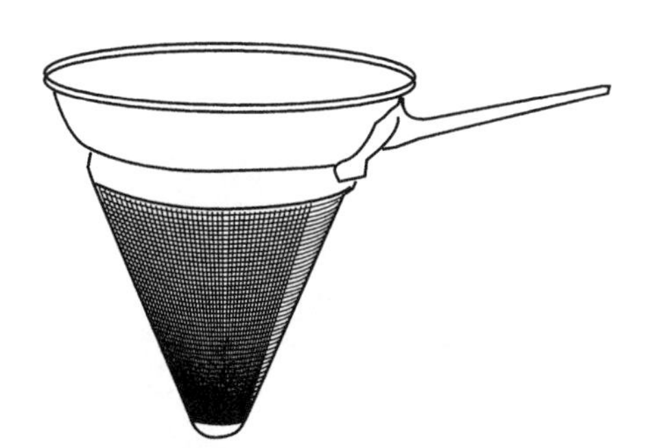

ILLUSTRATION 18-4 The China cap is a stainless steel strainer. When finer mesh is used, it is known as a chinois.

The colander is not really a hand tool, but more a member of the kitchen pots-and-pans family. Colanders usually have feet on the bottom to allow them to sit upright and handles for easy carrying and hanging. They may be made of aluminum or stainless steel (10 to 18 gauge) in capacities of 3 to 16 quarts. The perforations are larger than those of a strainer. Variations include:

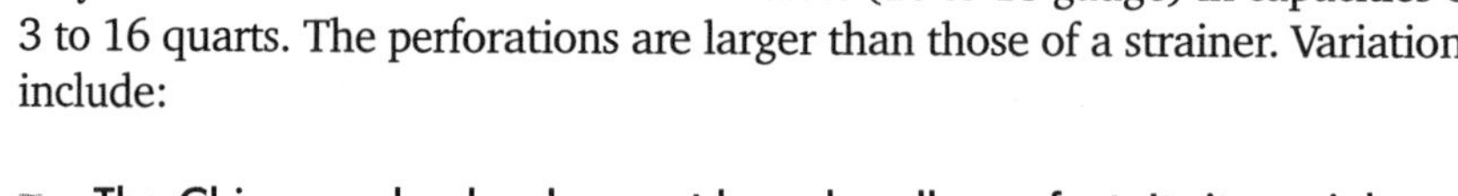

- The Chinese colander does not have handles or feet. It sits upright due to depressions pressed into its base.
- The Chinese food container is a related type of strainer used for washing and draining vegetables.

18-2 KITCHEN CUTLERY

Any good chef will have strong opinions about his or her kitchen knives, but they'll all agree on one thing: All cutlery is not created equal. In many kitchens, employees all share the knives and the unfortunate result is that no one in particular takes care of them. You'll notice, however, that the more culinary training a chef has, the more likely he or she is to invest in a personal set of knives, carried to and from work in a protective bag. This fairly recent development fosters pride in the tools of the trade and ensures that the correct knife is close at hand, sharpened, and ready to go when it's needed.

Your menu will determine the types of knives you'll need to purchase. It would be impractical to cover every little detail about kitchen knives in this text, but in the next few pages, you'll learn about the basic types of knives, knife construction, and care.

Knife Construction

The two most popular knife manufacturing techniques are ***forging*** and ***stamping.*** Forged blades are formed when steel is heated, then roughly shaped and compressed under great pressure. Then the blades are put through a number of honing and grinding processes. Forging is more expensive than stamping, and its proponents claim forged knives are sturdier.

Stamping means a single sheet of flat steel is imprinted with dies or molds, to make different shapes and sizes of blades. Several blades can be stamped at once, then individually honed and sharpened. The newest knife technology cuts the sheet of steel into blade shapes with a computer-guided laser beam. Because laser cutting is more precise than stamping, less steel is wasted in the manufacturing process.

A third and relatively new manufacturing technique creates a ***ceramic knife*** from zirconium carbide, in combination with aluminum oxide or heat-resistant zirconium oxide. Called ***hot isostatic pressing,*** the material is molded and fired all at once. The resulting blade is black or gray in color, and is known for remaining sharp for a long time. The downside of ceramic knives is that they are breakable. They also require a special tool to sharpen them, which we'll discuss in a moment.

The *blade* of the knife is usually made of high-carbon stainless steel; sometimes, it is just stainless steel or carbon steel. In the past, carbon steel was popular because it is easy to sharpen; however, blades made with carbon steel lose their edges quickly and need to be sharpened often. They also discolor (darken) in contact with high-acid foods such as tomatoes, onions, or oranges. Carbon steel will pit and rust, so it must be washed and thoroughly dried after each use.

Stainless steel doesn't have rust and discoloration problems, but it is harder to sharpen a stainless steel blade properly. The good news is, once you get it sharp, it stays sharp longer. Stainless steel is a stronger metal than carbon steel.

The most recent cutlery technology combines the advantages of both carbon and stainless steel. The *high-carbon stainless steel* blade keeps its sharp edge without rust or discoloration. Alloys with exotic-sounding names are part of each manufacturer's mixture: molybdenum (for heat resistance and strength); chromium (for hardness and corrosion resistance); manganese (for strength and durability); and vanadium (for flexibility), to name a few.

Knives have several different important parts, most of which are shown in Illustration 18-5. They are:

- The *tip* (the pointed end of the blade)
- The *cutting edge*
- The ***spine*** (the unsharpened edge of the blade, opposite the cutting edge)
- The ***heel*** (the end of the knife handle)
- The ***bolster*** (the thicker place at the "bottom" of the blade, where it meets the knife handle)
- The ***tang*** (the extension of the blade that is embedded in the knife handle; not seen in the illustration)
- The *rivets* (metal bolts that hold the tang and the handle together)
- The *handle*

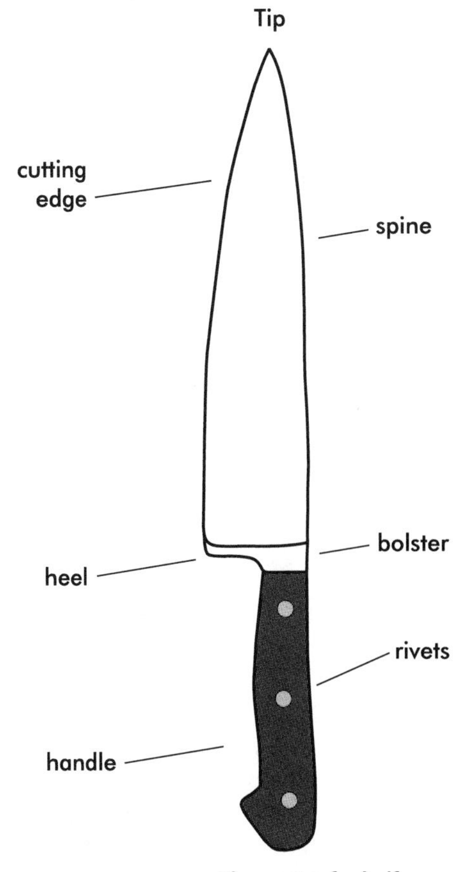

ILLUSTRATION 18-5 **The parts of a knife.**

The professional kitchen knife blade should always be ***taper ground.*** This means the entire blade is made from a single piece of steel, ground so that it tapers smoothly from the spine to the cutting edge. You may also hear the term ***hollow-ground blade***; this means the blade is made from two pieces of steel welded together, with beveled or fluted edges. Its primary selling point is its very sharp edge, but with the extraordinary use it will get in a foodservice kitchen, the hollow-ground blade lacks balance and has a much shorter life.

The bolster is also known as a *collar* or *shank.* Not all knives have bolsters, but all well-made ones do. For longest useful life, the bolster should be an extension of the blade, not a separate piece of metal attached to the handle. The purpose of the bolster is to provide more weight to the knife, improving its balance, but it is also important that the design or shape of this area not harbor bacteria by being difficult to clean.

Also important for heavy-duty commercial use is that the tang run the entire length of the handle. You'll see plenty of bargain-priced knives with partial tangs or so-called ***rat-tail tangs,*** in which the tang runs the length of the handle but is tapered to be much thinner than the spine of the blade. Neither of these will hold up to continuous use.

Knife handles are made of a number of different substances, including heavy-duty plastic or wood, unfinished or with a painted or lacquered finish. Some sanitation codes require that plastic handles be used in meat cutting, because they are considered easier to sanitize than wooden handles. The newest and most ergonomic knife handle is made of rubberized polypropylene, for a nonslip grip and a knife that conforms better to the hand.

In some cases, the appearance of the knife is a consideration in choosing a handle: A sleek wooden handle looks better carving roasts on a fancy serving line, for example, than a plastic handle. The highest-quality wooden handles are made of rosewood. Because it is extremely hard and has no grain, it resists splitting and cracking. Many wooden handles are smaller in diameter than plastic ones, so they often feel more comfortable to female chefs or others with comparatively small hands. A note about caring for unfinished rosewood handles: They can soak up water and become stained with constant use and cleaning. Overall, painted or lacquered handles are much more durable.

The rivets that fasten the handle to the tang should be completely smooth and flush with the surface of the handle. If they're not, they can irritate the user's hand and also create a sanitation hazard by trapping microorganisms.

Among your most important safety considerations in knife selection is its comfort—quite simply, the way it feels in your hand. A knife's weight should be balanced between the handle and the blade. If it's not distributed evenly, the knife can become cumbersome to use, or even dangerous. The handle should provide a comfortable grip, which will result in accuracy and speed.

The repetitive motions of knife use can easily lead to inflammation of the hand and wrist, often referred to as carpal tunnel syndrome, a costly worker's compensation issue in some businesses. To that end, several manufacturers now offer knives with ergonomically designed handles to minimize this risk. One is an angled handle. It may feel a little odd at first, but it is designed to allow the bones in the hand and wrist to align more naturally than they do with a straight, horizontal knife. Another improvement is the easy-grip handle, which allows the user to grip the knife handle with less pressure and still maintain control of the blade. There are even knives designed for left-handed persons. So consider these options, especially if your kitchen will require certain workers to perform repetitive tasks for long time periods.

Types of Knives

The most basic kitchen knife goes by a number of different names: the ***chef's knife,*** *French knife,* or *cook's knife.* They all refer to the 8- to 12-inch blade used for most chopping, slicing, and dicing tasks. A laminated wood handle is recommended for this knife, which takes a lot of abuse even in a home kitchen, but color-coded heavy-duty plastic handles are also available.

The ***utility knife*** or *salad knife* is a smaller (6- to 9-inch) version of the chef's knife, used mostly for pantry work such as preparing salad greens or slicing fruit. Its stainless steel blade should have a scalloped cutting edge and a rosewood handle with a bolster.

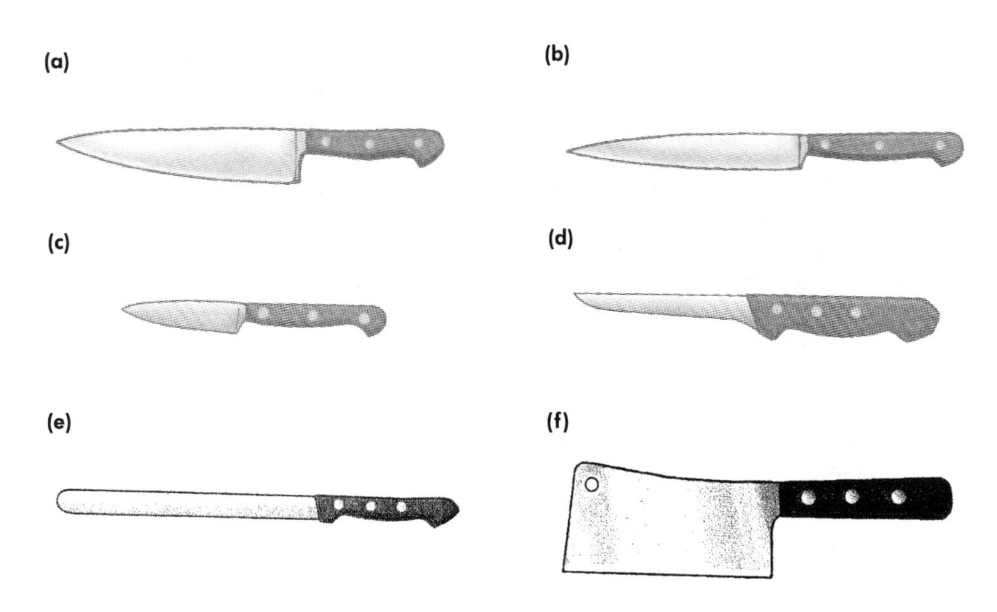

ILLUSTRATION 18–6 **(a) Chef's knife (b) utility knife (c) paring knife (d) boning knife (e) slicer (f) cleaver.**

SOURCE: *PROFESSIONAL COOKING,* FIFTH EDITION BY WAYNE GISSLEN. THIS MATERIAL IS USED BY PERMISSION OF JOHN WILEY & SONS, INC. © JOHN WILEY & SONS, INC., 2003.

The ***paring knife*** is an even shorter blade (3 to 4 inches in length) with a sharp tip that may be pointed or slightly curved. (The curved point is called a *tourne.*) This knife is used to pare and trim fruits and vegetables. It's in big demand in most kitchens and, because it is small, it is the knife most often misplaced—so order extras! In some kitchens, while no expense is spared on other knives, inexpensive paring knives work best because they have to be replaced so often.

The ***boning knife*** is used to separate raw meat from the bone. The blade is thinner than the other knives we've mentioned; for most meats, it should be stiff and pointed, but, for deboning poultry and/or fish, choose a more flexible blade. The stiff knife will be 5⁄8 inch wide; the all-purpose, more flexible model will be 7⁄8 inch wide. The length of the boning knife blade is 5 to 6 inches. Some manufacturers also offer a slim, curved blade in addition to the straight blade.

The ***slicer*** is a long, slender knife (12 to 14 inches) with a flexible blade that is used to slice cooked meats. You can order it with either a rounded or a pointed tip, as well as a solid or serrated edge. High-carbon stainless steel is the recommended material for this hard-working blade.

The ***cleaver*** is used for heavy-duty chopping, primarily of meats, and should be sturdy and hefty enough to cut through bones. A typical cleaver will weigh from 3⁄4 to 1½ pounds. The blade is rectangular in shape, 6 to 7 inches in length, and about 4 inches wide.

That is the basic selection, but one look at a knife catalog and you'll see there are many more to choose from—scimitars for meat-cutting, small oyster and clam knives, and so forth. A catalog or two from major manufacturers should always be on hand for the person who does the knife buying in your operation.

When professional cutlery is purchased, it's important to plan first what cutting tasks will be performed and by how many persons at the same time; how often these tasks will be done; what types of products will be cut (whole cuts of meat, for example); and whether specialty tools will be needed for tasks such as pizza slicing, oyster shucking, or meat carving. We'll discuss knife sharpening in just a moment.

Cutting Boards

The surface on which cutting is done is almost as important as the knife itself. A good cutting surface makes the job easier and lessens wear on the knife. Most cutting will be done on a cutting board, either wooden or plastic. Although wooden boards are used in many situations, they are frowned upon by health inspectors. The modern preference is the plastic board, which can be soaked in sanitizing solution or run through the dishwasher. New-generation cutting boards made of resin are even harder and more heat-resistant than NSFI-approved plastic boards, and another excellent option.

If not cleaned correctly, any type of cutting board can pose a health hazard. Over time, they all develop grooves, nicks, and pits from the cutting process, in which bacteria can grow and contaminate food. And the tendency in a busy kitchen is to use the board, hurriedly wipe the surface, then use it again for some other type of food.

There's been a debate in recent years about whether putting a cutting board in the microwave oven will heat it sufficiently to destroy bacteria, but experiments with this technique have resulted in steam buildup, which turned some hot cutting boards into fire hazards! It's much less risky to combine 1 cup of liquid beach with 1 gallon of water and submerge the board in this solution. The solution can also be used to wipe the board after each use.

Restaurant owners can minimize the safety hazards with another simple system, which starts by having more than enough cutting boards on hand. This lessens the chance that a busy worker will reuse the same board for different foods. Many establishments use color-coded plastic boards, matched to the types of food being cut: red for meat, pink for poultry, green for lettuce, and so on. Plastic cutting boards come not only in different colors, but in different thicknesses, too.

Knife Sharpening

Regular, correct sharpening of knives actually prevents injuries. Think about it—it takes a lot more work to use a dull knife than a sharp one. Caring for knives is an ongoing process, which should be entrusted to employees who will do it often and well. Many people assume that the tool called a ***steel*** is what sharpens knives, but this is not true. The steel is a long (10- to 14-inch), slim tool that looks a little bit like a bayonet. It is made of steel, electroplated with a hard nickel coating. Here's how it works: Tiny, industrial-grade diamonds are embedded in the nickel, giving it a slightly gritty surface. The idea is to scrape the knife blade quickly back and forth across the steel. This motion keeps the blade in alignment. It does not "sharpen" the knife, but helps maintain the sharpness of the cutting edge. You place one hand on the handle of the steel and run the knife across it with the other hand. There's a protective hand guard separating the handle from the steel itself. Using a magnetized steel will prevent microscopic dust particles from clinging to the newly sharpened knives.

Knives may also be sharpened with ***sharpening stones.*** The blade is passed over the surface of the stone at an angle, and the coarseness of the stone (also called its ***grit***) abrades the cutting edge to sharpen it. Sharpening stones are made in a variety of grits—begin by using the coarsest one, then moving on to the finer surfaces. Hold the knife blade at a 20-degree angle to the stone's surface and draw the entire blade across the stone. Stones may be used dry or may be moistened first with mineral oil. A favorite is the triple-faced stone, mounted on a rotating frame so it can be locked into position for use. You can see a steel in Illustration 18-7.

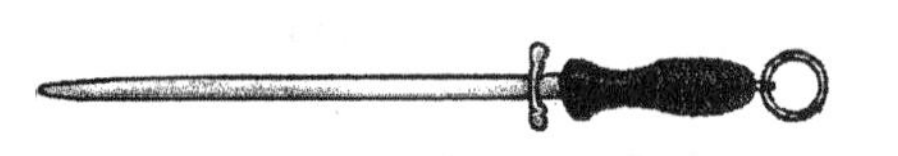

ILLUSTRATION 18-7 Steels, as well as sharpening stones, are used regularly to keep knives at their best.

SOURCE: *PROFESSIONAL COOKING*, FIFTH EDITION BY WAYNE GISSLEN. THIS MATERIAL IS USED BY PERMISSION OF JOHN WILEY & SONS, INC. © JOHN WILEY & SONS, INC., 2003.

Knives can also be sharpened with a *rotating wheel and belt* or a *diamond hone and knife sharpener.* In the latter case, roller grids precisely control the sharpening angle. Small diamond abrasives remove the old, weak edge and create sharp bevels there. Then the blade goes through a second process to hone the bevels and polish the edge to razor sharpness. The diamond hone tool weighs about 3 pounds.

The ceramic knife requires a ***ceramic steel*** to do the sharpening. It is made of aluminum oxide, a strong compound found in some precious stones.

18-3 MEASURING TOOLS

If cooking is art, it's also at least half science, and the scientific part includes accurate measurement of ingredients. Measuring requires several different types of tools, including scales, volume measures, ladles, and scoops. Don't forget food safety, with thermometer probes and other temperature-testing devices.

Kitchen Scales

The scales discussed in Chapter 10 were the types used to weigh bulk materials in your receiving and loading areas. This discussion focuses on scales used in the cooking process, to weigh ingredients and check portion sizes. No matter what style of scale you choose, consider a few important points as you shop:

CAPACITY. Do you want scales that can weigh a whole chicken or just a chicken breast? Generally, choose the larger capacity, since it will be more useful in the long run.

DESIGN. Scales with a large area or basket for the items to be measured are preferable, or those that allow you to use a bowl when needed. Also, is it easy to clean?

PRECISION. You can buy scales that measure as precisely as ¼ ounce or 1⁄10 of a pound. Again, the more precise it is, the more useful it will be. The greater their precision and accuracy, the more expensive scales are.

ACCURACY. Are the scales easy to calibrate, and will someone on your staff do this regularly? The greater the accuracy, the better your portion control and cost control.

READABILITY. Some scales are easier to read and, therefore, better suited to precise measurement. Digital scales are the easiest to read, but also the most expensive.

Now, here's a quick rundown of what's available on the market.

The spring platform scale has a small stainless steel platform (4 by 6 inches) on which food is placed, with spring and chassis below. A circular dial on the chassis tells you the weight of the food. Choose a model with an adjustable dial that allows you to "zero out" the weight of the container and determine the true weight of the food. Spring platforms come in two basic sizes: those that can weigh up to 32 ounces in ¼-ounce increments, or up to 50 ounces in 4-ounce increments.

The digital portion scale is battery operated. Its stainless steel platform is 7 by 8½ inches, and its chassis is only about 3 inches tall. It measures up to 6 pounds in ½-ounce increments, announcing its findings in 1-inch numbers on a liquid crystal display (LCD) window. A push button on the face of the scale can "zero out" the container weight. To save battery power, the scale turns itself off automatically if the platform remains empty for more than three minutes. In supermarkets and delis, some models can print price and weight labels.

Anyone who bakes from scratch will find the baker's dough scale indispensable. It has a large scoop for holding flour, sugar, and other common baking ingredients. Stainless steel scoops last longer than plastic ones, which chip with use. The scoop is filled, then rested on one side of the scale; counterweights of 1 to 4 pounds are placed on the other side. There is also a sliding pose, which can be adjusted in ¼-ounce increments.

Costly mistakes are made when the correct type of scale is not used. Think of it as a bar owner who allows the bartenders to "free-pour" alcohol instead of measuring it, resulting in lower profits. Unfortunately, this point is overlooked in many kitchens. Sometimes, an owner purchases only one scale when at least two are needed: one in the hot-food area and one in the cold-food area.

Volume Measures

You'll need measuring cups and spoons. For liquids, the cups resemble pitchers, with a raised lip to pour ingredients. These can also be used for dry ingredients, but there are separate dry measuring cups. In many instances, it is probably smarter to weigh things like flour and sugar.

Volume measures are available in a variety of sizes, from sets of measuring cups like you find in home kitchens, to quart-sized sets, with containers that measure from ½ quart to 4 quarts (1 gallon). Commercial-type measures are made from seamless, heavy-duty aluminum. Keep them from getting dented, since this may distort the capacity of the container. Avoid using glass measures and you'll avoid the inevitable breakage.

Ladles

This kitchen hand tool is most often used to transfer liquids between containers, but it is also a measuring tool. Let's look at the value of the ladle to kitchen efficiency. An employee might portion 8 ounces of soup into a 12-ounce bowl with two dips of a 4-ounce ladle or with one dip of an 8-ounce ladle. Simple enough, right? Well, the worker who serves, say, 200 of these bowls per day wastes 10 seconds or so per serving by having to dip the smaller ladle twice each time. This means 33 extra minutes per day is spent ladling soup, which could be spent more productively.

In a self-service situation, an oversized ladle in the soup can mean customers who, understandably, help themselves to 10 ounces but pay for only 6 or 8 ounces. Bottom line: A properly sized ladle helps control both food and labor costs.

Most ladles are made of 18-gauge steel, and the capacity of their bowl is stamped on the handle. Ladles begin at 1 ounce in size and range all the way up to 24 and 72 ounces. There are even 1-quart and 40-ounce capacities, with lesser volumes marked on the side. The lengths of their handles vary from 9 to 18 inches. The NSF has sanitation and safety standards for ladles, including one-piece bowls with no welds or crevices and grooved handles for added strength. You can also order plastic-covered handles, which stay cool even when used with liquids up to 150 degrees Fahrenheit and are dishwasher safe.

Scoops

There are two types of tools commonly known as scoops. One is a rounded spoonlike instrument used primarily for dishing ice cream. The other is much larger, and looks more like a three-sided trough with an opening on the fourth side. This type of scoop is used for handling dry ingredients or ice. Let's discuss the ice cream scoop first.

The ice cream scoop is also sometimes called a ***dipper*** or *disher.* The bowl is made of stainless steel; the handle of sturdy plastic. The bowl usually includes a squeeze mechanism used to push the ice cream out and into a cone or dish. Bakers also like using these to portion batter for muffins and cupcakes. Sometimes you squeeze the mechanism in your hand; sometimes with just the thumb. (The latter type is less expensive, but not well-suited for big jobs—the repetitive motion can be uncomfortable.) The capacity of the bowl is printed on the scoop, expressed not as ounces but as the approximate number of scoops you can get out of 1 quart of ice cream using this particular scoop. The most common dishers are numbers 12, 16, and 20, although there are several others. Some manufacturers color-code the ends of the plastic handles to help identify different-sized scoops.

A variation of the ice cream scoop is the ***portion server.*** This is exactly the same tool as the spoodle described earlier in this chapter, made of 18-gauge steel with a plastic-covered handle, also color-coded for capacity. Portion servers range in size from 2 to 8 ounces and come with perforated bowls or solid bowls.

Scoops used for dry ingredients or ice should be made of stainless steel or heavy cast aluminum, with a handle that is ribbed for easy gripping. In some cities, scoops cannot rest directly on the product when not in use—that is, you can't store the flour scoop in the flour bin or the ice scoop inside the ice machine, but must provide a pan outside the container that can be kept clean.

Scoops vary in size from 4 ounces (perfect for a bar, because that's about as much ice as you'd put in an average drink) to 84 ounces (well suited for loading buckets of ice to stock a salad bar). The physical size of the scoop will range from almost 8 inches long and about 1 inch deep to 12 inches long and about 6 inches deep.

Thermometers

HACCP compliance requires constant vigilance to guard against food-borne illnesses, and temperature control is a critical part of this process. When cooking hamburgers, for instance, you prevent the presence of deadly *Escherichia coli (E. coli O:157)* bacteria by cooking the meat at 155 degrees Fahrenheit for at least one minute. Refrigerators, freezers, fryers, and ovens all need to be set at the correct temperatures to do their jobs.

Luckily, temperature-sensing technology has improved greatly in the last few years. Many thermometers are small and easy to handle. They have digital faces to take the guesswork out of reading them accurately.

There are several different types of thermometers you should know about, and you'll be using most of them daily in foodservice applications. (Illustration 18-8).

(a)

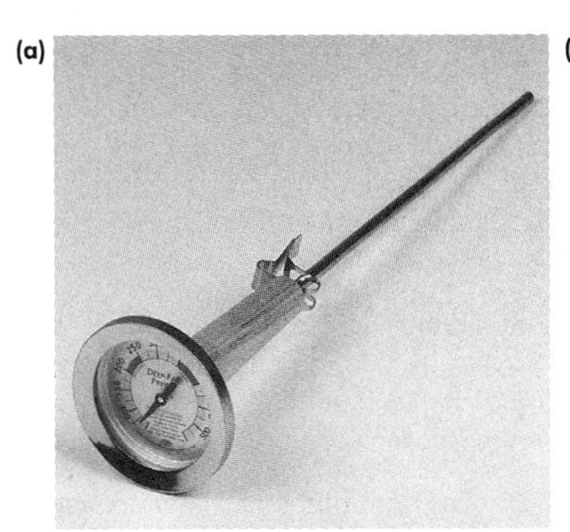

(b)

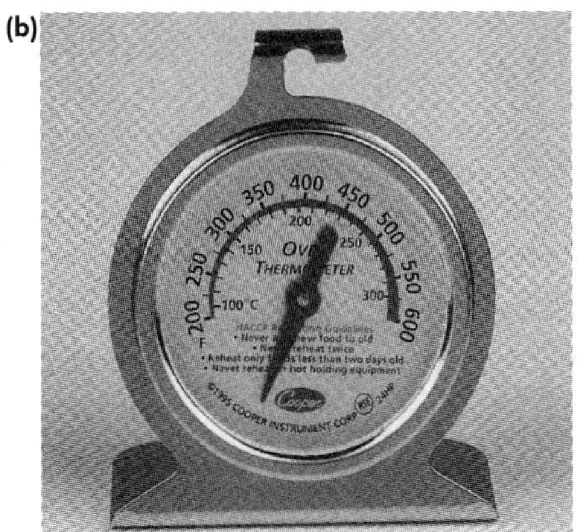

(c)

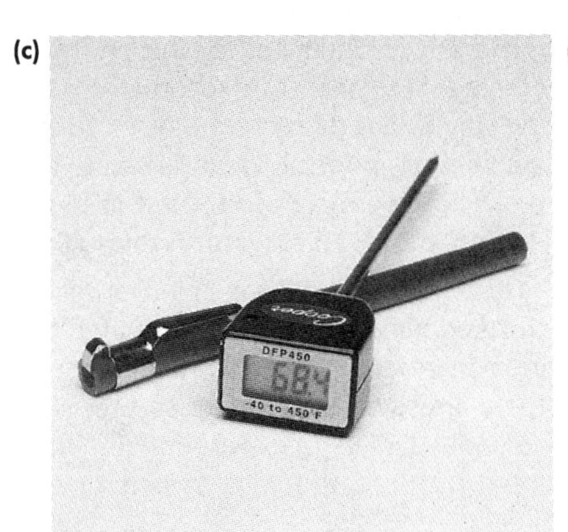

(d)

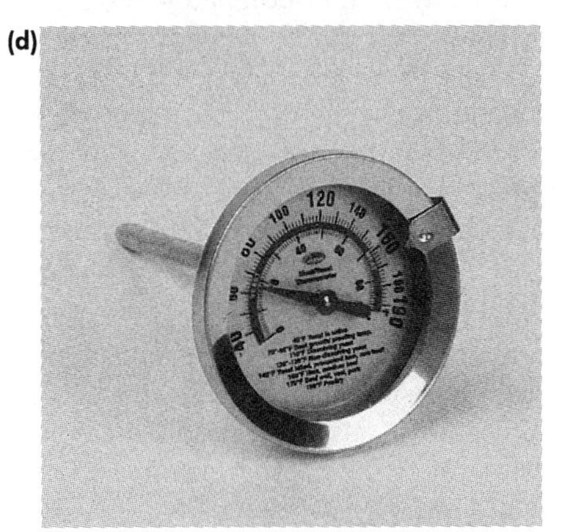

(e)

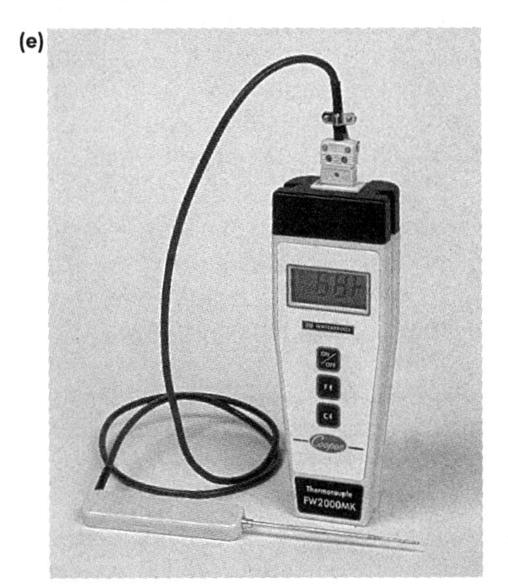

ILLUSTRATION 18-8A-E **The most common types of temperature-measuring devices used in food preparation: (a) stem-type or bimetallic thermometer; (b) oven thermometer; (c) pocket-size infrared thermometer; (d) meat thermometer; (e) waterproof thermocouple (Courtesy of Cooper Instrument Corporation, Middlefield, Connecticut).**

Bimetallic Thermometer. Common names for this device are the *stem-type* or *stemmed, portable, bimetal* and *testing thermometer.* It has a long, somewhat sharp end that is pushed into the center of a roast, for instance; the dial face at the other end will register the temperature. For most kitchen purposes, a 5-inch stem on the thermometer is sufficient, but longer ones may be needed for use in larger food containers. There are also convenient, 3-inch pocket sizes. All are water-resistant.

The typical thermometer, either dial-face or digital readout, is accurate to within plus or minus 2 degrees, but only if used correctly. The entire sensing area, not just the tip, should be inserted into the deepest part of the food. Most stems have a tiny indentation or dimple that marks the sensing area.

The bimetallic thermometer is battery operated, using a button-sized watch battery, and has a "power off" feature to save batteries when it's not in use. And for use in freezers, on grills, and in deep fat fryers, heavy-duty models are available.

Thermocouple. The ***thermocouple*** is a battery-operated hand-held or wall-mounted meter. It is connected by a flexible cord to a probe that can read the temperature of anything it touches, from 0 to 500 degrees Fahrenheit. Since the sensor is located in the tip of the probe, there is no need to insert it very far into the food. Thermocouples are required in some kitchens for temperature checks of foods that are thin, like burgers and chicken breasts. Different types of probes accomplish different tasks:

- The penetration probe is used to measure the internal temperatures of foods.
- The immersion probe is for measuring liquids, from frying oil to sauce or soup.
- The surface probe can be used for taking surface temperatures of appliances.
- The air probe measures the air temperature in ovens or refrigerators.

Most thermocouple meters can be used with probes from different manufacturers. Inside the probe are two wires, welded or crimped together at the tip. Heating this junction causes thermoelectric energy, which is conducted up the wires to another junction at the other end of the probe. The signal travels through a cord to the meter, which displays the voltage as a temperature reading on a digital display.

Depending on the type of metal used in the junctions, thermocouples can read extremely high temperatures—up to 1800 degrees Fahrenheit in scientific models! There are nine basic types of thermocouples, each identified with a one-letter name. The type most often used in foodservice is "K." Its price tag is between $100 and $200—compared to less than $10 for many stem-type thermometers.

Thermistor. Like its cousin the thermocoupler, the ***thermistor*** (short for "thermo resistor") is a type of semiconductor. Inside its probe, metal oxides are formed into a tiny bead and coated with glass or epoxy. The resistance of the metal drops as its temperature increases. Thermistors are far more sensitive than thermocouplers, useful in laboratories because their pin-sized sensors are so precise. But the range of temperatures they can detect is smaller, only up to about 150 degrees Fahrenheit, which limits their use in foodservice. However, thermistors are often used inside data-logger systems (see "Specialty Thermometers" later in this chapter.)

Infrared Thermometer. Another recent development is the ***infrared thermometer,*** also called a *noncontact thermometer* because it doesn't even have to touch the food to read its temperature. Its use is based on the principle that all objects emit infrared energy, and the hotter they get, the more energy they emit. You point it at an object, squeeze a trigger, and the instrument's internal optics (sometimes a laser) collect the energy and focus it on a detector, which displays it as a temperature reading. The process only takes half a second.

Using an infrared thermometer is easy, but also tricky. Plastic film and reflective surfaces such as glass or tinfoil may interfere with its operation, giving you the temperature of the glass or metal instead of the food. Soup or stew will have to be stirred first, or you'll get a reading of only its surface temperature. It's smart to take two temperature readings—take a surface temperature reading, then stir and take a second reading.

Infrared thermometers run on batteries and cost $200 or more. They are handy, but they can't really replace the accuracy of an internal temperature probe, so they are used most often for quick scans of items on salad bars or buffet lines, chores that would otherwise require sanitizing a probe-type thermometer between each dish.

Environmental Thermometers. These are not meant to measure food temperatures, but the temperature of the environment in which they are placed. They're sometimes called *equipment thermometers.*

For refrigerated and dry storage areas, a round or rectangular dial thermometer can be mounted inside the door or can stand on a shelf. These cold-temperature thermometers measure from minus 40 to 80 degrees Fahrenheit and, in most cases, also show temperatures in degrees Celsius. You can also specify, with most new refrigerators and freezers, that the thermometer be wired to read the inside temperature, but be displayed on the outside of the door for the most accurate reading. Some are combination temperature and humidity sensors.

The *liquid crystal thermometer* is an inexpensive choice for coolers. It isn't battery-operated, but filled with chemical crystals that react to cold temperatures, registering a color change instead of a digital readout. These are also called *time-temperature indicators (TTIs)* and can be useful when transporting foods for off-site catering.

Oven thermometers will verify that your oven is heating accurately. Two variations on this idea: the thermometer with a clip-on heat probe, which is useful for testing the temperatures of reach-in appliances; and the surface-temperature probe, which identifies hot spots on griddles or poorly functioning burners or heating elements.

Dish machines also require thermometers to check water temperatures.

Specialty Thermometers. Candy thermometers are used for making candy and jellies, and also in deep-fat fryers. The intense heat of these processes requires a thermometer that can read up to 400 degrees Fahrenheit. The infrared thermometer is taking the place of the candy thermometer in many cases, since it's less messy.

Most specialty thermometers are distinguished primarily by the temperature ranges they cover. For example:

- Sauce thermometers –35 to 180 degrees Fahrenheit
- Yeast/dough thermometers –40 to 160 degrees Fahrenheit
- Coffee thermometers (for measuring the temperature of frothed milk for lattes) –0 to 220 degrees Fahrenheit
- Kettle thermometers (have longer stems) –50 to 550 degrees Fahrenheit

Disposable thermometers aren't designed to read precise degrees, but they have a sensor spot that changes color to indicate that a food item is below or above an ideal temperature reading. They come with a sensor spot of either 140 or 160 degrees Fahrenheit. They are handy for testing dish machine water temperatures, dishes on buffet lines, and individual portions of items like hamburgers.

The cook-chill system adopted by many high-volume operations requires close control over both cooking time and temperature. Thermometers with heat probes continually monitor both the cooking and the cooling process. There are specific *cooling thermometers,* which can signal a temperature drop from 165 to 40 degrees Fahrenheit within the specified time to meet food safety guidelines. Systems called ***data-loggers*** are used to verify product safety during meat-smoking, hot- or cold-holding, transport, or storage. They have the ability to keep a running chronological temperature check and upload it to a computer for printout.

Food safety isn't the only reason thermometers and temperature probes should be readily available to kitchen personnel. The simple failure to use a meat thermometer when cooking a roast, for example, can mean the roast overcooks by 1 to 3 degrees. Just this seemingly tiny temperature difference can result in a decreased yield of 10 to 15 percent from that one cut of meat.

Use and Care of Thermometers. Any type of thermometer will only work well if it is properly sanitized, properly calibrated, and used correctly. Here are a few tips:

- Use the right thermometer for the job. A meat thermometer (with a readout that starts at 120 degrees Fahrenheit) is not the correct tool for measuring the temperature of a bag of solid-pack spinach to be sure it's still frozen. A bimetallic thermometer with a 2-inch sensing area isn't going to work on items like hamburgers or eggs, but a thermocouple with a tiny probe tip will.
- No matter what the manufacturers' claims about how quickly the thermometer works, wait about 15 seconds after you're inserted a stem or probe before recording the temperature. This ensures the thermometer has had a chance to get a good, steady reading.

- Thermometers should be washed, rinsed, sanitized, and air-dried between uses. Soaking it in any sanitizing solution that's safe for food-contact surfaces will work. If you're testing both raw and ready-to-eat foods with the same thermometer, it must be washed and dried in this way. Don't forget to sanitize its storage case as well.
- If you're testing only cooked foods, or only raw foods, it's okay to simply swab the food contact surface of the thermometer with rubbing alcohol between uses. Give the alcohol enough time to completely evaporate before reuse.
- The old "shake-down" mercury-filled type of glass thermometer is absolutely never permitted for taking temperatures directly in hot or cold foods because of the danger of breakage and mercury poisoning.
- Infrared thermometers work best when they are held as close to the product being measured as possible, without touching it.
- If you're going to take food temperatures, have a simple system to do it at regular intervals, and a form that employees can use to record the information, posted on clipboards throughout the kitchen and storage areas. Make sure everyone is trained to monitor temperatures and understands the importance of these tasks as well as the procedures.

With constant use, or if they are dropped, stem-type thermometers often lose their correct calibration, making their readings slightly off. When it comes to food safety, even a couple of degrees are critical and luckily, this is an easy problem to correct. There are two ways to recalibrate, the ice point method and the boiling point method.

Ice Point Method. Pack a 10-ounce water glass with clean, crushed ice, then add water. Let it sit for about five minutes, until the water is cold and slushy. Place the thermometer probe or stem into the icy water until the temperature reading stabilizes, which should take no more than 30 seconds. Don't let it rest on the sides or bottom of the container. Then, following the manufacturer's instructions, take the thermometer out of the water and adjust its calibration nut (located behind the dial) until its temperature reads 32 degrees Fahrenheit. Check your progress by returning the thermometer to the glass and seeing if it registers 32 degrees Fahrenheit.

An infrared thermometer can also be checked using ice water. Stir the ice water and, from a distance of about 12 inches from the water, press the trigger to display the temperature. If the thermometer is working properly, the readout should be 32 degrees Fahrenheit.

Boiling Point Method. This is not quite as accurate a method as the ice point, but it's the only way to recalibrate a maximum-registering thermometer. Put a pan of clean water on the range top and let it come to a full, rolling boil before inserting the thermostat. If it's a stem-type thermometer, hold the tip about two inches from the bottom of the pan; if it's an immersion probe, immerse it completely in the water. When it has reached a stable temperature reading, adjust the thermometer to read 212 degrees Fahrenheit by holding the calibration nut with a wrench and turning the head or face of the thermometer until it reads correctly.

Remember, water boils at slightly different temperatures depending on altitude, so your thermostat's absolute accuracy will be based on that number. The rule is that its boiling point decreases about 1 degree Fahrenheit for each 550 feet above sea level.

18-4 POTS AND PANS

There are two types of cookware: range-top cookery and oven cookery. Together, they represent the highest-priced and largest variety of smallware in a restaurant kitchen. These are heavy-duty items, made of metal. You will occasionally find glass, ceramic, or terra-cotta cookware in a commercial kitchen, but only for specialty uses—and with the chef's admission that there's a lot of breakage and chipping.

Before we continue, let's talk about the difference between pots and pans. To the average person, this isn't a critical issue, but it is to the person managing a professional kitchen. Consider a 9-quart saucepan and a 9-quart saucepot, as shown in Illustration 18-9. The saucepan, with its single, extended handle, is easy to lift and pour from. The long handle keeps the cook's hand away from the hot food inside, helping to prevent burns. The saucepan is best suited to range top cooking on the front burners, where cooks need to tend and periodically turn the food being prepared. However, also note that the long handle might be a "traffic hazard" in a busy kitchen, if it sticks out into an aisle.

The saucepot is harder to pour from, but it is well suited to slow-cooking sauces on the back burner of a range top. Can you see why both styles are needed for maximum kitchen efficiency?

Whether it's a pot or pan, the metal it is made of should distribute heat evenly and uniformly. Otherwise, the cookware will develop hot spots, which cause food to scorch and burn in spots. Good chefs can compensate for an inferior-quality pan, but why should they have to? Two basic factors govern a metal's ability to cook evenly: the type of metal and its thickness. The thicker a pot or pan is, especially the bottom, the better it holds heat. Conductivity—how fast a metal heats up and holds that heat—varies with different metals.

Manufacturers often bond thin sheets of different metals together, getting the best of both to form a pan or pot. This process is called ***cladding.*** Often, cladded cookware is copper on the outside (for maximum heat conductivity) and stainless steel on the inside (for sturdiness and stain resistance). "Triple-cladded" cookware has two sheets of stainless steel—outside and inside—with a carbon steel core between them. The most common metals used in cookware are:

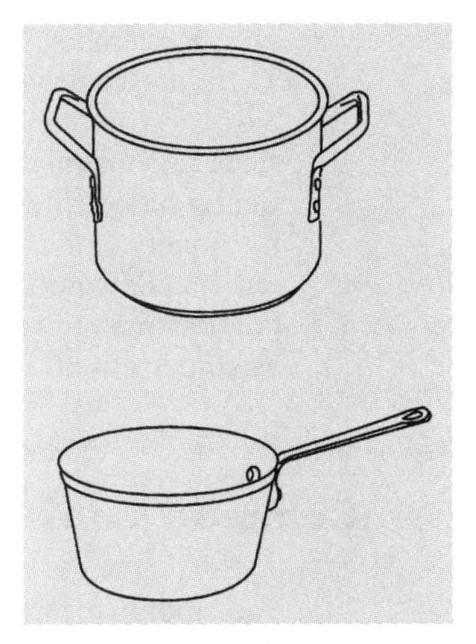

ILLUSTRATION 18-9 **The differences between a saucepot (top) and a saucepan (bottom) are obvious when viewed together.**

ALUMINUM. This metal is lightweight and yet it is a good conductor, so it is a very popular choice for pots and pans. However, as metals go, it is fairly soft, which makes it easy to scratch or dent. Also, aluminum should not be used for lengthy cooking or storage of foods with a high acid content. These foods may react by taking on a slightly metallic taste or, in the case of pale-colored foods such as béchamel sauce or cream of mushroom soup, aluminum may discolor them. Most often in commercial cooking, aluminum is used to coat a stainless steel pan. This cladded combination is known as ***aluminized steel*** or ***tinned steel.***

COPPER. This metal heats quickly and evenly and cools fast, too. It is the best heat conductor, but it is also the most expensive type of cookware. In addition to being pricey, it is heavy. Like aluminum, copper reacts unfavorably to some foods, so it is usually lined with tin or stainless steel. Tin is soft and easily scratched.

STAINLESS STEEL. By itself, this metal is not the best choice for cookware. A poor conductor of heat, food scorches easily in stainless steel and has to be watched more carefully. A better choice is the cladded stainless steel pan with a layer of either copper or aluminum bonded to the (exterior) bottom, which combines conductivity with the durability that is stainless steel's best asset. Unless it has this bottom coating, stainless steel is best suited for low-temperature duties such as keeping foods warm (steam table pans) or storage. One important advantage is that foods don't have chemical reactions to stainless steel.

CAST IRON. Used primarily to make skillets and heavy griddles, cast iron distributes heat evenly and holds its temperature well. However, it cracks easily and rusts very quickly unless it is kept properly conditioned and dry when not in use. Sometimes cast iron is cladded with a decorative exterior of porcelain enamel.

NONSTICK COATINGS. You may already be familiar with these synthetic layers that are applied to the interiors of some pots and pans, with names like Teflon and Silverstone. These coatings provide a nonreactive finish that prevents foods from sticking to the pan, allowing the cook to sauté with little or no added oil. The important thing to note about coated pans is that great care must be taken to keep the coating intact. Don't use metal spoons or spatulas, which can nick the coating, and dish room employees must be trained never to clean them with abrasive scouring pads, steel wool, or gritty cleaners. In short, coated pans often do not stand up to the rigors of a busy commercial kitchen. However, recent technology has introduced cookware with an incredibly thin layer of stainless steel over the traditional nonstick coating, to seal it more securely to the pan.

Remember, your cookware choices will be different of necessity if you're using induction appliances. There are cookware lines advertised specifically as "induction-ready," and manufacturers of induction burners recommend a cladded combination of copper or aluminum (for conducting heat) and carbon steel (for magnetic properties.) Interestingly, when Pacific Gas and Electric's Food Service Technology Center tested three different induction cooktops with 14 different pans in its research laboratory in 2001, the quality of the cooktop had more to do with the results than the pan. A "value-priced" cooktop that put out 1.5 kilowatts didn't have the same results as the "best-quality" model that with the same kilowatt output. Between the three units and the various sizes of pans, the heat transfer sometimes dropped as much as 40 percent off the manufacturer's rated input of the unit.

Common Cookware Types

Saucepots are large, round vessels with straight sides and two welded loop handles at opposite sides of the pot. Saucepots are shallower than stockpots, and are used to make soups, sauces, and other liquid foods. Its shallowness makes it easier to stir the contents of a saucepot. It is made of aluminum, and the best ones have a double-thick, 8-gauge rim and bottom. Saucepots range in capacity from 5 to 60 quarts. The smallest are about 8 inches wide and 5 inches deep; the largest, 20 inches wide and 11 inches deep.

Never buy a saucepot without a matching lid. In commercial cookware, lids or covers often must be ordered separately. Be sure the lid you choose from a cookware catalog matches the item number of the exact pot or pan you intend to use it with. Lids made of heavy-duty aluminum will hold their shape best. And remember, covers measure slightly larger in diameter than the pots they are designed to fit, from 6 inches to almost 22 inches. Check the catalog numbers carefully for the correct fit.

You can also order other accessories for your saucepot, to adapt it to special duties such as pasta and vegetable cooking. Baskets come in various sizes and attach to the saucepot's rim, allowing the user to immerse and/or drain foods.

Stockpots look like taller versions of saucepots. They are used for preparing stocks and simmering large quantities of liquids. In fact, some have spigots to remove liquid without lifting the heavy pot, and vegetable screens (that clog easily, unfortunately) to filter the liquid as it flows out. The basic stockpot is made of ¼-inch-thick aluminum, with straight sides and a rim that is twice as thick as the sides. Using 8- to 10-gauge aluminum makes the pot a bit lighter. Some manufacturers make stainless steel stockpots with aluminum bases. Capacities range from 8 quarts to a whopping 140 quarts; depths from 6 to 22 inches; diameters from 8½ to 21¾ inches.

Match the circumference of your stockpot to the size of the burners you'll be using and, once again, match the cover to the pot by coordinating the item numbers if you order by catalog. Hollow handles on stockpots stay cooler, making them easier to move and carry.

Double boilers are used to cook anything prone to scorching. Some chefs also like them because they cook slowly, eliminating the need for constant stirring of delicate items such as sauces, puddings, and pie fillings. The double boiler consists of two pots: a base pot on the bottom and an insert pot that nests inside the base. Boiling water in the base gently heats the contents of the insert pot above it, so the food in the insert pot doesn't come into direct contact with the heat of the burner (see Illustration 18-10). Double boilers range in size from 4-quart inserts with 6-quart bases to 32-quart inserts and 40-quart bases. They're made of either aluminum or stainless steel.

The ***brazier*** is a wide, shallow version of the saucepot, used to braise, brown, and stew meats. Also called a *rondeau,* it is made of aluminum, with wide, sturdy handles for easy lifting. Braziers measure from 11 to 20 inches in diameter, and all are less than 6 inches deep. Order a matching cover for your brazier.

Saucepans are cousins to saucepots, with these basic differences: The saucepan is shallower and has one long handle (making it a pan) instead of the two loop handles on a pot. A saucepan may have straight or tapered sides. If the sides are slightly rounded where they meet the bottom of the saucepan, stirring will be easier. Saucepans are made of 4- to 8-gauge aluminum or stainless steel. You'll want a variety of sizes, from 1½ to 14 quarts.

ILLUSTRATION 18-10 **The double boiler is a stockpot with an insert that fits inside.**
COURTESY OF VOLLRATH COMPANY, LLC, SHEBOYGAN, WISCONSIN.

Sauté pans have almost as many names as the chef's knife! Many people refer to them as *frypans,* but there are other names for different shapes and uses. Sauté pans may have straight sides or sides that slope gently inward from the rim to the bottom. The straight-sided sauté pan is also called a ***satoir***; the slope-sided sauté pan is also called a ***sauteuse.*** To further confuse the novice, either of them may be referred to as an *omelet pan* or *crepe pan.* Finally, the *cast-iron skillet* is also a type of sauté pan. All of them are used to sauté, brown, braise, and fry. The broad surface area of the pan speeds the cooking of sauces or other liquids that require a rapid reduction in volume They are typically 8 to 12 inches in diameter, but can be as big as 20 inches in diameter.

Both the satoir and sauteuse are shown in Illustration 18-11. Following are a few features of each type of sauté pan.

Satoir. The classic, traditional sauté pan, the satoir is thicker and heavier than its counterparts. It has straight sides, keeping juices in contact with meats as they are stewed or braised. Its large, flat bottom and uniform heat distribution make it ideal for range top browning and simmering. The recommended metal for this type of pan is 8-gauge aluminum, although you can order equally fine stainless steel pans with aluminum bases and sides. Capacities range from 2 to 8 quarts. The satoir usually has a 9-inch handle but can be ordered with a 12-inch handle or two loop handles at each side. Be sure any handles are secured to the sides with three rivets and that the pan can be used in the oven as well as on the range top.

Sauteuse. This shallow skillet, traditionally known as a frypan, has gently sloping sides and a single long handle. Its angled sides allow moisture to dissipate more quickly from food, and enable a proficient chef to flip and toss its contents without using a spatula. Diameters range from 7 to 14 inches at the (wider) top of the pan and from 4¾ to 11 inches at the (narrower) base. Eight-gauge aluminum or stainless steel are your choices of material, with nonstick coatings now widely offered on this type of sauté pan. Matching lids are a requirement; some pans also have silicon rubber sleeves that fit the handles. The sleeves make the pan easier to grip, and they are removable if it's used under a broiler.

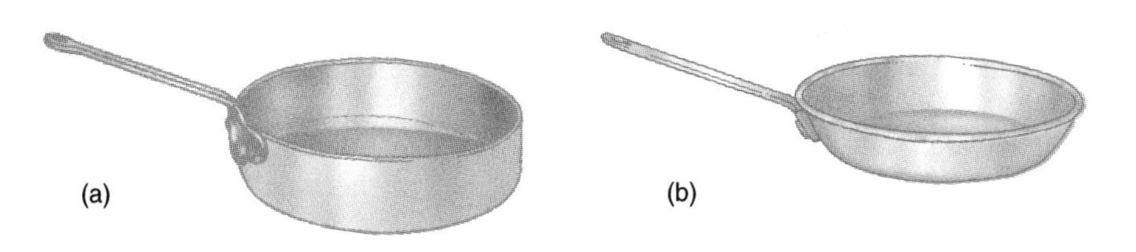

ILLUSTRATION 18-11 **a, b Sauté pans come in a variety of sizes and depths. (a) Satoirs are straight-sided; (b) sauteuses have sloping sides.**
SOURCE: *PROFESSIONAL COOKING,* FIFTH EDITION BY WAYNE GISSLEN. THIS MATERIAL IS USED BY PERMISSION OF JOHN WILEY & SONS, INC. © JOHN WILEY & SONS, INC., 2003.

There are several handy variations of sauté pans for special uses. They include:

- The *carbon steel frypan,* made of heavy-duty steel, which heats particularly fast.
- The *stir-fry pan,* commonly known as a *wok,* a wide, bowl-shaped aluminum pan used in Oriental cooking.
- The *paella pan,* named for the famous traditional Spanish seafood and rice dish, is a shallow, thin pan made of tinned sheet metal. Part of the "personality" of this pan is that it blackens with use.
- The *induction cooking frypan,* also made of carbon steel, especially to be used on the magnetic field heating elements of induction cooktops.
- The *cast-iron skillet,* a heavy, thick-bottomed frypan used for pan frying with steady, even heat. Also known as a ***griswold,*** this pan has a single, short iron handle that should be handled with care. Cast-iron skillets can be used on the range top or in the oven.
- The *omelet/crepe* pan is a shallow skillet with short, sloping sides made of rolled steel, which heats up fast. A good breakfast chef guards these zealously and allows no food other than eggs to be cooked in them!

Your menu and kitchen output will determine your sauté pan needs. The chef's personal preference will play an important role. Some cooks stick with a couple of pan sizes they feel are comfortable and versatile; others demand different sizes for different foods and/or styles of preparation.

Oven Cookware

The pans used for roasting and baking in ovens are made of the same basic materials as range top cookware, but in oven cookery, we may also find glazed and unglazed earthenware, glass, and ceramic dishes. Oven heat is typically less intense than range top heat, so these less sturdy materials usually don't crack or shatter due to temperature extremes. It is important to protect them, however, by avoiding things like submerging them in water while they're still hot from the oven. Quick temperature changes may damage them.

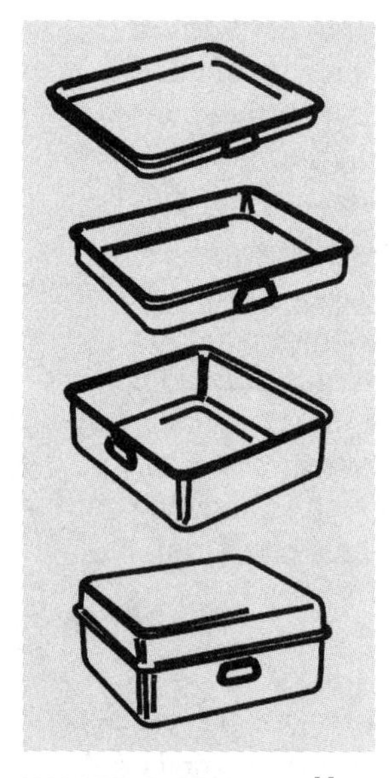

ILLUSTRATION 18-12 Versatile roast-and-bake pans can also be used for food storage. They are usually made of aluminum.

Common Ovenware Types

The *roasting pan* is an essential part of any kitchen arsenal. It is a rectangular pan with sides from 4 to 9 inches tall. It typically has handles on both ends, making it easier to carry when it is full. The deeper the roasting pan, the easier it is to remove from the oven without spilling the contents. Made of heavy (10- to 14-gauge) aluminum, roasting pans are used primarily for roasting meats and poultry (see Illustration 18-12).

You'll definitely want to order a cover for your roasting pan. Keeping the cover on during the cooking process circulates moisture around the food, resulting in juicier meat with less shrinkage. Remove the cover during the last 15 to 20 minutes of cooking if you want the food to brown before serving. One advantage of a roasting pan cover is that it is deep enough and large enough to turn over and use as a separate pan, for open-pan roasting.

A variation of the roasting pan is found in large-volume operations such as military mess halls. The *heavy-duty roasting pan* is exactly what its name indicates: It is made of thicker, 48-gauge metal, and it is bigger than the largest standard roasting pan, almost 22 by 18 inches.

Roast-and-bake pans are designed for versatility by their manufacturers to fit the standard baker's rack. They're handy for many foods, but, of necessity, they must be rather short, only 2 to 3½ inches deep, with capacities from 5½ to 16 quarts.

Some kitchens use steam table or hotel pans for roasting, but this isn't such a good idea. If you see these types of pans warped and discolored, you can bet they were used randomly for roasting. In fact, they can be used for baking, but only for certain foods, such as lasagna, which completely fill the pan. Otherwise, use them only for gently reheating foods that will go directly from oven to steam table or other hot holding unit.

The ***sheet pan*** is another rectangular pan, even more shallow than the roast-and-bake pan. Heavier-weight sheet pans are used for baking; lighter-weight ones, for product storage. Both types are available in full size (18 by 26 inches) or half size (18 by 13 inches). Their sides are only 1 inch high. Half-sized pans are often used for displaying baked goods in pastry shops. They fit the standard baker's rack, and they all have tapered edges, making them easy to stack. Most sheet pans are made of aluminum, light-weight 12 gauge or heavier weight 18 gauge.

Select your sheet pans to match the types of products to be baked. Brownies, for example, will cook evenly on nearly any type of sheet pan surface, but some chocolate chip cookies will burn on the bottoms before the tops are cooked if they are baked on a pan that's too delicate. Nonstick finishes are available on sheet pans, but they're only recommended for cookies, rolls, or croissants—not for bar-type cookies that must be cut into pieces while still in the pan. Instead, try anodized (coated metal) sheet pans, because their surfaces resist stains and do not scratch easily. Finally, whether the sheet pan's surface is dull or shiny, dark or light may have an impact on the final product. Bright, shiny finishes produce the best jelly rolls and sheet cakes.

Ask kitchen personnel to store sheet pans carefully, keeping the lighter ones separate from the heavier ones, to ensure the right pan for the job is always available.

Specialty Bakeware

In baking, there's a whole array of specialty pans to choose from: madeleine pans, muffin tins, pie pans, nested cake pans, pizza pans, springform pans, tube pans, quiche pans, and more. As always, let your menu be your guide. Here are notes about a few of the most popular specialty items.

A ***springform pan*** is used to bake cheesecakes or other dense cakes that need special handling to be unmolded. The bottom of the pan can be removed by releasing a spring or clamp on the side of the pan. The pan works like a mold, holding the cake together as it sets up. Springform pans vary in diameter from 8 to 12 inches, and they're about 3 inches tall.

Muffin tins come in 6-, 12-, and 24-cup sizes, and you can choose a variety of capacities, starting at 1 ounce per cup. Nonstick coatings on these pans are helpful so you don't always have to use paper baking cups to prevent sticking.

Bread pans come in both rectangular and round shapes. The round ones are sometimes called dough retarding pans. They come in a range of sizes, most designed to hold either 1 or 1½ pounds of dough.

Perhaps the newest and coolest baking pan innovation is known by their trademarked names as Flexipan® and Silpat® baking mats, both made since the 1980s by the French-based company Demarle, Inc., but fairly new to the U.S. (The U.S. website is www.demarleusa.com.) The company makes molds out of knitted glass fabric, coated with nonstick silicone. After baking, the pan can be literally peeled away from the food, then washed and reused. The mat is simply laid over a standard baking sheet instead of having to grease and flour it. Some recipes have to be adjusted slightly, to bake at lower temperatures for longer time periods for optimum results; and the baking mats, although flexible, cannot be folded for storage. But for pastry chefs who are tired of hand-cutting parchment paper, these products are certainly an interesting option.

Pizza-Making Supplies

If pizza is on your menu you'll have need for quite a few unique baking items. *Pizza pans* are round, made from 14-gauge aluminum, generally anodized (coated with a harder finish), which makes the aluminum appear darker in color. Diameters vary from 6 to 28 inches, or even larger, with the bottom of the pan slightly smaller than the top rim. Pizza pans have tapered sides of ¼ to ½ inch in height. The anodized finish won't flake off, but aluminum is a soft metal and with repeated use and scouring, the coating will thin. For this reason, it's important to "season" the pan properly. Before you use it, wash it in warm, soapy water and dry it thoroughly. Brush a light coating of cooking oil on the entire pan and allow it to soak in for 30 to 60 minutes. Then bake the empty pan at regular time and temperature, as though it had a pizza on it. After baking, don't wash it again; just wipe it off. And any time you wash it, oil it again. Keeping a light oil film on the pan will keep the pizzas from sticking to it. Don't put your pizza pans in the dishwasher, which will erode their anodized coating.

You can also purchase tin-plated pizza pans. Like aluminum pans, these also need to be seasoned with oil. Some new pans may need a second or third coating before they're fully seasoned. Tin pans will darken with use, and they'll rust quickly if refrigerated, soaked in water, or left unseasoned. They should be completely dried after every washing. Popular pizza pan variations include the *deep-dish pan,* with high, 2-inch sides for thick-crust pizzas. It's made of aluminum or black steel. The bottom of the *perforated pizza pan* contains holes, which allow steam to escape as the dough bakes. Perforated pans work well when baking frozen pizzas; they are often also dishwasher safe.

Here are some other accoutrements. ***Dough boxes*** are used to store balls of pizza dough until it is needed for baking. A good dough box should be able to keep the dough from drying out whether it is frozen, thawing, or ready to use. It measures about 18 by 26 inches in diameter, and several boxes can be stacked to save space.

A ***docker*** is a tool for poking tiny holes in dough for pizza, focaccia, and the like just before it's ready to go into the oven, to keep air bubbles from ruining the nice, smooth crust. It's a round barrel, either metal or plastic, with small stainless steel pins protruding from it. If the pizza dough gets a big air bubble in it anyway, there are aluminum *bubble poppers* to remedy that. A little hook on a long metal arm does the job.

The long-handled pan with a flat surface used to place pizza directly onto a hot hearth is called a ***peel.*** Its flat surface may be made of wood or aluminum; the square surface can be as small as 7 inches or as large as 18 inches, and the handle can be as short as 9 inches or as long as 38 inches. Between uses, the pizza oven is swept out with a long-handled brush.

18-5 SERVING AND HOLDING CONTAINERS

Because they are so versatile, you'll probably need to have plenty of ***steam table pans*** in your kitchen. They are so common that they go by several different names, including *hotel pan, counter pan,* or service pan. Some are also nicknamed *Number 200s* (2 inches deep) or *Number 600s* (6 inches deep). These rectangular stainless steel pans can be found in almost every area of the kitchen, from preparation to freezer to serving line. The standard size is 12 by 20 inches, but they come in half-pan (12 by 10 inches), third-pan (12 by 6½ inches), and fourth-pan (6½ by 10 inches) sizes, with depths ranging from 1¼ to 6 inches (see Illustration 18-13). Steam table pans are made of 20- to 24-gauge stainless steel, with notched lids that fit securely. The handles on the lids are indented so the covered pans can be stacked conveniently. For serving, you can also order high, dome-shaped covers that permit the user to open only half of the cover. Stainless steel adaptor plates convert the standard rectangular openings in steam tables to accommodate other shapes of containers.

If you're using these pans for steam cooking, also consider purchasing adaptor plates. Setting the food on such a "shelf" allows for free circulation of steam. The *egg poacher* is one type of adaptor plate, with holes that each fit a single aluminum poaching cup. The adaptor plate does double-duty on a buffet or serving line, where it can be used to hold full juice glasses upright.

A couple of common kitchen containers often do double-duty for display and storage. They are the half-sized sheet pan and the mixing bowl. If bowls of large capacity will be used for serving—let's say, bowls over 30 quarts—consider using a mobile bowl stand, a wheeled cart onto which the full bowl can be placed. The casters on the stand should have locking brakes, and you'll probably want the convenience of tray shelves that slide in and out for storage beneath the bowl. To ensure stability, only use the mobile bowl stand with a bowl that has a fairly wide, flat bottom.

ILLUSTRATION 18–13
Steam table pans should be purchased with matching lids.
COURTESY OF VOLLRATH COMPANY, LLC, SHEBOYGAN, WISCONSIN.

18-6 CARING FOR SMALLWARE

We've already mentioned care and cleaning of some smallware items, but there are a few general rules to cover. Like anything else, your cookware and utensils will last longer and give better results if you take care of them. This means:

- Avoid using the dishwasher and/or caustic detergents to clean your pots and pans. Wash them by hand, with cleaners and polishes formulated for the material your pots are made of. There are specialty products for the care of aluminum, copper, cast iron, and so on.
- Empty and clean cookware as soon as possible after cooking. Don't let food sit in it. Cookware is designed for cooking food, not storing it. In some cases, prolonged contact with acidic foods will pit or stain the surface; liquids left too long in a pan may cause permanent discoloration.
- Hang your cookware when not in use or put it on a pot rack. Don't stack it on shelves unless storage space is so limited that you have no choice. If this is the case, it should be stacked very carefully to prevent dents and scratches. If your establishment is located near salt water—an island, seaport, or waterfront—special care must be taken to provide enclosed storage space. Cookware can be damaged by prolonged exposure to the salty air.
- Hang knives on magnetic strips mounted on the wall near where they will be used. If they must be stored lying in drawers, slip the blades into protective sheaths first. Most chefs also insist that their "good knives" be hand-washed.
- Keep empty cookware off the range top and/or out of the oven. Warping is a common problem that occurs when empty cookware is exposed to high heat. If you insist (as some chefs do) on keeping a sauté pan always ready on the range top, at least keep it on low heat until it's needed.
- When rivets and/or handles become loose, it's smart to replace that piece of cookware. In purchasing pans, make sure the handles are secured by a minimum of three large rivets. Cheaper handles are made, but trying to save money by cutting corners in this regard is false economy. And each time you use a pan, pick it up and do a quick inspection: Check the rivets and make sure the bottom of the pan sits perfectly flat on the range top or counter.
- Preserve your cooking surfaces by ***seasoning*** them. Seasoning a sauté pan is different from seasoning a piece of ovenware, such as a pizza pan. To season a new range-top pan, clean and dry it. Then heat it until it begins to smoke. At this point, add enough table salt to cover the bottom of the pan. Roast the salt until it's light brown, then remove the pan from the heat and empty it. Wash the cookware by hand after each use and reseason it if you ever put it in the dishwasher.
- Finally, go easy on your cookware when using metal utensils. Pans with nonstick surfaces don't need seasoning, but they do need special care not to become nicked or scratched. In fact, any pot or pan can become damaged if it's slammed enough times with a spatula.

Every manufacturer and many culinary associations have guidelines and tables that list smallware requirements for different types of foodservice businesses. There is no ideal smallware package, because no two operations are alike. To estimate your needs before you buy, break them down into what might be needed for each separate section or station in your kitchen. As you'll soon see, "smallware" is no small task!

SUMMARY

Always match the proper utensil to the task at hand. In some cases, it wastes time to do the job with the wrong utensil or to do chores like slicing by hand when a mechanized slicer can do it faster and with less waste. For small items such as spoons, whips, spatulas, ladles, and forks, look for sturdy construction and heat-resistant handles.

Good knives are absolutely essential in a kitchen, but all too often they are not correctly stored or sharpened. Knives are made by forging or stamping. The latest knife-making technology cuts the steel into blade shapes with a computer-guided laser beam. Knife handles are made of heavy-duty plastic or wood. This chapter acquainted you with the names of and uses for common kitchen knives.

The cutting surface is almost as important as the knife you use; health inspectors often prefer plastic cutting boards to wooden ones because they're easier to sanitize, and a new generation of bacteria-resistant resins are also being used to make cutting boards. Most kitchens now use some sort of color-coding system—different colors of cutting boards, knife handles, and the like—for different types of food.

Keeping foods at their proper temperatures is one of the most critical safety considerations. For this, you will need a variety of thermometers. Some are designed to measure a food's internal temperature; others to spot-check surface temperatures; still others to track environment temperatures inside ovens, walk-ins, and so on. For hand use, the most popular are the basic stem-type bimetallic thermometer; the thermocouple, a thermoelectric sensing device which can read the temperature of anything it touches; and the infrared thermometer, which can read a temperature by being pointed at an object without even touching it.

No kitchen would be complete without a wide variety of pots and pans. There are two types of cookware—for the range top and for the oven—and they come in a wide range of shapes and sizes. Always purchase them with matching lids. The most common metals used to make cookware are aluminum, copper, stainless steel, and cast iron. There is a specialty pot or pan for almost every use, including new flexible bakeware made of woven glass that molds to a shape and can be peeled off the finished food and reused, right out of the oven.

Your cookware will last longer if you take care of it. This means washing it as soon as possible after use; using cookware for cooking, not storing, food; hanging things up whenever possible instead of stacking them; and using utensils carefully so as not to nick or scratch the surfaces of pots and pans.

STUDY QUESTIONS

1. Why would you use a spoodle instead of a spoon?
2. Explain three things that would indicate if the kitchen knife you're about to buy is of top quality.
3. What is cladding? Is it important? Why or why not?
4. List three of the five major considerations when buying kitchen scales, and explain why each is important.
5. What's the difference between a ladle and a scoop? Why would you use each of them?
6. What types of thermometers should most kitchens have? Where should they be used?
7. Explain the advantages and disadvantages of using cookware with a nonstick coating.
8. What is the difference between a satoir and a sauteuse?
9. Why would you choose to order different types (light and heavy) and sizes of sheet pans?
10. Briefly explain the process used to season a new sauté pan.

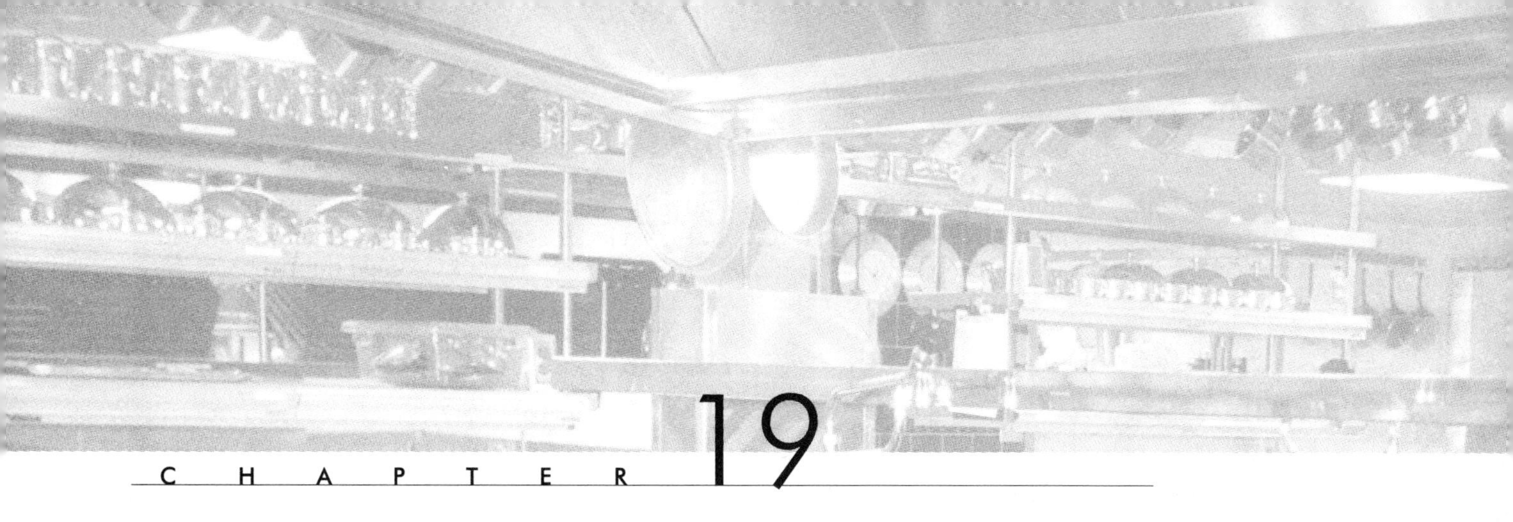

CHAPTER 19

TABLEWARE

INTRODUCTION AND LEARNING OBJECTIVES

Tabletop is the foodservice industry's term for the various supplies that make up a table setting: flatware, glassware, and plates; tablecloths and napkins; condiment containers; and other dining accessories that sit on a table. Together, they do quite a bit to contribute to the personality—and, ultimately, to the success—of your restaurant. As you select the colors, patterns, and sizes of your tableware, you must think about the concept of your restaurant, the menu items and price range, the chef's preferences, your budget, your target customer, and more.

The investment you'll make in tableware is considerable, partly because it never seems to end. In the everyday bustle of restaurant life, plates chip, glasses break, flatware disappears or is damaged. Experts estimate that most businesses replace an average of 20 percent of their original tableware purchase per year; in high-volume operations, the estimates are as high as 80 percent!

So how do you choose items that are sturdy, stylish, *and* affordable? In this chapter, we'll discuss each tabletop component: how it is made; shapes, sizes, and selection criteria; care and cleaning tips; and guidelines for how much inventory to have on hand. This includes:

- Plateware
- Glassware
- Flatware

First, let's begin by discussing the basic functions of the tabletop components and their importance to your foodservice setting. The most obvious, of course, is their use as tools and utensils for the dining customer. Your plates, glasses, and silverware also help display foods and beverages to your best advantage. This display function is just as important to a restaurant as the artfully decorated display windows of a department store; it adds to the ambience and attractiveness of the place.

You could also think of tableware as having a point-of-sale function. It reinforces your theme or concept, as well as any advertising or promotion you may have done, and it tailors diners' expectations of what's to come. This marketing function is particularly critical because it can so easily backfire: Something as simple as a lipstick smudge on a glass can indicate a lack of attention to detail that will affect the customer's mood or opinion about the entire dining experience, no matter how good the food or service.

19-1 DEFINITIONS OF TERMS

At this point, let's define a few of the terms you'll see often in this chapter.

Plateware refers to the dishes used to serve food: dinner plates, salad plates, bread plates, saucers, etc. In some cases. ***Holloware*** is the general term for dishes that aren't flat: bowls, pitchers, and cups. The whole lot may be referred to as *dinnerware* or *tableware.*

Glassware refers to containers used for serving water or beverages. Different shapes or sizes of glassware have different names, including *tumblers, stemware, mugs,* or *footed glasses.* In institutional or very casual settings, these are often made of plastic instead of glass; in more expensive eateries, they may be made of ***crystal.***

Flatware means forks, knives, and spoons, in all their various sizes. Generally made of stainless steel, they may also be silver plated and, on rare occasions, true silver. They're also called *silverware* or *utensils.*

A couple of notes to help avoid confusion. In promotional materials from various manufacturers, the term "holloware" may also be used to refer to silver-plated knives with hollow handles; and the term "flatware" is also used to refer to plates, because they're flat! The context of the information, and the type of manufacturer, should make it obvious which type of item fits the definition in a particular catalog or brochure.

A ***place setting*** is one person's set of the preceding items, arranged on the table. A ***table setting*** is the place setting, plus the placement or tablecloth and napkin. And ***tabletop*** is the basic table setting, plus a collection of accessories (salt and pepper shakers, candleholder, bud vase, sugar and cream containers, additional wineglasses, and so on). The contents of any tabletop vary drastically with the type of restaurant and type of cuisine. What can you tell about the restaurants from the tabletops pictured in Illustration 19-1?

19-2 TABLETOPS AND TRENDS

The most important part of any tabletop is the 24-inch space directly in front of the guest when he or she sits down. Anything you can do to make that space more visually appealing will go a long way toward enhancing the dining experience. First, however, you must develop a tabletop philosophy, deciding what it is you want to convey. This message should be carefully considered and completely intentional.

It is not just the way the tabletop looks that makes an impression. Dining is a very tactile experience—guests will touch their plates, cups, and utensils—and the feeling is completely different when you pick up a thick mug or a delicate china teacup. Some operations also select different types of table settings for different needs: coordinating or complementary patterns to help distinguish between lunch and dinner service; dining room, banquet, or room service; formal or informal events.

As restaurant owners and chefs try to gain an edge in a fiercely competitive industry, more of them are developing "signature" dishes and serving them with distinctive—albeit unlikely—touches to add more panache to the presentation. For instance, at the time this was written, a Florida eatery called Barton G The Restaurant serves fried chicken in large ceramic roosters, and swordfish on actual swords, four feet in length. Tru, a Chicago restaurant, serves ceviche (raw, marinated fish) in a cone-shaped glass, perched atop a small fishbowl with a live fish inside. At Famous Dave's Barbecue restaurants based in Eden Prairie, Minnesota, guests can order a barbecue platter for four served on an upturned trash can lid. Whimsical? You bet! We've heard of foods being served in cans, jars, hubcaps, flowerpots, egg cartons, and more. And why not, if it's fun and creates the kind of attention you're looking for?

Some restaurants attempt to upgrade and refresh their look by redesigning their tabletop. It's a relatively inexpensive way to redecorate. Colorful new linens or unique dessert plates may be all that's needed to perk up tired place settings with minimal cash outlay.

ILLUSTRATION 19–1 These tabletop designs were created by Mike Fleming of U.S. Foodservice.
COURTESY OF JOHN RIZZO, PORTLAND, OREGON.

However, don't mindlessly follow tabletop "fashion"—and yes, there are as many trends in this area as in any other design industry. The search for uniqueness leads many restaurateurs to hire designers who create so-called "custom" tabletops. Although they probably do a fine job, temper your enthusiasm based on the following considerations:

- Your budget. Will you have high enough volume and check averages to justify the higher cost of a custom-designed place setting? First-time restaurateurs are probably better off choosing existing plateware patterns and conserving their precious start-up funds.
- Designs may be stunning on paper, but how do they actually look on a plate? Technology has improved the design process, but much is still done by hand. Some colors are not faithfully reproduced on the clay that becomes the plate; the manufacturing process that produces

red requires lead and cadmium, both food safety concerns. And decals may not withstand the rigors of mechanical dishwashing. You certainly want to minimize the amount of time-consuming hand washing of dishes.

- How does food look on the plates? If you want your cuisine to stand out, a simpler plate is often better. Same with wineglasses: Wine lovers know that colored or busily patterned goblets don't allow you to appreciate the true color of the wine. A colored stem is fine; a colored bowl on the glass is not.
- Finally, custom-designed tabletop items may take as long as three months for delivery. Order early and prepare to wait. And, when you need to replenish your supply of a particular plate or pattern, will it be readily available?

In his handy treatise *Tabletop Presentations,* Irving J. Mills also cautions the prospective restaurant owner about short-lived fads. Instead, he suggests the following tips for tableware selection:

- Appearance and durability should be given equal weight in making your decision.
- Utility and durability are more important in your choice of basic tableware than they are in selecting accessory items. This is primarily because dishes are removed from the table and washed after each turnover, so they probably need to be sturdier than items that remain on the table for longer time periods.
- Don't crowd the tabletop with too many items. Decide how much is necessary and how much is clutter.
- If you see your tabletops as a stage designer would look at the set of an upcoming play, you'll learn to think of them as props, and even scenery. Match the "scenery" to the elaborateness of the production.

To Mills's advice, let's add that (as you're "decorating your set") you may hear conflicting opinions about how to coordinate the variety of colors and textures. Most experts frown on the idea of mixing and matching pieces that don't normally go together; they claim the table looks like a maze or quilt, which sends a message to the guest that the restaurant has not shown proper attention to detail. However, others insist there are classy and coordinated ways to combine such diverse substances as metal and china, or china and glass, to deliver an interesting and powerful statement to diners. Generally, the suggestions and ideas that are common in home dining rooms simply don't work in the foodservice industry. Items such as delicate crystal glasses, real silver, and high-dollar accessories are sometimes found in equally high-dollar restaurants, but, for the most part, these items don't hold up to the rigors of commercial dining.

In recent years, tabletop trends have included some fun and interesting styles and innovations, such as:

- Oversized white plates with brightly colored accents, like napkins and side-dish plates.
- Decorated plate rims, which you'll read more about in a moment. This is a bit more labor intensive, and can only be accomplished with plain white plates.
- Stacking plates of different shapes, sizes, and colors—a simple white dinner plate with a bright, triangular salad plate on top, for instance.
- Glass plates (bright colors for casual dining; clear and etched for fine dining) instead of china.
- Interesting, geometric shapes for plates—squares, triangles, rectangles, and so on.
- European-sized flatware, which is larger and heavier than American-sized.
- Oversized Japanese soupspoons to hold small dollops of condiments served with appetizers.
- Earth tones for table linens, which represent warmth and comfort and don't show spills.
- Large or unusual beer glasses to showcase high-end microbrews.
- Classy or whimsical individual teapots, as restaurateurs learn to cater to the growing (and profitable) market of hot-tea drinkers.
- Brightly colored plastic water pitchers for wait staff to use instead of clear plastic or heavy glass.
- The use of quick-ship services that allow you to reorder items quickly without having to stock lots of inventory.

And what's out of fashion? Table tents and ashtrays.

19-3 PLATEWARE

You might think of plateware as "packaging" for food, but it is a lot more than that. The dinner plate and its accessories are the focal points of the entire tabletop, contributing to the decor of the eatery and helping to deliver the promise of a first-rate meal. In fact, the "right" plateware can actually build sales by up to 25 percent, without any other operational change! It is capable of adding color and drama to the dining experience.

On a daily basis, plateware must be a priority. You should supervise how it is handled, from storage to the plating process, to serving, to its removal from the table, to the dishwashing process, and to its return to storage. What you'll find is that most foodservice venues are very hard on their plateware. The National Restaurant Association estimates the useful life of a typical restaurant plate is only twelve months. So before you choose (or change) yours, completely answer each of these questions:

1. What are the style, pattern, and price of the current plateware being used?
2. What are your immediate competitors using?
3. What is your overall concept? Include theme, decor, atmosphere, lighting levels, and typical prices.
4. What are the sizes and types of tables in your dining room?
5. What types of food will you serve, and how large are the portions? Do you intend to serve main courses and side dishes on separate plates or on single, large dinner plates? (Your chef will probably have some strong preferences.)
6. Will your customers prefer a thick, solid feel to their dinnerware, or is your concept better suited to the lighter weight of delicate china?
7. Will dishes be expected to sit for long periods of time in conditions that are very hot (under the salamander) or very cold (on ice at the salad bar)? Will they be used in a microwave, or a food finisher?
8. At business peaks, how many meals do you expect to serve? Are there additional catering or banquet functions off site that will require more plateware?
9. What is the capacity of your dishwashing system?
10. Finally, how and where will dishes be stored? What is the capacity of your storage area?

After all that is determined, it may be a relief to hear that plain white plateware never seems to go out of style. At this writing, oversized white plates are popular, yet so are plates that are primarily white but decorated with a bold stroke of embossed color, a logo, or an accent pattern. Other trends you may notice:

FEWER PIECES. Cost pressures are slowly eliminating items such as candleholders, butter plates, and fruit plates.

THE BIGGER, THE BETTER. Dinner plates come in sizes from 9 to 14 inches. The standard is a 10-inch dinner plate, but large plates with wide rims give the chef more of a "canvas" on which to be artistic, while offering the consumer the perception of better value. It's not uncommon to see an 11- or 12-inch plate or a wide, 11- or 12-inch bowl for salads and pastas.

DECORATED RIMS. As long as chefs have that wide rim to work with, you'll find it sprinkled with parsley or paprika, drizzled with chocolate, or painted with a colorful sauce. Decorating the outer edges of a plate can add fun and drama, enhancing the meal presentation and controlling the eye of the diner. As a practical matter, it can also divert attention from a smaller portion size.

BRIGHTLY COLORED PLATES. They're not the norm, but we've already mentioned the bright colors and wild shapes being used in a few dining establishments. Even the trendsetters, however, must consider the same timeless selection criteria: durability, availability, and temperature retention.

How Plateware Is Made

A plate is not just a flat surface! Illustration 19-2 shows its basic parts. Knowing these, and the technical names for the types of plateware, should help you distinguish between the various offerings of manufacturers. Most plateware is ***ceramic,*** which means it is made of clay that has been baked. The ceramic used to create foodservice plates is called *vitrified china.* It is high-quality, highly refined clay that is mixed with water to form a creamy liquid called ***slip*** and poured into a mold. The slip clay is heated intensely (more than 2200 degrees Fahrenheit) until it fuses into a solid, composite mass as the silica in the clay turns to glass. This heating process, called *vitrification,* hardens the china, making it strong and nonpermeable and protecting it from moisture and food stains.

When the manufacturer adds ground bone or calcium phosphate to the clay, the resulting china has a finer, more delicate texture and a translucent (semitransparent) appearance. It is known as *fine china, bone china,* or *English china,* the latter because it was first manufactured in England. Today, however, Japan and the U.S. are also major china suppliers. Despite its delicate appearance, fine china is actually quite strong and sturdy when well made.

Here's the basic plate-making process: The wet clay is shaped and dried, then baked at high temperatures to create a dull, rough-looking material known as ***bisque.*** If the plate is to be decorated, it is done at this time. Whether or not it is painted or embossed, a ***glaze*** (or *miffle*) is then sprayed on the plate, and the plate is baked a second time. The second baking fuses the glaze onto the bisque, creating a protective covering that completely seals the bisque from contact with food. The glaze should be even and unpitted, giving the plate an attractive luster as well as protection. Often, the ***foot*** (the ridge on the bottom of the plate, on which it sits) is not glazed. When the foot is unglazed, it is called a ***one-fired plate.*** When the foot is glazed, it is called a ***two-fired plate.*** Two-fired plates are more expensive, but they can be stacked without scratching one another. The third alternative is called a ***polished plate.*** The foot of this one-fired plate has been smoothed and polished.

There is an ongoing debate among dinnerware manufacturers about the merits of one-fired versus two-fired plates. The one-fired process takes two days to complete; the two-fired process takes four days. The Center for Advanced Ceramics Technology at Alfred University in Alfred, New York, says the strength of the plate is based on the temperature and length of time it is fired, not how many times it is fired.

Correctly made, a finished plate should be able to withstand sharp knocks and raps (within reason) and temperature extremes (also within reason). Of course, its overall strength will depend on factors such as the quality and purity of the original clay, a lack of air bubbles (which weaken the clay) during processing, the thickness of the plate (doubling its weight increases its strength by 75 percent), and whether it has a rolled edge or an extra-thick rim. Dishes with scalloped edges are also slightly stronger than those with plain edges. Experts insist there is no such thing as unbreakable dinnerware and no reputable manufacturer would make that claim. The best you can hope for is "break-resistant."

Hotelware is a term you may hear for china dinnerware made for use in hotels and foodservice. Technically it is china, but it doesn't have the delicate look or feel because it is made for heavy use. ***Stoneware*** is chip-resistant and has a natural beige or gray color. It is made with a different grade of clay and only vitrified about 5 percent, with additives mixed in to make it nonporous. ***Earthenware*** is fired at lower temperatures than either stoneware or china, making it heavier, more porous, and opaque instead of translucent. It is the least expensive of the plateware types, but sturdy enough for casual, everyday use.

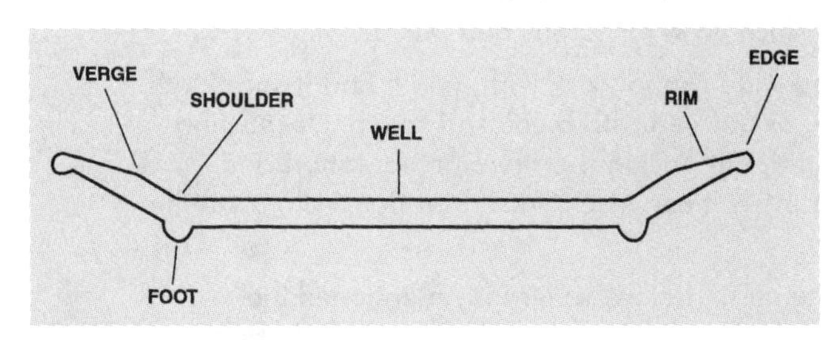

ILLUSTRATION 19-2 **The parts of a plate.**

SOURCE: ADAPTED FROM IRVING J. MILLS, *TABLETOP PRESENTATIONS: A GUIDE FOR THE FOODSERVICE PROFESSIONAL,* JOHN WILEY & SONS, INC.

There is one more popular plate type, and that is the metal plate. Used for steaks, fajitas, and other broiled items, these are made of aluminum or stainless steel, often in a rectangular shape to fit on wooden bases so servers can pick them up. Aluminum costs less, but both materials can be scratched by knives. The aluminum plate also begins to soften when it reaches about 1000 degrees Fahrenheit—and yes, that can happen if it is accidentally left in the oven or on a range top. The obvious advantages of the metal plate include the following: It holds heat better and longer than fired plateware, and it adds the "sizzle," literally, to serving a good steak.

Shapes and Sizes of Plateware

There are seven broad categories of plate shapes, and, within each category, there are design options. A few of them are shown in Illustration 19-3. Often, the shape and size of a plate is determined by the size of its rim: wide, medium, or narrow. *Wide rims* create an elegant, spacious palette for food presentation. It's one of the ironies of foodservice that the largest plates are often used to show off the smallest, most "gourmet" portions. If you like the idea of using wide-rimmed plates, remember you'll need plenty of table space to accommodate them.

Medium rims are industry standards, because they're so versatile, and they allow for flexibility in portion control. A cousin of the medium-rimmed plate is the ***rolled edge,*** a plate made with a thicker, rounded edge for maximum durability. Plates with *narrow rims* provide maximum space to pile on the food! They have a casual look, and are perfect for small table sizes.

Plates with *scalloped edges* resist chips and cracks better than plates with plain edges. However, they have a somewhat "traditional" look that is not quite right for every establishment.

So-called ***coupe-style*** plates have no rims at all. Their sleek, contemporary look offers maximum surface area for food presentation. Finally, the *square-shaped* plate is recommended for smaller tables. It should always have a rolled edge.

The ***pitch*** of the plate is a term that refers to how flat (or rounded) it is. This factor is important for presentation of the food, because it affects how sauces will spread on the plate.

Manufacturers most often list the sizes of their plates, platters, and saucers by diameter, but occasionally you'll find them listed by the diameter of the ***well*** (the inside portion of the plate, excluding the rim). To avoid confusion when ordering, give the total diameter of the plate and also specify the width of the rim.

We include cups and bowls here because they are made in the same way as plates. Cups and bowls are sized strictly by the volume they hold, and there are no standard sizes, particularly with mugs.

A cup is most likely to be identified by its shape. The *ovoid shape* is the cup most likely to be used with a saucer; there are also *conic cups, stack cups, Irish coffee cups,* and mugs (see Illustration 19-4). The latest trend is the wide-mouthed French coffee cup, which holds up to 12 ounces of liquid.

Plateware Care and Cleaning

Let's follow a typical place setting as it makes its daily cycle from storage to use to cleanup and back to storage. If plates are stored for more than a week between uses, protect them with a clean cloth or plastic cover or, at the very least, store them in cabinets with doors. Plateware components should always be stored by size—plates in stacks, cups in racks. Don't stack the cups inside each other, which is a common cause of chipping and breakage. Even if plates are stackable without scratching each other, they should never be stacked more than 12 inches high. Storage shelves should be made of high-grade, nonmagnetic stainless steel, to keep dishes free of dust and unsightly metal marks.

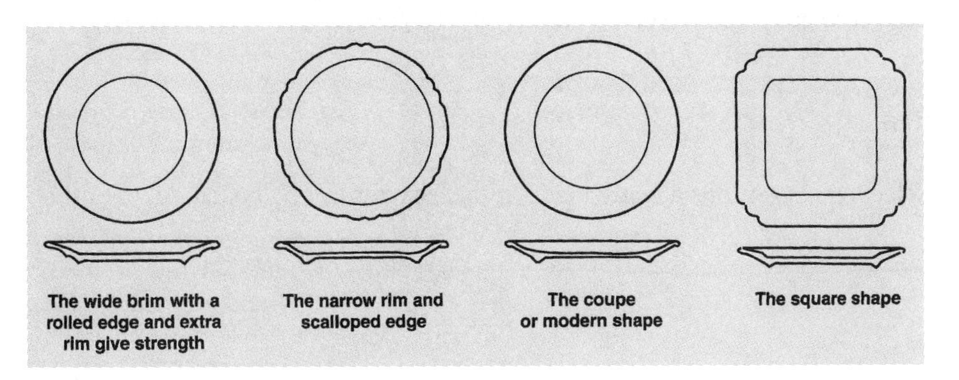

ILLUSTRATION 19-3 Common plateware shapes.

ILLUSTRATION 19-4 Common cup shapes.

Storage is especially critical in the soiled-dish area, where the American Restaurant China Council estimates 75 to 80 percent of all dish breakage occurs. China in active use should be stored at or below the food plating area; more breakage seems to occur when employees must reach into overhead storage to bring plates down.

In the dining area, teach your wait staff to load trays properly; the typical tendency is to overload. When stainless steel or silver-plated covers are used in banquet situations, the cover should fit correctly to minimize scratches and chips on plates. And, when clearing tables, busboys should be taught not to overload the bins, also called ***bus boxes.*** Make sure yours are plastic or plastic-coated. Soiled plates should be placed upside down in the bus box, with cups, bowls, and glasses on top. Don't shove the plates together on their sides, like so many books on a shelf; the unglazed footing of one plate can easily scratch the front of the next.

Now, we enter the dish room. Here, one of the most frequent dish-handling mistakes occurs, when one piece of tableware is casually used to scrape food off another. This is among the most common ways things get broken. Avoid this by providing plenty of proper scraping implements and lots of dish table room for scraping, stacking, and racking so all dishes can be correctly sorted and rinsed. This means never using metal scouring pads to scrub off baked-on food, or metal utensils to scrape it off. These actions simply damage the glaze, when most commercial dishwashers can remove the food without the extra scrubbing. Before washing, rinse and place similarly sized plates in stacks of no more than 12 inches high (a practice also called ***decoying***). It's important that soiled dishes be washed within 30 to 40 minutes after this first rinse. Coffee, tea, and some acidic foods can cause staining if allowed to sit for longer time periods.

Check the dishwasher curtains regularly to be sure they're clean. Also check the water pressure and temperature, spray pattern, and length of each phase of the dishwasher cycle, as described in Chapter 16. Be especially careful not to prerinse soiled dishes with water hotter than 110 to 120 degrees Fahrenheit; higher temperatures tend to stain china, and bake food particles onto the plates. If dishes continue to show persistent stains, you may have a water quality problem, from hard water or too much iron content.

Clean, wet dishes can be air dried in a well-ventilated area as they come out of the dishwasher. A rinse additive may be used in the rinse cycle to speed the drying process. China should never be stacked right out of the dishwasher, as it is more likely to be scratched or pitted when it is hot and wet. Tilt the dish racks slightly to drain water from the feet of the dishes, and handle clean dishes only at the edges. In fact, the use of rubber gloves is preferable. Position dish carts or self-leveling carts at the end of the dish machine line for easy movement from dish room to storage area.

China can be overworked—that is, used too often—when inventory is inadequate. This will result in more stains and discoloration and more breakage. And, as one manufacturer puts it, "Dishes don't break . . . they are broken!" Take good care of yours, and they'll take good care of your customers.

Purchasing Plateware

Most of the plateware selection suggestions already made in this chapter can be boiled down to three main categories:

- Cost
- Availability
- Durability

FOODSERVICE EQUIPMENT

MANAGING METAL MARKING

Many operators may think metal marking is an unavoidable part of owning china. However, these unsightly marks can be eliminated. Brian Joyce, a principal at Joyce Pacific Tableware in Novato, California, says metal marking can be prevented, as well as removed. He says operators should note the following tips to keep their plates looking as new as the day they received them.

- Eliminate metal-to-china contact wherever possible.
- Keep kitchen surfaces that contact china as clean as possible.
- Maintain the foot of the china; this area can be cleaned fairly abrasively.
- Maintain the surface; don't let it get so bad that it can't be recovered.
- Never use metal pads or scrubbers on the china.
- Utilize rubber mats, plastic mats, and chemicals.

To rid china of existing metal markings, operators can try using nonabrasive cleansers. As a last resort, operators can try using bleach or other abrasive cleaning products. Chemical providers can also be a great source for information on ridding plates of metal marks.

Source: ***Foodservice Equipment and Supplies,*** **a division of Reed Business Information, Oak Brook, Illinois, May 1998.**

Because the total amount you'll spend on plateware seems small compared to other, large equipment investments, you may be tempted to spend a little more than you'd originally budgeted. So here's another reminder: This is not a fixed cost. It is an ongoing expense. Replacement costs should be a line item on a restaurant's income statement, shown as a percentage of sales.

There are four different "grades" of tableware, listed here from least to most expensive:

- *Commodity* means simple, unadorned (or minimally decorated) pieces, best suited for high-volume serving.
- *Stock* means a popular pattern, either inexpensive or top-dollar, that is such a staple of the manufacturer that it is always kept in stock. *Open stock* means the dishes can be bought by the piece instead of in full place settings or with large, minimum orders.
- *Custom* means dinnerware that is made to order for a particular customer, perhaps with a logo or special design.
- *Exclusive* means tableware that is made to the specifications of one distributor, and available only through a single manufacturer.

Availability becomes an issue if you choose anything other than an *open stock* pattern. Remember that you can keep your existing inventory of dishes (and, therefore, your initial cash outlay) low if you know you can always place an order "as needed" and have it shipped to you within days. Try to deal with manufacturers or distributors who can promise to keep your particular patterns in stock for three to five years.

Durability isn't just a cost factor; it's a safety issue. There's no universal quality standard for plateware, but ask as much as you can about the manufacturing process and any safety features: Is it shatterproof? Does it have a finish or edge that makes handling or stacking easier? And so on.

Most manufacturers will offer basic inventory tables, which are to be used as ordering guidelines; your needs will vary depending on the factors we've already mentioned. Oneida Foodservice Ltd. has replaced a full-sized inventory table with a fairly simple formula, shown in Table 19-1. Just multiply to compute your own plateware needs.

TABLE 19–1

GUIDELINESFOR ORDERING PLATEWARE

Multiply the number of seats in your restaurant by the following factors and divide by 12:

Type of Operation	Cups	Saucers	Fruits	Grape-fruits	Bouillons	Bread and Butter Plates	Salad/ Dessert Plates	Dinner Plates	Platters	Bowls	Serving Items
Dining room	3.5	2	3.5	1.5	1.5	3	2.5	2.5	2	1	0.5
Coffee shop	3	1.5	3	1.5	1.5	2.5	2	2.5	1.5	—	—
Health care	2.5	2.5	2.5	2.5	2.5	—	2.5	2.5	—	—	—
College/ university	5	3	4	—	2	—	5	3	—	3	—

For example, if your operation has a dining room that seats 100, we recommend the following initial order:

	Cups	Saucers	Fruits	Grape-fruits	Bouillons	Bread and Butter Plates	Salad/ Dessert Plates	Dinner Plates	Platters	Bowls	Serving Items	Total Order
Quantity (dozens)	30	17	30	13	13	25	21	21	17	9	5	201

Source: Oneida Foodservice, Ltd., Oneida, New York.

19-4 GLASSWARE

Paper and plastic cups may be practical for quick-service dining, but even the most casual sit-down restaurants today serve beverages in glass, to subtly position themselves more as "full service" than "fast food." Glassware is an important decorative component of the tabletop; it helps set a mood and deliver a promise. Because they sit vertical to the tabletop, glasses are noticed immediately, even reflecting the light and color of the room itself. The right glassware can add profitability to your beverage service, whether you serve an expensive wine or a humble glass of iced tea. This is because the right glass adds visual appeal to the beverage, which increases the guest's perceived value of the drink. In short, the better it looks, the easier it is for the guest to happily pay a higher price without your having to increase the size of the drink. Also, imaginative styles and uses of glassware help whet the appetite: Serving shrimp cocktail, sorbet, or a dessert in elegant stemware may prompt an impulse buy as the customer notices how good it looks at the next table.

Of course, you can probably be *too* inventive! When it comes to glassware, customers have some distinct preferences and expectations. A martini in a beer mug? Milk in a champagne flute? There are limits to what "works" and what doesn't. Fortunately, there's enough leeway so that you can be creative without looking silly.

How Glassware Is Made

Commercial glassware is manufactured in two ways: blowing or pressing. The molten glass is either blown into its final shape by introducing air into it, or pressed into a mold to create its shape. Most commercial glasses are *pressware,* especially if they feature handles or designs, since greater detail can be achieved by pressing.

Glass is made of very fine sand (called ***silica***) that is mixed with soda, lime, and ***cullet*** (reused broken glass bits) and fired at very high temperatures, nearly 1500 degrees Fahrenheit. The molten glass is blown or pressed into shape when it's at least 1250 degrees Fahrenheit. In lower-cost *pressware,* this usually results in a very clear, faintly raised line on each glass that indicates where the two halves of the mold came together. The most delicate glasses are created by glassblowers, who use their mouths and hand tools to introduce air into still-warm glass and bend it to fit their artistic aims. Machines are now used to blow larger volumes of high-quality commercial glassware.

After glass is shaped, it is put into a warm oven to cool slowly, a process called ***annealing.*** This slow cooling period strengthens and stabilizes the glass and removes stress points that may have developed during shaping.

Manufacturers also do other things to ensure the glass is strong and durable. The shape and thickness of the glass have a lot to do with its resiliency. A curved or barrel-shaped glass is stronger than a straight-sided one; a thick, short, fluted stem is more durable than a tall, thin one. You may also notice extra thickness at spots of greatest stress. Ribbed or swirled patterns, or a rolled edge at the rim of the glass, all indicate extra thickness where it is needed most.

After annealing, some glassmakers add another step called ***tempering.*** The cooled glass is reheated, almost to its melting point, and then exposed to a quick blast of cold air. The cold air slightly shrinks the surface of the glass, while the interior of it remains boiling hot. Oddly enough, this builds the resistance of the glass to temperature extremes. The term *fully tempered* means the entire glass underwent this process, but sometimes only the rim receives this extra treatment. Most stemware is fully tempered; most tumblers are *rim tempered.*

A final way strength is added to glass is to mix chemical compounds in during the glassmaking process. Popular examples of this method (with their "secret formulas" closely guarded) include the brand-named Corningware, Pyroceram, and Pyrex.

Other types of glass are known not for their strength, but for their delicacy. ***Crystal*** or *rock crystal* is created by adding lead oxide and potassium silicate to the regular glassmaking process. And ***cut glass*** is high-quality crystal with a design etched onto the glass. The etching is done by careful sandblasting or by brushing on fluorine, a highly corrosive acid. Glasses can also be decorated either with decals or by silk-screening on enamel-based paints.

Shapes and Sizes of Glassware

Like other kinds of dishes, you'll select glassware based on its cost, availability, and durability. However, attractiveness must also rate among your priorities. Look for glasses that are clear and bright, without specks or bubbles to mar their shiny finish. Rounded corners are desirable simply because they are the easiest to clean. The best way to make your choice is not from a catalog, but by picking up the glass and deciding how it feels. Is there enough weight to the bottom to keep it from tipping? Can you grip it securely? Does it balance well when you set it down?

You'll have a lot to choose from, but most glasses fall into one of several broad categories (see Illustration 19-5):

CYLINDRICAL. Also known as a *tumbler,* this is a straight-sided glass with a flat, stable (usually thick) bottom. It is used for water, milk, and (in larger sizes) iced tea.

STEMWARE. So named because it has a thin stem that separates the bottom from the bowl of the glass, it is most often used for wine. Some bars also use stemware to serve cocktails.

FOOTED GLASS. This looks like a cross between a tumbler and a wineglass. It has a disk-shaped base, either attached directly to the bowl or with a very short stem. It is considered sturdier than stemware and more upscale than a tumbler.

MUG. Also known as a *stein,* this large, heavy glass with a side handle is used primarily in casual dining for beer, soda, and iced tea.

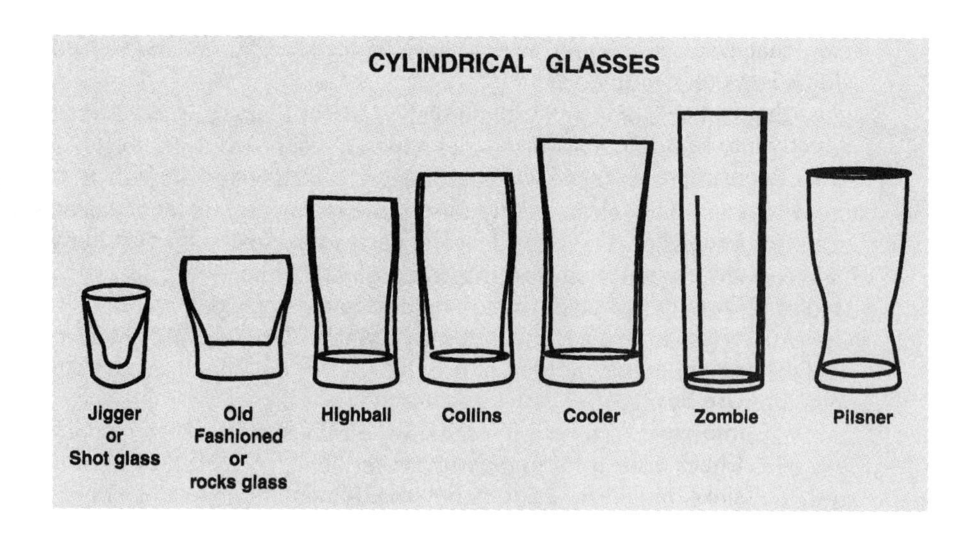

ILLUSTRATION 19-5 There are almost as many shapes and sizes of glassware as there are restaurant concepts.

SOURCE: COSTAS KATSIGRIS, MARY PORTER, AND CHRIS THOMAS, *THE BAR AND BEVERAGE BOOK*, 3RD ED., JOHN WILEY & SONS, INC., 2003.

In addition to the basic types mentioned previously, there is an almost overwhelming number of specialty glasses on the market. Your selection process should depend partly on whether you are serving alcoholic or nonalcoholic beverages and how brisk your service is. Generally, the faster the pace, the sturdier you want your glassware to be, since some breakage is inevitable. Here are a few of the possibilities you'll be faced with:

For trendy microbrewed beers or ales, you might choose an *English pint* glass or a slim *German pilsner* glass. For most cocktails, consider the size of drink you want to serve: 4-ounce or 6-ounce? Bulbous *brandy snifters* come in capacities from 10 to 18 ounces—of course, they're not filled to capacity. They are also used to serve ports and some dessert wines, or desserts! These after-dinner treats may also be showcased in 3- to 4-ounce *liqueur glasses. Highball glasses,* never less than 8 ounces in capacity, are versatile, straight-sided tumblers with heavy bases. And *old-fashioned glasses* (named for the cocktail, not for bygone days) are squat, footed glasses of 5- to 8-ounce capacity.

As if all that doesn't provide you with enough decisions to make, just wait until you start looking at wineglasses. There is general agreement that wine should be served in stemware, although, in recent years, some more casual restaurant chains have started using 7-ounce tumblers for their house wines.

There's running controversy in the industry about having one all-purpose wineglass or different types of glasses appropriate for reds, whites, champagnes, and dessert wines. Perhaps a theme restaurant with a small wine list can do well with an all-purpose, 8- to 10-ounce glass for a 5- or 6-ounce serving. However, if yours is a trendy wine bar or an upscale restaurant, you will certainly want to invest in a variety of glasses to suit the offerings: wide-mouthed, large-bowl stemware for reds; smaller-bowl stemware for whites; elegant, narrow ***flutes*** to preserve the effervescence of sparkling wines; and a variety of snifters and liqueur glasses for after-dinner pours.

One interesting trend in fine dining is to showcase all wines (except sparkling) in oversized, stemmed glasses. The clarity and thinness of the glass, its large capacity (although it is never filled more than halfway), and no top rim are the hallmarks of this type of glassware. Only 5 or 6 ounces of wine are poured into its 11- to 13-ounce bowl, which allows the bouquet (aroma) of the wine to develop in the glass, with plenty of room to swirl it and look at it without spilling. The glasses are elegant but pricey, costing up to $35 a stem. It is wise to have one restaurant staff member (perhaps the sommelier, if you have one) be responsible for the care of this glassware.

Glassware Care and Cleaning

A continuous training program for people handling glassware is a must for both restaurants and bars. Improved handling means less breakage, higher productivity, and less likelihood of accidents and injuries. Glass breakage occurs for two major reasons. ***Mechanical impact*** results when the glass hits another object, however slightly or accidentally. This contact can cause tiny abrasions, not even visible at first, that weaken the glass and make it susceptible to breakage with the next impact. And ***thermal shock*** is a quick, intense temperature change that can cause enough stress on the glass to break it. A glass that comes out of the dishwasher steaming hot, for instance, should not be filled with ice and put directly into service. The thicker the glass, the more time it needs to reach room temperature again.

To ensure a longer, safer life for your hard-working army of glasses, here are a few tips:

1. Keep enough inventory on hand that you're not constantly faced with using glasses that are still hot from the dish area.
2. Organize incoming soiled dishes so that busers can place glasses, plates, and flatware in separate areas. They can probably load glasses directly onto divided racks appropriate for that size of glass.
3. Use bus tubs that have separate flatware baskets, so busers won't be tempted to pile flatware into empty glasses.
4. Check dishwasher temperatures before every shift. And, in hand-washing situations like bar sinks, regularly replace worn-out cleaning brushes.
5. Use plastic scoops in ice bins; metal scoops can more easily chip the rim of a glass. And never, *never* scoop ice with the glass itself!
6. Don't pour hot water in an ice-cold glass or put ice in a hot glass. When pouring hot drinks, preheat the glass by running it briefly under warm water.
7. Never pick up multiple glasses at a time, "bouquet" style, in one hand. In fact, avoid glass-to-glass contact whenever possible. Unload dish racks or bus tubs one piece at a time, not in stacks.
8. ***Pyramiding*** is the best way to store glassware (see Illustration 19-6) if you do not have sufficient dish racks for the particular type of glass or cup.

ILLUSTRATION 19-6 **Stacking trays of glasses in a "pyramid" shape allows more storage without damaging the glasses.**

Sparkling clean glasses say more about your operation than almost any type of advertising you can pay for. In most situations, it's okay to run glassware through the same dishwashers as your plateware and flatware. However, if you have a constant need for quick turnaround of your glassware supply in very busy surroundings, you might consider a separate washing area just for glasses. Restaurants that do a high-dollar fine wine business sometimes wash wineglasses separately, in superheated water with no detergent. This avoids the problem of soap residue and spotting on the glasses.

For regular glasses (not wineglasses), mechanical or manually operated brushes can be used to scrub the insides with detergent, even before the final wash cycle. During the wash itself, water and more detergent are forced onto the glassware; then a 180-degree-Fahrenheit rinse provides the sanitizing process. Most professionals suggest a wetting agent in the detergent, which allows the rinse water to run freely from the glasses and prevents spotting. Air drying is recommended; towel drying is inefficient and is likely to leave bits of towel fuzz on the clean glasses, as well as a bit of residue from the laundry soap the towels were washed in. However tiny the amount, it has a negative effect on carbonated beverages: Soda, beer, and champagne will quickly lose their fizz when they come into contact with even a hint of detergent.

Over time, glassware may develop a cloudy look, which is not easily removed with brushing or normal detergents. This film can be the result of hard-water minerals or a combination of food protein and detergent. Beer glasses seem to be especially susceptible to this buildup. Special detergents and more brushing will usually remove it.

After they're washed and dried, storing the glasses in glass racks will save time and handling. A major cause of breakage is storing glassware in the incorrect type of rack. The only exception to rack storage is with bar glasses, which may be displayed at the back bar or in special overhead racks built onto the ceilings or cabinets.

Purchasing Glassware

The same rules apply for glassware as for purchasing other components of your place settings: Order more inventory than you need. When you have extras on hand, you are not tempted to use hot, wet glasses right out of the dishwasher and risk thermal shock by filling them with ice or chilled beverages. Choose from existing patterns that will be readily available in months and years to come. And, as you've read before in this chapter, never order glassware from a catalog or mail-order source until you've had a chance to feel, and even drink from, the glass. It's the only way to gauge how comfortably it will fit with your menu, plateware, and the overall style of your restaurant.

Here, we'll end the glassware discussion by reproducing a table that lists guideline amounts for your opening glassware order (see Table 19-2).

TABLE 19–2

GLASSWARE REQUIREMENTS

	Item	Glassware Required Based on Seating Capacity		
		Seating 100	Seating 200	Seating 300
Coffee Shop	5 oz. juice	12 dz.	24 dz.	36 dz.
	10 oz. water	24 dz.	36 dz.	72 dz.
	10 oz. iced tea	12 dz.	24 dz.	36 dz.
	Sugar packet holder	10 dz.	20 dz.	30 dz.
	Salt and pepper	3 dz.	6 dz.	9 dz.
Dining Room	5 oz. juice	12 dz.	24 dz.	36 dz.
	12 oz. iced tea	12 dz.	24 dz.	36 dz.
	10 oz. water goblet	24 dz.	48 dz.	72 dz.
	5½ oz. sherbet	12 dz.	24 dz.	36 dz.
	6 oz. flute	12 dz.	24 dz.	36 dz.
	Sugar packet holder	10 dz.	20 dz.	30 dz.
	Salt and pepper	3 dz.	6 dz.	9 dz.
Banquet	5 oz. juice	12 dz.	24 dz.	36 dz.
	12 oz. iced tea	12 dz.	24 dz.	36 dz.
	10 oz. water goblet	24 dz.	48 dz.	72 dz.
	5½ oz. sherbet	12 dz.	24 dz.	36 dz.
	6 oz. flute	12 dz.	24 dz.	36 dz.
	8½ oz. wine	12 dz.	24 dz.	36 dz.
	Sugar packet holder	10 dz.	20 dz.	30 dz.
	Salt and pepper	3 dz.	6 dz.	9 dz.
Room Service	8 oz. room tumbler	Minimum 2 to a room		

Source: Libbey, Inc., Toledo, Ohio.

19-5 FLATWARE

No matter how popular finger foods become, flatware will still be an absolute necessity for eating and a timeless component of the tabletop. They may not add the color of plateware or the sparkle of glassware, but utensils have an inherent appeal nonetheless, because they are associated with precious metals. Think about it! At most tables, utensils—with the notable exception of plastic forks and spoons—are commonly referred to as "silverware." And even the most plain, functional flatware includes an element of design. It has a look, and a feel, that should be carefully chosen to reflect the theme of the restaurant and the type of dining experience. By giving proper attention to the size and weight of the flatware, you can subtly convey an impression of quality.

Weight, in fact, appears to be the most distinctive characteristic of flatware in the eyes of the guest. It is an assumption rooted in history, because utensils really were made of gold and/or silver in ancient times, and the higher your standing in the community, the heavier your eating utensils.

Flatware was first introduced in the West by Crusaders returning from their military expeditions to the Holy Land. Later, traders brought flatware to Venice and Genoa, where the wealthy aristocrats of the day snapped it up. An Italian, Catherine de Medici, introduced the idea of dining with utensils to the French court when she married King Henry II and brought her silverware with her. The British were reportedly slow to accept the custom, preferring to eat with their hands. In many cultures, the right hand is still considered the most useful and proper dining tool—but not in *this* chapter!

Today, oversized flatware is often known as ***European-sized,*** and it can make an opulent statement on some tabletops. However, drama and style may be enhanced with the addition of a single accessory piece selected for a specific course, such as sauce spoon, steak knife, or salad fork.

How Flatware Is Made

Foodservice flatware is generally either *silver plate* or *stainless steel.* Solid silver is impractical in heavy-use settings like restaurants, not only because it's expensive, but because it is soft and can be easily scratched or bent.

Silver plate is the choice in upscale clubs and restaurants. It begins when a utensil is crafted from a ***base metal.*** Today, this base metal is an ***alloy,*** or mixture, of nickel, brass, copper, zinc, and/or stainless steel melted together at high heat. The piece made of alloy is called a ***blank.*** All alloys are not created equal: Those without enough nickel are too soft; those with too much iron can rust easily, and so on.

A thin coat of silver is ***plated,*** or fused, onto the blank using an electric current. The silver plate will be identified by the thickness of the plating in ***mils.*** Each mil is one-thousandth of an inch. After plating, the piece is finished, or given a texture, and polished. Listed from most opaque to clearest, the types of ***finishes*** are *satin, butler, bright,* and *mirror.* (The polishes usually have numbers instead of names.) The satin or soft butler finishes are the most popular in commercial use, but sometimes, a single soupspoon will have two: a mirror finish to make the bowl especially shiny and a stain finish for the handle.

Custom patterns or logos can be stamped right onto the metal. There'll be an extra charge for making the die and for the stamping process.

Silver-plated knives with hollow handles instead of solid ones are called ***holloware.*** Many prefer holloware for foodservice use because it is lightweight and easy to handle, but others feel solid-handled knives are more durable, especially when blade and handle are forged from a single piece of metal. Some knives consist of a stainless steel blade, welded onto a silver-plated handle. When choosing knives, an indication of quality is the point at which the blade and handle are welded together—potentially the weakest part of the knife. Check it, feel it, and ask about it.

Silver-plated flatware will naturally wear down at points of use, such as the tines of forks or the backs of spoon bowls, so better-quality utensils are sometimes brushed, sprayed, or dipped in extra silver at these points. The process is known as ***overlaying.*** The amount of silver that's used is measured in terms of how many troy ounces of silver have been plated onto a gross of standard-sized teaspoons. (There are 144 teaspoons in a gross.) This is all pretty confusing, so take a look at Table 19-3, which lists the common names and numbers of ounces of silver in different "grades" of silver plate.

TABLE 19–3

GRADES OF SILVERPLATED FLATWARE

Common Name	Ounces of Silver per Gross
Half standard	2½ ounces
Banquet	4 ounces
Standard	5 ounces
Triple	6 ounces
Extra heavy	8 ounces

An order of new flatware should be examined thoroughly for flaws. The pieces should have a smooth, even coating of silver so that no thin spots show through with wear.

The weight of a fork or spoon is not necessarily an indication of its strength or hardness—in fact, the quality of the alloy used to make the blank is the most important factor in determining durability. By the time you've received your shipment, you've got to take the manufacturer's word for that, but there are a few tricks you can use to check for overall quality:

- Put pressure on the bowl of the spoon where it joins the handle, to see how easily you can bend it.
- Put a fork tine down on a solid tabletop and try to bend it by pushing it down.
- Closely examine the solder between the knife handle and blade, to see if it is easily bent or bowed.

The one substance that will outlast silver plate of any weight is stainless steel. This is the material of choice for most foodservice flatware because it requires so much less care than silver plate: It's hard to scratch, dent, or stain; doesn't need polish; doesn't rust or tarnish; and doesn't need replating. However, like silver plate, stainless steel utensils are only as good as the alloys they are made of.

The accepted standard is nicknamed "18/8," which is short for 18 percent chrome, 8 percent nickel, and 74 percent steel. Chrome adds softness and luster; nickel and steel provide the hardness. Together, the mixture has a natural resistance to food acids and cleaning chemicals, but still offers a warm, silver-like appearance. A more expensive option is the "18/10" blend, which includes slightly more nickel. There are also alloys of stainless steel and chrome without any nickel. They are referred to as *18% chrome stainless* and *13% chrome stainless.* Finally, alloys known as "410" and "430" are used to make less expensive flatware.

Better-quality flatware is *graded,* or shaped so that the stress points of the utensil are made slightly thicker. Graded flatware also looks nicer than lower-quality flatware, which is stamped out of metal sheets in uniform thickness.

Flatware Care and Cleaning

One of the biggest problems with utensils is that they disappear so easily. Employee theft is certainly part of this, but carelessness is equally to blame. Several manufacturers have created flatware retrieval devices that can be placed over a trash can. They simply look like a lid with a wide slot or opening in them, but the models we've seen contain magnets that catch falling flatware and hold it at the top of the container so it doesn't end up in the trash (by KatchAll Industries International of Cincinnati, Ohio); or a small, battery-operated metal detector that sounds an alarm when silverware passes through the slot. The latter, called a Silver Chute (by Golden West Sales of Huntington Beach, California) can be ordered with a wide rubber squeegee attachment to scrape plates as well.

Both stainless steel and silver-plated flatware will show film, stains, rust, or corrosion unless care is taken to store and clean them properly. Soiled flatware should be removed from tables as soon as possible, and it should be presoaked before going through the dishwashing cycle. The benefits of presoaking are threefold:

1. Wetting utensils makes it easier to remove caked-on food particles. Add a small amount of detergent to the soak water to increase its effectiveness.
2. Presoaking minimizes the utensil's contact with acidic foods, eggs, and other substances that may tarnish or corrode it.
3. When using silver plate, the addition of a small sheet of aluminum foil to the bottom of the soaking tub prompts an interesting electrolyte reaction: The tarnish leaves the silver and adheres to the aluminum. This *does not work* with stainless steel flatware, however, and might even damage it. Try it only with silver-plated utensils.

When it comes to presoaking, your flatware can get too much of a good thing. Limit soak times to half an hour at most, and use a mild alkaline detergent in low concentration—only 1 ounce per gallon of water—in the soaking tub. From here, the flatware should go immediately into the dishwasher. To allow it to dry would defeat the purpose of presoaking.

Some foodservice businesses buy flatware washers, special machines that can clean and sanitize utensils in 90 seconds. These combine mechanical and hydraulic action to "lift" and separate the flatware during washing and rinsing, no matter how tightly it is packed into the perforated wash baskets. A photo for a style of flatware washers is shown as Illustration 19-7. These particular models use the rinse water from the previous cycle as the wash water for the next cycle, saving water and energy.

ILLUSTRATION 19-7 **Flatware washers are specialty machines for fast washing of silverware.**
COURTESY OF INTEDGE INDUSTRIES, INC., WOODRUFF, SOUTH CAROLINA.

In the dish room, never use steel wool or other metal-based scouring pads to remove caked-on foods from utensils. These can easily scratch and mar the mirror finish. If it's necessary to wipe them off, do so with a cloth or plastic pad. The other dish room enemy of flatware is, of course, the garbage disposer unit. Keep it out of the disposer, and both will last longer.

Pack the flatware loosely in its perforated baskets and arrange it so that water can reach all parts of the utensils. Turn forks and spoons up, but keep knives pointed downward as a safety precaution. Segregating them only makes it easier for them to nest together and not get fully clean, so mix forks, spoons, and knives together at random. Temperatures for proper flatware washing should be 140 to 160 degrees Fahrenheit for the wash cycle, with a final rise of 180 degrees Fahrenheit. More temperature guidelines are available in Chapter 16.

There are several additions to the wash cycle other than the detergent. A wetting agent in rinse water will help water run off dishes more thoroughly, preventing unsightly spotting. In areas where water is "hard" (has a high mineral content), you may need a water softener to prevent spotting. Finally, flatware should always be air dried, not towel dried, for maximum sanitation.

Silver plate must occasionally be detarnished, usually with a chemical compound that is rubbed onto the utensils. Rinse carefully after removing tarnish, because the compound contains chloride and will actually corrode the silver if left on too long. Or make your own detarnishing solution by dissolving 1 tablespoon of salt and 1 tablespoon of baking soda in 1 gallon of boiling water. Put this solution into a soaking tub with a small sheet of aluminum foil at the bottom. Make sure all the utensils come into contact with both the solution and the foil. It should take no longer than 30 seconds to do the job.

You may also purchase a ***burnisher***, a machine that uses tiny, vibrating stainless steel balls to polish silver plate by removing some of the surface scratches and leaving it with a fine, soft sheen. Over time, its proponents say burnishing also helps utensils become harder and more durable. Different sizes of burnishers can accept up to 300 pieces of mixed flatware at one time, and the cycle lasts

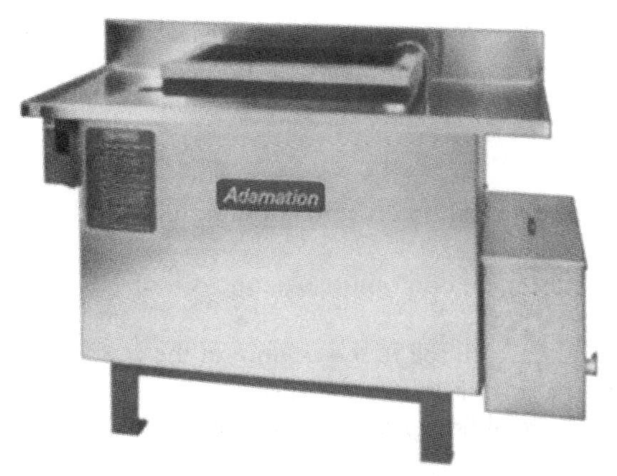

ILLUSTRATION 19-8 Silver burnishers come in many shapes and sizes. Vibrator motion burnishers (shown here) polish large pieces such as trays, pitchers, bowls, and coffee pots. Rotary burnishers (not shown) are used to polish flatware.

COURTESY OF ADAMATION, INC., NEWTON, MASSACHUSETTS.

about five minutes. Burnishers require both water and electricity to operate (see Illustration 19-8). Where space is tight, they can be rolled into an area on casters.

The appearance of stainless steel flatware can be improved by soaking it in a solution of one part vinegar to three parts water, then rinsing and air drying it. This usually removes hard-water film and/or mineral deposits. Like silver plate, stainless steel should be presoaked promptly after use.

Purchasing Flatware

The standard flatware pattern consists of eight items, although many are not used in casual-dining settings. They are the teaspoon; the (larger, oval) dessert spoon (sometimes inaccurately called a "tablespoon"); the (rounded) soupspoon; the (tall) iced-tea spoon; the dinner fork; the salad fork; the cocktail fork; and the dinner knife. Some patterns contain as many as 20 items, including such specialty utensils as short, round-tipped butter knives; steak knives with serrated blades; the fish fork, with wider prongs; and the fish knife, which looks like a larger butter knife. As always, your menu determines your flatware needs (see Illustration 19-9 and Table 19-4).

Always ask the manufacturer or sales representative to show you samples of flatware. This is the type of thing you should never purchase by catalog without seeing and handling it first. Besides the basic look and feel of each piece, a quick quality check should reveal:

- **Forks with smooth, rounded edges. The back of the fork handle should be slightly concave. Feel the tines or prongs to see how solid and sturdy they are. One of the first things to go "wrong" on utensils is that fork tines get bent.**

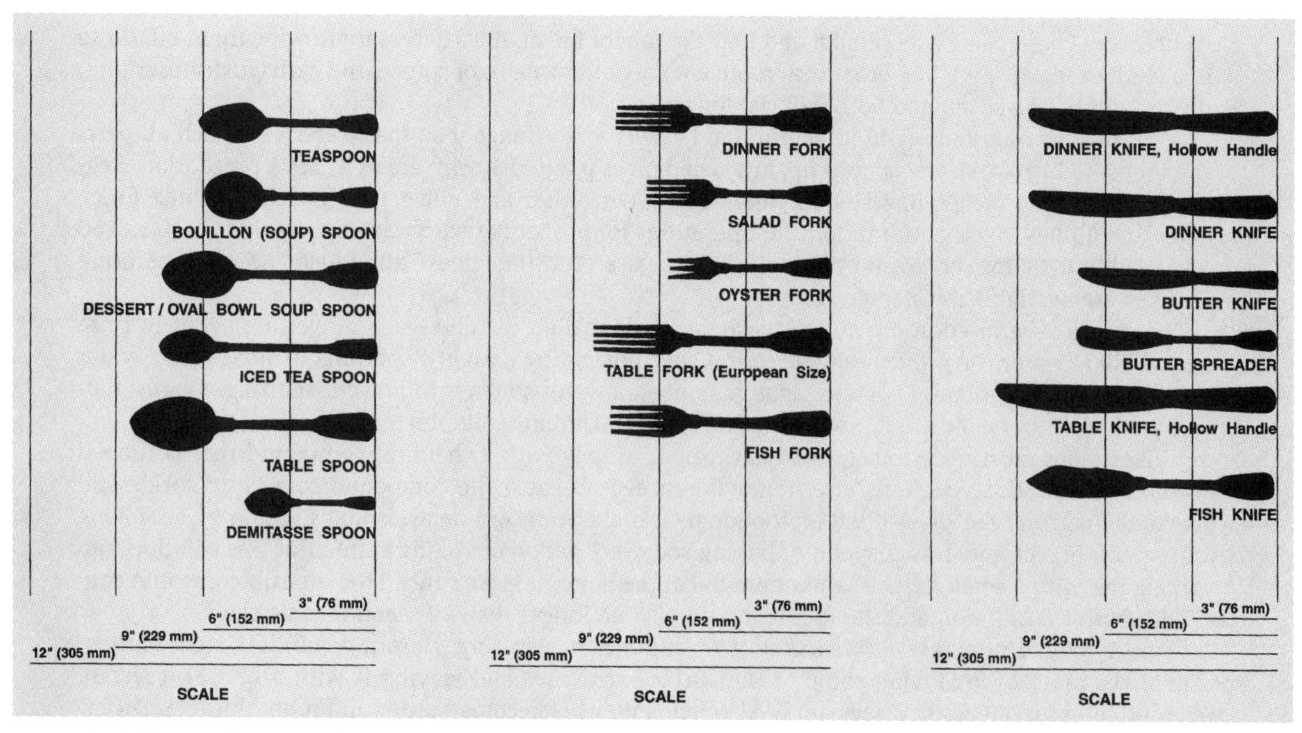

ILLUSTRATION 19-9 The most common sizes of flatware.

SOURCE: ADAPTED FROM DELCO TABLEWARE INTERNATIONAL, PORT WASHINGTON, NEW YORK.

TABLE 19–4

SUGGESTED FLATWARE REQUIREMENTS

	Requirements per Seat		
	Hotel Dining Rooms	Average Good Restaurant or Coffee Shop	Smaller Operation
Teaspoon	5	4	4
Dessert/oval bowl soup spoon	2	2	2
Tablespoon	¼	¼	—
Iced-tea spoon	1½	1½	1
Demitasse or after-dinner coffee spoon	2	—	—
Bouillon spoon	2	2	2
Dinner fork	3	3	3
Salad fork	1½	1½	—
Oyster fork	1½	1½	1½
Dinner knife	2	2	2
Butter spreader/butter knife	1½	1	—
Steak knife	1½	1	—

Special consideration should be given to the following factors: average seating capacity, maximum seating capacity, turnover during rush hours, and special food features (seafood, steaks, etc.).

Source: *Foodservice Equipment and Supply Specialist.*

- Knives with smooth, rounded edges. Solid-handle knives should be one piece; holloware handles should be securely welded to the blade. Cut meat with them; press the sample firmly down on something solid. The blade should not bend.
- All edges of spoon bowls should be the same. The back of the spoon handle should be slightly concave. Try bending it, as bent spoon bowls are another common flatware problem with heavy use.
- For commercial use, a satin finish (which may be called a number 7 polish) should be used, except for spoon bowls. These should always be shined to a mirror finish.

When it comes to designs and patterns, you'll be amazed at how much the industry has to offer. Pare down your choices based on the size and style of your plateware, because the two must complement each other.

Also, think about the uses of your flatware. In the most contemporary dining situations, guests tend to use the American exchange. That is, they transfer their fork from one hand (as they use it to hold something they're cutting) to the other hand (as they eat). Oversized European flatware may not be the most comfortable choice for them.

In casual settings where "grazing" or appetizers are the focus, smaller airline-sized flatware may be perfect. Not everyone needs butter knives, fish forks, or iced-tea spoons. Decide what is truly necessary based on your menu, and stick to your list.

In your attempts to bargain-hunt, resist the urge to buy what may seem like identical patterns and styles from different manufacturers. The weight or finish will be just different enough that they won't really match the other pieces when placed side by side. And remember, the more ornate the design, the fewer scratches will show.

Finally, if you customize your flatware with a logo or initials (called *cresting*), consider customizing your plates as well. Flatware alone cannot make the expected impact.

SUMMARY

The flatware, plates, glassware, linens, and accessories that sit on the dining table are known as tabletop. Your tabletop is a prime component of the personality of your restaurant, and the investment you will make in these components is considerable. Most businesses replace about 20 percent of their tableware each year, the result of breakage or theft. The most important part of the tabletop is the 24-inch space directly in front of the guest, and the items there comprise the table setting.

As you choose your plates, glassware, and flatware, think about how the items will feel in the guest's hand or mouth, not just how they look. Durability is another critical factor. Consider your budget, and how the food will look on the plate. Custom-designed plates are impressive, but can you reorder in a hurry when you need more of them? The oversized white plate seems to be the most functional in foodservice settings, and even this has half a dozen different types of styles, sizes of rims, and so on. Many restaurants choose brighter accent colors and/or patterns for linens, accessories (bread containers, salt and pepper shakers, and the like), and side plates for appetizers, salads, and desserts. Being fun and whimsical and using unusual items as serving pieces is great for tabletop design, but only if it truly fits the concept and types of food you'll be serving.

To minimize breakage, teach employees how to properly clear tables, load bus boxes and dish racks, and scrape waste off of dishes with the correct implements (*not* with another dish!). Most breakage occurs in the dish room.

STUDY QUESTIONS

1. Name and describe three different functions of restaurant tableware.
2. What is vitrification, and why is it important?
3. List the pieces of plates, glasses, and flatware that you would require for a single place setting in an elegant, four-star restaurant.
4. How should plateware be stored, and why? Give at least two tips.
5. Why should you select glassware that has been tempered? What's the difference between tempered and fully tempered?
6. List at least two precautions that can be taken to prevent thermal shock in glassware.
7. Why would you choose stainless steel flatware for your dining room? Why would you choose silver plate?
8. What are the guidelines for when and how to presoak flatware?
9. What is the American exchange, and what impact might it have on your flatware selection?
10. Now that you've read the pros and cons of using a designer to create custom table settings, would you hire one or would you do it yourself? Why?

A Conversation with . . .

Mike Fleming

TABLETOP DESIGNER, RYKOFF AND COMPANY
PORTLAND, OREGON

Mike Fleming loves his work. With a background in sales—everything from stocks and securities to restaurant equipment—he now specializes in tabletop presentation and sells dishes, glassware, and flatware to foodservice establishments around the Western United States for Rykoff and Company. He says his job is a perfect blend of his natural artistic tendencies, his sales acumen, and his love of good food. Mike's favorite recent clients have been the foodservice departments of U.S. national parks, like Crater Lake, Oregon, where he designed a commemorative dinner plate for the park's anniversary.

Q: How did you get started in this line of work?

A: As a restaurant equipment salesperson, I was doing contract and design work, literally building kitchens from the ground up. I got hooked up with a couple of representatives of china manufacturers who appreciated my ability to sell myself, and they helped me learn about the china.

Q: What's the most fun part of the job?

A: The best part of what I do is work with chefs. I look at a real chef as an artist, an absolute artist. And so many of them do not get an opportunity to express themselves on their tabletop. They do the food preparation and the color, but they never really have the canvas to put it on. So what I do is enable them to enhance their food product, and I give them a canvas to put their artistic work on.

Q: Do they have some strong ideas about how tabletops should look, or are they not taught that kind of thing?

A: For a chef, a lot of it is on-the-job training. I know there are general courses that include some tabletop information, but there are always so many new trends and ideas.

If it's a high-end restaurant, I start out by selling them a handblown glass product. It doesn't have to be a real expensive one, but it does have to be handblown. You don't want lead crystal, it's generally too soft, but I start out with a 21- to 24-ounce glass that they use for red wines. I work with the chef or owner's attitude about profit making, and with a 24-ounce glass, by filling it one-quarter or so, four people will drink a whole bottle. So then to have a second glass, they have to order a second bottle. This elevates the profit, elevates the tip, elevates everything. And then I'll use maybe a 10-ounce glass for the whites. I urge people to always think about the potential profit behind their decisions, whether it's the size of wineglass or coffee mug or dessert plate.

Q: What is the most ignored part of the tabletop that you feel restaurants should pay more attention to?

A: There are several things. First, salt and pepper shakers could be more fun than they are in most places. Of course, at real high-end restaurants there's no salt or pepper on the table because if the chefs do it right, you don't need to season the food.

But the most important things on the table are what the customers will feel in their mouths: the rim of the glass, the tines of the fork. If I'm gonna pay $17 to $23 per entrée for my dinner, I'm going to want to drink out of a wineglass that feels very

good to my mouth, with a fine, sheer rim instead of a large, clunky edge.

The flatware should feel good. You can always tell inexpensive flatware because it has very sharp tines, rough and abrasive. I'm not saying there isn't a market for inexpensive products—and they're not "cheap," they're inexpensive. It depends on your situation and the atmosphere you're trying to create.

Another little thing most people don't think about is how to serve sugar. You see a lot of standard sugar-packet holders, the rectangular ones you can get in glass or ceramic. Some restaurants use a little silver-plated one, or a wire basket, and in your mom-and-pop places you'll want a sugar shaker. This is not only sanitary, it's a way of portion control and it keeps the health inspector off your back.

Q: What is the priority when you're just opening a restaurant, and you know you can't afford top of the line for everything? What tabletop item should you splurge on, in your opinion, above anything else?

A: My main thing would be the plateware. When you look at a painting, you look right at the center of it. The china is the center of your place setting. A lot of chefs choose plain white because of the garnishes and sauces they use. Imagine how a beautiful red tomato looks on a white plate, or the rich burgundy color of pomegranate juice.

Q: Tell us a little bit about the different kinds of plates available.

A: If your problem is food that comes out of the kitchen hot but is too cold by the time it gets to the table, you should be looking at your choice of plates. Thick ceramic products retain the heat longer.

I also caution restaurateurs not to choose an overbearing plate with so much color that the food gets lost on it. The customer doesn't get the excitement of looking at what you've created, seeing the shrimp and the greens and the chives. Use specialty plates for signature dishes, like the really fun dessert plates, but otherwise stick with something plain white and classy.

You know, there are even different shades of white—it's not just plain white plates. You have a creamy or off-white color, which is what we typically think of as a white plate. Then there's a brighter white called "American white" in the industry, like a white piece of paper. And there's a "European white," that's so white that it's almost blue or gray/blue.

You can also have a custom plate designed for you, with your picture on it, your logo, or just about anything you want.

Those can really be fun projects to work on.

Q: You work with so many restaurants. Do you notice anything special about the ones that are really successful?

A: Yes. Their wait staffs are extremely knowledgeable about the products. They have tasted the daily special, so when the customer asks about it, the explanation is informed and enthusiastic. And the wait staff itself—they're hustlers, they're educators. They want people to know about the food. They're proud to serve it and talk about it. You know, in a really fine dining environment, a waiter can make $50,000 a year and work his own hours, on his own terms. I don't think you can ever go wrong with more staff training.

Q: Let's talk for a moment about care of your tabletop items.

A: Glasses and china do not break; they are broken. So it's important to subdivide your busing. Don't put your wineglasses on the bottom and platters on the top of your bus tubs. That's stupid! Just running your cart across the floor, the vibration will break them. Have a separate glass rack on the cart, and put your glasses on that. Only your plateware should go in the bus tub. You should also have a refuse container on your bus cart, and a light presoak for your flatware.

Don't bunch your flatware too tightly to be washed. Especially with a low-temp machine, if they don't have a chance to properly air-dry, detergent could remain on the utensils and even cause a chemical reaction that eats through the finish of the flatware. Monitor how the flatware is bused, too, because often it is used to scrape leftovers off of plates and down it goes, into the trash. I also joke with people that once all your employees have a five-piece place setting at home, your pilferage rate will drop!

For high-end glassware, don't wash it in an open rack; it just bangs together. Get compartmentalized racks so each glass fits into its own compartment. These are easier to stack, too, and you can store the clean glasses right in the rack until you use them again.

The less you handle the glasses, the better off you are. Remember, table to rack and rack to table, and that will eliminate 90 percent of your breakage, right there.

CHAPTER 20

LINENS AND TABLE COVERINGS

INTRODUCTION AND LEARNING OBJECTIVES

Table surfaces of wood, brushed metal, and even leather are chic today. But there's nothing quite like a freshly starched tablecloth and attractively folded napkin to set, or change, the mood of a dining space. They can add excitement, elegance, and even nostalgia to the dining experience.

Whether you rent or purchase them, table linens should never be more than 1.5 percent of your budget. However, it's a very important percentage point, since it does so much for your image. Table linens enable us to include splashes of color that enhance the rest of the restaurant's decor. And, like any other component of the tabletop, there's a lot more to linen selection than meets the eye.

In this chapter, we will discuss:

- Types and sizes of tablecloths and napkins
- How different types of fabrics are made
- Selection guidelines
- Alternatives to fabric
- Options to rent or purchase table linens
- Care and laundering of your table linens

20-1 LINENS IN THE DINING AREA

Unlike other tabletop components—flatware, plates, or glassware—table linens are one- dimensional. Think of them in the same way an art gallery uses its wall space, to highlight the artwork displayed on it. In some cases, the wall simply enhances the display; in others, the wall actually becomes part of the display.

The foodservice term for table linens is ***napery.*** In addition to providing color, napery also offers the greatest "comfort level" of any tabletop item. Tablecloths and napkins are the only items that don't

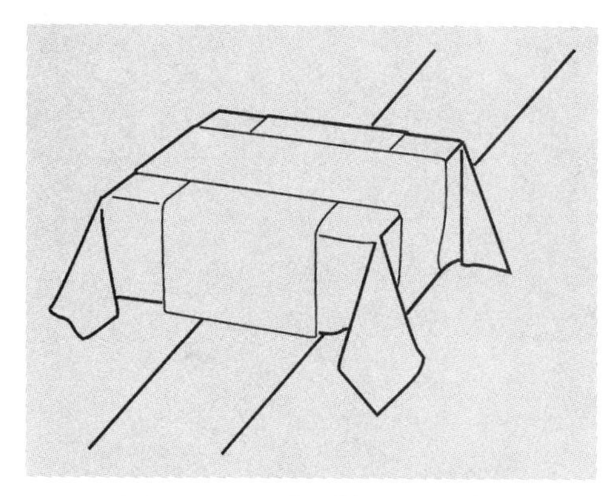

ILLUSTRATION 20-1 A table runner is so named because it "runs" the length of the table.

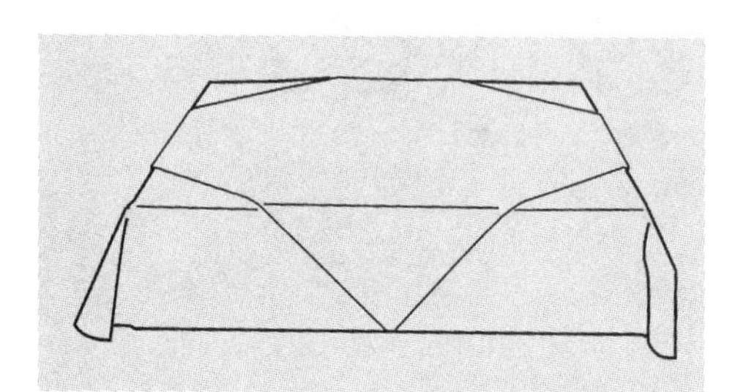

ILLUSTRATION 20-2 An overlay is the term for one tablecloth layered over another.

have hard surfaces, and they should feel good, as well as look good. We'll discuss fabric choices later in this chapter.

One manufacturer of tablecloths and napkins suggests that five basic table colors—white, red, hunter green, royal blue, and rose petal pink—can be used by any restaurant to create festive moods year-round for any holiday theme. Some common examples are listed below. Can you think of others?

- Red, White, and Blue. Any patriotic holiday
- Red and Green. Christmas and New Year's Day
- Red, White, and Green. Cinco de Mayo or other Hispanic festival days; Italian holidays or events
- Red or Pink and White. Valentine's Day
- Pink and White. Weddings, Mother's Day
- Green. St. Patrick's Day, Thanksgiving

No matter what color or fabric is used, the linens won't have the desired effect if they are not spotlessly clean. Clean napery fulfills the restaurant's promise of a safe, clean place to eat. Elsewhere in this chapter, we'll discuss laundering requirements to keep your table linens looking their best.

There are only a few types of napery to become familiar with. Besides the basic tablecloth and napkin, there are placemats and ***table runners.*** Runners are lengths of fabric that are placed down the center of a table (often over the tablecloth) to introduce a complementary print or accent color to the tabletop. They are sometimes known as ***deco-mats*** (see Illustration 20-1).

Some restaurateurs use two tablecloths on each table, putting on a larger cloth first, then covering it with a second, smaller tablecloth. The smaller cloth, called a *top* or ***overlay,*** is changed more often than the larger one. In busy restaurants, using tops makes busing the table easier and less noisy, while enabling you to introduce an accent color with the top (see Illustration 20-2). Often, though, tops and the cloths beneath them are both the same color.

20-2 TEXTILE FIBERS AND CHOICES

As a foodservice operator, you will have so many choices of materials and colors that it helps to have some basic knowledge about the fabrics from which napery is made. Even if you decide to rent instead of buy your linens, you'll get better service from the rental company (and you'll be able to tell your staff how to care for the items) if you know something about how they are made.

Textile fibers may be natural or synthetic, but they all have a base that is either *cellulose* (plant) or *protein* (animal/insect). Examples of cellulose-based fibers are cotton, flax, and hemp, which begin their life cycle as plants. Examples of protein-based fibers are silk, wool, or other cloths that begin as part of an animal or insect. The two types of fiber react differently to heat, moisture, dye, and other variables that govern their suitability for foodservice use.

Cotton begins as the puffy, fibrous material that surrounds the boll (or seedpod) of the cotton plant. It is thin and twists easily, so many small cotton fibers are woven together to make long, strong strands that can then be woven into cloth. Cotton is relatively inexpensive. As a fabric, it is durable and absorbent; in fact, it actually becomes stronger when it gets wet. It is also flexible and does not attract static electricity, and both qualities make it an attractive choice for draping tables (see Table 20-1). On the negative side, cotton shrinks and stains rather easily. Fortunately, modern technology allows us to treat today's cotton fabric so it resists both these tendencies. The fancy term for making cloth shrinkproof is to make it ***dimensionally stable.***

TABLE 20-1

100 PERCENT COTTON NAPERY AT A GLANCE

Advantages	Disadvantages
Absorbs spills	Can be damaged or destroyed by:
Breathable	Acids
Easy to wash	Alkalis
Feels good	Heat
	Bleach
	Mildew
	Can shrink
	Colors can fade

Linen is a *bast* fiber, meaning it comes from the inner portion of a plant stem. This particular plant is *flax,* and it yields long fibers (up to 3 feet in length), as well as short ones. The finest linen comes from the longest fibers. Linen, like cotton, absorbs moisture well and is stronger wet than dry. It is two to three times stronger than cotton and good-quality linen does not shed lint like cotton often does. However, linen does not retain its strength under the rigors of laundering. Bleaches and harsh detergents are more likely to damage linens than cottons.

Linen has a stiff, crisp feeling because the fibers used to make it are not very flexible. It wrinkles and creases easily, but it is also nonstatic and holds dye well.

There are other cellulose-based fibers, such as ramie, jute, and hemp, but they are not widely used as table coverings. Nor are protein-based fabrics, such as wool (from animals) or silk (from insects). They just wouldn't hold up to the heavy use they'd receive in a restaurant.

However, there are several synthetic fibers in common use in foodservice settings. These include rayon, acetate, nylon, polyester, and combinations thereof.

Rayon is a cellulose-based fabric, made from bits of wood, cotton, and other cellulose products. *Acetate* begins its life as the same basic cellulose mixture, but different chemicals are added to each mixture in the manufacturing process. Both rayon and acetate are flammable unless specifically fireproofed. Both fabrics are low in cost.

Second only to cotton, rayon is one of the world's most popular fabrics. It dyes well and can be made shrink resistant; it also absorbs moisture and can be laundered or dry-cleaned with equally good results. However, rayon lacks strength and resiliency and is easily attacked by acids and/or alkalies.

Acetates also launder well and can be bleached without problems. Acetate fabrics have soft, luxurious textures and good draping qualities. They don't hold dyes well, though, and, unless specially treated, they build up static electricity. They must be ironed at low temperatures, too, because they'll glaze, stick, or even melt under an iron that's 300 degrees Fahrenheit or hotter.

Nylon is a nitrogen compound, a mixture of several nitrogen-based chemicals. Nylon has the disadvantage of not absorbing moisture well, and it tends to build up static electricity unless it is treated with an antistatic coating. Although nylon is strong and durable, it will yellow with repeated washings. Nylon must also be washed, dried, and pressed at moderate heat settings.

Polyester is a current favorite of both manufacturers and foodservice professionals. It's a synthetic concoction of glycerin and a variety of acids and goes by a number of brand names: Dacron, Vycron, and so on. Polyester is strong, easy to dye, easy to wash, dries quickly, resists wrinkles, and doesn't shrink or fade (see Table 20-2). Some people don't like the "slick" feeling of polyester, which tends to slide off the customer's lap and onto the floor. From a budget standpoint, however, polyester lasts much longer than cotton. An odd characteristic of polyester is that it's not very water absorbent, but it soaks up oil quickly.

TABLE 20–2

POLYESTER NAPERY AT A GLANCE

Advantages	Disadvantages
Strong	Low absorbency of water
Colorfast	Holds oily solids
Easy to wash and dry	Slick finish
Resists damage by acids, bleach, mildew	Not easy to starch
Little or no shrinkage	

Polyester fabrics must be specially treated to prevent stains from typical foodservice sources such as ketchup, iced tea, and lipstick. And an environmental note: Because polyester is not biodegradable, it may not be allowed in your city's landfill. Check before you buy.

In their endless quest for customer satisfaction, manufacturers are always tinkering with their synthetic fabrics, trying to come up with combinations that work better than cloths made from individual fiber types. To date, the most popular amalgam is 50 percent cotton, 50 percent polyester, commonly known as the ***50/50 blend.*** Restaurateurs favor blends because they are such faithful reproductions of the textures of all-natural fabrics, but they don't stain as easily. The best of both worlds? Well, yes and no. The 50/50 blend is certainly stronger than cotton, more colorfast, and less likely to shrink; it also absorbs spills as well as cotton. However, its polyester content means it absorbs oily spills, too, which makes for some nasty stains if they're not treated promptly. Heat, bleach, and mildew can wreak just as much havoc on a 50/50 blend as on 100 percent cotton. Experts agree the 50/50 blend offers a lot of convenience, but, overall, you end up with a product that wears and looks more like polyester than like cotton.

How Fabrics Are Woven

Fibers are combined to make *yarn* by a process known as *spinning*; then the strands of yarn are woven together to make fabric. You can usually tell how a fabric is woven by how it feels; different textures are created by weaving the yarn (which may be smooth, coarse, or nubby) tightly or loosely.

For most tabletop fabrics, a plain, unadorned texture known as a ***satin weave*** is used. Most white tablecloths and napkins are made of satin weave fabrics.

Damask weaves are often used for napkins. A damask fabric contains an elaborate design on both sides, making it look rich and lustrous. The front of the fabric is known as the *weave brocade*; the back side, showing the opposite of the front design, is called the *fill* side. The word "damask" comes from the city of Damascus, the capital of Syria. In ancient times, artisans there were known for their intricate inlaid designs on everything from artwork to architecture.

Today, it's sometimes called *Jacquard damask,* named after the inventor of the Jacquard loom used to weave a particular style of damask cloth. *Double damask* means two thicknesses of cloth are woven together, making the final fabric heavier and more durable.

Momie cloth is the name for a weave that looks like a cross-stitch or herringbone pattern; it has found favor in foodservice because of its durability. Momie cloth is usually either 100 percent cotton or a 50/50 blend. *Oxford cloth* is similar, comes in a plain or basket weave, and is known for its softness (think Oxford shirts).

After weaving, ***mercerizing*** is a common process used to make cotton look glossy and more lustrous. The fabric is stretched and coated with a caustic soda solution, then allowed to dry. Mercerizing (named after the English textile merchant who invented the process in the 1800s) enables the fabric to accept dye more easily, while providing some protection against dirt and stains.

Functional but Fashionable

A study done for the table linen industry about a decade ago indicated a couple of statistics that are probably timeless: 93 percent of all diners notice whether a restaurant's napkins are cloth or paper, and 78 percent prefer cloth, saying it "upgrades" their expectations of the dining experience. Indeed, the types and styles of napery you use will convey the degree of sophistication of your dining establishment. So consider the following factors as you look at samples and talk with salespeople.

Fashion. In some ways, table fashions mirror clothing fashions. Although cotton is a popular choice, today's fast-paced lifestyle often requires the durability and convenience of synthetics. Always examine potential linens with other tabletop components. At this writing, placemats and table runners are "hot" accessories, and it's no longer chic to simply roll up a napkin into a tube when you could make it into a more interesting shape, or stuff it attractively into an empty glass to add height to a table setting. Also consider the overall impact of the colors on the entire room, not just on a single table.

In choice of color, remember that tablecloths in a warm, fall hue may look dramatically out of place next spring; or that the Southwestern print tops you're ready to order won't fit if you're not serving that type of food six months from now.

Budget. If your tables are showing wear and tear, you can buy new tables or cover the old ones. Yes, purchasing tablecloths is less expensive in the short run. However, from then on, you must factor in the costs of cleaning, repair, and replacement. When it comes to napkins, an Illinois Restaurant Association survey found the cost difference between using cloth and paper is only 2 to 3 cents more per meal for cloth, since diners tend to use two to three paper napkins, but only one cloth napkin.

Environment. Many foodservice operations use fabric napery simply because their customers think using throwaway paper products is wasteful. Others spurn polyester because it is not biodegradable. (Researchers have found ways to recycle it, however.) Environmental issues such as recycling and waste disposal are important to customers and should be to you in your decision making.

Tactile Sensations. Multiple surveys indicate that restaurant patrons simply prefer natural fibers for napkins. They say they "feel better" than synthetics. The tactile quality of a fabric is called its ***hand*** in the linen industry. However, consider carefully what sensation you wish to convey. High-dollar, formal restaurants want the feeling of crisply starched linen, while a more casual place can choose a softer material.

Some restaurants use padded tables, which gives a nice, soft feeling but can cause stemware to topple more easily. Padded tables cannot be used without tablecloths.

Lifestyle. Although casual dining seems relaxed, do not confuse it with a lack of sophistication. The use of nice linens can add just the right touch of class to a menu of upscale burgers and microbrewed beers. Diners don't expect mediocrity when they dine out, and your tabletops should exceed their expectations.

On the other hand, decide whether your target clientele will "accept" a tablecloth. In their minds, it may mean they "have to" dress up when they don't really want to.

Menu. Family restaurants or barbecue restaurants will be harder on their linens overall. Is it worth it to buy table linens if you'll always be stain treating them? The Sticky Fingers Restaurant Group, a chain of barbecue eateries based in Mt. Pleasant, South Carolina, uses dark green hand towels instead of napkins. Another option is a heavier, higher-end type of paper napkin.

Technology. Keep up with the market, long after you've made your initial purchases. This is a very competitive field and manufacturers are continually improving the quality of tabletop fabrics. Ask to see new samples now and then. Ask about new developments that make fabrics more absorbent, or more stain resistant.

20-3 DETERMINING NAPERY NEEDS

How many tablecloths? How many napkins? These are not easy decisions. Too much inventory ties up capital; too little necessitates frequent washing, which means excessive wear, which means more expense replacing napery!

To avoid this vicious cycle, you've got to do some homework. There are several different options for obtaining and laundering your table linens:

1. Rent them from a reliable rental service.
2. Purchase the fabric yourself, have it cut and hemmed to the sizes you need, and hire a professional laundry to wash and iron it on a regular basis for a fixed price per pound. There should be regularly scheduled pickup and delivery times. In a written contract, spell out the consequences if these are not met or if the laundry loses or damages any of your napery.
3. Buy your own and wash it yourself. A linen inventory requires some ongoing maintenance and a budget to replenish it as items wear out. However, by laundering it yourself, the experts suggest you can save 50 percent over what it costs to rent table linens. If there is space available, perhaps the most economical alternative is to have your own on-premise laundry room.
4. Buy your own and hire your mom to do the wash. (Just kidding . . .)

We'll talk more about linen-rental companies and the on-site laundry option in a moment. First, you've got to select the napery you need. Questions of quality, size, and color can only be answered after you have made a battery of other decisions about your restaurant's concept, design, price range, and menu. If nothing else, you can't pick out tablecloths unless you know the sizes and shapes of your tables.

If yours is a completely casual or fast-food venue, you'll probably stick with paper products. If it's a franchise operation, the size, style, color, and imprinting of napkins or placemats may be dictated by your parent company. For other types of restaurants, however, cloth should at least be considered.

When you buy your own linens, most napkins and tablecloths can be cut to order. However, as shown in Table 20-3, they do shrink slightly when laundered for the first time, depending of course on fabric type.

What does it matter if your guest puts a 16-inch or a 20-inch square of cloth on his or her lap? If you're going to use certain types of classy-looking napkin folds or if you want to give the appearance of luxury (like an oversized towel in a nice hotel), you'll choose a bigger napkin.

For tablecloths, choose the correct size by adding 8 to 12 inches to the size of your tabletop. Twelve inches is the usual amount of space between the table and the seat of the chair. The amount of tablecloth that hangs over the side of the table is called the ***drop.*** The drop should not interfere with the guest's comfort getting into and out of a chair or booth, but it should be long enough to completely cover all ends of the table and to drape attractively.

Illustration 20-3 shows tablecloth size options, and Table 20-4 lists standard tablecloth sizes for almost every common table shape and size, both finished (newly cut and hemmed) and laundered (actual size after first wash), as well as the number of inches for the drop.

TABLE 20–3

NAPKIN SIZES, BEFORE AND AFTER LAUNDERING

Cut Size (inches)	Laundered Size (inches)
13 × 18	11 × 16
16 × 16	14 × 14
16 × 23	14 × 21
18 × 18	16 × 16
21 × 21	19 × 19
22 × 22	20 × 20
24 × 24	22 × 22

For banquet tables, one option that may be less expensive than buying finished tablecloths is to buy fabrics in standard widths (54-, 64-, 72-, or 90-inch) and have them cut and hemmed to fit 6- to 8-foot tables.

Even more economical, purchase your tables in small, standard sizes. You can always accommodate larger parties by moving tables together and covering them with smaller tablecloths, eliminating the need for larger ones.

Now, let's discuss quantities. A reliable guideline is the "rule of three:" For every cloth on a table, there should be a clean cloth in inventory awaiting use and a third one in the laundry. To determine your own target number, begin by multiplying your total number of tables by your turnover rate. For example, a restaurant with 40 tables "turns" them (has them occupied by a new set of customers) three times a day:

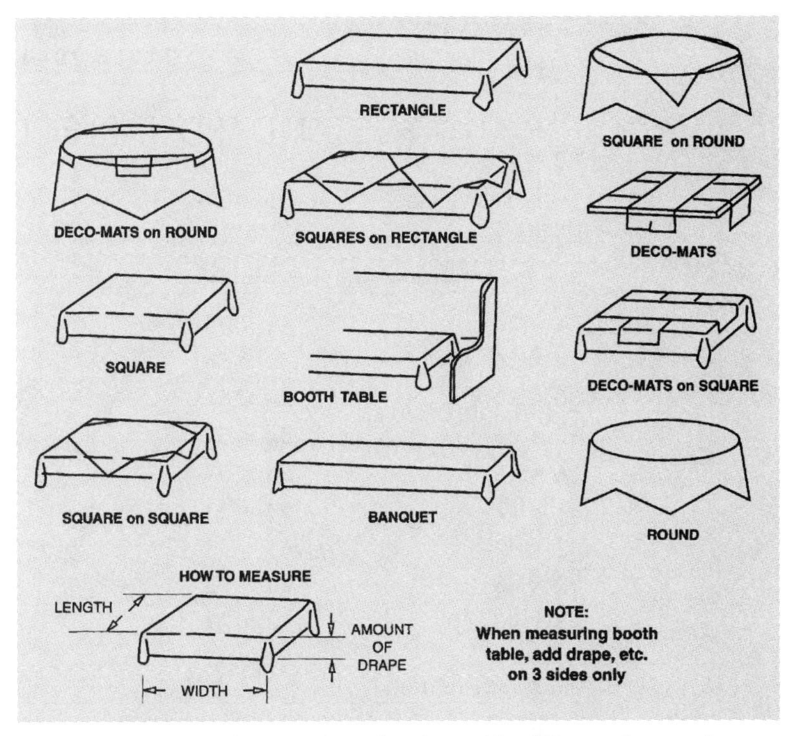

ILLUSTRATION 20-3 **There are lots of options with table coverings, as long as you measure correctly before buying.**

40 (tables) × 3 (turns per day) = 120

Then, multiply that total by our "rule of three" (one on the table, one in inventory, one in the laundry), and you'll soon see why your mom may not want to handle your restaurant's laundry duties:

120 (tablecloths) × 3 (rule of three) = 360 tablecloths

For napkin quantities, the same formula works if, instead of the number of tables, you use the number of chairs. Let's say our 40-table dining room has four chairs per table:

40 (tables) × 4 (chairs) = 160 chairs

and

160 (chairs) × 3 (turns per day) = 480 napkins

and

480 (napkins) × 3 (rule of three) = 1440 napkins

Another common guideline suggests an inventory of three or four tablecloths per seat and nine to 12 napkins per seat. If your tablecloths are used with tops, count on having four tablecloths and 12 tops per table.

These quantities are recommended when 24-hour laundry service is available—that is, when laundry will be delivered each day for use the following day. On weekends, or perhaps during holidays, you'll need to plan ahead and stock as many as three days' linens in advance.

Renting Linens

Working with a linen-rental company is another opportunity to build a good business relationship. You can rent more than table linens from these firms. Most also offer kitchen towels, aprons, and uniforms; and it's nice to have a supplier for special occasions when you need table skirting and different colors or types of linens that you wouldn't otherwise have in inventory.

TABLE 20–4

STANDARD TABLECLOTH SIZES

Table Size (inches)	Finished Size (inches)	Laundered Size (inches)	Inches of Drop (inches)
For Square or Rectangular Tables			
24 × 24	45 × 45	43 × 43	9 × 9
24 × 30	45 × 54	43 × 52	9 × 11
30 × 30	54 × 54	52 × 52	11 × 11
30 × 36	54 × 54	52 × 52	11 × 8
36 × 36	54 × 54	52 × 52	8 × 8
36 × 42	54 × 63	52 × 60	8 × 9
42 × 42	63 × 63	60 × 60	9 × 9
42 × 48	63 × 72	60 × 69	9 × 10
48 × 48	72 × 72	69 × 69	11 × 10
For Booth Tables			
24 × 24	45 × 54	43 × 52	9 × 10
30 × 42	54 × 54	52 × 52	11 × 10
For Banquet Tables			
30 × 72	54 × 96	52 × 91	11 × 9
30 × 84	54 × 108	52 × 103	11 × 9
30 × 96	54 × 120	52 × 114	11 × 9
For Convertible or Round Tables			
36-inch square to 52-inch round	72 × 72	69 × 69	8 inches square
60-inch round	82 × 82	79 × 79	9 inches square

Source: Riegel Consumer Products Division, Mount Vernon Mills, Inc., Johnston, South Carolina.

A big part of shopping for this service is checking the company's references. Ask other customers about the kind of service they provide. Do they really deliver when they say they will? How often is the delivery late? How often is the quantity short? Most important, how do they handle emergencies—that unexpectedly busy weekend when you need more, and you need it now? Will they make special deliveries at no extra charge?

Be wary of the linen-rental company that proposes charging you a flat fee by the week for a maximum number of linens. What this really means is that, each and every week, you're charged to keep the same 400 tablecloths in stock, even if you only used 150 of them. Always arrange to pay based on what you actually use, for a fixed price per use. In the contract, a price should be quoted per napkin and per tablecloth, depending on the sizes of napkins and/or tablecloths. Some colors or fabric blends may be more expensive than others. All of these variables should be listed in writing, along with:

- Specific quantities to be delivered
- Times and days for pickup of soiled linens and delivery of fresh ones
- Your recourse, if anything goes wrong

It's tempting to accept incoming orders without the hassle of counting them, since they are neatly bundled for storage, but they should be counted if possible. Be sure to note if you're short a few napkins, and set aside any that are stained or damaged, and then ask for credit for these pieces.

The linen company contract usually includes a replacement charge for each item that is damaged. Monitor these "replacement charges" carefully. It's not out of line to ask to keep the damaged items. After all, you're paying for them outright when you pay that replacement charge. Also ask the rental service what happens (charges or penalties) if linens disappear, because pilfering is an unfortunate reality of the restaurant business.

When interviewing both linen-rental and laundry-service companies, look for added value. Will they come in periodically and help train your employees about linen care, or offer stain treatment guidelines and products? Again, these are long-term partnerships. Make sure you are getting that "partnership feeling" from these important suppliers.

Fun with Folds

If you're going to buy or rent hundreds of napkins, you might as well have some fun with them. The way napkins are folded says a lot about the attention paid to detail in your dining room. Most are quite simple, for the elegant effect they can have. Try different folds on your tabletop to see which ones look the best (see Illustration 20-4).

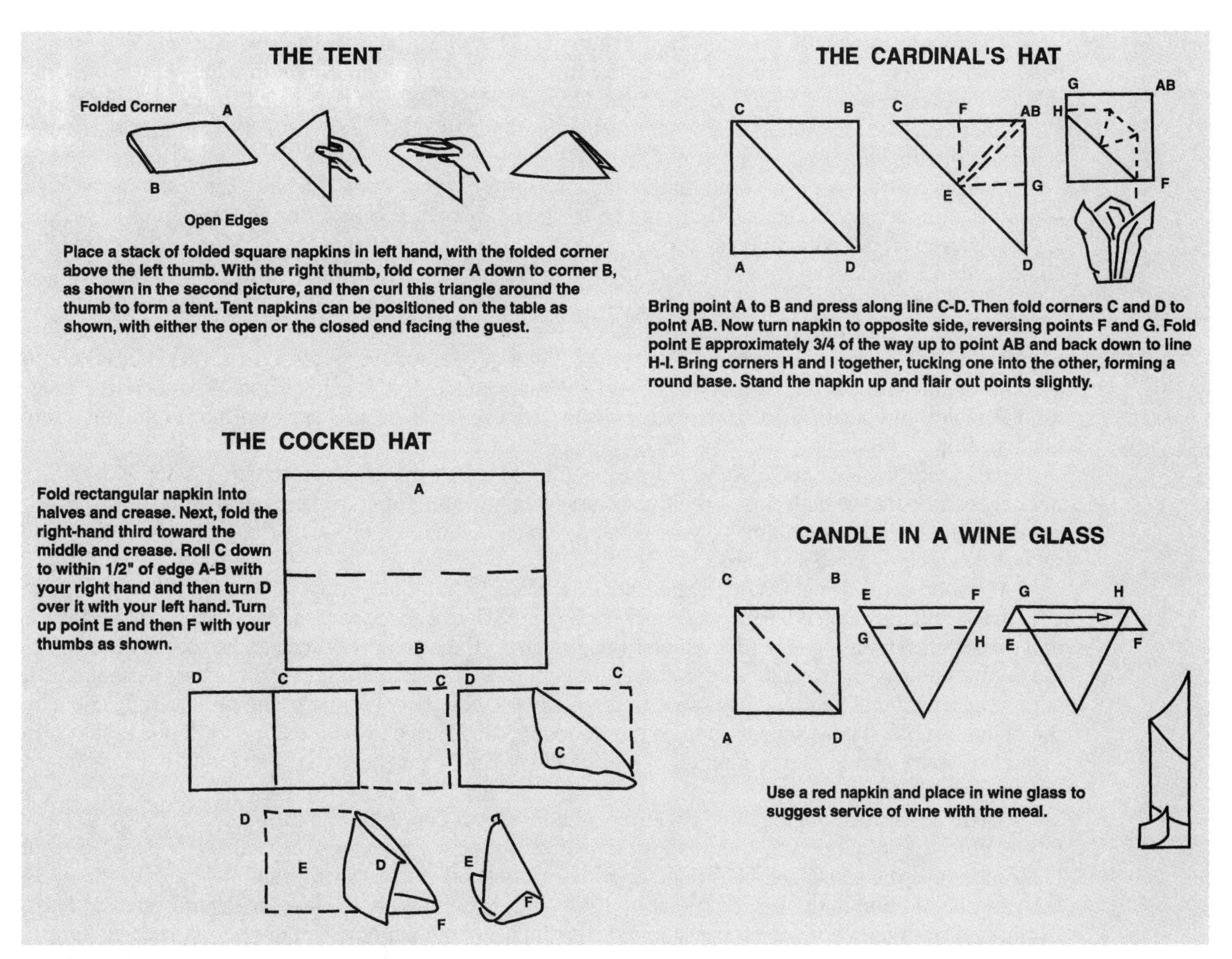

ILLUSTRATION 20-4 **Napkin-folding techniques.**

SOURCE: *CORNELL UNIVERSITY HOTEL AND RESTAURANT ADMINISTRATION QUARTERLY,* COPYRIGHT © CORNELL UNIVERSITY. ALL RIGHTS RESERVED. REPRINTED WITH PERMISSION.

20-4 ALTERNATIVES TO FABRICS

There's nothing wrong with using paper products. Many family restaurants bedeck their tables with white butcher paper and furnish crayons so the kids can draw on it. Even more eateries use paper napkins. Those who prefer paper say it is:

- Easy to use and discard
- Easy to store and takes up less space than cloth
- Safe and sanitary
- An additional way to advertise when imprinted with name or logo
- Less expensive than cloth

Paper napkins come in a variety of sizes. It seems the most difficult thing to control, especially in a self-service or takeout setting such as a cafeteria or fast-food restaurant, is that customers use so many napkins. The tendency is always to take more than they need. Some restaurants use paper napkins for breakfast and lunch service, then switch to linens for a more upscale mood at dinner.

The casual, theme restaurants were the first to eliminate the tablecloth altogether. The marketing implication is clear: to position themselves somewhere between fast-food and white-tablecloth dining. A decorative table can be attractive with no cover at all, and the cleanup is quicker and easier, too. Also, the table itself can be used as a marketing tool, to mirror the theme of the eatery.

The trend begins with the idea that, like the walls of your dining area, table surfaces can also be decorated to embellish the atmosphere. Colorful tiles, postcards, sports memorabilia—just about anything whimsical may be embedded into the tabletop itself, which is then sealed to prevent stains and water damage. The tabletop itself may be made of ceramic tile, although this is unadvisable because it can crack or chip with use.

One successful example of alternative tabletops is California Pizza Kitchen. The restaurant's walls feature custom-designed tiles, but using tile on tabletops was not practical. Instead, tabletops and countertops were surfaced with DuPont Corian, with random patterns of tiny, multicolored epoxy triangles embedded in a solid color. It's a fun look, like confetti scattered across these surfaces, and helps set the stylish, casual mood for these upscale pizzerias.

The use of glass tops over tables is another alternative. A tablecloth may still be used, but the glass fits over it exactly, minimizing spills and stains. Cleanup of the table is certainly easier, but take into account the increased noise level as dishes and flatware clank onto the hard surface. If you choose glass-topped tables, buy heavy plate glass with polished edges. Sharp corners snag on fabrics and are more likely to chip.

An interesting advantage of the glass-topped table is that it allows you to experiment with fabrics not necessarily used as tablecloths; the rich colors and patterns of upholstery fabrics, for instance, which are dry-cleaned about three times a year instead of being laundered. Because the glass covers the table surface, your cloth choices are much wider.

Yet another easy-maintenance alternative is the vinyl table covering. A good, commercial-grade vinyl has a soft, cushioned backing to muffle noise, and wipes clean with a damp cloth. It comes in several thicknesses, from 4-gauge (the thinnest) to 10-gauge (the heaviest). Vinyl can be used for upholstering booths and chairs, too, and is appropriate for fast-food, casual, and outdoor dining establishments. If your vinyl covers will be used on outdoor umbrella tables, be sure you order them with a strong stitch on the umbrella hole. If not, you will soon be reordering, because just a few windstorms will pull at the cloth enough to tear it around the opening. Like fabrics, vinyls can be purchased as yard goods and customized to fit your table sizes.

There are also laminated table coverings—choose from hundreds of existing designs, or select any fabric, purchase it by the roll, and have it laminated, cut, and sewn to fit your tables. Laminated cloth looks more upscale and "linen-like" than vinyl. The initial cost of laminating can be daunting—$6 or $7 per yard, in addition to the cost of the fabric itself—but remember, it will last for several years, at least. Laminated cloth does not have the cushioned backing like vinyl, but you can order inexpensive, nonslip padding to use beneath the cloth. To see some of the hundreds of designs and styles available in vinyls and laminates, check out the websites of two of the largest manufacturers: Americo, Inc. (americo-inc.com) and Marko International (marko.com).

20-5 CARE AND CLEANING OF NAPERY

Unfortunately, the people who work for you may actually work against you in your crusade for great-looking table linens. This is the real world, and napery is likely to be used as everything from a dishrag to a potholder as it makes its way from tabletop to laundry and back again.

Research by the American Hotel–Motel Association indicates tablecloths and napkins are washed an average of 100 times before they are discarded as ***rag-outs,*** the informal term for making them into rags. However, they would last longer if they were not abused by foodservice staffers. Linen replacement costs the typical restaurant from 13 to 25 percent of its laundry budget, so it is worth your while to convince employees that the sole purpose of a napkin is to place on the dining table. Don't wipe with it, use it to grab hot pans or dishes, or wrap the flatware in it when tables are being cleared. (Flatware is often sent accidentally to the laundry and rarely returns.)

There should be a training process for handling linens in your business. It should include:

- Educating employees about the cost of linens and the importance of handling them correctly.
- A system for sorting soiled linens by item and color into different hampers. This applies whether you do your own laundry or use a rental service.
- Procedures, supplies, and a specific place to pretreat stains before tossing items into the hampers.
- A place to put damaged linens that must be inspected and/or returned to the rental company.
- Some control of who has access to the linen supply to prevent theft, including a sign-out or check-out system for obtaining linens.

In addition to training of your wait staff and busers, there are other factors to consider in your choice of linens:

COLOR RETENTION. You'll want some assurance from your salesperson that the color will remain consistent, no matter how many times the item is washed. This is also known as being ***colorfast.***

SOIL/STAIN RELEASE. When properly laundered, all of the typical restaurant stains and odors should wash out easily. The now-famous Scotchguard process is advisable for many types of fabric. It coats the fibers to prevent even the toughest food stains, and enables you to wash the items successfully without heavy-duty laundry equipment.

ABSORBENCY. No matter how crisply it is starched and ironed, the fabric should maintain its ability to soak up liquids.

FINISH. The surface of a napkin should cling to a customer's lap without sliding off; permit fancy folds; and feel natural and somewhat soft, even when it's been starched. Remember, in the industry, the feel of a fabric is called its ***hand.***

MINIMAL SHRINKAGE. After their initial shrinkage, both napkins and tablecloths should maintain their basic square shape. The latter should retain its ability to drape the table attractively.

LACK OF LINT. Who wants to stand up from an important business lunch to find bits of white stuff all over that nice dark suit? Lower grades of fabric break down more easily in the wash process and produce more lint. Choose fabrics that won't shed.

An important part of the staff training process mentioned earlier is treatment of problem stains immediately after a meal, before they can "set." Some restaurants have a policy of putting a loose knot in a stained or soiled item (like a napkin), or in a separate plastic bag (like a tablecloth), to signal that it needs special attention. This also keeps it from staining other items when placed in the same laundry bin or hamper.

Your laundry service may recommend a professional-strength stain remover that is safe for the kinds of fabrics you're using. There are also some tried-and-true "restaurant remedies:"

- Berry and other fruit stains, or red wine spills, can be sprinkled with salt, or saturated with club soda, or both.
- Mildew or rust spots respond to repeated applications of a paste of lemon juice and salt. (Rinse completely before reapplying the paste.)
- Coffee or tea stains and iron scorch marks can be lifted out of white linens by soaking them in a solution of one part borax to six parts water.
- Similar coffee-type stains that have been sitting long enough to dry onto the fabric can disappear when rubbed gently with a mixture of glycerin and water.

Just remember that, if you're going to soak something to remove a stain, don't soak it for more than fifteen minutes. Any longer and you're just soaking the fabric in the dirty water. Your stain may have come out, but the whole napkin may look dingy from then on.

If your restaurant has its own linens and laundry, give someone the responsibility of checking it periodically for wear and tear. Rips only get bigger in the wash, so they should be mended before laundering.

There's a whole list of laundry "dos" and "don'ts." Your linen supplier probably has suggestions, and so will the instructions that come with your commercial washers and dryers. Here are a few of the top tips from foodservice operators and commercial laundries with restaurant clients:

- Sort and wash colors separately four to five times before washing them with whites or other colors.
- The minimum wash temperature should be 140 degrees Fahrenheit, but no higher than 160 degrees Fahrenheit. Lower water temperature won't sufficiently clean; hotter water will fade fabrics faster.
- Set the machine for the proper water levels for your wash load.
- Use only detergents that are labeled correctly, use the right amounts, and add them at the right times.
- Avoid overloading the dryer. Load dryers to 60 percent of capacity when you will be tumble-drying without ironing afterwards; 80 percent of capacity when planning to iron the fabric.
- If your dryer has a cool-down cycle, use it. It's best to dry fabrics completely and gradually cool them to about 100 degrees Fahrenheit before they are removed from the dryer.
- All napery should be ironed while it's still damp. Especially if you're going to starch an item, pass up the dryer altogether and iron immediately after washing.
- Iron table linens at temperatures between 315 and 350 degrees Fahrenheit; lower isn't effective and higher will damage them.
- Over-starching linens will make them more fragile.
- If you're not ironing them, remove them promptly and fold or drape immediately. Avoid mildew by storing clean tablecloths on coat hangers with plenty of air circulation, instead of folding and stacking them.
- Fine linen needs time to "rest" after being cleaned, to restore a bit of its natural softness. Newly washed, dried, and folded linens should be stored at least 24 hours before being returned to service.

Milliken and Company, a leading manufacturer of table linens, has graciously permitted us to reproduce its table of the most common laundry situations you may face. You'll want to refer to this Napery Troubleshooting Guide again and again.

Doing restaurant laundry is not as simple or mindless a process as you might assume. Another leading table linen manufacturer, Artex International, Inc., suggests you perform a few simple tests on samples of your selected fabric. By determining the right detergents and temperatures to use, you will get the maximum useful life out of your napkins and tablecloths. These tests include bleaching samples of the

FOODSERVICE EQUIPMENT

NAPERY TROUBLESHOOTING GUIDE

Problem	Cause	Solution
Discoloration	Bleach on colors	Do not use bleach on colors.
	Residual dye transfer	Prewash colors separately on first wash to prevent residual dye transfer. Sort napery into recommended color groupings for subsequent washings. Always wash whites separately.
	Soil redeposition	Reclaim napery with soil redeposition by using increased temperature and supplies. Prevent re-deposition by adjusting the formula for soil level.
	Chemical reaction	Avoid chemical discoloration by thoroughly rinsing all chemicals out of the napery before drying or finishing.
	Yellowed whites	Chlorine bleach will not damage VISA[a] fabric, but residual bleach on white napery can cause yellowing if the chlorine is not neutralized before exposure to heat. Avoid yellowing by using an antichlor in the second rinse after bleaching with chlorine.
	Incorrect ordering	Order napery colors by the four-digit code number or the distributor color codes to avoid confusing similar colors.
	Glazing	Keep ironer chest temperatures under 350°F and use correct roll pressures to prevent glazing.
Waterproofing	Fabric softeners	Fabric softeners prevent VISA napery from absorbing liquids. Do not use fabric softeners on VISA napery.
	Tallow soap	Do not use tallow soaps on VISA napery. Instead, use built detergents and surfactants.
	Mildeweides	To help minimize mildew growth on VISA napery, use only mildeweides that have no quaternary ammonium base.
	Washing with cotton	Wash VISA napery with other 100% synthetic fabrics. Do not wash with cotton or polycotton blends.
	Soil redeposition	See Discoloration—Soil redeposition.
	PVAc buildup	Reclaim, then reformulate the starch ratio: four parts natural starch to one part PVAc.
Static	Overdrying	Reduce extraction or conditioning, and cover wet work to maintain 20–25% moisture retention in napery before ironing.
	Incorrect grounding	To effectively ground equipment, sink a 6′ steel rod in the ground and attach grounding straps. Also, use static bars on folders and conveyors.
	Friction	Eliminate friction from goods slipping on the belts by synchronizing the speeds of adjacent belts.
	Folders	Check folder adjustments and/or use more starch on napery.

Continued

Problem	Cause	Solution
	Low humidity	Low relative humidity can cause goods to stick due to static electricity. On particularly dry days, a humidifier may be needed.
Stains	Permanent: Bleach spots/ cleaners carbon/metal heat set food/cement	For all permanent stains, try reclaiming with more supplies and higher temperatures to reduce stain visibility. Napery with noticeable stains should be ragged, overdyed by a qualified dye house, or cut into smaller pieces.
	Removable: Blood	Use a warm water flush, then normal washing. Do not use hot water on protein soils.
	Fats/corn oil salad dressing	Use solvated surfactants to boost regular formula.
	Motor oil/grease	Use solvated surfactants to boost regular formula.
	Lipstick/candle wax	Use solvated surfactants and/or higher temperatures.
	Rust	Use an oxalic acid prewash or rust-removing sours.
	Mildew	Use chlorine bleach on whites. As a last resort, use 1% available chlorine bleach at 1–2 quarts/CWT to remove mildew from colored napery. This will cause some color deterioration. A mildeweide with no quaternary ammonium base should be used to minimize mildew growth.
Ironing rejects	Dirty ironer	First, do a thorough downtime cleaning. Then follow up with regular cleaning and maintenance checks. Do not overwax the ironer to avoid wax buildup.
	Roll pressure	Do the paper test on the first ironer roll. Correct uneven or incorrect pressure.
	Side-to-side pressure	Check bearings, individual roll pressures, and the pillow blocks.
	Drafting	Check the circumference of each roll with adding machine tape to determine if there is appropriately increasing diameter from front to back. If not, replace the ironer padding.
	Incorrect feeding	Carefully instruct all personnel on correct feeding procedures.
	Cold chest	Maintain a minimum chest temperature of 310°F.
	Warped chest	A warped chest must be replaced.
	Over- or undersoured napery	Adjust the amount of sour to maintain 5.5–6.5 pH.
	Ironer tapes	Be sure there are two ironer tapes per lane for napkins. Tapes should be around the ironer rolls and tension bars only, not around the finger roll.
	Residual chemicals	Rinse goods thoroughly, then sour to neutralize rinse water alkalinity. Also, thoroughly clean the ironer chests and roll pad covers to remove chemical buildup.
	Goods too wet	Increase extraction or conditioning if napery is too wet going into the ironer. Goods should feel damp at the recommended 20–25% moisture retention.

Problem	Cause	Solution
	Roll motion and covers	Correct rough roll motion and loose or rough roll covers. Also check for excess wear on covers, belts, and aprons.
	Static	See Static—Grounding.
	Glazing	Keep ironer temperature under 350°F and maintain 20–25% moisture in napery to avoid glazing.
Customers picks and snags	Rough shelves and table corners	Eliminate rough spots and protruding nails on shelves. Tape table corners.
	Personnel	Educate personnel on correct handling techniques.
Laundry picks and snags	Burrs and sharp edges	Check for burrs and sharp edges on machinery and handling equipment by using a wet VISA napkin. Do not use staples to fasten ironer tapes, and check for loose or broken wires on feed and exit apron connectors.
	Washing with tableware	Be sure all tableware and other foreign objects are removed before washing.
Starch/sizing	Too stiff: PVAc buildup or excessive starch	Reclaim with additional alkali and more heat. Then reformulate four parts natural starch to one part PVAc.
	Too limp: Water level	Use lowest available water level for optimum starch penetration. Actual level varies by washer.
	Sour	Starching results are best at pH levels between 5.5–6.5. Add sour at least two minutes before starch to allow even distribution.
	Temperature	Maintain bath temperatures between 90 and 100°F for starch.
	Supplies	Check with your Milliken Technical Services Representative for the recommended amounts and ratios of supplies for each type of starch or sizing material.
	Load size	Starch penetration is limited when the washer is overloaded. Use the following clean dry weight capacities as a guideline for load size: full drop: 90%, split pocket: 75%, Y pocket: 65%.
	Overdrying	Too much extraction or conditioning causes starch to be lost. Maintain 20–25% moisture retention in napery before ironing.
	Time	Allow at least eight minutes starch time for even penetration of starch.
	Inadequate cleaning	Reformulate washing process to ensure thorough cleaning of napery so starch can adhere to the fabric.
	Personnel	Carefully instruct all personnel on correct starching procedures to insure consistency from load to load.
Wrinkles	Thermal shock	Thermal shock wrinkles occur when napery is exposed to sudden changes in temperature. Avoid thermal shock by tempering cold water in the winter. Then reduce the water temperature in 15° increments to 100°F before extracting.

Continued

Problem	Cause	Solution
	Extraction	Reduce pressure, RPM, or time during extraction.
	Insufficient cool down	Cool to a temperature of 100°F or less before extracting or removing from washer or dryer.
	Hot spots on dryer	Be sure gas flame is not impinging on the dryer basket.
	Malfunction	Inspect all machinery and maintain on a regular schedule.
	Overloading	Washer capacity should not exceed 90% of clean dry weight for full drop machines, 75% for split pocket, and 65% for Y pocket. Tumbler loads should be 50%. Also, do not leave carts or slings overloaded for extended periods.
	Folder stacks	Reduce the size of napery stacks on the folder or increase airflow to cool the napery before stacking.
	Storage	Fold napery correctly before storing, and allow adequate storage space to prevent wrinkling.
Folder rejects	Settings	Check manufacturer's recommendations for correct settings.
	Slippage	Inspect and maintain gears, belts, and conveyors at the apron/conveyor junction. Also, try slowing down the conveyors or using more starch.
	Uneven folds	Adjust folder alignment to manufacturer's specifications and repair or replace worn belts.
	Belt angle	Reduce the incline if the conveyor belt angle is too sharp.
	Static	See Static—Grounding, and check the speeds of adjacent surfaces.
	Dirty folder	Clean each folder and folder belt with an air hose as needed.
	Personnel	Carefully instruct all personnel on correct feeding techniques and lane alignment.
Customer abuse	Excessive heat	Explain to the customer that excessive heat such as a hot grill will damage linens.
	Incorrect storage	Set up storage for both clean and soiled napery in a convenient place. Check to be sure correct procedures are being followed.
	Soil segregation	Advise the customer not to mix soiled napery with bleach rags or bar wipes.
	Incorrect usage	Napery should not be used as a grease rag or bar wipe. Offer the customer appropriate items for these applications.

[a]VISA is a registered trademark of Milliken and Company for fabrics.

Source: Milliken and Company, Spartanburg, South Carolina.

fabric, first with only a mild concentration of bleach, and then with a high bleach concentration, to see how the fabric reacts. This should help determine the amount of bleach to use. Bleach does not technically remove stains; it simply oxidizes them to appear colorless. It should always be used sparingly, and with caution. You'll also want to check with the napery manufacturer to see which type of bleach is recommended, if any, depending on the type and color of fabric: chlorine, sodium hypochlorite, or sodium peroxide.

Other tests can indicate the fluidity of the fabric (how much a single wash can damage it); the redeposition rate (how much dirt or lint is "redeposited" on fabric that is improperly washed—not enough detergent, overloading the machine, and so on); and the effects of temperature and water acidity or alkalinity.

An alkalinity check of your water is especially important, because this will affect how well the soap dissolves in the water and how thoroughly the fabrics can be rinsed. The pH level of the wash water should never be lower than 7. You can purchase heat-sensitive strips to measure the temperature of your wash water within 10 degrees, or simply stop the machine and use a thermometer.

Water softeners and filters can help enormously with sediment build-up in appliance water lines, and also with rust-related stains. Rust is dissolved iron in tiny particles that are difficult to filter out, and can make light-colored fabrics look dull over time. In areas where rust is an ongoing problem, commercial laundries use a product known as a "sour" in the rinse water, to dislodge rust particles.

Yet another type of laundry technology "zaps" linens clean without heat, detergents, or bleach. GuestCare, Inc., in Dallas, Texas, has discovered a way to harness ozone gas for cleaning purposes. Hotels that have field-tested the system claim their laundry bills were cut by 60 percent.

20-6 AN ON-PREMISE LAUNDRY

Earlier in this chapter, you read about the possibility of handling your own laundry in-house. Indeed, the most economical long-term method of dealing with those mountains of linens is the on-premise laundry.

Who benefits from this solution? Think high volume—hotels and motels, hospitals, and colleges—because they also wash bed linens. Athletic clubs and country clubs may have sufficient laundry volume, between dining linens and towels. In fact, any restaurant or catering facility that produces more than 500 soiled napkins and 100 soiled tablecloths per day is probably a candidate for an on-premise laundry. One benefit of doing it yourself is that you'll never get caught short on weekends or holidays. However, operations with lower counts should carefully weigh the costs of owning versus renting from a reliable linen supply firm. In estimating your laundry volume, don't forget to include kitchen and bar towels, aprons, and uniforms, if applicable.

The decision to install a laundry room is a big investment of money, as well as space. Once it's in place, it will cost a lot to run (electricity) and a lot to maintain (repairs). So weigh these factors carefully:

- Initial cost of the equipment
- Installation charges (plumbing, electrical)
- Utility costs (electricity, water)
- Ongoing supply costs (detergents, bleach, and the like)
- Repair and maintenance costs
- Cost of hiring or training someone to run and maintain the facility and troubleshoot when necessary
- Cost of purchasing all napery
- Costs to repair napery when necessary

The development of wrinkle-free (more accurately, "noniron") fabrics has prompted more foodservice managers, especially in casual establishments, to consider on-premise laundries. They can be run by less-skilled personnel, don't use as much hot water, and the linens themselves are less expensive than higher-maintenance cottons. If you decide to move in this direction, your three major considerations should be economy of operation; consistent quality of the finished linens; and selection of the right person to accomplish these goals.

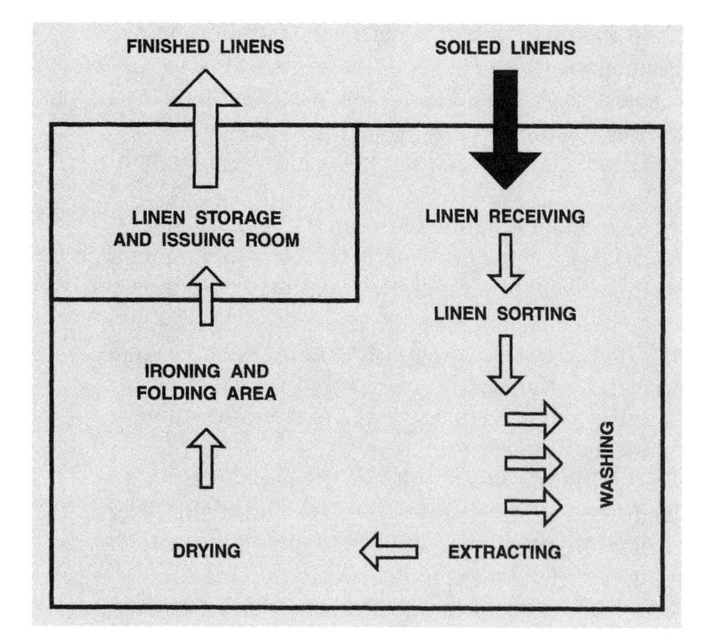

ILLUSTRATION 20-5 A sample laundry room layout.
SOURCE: FRANK BORSENIK AND ALAN STUTTS, *THE MANAGEMENT OF MAINTENANCE AND ENGINEERING SYSTEMS IN THE HOSPITALITY INDUSTRY,* 4TH ED., JOHN WILEY & SONS, INC.

And where to put this fabulous new facility in your already cramped back-of-house area? Give careful thought to the location, size, and construction of a laundry room. Unlike the washer and dryer tucked neatly into a pantry or large closet at home, this laundry will need space for: worktables for receiving and sorting; the machines; an extraction area (moving items from washer to dryer, which, in very large operations, can be done mechanically); ironing and folding; linen storage, and checkout (see Illustration 20-5).

The size of your laundry facility will be determined by your volume of business, number of seats, types of table covers, and more. For discussion purposes, let's examine the needs of a restaurant that soils 500 napkins and 100 tablecloths per day. Here, one washer-extractor and one dryer could be installed in an area as small as 120 square feet. But where would cleaning supplies be stored? Where would ironing be done, and clean linens be stored? You must make room for at least two worktables, as well as the machines themselves, and the best way to store tablecloths is on hangers in an area with good air circulation to prevent molding. Soiled linens must also be stored somewhere until washing, and hampers should be out of sight of guests.

Your laundry should be located near the receiving area at the back of the house, preferably on an outside wall of the building to provide venting for the dryers. Modern technology also allows for design of the equipment exit vents so that the hot air passes through a heat exchanger and is reclaimed. Some commercial dryers have heat reclaimers built in.

A ground-floor location is ideal because, of course, laundry is heavy. The walls and floors of this area should be made of durable, moisture-resistant material, highly insulated to keep noise to a minimum. You don't need windows in the laundry area, because wall space is better used for storage shelves and bins. However, in some cities they are required, so take care to place them where glare won't be a problem for employees. Place them high up on the walls, and use wireglass (sheet glass with wire netting embedded in it) to prevent breakage. Doors should be large enough to get the washers and dryers into the room in the first place, and to enter and exit easily with utility carts or baskets in tow. Windows in this set of swinging doors should provide a clear view to prevent accidents; bumper guards on the doors can be placed so the utility cart can be used to shove the door open without scratching or denting either door or cart. Door thresholds should be flush with the floor so there are no bumps to clatter over.

A 9- to 11-foot ceiling is recommended, and sound absorption is especially important. The floors should be level concrete slabs, capable of supporting the heavy cleaning machines. Have floor drains installed near the washer sites, and the floors sloped about ½ inch per foot toward the drains to facilitate runoff in case of water overflow or backup. Other than the areas directly around the drain, no part of the floor should be a low spot where water could pool. Concrete is the most suitable floor material, but for employees' comfort and safety, it should be covered with mats made of nonskid rubber or other synthetic material wherever people will be walking or standing.

These facilities require lots of utilities. You'll need hot water, cold water, steam vents, and gas and/or electricity, as well as large (6-inch) drains and 1-inch water lines. Preplanning is essential to accommodate all these requirements. Your appliances will probably require electrical outlets for both 115- and 220-volt use, and at least 100 psi of steam pressure. Ceiling light fixtures should provide at least 40 to 50 foot-candles of illumination, and all fixtures should be vapor-proof so they're not affected by the humidity that is a natural byproduct of doing laundry. Both humidity and temperature can be controlled by exhausting the machines to the outside and installing exhaust fans within the room.

Of all the utilities you'll juggle, water is among the most important. Hard water is undesirable for an on-premise laundry. It contains mineral salts, which can mix with the alkaline cleaning chemicals to become a sticky mess known as ***soap curd.*** Soap curd looks just like it sounds; it sticks to fabric, making it so stiff and unsightly that it can cut its useful life by half. It may also discolor some fabrics. It is such a potential problem that both hot and cold water should be softened before use in a commercial laundry. Softening should be considered when the hardness of your incoming water supply is rated between 5 and 10 grains, or 171 parts per million of solids. Consult your local water department when making this decision.

Buying and Using Laundry Equipment

The appliances you buy should be able to handle both lightly and heavily soiled laundry. Shop for washers and dryers with a variety of cycles and temperatures to launder the wide range of fabrics you will be using. Make sure the interiors are large enough to accommodate your bulkiest items and largest loads. Look for large door openings (for easier loading and unloading) at convenient heights (to prevent back strain). Today's commercial washers and dryers are highly computerized, with microprocessor controls instead of electrical contacts and wiring. They're built to save energy, and they can often be preprogrammed.

You might consider purchasing two smaller washing machines instead of a single, big one. It will take less time to accumulate enough laundry for a full load, and you'll always have a backup in case one machine needs to be serviced. You can handle smaller loads or odd lots more efficiently and/or wash two types of items simultaneously. Again, the average home-use washer is simply not built for the wear and tear of restaurant use.

The commercial washer even does its job differently than a home machine. Instead of a central agitator, a commercial machine launders by lifting and tumbling the linens in its internal cavity, which is called a cylinder. If the cylinder is small, the lifting and tumbling action is limited—not good for getting things really clean.

Often, a commercial washing machine is installed by being bolted to the floor. This keeps it firmly in place. But the latest commercial laundry technology is a ***floating suspension washer,*** also called a *soft-mount* washer. It does not have to be bolted down, since it does not "vibrate" during the spin cycles like a regular washer. It was created primarily for hotel use, so that washers on guest floors didn't create annoying noise and vibration when they were running, but it's a useful option if space is tight and vibration or unbalanced loads are problems. Floating suspension washers also spin laundry much faster than their regular commercial counterparts, running at higher revolutions per minute (rpms). The faster a machine's spin cycle, the less moisture in the items that emerge from the washer. That means less drying and ironing time.

Another way to accomplish this is with an ***extractor.*** As its name suggests, this machine extracts moisture from wet fabrics by spinning them around under high-speed centrifugal force—so fast that the wet fabrics weigh only 1½ times what they would if they were completely dry. Commercial *washer-extractor* combinations are available, which require less space and eliminate the step of having to remove wet laundry from the washer and place it in the extractor. A washer-extractor with a 35-pound capacity will take about 25 minutes to complete both wash and extract cycles (see Illustration 20-6). Some of these machines also have a cool-down cycle, during which cold water is slowly injected into the wash to prevent non-iron fabrics from wrinkling. The capacities of these machines vary from the 35-pound model (about the same size as a home washer) up to the 700-pound model for large hotels and hospitals. Use Table 20-5 to see which capacity is correct for your estimated amounts of napery.

ILLUSTRATION 20-6 **Choose a washer-extractor combination based on how many pounds of laundry it can handle in one load.**

COURTESY OF SPEED QUEEN, INC., RIPON, WISCONSIN.

TABLE 20–5

WASHER-EXTRACTOR CAPACITIES

Size	54-inch Tablecloths	20-inch Napkins
35-pound model	37	200
50-pound model	54	290
75-pound model	81	440
95-pound model	102	560

When you see the volume of linens most restaurants use, it should become clear that there isn't a soul who would consider pressing it by hand. That's why there are commercial ***ironers*** and ***folders.*** Most napery can be sent directly from the extractor to the ironer. The ironer gives napkins and tablecloths a crisp, finished look. Often, even no-iron fabrics are sent through the ironer to improve their appearance. The ironer/folder step is sometimes called ***finishing.***

The ironer may run on gas, steam, or electricity, but the principle is the same: It is a continuous-feed machine with rollers. Linens are sent separately through the row of rollers, emerging warm and neatly pressed on the other side. For tablecloths and napkins, the standard finishing temperature is about 338 degrees Fahrenheit, but this temperature can (and should) be adjusted when fabric instructions call for it. If some types of linens are ironed too hot, they develop a shiny, unattractive glaze.

An automatic folder takes over when pressing is finished. The linens are fed directly from one machine to the next. The folder acts as an extra pair of hands, holding the laundry, then blasting it with air to make sharp creases. Modern folders have computer chips in them, to calculate the specific points at which folds should be made and time the air blasts accordingly. Today's folders will even stack and count the finished linens for you!

ILLUSTRATION 20-7 The drying tumbler you choose should be larger in capacity than the washer-extractor to be used with it.

COURTESY OF SPEED QUEEN, INC., RIPON, WISCONSIN

Ironers are identified based on how many pounds they can handle per hour; the smallest models range from 40 to 100 pounds per hour. Be sure you purchase an ironer that is large enough to handle your widest tablecloth, and consider a dual-finish machine, which irons the fabric on both sides at once.

Not all your laundry will be extracted, ironed, and folded, so you'll also need at least one dryer. Commercial dryers are also known as ***tumblers.*** The tumbler should have about 25 percent more capacity than your washer and/or extractor; that is, pair a 35-pound washer with a 50-pound dryer; or 100-pound washer with a 125-pound dryer. In fact, you should probably have more dryers than you have washers, so multiple loads can be drying simultaneously. Dryers should be located close to washers and extractors, to minimize employees' having to lug loads across the room (see Illustration 20-7). Today's commercial dryers have more efficient heating elements and new lint removal methods. Some feature fire suppression systems, in case of overheating.

Just toss in the kitchen towels and come back when they're dry? Hardly. There's a lot more to it than that. Mike Diedling, of *Laundry News* magazine, has agreed to share his "top ten" most common mistakes and misconceptions about using commercial dryers.

Finally, worktables with casters are the most convenient spots for folding the items that don't receive automatic folder treatment. These tables can be moved as needed. You'll also need plenty of baskets and hampers, each holding a minimum of 12 to 16 bushels. There should also

FOODSERVICE EQUIPMENT

THE TEN MOST COMMON DRYING MISTAKES

Although these tips were written with large laundries in mind, many of them can also help managers of mid-size and small laundries.

1. **Not loading the dryer to full capacity.** Underloading is the most common and most costly mistake made in institutional laundries. Dryers with microprocessors can eliminate this problem with "small load" dry cycles for each fabric classification.
2. **Allowing laundry to dry too long.** Besides wasting energy and production time, overdrying creates friction, which wears fabrics out more quickly.
3. **Not running the dryer at constant production rates.** It's a waste of production time to let employees dictate when a dryer is loaded. Often the machine sits idle after a load is dried because employees are engaged in other tasks. There are automatic or semi-automatic dryer loading and unloading possibilities for almost any laundry.
4. **Taking too long to load or unload machines.** Besides wasting time, this allows the dryer's cylinder to cool off between loads, resulting in the dryer having to work harder to bring the temperature back up for the next batch.
5. **Too much heat at the end of the load.** The most effective means of drying is to apply the greatest heat to laundry at the beginning of cycles. It is best to restrict heat toward the end, when goods are most susceptible to damage. Dryers with microprocessors can do this easily.
6. **Not enough cool-down.** If dried laundry is not cooled down enough during the dryer's cool-down cycle, at worst there is a chance that it may smolder—or even catch fire! At best, wrinkles are set into the fabric, which means employees must spend more time in the finishing area trying to get them out. Sometimes the laundry must be rewashed. Ideally, the cool-down cycle should occur when the laundry reaches a certain temperature, not at a preselected time. Microprocessors make this possible.
7. **Not keeping filters clean enough to allow the dryer to operate at optimum levels.** Although it should be a given that laundries follow the manufacturer's recommended filter-cleaning schedule, it's amazing how few do.
8. **Not maintaining the dryer's seal.** A dryer's efficiency is directly related to the condition of its seal, since it keeps cold air out and hot air inside the basket so it can flow through the laundry. Most seals wear out quickly since they are subjected to the reverse side of the basket's perforations (a cheese-grater effect). A worn-out seal means the dryer has to work harder and longer to dry goods. You can bypass this problem entirely with new dryers, which have an unusual placement of the seal on a smooth band around the basket.
9. **Not moving laundry to finishing stations quickly.** Optimum use of ironers, for example, relies on a specific amount of moisture in the goods before ironing. Allowing goods to sit too long makes them wrinkle; sometimes rewashing is the only cure. Besides improving scheduling, consider installing a conveyor or some other automated means of transport between dryers and finishing stations.
10. **Not keeping accurate production and cost figures for drying.** To improve anything, you need to know what you're starting with—how much time and fuel are projected for each classification versus actual numbers. New personal computer systems can provide this information in summarized reports.

Source: Mike Diedling. "Avoiding Ten Most Common Errors in Drying Reduces Fuel Costs in Institutional Laundries," *Laundry News*.

be ample cupboard and bin space to store the full complement of detergents, stain removers, bleaches, and fabric softeners you will need.

The Art and Science of Washing

You probably didn't come to school to learn how to do laundry. (Then again, it couldn't hurt!) And, as you've seen in this chapter, foodservice linens are not just ordinary wash. They're a major investment. In your laundry room, you'll need to sort and, in some instances, weigh your soiled napery.

Sorting is done based on the type of fabric, its color, and its weave, as well as what type of item it is. You may also need to presort linens that have a particular type of stain—grease, red wine, or lipstick, for example. These may require presoaking, treatment with a special solvent, a longer wash time, and/or higher water temperature. You'll separate natural from synthetic fibers and colors from whites. You don't want stained laundry to sit overnight before washing without, at least, having stain spots pretreated.

How does washing work? Laundry is placed into a rotating cylinder dotted with holes. The cylinder tumbles the items around, forcing water through them and extracting soil from them. In commercial machines, there can be up to four wash cycles (which each last from 5 to 20 minutes) and four rinse cycles (which each last 3 to 5 minutes). Using too much detergent is as bad as using too little. Automatic dispensers—for detergents, fabric softeners, and whiteners—can eliminate this problem. Some dispensers mix the product with a little water before adding it to the wash load, so nothing fully concentrated "hits" the fabric and discolors it.

The extraction or spin cycle follows washing. Centrifugal force flings excess water out of the fabrics and dries them as much as possible. In commercial machines, this process should remove enough moisture so that napkins and tablecloths can go directly to the ironer/folder. If not, it's on to the extractor, which uses more centrifugal force for moisture removal; and then to the dryer, another perforated cylinder in which warm air passes through the items as they are tumbled about.

Once your purchasing and planning decisions are made, running your own laundry is a relatively simple process. Always remember, however, it has lasting implications for your inventory, budget, and the overall image you want your restaurant to project.

SUMMARY

Table linens allow you to add class and color to your restaurant's decor. You might think of them as the canvas on which the rest of the "artwork" is displayed on your tables. The foodservice term for table linens is *napery.* They may be made of cotton, linen, rayon, acetate, nylon, or polyester. Each fabric has advantages and disadvantages, but you'll want napery that is colorfast, stain resistant, easy to launder, and shrink-proof. Your budget, current fashion, the way the fabric feels, the menu, and style of the restaurant will impact your selection.

You must also decide whether you want to buy and launder your own linens or rent them from a linen-rental company. How many will you need? The "rule of three" applies in most cases: For every cloth on a table, there should be one in the laundry and another ready to use. For napkins, three per chair should be sufficient.

Employees should be trained in the proper care of linens. Your operation should include a set of procedures for sorting and pretreating stains; a way to secure inventory so that linens don't disappear; and a process for checking incoming linens (from laundries or rental companies) so you aren't charged for missing or damaged ones.

For casual dining, paper products, vinyl or laminated cloths, and tile-topped tables are popular alternatives. The use of a glass top over the table will help keep linens looking good.

If you launder your own napery, it is helpful to test the linens first, with different types of detergent and bleach concentrations, to see how the fabric will react. Test the water for alkalinity and rust content that may dull the fabrics or compromise the effectiveness of the detergents; incoming water may have to be filtered for the best results. The linen manufacturer or appliance manufacturers may have recommendations.

An on-premise laundry room must include washers and dryers and/or washer-extractor combinations, to wash and then wring out almost all the moisture in the fabric; plus work tables, ironing and folding space, and storage areas for linens and laundry supplies. Floors should be level and include drains; vents will be necessary to remove moisture from the room.

STUDY QUESTIONS

1. Why would you use a top in addition to a regular tablecloth?
2. Discuss the advantages and disadvantages of using blended fabrics as table linens.
3. Why would you rent your tablecloths and napkins from a rental service instead of buying your own?
4. How do you select the correct size of tablecloth for a table?
5. Explain the "rule of three." Do you agree with it, or not?
6. Your restaurant has 85 tables, and is open for lunch and dinner six days a week. How many tablecloths and napkins do you need?
7. What are three alternatives to covering tables with the standard fabric tablecloth?
8. What are the advantages to having an on-premise laundry at a foodservice facility?
9. What is an extractor, and why might you need one?
10. List three common mistakes made when drying fabrics in commercial tumblers, and explain why they are mistakes.

GLOSSARY

Absorption ratio The percentage of water a particular type or amount of flour can hold as it is made into a dough.

Acceptable pressure drop A formula for calculating the effectiveness of a steam system that takes into account the lengths of pipe and the natural pressure drop that occurs as steam is piped from one place to another.

Adaptive reuse Refurbishing or rehabilitating an old or historic building.

Aggregate A substance added to flooring material to make it more slip-resistant. It might be sand, clay, tiny bits of gravel or metal, Carborundum™ chips or silicone carbide.

Aggregator In markets where the sale of electricity has been deregulated, this is a broker who researches rates and terms of different providers and negotiates to buy power on behalf of a group of customers.

Agitator kettle A special steam-jacketed kettle that includes an agitator or mixing arm for stirring while cooking. Also called a *mixer kettle.*

Agitators The beaters on a mixer.

Air door A curtain of air circulated across the opening of an oven instead of a solid door. The air curtain prevents heat from escaping. Also called an *air deck,* this feature is most often found on pizza ovens.

Airpot An insulated thermal container with a pump, used to store hot beverages such as coffee.

Air shutter A device in a gas range top burner that can be adjusted to control the amount of air that comes into the burner.

À la carte Items cooked individually instead of in batches. Also called *to-order.*

Allowable pressure drop The amount of pressure lost as steam is piped from its source to the equipment it powers or heats.

Alloy A mixture of metals used to create flatware.

Alternating current (AC) An electric current that reverses its flow at regular intervals as it cycles. Alternating current can be transmitted over long distances with minimal loss of strength and without overheating wires and appliances.

Aluminized steel A stainless steel pot or pan coated with aluminum Also called *tinned steel.*

Ambience A feeling or mood associated with a particular place or thing. See also **Atmosphere.**

Amp An ampere; the term for how much electric current flows through a circuit.

Annealing A slow cooling period for newly fired glassware to strengthen and stabilize it.

Antimicrobial A property or substance that naturally retards bacteria growth, used in manufacturing items like cutting boards, slicer blades, ice machines, and protective gloves.

AREP Abbreviation for "Affiliate Retail Electric Provider," a company affiliated with an electric utility that sells power to end users and provides customer service and billing functions.

Ash dolly A metal tray on wheels used for storing and transporting ashes from a wood- or charcoal-burning fire.

Assembly line flow See **Straight line.**

Atmosphere The overall mood of a restaurant, determined by both practical and aesthetic elements such as lighting, artwork, noise levels, aromas, and temperature. See also **ambience.**

Atmospheric burner An open gas burner that uses air from the surrounding atmosphere to mix with the gas and ignite the flame. Refers to gas burners on ranges, and in fryers and broilers.

Atmospheric vacuum breaker A small shutoff valve on a water line that allows it to drain completely after a faucet is shut off.

Augur A flexible metal rod used to clean out plugged drains. Also called a *snake.*

Average check The average amount each guest will spend for a meal at a restaurant.

Back bar The rear structure of the bar behind the bartender, where glasses, liquor, and other equipment are stored.

Backflow Water that flows back or returns to its source, usually because of pressure differences. Because backflow can contaminate fresh incoming water, a backflow preventer may be installed to avoid this problem.

Bain marie A piece of equipment that holds heated or cooked foods and keeps them warm using steam or hot water.

Balanced system A series of ducts, hoods, and fans designed to work together to circulate fresh air, remove stale air, and prevent drafts in a building.

Ballast A device that limits incoming electric current into a fluorescent light fixture and acts as a starting mechanism for that fixture. There are magnetic and electronic ballasts.

Ballast factor The ratio of the light output of a fixture with a particular ballast to the maximum potential output of that fixture under perfect conditions.

Banquet kitchen A separate kitchen facility specifically used for banquets or on-premise catering. Also called a *service kitchen.*

Banquette An upholstered couch or booth fixed to a wall, with a table placed in front of it.

Bar die The vertical front of a bar, which separates the customers from the back bar area.

Barista A person who specializes in making coffee drinks using an espresso machine.

Bar mixer A blender.

Base The area beneath a table where its legs are. The base should allow customers a comfortable amount of legroom. Popular options include the spider base (with four legs) and the cylindrical mushroom base.

Base metal The alloy or mixture of metals used to create flatware.

Base rate The amount per square foot that is charged to rent space.

Basket rack A place to hang multiple fryer baskets.

Bast A fiber made from the inner portion of a plant stem. Flax, for instance, is a bast fiber used to make cloth.

Batch cooking Preparing numerous servings of food at the same time; cooking in quantity.

Battery A solid line formed by fitting together heavy-duty range tops in almost any combination of grate tops, hot tops, or griddles. Also called a *hot line.*

Beer box A refrigerated compartment designed to hold a keg of beer as part of a draft-beer system. Also known as a *tap box.*

Belly bar A front rail on a range that keeps cooks from touching the hot surface if they lean on the range as they work.

Benches Special stainless steel tables for tabletop appliances with a shelf beneath.

Bench scraper A stiff metal or plastic blade on a wooden handle, used for scraping and cutting dough.

Biofilm The more technical term for slime, caused by build-up of bacteria, fungi, and other microbes in moist areas.

Bisque Plateware in very rough form, made of wet clay that has been shaped and dried and is ready to be decorated.

Blank A newly formed utensil, made of a metal alloy, that has not yet been silver plated.

Blast chiller A machine that looks like a refrigerator, available in reach-in and roll-in models used for blast chilling.

Blast chilling A type of quick chilling that circulates cold air at high velocity around pans of food to cool it quickly.

Blower A fan that circulates air in a steamer or oven compartment.

Blower-dryer An electric or steam-heated compartment in a dishwasher, used to speed up the dish drying process.

Boiler A small steam generator that boils water to be made into steam.

Bolster On a knife, the point at which the blade meets the handle. Also called a *collar* or *shank.*

Boning knife A knife with a thin blade, used to bone meats.

Booster heater A heated tank that can increase ("boost") the temperature of water before it enters an appliance. Used in conjunction with a regular water heater when hotter temperatures are needed, as in dishwashing.

Bottle wells Indentations in a bar (in the under bar area) meant for holding bottles of liquor.

Bowl adapter An adapter used to fit large mixers with smaller bowls.

Bowl dolly A rolling platform that allows a mixer bowl to be rolled instead of lifted. Also called a *bowl truck.*

Bowl guard Heavy wire shield that fits over mixing bowls to prevent injury to workers' hands while the mixer is running.

Bowl jacket An accessory for mixing bowls that can be filled with hot or cold water to keep food inside a mixing bowl at a certain temperature. The bowl fits into the bowl jacket.

Brand name The name of a company that manufactures a certain commodity.

Brass rail The metal footrest that runs the length of the bar, about one foot off the floor.

Brazier A wide, shallow saucepot used to braise, brown, and stew meats. Also called a *rondeau.*

Brick oven One name for a wood-fired or wood-burning oven.

British thermal unit (Btu) A standard measurement of energy use: 1 Btu is the amount of heat needed to raise the temperature of 1 pound of water 1 degree Fahrenheit.

Broiler/griddle A dual-purpose cooking appliance that uses tubular burners for the griddle and infrared bulbs for the broiler.

Bubble diagram A drawing to help organize the layout of work stations or duties. A circle ("bubble") represents each station or duty; intersecting circles indicate areas that are related.

Building-related illness (BRI) A medical condition in which 20 percent or more of a building's occupants experience similar symptoms of discomfort that disappear when they leave the building. Also called *sick building syndrome.*

Build-out allowance See **finish-out allowance.**

Burner ports The pattern of holes in a gas burner where the flames come out.

Burner tube The section of a gas burner where gas mixes with air and ignites. Also called a *mixer tube.*

Burn-in A procedure used with recently purchased cooking equipment to prepare it for regular use. The burners and/or ovens are turned on and heated up to burn off any protective coating.

Burning speed The speed at which a flame shoots through the mixture of air and gas in a gas burner.

Burnisher A machine that polishes silver plate by removing some of its surface scratches.

Bus boxes The plastic bins used to clear soiled dishes from tables.

Bypass line A cafeteria layout in which foods are grouped into sections with indentations of space between them. Customers can line up at the sections they want and bypass the others. Also called a *shopping center* layout.

Cable Several electrical wires bound together and insulated.

Cappuccino Espresso coffee topped with foamed milk.

Capture velocity The ability or strength of an exhaust hood to pull stale air up and away from the cooking line.

Carbon soot A black buildup that coats and clogs the parts of a gas burner, created by a poorly adjusted burner.

Casters Wheels on racks, carts, or pieces of heavy equipment that allow them to be rolled from place to place.

Cathodic protector A metal device mounted on the inside of a steam generator, suspended in the water, which works to control corrosion.

Ceiling-type refrigeration A single fan mounted on the ceiling inside a refrigerator, which circulates cold air.

CE Marking The logo affixed to equipment, toys, and medical devices sold in the nations of the European Economic Area, denoting that the product has met minimum health, safety, and environmental protection standards.

Ceramic Clay that has been baked; the substance used to make most plateware.

Ceramic knife A knife made from zirconium carbide, in combination with aluminum oxide or heat-resistant zirconium oxide. Often gray or black in color; known for keeping its sharpness.

Ceramic steel A tool made of aluminum oxide, used to sharpen ceramic knives.

Chan A wooden spatula used in Asian countries for stir-frying and wok cooking.

Charbroiling Cooking foods using flames, smoke, and radiant heat.

Cheesemelter A specialty type of broiler used primarily to melt cheese and heat prepared foods.

Chef's knife An 8- to 12-inch knife blade used for most chopping, slicing, and dicing tasks. Also called *cook's knife* or *French knife.*

China cap A metal mesh cone with a handle used as a strainer; a very fine mesh China cap is called a *chinois.*

Chinese range A specialty range also called a *wok range* or *Oriental range.* The rings of its recessed, circular burners can be adjusted to accommodate large woks for stir-frying under very high heat conditions.

Chromaticity The warmth or coolness of light, measured in degrees Kelvin.

Circuit The complete path of an electrical current, from its source through wiring and into an appliance or socket. A *closed circuit* is complete and working; an *open circuit* has been interrupted to stop the power flow.

Circuit breaker A switch that automatically trips into the "off" position when it begins to receive a higher load of electricity than its capacity.

Cladding The process of bonding thin sheets of different metals together when making cookware, to use the best characteristics of each. Such a pot or pan is referred to as *cladded.*

Clamshell griddle A griddle with a heated hood that raises and lowers, cooking food on two sides at once. Electric clamshells are called *contact grills.*

Classroom-style seating A banquet table setup in which people sit on only one side of the table.

Clean-air vacuum A vacuum in which incoming dust is filtered into a dust bag before entering the motor.

Clean steam Pure steam, not contaminated by chemicals.

Cleaver A heavy, sturdy knife with a square-shaped blade used for chopping meat.

Cobra A multilined hose with a dispensing head to serve soft drinks and water; a necessary part of a bar.

Coil A piece of coiled copper or stainless steel tubing found in appliances.

Colander A perforated bowl used to wash and rinse foods.

Cold plate A chilled metal sheet at the bottom of an ice bin, which helps to chill the lines of an automatic dispensing system for liquids.

Cold zone In a fryer kettle, a space about 2 inches deep located between the heating elements and the bottom of the kettle, where crumbs and debris fall during the frying process.

Colorfast The ability of fabric to retain its color without spotting or fading after multiple washes.

Color rendering index (CRI) A numerical scale from 0 to 100 that indicates how bright a color appears based on how much light is shining on it.

Color temperatures A scientific comparison of colors based on their warmth or coolness, measured in kelvins. Whether a light source appears warm or cool is its *correlated color temperature* (CCT).

Comal A broiler used primarily in restaurants for display cooking of fajitas.

Combi Short for "combination," this term refers to a multifunctional cooking unit such as an oven/steamer.

Combination steamer A steamer that can be adjusted to be either pressurized or pressureless.

Combi-wave A small oven that combines microwave and convection cooking techniques.

Combustion The burning of trash to minimize waste.

Combustion efficiency An estimate of how well a gas-powered appliance uses its energy.

Comfort zone A combination of external temperature and humidity level that makes people comfortable.

Compartment steamer Another term for a pressure steamer.

Composting The natural breakdown of organic materials over time, turning them into a soil enhancer that is useful in gardening and farming.

Compressor The part of an air-conditioning or refrigeration system that acts as a pump, impacting the temperature and pressure of refrigerant gas.

Concept The idea for a restaurant, which encompasses menu, theme, decor, and other factors that create an image in the minds of customers.

Concept development The process of identifying, defining, and collecting ideas to create an image for a new business or product.

Condensation In a system, the steam that cools enough to become liquid again and must be drained or removed from the system.

Condenser The part of an air-conditioning or refrigeration system where the refrigerant releases its heat and becomes liquid again. A condenser may be filled with either air or water.

Conduction The simplest form of heat transfer, when heat moves directly from one item to another.

Conduction shelves Shelves that contain a solution of water and antifreeze to cool food rapidly.

Conductor Another name for an electrical wire; a material or object that allows electricity to flow easily through it.

Conduit Wires encased in rigid steel pipes.

Constant multiplier The number used by a water meter reader to multiply the readings from a meter for an accurate reading of how much water was used.

Continuous cooker A rotisserie oven that contains a vertical, moving ladder with product on each "rung."

Continuous toaster An appliance for high-volume toasting; food is fed into the heated cabinet on a conveyor belt. Also called a *conveyor toaster.*

Controlled evaporation cooker A type of fryer that uses both heat and steam to hold cooked food at its peak quality. Also called a *CVAP* or *Collectramat.*

Convection Cooking with hot air that is circulated by fans in the oven cavity, providing excellent heat transfer.

Convenience oriented A restaurant that relies primarily on customers very close to its location for business. A fast-food restaurant is usually convenience-oriented.

Cook-chill The process of cooking food in quantity, then rapidly chilling it.

Cook-and-hold oven A low-temperature roasting appliance that cooks foods slowly, then keeps them warm until serving.

Cook-cool tank A tank used to cook large cuts of meat or other foods not suited to kettle cooking. Also called a *chiller-cook tank* or a *cook-chill tank.*

Cook-to-inventory Another term for the **cook-chill** food preparation method.

Cooking suite A freestanding, custom-built battery of kitchen appliances that is installed as an "island," rather than against a wall. Also called a cooking island.

Cooking zone The area in a fryer kettle where hot oil circulates and frying takes place.

Coupe style A plate made without a pronounced outer rim.

Cowl A hood-shaped cover fastened to the end of a conveyor dishwasher that is connected to an exhaust system. Makes the system more efficient by removing moisture quickly.

Cradle A device to transport beer kegs or other drum-shaped objects and hold them steady as they roll.

Critical control points (CCPs) Any place or procedure in a foodservice operation where loss of control may result in a health risk.

Crumb tray Located at the bottom of a fryer kettle, this removable plate captures food particles until they can be discarded.

Crystal Glass created by adding lead oxide and potassium silicate to the regular glassmaking process. Also called *rock crystal.*

Cullet Bits of broken glass melted with other substances to form new glass.

Curved unloader An addition to a dish room that helps push finished dish racks off a conveyor belt.

Cut glass High-quality crystal with a design etched on the glass.

Cutter mixer A combination mixer and blender with attachments that can be used to mix or mash foods.

Cycling thermostat A temperature monitoring device on fryers that regulates heat up to 400 degrees Fahrenheit.

Damask weave Fabric with an elaborate design woven onto both sides. Often used for napkins. The front side of Damask cloth is known as its *weave brocade*; the back is called the *fill* side.

Damper A valve or plate that regulates airflow inside the ductwork of an HVAC system.

Damp-labeled The term for a lamp or light bulb that can withstand humidity, designed for use in areas like walk-in refrigerators and dishrooms.

Data-logger A computerized system for measuring both time and temperature, recording and storing the data chronologically to track food safety.

Deck The bottom or floor of an oven cavity. In pizza ovens, called a *hearth.*

Deck oven An oven that has more than one cavity and a separate set of controls for each cavity. Also called a *stack oven.*

Deco-mat A table runner.

Decoying The scraping and stacking of dirty dishes by wait staff or bus persons, to prepare them for washing.

Dehumidifier An appliance needed in some HVAC systems to remove excess moisture from the air.

Demand charge An electric bill based on the highest amount of power that is used during its busiest (peak) time period. Also called a *capacity charge.*

Demand meter A device that keeps track of the maximum amount of power used by a business during its busiest times, in a fixed number of minutes.

Demographics The characteristics of potential customers; age, sex, income, education levels, and so on.

Design In a space plan, the definitions of sizes, shapes, styles, and decorations of the facility and furnishings.

Design program A document that includes all the criteria for a business' design, including equipment and flow patterns.

Dessicant A substance used as a drying agent in a heating and air conditioning system to reduce the humidity of incoming air. Silica gel or titanium silicate are commonly used dessicants.

Destination oriented A business that attracts guests who don't necessarily live or work nearby, because of its unique attributes or ambience. A fancy restaurant where people might go to celebrate a birthday or anniversary is an example of a destination-oriented restaurant.

Deuce A table that seats two people. Also called a *two-top.*

Diffuser Another term for a vent in an HVAC system.

Digital meter A modern type of electric meter that has a digital readout instead of a series of dials to be read.

Dimensionally stable Cloth that is shrink-proof.

Dipper An ice cream scoop. Also called a *disher* or *portion server.*

Direct current (DC) This type of electricity has a constant flow of amps and volts. Not widely used in the United States, direct current cannot travel through power lines for long distances and can sometimes overheat appliances.

Direct lighting Light that is aimed at a specific place or object to accent it.

Discharge system The drainage system plumbed so that liquids from kitchen, sinks, and restrooms flow into it.

Dish table A metal table, often with shelves underneath or glass racks above. Used as landing space in a dish room.

Dishwashing The process by which glasses, dishes, and flatware are cleaned using hot water, detergent, and motion.

Display kitchen A kitchen where much of the cooking and food preparation takes place in view of the customers.

Docker A handheld implement that looks like a round barrel on a stick. The "barrel" contains pins to poke tiny holes in dough before it is baked. Also called a dough docker.

Dolly See **hand truck.**

Dome strainer A stopper for sinks, perforated to prevent bits of solids from going down the drain. Also called a *sediment bucket.*

Double boiler A base pot that holds water and an insert pot that nests inside the base. Water is boiled gently in the base pot to heat delicate foods that might scorch if exposed to direct heat.

Double broiler A broiler with two separate heat controls and cooking areas.

Dough box An insulated container used to keep pizza dough moist while it is stored until use.

Downlighting Positioning a light fixture to provide direct lighting to the area directly beneath the fixture.

Drawer warmer A heated cabinet with one or more drawers that slide open to load or remove food.

Drift The term used to describe an electronic or digital scale that doesn't hold a steady readout when weighing an object—the numbers shown on the digital readout "drift" up or down. This may indicate a temperature problem, or a source of static electricity that is interfering with the scale's balance.

Drip pan A removable tray that catches grease drippings and crumbs at the bottom of a broiler. Also called a *drip shield.*

Drop The amount of tablecloth that hangs over the side of the table to drape it attractively without getting in the way of the guests.

Drop-in burner Modular cooking surfaces that come in configurations of one or two burners that "drop in" to the range top, where they are connected to a power source.

Dry steam Steam with a small amount of superheat in it.

Dry storage Storage for canned goods, foods, paper products, and other items that don't require refrigeration.

Dual voltage Large, commercial appliances sometimes require more than one type of electric current: one for its motor, and a more powerful voltage for its heating element, for example.

Duct-type refrigeration A system inside a refrigerator that circulates cold air between a forced-air unit on the ceiling of the interior cabinet and a series of small air ducts on the back wall of the cabinet.

Dump pan A pan into which food is unloaded after leaving a fryer.

Dunnage racks Racks for storage of bulky items such as bags and cases, which would otherwise take up valuable shelf space.

Earthenware Sturdy, inexpensive plateware that is opaque and somewhat porous because it is fired at lower temperatures than either stoneware or china.

Effective length The actual length of a pipe when the lengths of pipe fittings are added to it.

Efficacy The ability of a lighting system to convert electricity into light.

E-lamp A long-lasting bulb that uses a high-frequency radio signal instead of a filament to produce light.

Electric discharge lamp A fixture that generates light by passing an electric arc through a sealed space filled with a special mixture of gases. Sometimes called *gaseous discharge lamps,* these include mercury vapor, fluorescent, halide, and sodium lamps.

Electric service entrance The physical point at which a building is connected to its source of electricity.

Elliptical reflector (ER) lamp A specially shaped bulb for recessed light fixtures.

End cap In a strip shopping center, the part of the building at either end.

Energy audit An analysis of a building's utility use, with suggestions for saving energy and money, performed by a utility company or consultant. A basic audit is called a *walk-through*; a more detailed one is called an *analysis audit.*

Energy charge The amount of money a business pays for the total amount of electricity it uses during a single billing cycle. Also called a *consumption charge.*

Energy-efficient rating (EER) An estimate on air-conditioning systems of how much power a system will use.

Energy index The total amount of energy used by a business, calculated by reducing each type of fuel to Btus and adding them together.

Environment The conditions in a restaurant that make guests feel comfortable and welcome; or that give employees a safe and productive work setting.

Equipment key A diagram on paper that sketches out the locations of your equipment (in a kitchen) or your tables and booths (in a dining area). Also called an *equipment schedule.*

Equipment specifications Concise written statements about a piece of equipment that state exactly what is needed so potential sellers can supply what the customer wants. Nicknamed *specs* or *spec sheets.*

Equivalent length An amount of space added to pipe calculations for the pipe fittings, elbow joints, and the like.

Ergonomics The science of studying the characteristics of people and arranging their activities to be done in the safest and most efficient manner.

Escalation clause The stipulation in a building lease that the owner may increase the rent after a certain period of time.

Espresso Strong coffee in small, brewed-to-order servings. The word means "to press" or "to press out" in Italian.

European sized Oversized flatware.

Evaporator A series of coils in a refrigeration system filled with liquid refrigerant. The refrigerant evaporates inside the coils, creating gas vapor.

Even-heat plate Another term for a uniform heat top or "hot top" on a commercial range.

Expediter A person hired in busy kitchens to act as a liaison between kitchen staff and wait staff, to facilitate ordering and delivery of finished plates of food. Also called a *wheel person* or *ticket person.*

Extinction pop A type of flashback that occurs when a gas burner is turned off. It is a popping sound that sometimes accidentally extinguishes the pilot light flame.

Extraction A method of kitchen grease removal in which exhaust air moves through a series of specially designed baffles, throwing off the grease particles by centrifugal force.

Extractor In a commercial laundry, a machine that takes most of the moisture out of newly washed, wet fabrics by spinning them around under high-speed centrifugal force.

Fast casual A restaurant market niche which is similar to "fast food" or quick-service restaurants, in that it does not offer table service, but offers somewhat higher food quality and atmosphere. Also known as quick-casual.

Fat filter A filtering device inside a combi oven that traps airborne grease and food odors. It can be removed for cleaning.

Fat turnover In frying, the amount of fat (cooking oil) used each day compared to the total capacity of the fryer kettle.

Feasibility study A written document prepared in the planning stages of a new business, which compiles the research done to justify opening the business. See also **financial feasibility study** and **market feasibility study**.

Feed grip The part of a slicer that holds the food steady as it is pushed toward the spinning blade.

Field convertible Refers to a dishwasher that can be adjusted to sanitize dishes using either chemicals or hot water.

50-50 blend A fabric made of half cotton and half polyester that is often used to make table linens.

Filler cap A blender cover that is made of two pieces so one can be lifted for adding ingredients during blending.

Filtration system The equipment used to clean debris out of frying oil.

Financial feasibility study The written document that outlines income and expenses necessary to start a new business.

Finger panels Metal panels perforated with small holes through which hot air is forced in an impinger/conveyor oven.

Finish **1.** The amount of polish or shiny surface on stainless steel equipment. In equipment specifications, finish is measured on a scale from 1 to 7; 1, meaning rough, to 7, meaning mirrorlike. **2.** The surface of a piece of silver plate. Popular finishes have names, and, listed from most opaque to most shiny, they are satin, butler, bright, and mirror.

Finishing The ironing and folding process for newly laundered tablecloths and napkins.

Finish-out allowance The amount of money set aside in a building or remodeling project for the new occupant of the space to finish it instead of the contractor. The finish-out allowance in a restaurant is often used for paint and fixtures.

Fins In a refrigeration system, the metal plates that surround the copper coils of the evaporator.

Firing rate The amount of power in a gas-fired steam generator, expressed in Btus.

Flaker An ice machine that produces soft, snowlike ice beads used to keep foods or wine cold.

Flame lift A potentially dangerous condition where some of the flames on a gas burner lift and drop without being manually adjusted.

Flame rollout A serious gas burner fire hazard that occurs when flames shoot from the burner's combustion chamber instead of through the burner ports.

Flame spreader The walls of a commercial gas oven that distribute heat evenly through the cavity.

Flame stability A clear, blue-colored ring of gas flames with a firm center cone. This means the correct mixture of air and gas is being used.

Flashback The roar or "whoosh" a gas burner emits as it is lit, which occurs because the burning speed is faster than the gas flow.

Flash steamer A countertop steamer used to heat sandwiches and pastry products quickly without making them soggy.

Flat truck A rolling cart mechanism used in storage and warehouse areas to transport cases of product.

Flatware Utensils that guests use to eat. Often called *silverware,* even if they're not silver.

Flight-type dishwasher A dish machine that uses a conveyor belt to move racks of dishes in a straight line through the wash, dry, and sanitize process. May also be called a *rackless conveyor* or *belt conveyor,* depending on the features of the conveyor belt.

Floating flames Flames that don't assume a normal, conical shape. Indicates the need to adjust a gas burner.

Floating interest rate The amount of interest to be paid on a business loan; usually one or two points higher than a bank's prime lending rate.

Floating rod A type of broiler grate, chrome plated and warp-proof. Also called a *free-floating grate.*

Floating suspension washer A type of commercial washing machine that spins laundry without as much external vibration as a regular washer, so it does not have to be bolted to the floor for installation. Also called a *soft-mount washer.*

Flow (or flow pattern) Like the traffic pattern of a neighborhood or intersection, this is the method and routes used to deliver food and beverages to customers.

Fluorescent lamp A long-lasting type of electric discharge lamp. The bulb is tube shaped and contains mercury vapor. The smallest bulbs are called *compact fluorescent lamps* (CFLs).

Flute A tall, narrow glass used to serve Champagne.

Folder A machine that folds table linens.

Food finisher A type of fast-cooking equipment used to reheat or "finish" foods that require quick browning, toasting, crisping, and so on before serving.

Food mill A hand-cranked appliance that can be used to strain, grind, or puree foods by forcing them through a perforated disk.

Food slicer An apparatus used in slicing meats and cheeses. Provides more portion control than a knife.

Food waste disposer Also known as a garbage disposal. Installed in kitchen sinks to grind up and dispose of food waste.

Foot The raised ridge located at the bottom of a plate, on which the plate sits.

Foot-candle A measurement of light; the light level of 1 lumen of light on 1 square foot of space.

Foot treadle A pedal or lever you can press with your foot, which opens the refrigerator door.

Forging A way to make knives by heating steel, then shaping it and compressing it under pressure.

Four-top A table that seats four persons.

Free-floating grate See **floating rod.**

French press A type of coffee pot. A glass carafe where ground beans and water are combined and left to steep; the beans are then forced to the bottom of the pot (manually, with a plunger) when the coffee is ready to drink.

French whip A standard kitchen whip used for mixing air into soft foods such as mashed potatoes.

Fryer basket A perforated wire mesh container that is filled with food to be lowered into the hot oil of a frying kettle.

Fryer cabinet The metal cabinet that houses a fryer kettle and, in some cases, its filtration system.

Fryer screen A perforated wire mesh panel that separates the cold zone at the bottom of a fryer kettle from the rest of the oil.

Frying kettle The receptacle in a fryer where the oil is poured to be heated. Also called a *frying bin.*

Fry powder A substance that can be added to frying oil to minimize oxidation and improve its useful life.

Fry station An area of a foodservice kitchen where fried foods (especially french fries) are prepared for serving. May include landing or storage space, fryers and filtering systems, food warmers, basket racks, and a bagging area.

Fry-top range Another name for a griddle with an oven installed below it.

Fuel adjustment charges Fluctuations in power costs that a utility is authorized to pass on to its customers. Also called *energy cost adjustments.*

Fuse A load-limiting device that can automatically interrupt electricity in a circuit if an overload condition exists.

Garde manger A term for the cold-food preparation area of a commercial kitchen; the proper (and fancier) European term for pantry.

Gas-assist oven A wood-fired oven that is equipped to use natural gas to start its flames. Also called a wood/gas combination.

Gas connector The heavy-duty brass or stainless steel tube used to attach a gas-powered appliance to a gas source.

Gauge A measurement of the thickness of metal. The lower the number, the thicker the metal.

Geoexchange technology See **ground source heat pump.**

Glare zone The angle at which most types of lighting cause reflective glare or are uncomfortable to look at.

Glassware The wide variety of containers used to serve water or beverages, such as mugs, tumblers, and stemware.

Glaze A protective coating sprayed on plateware during the manufacturing process. Also called *miffle.*

Glide A self-leveling device on the leg mechanism of a banquet table.

Grade The three-digit number used to classify stainless steel by its chemical composition. Grades 304 and 420 are the ones most often used in foodservice.

Grate top An open gas burner. Also called a *graduated heat top.*

Gravity feed machine A soft-serve ice cream machine where the mix is loaded into a hopper and flows as needed into a cylinder below, where it is frozen, scraped out of the cylinder, and dispensed.

Gray water Water that is unfit for drinking but may be used for other, less critical situations where purity is not an issue.

Grease guards Devices installed at the base of an exhaust fan to catch grease so it does not settle on the roof of a building and create a fire hazard.

Grease trap An underground site where water and waste output drains. Its role is to prevent the grease and solids from entering the sewer system, and it must be cleaned regularly by professionals. Also called a *grease interceptor.*

Grease tray A receptacle for catching grease runoff in the cooking process, or for aiming it into a drip pan.

Grid assembly A metal grill on which food is placed to be broiled. It rolls in and out of the broiler smoothly and its proximity to the heat source can be adjusted.

Griddle A flat, heated cooking surface made of a thick steel plate.

Griddle stone A pumice stone used to scrape and clean a griddle. Also called a *pumice* or *brick.*

Grill Food is placed over this grid of metal bars, which is set over a heat source for cooking.

Grille A cover on a vent that can be manually opened and closed to regulate airflow. Also called a *supply grille.*

Griswold A nickname for a cast iron skillet.

Grit The amount of coarseness of a knife sharpening stone.

Ground loops The underground lines or piping in a ground source heat pump system.

Ground source heat pump An HVAC system that operates on a simple electric compressor, with its lines buried underground and no condenser. The ambient ground temperature keeps the lines at moderate temperatures, meaning the system uses less energy to heat and/or cool the fluid running through them. Also known as **geoexchange technology**.

Gun The dispensing head of a multiline hose (a *cobra*), which can dispense as many as eight beverages from a single head.

HACCP Stands for Hazard Analysis of Critical Control Points. A multiple-step method to identify and correct potential health and safety hazards in foodservice.

Hand The linen industry term for the tactile quality of a fabric; how it feels to the user.

Hand truck Also called a *dolly,* a small sturdy platform on two wheels that is designed to support and move heavy items such as cases of canned goods.

Hard-products slicer A blade for processing foods such as carrots, potatoes, and cheese.

Harmonics A distortion of an electric power frequency that can create interference on power circuits.

Head space The amount of room left between the top rim of a steam-jacketed kettle and the food inside it.

Heat exchanger A device that takes steam from one source, circulates it through a series of coils to clean it, and pipes it to another appliance.

Heating plant The part of an HVAC system where fuel is consumed and heat is produced.

Heating system The ductwork and vents of an HVAC system that distribute and control heat.

Heat pipe exchanger Part of an HVAC system that evaporates and condenses fluid in a continuous loop to take heat from supply air to return air streams, without the use of electrical power.

Heat pump water heater (HPWH) A tank that uses waste heat from an air-conditioning system to heat water.

Heat recovery The process of capturing heat vented from one appliance that would otherwise be wasted, and piping or transferring it to another appliance for another use. Also called *heat reclamation* or *cogeneration.*

Heat recovery time The time it takes for frying oil to return to its optimum frying temperature after cold food is immersed in it to be fried.

Heat transfer fluids (HTF) A gas burner system with a series of pipes and heated fluid that runs through them. The pipes are hooked to several pieces of equipment, allowing the system to run more than one appliance simultaneously.

Heat transfer rate In steam cooking, a complex formula that measures (in Btus) how much energy it takes for heat to be transferred from one substance to another.

Heat zones Individually heated sections of a griddle, each with its own controls.

Heavy-duty range A range best suited for high-volume cooking.

Heel The back or end of a knife handle.

High-density storage Storage that can increase the overall capacity of an area by 30 to 40 percent.

High-limit thermostat A safety feature on fryers to detect overheating and turn the fryer off automatically.

Hob Another name for an electric induction range.

Holloware **1.** A general term for dishes that aren't flat, such as bowls, pitchers, and cups. **2.** A silver-plated knife with a hollow handle instead of a solid one.

Hollow-ground blade A knife blade made from two separate pieces of steel welded together.

Hollow-square system See **Scramble system**.

Hood An exhaust canopy located directly over cooking equipment that draws smoke, heat, and moisture away from the kitchen.

Hot-food table A table used to keep food warm by heating multiple internal compartments with hot air. Pans of food are placed in openings on top of the table. May be called a waterless hot-food table. Also see **hot well**.

Hot line The battery of kitchen appliances where cooking is done—ovens, ranges, steam jacketed kettles, and so on—located under an exhaust hood.

Hot isostatic pressing The manufacturing process for making ceramic knives, which involves simultaneous molding and heating of the blade.

Hot well The opening in a hot-food table or steam table into which a pan of food is placed; or the nickname for the table itself.

Hotelware A heavy-duty type of china dinnerware made for use in hotels and foodservice.

Hubbelite A sturdy, comfortable type of brick-red flooring made of cement, copper, limestone, magnesium, and a few other substances.

Humidifier An appliance that adds moisture to the air.

HVAC The abbreviation for a heating, ventilation, and air-conditioning system.

Hydrogenated A term to describe cooking oil that has been chemically combined with hydrogen molecules to make it more solid at room temperature and to increase its stability.

Ice bin An insulated storage container that holds ice produced by an ice maker.

Ice builder A refrigerated machine that supplies ice water to a tumble chiller.

Ice cuber An ice-making machine.

Ice dispenser A refrigeration unit that makes ice and dispenses it as needed with the push of a button or lever instead of being scooped from an ice bin.

Ice shock When food is cooked and then immediately plunged into ice water.

Illumination The effect achieved when light strikes a surface. Also called *illuminance.*

Image The way the public perceives your restaurant, based on the concept you have created.

Impinger/conveyor oven An oven that cooks by blasting high-velocity hot air onto food that moves through the cooking chamber on a conveyor belt.

Incandescent lamp A filament encased in a sealed glass bulb, commonly known as a light bulb.

Incomplete combustion A natural gas flame that does not use the correct mixture of air and gas. Flames are unstable and are tipped with yellow.

Indirect lighting Lights that wash a space with light instead of focusing directly on one object or spot. Minimizes shadows and is considered flattering.

Indirect waste An inch or more of space between a water pipe and its corresponding drain, which prevents contaminated water from flowing back into the freshwater supply.

Induction Cooking that creates heat when an electric current runs from the surface of the stove into a pot or pan, heating the food and not the pan. The induction process only works with ferrous metal, such as cast iron or stainless steel.

Infinite-heat knobs Heat controls that allow for small adjustments instead of just low, medium, and high settings.

Infrared Invisible rays that have a penetrating heating effect when used in cooking.

Infrared thermometer A heat measuring device that can read food temperature without having to touch the food, by detecting its infrared rays.

Ingredient room An area in some kitchens where everything needed for one recipe or task is organized, to be picked up and delivered to a specific work station.

Inlet chiller A small tank that collects water that would normally be discharged from an ice maker and recirculates it. The cold outgoing water chills fresh incoming water, allowing it to freeze more quickly for faster ice production.

Integral ballast lamps A lamp and ballast that are a single unit.

Inventory turnover rate The number of days your raw materials and supplies will be in your inventory, from the time they are delivered until the time you use them.

Ionizer The electrically charged field in an air cleaner. Air is forced through this field and into a series of metal plates that collect impurities and airborne particles.

Ironer A machine that irons damp linens.

Jet The stream of gas that flows into a gas burner.

Kelvin (K) A measurement of the warmth or coolness of color, based on the Celsius temperature scale.

Kickplate A protective metal plate covering the lowest edge of a door that is frequently "kicked" open by waiters with full trays, kitchen employees, etc. Also called a *scuff plate.*

Kilovolt-ampere (kVA) The power it takes to propel 1000 volts of electricity through a power line.

Kilowatt-hour (kWh) The time it takes to use 1000 watts of electrical power.

Kilowatt-hour meter A device that keeps track of the number of kilowatt-hours of electricity used in a building.

Lamp lumen depreciation The gradual reduction in the light output of a bulb as it ages.

Landing space Work surfaces for setting things down.

Lavatory A hand sink installed in a restroom.

Layer platform Commonly called a *flat truck,* it has four wheels and is strong enough to hold 10 to 15 cases of canned goods or other products.

Layout The detailed arrangement, often on paper, of a building and its floor and/or counter space.

Lazy steam Pressurized steam in a cooking cavity that works its way through food and cooks it from the outside in.

Lens A clear plastic screen or covering that shields a fluorescent bulb to diffuse glare.

Life-cycle costing A method for determining the true cost of an appliance by taking into account its power and water use, durability or expected useful life, ease of use and training, and potential repair and/or service costs.

Light-emitting diode (LED) A low-voltage, extremely energy-efficient light source that contains a microchip designed to release photons. LEDs are generally brighter and last longer than other, conventional light sources.

Light fixture The base and wiring that connects to an electric power source, and a reflective surface to direct the light of a bulb. Also called a *luminary.*

Lighting transition zone An area where people are moving between two different types of lighting, which gives their eyes a moment to adjust to the change.

Light loss factors Conditions that contribute to a light fixture putting out less light than it is capable of.

Liquefied natural gas (LNG) Gas that has been highly compressed under very cold conditions for storage.

Load calculation An expert estimate of the size of unit you will need to heat and cool a space.

Load cell A metal bar or beam inside a digital or electronic scale that bends when an object is being weighed on the scale. As the load cell bends, the strain gauges attached to it also bend, measuring the weight of the object by their resistance to the change.

Load factor The amount of power consumed, versus the maximum amount of power that could be consumed, over a stated time period. Expressed as a percentage, this figure allows a business to determine its overall level of energy efficiency.

Load profile A summary of how much power a building needs and what times of day its power usage is highest. The load profile is determined by tracking the daily load factor over weeks or months to get an accurate picture of power requirements.

Long-throw system In HVAC, a system in which air inlets are located far from the kitchen's exhaust hoods. The air must travel toward the hood to be exhausted, which allows for cross-ventilation.

Louver A small shade that can be adjusted to aim light to eliminate glare at certain angles.

Lowerator A spring-loaded plate holder that can be temperature controlled to provide preheated plates.

Low-velocity system A makeup air system that brings air in through the exhaust hood and vents it straight down across the front of the cooking equipment. Also called an *air curtain.*

Lumen A measurement of light output; the amount of light generated when 1 foot-candle of light shines from one source.

Lumen direct depreciation A reduction in a bulb's light output because dirt or grease has accumulated on the bulb.

Marche kitchen The European term for the system of guests ordering at a counter and having their food cooked to order as they wait.

Market feasibility study A written document that examines the competition, potential customers, and trade area where a new business plans to locate.

Masonry base The concrete base on which a heavy-duty appliance may be installed.

Mechanical impact Breakage that occurs when a glass hits another object.

Mechanical oven An oven used in large-volume operations where the food is in motion as it cooks inside the oven compartment. Depending on the type of motion, also called a revolving or rotary oven.

Mechanical pot/pan washer A special type of dish machine designed to wash pots and pans.

Mercerizing The process of stretching fabric and coating it with a caustic soda solution to make it look glossy and provide some stain protection.

Merchandising doors Full-length glass door panels on refrigeration units, often seen in supermarkets and convenience stores.

Mercury vapor lamp A type of bright light used for lighting streets and parking lots. Also called a *high-intensity discharge lamp.*

Metal halide lamp A variation of the mercury vapor lamp, with metallic halide gases inside to improve color and efficiency.

Metal sheath A type of heating element in a toaster.

Metering station The area of a kitchen where food that is being cook-chilled is sealed, weighed, and labeled; often part of the pump-fill station in the tumble-chilling process.

Methane The most commonly used form of natural gas.

Microwave An oven that uses electromagnetic waves to cook with radiant heat. Also see radiation.

Mils A measurement of the thickness of silver plating on flatware. One mil is 1/1000 of an inch.

Mission statement A written statement that expresses your concept and image, to be used as a guideline for creating a business.

Momie cloth A durable cloth with a subtle herringbone pattern woven into it, used to make table linens.

Mongolian barbecue A type of griddle used in restaurants that serve Mongolian-style cooking.

Monolithic floor A one-piece floor, poured all at once and made of epoxy, polyester, urethane, or cement.

Mullion-type refrigeration An airflow system in a refrigerator that takes in air above the top food storage shelf of the refrigerated cabinet and discharges it below the bottom shelf. Also called a *back wall* system.

Multistage flash distillation A method of sanitizing seawater by boiling and then condensing it.

Musculoskeletal disorders (MSDs) Chronic health conditions, such as carpal tunnel syndrome, back pain, and tendinitis, that may be caused or exacerbated by repetitive motion or unsafe work methods.

Napery Tablecloths and napkins.

Needle injection steamer A steamer that contains a series of needles that can be arranged in different patterns to heat food. When food is placed on the needles, steam is released through the needles directly into the food.

Negative air pressure zone A space in which more air is removed from the area than is piped into it.

Net sales The amount of money a business makes after all expenses have been subtracted.

Occupancy sensor A controller device that can turn on light fixtures when it detects motion in a room, and turn them off when the room is empty.

Offset spatula A hamburger turner.

Ohm A measurement of electrical resistance; whether a substance is a good or poor conductor of electricity.

Oil take-up The amount of oil that is absorbed into food products being fried.

One-fired plate A plate that is fired without its foot being glazed.

Opacity The degree to which smoke blocks the flow of light in an enclosed space.

Orifice The hole in a gas burner through which the gas flows.

Overbroiler A broiler with its heat source located above the food. Also called an *overhead* or *overfired broiler* or a *hotel broiler.*

Overlay The process of brushing or spraying on an extra coat of silver on the parts of a piece of flatware most likely to wear down with repeated use; or, in napery, a smaller, accent tablecloth that rests on top of a larger tablecloth.

Overrun The percentage of air forced into soft-serve ice cream by a soft-serve machine to create its soft, creamy consistency.

Oyster kettle A small steam-jacketed kettle, so named because it is often used to cook oysters.

Packaged air-conditioning system A self-contained HVAC system, in which heating and cooling functions use the same ductwork, dampers, and thermostat.

Palette knife A straight spatula with a flexible blade and rounded end, used to scrape bowls or spread icing.

Pallet A low, raised platform on which boxes or cases can be stacked and stored to keep them off the floor.

Pan glides Adjustable, easy-to-slide drawers or racks for placing serving pans inside refrigerators. Also called *pan slides.*

Pantry A dry storage room; the cold-food preparation area.

Parallel flow A pattern of food preparation in which two straight lines of workers perform their tasks. For instance, they may be working back to back or facing each other. Parallel flow is handy where space is tight.

Paring knife A short-bladed knife with a sharp tip, used to pare and trim fruits and vegetables. A slightly curved tip on some paring knives is called a *tourne.*

Pass-through refrigerator A type of refrigerator with two sets of doors, located opposite each other. Often placed between a kitchen and service area, so the wait staff doesn't have to enter the kitchen to get items out of the refrigerator.

Pass window An opening between kitchen and dining area to pass food back and forth conveniently without having the wait staff enter the kitchen.

Pasteurize The process of (at least partially) sterilizing or sanitizing a food or liquid by heating it to temperatures that will kill harmful microorganisms without chemically altering the food or liquid itself.

Payout period The number of years it will take to pay back a loan amount; in foodservice, usually 10 to 15 years.

Peak loading Energy that is used at peak demand times, usually during the daytime and early evening. Reducing peak loading can reduce your power bill.

Peak shaving Curtailing power use or shifting to another power source to avoid peak loading and the higher costs it entails.

Peel A long-handled pan with a flat surface, used to take pizzas in and out of a hot oven.

Percentage factor A percentage amount added to the base rate or monthly rent of a leased building. The business operator in that building must pay the building owner a percentage of his or her profit as part of the lease.

Photocells Light-sensitive timers that turn on outdoor lighting when the surroundings become dark.

Piano whip A kitchen whip with finer, rounder wires than the standard French whip. Used for whipping sauces.

Pitch The term for how flat (or rounded) the surface of a plate is.

Place setting One person's set of flatware, plates, and glasses, to be arranged on a dining table.

Plate **1.** The surface of a griddle. **2.** A term for the blade on a food processor.

Plated The process by which a thin coat of silver is fused onto a new piece of flatware using an electric current.

Plateware Dishes used to serve food and hot beverages. Also called *dinnerware* or *china.*

Polished plate A one-fired plate that has had its rough foot smoothed out without being glazed and fired a second time.

Polyvinyl chloride (PVC) The most common substance used to make heavy-duty plastic piping.

Pom tongs Tongs that are not joined together by a spring.

Positive air pressure zone An area where more air is supplied to it than removed from it.

Postmix system A carbonated beverage dispenser that operates using two chilled tubes (*lines*), one for syrup concentrate, and one for carbonated water.

Pouring station The central part of an under bar, the area used to add juices and carbonated beverages to drinks.

Power burners Gas burners on a range top that can supply twice the heat of a conventional gas burner. Power burners also use more fuel.

Power grid Another term for a building's electrical system.

Power factor (PF) A measurement of how efficiently an individual light fixture, or a whole business, uses power. Customers who have low power factors may be charged more for their electricity.

Power sink A motorized three-compartment sink that circulates water rapidly to wash pots and pans.

Premix system A carbonated beverage dispenser in which the syrup and carbonated water are purchased already mixed, then propelled up a single line into the dispensing head.

Pre-prep A work station where food is first taken from receiving or storage to be prepared for delivery to the various preparation areas of a kitchen. In pre-prep, crates of produce are opened, meats are cut into smaller pieces, and so on.

Pressure footprint The diagram of an HVAC system's positive and negative air pressure and direction of airflow.

Pressure fryer A specialty fryer that cooks food with a combination of hot oil and steam.

Prevailing market rate The typical cost of a similar building or space at the time a lease is negotiated.

Prewash option In a commercial dishwasher, a separate chamber that blasts dirty dishes with high-velocity water to remove debris. Also called a *power prewash.*

Primary air The rush of air that flows into a burner tube of a gas appliance along with the first jet of gas.

Prime costs The major foodservice expenses: food, beverage, and labor.

Product flow The pattern of moving food and supplies from their arrival at a receiving area, to preparation sites, and, finally, to the guests.

Production line The area of the kitchen where the food is cooked, plated, and garnished prior to serving. Also called a *hot line.*

Proofer A heated cabinet that permits proofing of breads. Also called a *proofing cabinet.*

Proofing Gently warming dough to allow it to rise before baking.

Protective lighting A light fixture or lamp that is specially coated to prevent any bits of glass or chemical from flying out in case of breakage. These have multiple uses in kitchens and food storage areas, and are now required in the U.S. by federal law.

Psi The abbreviation for *pounds per square inch,* a measurement of steam pressure.

Psychographics The attitudes and tastes of potential customers.

Pump-fill station An area that contains the equipment to transfer cooked foods directly from a kettle through tubing and into plastic storage casings. Part of the cook-chill process.

Pyramiding A method of stacking glasses for storage.

Quartz sheath A type of heating element in a toaster.

Quick-disconnect coupling A shutoff device that instantly stops the flow of gas to a gas-powered appliance. A safety feature of gas connections.

Quick-service restaurant (QSR) The restaurant industry terminology for what is commonly known as a "fast food" restaurant, with counter-style and drive-through service.

Raceway The long, horizontal section of a utility distribution system. It contains the outlets that hook the system to each piece of cooking equipment.

Rack conveyor A type of dishwasher in which racks of soiled dishes are placed on a conveyor belt that moves them through wash and rinse cycles.

Rack oven An oven that is tall and thin enough to accommodate whole rolling racks of food for cooking.

Radiant heat Heat transmitted by radiation instead of conduction or convection. The sun's rays are an example of radiant heat.

Radiation The term for cooking with radiant energy, which is heat in the form of electromagnetic waves, found in microwave ovens.

Rag-outs Discarded tablecloths and napkins recycled into rags.

Rail The recessed area on the top surface of a bar, nearest the bartender. Also called a *glass rail, drip rail,* or *spill trough.*

Range The appliance commonly known as a "stove" in home use. In foodservice, the range usually consists of the range top for surface cooking and the range oven for baking and roasting.

Ratchet clause A utility company may require this minimum charge for electric power, based on a customer's highest energy use period.

Rat-tail tang The part of a knife blade that runs the length of the handle but is tapered to be much thinner than the rest of the blade. Seen in inexpensive knives. See also **tang**.

Reach-in refrigerator A refrigerator with a door that pulls open to retrieve items from inside.

Reactive organic gases Cooking odors.

Receiving area The place where food and supplies are delivered to the building, often near the back door.

Recognition clause Part of a lease agreement stipulating that an existing business cannot be forced out, and may continue to lease under the terms of the existing agreement, if the original owner dies or sells to someone else.

Recycling The collection and separation of specific waste materials that can be processed and reused instead of being thrown away.

Reducing collar A device in a dishwasher that restricts the pressure of spraying water to protect dishes.

Reflector **1.** A shiny, custom-fit surface that fits between a bulb and fixture and mirrors light. **2.** A stainless steel or cast iron piece in a broiler that absorbs and concentrates heat from an electric or gas burner.

Refrigerant A substance, usually a gas, that is used in refrigeration.

Refrigerated storage Storage for items that must be kept chilled or frozen until used.

Refrigerated ton A measurement of the cooling capacity of an air-conditioning system; one refrigerated ton is the equivalent of how much energy it takes to melt 1 ton of ice at 32 degrees Fahrenheit in a 24-hour period.

Refrigeration circuit The machinery and system that make heat removal possible from a refrigerated space.

Refrigeration cycle The process of removing heat from a refrigerated space.

Registers Another term for the dials on a gas meter.

Regulation valve A valve used to decrease water pressure.

Reheater This part of an HVAC system heats and dries air that would otherwise be too moist to use in the ducts.

Relative humidity The percentage of moisture in a refrigerated space, which can impact the appearance and rate of deterioration of many foods.

REP Abbreviation for *Retail Electric Provider.* In markets where the sale of electricity has been deregulated, these are companies that purchase power from utilities and sell it to retail and residential customers.

Restaurant cluster An area of a city or neighborhood that contains a lot of restaurants and consequently becomes a popular destination for dining out.

Retarder A temperature-controlled cabinet that stops the proofing process for dough that has risen sufficiently.

Rethermalization Reheating.

Rethermalization cart A specialty appliance that brings pre-plated cold food up to proper serving temperature in less than one hour and can be rolled from place to place.

Return piping A set of pipes in a steam system through which condensation is removed.

Reverberant sound Noise that reflects or "bounces" off surfaces.

Reverse osmosis A method of distilling seawater by pumping it through a permeable membrane to remove minerals and contaminants.

Riser This word has different meanings in different situations. In this text, it refers to a folding platform used to elevate people or tables, usually in a banquet setting; or the tall, vertical posts on either end of a utility distribution system; or the height of a single step on a staircase.

Rolled edge A plate made with a thicker, rounded outer edge for more durability.

Roll-in refrigerator A refrigerator large enough to roll carts in and out of its refrigerated space.

Rollout Stale or contaminated air from above the cooking line that escapes into the rest of the kitchen. An exhaust system should prevent rollout.

Rotisserie An oven that contains rows of metal spikes (called *spits*). Meat placed on the spits rotates and roasts slowly as warm, moist air fills the oven.

Rough-in A preliminary diagram of where utility connections will be in a building.

R-value A measurement of the amount of heat resistance in a building's insulation.

Salamander A smaller version of a broiler that can be placed on a shelf or counter. Often used as a plate warmer.

Satellite brewer A 1.5-gallon coffee-maker that is easy to transport and can be connected to a small docking station that keeps the coffee warm until serving.

Satin finish A less glaring surface for stainless steel worktables and refrigerator doors that reduces glare.

Satin weave A simple texture used in making most tabletop fabrics.

Satoir A straight-sided sauté pan.

Saucepan A shallow pan with a single, long handle.

Saucepot A large, round pot with straight sides and two welded loop handles for easy lifting.

Sauté pan A versatile pan used for range top duties such as sautéing, browning, braising, and frying. Also called *frypan, omelet pan,* or *crepe pan.* See also **satoir** and **sauteuse**.

Sauteuse A sauté pan with gently sloping sides.

Scale A person's visual perception of the size of an object.

Scaling The buildup of lime or other chemical residue from direct contact with water.

Scoring Creating burn marks on broiled food by placing it on a grate. Also called *branding.*

Scramble system A self-service flow pattern for cafeterias that allows customers to wander between displays or stations without standing in one line. Also called a *free flow* or *hollow-square system.*

Scrap basket A perforated basket in a dish room sink or dishwasher, used to scrape food waste into or collect it during the washing process.

Scrap collector A perforated pot that sits in a sink and captures solid waste before it goes down the drain.

Scrapping area Part of the dish room or pot sink washing area in which dishes or pots and pans are prescrubbed or scraped by hand, to remove food waste scraps before insertion into a dish machine or pot washer.

Scrap screen A screen panel located inside a dish machine that collects food residue during the washing process.

Screed A sturdy strip applied to a floor, usually at the entrance to a walk-in refrigerator, to level the floor and more easily allow carts to be wheeled in and out.

Seasoning The process of cleaning, oiling, and heating a new pan to help lubricate and preserve its cooking surface.

Seat turnover The number of times a seat in the dining area is occupied during a mealtime. Also called *seat turn.*

Secondary air The air in the immediate atmosphere that surrounds a gas flame.

Service system The way food is delivered to the guests; by a wait staff, at a pickup window, or the like.

Sharpening stone A stone used to sharpen knives by passing the knife blade over it at an angle.

Sheet pan A flat, rectangular pan used for baking and food storage.

Short circuit An unsafe electrical flow that has been interrupted to follow an unplanned path; in HVAC terminology, an HVAC system that introduces fresh air through one duct in the exhaust hood. Also called a *short-cycle* system.

Side loader A table that can be fastened to a dish machine, which allows the operator to load dishes in a small space.

Sieve A utensil with a fine metal screen at the bottom, used for sifting flour or other fine, dry ingredients.

Silica Very fine sand used in the glassmaking process.

Single-phase current One electrical current that flows into an appliance. Single-phase power is usually required for small appliances.

Slicer A long, slender knife with a flexible blade, used to slice cooked meats.

Slip A mixture of clay and water that forms a creamy liquid poured into molds to make plates, bowls, cups, and the like. The type of clay used determines whether the finished product will be china, stoneware, and so on.

Sloping plate A griddle plate that is tilted slightly to allow grease to drain into a large, removable pan.

Slurry The mixture of solid waste and water that is created inside a waste pulper.

Smallware Pans, containers, and utensils used for a variety of kitchen duties, including hand tools, measuring tools, and cookware.

Smoke point The temperature at which a frying oil or shortening begins to smoke and take on a sharp, burnt smell.

Smoker cooker A type of oven that cooks using smoke. Its energy source can be wood, charcoal, and/or electricity.

Snap-action thermostat A temperature-regulating device that opens fully to permit maximum heating in appliances (such as deep-fat fryers) that require quick heat recovery.

Soap curd The sticky residue that results when laundry detergent combines with the chemicals in hard water.

Sodium pressure lamp A highly efficient but harsh light source used in areas like parking garages and hallways, where the color, temperature, and "eye appeal" of the light itself are not a concern.

Soft-mount washer See **floating suspension washer**.

Solar tube An aluminum tube installed in a ceiling to bring daylight into an interior room.

Sorbite A mineral product used to absorb humidity and odors inside a refrigerator.

Source reduction Making less waste or reducing the toxicity of the waste that is produced.

Sparkle The pleasant glittering effect of a light fixture.

Speed rail A place on a bar to store bottles that are used often to make drinks.

Spindle blender A small mixer that may have multiple heads. Used to make milkshakes and other blended drinks.

Spine The nonsharp edge of a knife blade.

Spiral mixer A mixer that works gently to knead dough.

Splash guard A 3- to 6-inch vertical piece of metal on a griddle that helps control grease spatters and prevents food from falling behind the griddle. Also called a *fence.*

Splash zone A surface that is subject to routine spills, splashes, and contamination. Also called a *splash contact surface.*

Split basket A half-sized basket for frying or cooking smaller portions.

Spoodle A kitchen utensil that combines the stirring ability of a spoon with the portion control of a ladle. Also called a *spoonladle, portioner,* or *spooner.*

Springform pan A baking pan with a spring or clamp on the side to make food removal easier. Most often used with cheesecakes.

Stack A vertical drain pipe in a plumbing system. Also called a *soil pipe.*

Stamping A knife-manufacturing technique in which more than one knife blade is cut out of a single sheet of steel.

Standard Another term for the tap or faucet of a beer dispensing system.

Stationary-rack dishwasher Another name for a single-tank, door-style dishwasher. Racks of dishes are loaded into the machine and sit still as they are washed instead of moving on a conveyor belt.

Steamer An appliance with one or more enclosed compartments that cooks food using steam.

Steam injector A method of shooting pressurized steam into an appliance to create heat.

Steam-jacketed kettle A large bowl within a bowl used for making soups, sauces, and stocks. Steam or water can be pumped into the space between the two bowls to cook, heat, or cool the food inside the kettle.

Steam table A table with openings to hold containers of cooked food over steam or hot water, to keep the food warm. Also called a *hot-food table.*

Steam table pans Rectangular stainless steel pans used for food storage and serving. They can be stacked, frozen, heated, and come in a variety of standard sizes. Also called *hotel pans, counter pans,* or *service pans.*

Steam trap A valve in a steam system that helps regulate the overall pressure of the system. There are two types of steam traps, the *inverted bucket* and the *thermodynamic disk.*

Steam tube oven A specialty oven usually used for baking, which contains hollow tubes placed above and below the cooking cavity that fill with super-heated steam to cook product.

Steel A long, slim tool made of steel that is used to keep knife blades in alignment as part of the sharpening process.

Stemware Glassware that has a thin stem separating the foot of the glass from the bowl, such as a wine goblet.

Step-rate A graduated schedule of costs for electrical power, depending on how many hours are used. The cost decreases as usage increases. Also called a *declining-block schedule.*

Stockpot A taller saucepot, used for preparing stocks and simmering large quantities of liquids.

Stomper A hard plastic or wooden device used to push meat into a grinder. Also called a *pusher.*

Stone hearth oven Another term for a wood-burning or wood-fired oven.

Stoneware Chip-resistant dinnerware in natural beige or gray colors, made the same way as china, but not translucent.

Straight line A type of flow pattern in which materials or people move steadily from one process to another in a straight line. The term is used to describe kitchen work stations or cafeteria customer lines. Also called *assembly line flow.*

Strainer A hand tool shaped like a cup or cone and made of mesh, through which liquids can be strained.

Strain gauge A measuring device attached to the hinges of a load cell (inside a digital or electronic scale) that bend when an object is being weighed on the scale. As the load cell bends, the strain gauges also bend, measuring the weight of the object by their resistance to the change.

Strip curtains Heavy plastic strips hanging directly inside a walk-in refrigerator door that help keep cold air from seeping out when the door is open.

Sump A floor drain with a large strainer to trap solids before they reach the drainage system.

Super-cooker A multitask oven designed for cooking very quickly using two different cooking technologies: microwave and convection, for instance, or microwave and heat impingement.

Superheated steam Water vapor that has been heated to a temperature above the boiling point of water.

Sustainability Building structures or planning businesses with conservation of natural resources in mind. Also called *green building.*

Sweep system A device that can be programmed to turn lights on or off sequentially as needed.

Table runner A length of fabric placed down the center of a table, often a complementary print or accent to a tablecloth.

Table setting The place setting, plus the placemat or tablecloth and napkin.

Tablet arm The small table attached to an individual chair for casual dining. This type of chair looks like a school desk.

Tabletop The foodservice term for the supplies that make up a table setting: flatware, dishes, glasses, table linens, condiment containers, and other dining accessories.

Tang The part of a knife blade that is embedded in its handle.

Taper ground A knife blade made from a single piece of steel, shaped so it tapers smoothly from spine to cutting edge.

Tariff analysis An overview report of the billing, rates, and rules of a utility company, which should be reviewed and revised as these rates and rules change. A tariff analysis can be done for individual buildings based on their history of energy use.

Temperature shock The term for an abrupt temperature change in a fryer kettle, caused by dropping a block of cold, solid shortening into hot frying oil.

Tempered A makeup air system that can either heat or cool incoming air.

Tempering Reheating and then quickly cooling newly made glass, which builds its resistance to temperature extremes. Some glasses are *fully tempered*; and others are only *rim tempered.*

Teppanyaki griddle An oversized griddle used in Japanese-style exhibition cooking, with customer seating around its diameter.

Theme The characteristics or qualities that make up the mood of a dining experience; for example, a restaurant that is decorated to resemble a medieval village.

Thermal capture system An exhaust hood that cycles air in the upper part of the hood without venting air into the kitchen. This type of system is sufficient for a small kitchen but not a large one.

Thermal head space The upper space in the interior cavity of a wood-fired oven. The more thermal headspace an oven has, the more heat stays inside the oven cavity instead of being vented away.

Thermal shock A quick, intense temperature change that creates enough stress on a glass to cause breakage.

Thermistor A very sensitive temperature-sensing device that uses metal oxides formed into a tiny bead to measure resistance when subjected to heat.

Thermoacoustics The use of sound waves to cool foods.

Thermocouple A pair of electrical wires joined together at one end and attached at the other end to an appliance.

Thermoelectric control A safety feature on a gas pilot light appliance that heats two metal wires and creates a very low electric voltage to hold the gas valve open.

Thermostat An automatic device for regulating temperature. On gas appliances, this device is a knob or dial called a *throttling control* or *modulating control.*

Three-phase current Three different streams of electrical current that peak in a regular, steady pattern. Three-phase power is usually required for large, heavy-duty appliances.

Throat The opening at the top of a food waste disposer, into which scraps are fed to be ground up.

Tilting braising pan A versatile cooking appliance made up of a flat, griddle bottom with vertical sides and a tilting body.

Time-of-day-option A billing option for electricity that allows a business to be charged for power use based on the time of day it is open, giving lower rates to businesses that use power at nonpeak times (nights, weekends).

Time-temperature indicator (TTI) A device filled with chemical crystals that are temperature-sensitive, used to verify safe temperatures during food holding, transport, or storage.

Tinned steel See **aluminized steel.**

Total harmonic distortion (THD) ratings A measurement of how likely a ballast is to distort the power line by creating electromagnetic interference.

Traffic flow The movement of guests and employees through the building. The ideal flow minimizes backtracking and crossing each other's paths, for safety and efficiency.

Trap A curved section of pipe in a discharge system that retains some water.

Triple net lease A lease price that includes a building's rent, taxes, and insurance in a single monthly payment.

Trough A recessed area that surrounds the griddle top on three sides, used to scrape grease and food bits into to be discarded. Also called a *gutter.*

Trunnion A device that allows heavy kitchen equipment to pivot for easy loading and unloading.

Tumble chiller A rotating drum full of cold water that gently kneads and chills packages of food as part of the cook-chill process.

Tumble chilling A method of cooling cooked food quickly by immersing the packaged food in cold liquid.

Tumbler A commercial laundry dryer.

Tuning The precise adjustment of lighting.

Two-fired plate A plate that has been fully glazed, including its foot, making it safer to stack without scratching other plates.

Under bar The part of the bar directly beneath the top surface, where supplies are stored out of view of customers.

Underbroiler A broiler with its heat source located under the food being cooked. Also called an *underfired* or *open-hearth broiler.*

Uniform charge A single rate for each hour of electricity used, found most often on a residential power bill.

Uniform heat top A flat heated surface on a rangetop (instead of individual burners) that allows the user to cook on the entire surface. Also called a *hot top* or *even-heat plate.*

Untempered A makeup air system that moves air without the capability to heat or cool it.

Upfeed system The natural pressure of water inside piping that allows it to move through a series of water lines, even to upper floors of a building.

Utility cart A rolling shelf unit most often used in restaurants to transport dishes and cleaning supplies.

Utility distribution system A central location for all necessary kitchen utility services, connected directly to the exhaust canopy. Contains a single connection for each utility, plus emergency shutoffs and inspection panels.

Utility knife A small version of the chef's knife, popular for pantry work. Also called a *salad knife.*

Vacuum tumbler A motorized machine that gently kneads salt-based marinades into meat, to coat the meat and seal in juices when it is cooked.

Vapor compression A popular type of commercial air conditioning in which a refrigerant gas is compressed into liquid form, then evaporates back to its gaseous state.

Vapor oven A type of combination oven/steamer that cooks food using low-temperature steam.

Variance An exception to a zoning ordinance, which must be granted by a city, county, or other governing body.

Ventilation To provide fresh or properly treated air into a room, and to send a corresponding amount of stale or used air out of the room.

Venting system The part of the waste water discharge system that prevents water from being siphoned from the grease trap. A series of vents equalize pressure in the system and circulate enough air to reduce pipe corrosion and remove odors.

Ventless fryer A fryer with an internal air filtration system that does not require venting waste air to the outside and, therefore, can be used without being placed under a ventilation hood.

Vertical cutter mixer (VCM) A kitchen appliance that combines mixing and blending functions with a variety of blades and a high mixing speed.

Vertical mixer A motorized appliance that mixes batters, dough, etc. Also called a *planetary mixer.*

Volt The force that pushes an ampere through an electrical wire.

Wallsaver legs Chair legs that extend slightly beyond the back of the chair. These prevent the back of the chair from hitting the wall behind it.

Warewashing The dishwashing process, including collection of the soiled dishes, scraping, rinsing, washing, sanitizing, and drying them.

Waste audit form A means of analyzing how much waste a business is creating, with suggestions for ways to reduce the amount.

Waste heat Warm air given off by appliances or air conditioners that may be harnessed for use.

Waste pulper A machine that can digest food waste and paper products, mincing it into bits and mixing it with water to create more compact amounts of waste to be discarded.

Watt The amount of power required for 1 amp of electricity to flow through a circuit at 1 volt of pressure.

Watt density The number of watts an appliance's heating element can produce per square inch, a measurement of its ability to provide uniform heating.

Wave guide A fan located in the top of a microwave oven cavity that propels the waves into the cavity.

Wave stirrers Devices in a microwave oven to distribute the waves throughout the cavity, to ensure even heating.

Welding Joining together two pieces of metal by heating them.

Well The inside portion of a plate, not including its rim.

Wetting action The ability of a dishwashing detergent to soften water and loosen food particles from dish surfaces.

Wide-area vacuum A two-motor vacuum cleaner with a nozzle width of up to 36 inches for use in large areas.

Wishbone-style legs Table legs that provide plenty of legroom for customers who sit at the table. Popular on oblong banquet tables.

Wood-fired oven An oven that burns wood as its heat source. It has many other common names, including: brick oven, hearth stone oven, and wood-burning oven.

Work center An area in which workers perform a specific task, such as tossing salads or garnishing plates.

Work section An overall area of the kitchen that encompasses several work centers, such as cooking or baking.

Zoned cooking In a conveyor oven or on a griddle, separate controls for different sections of the appliance.

INDEX